To David Bailey,

with gratitude

the

Nigel Dawson.

Feb. 1973.

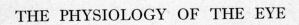

THE PHYSIOLOGY OF THE EYE

To
SAMUEL L. SALTZMAN

THE PHYSIOLOGY OF THE EYE

by

HUGH DAVSON, D.Sc.(Lond.)

Scientific Staff, Medical Research Council;
Honorary Research Associate and Fellow of University College,
London

THIRD EDITION

With 453 illustrations

ACADEMIC PRESS
New York

First Edition 1949
Reprinted 1950
Second Edition . . . 1963
Third Edition 1972

I.S.B.N. 0-12-206740-1
Library of Congress Catalog Card Number 76-181061

Printed in Great Britain

Published in the United States by
Academic Press, Inc., 111 Fifth Avenue
New York, N.Y. 10003

PREFACE

It was in 1948 that I called at the offices of Messrs. Churchill with a brown paper parcel containing the manuscript of *Physiology of the Eye*, and offered it to them for publication, and the first edition appeared in 1949. A second edition appeared in 1963, and in the preface to this I wrote: 'The task (of revision) has now been completed and has resulted in about a 50 per cent increase in size, an increase that is the necessary consequence of the tremendous advances made in those aspects of the physiology of the eye that are dealt with in Sections I and II. Thus, it is true to say that in 1948 our knowledge of the vegetative physiology and biochemistry of the eye was rudimentary in the extreme. Now, thanks to the establishment of numerous ophthalmological research organizations—mainly in the U.S.A.—it is safe to say that out knowledge in this branch of ocular science is abreast of our knowledge in comparable fields of physiology and biochemistry. Although, in 1948, the field covered by the title *Mechanism of Vision* rested on a much firmer basis of accurate scientific investigation, the striking improvements in photochemical and electrophysiological techniques that have since taken place have permitted just as rapid progress in elucidating basic problems. Consequently, because I have tried to maintain the same sort of treatment as I had in mind when I wrote the first edition of this book—namely, to present a simple yet thorough exposition of the fundamental principles of ocular physiology—I have had to double the amount of space devoted to these first two sections. The muscular aspects covered by Section III have been expanded a little to take account of the important advances made possible by the application of modern methods to the analysis of the eye movements, pupillary function, and so on. Sections IV and V rest on the firm theoretical and practical foundation laid by the work of Helmholtz and other classical research workers in the realm of physiological optics, so that they have been tampered with least'.

To this I need only add that further advances have required the addition of some 150 pages, and these have been fairly equally divided between the first four sections; perhaps the most striking feature of the progress made between this and the previous edition is the successful application of modern microelectrode techniques to the recording of activity at all stages in the visual pathway pioneered by Hubel and Wiesel; this has already necessitated a considerable expansion of Section IV by virtue of our increased understanding of the neuronal basis of higher interpretative processes in visual physiology.

London, 1971   HUGH DAVSON

ACKNOWLEDGEMENTS

IN preparing this new edition I have been helped in the way only an author can appreciate by the kindly and skilful assistance of the head librarian of the Thane Library at University College, Mr. C. F. A. Marmoy. In addition, it is a pleasure to acknowledge the secretarial assistance of Mrs Jane Barnett, and the co-operation of my publishers who, in the person of Mr. A. S. Knightley, have made the task of seeing the book through the press so agreeable.

To the authors and publishers from whose works illustrations have been reproduced, I convey my sincere thanks, particularly to Drs. de Robertis, Fine, Gavin, Ishikawa, Jakus, Kleifeld, Pappas, Pirie, Sjöstrand, Smelser, Tousimis, Lasansky, Wanko, Bloemendal, Tripathi, and Kayes, for generously providing originals of micrographs.

H. D.

CONTENTS

vii

ANATOMICAL INTRODUCTION

The Three Layers. The globe of the eye consists essentially of three coats enclosing the transparent refractive media. The outermost, protective tunic is made up of the *sclera* and *cornea*—the latter transparent; the middle coat is mainly vascular, consisting of the *choroid, ciliary body,* and *iris.* The innermost layer is the *retina,* containing the essential nervous elements responsible for vision—the *rods* and *cones;* it is continued forward over the ciliary body as the *ciliary epithelium.*

Dioptric Apparatus. The dioptric apparatus (Fig. 1) is made up of the transparent structures—the cornea, occupying the anterior sixth of the surface of

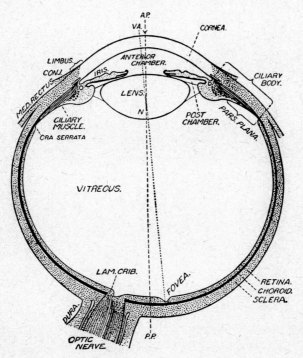

FIG. 1. Horizontal section of the eye. P.P.= Posterior pole; A.P. = Anterior pole; V.A. = Visual axis. (Wolff, *Anatomy of the Eye and Orbit.*)

the globe, and the *lens,* supported by the *zonule* which is itself attached to the ciliary body. The spaces within the eye are filled by a clear fluid, the *aqueous humour,* and a jelly, the *vitreous body.* The aqueous humour is contained in the *anterior* and *posterior chambers,* and the vitreous body in the large space behind the lens and ciliary body. The posterior chamber is the name given to the small space between the lens and iris.

The iris behaves as a diaphragm, modifying the amount of light entering the eye, whilst the ciliary body contains muscle fibres which, on contraction, increase the refractive power of the lens (accommodation). An image of external objects is formed, by means of the dioptric apparatus, on the retina, the more highly specialized portion of which is called the *fovea*.

Visual Pathway. The retina is largely made up of nervous tissue—it is an outgrowth of the central nervous system—and fibres carrying the responses to visual stimuli lead away in the *optic nerve* through a canal in the bony orbit, the *optic foramen* (Fig. 2); the visual impulses are conveyed through the optic

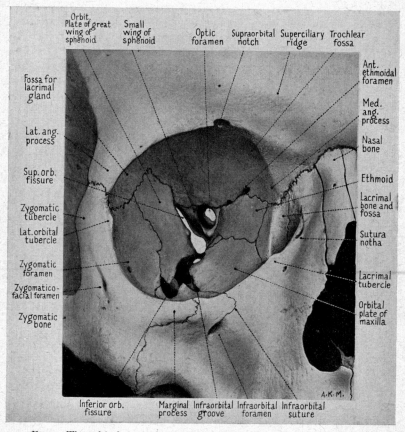

FIG. 2. The orbit from in front. (Wolff, *Anatomy of the Eye and Orbit.*)

nerve and tract to the lateral geniculate body and thence to the cerebral cortex; on their way, the fibres carrying messages from the medial, or nasal, half of the retina cross over in the *optic chiasma*, so that the lateral geniculate body of the left side, for example, receives fibres from the temporal half of the left retina and the nasal half of the right. The nerve trunks, proximal to the chiasma, are called the *optic tracts*. This partial decussation may be regarded as a development associated with binocular vision; it will be noted that the responses to a stimulus from any one part of the visual field are carried in the same optic tract (Fig. 3), and the necessary motor response, whereby both eyes are directed to the same point in the field, is probably simplified by this arrangement. It will be noted also that a right-sided event, i.e. a visual stimulus arising from a point in the right half of

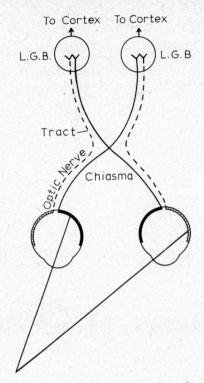

FIG. 3. The decussation of the optic pathway.

the field, is associated with impulses passing to the left cerebral hemisphere, an arrangement common to all peripheral stimuli.

Vascular Coat

The nutrition of the eye is taken care of largely by the capillaries of the vascular coat; let us examine this structure more closely.

Choroid. The choroid is essentially a layer of vascular tissue next to the retina; it is separated from this nervous tissue by two membranes—the structureless *membrane of Bruch* and the pigment epithelium. The retina comes to an end at the ciliary body, forming the *ora serrata* (Fig. 4), but the vascular coat continues into the ciliary body as one of its layers, the *vessel layer*, which is separated from the eye contents by membranes—the two layers of ciliary epithelium which, viewed embryologically, are the forward continuations of the retina and its pigment epithelium, and the *lamina vitrea*, which is the continuation forward of Bruch's membrane.

Ciliary Body and Iris. The ciliary body in antero-posterior section is triangular in shape (Fig. 5) and has a number of processes (seventy) to which the zonule is attached; viewed from behind, these processes appear as radial ridges to which the name *corona ciliaris* has been given (Fig. 4). The relationship of the iris to the ciliary body is seen in Fig. 5; the blood vessels supplying it belong to the same system as that supplying the ciliary body. Posteriorly, the stroma is

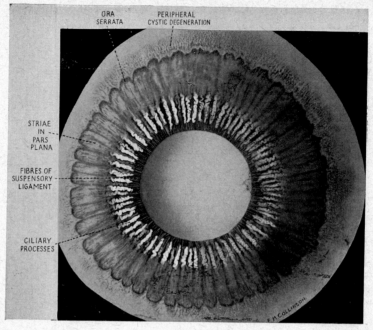

FIG. 4. Posterior view of the ciliary body showing ora serrata and corona ciliaris.
(Wolff, *Anatomy of the Eye and Orbit*.)

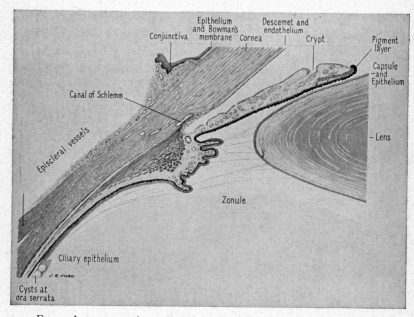

FIG. 5. Antero-posterior section through the anterior portion of the eye.
(Wolff, *Anatomy of the Eye and Orbit*.)

separated from the aqueous humour by the *posterior epithelium,* a double, heavily pigmented layer which may be viewed as a prolongation of the ciliary epithelial layers.

Nutrition. The vessels of the vascular coat nourish the internal structures of the eye; so far as the lens is concerned, this process must take place by diffusion of dissolved material from the capillaries through the aqueous humour and vitreous body. The inner (nearest the vitreous) nervous elements of the retina, however, are provided for by a functionally separate vascular system derived from the *central artery of the retina;* this artery, a branch of the *ophthalmic,* enters the globe with the optic nerve and it is its ramifications, together with the *retinal veins,* that give the fundus of the eye its characteristic appearance. The choroid, ciliary body, and iris are supplied by a separate system of arteries, also derived from the ophthalmic—the *ciliary system of arteries.*

The anterior portion of the sclera is covered by a mucous membrane, the *conjunctiva,* which is continued forward on to the inner surfaces of the lids, thus creating the *conjunctival sac.* The remainder of the sclera is enveloped by *Tenon's capsule.*

Muscle. Movements of the eye are executed by the contractions of the six extra-ocular muscles; the space between the globe and orbit being filled with the orbital fat, the movements of the eye are essentially rotations about a fixed point in space.

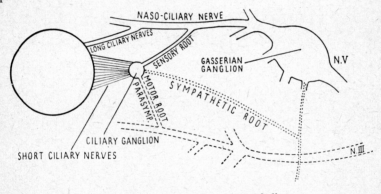

FIG. 6. Nervous supply to the eyeball.

Nerve Supply. The essentials of the nerve supply to the eyeball are indicated in Fig. 6; sensory impulses (excluding, of course, the visual ones) are conveyed through the *long* and *short ciliary nerves.* The long ciliary nerves are composed mainly of the axons of nerve cells in the *Gasserian ganglion—*the ganglion of the trigeminal (N. V); they convey impulses from the iris, ciliary body, and cornea. The short ciliary nerves also contain axons of the trigeminal; they pass through a ganglion in the orbit, the *ciliary ganglion,* into the *naso-ciliary nerve;* the fibres carry impulses from all parts of the eyeball, but chiefly from the cornea.

The voluntary motor nerve supply to the extraocular muscles is through the cranial nerves III (oculomotor), IV (trochlear) and VI (abducens). Parasympathetic motor fibres to the ciliary muscle and iris travel through the lower division of N. III as the *motor root* of the ciliary ganglion; post-ganglionic fibres to the muscles are contained in the short ciliary nerves.

Sympathetic fibres from the superior cervical ganglion enter the orbit as the sympathetic root of the ciliary ganglion and run in the short ciliary nerves to supply the vessels of the globe and the dilator fibres of the pupil. Other sympathetic fibres avoid the ciliary ganglion, passing through the Gasserian ganglion and entering the globe in the long ciliary nerves, whilst still others enter the globe in the adventitia of the ciliary arteries.*

* Ruskell (1970a, b) has described a second source of parasympathetic fibres to the eye; these are derived from the facial nerve (N. VII) as *rami orbitales* of the pterygopalatine ganglion, and they enter the eyeball mainly with the long ciliary nerves; nerve terminals on all ciliary arteries examined showed degenerative changes after damage to the pterygo-palatine ganglion.

The Vegetative Physiology and Biochemistry of the Eye

PART I

AQUEOUS HUMOUR AND THE INTRAOCULAR PRESSURE

THE aqueous humour is a transparent colourless fluid contained in the anterior and posterior chambers of the eye (Fig. 1, p. 1). By inserting a hypodermic needle through the cornea, the fluid may be more or less completely withdrawn, in which case it consists of the mixed fluids from the two chambers. As Seidel first showed, the posterior fluid may be withdrawn separately by inserting a needle through the sclera just behind the corneoscleral junction; because of the valve-like action of the iris, resting on the lens, the anterior fluid is restrained from passing backwards into the posterior chamber, so that subsequently another needle may be inserted into the anterior chamber and this anterior fluid may be withdrawn. In the cat Seidel found that the volume was some 14 per cent of the volume of the anterior fluid, whilst in the rabbit it is about 20 per cent, according to Copeland & Kinsey. The fluids from the two chambers are of very similar composition, so that we may consider the composition of the mixed fluid when discussing the relationships between aqueous humour and blood. Thanks largely to the studies of Kinsey, however, it is now known that there are characteristic differences between the two fluids, differences that reflect their histories, the posterior fluid having been formed first and therefore exposed to diffusional exchanges with the blood in the iris for a shorter time than the anterior fluid.

ANATOMICAL RELATIONSHIPS

The nature of the aqueous humour is of fundamental interest for several reasons. There is reason to believe that it is formed continuously in the posterior chamber by the ciliary body, and that it passes through the pupil into the anterior chamber, whence it is drained away into the venous system in the angle of the anterior chamber by way of Schlemm's canal. As a circulating fluid, therefore, it must be a medium whereby the lens and cornea receive their nutrient materials. Furthermore, because it is being continuously formed, the *intraocular pressure* is largely determined by the rate of formation, the ease with which it is drained away, and by the forces behind the process of formation. To understand the nature of the fluid we must consider its chemical composition and variations of this under experimental conditions, whilst to appreciate the drainage process we must examine the detailed micro-anatomy of the structures that are concerned in this process. Since, moreover, the aqueous humour cannot be considered in isolation from the vitreous body, the posterior portion of the eyeball is also of some interest.

The Vascular Coat

The choroid, ciliary body and iris may be regarded as a vascular coat—the *uvea*—sandwiched between the protective outer coat—cornea and sclera—and the inner neuroepithelial coat—the retina and its continuation forwards as the ciliary epithelium (Fig. 7). This vascular coat is made up of the ramifications of the ciliary system of arteries that penetrate the globe independently of the optic nerve.

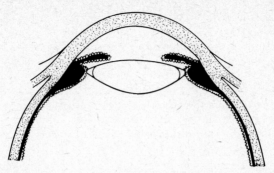

FIG. 7. Illustrating the uvea or vascular coat (black) sandwiched between the sclera on the outside and the retina, ciliary epithelium, etc., on the inside.

Ciliary Arteries. The posterior ciliary arteries arise as trunks from the ophthalmic artery (Fig. 8), dividing into some ten to twenty branches which pierce the globe around the optic nerve. Two of them, the long posterior ciliary arteries, run forward through the choroid to the ciliary body where they anastomose with the anterior ciliary arteries to form what has been incorrectly called the *major circle of the iris*; incorrectly, because the circle is actually in the ciliary body. From this circle, arteries run forward to supply the ciliary processes and iris, and backwards—as recurrent ciliary arteries—to contribute to the choroidal circulation.

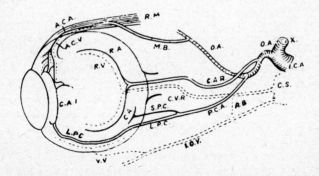

FIG. 8. Physiological plan of the circulation in man. O.A., ophthalmic artery; M.B., muscular branch; A.C.A., anterior ciliary artery; C.A.I., Circulosus arteriosus iridis major; L.P.C., long posterior ciliary artery; S.P.C., short posterior ciliary artery; P.C.A., posterior ciliary arteries; C.A.R., central artery of the retina; R.A., retinal artery; V.V., vortex vein; I.O.V., inferior orbital vein; A.B., anastomosing branch; C.S., cavernous sinus. (Duke-Elder, *Brit. J. Ophthal.*)

By the *choroid* we mean the vascular layer between the retina and sclera; it is made up of the ramifications of the recurrent ciliary arteries, just mentioned, and of the short posterior ciliary arteries, which enter the globe around the posterior pole. The arteries break up to form a well-defined capillary layer—the *choriocapillaris*—next to the retina, but separated form this by a transparent glassy membrane—the *membrane of Bruch* or *lamina vitrea*. This capillary layer is certainly responsible for nutrition of the outer layers of the retina—pigment epithelium, rods and cones, and bipolar cells—and may also contribute to that of the innermost layer.

Venous System. The venous return is by way of two systems; the anterior ciliary veins accompany the anterior ciliary arteries, whilst the vortex veins run independently of the arterial circulation; these last are four in number, and they drain all the blood from the choroid, the iris and the ciliary processes, whilst the anterior ciliary veins drain the blood from the ciliary muscle and the superficial plexuses, which we shall describe later. According to a study by Bill (1962) only 1 per cent of the venous return is by way of the anterior ciliary veins.

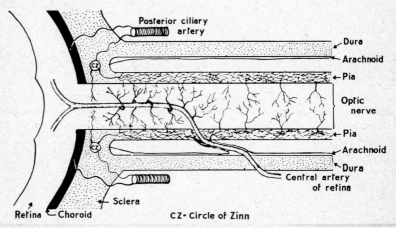

FIG. 9. Schematic section of optic nerve and eyeball showing course and branches of the central retinal artery. (Singh & Dass, *Brit. J. Ophthal.*)

Retinal Circulation. As indicated earlier, the innermost part of the retina is supplied directly, in man, by the ramifications of the central retinal artery, a branch of the ophthalmic that penetrates the meningeal sheaths of the optic nerve in its intraorbital course and breaks up into a series of branches on the inner layer of the retina, branches that give the fundus its characteristic appearance in the ophthalmoscope. As Fig. 9 indicates, the central artery gives off branches that supply the optic nerve*; moreover, the posterior ciliary arteries may send off branches in the region of the optic disk, forming a circle—*circle of Zinn or Haller*—which may form anastomoses with the branches of the central artery of the retina. To the extent that these anastomoses occur, the uveal and retinal vascular systems are not completely independent. Venous return from the retinal circulation is by way of the central retinal vein which, for a part of its course, runs in the subarachnoid space of the optic nerve and is thereby subjected to the intracranial pressure. For this reason, an elevated intracranial pressure may be manifest as an engorgement of the retinal veins.

So much for the general lay-out of the vascular circulation of the eye; as we

---

* Some authors, for example François & Neetens, speak of a *central artery of the optic nerve*, derived either as a branch of the central retinal artery or entering separately as a branch of the ophthalmic artery. Wolff describes them as *arteriae collaterales centralis retinae*. The literature relating to this and other points is well summarized in the careful studies of Singh & Dass (1960), whilst Hayreh (1969) has described the capillary supply to the optic disk; a large part of this—the prelaminar—is from cilio-retinal arteries.

have seen, the uveal circulation supplies the ciliary body and iris, the two vascular structures that come into close relationship with the aqueous humour. It is important now that we examine these structures in some detail.

Ciliary Body

In sagittal section this is a triangular body (Fig. 5, p. 4), whilst looked at from behind it appears as a ring, some 6 mm wide in man. Where it joins the retina—the *ora serrata*—it appears smooth to the naked eye, but farther forward the inner surface is ridged, owing to the presence of some seventy to eighty *ciliary processes*, and acquires the name of *corona ciliaris* (Fig. 4, p. 4). As the oblique section illustrated in Fig. 10 shows, the ciliary processes are villus-like

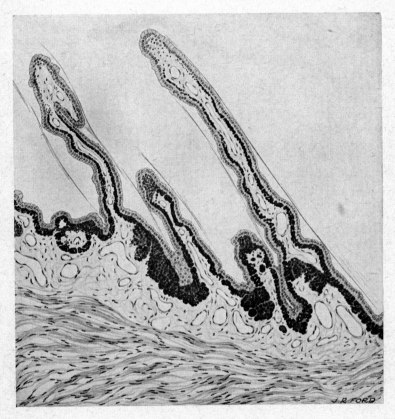

FIG. 10. Oblique section of the ciliary body.
(Wolff, *Anatomy of the Eye and Orbit.*)

projections of the main body, jutting into the posterior chamber. According to Baurmann, the total area of these processes, in man, is some 6 cm^2 and we may look upon this expansion as subserving the secretory activity of this part of the ciliary body, since there is good reason to believe that it is here that the aqueous humour is formed. The vessel-layer of the ciliary body is similar to the choroid; it extends into each ciliary process, so that this is probably the most heavily vascularized part of the eye. Each process contains a mass of capillaries so arranged that each comes into close relationship, at some point in its course, with the surface of the epithelium. As stated above, blood from the ciliary processes is carried away by the vortex veins. In the rabbit, the ciliary processes are of two kinds—*posterior principal processes* projecting from the posterior region, and the *iridial processes*, located more anteriorly, and projecting towards the iris. These last have no zonular connections and may be those that are most closely concerned in production of the aqueous humour since they are the only processes to show significant morphological changes after paracentesis (p. 40), according to Kozart (1968).

Ciliary Epithelium. The innermost coat, insulating the ciliary body from the aqueous humour and vitreous body, is the *ciliary epithelium*, a double layer of cells, the inner layer of which, i.e. the layer next to the aqueous humour or vitreous body, is non-pigmented, whilst the outer layer is heavily pigmented. These cells, which are considered to be responsible for the secretion of the aqueous humour, have been examined in the electron microscope by several workers; a striking feature is the interdigitation of the lateral surfaces of adjacent cells, and the basal infoldings (Fig. 11, Pl. I), which are characteristic features of secretory epithelia concerned with fluid transport. The relations of the two epithelial cell layers are of importance, since the secreted aqueous humour must be derived from the blood in the stroma of the ciliary body, and thus the transport must occur across both layers. This relation has been studied in the electron microscope by Bairati & Orzalesi (1966); the two layers, which face apex-to-apex, are united at frequent intervals by a variety of junctional complexes including zonulae occludentes, which would restrict diffusion through the intercellular space; the fusion of cell membranes brought about in this way would provide a relatively unrestricted passage of material from one cell layer to the other. Localized expansions of these intercellular spaces occur, and are called ciliary channels. Diffusion from stroma towards the aqueous humour is restricted by lateral zonulae occludentes between adjacent pigmented cells, but the spaces between the non-pigmented cells are apparently open.

The Iris

This is the most anterior part of the vascular coat; in a meridional section it is seen to consist of two main layers or *laminae*, separated by a less dense zone, the *cleft of Fuchs*. The posterior lamina contains the muscles of the iris, being covered posteriorly by two layers of densely pigmented cells, the innermost being the posterior epithelium of the iris which is continuous with the inner layer of the ciliary epithelium. The blood vessels, derived from the major circle of the iris (p. 9), run in the loose connective tissue of the laminae. Both arteries and veins run radially; venous return is by way of the vortex veins.

On its anterior surface the iris is in contact with the aqueous humour, so that it is of some interest to determine exactly how the fluid is separated from the tissue. It is customary to speak of an *anterior endothelium*, continuous with the corneal endothelium anteriorly and the posterior iris epithelium posteriorly, covering the anterior surface of the iris and thus insulating the aqueous humour from the underlying spongy tissue. The weight of recent evidence, however, bears against this so that it would seem that, in man at any rate, the aqueous humour has direct access to the stromal tissue.

PLATE I

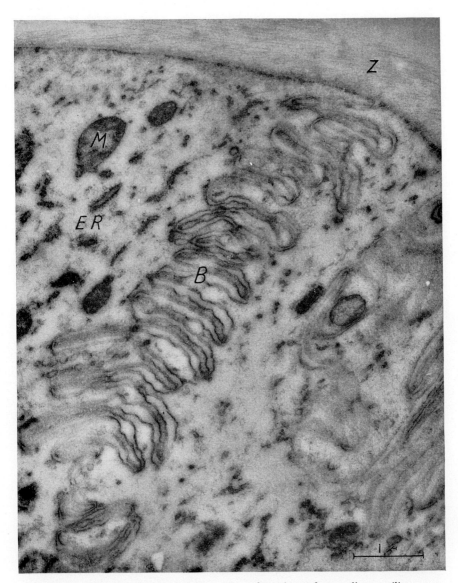

FIG. 11. Electron micrograph of a section of portions of two adjacent ciliary epithelial cells of the rabbit showing complex interdigitations (B) of their boundaries. Zonular fibres (Z) can be seen in close approximation to the surface of these cells. ER, endoplasmic reticulum; M, mitochondrion. Osmium fixation. × 19,000. (Pappas & Smelser, *Amer. J. Ophthal.*)

To face p. 12.

PLATE II

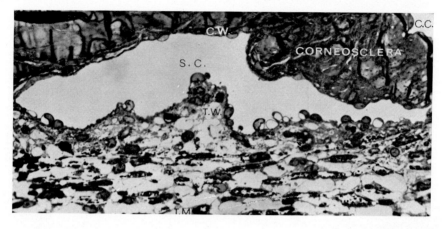

FIG. 14. Meridional section of Schlemm's canal and adjacent trabecular meshwork of rhesus monkey. Osmium fixed; toluidine blue stained. C.C., Collector channel; C.W., corneo-scleral wall of Schlemm's canal; S.C., Schlemm's canal; T.M. trabecular meshwork. Light micrograph. × 610. (Tripathi, *Exp. Eye Res.*)

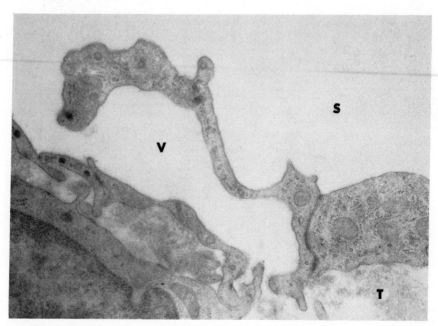

FIG. 24. Section through vacuole, V, of the endothelium lining Schlemm's canal, S. Note communication between Schlemm's canal and trabecular meshwork, T, through the vacuole. × 19,500. (Kayes, *Invest. Ophthal.*)

The Drainage Route

We have said that the aqueous humour is most probably secreted continuously by the cells of the ciliary epithelium; to make room for the new fluid there must obviously be a continuous escape, or drainage, and this occurs by way of the *canal of Schlemm*. This is a circular canal in the corneo-sclera of the limbus (Fig. 5, p. 4) which comes into relation with the aqueous humour on the one hand and the *intrascleral venous plexus* on the other, as illustrated by Figs. 12 and 13.

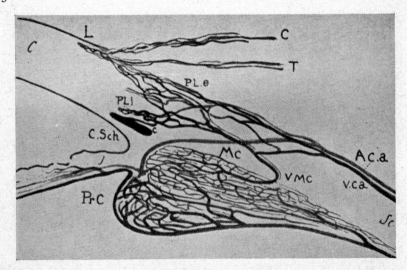

FIG. 12. Illustrating the canal of Schlemm and its relation to the plexuses of the anterior segment of the eye. C, conjunctival plexus; T, Tenon's capsule plexus; PLe, episcleral plexus; PLi, intrascleral plexus; Mc, ciliary muscle; VMc, vein of ciliary muscle; Ac.a, anterior ciliary artery; v.ca, anterior ciliary vein; Pr C, ciliary process; C, cornea; c, collector; L, limbus. (Maggiore, *Ann. Ottalm.*)

Plexuses of the Anterior Segment. The intrascleral plexus is largely made up of the ramifications of branches of the anterior ciliary veins, but, as Maggiore showed, there is also an arterial contribution, branches of the anterior ciliary arteries passing directly to the plexus and breaking up into fine vessels that come into direct continuity with the finer vessels of the plexus. It connects with the canal of Schlemm by some twenty fine vessels—*collectors*—which thus transport aqueous humour into the venous system. The blood from the plexus is drained away by large trunks into the *episcleral plexus;* from here large veins, also draining the more superficial plexuses—conjunctival and subconjunctival—carry the blood to the insertions of the rectus muscles in the sclera and accompany the muscular veins.

Canal of Schlemm and Trabecular Meshwork. The canal of Schlemm is usually represented as a ring lying in the corneo-sclera, which on cross-section (as in the meridional section of Fig. 13) appears as a circle or ellipse; its structure is, in fact, more complex, since it divides into several channels which, however, later reunite; hence the appearance of a cross-section varies with the position of the section. The canal is essentially an endothelium-lined vessel, similar to a

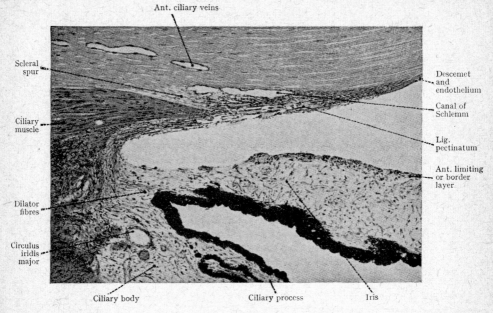

FIG. 13. Section of the angle of the anterior chamber.
(Wolff, *Anatomy of the Eye and Orbit.*)

delicate vein; on the outside it rests on the scleral tissue whilst on the inside, nearest the aqueous humour, it is covered by a meshwork of endothelium-covered trabeculae—the *trabecular meshwork* or *pectinate ligament* (Fig. 14, Pl. II). Aqueous humour, to enter the canal, must percolate between the trabeculae and finally cross the endothelial wall. On meridional section the meshwork appears as a series of meridionally orientated fibres (Fig. 13), but on tangential section it is

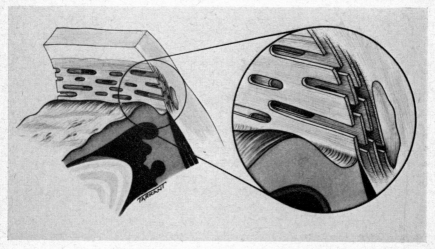

FIG. 15. Schematic illustration of lamellae of scleral meshwork.
(Ashton, Brini & Smith, *Brit. J. Ophthal.*)

seen to consist of a series of flat lamellae, piled one on top of the other, the fluid-filled spaces between these lamellae being connected by holes, as illustrated schematically in Fig. 15. The meshwork has been divided into three parts with characteristically different ultrastructures; the innermost portion (1 of Fig. 16) is the *uveal meshwork*; the trabeculae making up the lamellae here are finer than those of the outer or *corneoscleral meshwork*, whilst the meshes are larger. Each

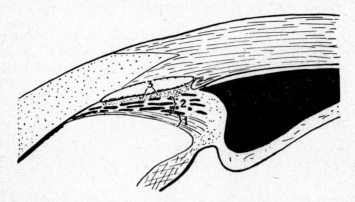

FIG. 16. Schematic drawing of the chamber angle in primates, showing the three distinct parts of the trabecular meshwork. 1, uveal meshwork; 2, trabeculum corneosclerale; 3, inner wall, or pore area. (Rohen, *Structure of the Eye*.)

lamella of the uveal meshwork is attached anteriorly to the corneal tissue at the end of Descemet's membrane, constituting *Schwalbe's ring*,* and posteriorly to the meridional fibres of the ciliary muscle. A pull of these fibres may well operate to open up the meshes of this trabecular structure, thereby favouring passage of fluid. The third portion, revealed by the electron-microscope, immediately adjacent to the canal, is called the *pore area* or *pore tissue* and is made up of endothelial cells with an intervening ground substance. Thus the inner wall of Schlemm's canal, facing the trabecular meshwork, consists of a double layer of endothelial cells, which are closely connected by tight junctions, whilst beneath this is the pore-tissue or what Rohen (1969) calls the "cribriform meshwork".

THE DRAINAGE PROCESS

Experimental Study. In 1921 Seidel established beyond reasonable doubt that the aqueous humour was continually formed and drained away. He inserted a needle, connected to a reservoir containing a solution of a dye, into the anterior chamber of a rabbit (Fig. 17). By lowering the reservoir below the intraocular pressure a little aqueous humour was drawn into the needle and then, when the reservoir was raised, this fluid, mixed with dye, was returned to the anterior chamber. Soon the dye mixed with the rest of the fluid. Seidel observed that the dye stained the blood vessels in the surface of the globe—episcleral vessels—indicating that aqueous humour was being carried into the venous system. This

* *Iris processes* are made up of an occasional lamina of the uveal meshwork that crosses posteriorly to attach to the root of the iris.

passage out would only occur, however, if the pressure in the eye, determined by the height of the reservoir, was at or above 15 mm Hg; when the pressure was reduced below this it was presumed that drainage of aqueous humour ceased. Since the pressure within the eye is normally above 15 mm Hg, we may conclude that fluid is being continuously lost to the blood; moreover, the loss presumably depends on the existence of a pressure-gradient between the aqueous humour and the venous system. Since the episcleral vessels are derived from the anterior ciliary system, we may conclude that the dye was not being absorbed from the iris, the venous return from which is by way of the vortex veins which were un-coloured in these experiments. More recent studies, in which isotope-labelled material is introduced into the aqueous humour and the venous blood collected from both ciliary and vortex veins, have confirmed that at least 99 per cent of drainage is by the anterior route (Bill, 1962).

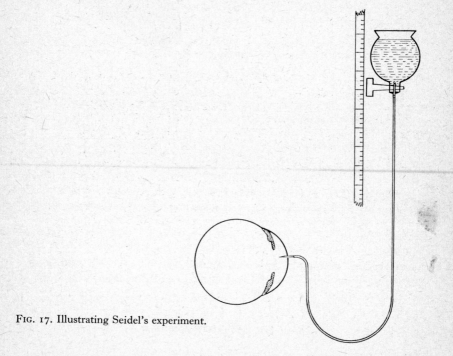

Fig. 17. Illustrating Seidel's experiment.

Aqueous Veins. Confirmation of Seidel's view was provided in a striking manner by Ascher in 1942; when examining the superficial vessels of the globe with the slit-lamp microscope, he observed what appeared to be empty veins but they turned out to be full of aqueous humour, and he called them *aqueous veins*. Usually one of these veins could be followed till it joined a blood vessel, in which event the contents of the two vessels, aqueous humour and blood, did not mix immediately but often ran in parallel streams forming a *laminated* aqueous vein, as illustrated schematically in Fig. 18. If the blood-vein beyond the junction was compressed, one of two things happened; either the blood drove the aqueous humour out of its channel or the aqueous humour drove the blood out.

The latter situation was described as the *positive glass-rod phenomenon*, whilst the former, with the blood driving aqueous humour out, was called the *negative glass-rod phenomenon*. These superficial vessels, full of aqueous humour, are probably derived from the deeper vessels of the intrascleral plexus, the pressure-relationships in the collectors from Schlemm's canal being such as to favour displacement of blood throughout the course of the fluid to the surface of the globe as illustrated by Fig. 19.

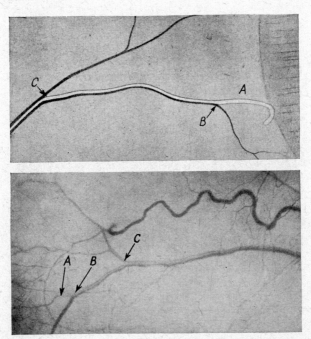

FIG. 18. Typical aqueous veins. *Above:* Sketch made by Ascher (*Amer. J. Ophthal.*) in his original description. *Below:* Untouched photo taken by Thomassen and Perkins. The aqueous vein emerges from the deeper tissue and is visible at A. At B it joins a blood-vein so that the latter has two strata. At C the stratified vessel joins another vein to give three laminae.

Goldmann proved conclusively that these vessels contained aqueous humour by injecting fluorescein intravenously. This substance passes very slowly from blood into aqueous humour so that one would not expect any vessel containing aqueous humour to fluoresce immediately after the injection, whilst vessels full of blood or plasma should so so; in fact only the blood vessels did fluoresce. Later, Ashton identified an aqueous vein in a human eye that was about to be enucleated; a wire was tied round the vein and, after enucleation, the canal of Schlemm was injected with Neoprene. After digesting the tissue away a cast of the canal and its connecting vessels remained, and examination showed that the aqueous vein did, indeed, take origin from the canal of Schlemm. Subsequent studies of Jocson & Sears (1968), employing a silicone vulcanizing fluid, have shown that the connections of the canal of Schlemm with the vascular system are of two main kinds, large (aqueous veins) to the episcleral vessels and much finer connections to the intrascleral vessels. In addition, Rohen (1969) has described

Fig. 19. Ascher's view of the origin
of an aqueous vein.

bridge-like channels running parallel with Schlemm's canal, which branch off
and later rejoin it; these make connections with the intrascleral plexus.*

PHYSIOLOGY OF THE VASCULAR SYSTEM

The vascular circulation of the eye has attracted a great deal of interest because
of its close relationship to the formation and drainage of aqueous humour. Space
will not permit a detailed description of this aspect of ocular physiology, and we
must be content with a few salient points.

Pulses. With the ophthalmoscope an arterial pulse may be observed in the
retinal vessels; if the intraocular pressure is increased by compressing the globe
the pulse may be exaggerated, since the collapse of the artery during diastole
becomes greater; if the pressure is raised sufficiently the pulse will cease, at
which point the pressure in the eye is equal to the systolic pressure in the artery
from which the retinal artery is derived, namely, the ophthalmic artery. The
retinal veins also show a pulse; it is best seen where the large veins lie on the
optic disk; it appears as a sudden emptying of this portion of the vein, progressing
from the central end towards the periphery followed by a pronounced dilatation,
beginning at the periphery and passing centrally. The pulse is not simply a
reflexion of events taking place in the right auricle, and is closely related to ventri-
cular systole. The rise in intraocular pressure associated with systole will tend to
compress the veins; and a collapse of the latter will take place where the pressure
is least, i.e. most centrally. The subsequent refilling of the vein will occur in
diastole, or late in systole, and will proceed in the reverse direction. The exact
time-relationships of the pulse were established cinematographically by Serr.

Pressures. These were studied experimentally by Seidel and later by Duke-
Elder; according to the latter author the following values (mm Hg) are associated
with an intraocular pressure of 20 mm Hg:

Ophthalmic artery	99·5	Retinal vein	22·0
Retinal arteries	75·5	Intrascleral vein	21·5
		Episcleral vein	13·0

* The outer wall of Schlemm's canal has been described by Rohen & Rentch (1969);
the tissue immediately adjacent to the endothelium is differentiated from the sclera proper;
the regions of confluence of the canal with the collector ducts may be regarded as out-
pouchings of this outer wall, whilst the septa that partially occlude the canal at various
points are to be regarded as extensions of the outer wall into the lumen. In old eyes there
is an accumulation of amorphous material in the outer wall, but its significance for drainage
is doubtful.

Of most importance are the values in the anterior ciliary or episcleral veins; these are probably not independent of the intraocular pressure, however; thus in human subjects Weigelin & Löhlein found the following empirical relationship: $P_{ev} = 0.48 \times P_{oc} + 3.1$, where P_{ev} is the episcleral venous pressure and P_{oc} the intraocular pressure. In animals, too, Macri (1961) found a linear relationship between the two.

THE INTRAOCULAR PRESSURE

On the basis of Seidel's studies, we may expect the flow of aqueous humour to be determined by the difference of pressure between the fluid within the eye—the *intraocular pressure*—and the blood within the episcleral venous system into which the fluid must ultimately flow to reach the surface of the globe. Thus, according to Seidel's formulation we should have:

$$\text{Flow} = (P_{oc} - P_{ev})/R$$

if the flow follows Poiseuille's Law, R being a resistance term determined by the frictional resistance through the trabecular meshwork and along the various vessels through which the fluid flows. To understand the dynamics of flow of aqueous humour, then, we must consider in some detail the nature of the intraocular pressure, P_{oc}, and its relationships with the other factors in the above equation, namely, the episcleral venous pressure, P_{ev}, and the resistance term, R.

Measurement of the Intraocular Pressure

On inserting a hypodermic needle into the anterior chamber the aqueous humour flows out because the pressure within is greater than atmospheric; we may define the intraocular pressure as that pressure required just to prevent the loss of fluid. Manometric methods have been developed that permit the measurement of this pressure with a minimal loss of fluid, this being necessary since, as we shall see, loss of fluid *per se* may upset the normal physiology of the eye.

Manometry. The general principle of the manometric methods employed for measuring the intraocular pressure is illustrated by Fig. 20. The fluid-filled chamber is connected to a reservoir, R, which is of variable height, and also to the hypodermic needle. The end of the chamber is covered by a membrane, and it is essentially the movements of the membrane, caused by changes in pressure, that are recorded. If the membrane consists of a latex skin, it may be made to move a small mirror in contact with it, so that movements of the membrane can be magnified into movements of a spot of light reflected from the mirror. Alternatively, the membrane may be made of metal and act as the plate of a condenser, as in the Sanborn electromanometer, or it may be connected to a transducer valve, i.e. a valve whose anode may be moved from outside, so that a change in its position is converted into a change of voltage. The reservoir serves to fill the system with a saline solution and may be employed to calibrate the system; thus, with the needle stopped and the tap open, raising and lowering the reservoir will establish known pressures in the chamber.

Tonometry. For studies on man, the introduction of a needle into the anterior chamber is rarely permissible, so that various *tonometers* have been developed permitting an indirect measure of the pressure within the eye. With the *impression*

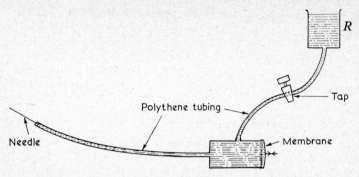

FIG. 20. Illustrating the general principle on which manometric measurement of the intraocular pressure is based. The reservoir serves to fill the system with saline. When the needle is in the anterior chamber, and the tap is closed, changes in pressure cause movements of the membrane.

type of tonometer, such as the Schiøtz instrument, the depth to which a weighted plunger applied to the cornea sinks into the eye is measured, whilst with the *applanation tonometer* the area of flattening of the cornea, when a metal surface is applied with a controlled force, is measured. The two principles are illustrated schematically in Fig. 21; the greater the intraocular pressure the smaller will be the depth of impression, or area of applanation.

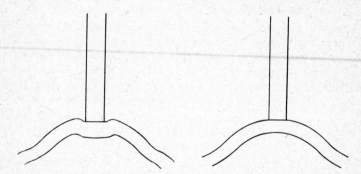

FIG. 21. Illustrating the principles of impression tonometry (*left*) and applanation tonometry (*right*).

Schiøtz Tonometer

So widespread is the use of this instrument, or its electronic adaptations, that it may profitably be described in some detail. An actual instrument is illustrated in Fig. 22, whilst the mechanical features are shown schematically in Fig. 23. The footplate, F, is curved to fit the average curvature of the human cornea. The weighted plunger, P, passes through the footplate; movements of this plunger operate on the hammer, H, which converts vertical movements into readings of the pointer on the scale. This scale reads from 0 to 20, the greater the indentation the lower the intraocular pressure and the *larger* the scale-reading. The frame, Y, serves to hold the instrument upright on the eye, and because the footplate can move freely within the cylindrical part of the frame, the actual weight resting on the eye is that of the plunger, footplate, hammer and scale (17·5 g), the frame being held by the observer. With the smallest weight on the plunger, the total effective thrust on the eye amounts to 5·5 g, and the readings made with this are referred to as the "5·5 g readings". By using heavier weights the thrust may be increased to 7·5 or 10 g, thereby allowing the instrument to give reasonable scale-readings at higher levels of intraocular pressure.

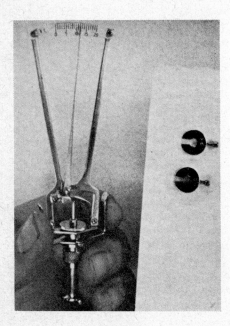
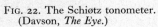

FIG. 22. The Schiøtz tonometer. (Davson, *The Eye.*)

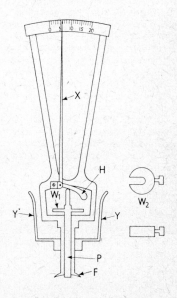

FIG. 23. Diagrammatic drawing of the Schiøtz tonometer. (Davson, *The Eye.*)

Calibration. To calibrate the tonometer, the scale-reading, R, must be converted to the actual pressure within the eye before the tonometer was applied, P_o. The detailed procedure has turned out to be very complex and not completely satisfying*; it is impossible to enter into details of the procedures here, so we must content ourselves with the general principles. When the tonometer is on the eye, the pressure within it obviously rises because of the weight of the instrument; and the pressure within the eye with the tonometer on it is referred to as P_t. P_t may be related to the scale-reading, R, by taking an excised eye, and inserting a needle connected to a reservoir and manometer into the anterior chamber. Different pressures may be established in the eye, and tonometer readings corresponding to these may be made, Friedenwald found the relationship:

$$W/P_t = a + bR$$

so that with a and b determined experimentally we have the relationship between P_t and scale-reading, R. To convert P_t to P_o, we require to know by how much the indentation, corresponding to the scale-reading, R, has raised the pressure; and

---

* The literature on calibration scales and ocular rigidity is quite large and cannot be summarized here; some key references are Friedenwald, 1957; McBain, 1960; Drance, 1960; Perkins & Gloster, 1957; Hetland-Eriksen, 1966).

this depends on the distensibility of the sclera, i.e. the extent to which a change in volume of the eye will raise the pressure. Thus, if we know the volume of displacement, V_c, associated with a given scale-reading, and if we can relate V_c to a change in pressure of the eye, we can finally establish a relationship between R and P_o. Friedenwald considered that there was a logarithmic relationship between pressure and volume of the globe, to give an equation:

$$\text{Log } P_2/P_1 = k(V_2 - V_1)$$

whence he defined a *rigidity coefficient*, k, characteristic of a given eye. Experimentally he measured the volumes of displacement, V_c, for different scale-readings, and making use of an average rigidity coefficient he was able to construct a table relating scale-readings to values of P_o, the intraocular pressure corresponding to the scale-reading. As subsequent work has shown (Perkins & Gloster; Macri, Wanko & Grimes; McBain), the weakness in this calibration is the assumption that the rigidity coefficient is a constant characteristic of each eye, i.e. that it is independent of pressure; in fact values ranging from 0·003 to 0·036 could be obtained in a given eye according to the intraocular pressure. Moreover, there is no doubt that the distensibility of the globe varies from one individual to another, so that unless this can be allowed for, it is not very sound to use a single calibration scale.*

Because of these difficulties in accurate calibration, the Schiøtz instrument is slowly being superseded by an applanation type of manometer, since with this instrument the deformation of the eye during applanation is very small, with the result that calibration is independent of scleral extensibility (Goldmann, 1955; Armaly, 1960).

APPLANATION TONOMETER

This measures, in effect, the force required to produce a fixed degree of applanation when a flat surface is pressed against the cornea. The earliest instrument (Maklow) consisted of a cylindrical piece of metal with a flat base weighing 5–15 g. A drop of a concentrated dye-solution was spread over the cornea, the instrument was allowed to rest a moment on the corneal surface and a print of the area of contact between cornea and metal was obtained by placing the foot of the instrument on a piece of paper. Goldmann's (1955) instrument is a refinement on this procedure, in which the degree of applanation is kept fixed and the instrument indicates the force required to produce this. There is no doubt that the calibration of this instrument involves far fewer uncertainties than does that of the Schiøtz indentation type. Other applanation instruments are the Mackay-Marg (1959) tonometer and the pneumatic applanation tonometer of Durham *et al.* (1965).

Because of the superiority of the Goldmann applanation tonometer to the indentation type, it seems reasonable to calibrate the Schiøtz tonometer by measuring pressure on the same eye with both instruments successively, taking care that the head is in the same position for both measurements. The most recent calibration carried out in this way has shown that the pressure deduced

---

* If the pressure is measured on the same eye with different weights on the tonometer, the same result should be read off the calibration scales. If this did not happen it was attributed to the circumstance that the ocular rigidity was abnormal in the eye being studied and Friedenwald, and later Moses & Becker, actually devised a nomogram and tables from which the ocular rigidity could be estimated from paired readings on the same eye using different weights. It will be clear, however, that the scales must be rigidly consistent within themselves if deviations from them are to be interpreted as the correlates of the mechanical properties of the eye rather than as errors in the scales themselves. Subsequent studies summarized by McBain (1960) have shown that they were not sufficiently consistent to warrant their use in this way.

from the Schiøtz measurement, using Friedenwald's calibration table, is lower than that measured by the Goldmann instrument (Anderson & Grant, 1970).

The Normal Intraocular Pressure

Clearly, the magnitude of the normal average intraocular pressure in man will depend on the validity of the calibration scale employed; early studies, using Schiøtz' original calibration, gave mean values in the neighbourhood of 20 mm Hg, but later studies in which more modern calibration scales have been employed, e.g. that of Leydhecker *et al.*, or in which a more accurate applanation technique was used (Goldmann), indicate a lower mean value in the range 15–16 mm Hg. In general, Leydhecker *et al.*'s figures, based on 13,861 apparently healthy eyes, suggest that 95·5 per cent of all healthy eyes have a pressure lying between 10·5 and 20·5 mm Hg. A pressure above 20·5 mm Hg may thus be a sign of abnormality. Between the ages of 10 and 70 there is no significant change in mean intraocular pressure, and no difference between the sexes is found. In newborn infants likewise the pressure falls within the normal adult range, the mean value found by Hörven being 16·5 mm Hg with limits of 10·9 and 24·2 mm Hg. In experimental animals the intraocular pressure may well be of the same order; thus in cats Davson & Spaziani reported a mean value of 16·5 mm Hg, whilst Langham found 20 mm Hg in this animal and 20·5 mm Hg in the rabbit.*

Diurnal Variation. In man the intraocular pressure shows a slight but significant diurnal variation, being highest in the early morning and lowest at about midnight. The range of fluctuation is small, 2–5 mm Hg. According to Ericson (1958) the cause is a diurnal variation in the rate of secretion of fluid, as measured by the suction-cup technique (p. 41).

Mechanical Factors Affecting the Intraocular Pressure

The simple relationship between flow-rate of aqueous humour, F, and the pressure-drop, mentioned above, may be rewritten:

$$P_{oc} = F \times R \times t + P_{ev}$$

where P_{oc} is the intraocular pressure, F is the volume of aqueous humour drained during time t, R is the resistance and P_{ev} is the episcleral venous pressure. With the aid of this relationship some of the factors determining the intraocular pressure may be easily appreciated. Thus an increase in resistance to flow will increase the intraocular pressure, provided that the rate of formation of fluid does not decrease in proportion. The factors determining the rate of production of fluid are not completely understood, but it seems safe to assume that this is independent of intraocular pressure over a fairly wide range, so that we shall not be far wrong if we conclude that intraocular pressure varies inversely with resistance to flow. Alterations in episcleral venous pressure, P_{ev}, may likewise be expected to be reflected in predictable alterations in intraocular pressure.

Tonographic Measurement of Resistance. The essential principle on which this measurement is based is to place a weight on the cornea; this raises the intraocular pressure and causes an increased flow of aqueous humour out

---

* Manometric methods usually demand the use of a general anaesthetic, so that the figures given may not be true of the normal unanaesthetized animal; according to Sears (1960) and Kupfer (1961), the normal unanaesthetized value for rabbits is 19 mm Hg; with urethane it was 16·8 mm Hg, with paraldehyde, 20·4 and aprobarbital, 17 mm Hg.

of the eye. Because of this increased flow, the intraocular pressure falls eventually back to its original value, and the rate of return enables the computation of the resistance to flow. Thus:

$$F_o = \frac{P_{oc}-P_{ev}}{R}(t_2-t_1)$$

F_o being the normal volume flow, without the weight on the eye, during the period t_2-t_1.

With the weight on the eye we have:

$$F_t = \frac{\frac{1}{2}(P_{t_2}-P_{t_1})-P_{ev}}{R}(t_2-t_1)$$

Where the average intraocular pressure during the period t_2-t_1 with the weight on the eye is taken as: $\frac{1}{2}(P_{t_2}-P_{t_1})$.

The extra flow,

$$F_t-F_o = \left\{\frac{\frac{1}{2}(P_{t_2}-P_{t_1})-P_{ev}}{R} - \frac{P_{oc}-P_{ev}}{R}\right\}(t_2-t_1)$$

$$= \{\tfrac{1}{2}(P_{t_2}-P_{t_1})-P_{oc}\}\frac{t_2-t_1}{R}$$

$$\text{or } \Delta F = \frac{t_2-t_1}{R}.\Delta P$$

$$\text{or } \Delta F = (t_2-t_1).C.\Delta P$$

where C is the reciprocal of R and may be called the *facility of outflow.*

Experimentally the weight is applied to the eye in the form of an electronic tonometer so that it serves also to measure the intraocular pressure when appropriately calibrated. Since, moreover, the calibration of the tonometer had already given the change in volume of the eye associated with changes in P_t, ΔF was also known. Hence by measuring the change in P_t, ΔF was found and these two quantities could be inserted into the equation to give C, the facility of outflow. The average value found by Grant, who first described the technique, was 0.24 mm^3 . min^{-1} . mm Hg^{-1}, i.e. the flow when the difference of pressure was 1 mm Hg was 0.24 mm^3 per minute. Measurements on excised human eyes gave comparable results; in this case fluid was passed into the eye from a reservoir and the rate of movement measured at different pressures.

Morphological Basis of Drainage. The connections between the canal of Schlemm and the vascular system have been well established since the work of Maggiore; the mechanism by which fluid passes from the trabecular meshwork into the canal is not so clear. McEwan (1958) showed on hydrodynamic principles that the flow could be accounted for on the basis of a series of pores in the inner wall of the canal, and recent electron-microscopical studies have lent support to this view. Holmberg (1959) made serial sections of the trabecular tissue immediately adjacent to the canal and found channels apparently ending in vacuoles enclosed within the endothelial cells of the canal, and he considered that these vacuoles opened into the canal. Subsequent electron-microscopical studies of Kayes and Tripathi seem to confirm that the vacuoles, which may be as large as

2–5 μ and appear as spaces bounded by a thin layer of endothelial cytoplasm, open into both Schlemm's canal and the trabecular spaces. By studying serial sections it was possible to show that a vacuole made connections with both spaces and thus acted as a channel, there being some 300–400 per millimetre of canal (Fig. 24 Pl. II). We may note that Tripathi (1971) has suggested that the channel is just one stage in a cycle of vacuolization, as illustrated by Fig. 25, in which event it is possible to envisage changes in the resistance to outflow in accordance with changes in the state of vacuolization. By employing the scanning electron micro-scope, the surface appearance of the wall of the canal can be examined; most recently Bill has employed this device and shown that bulges in the surface correspond to the nuclei and associated vacuoles of the endothelial cells; some of theses bulges and holes were 0·3–2 μ in diameter; if erythrocytes had been injected into the anterior chamber, these could be seen blocking the holes.

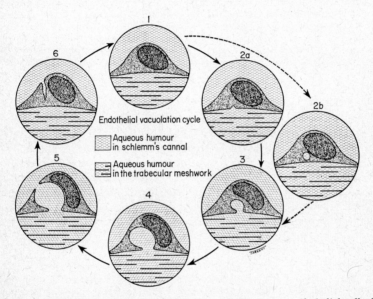

FIG. 25. Illustrating hypothetical cycle of vacuolation in an endothelial cell of Schlemm's canal resulting in creation of a temporary channel communicating between the canal and the trabecular meshwork. (Tripathi. *Exp. Eye Res.*)

EFFECT OF HYALURONIDASE

Bárány & Scotchbrook (1954) showed that the facility of outflow through the perfused rabbit eye was increased by adding hyaluronidase to the medium, suggesting that a factor in resistance to flow was the mucopolysaccharide matrix to the trabecular meshwork, but it would seem from the study of Melton & De Ville (1960) that this does not happen in the cat and dog. In this connection the possible role of plasmin, the enzyme that is formed from the precursor, plas-minogen, and that hydrolyses fibrin, is of some interest. This occurs in tissues containing narrow drainage channels, such as the kidney or the tear drainage mechanism, and histochemically Pandolfi & Kwaan (1967) have demonstrated that the endothelium of the canal of Schlemm has considerable fibrinolytic activity, presumably because it can activate plasminogen to plasmin. When monkey eyes were perfused with porcine plasmin a marked decrease in resistance occurred, but not in rabbit eyes (Pandolfi, 1967).

GLAUCOMATOUS CHANGES IN RESISTANCE

It is generally agreed that the rise in intraocular pressure in glaucoma is due to an increase in resistance to outflow; this may be due to obvious mechanical factors, as in the so-called angle-closure glaucoma, secondary glaucomas due to luxation of the lens, etc. The condition of "open-angle glaucoma", where an obvious mechanical interference with drainage is not present, doubtless involves more subtle changes in the resistance to flow through the trabecular meshwork and Schlemm's canal. In 1960 Theobald summarized the histological findings in this condition; these include replacement of the trabeculae with amorphous material, degeneration of collagenous tissue with proliferation of the epithelium of Schlemm's canal. Rohen & Straub (1967) have described a "hyalinization" of the corneo-scleral trabeculum, whilst Tripathi (1969) states that the number of vacuoles in the trabecular wall of Schlemm's canal is markedly reduced in chronic simple glaucoma. Rather similar changes in the cribriform meshwork, or pore tissue, are observed in senile eyes (Rohen & Lütjen, 1968). Yamashida & Rosen (1965) examined in the electron microscope the trabecular meshwork and canal of Schlemm of an eye excised with open-angle glaucoma secondary to haemorrhages; the endothelial meshwork was markedly condensed and pores in the canal were absent.

ESCAPE OF PARTICULATE MATTER

Particulate matter of diameter around $1 \cdot 5 - 3 \cdot 0 \ \mu$, when injected into the anterior chamber, escapes, presumably by way of Schlemm's canal; and this supports the notion of large pores in its wall; even erythrocytes, introduced into the anterior chamber with care not to alter the intraocular pressure, disappear and may be recovered in the blood when appropriately labelled; the rate of escape is not so rapid as that of serum albumin, showing that there is some obstruction to passage; moreover, if the cells are left in the anterior chamber and the eye is allowed to adopt its true pressure, this rises due to a partial obstruction of the outflow routes (Bill, 1968).*

SUBSIDIARY DRAINAGE MECHANISM

Bill (1964) deduced from his studies on the passage of albumin from blood into the uveal tissue that there must be some route for drainage additional to that provided by Schlemm's canal. By perfusing the anterior chamber of the monkey's eye with a mock aqueous humour containing labelled albumin, he was able to divide the losses of protein into two components, the one appearing in the blood and doubtless having] been drained through Schlemm's canal, and the other that remained in the uveal tissue, having presumably passed directly into the tissue of the iris; the average "uveo-scleral drainage" was $0 \cdot 44 \ \mu$l/min compared with a conventional drainage of $0 \cdot 80 \ \mu$l/min; thus the rate of formation of aqueous humour, the sum of these, was $1 \cdot 24 \ \mu$l/min. In the cat, the rate of secretion was $14 \cdot 4 \ \mu$l/min and only 3 per cent of this passed by uveoscleral routes. The passage of material was followed by Fowlks & Havener, who identified it in the tissue

---

* The actual locus of the main resistance to flow has been implicitly accepted as the trabecular meshwork and inner surface of the canal of Schlemm; in monkeys, Perkins found the pressures in the anterior chamber and the canal of Schlemm to be about equal, suggesting little or no resistance to flow into the canal, so that resistance was due to flow along the collector channels. Trabeculotomy, also, failed to affect resistance appreciably in monkeys (Edwards, Hallman & Perkins, 1967), although Grant (1958, 1963) found the opposite in the rabbit and man. The effects of ciliary muscle contraction seem to be so definite, however, that it is difficult to believe that the scleral meshwork is not the main site of resistance.

spaces of the ciliary body posterior to the processes, and it was concluded that it entered the ciliary cleft whence it passed into the perivascular spaces of the choroid and ciliary body, and thence to the suprachoroid.

Venous Pressure. In so far as the aqueous humour must drain into the episcleral venous system, either by way of the intrascleral plexus or more directly along the aqueous veins, the intraocular pressure will vary directly with the venous pressure in accordance with the simple equation given above. Certainly in man there is a good correlation between intraocular pressure and P_{ev}, whilst in cats Macri showed the same thing when he varied the intraocular pressure experimentally. Acute changes of venous pressure will influence the intraocular pressure in a twofold manner. First, by affecting drainage as just indicated, and second, by a direct transmission of the changed venous pressure across the easily distensible walls of the intraocular veins. The expansion of the veins will be eventually compensated by a loss of aqueous humour from the eye, so that the rise due to this cause may be more or less completely compensated, but the effect on drainage should last as long as P_{ev} is elevated. Experimentally the effects of changed venous pressure may be demonstrated by administering amyl nitrite; the peripheral vasodilatation causes a general increase in venous pressure and leads to a considerable rise in intraocular pressure in spite of a lowered arterial pressure. Again, ligation of the vortex and anterior ciliary veins causes large rises in intraocular pressure, as we should expect. Finally, destruction of the aqueous veins, by restricting the outlets of aqueous humour, may cause quite considerable rises in intraocular pressure in rabbits, according to Huggert.

Arterial Pressure. The arterial pressure can influence the intraocular pressure through its tendency to expand the intraocular arteries; since these are not easily distensible, however, relatively large changes in arterial pressure will be required to produce measurable changes in intraocular pressure. That the arterial pressure does affect the intraocular pressure in this way is manifest in the *pulse* shown in records of the intraocular pressure; the amplitude of this pulse is small, about 1 mm Hg, and coincides with cardiac systole. Thus a change of about 20 mm Hg, which is the pulse-pressure in an intraocular artery according to Duke-Elder, causes a change of only 1 mm Hg in the intraocular pressure.

In the rabbit, ligation of one common carotid artery lowers the intracranial arterial pressure on the side of the ligation, whilst the pressure on the other side rises above normal; hence we may study the effects of lowered arterial pressure on one eye, using the other as a control. Wessely in 1908 found a small lowering of the intraocular pressure on the ligated side and Bárány found that after 24 hours the pressure had returned to normal. The immediate effect of the ligation is undoubtedly the result of the fall in arterial pressure which leads to a reduced volume of blood in the eye. As Fig. 26 shows, there is a tendency to compensate for this since alternate clamping and unclamping of the artery leads to an overshoot on release of the clamp, suggesting that there has been some vasodilatation which would partly compensate for the reduced flow caused by the fall in pressure. The compensation is not complete, however, since Linnér found a 19 per cent decrease in the rate at which blood escaped from a cut vortex vein on the side of the ligature.

Rate of Flow of Aqueous Humour. Other things being equal, we may expect changes in rate of secretion of fluid to produce changes in intraocular pressure in the same sense, but the effects are not likely to be large. Thus, if the intraocular pressure, P_{oc}, is 16 mm Hg, and the episcleral venous pressure, P_{ev}, is 12 mm Hg, the pressure drop, $P_{oc}-P_{ev}$, is 4 mm Hg. If the rate of flow is doubled the pressure-drop must be doubled too, in which case it becomes 8 mm Hg, and if P_{ev} remains the same this means that the intraocular pressure, P_{oc}, rises from 16 to 20 mm Hg. Thus doubling the rate of secretion causes only a 25 per cent increase in intraocular pressure. In practice it is not easy to vary the rate of secretion; as we shall see, the carbonic anhydrase inhibitor, Diamox, reduces the rate and this is accompanied by a fall in intraocular pressure; the same is true of digitalis compounds, according to Simon, Bonting & Hawkins.

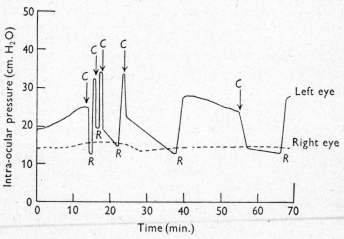

FIG. 26. Intraocular pressure of the rabbit. At points marked C the left common carotid artery was clamped, and at points marked R it was released. (Davson & Matchett, *J. Physiol.*)

By injecting saline continuously into the anterior chamber we may experimentally increase the flow of fluid through the eye and test the applicability of the simple equation relating flow-rate to pressure-drop; Langham found that the equation did not hold, the rise in pressure for a given increase in flow being smaller than predicted, and his results suggested some homeostatic mechanism that decreased the resistance to flow as the intraocular pressure rose. Section of the cervical sympathetic, or ligation of the common carotid, caused the relationship between flow and pressure to become linear, i.e. to follow the simple equation:

$$\text{Flow} \times \text{Resistance} = \text{Pressure Drop}$$

and this suggested that the adaptation was largely vascular. The situation is apparently complicated, however, by the effects of pressure on the rate of production of aqueous humour. Thus Bill & Bárány (1966) found a linear relation between flow-rate and pressure, but they also showed that the rate of secretion diminished with higher pressures, so that the linear relation found between flow-rate out of the cannula and pressure means that the actual flow through Schlemm's canal was not linearly related to pressure, a rise in pressure producing too small a flow, and thus suggesting an increased resistance at the higher pressure.

THE AQUEOUS HUMOUR

It has become clear that the intraocular pressure depends primarily on the rate of secretion of aqueous humour, so that factors influencing this are of vital importance in understanding the physiology of the intraocular pressure. We may therefore pass to a consideration of the chemistry of the fluid and the dynamics of exchanges between it and the blood. These aspects, although apparently academic, are important for the full understanding of the experimental approaches to measuring the rate of secretion of the fluid and to controlling this.

Chemical Composition

The most obvious chemical difference between aqueous humour and blood plasma is to be found in the protein contents of the two fluids; in the plasma it is of the order of 6–7 g/100 ml, whereas in the aqueous humour it is only 5–15 mg/100 ml in man and some 50 mg/100 ml in the rabbit. By concentrating the aqueous humour and then causing the proteins to move in an electrical field—*electrophoresis*—the individual proteins may be separated and identified. Such studies have shown that, although the absolute concentrations are so low, all the plasma proteins are present in the aqueous humour, so that it is reasonable to assume that they are derived from the plasma rather than synthesized during the secretory process.

TABLE I

Chemical Composition of Aqueous Humour and Blood Plasma of the Rabbit
(after Davson, 1969)

	Aqueous Humour	Plasma
Na	143·5	151·5
K	5·25	5·5
Ca	1·7	2·6
Mg	0·78	1·0
Cl	109·5	108·0
HCO_3	33·6	27·4
Lactate	7·4	4·3
Pyruvate	0·66	0·22
Ascorbate	0·96	0·02
Urea	7·0	9·1
Reducing value (as glucose)	6·9	8·3
Amino acids	0·17	0·12

Concentrations are expressed as millimoles per kilogramme of water.

Non-Colloidal Constituents. As Table 1 shows, the crystalloidal composition of the aqueous humour is similar in its broad outlines to that of plasma, but there are some striking differences; thus the concentrations of ascorbate (vitamin C), pyruvate and lactate are much higher than in plasma, whilst those of urea and glucose are much less.

GIBBS-DONNAN EQUILIBRIUM

The similarities between the two fluids led some earlier investigators to assume that the aqueous humour was derived from the blood plasma by a process of filtration from the capillaries of the ciliary body and iris; i.e. it was assumed that the aqueous humour was similar in origin and composition to the extracellular fluid of tissues such as skeletal muscle. If this were true, however, the ionic distributions would have to accord with the requirements of the Gibbs-Donnan equilibrium, according to which the concentrations of positive ions, e.g. Na^+ and K^+, would be greater in the plasma than in the aqueous humour, the ratio: $[Na]_{Aq}/[Na]_{Pl}$ being less than unity, whilst the reverse would hold with negative ions such as Cl^- and HCO_3^-. Moreover, the actual values of the ratios: *Concn. in Aqueous Humour/Concn. in Plasma* would have to conform to those found experimentally when plasma was dialysed against its own ultrafiltrate. In fact, however, as Table 2 shows, the ratios show considerable deviations from what is demanded

TABLE 2

Values of the Two Distribution Ratios; Concn. in Aqueous/Concn. in Plasma *and* Concn. in Dialysate/Concn. in Plasma. (*Davson*, Physiology of the Ocular and Cerebrospinal Fluids.)

	Concn. in Aqueous / Concn. in Plasma	Concn. in Dialysate / Concn. in Plasma
Na	0·96	0·945
K	0·955	0·96
Mg	0·78	0·80
Ca	0·58	0·65
Cl	1·015	1·04
HCO₃	1·26	1·04
Urea	0·87	1·00

of a plasma filtrate, and the deviations are not attributable to experimental errors. Thus with a plasma filtrate the ratio: $[Na]_{Aq}/[Na]_{Pl}$ should be 0·945 whilst it is in fact 0·96, indicating that there is an excess of 1·5 per cent of sodium in the aqueous humour. More striking is the discrepancy of the bicarbonate ratio, which in the rabbit is 1·26 compared with a value of 1·04 demanded by the Gibbs-Donnan equilibrium.

GLUCOSE

The concentration of glucose in the aqueous humour is less than in plasma, and the concentration in the vitreous body is even lower; these differences have been ascribed to consumption within the eye by the lens, cornea and retina; this utilization may be simply demonstrated by incubating the excised eye at 37°C; within two hours all the glucose has disappeared (Bito & Salvador, 1970). It is possible that a further factor is a low concentration in the primary secretion, since the concentration of glucose in the aqueous humour of cataractous eyes is not significantly lower than normal (Pohjola, 1966).

AMINO ACIDS

The relative concentrations of amino acid in the plasma aqueous humour and vitreous body depend on the particular amino acid, so that the concentration in aqueous humour may be less than, equal to, or greater than that in plasma (Fig. 27); the concentration in the vitreous body is always very much less than in either of the other fluids (Reddy, Rosenberg & Kinsey, 1961; Bito *et al.*,

1965). The low values in the vitreous body are not due to consumption within the eye since a non-metabolized acid, cycloleucine, shows a steady-state concentration in the vitreous body only 10 per cent of that in plasma after intravenous injection (Reddy & Kinsey, 1963), so that an active removal of amino acid by the retina has been postulated. When the dead eye is incubated, there is a large rise in concentration of amino acids in the vitreous body, due to escape from the lens, which normally accumulates these from the aqueous humour (Bito & Salvador, 1970).

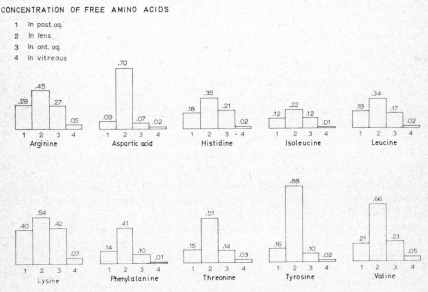

CONCENTRATION OF FREE AMINO ACIDS

1 In post. aq.
2 In lens
3 In ant. aq.
4 In vitreous

FIG. 27. Relative concentrations of free amino acids in lens, aqueous and vitreous humours of the rabbit eye. (After Reddy & Kinsey. *Invest. Ophthal.*)

ASCORBIC ACID

The concentration of ascorbic acid in the rabbit's aqueous humour is some 18 times that in the plasma; the degree of accumulation varies with the species, being negligible in the rat and galago, but high in man, monkey, horse, guinea pig, sheep, and pig. As Kinsey (1947) showed, when the plasma level is raised, the concentration in the aqueous humour rises until a "saturation level" is reached in the plasma, of about 3 mg/100 ml when the concentration in the aqueous humour reaches some 50 mg/100 ml. Raising the plasma level beyond the saturation level causes no further rise in concentration in the aqueous humour. According to Linnér (1952) the ability of the eye to maintain this concentration of ascorbic acid depends critically on the blood-flow, so that unilateral carotid ligation causes a 17 per cent fall in concentration on the ligatured side. Hence, under non-saturation conditions, the concentration of ascorbic acid in the aqueous humour is some measure of the rate of blood-flow through the ciliary body. When the isolated ciliary body is placed in a medium containing ascorbic acid, the tissue accumulates this to concentrations considerably higher than in the medium; D-isoascorbate competed for accumulation but glucoascorbate, containing an additional CHOH-group, had no effect (Becker, 1967). The isolated retina also accumulates ascorbic acid (Heath & Fiddick, 1966); this is true of the rat, although in this species the ocular fluids contain a very low concentration.

Aqueous as a Secretion. Studies along these lines have left the modern investigator in no doubt that the fluid could not have been formed by such a simple process as filtration; in consequence the aqueous humour is called a *secretion*. What is implied by this is, essentially, that in order that it should be formed from the blood plasma, something more than a mechanical filtration process that "skims" off the proteins is necessary; the cells of the ciliary epithelium must perform work—what is called *osmotic work*—in order to produce a fluid like the aqueous humour, which is not only different from plasma but also from a filtrate of plasma or dialysate. How this happens is still a closed book; presumably the first step is the production of a plasma filtrate in the stroma of the ciliary processes; this is prevented from passing into the posterior chamber by the tightly packed cells of the ciliary epithelium. We may assume that next these cells abstract from this fluid the various ions and molecules that are ultimately secreted, and together with water they drive them into the posterior chamber. In making the selection, certain ions and molecules may be transferred across the cells to give higher concentrations in the secretion than in the original plasma filtrate, *e.g.* bicarbonate and ascorbate, whilst others are transferred at lower concentrations, e.g. calcium, urea, and so on.

Alterations of the Primary Secretion. Once formed, the composition of the fluid may well be modified by metabolic and other events taking place during its sojourn in the posterior and anterior chambers; thus the concentration of glucose could be low because the lens and cornea actually utilize it, whilst the relatively high concentrations of lactate and pyruvate could be the consequence of glycolytic activity by these structures. That this is probably true is shown by comparing the vitreous body with the aqueous humour; the vitreous body, being exposed to both lens and retina, might be expected to have an even lower concentration of glucose than that in the aqueous humour, and higher concentrations of lactate and pyruvate. In fact this is true. Moreover, if lactate and pyruvate are produced in this way, being stronger acids than carbonic acid, they will have decomposed some of the bicarbonate entering in the original secretion, so that the original concentration of bicarbonate must have been higher (and the chloride concentration lower) than that actually found, and recent studies of Kinsey & Reddy have shown that this is indeed true.

Species Variations. The high concentration of bicarbonate in the aqueous humour shown in Table 1 is not a general characteristic common to all species; thus in horse, man, cat, rabbit and guinea pig the ratios: *Concn. in Aqueous Humour/Concn. in Plasma* are 0·82, 0·93, 1·27, 1·28 and 1·35 respectively. Hence in the horse and man the concentrations of bicarbonate are *less* than what would be expected of a filtrate of plasma. Study of the chloride composition shows an inverse relationship, as we might expect, the aqueous humours with high bicarbonates having low chlorides, and *vice versa*. It would seem that the secretion of a fluid with a high concentration of bicarbonate is concerned with the buffering needs of the eye, the bicarbonate being required to neutralize the large amounts of lactic and pyruvic acids formed by lens, retina and cornea. In animals like the rabbit and guinea pig the lens occupies a large proportion of the intraocular contents and so the buffering requirements are high; by contrast, in the horse and man the lens occupies a much smaller percentage of the contents (Davson & Luck). Wide species fluctuations in the concentration of ascorbate in the aqueous humour are also found; thus in the horse, man, rabbit, dog, cat and rat the concentrations are about 25, 20, 25, 5, 1 and 0 mg/100 ml respectively, although the concentrations in the plasma are all of the same order, namely about 1 mg/100 ml. Unfortunately the significance of ascorbate in the eye is not yet understood, so that the reason why the cat, dog and rat can manage with such low concentrations, whilst the rabbit and other species require high concentrations, is not evident. The suggestion that the lens actually synthesizes ascorbate and pours it into the aqueous humour has been disproved, at any rate as a significant source of the vitamin.

THE BLOOD-AQUEOUS BARRIER

Our knowledge of the relationships between aqueous humour and plasma has been furthered by a study of the exchange of materials between the two fluids, studies of what has come to be known as the *blood-aqueous barrier*. If various substances are injected into the blood so as to establish and maintain a constant concentration in the plasma, then the rates at which the concentrations in the aqueous humour approach that of the plasma are usually very much less than occurs in such tissues of the body as skeletal muscle, liver, etc. It is because of this that we have come to speak of the "blood-aqueous barrier", meaning that substances encounter difficulty in passing from the one fluid to the other.

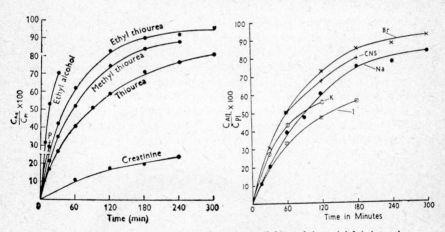

FIG. 28. Penetration of some non-electrolytes (*left*) and ions (*right*) into the aqueous humour of the rabbit. The concentration of the given substance in the plasma (C_{Pl}) was maintained constant, and the concentration in the aqueous humour (C_{Aq}) was determined at different times. (Davson, *Physiology of the Ocular and Cerebrospinal Fluids.*)

Experimental Studies

Fig. 28 illustrates the penetration of some substances into the anterior chamber of the rabbit. In general, the substances fall into three categories. First there are the very large molecules like the plasma proteins, dextrans and inulin which penetrate so slowly as to give the impression of an absolute barrier, although studies with isotopically labelled proteins leave us in no doubt that there is some, but very slow, penetration. Because the aqueous humour is constantly being drained away the concentration of these large molecules in the aqueous humour never rises very high and we may speak of a *steady-state* in which the ratio of concentrations is exceedingly low. This of course accounts for the low concentrations of plasma proteins in the aqueous humour. The second category of substance has a much smaller molecule, or may be an ion; the molecule is water-soluble, like urea, creatinine, *p*-aminohippurate, sucrose and so on. Penetration is considerably faster, usually, than that of the proteins. The third group are the lipid-soluble molecules which penetrate very rapidly; these include ethyl alcohol, various substituted thioureas, many sulphonamides. In general, the rate of penetration increases with increasing lipid-solubility, as measured by the oil-water partition coefficient. Water, although not strongly lipid-soluble, belongs to the category of rapidly penetrating substances. Again, the sugars such as glucose and galactose are insoluble in lipid solvents but they also penetrate the barrier rapidly.

Mechanism of the Barrier

We may imagine that the cells of the ciliary epithelium continually secrete a fluid, which we may call the *primary aqueous humour*, that is emptied into the posterior chamber and carried through the pupil into the anterior chamber to be finally drained away out of the eye. If a foreign substance, or the isotope of a normally occurring constituent of the aqueous humour, e.g. ^{24}Na, is injected into the blood it presumably equilibrates rapidly with the extracellular fluid in the ciliary processes, and thus gains ready access to the cells of the ciliary epithelium. If the substance penetrates these cells easily, then we may expect it to appear in the secretion produced by them relatively rapidly. The primary aqueous humour will therefore contain the substance in high concentration. This circumstance will favour penetration into the aqueous humour as a whole. On mixing with the fluid already present in the posterior chamber, the primary fluid will be diluted, so that it will take time for the aqueous humour as a whole to reach the concentration in the plasma, in the same way that if we were to add a coloured solution to a tank of water, removing fluid from the tank as fast as the coloured

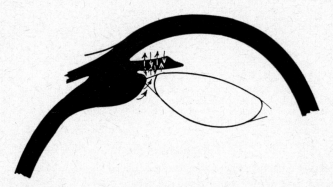

Fig. 29. Illustrating the diffusional exchanges between aqueous humour in posterior and anterior chambers on the one hand, and the blood plasma in the iris, on the other.

fluid was added, in time the concentration of dye in the tank would come up to that in the coloured solution, but the process would take time and be determined by the rate of flow.

Diffusion from Iris. While in the posterior and anterior chambers, however, the aqueous humour is subjected to diffusional exchanges with the blood in the iris, as illustrated in Fig. 29, so that if these exchanges are significant the rate at which the foreign substance comes into equilibrium with the aqueous humour is increased. There is no doubt that lipid-solubility is an important factor in determining the rapidity of these diffusional exchanges. The anterior surface of the iris is not covered by an intact epithelial or endothelial layer, so that restraint on diffusion from blood into the anterior chamber must depend on the capillaries of this tissue. According to Vegge & Ringvold (1969) the endothelial intercellular junctions are closed by zonulae occludentes a circumstance that would considerably reduce their permeability. These capillaries are in marked contrast with those of the ciliary body which have large fenestrations, i.e. attenuations of the endothelial cytoplasm that probably favour permeability; they have no zonulae occludentes between endothelial cells.*

Large Molecules

On the basis of this scheme, we may interpret the blood-aqueous barrier phenomena qualitatively. Thus large lipid-insoluble molecules such as the plasma proteins will probably be held back partially by the capillaries in the ciliary processes, so that only small amounts will be available to the epithelial cells; penetration into these cells, moreover, will be greatly restricted so that the primary secretion may contain only minimal amounts. During the fluid's sojourn in the posterior and anterior chambers some may diffuse from the iris, but these amounts will be small because of the obstructions that the large molecules will meet in the form of the epithelial covering of the posterior surface of the iris. Because of the sluggishness of all these processes, the concentration in the aqueous humour will never build up to that pertaining in the plasma; in fact, as indicated above, the concentration-ratio never rises above about 1/200 for the plasma proteins.

Ions and Lipid-Soluble Molecules

With substances such as ^{24}Na, the evidence suggests that the primary secretion contains this ion in the same concentration as in the plasma, whilst the contributions of diffusion from the iris are also significant. The over-all rate of penetration into the aqueous humour is thus considerably greater than that of the proteins. With ethyl alcohol penetration is altogether more rapid because of the lipid-solubility of the molecules; this permits of very rapid exchanges across the iris.

Blood Vitreous Barrier. Early studies on the kinetics of the blood-aqueous barrier tended to ignore the influence of the vitreous body on the penetration of material into the aqueous humour. In general, the vitreous body comes into equilibrium with the blood much more slowly than the aqueous humour and we may speak of a *blood-vitreous barrier*. The vitreous body probably receives

* We must note, however, that the iris capillaries are different from those of the retina in so far as they take up aminoacridine dyes; according to Rodriguez-Peralta (1962) the failure of retinal and brain capillaries to stain is symptomatic of a barrier to diffusion.

material from the choroidal and retinal circulations, but because, with many substances, equilibrium with plasma takes a long time to be achieved, it follows that the vitreous body must also receive material from the aqueous humour in the posterior chamber, the concentration tending to be higher in this fluid. The blood vitreous barrier is thus complex, and may be illustrated by Fig. 30.

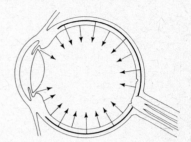

FIG. 30. The blood-vitreous barrier.

Since the vitreous body is exposed to diffusional exchanges with blood in the retinal circulation, as well as the choroidal, we must seek some reason for the slow rates of exchange of this body with blood. The pigment epithelium doubtless restricts passage from the choroid, and it is probably the presence of tight junctions that seal the intercellular clefts between endothelial cells of the retinal capillaries that acts as the restraint to permeability here (Reese & Karnovsky, 1967). The capillaries of the iris were in marked contrast to the retinal capillaries with only discontinuous regions of occlusion. So far as the restraint to passage from choroidal capillaries into the vitreous is concerned, it seems most likely that it is the pigment epithelium that constitutes the main site of resistance, the choroidal capillaries being highly permeable. Certainly treatment of rabbits with iodate, which has a specific effect in damaging the cells of the pigment epithelium (Noell, 1958), causes a marked decrease in electrical resistance across the retina (Noell, 1963) and an increased permeability of the blood-vitreous barrier (Davson & Hollingsworth, 1971).

DIFFUSION INTO VITREOUS BODY

A study of the penetration of, say, ^{24}Na into the aqueous humour must take account of the circumstance that quite a considerable fraction of the ^{24}Na passing from plasma to posterior chamber diffuses back into the vitreous body, instead of being carried into the anterior chamber. The extent to which this occurs depends largely on the relative rates at which the vitreous body and aqueous humour come into equilibrium with the plasma. With lipid-soluble substances like ethyl alcohol or ethyl thiourea the discrepancy is small, and the loss of material from the posterior chamber is not serious.* With ^{24}Na and various other ions the losses are very significant.

Mathematical Analysis. In consequence, the mathematical analysis of the process of equilibration becomes highly complex; moreover, the experimental study demands that we measure penetration into both anterior and posterior chambers separately, as well as into the vitreous body and sometimes the lens.

---

* With lipid-soluble substances, however, the penetration of material into the lens becomes a significant factor; in an animal like the rabbit, where the volume of the lens is about twice that of the aqueous humour, this can be a very serious factor.

It is not feasible to enter into the details of the mathematical analysis, which has been so elegantly carried out by Kinsey & Palm and Friedenwald & Becker. Fig. 31 shows the experimental basis for one such analysis, namely the penetration of [24]Na into posterior and anterior chambers. As we should expect, the posterior chamber tends to come into equilibrium more rapidly than the anterior chamber; the interesting feature is the crossing over of the curves, so that at the later stages the anterior chamber is ahead; this is because the posterior chamber is losing material to the vitreous body. A similar situation is found with chloride, studied with the isotope [36]Cl. The value of the mathematical analysis of the process of penetration is that it enables one to compute the probable character of the aqueous humour as it is primarily secreted (Kinsey & Reddy, 1959).

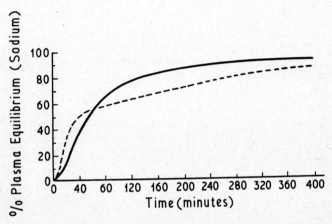

FIG. 31. The penetration of isotopically labelled sodium into the posterior (broken line) and anterior chambers (full line) of the rabbit. (After Kinsey & Palm, *Arch. Ophthal.*)

Posterior and Anterior Chambers

In describing the chemistry of the aqueous humour we considered simply the fluid obtained by inserting a needle into the anterior chamber and withdrawing all the fluid; the aqueous humour so studied was a mixture of posterior and anterior fluids. When the fluids are withdrawn separately and analysed, their comparative chemistry reveals, in some measure, their different histories. Thus

TABLE 3

Relative Compositions of Plasma, Aqueous Humours and Vitreous Body in the Rabbit.
(Reddy & Kinsey, 1960)

	Na	Cl	Total CO_2	Ascorbate	Lactate	Glucose
Plasma	143.0	109.0	20.6	0.04	10.3	5.7
Anterior Aqueous	138.0	101.0	30.2	1.11	9.3	5.4
Posterior Aqueous	136.0	96.5	37.5	1.38	9.9	5.6
Vitreous	134.0	105.0	26.0	0.46	12.0	3.0

Concentrations are expressed in millimoles per kilogramme of water.

the bicarbonate concentration in the posterior fluid is higher than that in the anterior fluid, in the rabbit, and this is because the freshly secreted fluid has a much higher concentration than in the plasma; while in the posterior chamber, some diffuses back to the vitreous body and some is decomposed by lactic and pyruvic acids; in the anterior chamber more is decomposed by the acids formed by lens and cornea, and some diffuses back into the blood in the iris. In a similar way we can account for the circumstance that the ascorbate concentration is higher in the posterior fluid; here, the diffusion from the anterior chamber into the blood in the iris is the determining factor. With chloride the concentration is less in the posterior fluid; this is because the freshly secreted fluid has a relatively low concentration of this ion; while in the anterior chamber, chloride diffuses from the blood into the anterior fluid. Some analyses, including the vitreous body, are shown in Table 3.

TABLE 4

Concentrations of certain ions (mequiv/kg H_2O) *in ocular fluids and plasma dialysate.*
(After Bito, 1970)

Species	Ion	Plasma Dialysate	Aqueous Ant.	Aqueous Post.	Vitreous Ant.	Vitreous Post.
Dog	K	4·06	4·74	5·32	5·24	5·11
Cat	K	4·53	4·15	4·70	5·15	4·99
Rabbit	K	4·37	4·26	4·69	4·69	4·62
Dog	Mg	1·18	1·07	1·06	1·28	1·38
Cat	Mg	1·26	0·99	0·89	1·01	1·08
Rabbit	Mg	1·49	1·41	1·50	1·99	2·16
Dog	Ca	3·05	2·96	2·92	3·34	3·56
Cat	Ca	3·02	2·70	2·78	3·04	3·22
Rabbit	Ca	4·12	3·48	3·48	3·90	3·94

Potassium, Magnesium and Calcium. The relative concentrations of these ions in plasma and ocular fluids have not received a great deal of attention, due, presumably, to difficulties in chemical analysis. The matter is certainly of great interest since the concentrations of these ions are very important for the functioning of the central nervous system, and they are maintained in the cerebrospinal fluid and extracellular fluid of the central nervous tissue at levels that are different from those of a plasma filtrate. Since the retina is a part of the central nervous system, and since it is exposed to the vitreous body, the concentrations of these ions in the vitreous body are of great interest; since, moreover, the vitreous body is in free communication with the aqueous humour in the posterior chamber, the concentrations in the aqueous humour are also relevant. The situation is complicated by species variability; for example, the concentration of K^+ in the anterior aqueous humour of the dog is considerably higher than that in a plasma dialysate, whereas in the cat and rabbit it is less. Some figures from a recent paper by Bito (1970) have been collected in Table 4. Perhaps the most striking feature is the existence of quite large differences in concentration between the anterior and posterior halves of the vitreous body; such differences indicate that the retina+choroid is acting as either a source or sink for the ion. Thus, in the dog, the concentration of K^+ in the posterior half is less than in the anterior half, and this is less than in the posterior aqueous humour; these gradients could be due to the retina+choroid acting as a sink for K^+, secreted by the ciliary body at a higher concentration than in a plasma filtrate. A further complicating factor is the possibility that the lens may be actively taking up K^+ at its anterior surface and releasing it at its posterior surface (Bito & Davson, 1964). The gradients with Mg^{++}, on the other hand, suggest that the retina+choroid is acting as a source for this ion.

Water and Osmotic Exchanges

The aqueous humour and blood plasma are approximately isosmolar, with the possibility that the aqueous humour is slightly hyperosmolar. Thus exchanges of water, apart from those concerned in the formation of the fluid, are normally negligible. When the osmolarity of the blood is altered experimentally, for example by intravenous injection of a hypertonic salt solution, then there is a rapid adjustment of the osmolarity of the aqueous humour, indicating that the barrier to water is relatively slight by comparison with the barrier to salts; and this is confirmed by direct studies with labelled water. The loss of water resulting from the hypertonic injection brings about a fall in the intraocular pressure, and the effectiveness of various solutes in reducing the intraocular pressure is presumably related to the ease with which they may cross the blood-aqueous and blood-vitreous barriers. Thus, if the solute penetrates rapidly into the fluids a difference of osmotic pressure cannot be maintained for any length of time and so the loss of water will be small. For this reason sucrose or sorbitol is effective. The special effectiveness of urea, which may be used clinically to reduce intraocular pressure temporarily, is due to a relatively slow rate of penetration of the barriers, but also to the circumstance that it is not rapidly excreted by the kidneys, so that a high level may be maintained in the plasma for some time; finally, its molecular weight is low, so that from an osmotic point of view a 10 per cent solution of urea is as effective as a 60 per cent sucrose solution.

Breakdown of the Blood-Aqueous Barrier

Under certain pathological or experimental conditions, the blood-aqueous barrier breaks down, so that substances such as proteins, that are normally almost completely excluded from penetration, now appear in the aqueous humour in measurable amounts. Thus, on treatment of the eye with nitrogen mustard, there is a large increase in the protein content of the aqueous humour, and this is easily revealed in the slit-lamp microscope as a pronounced Tyndall beam. It can be shown, too, that the rates of penetration of smaller molecules, such as sucrose or *p*-aminohippurate, are increased greatly. Presumably the general vascular engorgement causes an increased escape of proteins from the capillaries of the ciliary body and iris, and these find their way into the posterior chamber, or more directly into the anterior chamber from the iris.

Fluorescein Test. The dye, fluorescein, normally penetrates the blood-aqueous barrier very slowly, but to a measurable extent because, in the intact eye, it may be detected by the fluorescence it causes in the slit-lamp beam. When the blood-aqueous barrier is broken down, the rate of appearance of fluorescein increases, and the change in rate, measured by some simple fluorimetric device such as that devised by Amsler & Huber or Langham & Wybar, is used as a measure of the breakdown. Clinically this has been found useful in the diagnosis of uveitis, an inflammatory condition associated with breakdown of the barrier.*

Paracentesis. When aqueous humour is withdrawn from the eye in appreciable quantities, the fluid is re-formed rapidly but it is now no longer normal, the concentration of proteins being raised. The fluid is described as *plasmoid*; the breakdown of the barrier is probably caused by vascular engorgement, resulting from the sudden fall in intraocular pressure, since it can be prevented by stimulation of the sympathetic trunk or by application of adrenaline to the eye, according to Wessely. The re-formed fluid seems to be derived both from the ciliary body, presumably by the bursting of Greeff cysts, and also from the iris. The importance of the iris was shown by Scheie, Moore & Alder who observed that, in the completely iridectomized eye of the cat, the fluid, re-formed after paracentesis, contained only about a half as much protein as the re-formed fluid in the other, normal, eye.

ELECTRON-MICROSCOPY

Smelser & Pei (1965) and Bairati & Orzalesi (1966) examined the electron-microscopical appearance of the ciliary epithelium after paracentesis of the rabbit; the changes seemed to be confined to the pigmented layer, the intercellular spaces of which being dilated to form large vacuoles or cysts so that the pigmented cells were converted into slender cytoplasmic processes separated by large cavities filled with finely granular material; in spite of this, the cells retained their attachments to the basement membrane on the stroma side and to the non-pigmented cells on the other. In man, where the effects of paracentesis are not so large, the intercellular spaces showed only slight dilatations.

When particulate matter was in the blood before paracentesis, this apparently escaped mainly through the enlarged intercellular clefts; in particular Smelser & Pei (1965) noted that the interdigitations of the cells were reduced and the particulate matter became concentrated in the enlarged intercellular spaces. The basement membranes of epithelial and capillary endothelial cells offered no absolute restriction to movement of particles.

RATE OF FLOW OF AQUEOUS HUMOUR

The measurement of the rate of production (or drainage) of aqueous humour is of great interest for the study of the physiology and pathology of the eye, but unfortunately the measurement is not easy. Thus, simply placing a needle in the

---

*A few minutes after an intravenous injection of fluorescein, a line of green fluorescence may be seen along the vertical meridian of the cornea. This is called the *Ehrlich line*, and it results from the thermal currents in the anterior chamber. The rate of appearance of the Ehrlich line was used as a rough measure of rate of flow of aqueous humour, on the assumption that the fluorescein passed first into the posterior chamber. More recently the "fluorescein appearance time", determined with the slit-lamp microscope, has been recommended as a measure of rate of flow of aqueous humour, but the situation with regard to the mechanism of penetration of fluorescein into the aqueous humour is complex; thus the initial appearance seems definitely to be due to diffusion across the anterior surface of the iris (Slezak, 1969), a measurable concentration in the posterior chamber only building up later. There seems no doubt, moreover, from the work of Cunha-Vaz & Maurice (1967) that fluorescein is actively transported out of the eye, certainly by the retinal blood vessels and possibly, also, by the posterior epithelium of the iris. Thus appearance of fluorescein in the anterior chamber is determined primarily by diffusion from the surface of the iris rather than by the rate of secretion of aqueous humour.

eye and collecting the fluid dropping out would cause a breakdown of the barrier, and the rate of production of the plasmoid aqueous humour would not be a measure of the normal rate of production. In consequence, essentially indirect methods, largely based on studies of the kinetics of penetration and escape of substances into and out of the eye, have been employed. Space will not allow of a description of the methods. In general, the principle on which most are based is the introduction of some material into the aqueous humour and to measure its rate of disappearance from the anterior chamber. The substance is chosen such that the rate of escape by simple diffusion into the blood stream, across the iris, is negligible by comparison with the escape in the bulk-flow of fluid through Schlemm's canal, for example, ^{131}I-labelled serum albumin, ^{14}C-labelled sucrose or inulin, and so on.

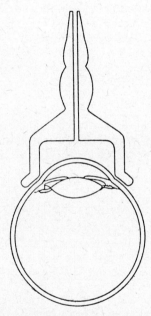

FIG. 32. The suction-cup applied perilimbally with the contact surface blocking the anterior outflow passages of the aqueous humour (Ericson. *Acta Ophthal.*)

Introduction of Marker. The main experimental difficulty is the introduction of the marker into the aqueous humour without disturbing the blood-aqueous barrier; and this is extremely difficult. Maurice applied a solution of fluorescein onto the cornea through which it diffused to establish a measurable concentration in the aqueous humour; the subsequent changes in concentration were measured with a slit-lamp fluorimeter and, with the aid of suitable equations, a flow-rate could be determined. The method is applicable to man. In animals Bill (1967), for example, has introduced needles into the anterior chamber and replaced the aqueous humour by an artificial one containing the labelled marker; by an ingenious arrangement of syringes this could be done without lowering the intraocular pressure. Oppelt (1967) has introduced cannulae into posterior and anterior chambers and perfused an artificial fluid through the system, measuring the dilution of the marker due to formation of new fluid; under these conditions, however, the eye was not normal since the perfusate contained a high concentration of protein.

Estimated Values. Estimates based on these principles agree on a value of some 1 to $1\frac{1}{2}$ per cent of the total volume per minute. Thus, in the cat, with a volume of about 1 ml, this would correspond to 10–15 μl/min, whilst in the rabbit, with a volume of about 0·35 ml, the absolute flow-rate would be less, namely 3·5–5·2 μl/min.

Tonography. If the resistance to outflow, or its reciprocal, the facility of outflow, has been determined by tonography, then Grant showed that the actual rate of flow, F, may be deduced from the relationship:

$$F \times R = P_{oc} - P_{ev}$$

provided the intraocular pressure, P_{oc}, and the episcleral venous pressure, P_{ev}, are known. Thus, with a resistance of 4 mm Hg . mm^{-3} . min and values of 16 and 12 mm Hg for P_{oc} and P_{ev} respectively, the flow-rate would be 1 mm^3/min (1 μl/min). Unfortunately, the value depends so much on the figure chosen for P_{ev} that, as a scientific estimate of rate of production of fluid, it is of little use; clinically, however, it has led to some useful information.

Pressure-Cup Technique. The outflow of aqueous humour may be blocked by applying a uniform positive pressure on the sclera at the limbus in such a way as to compress the drainage channels. The technique was developed by Ericson (1958) to permit an estimate of the rate of flow of aqueous humour, since, if rate of secretion is unaffected by this treatment, the rise in pressure, and calculated rise in intraocular volume, at the end of, say, 15 minutes' occlusion of the drainage channels should give the rate of formation. The blockage of outflow is achieved by placing a suction-cup on the globe (Fig. 32) fitting snugly round the limbus; a negative pressure in the cup leads to a positive pressure on the rim; thus a negative pressure of 50 mm Hg caused a positive pressure of some 29 mm Hg, which was adequate to block both the epi- and intra-scleral vessels. Langham (1963) obtained flow-rates between 0·8 and 1·9 μl/min in six human subjects, comparing with a mean of 0·85 μl/min found by Galin *et al.* (1961).

EFFECTS OF NERVOUS SYSTEM AND OTHER INFLUENCES ON SECRETION AND PRESSURE

Introduction

On theoretical grounds, nervous mechanisms may be expected to influence the steady-state intraocular pressure primarily by affecting either the resistance to outflow or the rate of secretion of aqueous humour. Temporary, non-steady-state, changes can be brought about by alterations in the volume of blood in the eye or mechanical compression brought about by the extraocular musculature. A direct influence of the nervous system on the secretory process is a possibility, whilst inhibition by drugs is well proven, as we shall see; alternatively, the secretory process may be limited by blood-flow in the ciliary body, in which event secretion could be controlled by an influence on the vasculature. The resistance of the angle of the anterior chamber could well be affected by the ciliary muscle, as originally suggested by Fortin, since the trabecular meshwork is the insertion of the meridional fibres; a contraction of these fibres should open up the meshwork and promote percolation of fluid through the holes in the lamellae. It seems unlikely that the nervous system can directly affect the meshwork, although the studies of the effects of ganglionectomy, to be described below, do suggest this; such an effect could be brought about by a shrinkage or swelling of the endothelial cells lining the trabeculae and possibly through an alteration of the size of the vacuolar system in the inner wall of the canal of Schlemm. Finally, the pressure within the blood vessels into which the aqueous humour drains is under nervous control.

Effects of Nervous Activity

Sympathetic System. Stimulation of the peripheral end of the cut cervical sympathetic trunk in the rabbit causes a constriction of the uveal vessels associated with a marked fall in the intraocular pressure. Section of the sympathetic, or extirpation of the superior cervical ganglion has, in general, remarkably little immediate effect on the normal intraocular pressure (Greaves & Perkins; Langham & Taylor), although occasionally quite large rises, lasting for thirty minutes, can be observed (Davson & Matchett). Twenty-four hours after ganglionectomy the intraocular pressure is definitely lower than normal, and a study of the dynamics of flow of aqueous humour suggested that this was due to a diminished outflow resistance, since P_{ev} was said to be unaffected by the procedure.

GANGLIONECTOMY

The "ganglionectomy-effect" has been the subject of a great deal of investigation largely in the laboratories of Langham and Bárány. The effect is definitely due to destruction of the postganglionic neurones, since preganglionic sympathotomy is without effect (Langham & Fraser, 1966). It is well known that denervation of a tissue causes hypersensitivity, and it was argued by Sears & Bárány that the resistance to flow was governed by an adrenergic mechanism which became hypersensitive after denervation, whilst the liberation of catecholamines from the degenerating terminals produced the adrenergic effect. The slow release of the catecholamines would account for the slow onset of the change, and the subsequent return of the intraocular pressure to normal would be accounted for by the eventual release of all stored catecholamines. The effects of reserpine and adrenergic blocking agents generally confirmed the hypothesis, although Langham & Rosenthal's observation that prolonged stimulation of the sympathetic had no effect on outflow resistance is in conflict with the theory. The effects of ganglionectomy are reflected in a hypersensitivity of the pupil, so that this dilates; in Fig. 33 from Treister & Bárány (1970) the mydriasis and rise in intraocular pressure, expressed as percentages of their peak values, have been plotted against time after ganglionectomy in the rabbit; on the same graph are the results of Sears & Gillis (1967) on the appearance of labelled norepinephrine in the anterior chamber; the correlations are striking, whilst the lag in the rise in intraocular pressure is significant and requires explanation.

In the vervet monkey Bárány (1968) found an increased facility with topically applied epinephrine. In this connection we may note that topically applied adrenaline and noradrenaline are used in the treatment of glaucoma, and the effects seem to be brought about by a decrease in the resistance to outflow; there seems to be a short-term effect, probably equivalent to the ganglionectomy effect, and a long-term effect which may be the result of metabolic changes leading to an altered composition of the meshwork (Sears, 1966). The strong vascular effects of adrenergic drugs, such as epinephrine, probably influence the blood supply to the ciliary processes, and if this is critical for rate of secretion of aqueous humour, the drugs could reduce this; thus an intravitreal injection of angiotensin lowers intraocular pressure without affecting resistance (Eakins, 1964).* Anselmi, Bron & Maurice (1968) using the rate of clearance of fluorescein, introduced through the cornea, as a measure of aqueous flow, found that guanethidine, which acts as a sympatholytic substance, caused a decrease in rate of secretion, lowering the intraocular pressure.

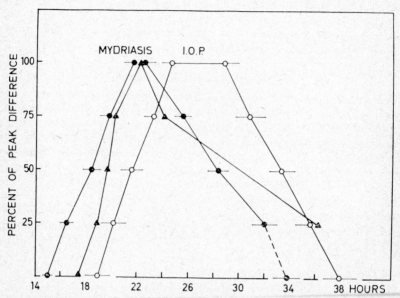

FIG. 33. Illustrating the post-ganglionectomy effects with respec to mydriasis and intraocular pressure. The line of black triangles represents the results of Sears & Gillis on the appearance of tritiated norepinephrine in the anterior chamber. (Treister & Bárány. *Invest. Ophthal.*)

Parasympathetic. Armaly (1959) obtained a definite lowering of the intra-ocular pressure on stimulating the parasympathetic root of the ciliary ganglion (Fig. 6, p. 5) by placing electrodes in the orbit. An initial rapid fall was probably caused by a reduction in ciliary blood volume, whereas a more prolonged action could have been due to the opening up of the meshwork resulting from con-traction of the ciliary muscle. At any rate, tonography (p. 42) suggested a decreased resistance to outflow. Drugs that mimic the effects of parasympathetic stimulation, e.g. pilocarpine and eserine, are used in the treatment of glaucoma, and their hypotensive effect has been regarded as largely exerted through the pull of the ciliary muscle on the trabecular meshwork. Atropine, the cholinergic blocking agent, may raise the pressure in glaucoma. In the normal human eye the

* It is impossible to give a fair and balanced account, in a short space, of the studies on adrenergic mechanisms in aqueous humour dynamics; the situation is complicated by the possibility that there are antagonistic α- and β-actions, as typified by the actions of noradrenaline and isoproterenol respectively, the one possibly acting mainly on secretion-rate and the other on resistance. A further complication is the use of different species, usually the rabbit and monkey; and finally the effects are often so small as to be beyond the resolution of the methods of measurement. The most recent paper on the subject that I have seen is that of Bill (1970), which should be consulted both for the review of the literature and the conclusion: "... adrenergic drugs have such complex effects in the eye that the effect on intraocular pressure can be expected to vary even qualitatively from one individual to another, and in the individual eye the mode of administration may play a role."

effects of pilocarpine and atropine are usually negligible, but this does not mean that the drugs are without effects on all the parameters that go to determine the intraocular pressure.

RESISTANCE TO OUTFLOW

Thus there seems little doubt that pilocarpine causes a decreased resistance to outflow, and Bárány (1967) has shown that the effect is so rapid that it occurs before sufficient of the drug has reached the aqueous humour in sufficient concentration to exert any direct effect on the meshwork; its action is therefore on the ciliary body and/or iris; and we may presume that it is exerted through the ciliary muscle. The effects obtained were just as large as any obtained through intracameral injection, so it is unlikely that the drug influences the trabecular meshwork directly. Direct stimulation of the oculomotor nerve as it emerges from the brain-stem in the monkey causes a fall in resistance to outflow, whilst cutting the nerve unilaterally causes the resistance to be higher on the same side (Törnqvist, 1970). Some of the complexities in the study of the effects of drugs on intraocular pressure are well revealed by the studies of Bill & Wålinder (1966), who showed that, in macaque monkeys, pilocarpine increased the intraocular pressure; this was associated with a fall in rate of secretion of aqueous humour from a control value of $1 \cdot 60 \ \mu l/min$ to $1 \cdot 05 \ \mu l/min$. The reason for the rise in pressure was an almost complete blockage of the uveo-scleral drainage route, so that drainage was confined to the canal of Schlemm.

Trigeminal. Perkins (1957) has shown that stimulation of the trigeminal nerve (N. V), especially if brought about mechanically, causes a marked rise in intraocular pressure accompanied by a dilatation of the intraocular blood vessels and a breakdown of the blood-aqueous barrier (p. 39). The effect on intraocular pressure is probably entirely due to the vascular engorgement and is reminiscent of the large rises obtained by irritative lesions.

Facial Nerve. Gloster (1961) has shown that stimulation of N. VII causes a rise in intraocular pressure; it is not merely due to a contraction of the extraocular muscles since gallamine did not abolish the effect. Cutting the nerve of the pterygoid canal causes a fall in intraocular pressure in the monkey (Ruskell, 1970b). This effect is consistent with Ruskell's (1970a) description of *rami orbitales* of the pterygopalatine ganglion containing parasympathetic fibres of N. VII; these supply the ciliary arteries, and if they are antagonistic to the sympathetic supply, the rise in intraocular pressure obtained by stimulation could be due to vascular dilatation.

Afferent Impulses. If there is a "centre" controlling the intraocular pressure, we may expect to find evidence of a sensory system capable of keeping the centre informed of the state of the intraocular pressure at any moment. Studies from the laboratories of Vrábec and v. Sallmann have shown that the trabecular meshwork is richly provided with fine nerve fibres, some of which may be sensory, whilst action potentials in the ciliary nerves, in response to alterations in the intraocular pressure have also been described. Nevertheless, as Perkins (1961) has shown, the evidence for sensory fibres, specifically responding to alterations in intraocular pressure, is still not completely satisfactory.

Hypothalamic Centre. Attempts to locate a centre controlling the intraocular pressure have been made by v. Sallmann & Loewenstein and Gloster & Greaves, by inserting stereotactically controlled stimulating electrodes into different parts of the brain. The parts of the brain from which effects on intraocular pressure could be obtained were in the hypothalamus, but since this region contains many centres controlling the autonomic system generally, it is important to distinguish primary effects on the intraocular pressure from the secondary consequences of stimulating these autonomic centres. Thus many stimulated points gave rise to rises or falls in intraocular pressure that were obviously associated with corresponding rises or falls of arterial pressure. Nevertheless, other points could be found where a rise or fall of intraocular pressure, independent of general vascular changes, resulted from stimulation; the falls in intraocular pressure were probably mediated by the sympathetic, since they could be prevented by section of the sympathetic trunk. It is possible, then, that there is a co-ordinating centre in the hypothalamus capable of influencing the intraocular pressure, and subject to afferent impulses carried in the long ciliary nerves. The evidence for this is not completely convincing, however, and we must bear in mind the possibility that there is no central control, local homeostatic mechanisms being adequate to keep the intraocular pressure within a narrow range.

Effects of Drugs

Diamox. There is a great deal of evidence suggesting that Diamox (acetazole-amide) inhibits the process of secretion of aqueous humour and in so doing produces a striking fall in intraocular pressure. We have seen that in order to produce a fall in intraocular pressure by reducing the flow of aqueous humour this reduction must be very high, and it is not surprising that it has been estimated that production of fluid is reduced by as much as 66 per cent after administration of the drug (Becker & Constant, 1955). As to whether the fall in intraocular pressure is entirely determined by the reduced rate of secretion is not completely certain; thus in the cat doses that reduce the intraocular pressure apparently do not reduce rate of secretion significantly; only when the doses have been increased well beyond this is inhibition observed (Macri *et al.*, 1965).*

Cardiac Glycosides. It is well established that many secretory processes, involving the active transport of sodium, are inhibited by ouabain and other cardiac glycosides and aglycones. The aqueous humour is no exception (Simon, Bonting & Hawkins, 1962). Oppelt & White (1968), using their perfusion technique in cats, obtained a 40 per cent inhibition, but an intracameral application of the drug was without effect, presumably because an adequate concentration cannot be built up in the cilary processes by this route. When an intravitreal injection was given, Bonting & Becker (1964) found a fall in intraocular pressure which correlated well with the course of inhibition of the Na-K-activated ATPase in the ciliary body.

Nitrogen Mustard and DFP. Nitrogen mustard applied to the cornea causes a congestion of the small vessels of the conjunctiva, iris and ciliary body; the result is a phenomenal rise in intraocular pressure (Fig. 34). This is associated with the appearance of protein in the aqueous humour, i.e. with a breakdown of the blood-aqueous barrier (p. 39). Smaller, but significant increases in intraocular pressure can be produced by the anticholinesterase DFP, whilst smaller effects still are

---

* More recently, however, Oppelt (1967), using a perfusion technique from posterior to anterior chamber, found a 45 per cent reduction in secretion rate in the cat with a dose as low as 10 mg/kg. In the vervet monkey, Wålinder & Bill (1969) found a 30–80 per cent reduction in rate of secretion.

caused by eserine and pilocarpine. Here the action is probably mediated by the parasympathetic, which presumably has a vasodilator action on the small blood vessels of the uvea. The rises in intraocular pressure obviously belong to the same class as the responses to contusion of the eyeball, irritation of the conjunctiva, scratching the iris, and so on. It is interesting that the drug, *polyphloretin*, known to reduce serous exudates, is able to counteract the action of nitrogen mustard (Cole, 1961).

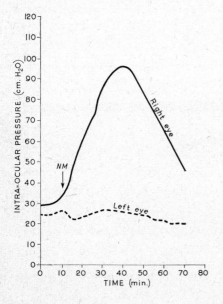

Fig. 34. The effect of instillation of nitrogen mustard (NM) into the right eye of the rabbit on its intraocular pressure. (Davson & Matchett, *J. Physiol.*)

Prostaglandins. The large rise in intraocular pressure resulting from treatment with nitrogen mustard is associated with exudation of protein into the aqueous humour and a powerful miosis. This triad is a general feature of the breakdown of the blood-aqueous barrier, e.g. that due to irritation of the iris by stroking (Duke-Elder & Duke-Elder, 1931), by injury to the lens (Davson & Matchett, 1951), by paracentesis, and so on. Ambache (1957) extracted a substance from the iris that he called *irin*; this seemed to be responsible for the pupillary constriction that followed stroking the iris or mechanical stimulation of the trigeminal (Maurice, 1953); and in a later study Ambache, Kavanagh & Whiting (1965) showed that the effects were not due to cholinergic mechanisms, since they were not blocked by atropine, nor yet was histamine or 5–HT responsible since mepyramine and LSD did not block. By perfusing the anterior chamber he was able to obtain material—irin— that caused contraction of smooth muscle; increased amounts of the irin were obtained by collapse of the anterior chamber, stroking the iris, or moving the lens. Chromatographic analysis of irin indicated that it consisted of one or more fatty acids of the prostanoic acid class called *prostaglandins* (Ambache *et al.*, 1966). Waitzman & King (1967) injected prostaglandins E_1 and E_2 into the anterior chamber of the rabbit's eye and obtained a sustained rise in intraocular pressure accompanied by a contracted pupil; the effect on facility of outflow was slight, so they concluded that the influence was on the production of fluid; these authors stated that the protein content of the aqueous humour was not raised significantly, but Beitch & Eakins (1969) found a considerable increase; Fig. 35 shows the course of the rise in intraocular pressure, the course being similar to that caused by nitrogen mustard; like Davson & Quilliam (1947), these authors found that application to one eye could have a contralateral effect on the other eye. As with nitrogen mustard, too, the effect of prostaglandin could be blocked by close arterial injection of poly-phloretin. Thus it would seem that the triad of increased intraocular pressure, plasmoid aqueous humour and pupillary constriction caused by various irritative insults, including the mechanical stimulation of the trigeminal nerve (Perkins, 1957), may best be accounted for by the liberation of prostaglandins, the irin of Ambache.*

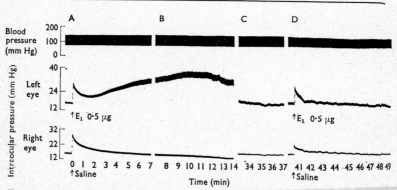

FIG. 35. Effect of prostaglandin PGE$_1$, on intraocular pressure of rabbit. Right eye is the control. A. Injection of 0·5 μg of PGE$_1$ into the anterior chamber of the left eye. B shows peak response 9–12 min after the injection. C shows return to normal after about 30 min and D shows failure of eye to respond to a second injection. (Beitch & Eakins. *Brit. J. Pharmacol.*)

Corticosteroids. François (1954) observed that the prolonged use of topical cortisone in the treatment of uveitis sometimes led to a rise in intraocular pressure; patients undergoing systemic corticosteroid therapy for rheumatic conditions also showed raised intraocular pressure. Bernstein & Schwartz (1962) showed that the cause was a decreased facility of outflow, which was reversible when the dose of corticosteroid was reduced. In human subjects, Anselmi, Bron & Maurice (1968) found no change in rate of secretion, although the intraocular pressure was high. In experimental animals, such as the rabbit, results have been contradictory (see, for example, Armaly, 1964, whose results were negative, and Jackson & Waitzman, 1965, with positive, but complex, results); the study of Oppelt, White & Halpert (1969), employing the posterior-anterior chamber perfusion technique in cats, indicated a dose-related fall in rate of secretion after intravenous hydrocortisone; associated with this, there was a 23 per cent decrease in facility of outflow, and presumably it would be this that causes the rise in intraocular pressure if results on cats may be transposed to man.

* Waitzman (1970) has studied the interaction between prostaglandin E$_1$ and norepinephrine with respect to intraocular pressure and pupil; there seems to be an antagonism between the two, presumably the α-adrenergic activity being inhibited by the prostaglandin; certainly Davson & Huber (1950) were able to block or delay the effects of nitrogen mustard whilst long ago Wessely (1900) showed that adrenaline prevented the rise in protein content of the aqueous humour after paracentesis.

Hypothermia

We may mention the effects of hypothermia on the production of aqueous humour; here we are reducing the rate of metabolism and therefore, presumably, reducing the amount of energy available for the production of aqueous humour. In fact, there is a considerable fall in intraocular pressure which is probably due, in part at any rate, to a reduced rate of production of fluid (Pollack, Becker & Constant); and it is interesting that the cells of the ciliary epithelium show the same accumulation of cytoplasmic vesicles as was seen after Diamox (Holmberg & Becker). According to Davson & Spaziani, the reduced rate of secretion could well be due, not so much to the lowered body-temperature, as such, but to the decrease in the blood supply to the ciliary body that must accompany the great reduction in arterial pressure that takes place in hypothermia.

The Effects of Pressure

It has been briefly indicated that Bill & Bárány (1966) found that a rise in pressure, artificially induced, caused a reduction in the computed rate of secretion of aqueous humour. This has been confirmed in the monkey by Bill (1968) who caused a hypertension by intracameral injection of erythrocytes; the suppression corresponded to 0·06 μl/min/mm Hg rise in pressure; thus with a normal flow of 1·24 μl/min this means that a rise of some 20 mm Hg would suppress aqueous humour formation altogether; since, in glaucoma, pressures much higher than this are reached, it seems unlikely that suppression would occur at only some 30 mm Hg. In general, secretory processes are able to work against very high pressures, so that this reduction in rate of formation may well indicate that there is a significant contribution of an ultrafiltrate from plasma to the total formation of fluid; such a contribution had been suggested by the work of Davson (1953) which showed that large water-soluble molecules, such as raffinose, entered the eye by a flow mechanism rather than by simple diffusion. That the pressure-suppressed inflow of aqueous humour is, indeed, formed by filtration is made very likely by Macri's (1967) finding that, when he raised the intraocular pressure by occluding anterior ciliary veins, there was no suppression; in this case the transmural pressure-difference across the blood capillaries was probably not decreased, whereas when the intraocular pressure was raised by increasing the pressure-head of saline during perfusion, there was a suppression of inflow of aqueous humour, as revealed by the phenomenon of pseudofacility.

Pseudofacility. If raising the intraocular pressure reduces the production of fluid, this means that, when facility is determined by measuring the flow of fluid from a cannula in the anterior chamber under an applied pressure, a reduction in fluid production caused by the raised pressure will make room for some fluid from the reservoir and so give too high a measure of the extra flow through Schlemm's canal; the extra fluid entering for a given rise in pressure can be expressed as a "pseudofacility" (Bill & Bárány, 1966). In rhesus monkeys Brubaker & Kupfer (1966) deduced a value of 0·19 μl min^{-1} mm Hg^{-1} compared with a total facility of 0·62 μl min^{-1} mm Hg^{-1}. In general, when different species are compared, the pseudofacility is some 11 to 20 per cent of the total (Brubaker, 1970). The outflow through "conventional channels" may be measured in the experimental animal by the escape of labelled protein or inulin into the blood.

Effects of Blood Osmolality

Aqueous humour and blood plasma are approximately isosmolal, and this is to be expected in view of the rapid exchanges of water across the blood-ocular fluid barriers (Kinsey *et al.*, 1942) compared with the much slower exchanges of solutes. Raising the osmolality of blood artificially causes a fall in intraocular pressure because of the osmotic outflux of water (Davson & Thomassen, 1950; Auricchio & Bárány, 1959); clinically an acute fall in intraocular pressure may be induced by oral urea, this being chosen because it is not rapidly excreted from the blood and its passage across the blood-ocular barriers is slow; mannitol is less effective presumably because its renal excretion is greater (Galin, Davidson & Pasmanik, 1963).

THE SECRETORY PROCESS

We may conclude this account of the physiology of the aqueous humour by a few remarks on the nature of the secretory process in the ciliary epithelium. The secretion of fluid in other systems is usually associated with an active transport of Na^+ through the epithelial cells of the secreting tissue in the direction of fluid movement; the passage of anions may be linked to this either actively or passively. The increased osmolality on the side of the epithelium to which the Na^+ has been transported causes the flow of water. Evidence for the location of a "sodium-pump" on the epithelial cells is provided by the presence of an ATPase on their membranes that requires, besides Mg^{++}, Na^+ and K^+ for its activity. Kaye & Pappas (1965) showed that this ATPase activity was confined to the regions of lateral interdigitations of the non-pigmented cells of the ciliary epithelium. Metabolic evidence also favours the non-pigmented layer as the site of secretion; thus Cameron & Cole (1963) found a high concentration of succinoxidase activity in this layer.

Cardiac Glycosides. The cardiac glycosides, digitalis, ouabain, etc. specifically inhibit the ATPase concerned in active transport processes; we have seen that they are able to reduce the rate of formation of the aqueous humour (Simon, Bonting & Hawkins, 1962) after intravenous injection into cats. Interestingly an intracameral injection was ineffectual (Oppelt & White, 1968). Ouabain also inhibited the metabolic activity of the ciliary epithelium; thus oxidative activity was inhibited by 66 per cent and aerobic lactate formation by 63 per cent (Riley, 1966).

Dinitrophenol. A less specific way of interfering with the formation of aqueous humour is to inhibit the metabolism that produces the energy for the process; in doing this, of course, one tends to kill the animal because one inhibits other vital activities. For this reason local application is necessary. With the eye, this can be achieved by injection of a poison, dinitrophenol (DNP), into the internal carotid artery or a branch of this. Cole showed that the transport of sodium into the aqueous humour was reduced by this drug.

Diamox. The decrease in intraocular pressure caused by administration of Diamox has attracted a great deal of interest because of its great clinical value; from the scientific aspect, too, the discovery was of fundamental interest since it suggested that at one stage in the production of aqueous humour the conversion of carbon dioxide to bicarbonate-ion is an important step. In animals like the rabbit and guinea pig, which produce an aqueous humour with a high concentration of bicarbonate, Diamox certainly causes a reduction in the concentration of this ion. In those species where the aqueous humour has a relatively low concentration of bicarbonate, Diamox seems to attack the chloride concentration, reducing this. It seems, therefore, that the drug interferes with what we may call an anion-secreting mechanism. It is presumably because of the interference with this that the fluid can no longer be formed at its normal rate.

Ciliary Body Potentials. The active transport of the Na^+ -ion by frog-skin causes the development of a difference of potential across the skin, the side to which transport is directed being positive. If electrodes are placed in the aqueous humour and in the jugular vein a potential is indeed observed, the aqueous humour being some 10 mV positive in relation to the blood (Lehmann & Meesmann; Cole, 1961) whilst if the ciliary body is stripped away from the eye and maintained in an artificial medium, a difference of potential may also be observed. By studying the effects of poisons, and of artificial variations of the potential by means of an applied electromotive force, Cole has made it extremely probable that the potential is in some way a measure, or at least a sign, of the active transport of sodium. It must be appreciated, however, that the chemical studies of the aqueous humour described earlier have made it very likely that *anions*, either chloride or bicarbonate or both, are transferred against gradients of electrochemical potential, i.e. are actively transported, so that the observed potential is doubtless the sign of both processes, namely transport of positive and negative ions. This may be why the potential is so small, by comparison with the 50 or more millivolts observed with the frog skin where only positive ions seem to be actively transported; presumably the potential created by the transport of positive ions by the ciliary epithelium is largely balanced by an opposite potential associated with transport of negative ions.[*]

ACTIVE TRANSPORT OF CHLORIDE

Evidence favouring an active transport of Cl^- by the rabbit's ciliary body has recently been provided by Holland & Stockwell (1967)[†] in the cat and by Cole (1969) in the rabbit; in general, the active transport of Na^+ dominates the picture, and it may well be that the variations in chloride concentration found in the aqueous humours of different species are a reflection of the different degrees to which chloride is actively transported.

In Vitro Secretion. Berggren (1964) spread the ciliary body over a cylinder of

---

[*] The potentials within the epithelial cells have been measured by Berggren (1960); the unpigmented cells have a "resting potential" of -30 mV in relation to the aqueous humour, whilst the pigmented cells have one of 60 mV. Similar results were obtained by Miller & Constant (1960). We shall see that the electroretinogram is a measure of changes in potential between inside and outside of the eye; i.e. they are measurements of electrical changes induced by light, superimposed on the aqueous humour potential, or *d.c.-potential* as it is called in this context. The origin of the d.c.-potential is attributed to current-flow across the receptors; if this is true the effects of the secretory process on the potential may well be obscured or they may be non-existent.

[†] More recently Holland & Gipson (1970) and Holland (1970) have made Ussing-type experiments with the cat's isolated ciliary body preparation, and have shown that a large part of the short-circuit current is accounted for by an active transport of chloride; the steady-state potential across the preparation was such that the aqueous humour would have been negative in respect to the sclera, by contrast with the rabbit where Cole (1969) found the reverse sign; as Holland (1970) points out, the species differences in the chloride-distributions between plasma and aqueous humour, described by Davson, Matchett & Roberts (1952), doubtless reflect different degrees of active transport of this ion.

Perspex in such a way that the processes faced outwards and the whole was immersed in a bath. Microscopical examination of the processes showed that they shrank as a result of secreting aqueous humour, and it was possible to study the effects of metabolic poisons on the process.

Absorption from the Eye. It is common to think of the ciliary processes as organs for the production of secretion, in the sense of transporting material from blood to posterior chamber. It is therefore of considerable interest to discover that the ciliary processes are capable of actively removing certain substances from the aqueous humour into the blood. Thus Forbes & Becker showed that, if diodrast was injected into the vitreous body, it disappeared from this into the blood but did not appear in the aqueous humour, being apparently actively removed by the ciliary epithelium; studies on the isolated ciliary body confirmed that the epithelium was indeed able to accumulate diodrast and also iodide. Cunha-Vaz & Maurice (1967) have described an active transport of fluorescein from the vitreous body after injecting it into the vitreous body; the rate of loss was far too great to be accounted for by passive diffusion into the posterior chamber, unless the process had been previously inhibited by probenecid; and a study of the concentrations in different parts of the vitreous body indicated removal by both retina and anterior uvea. An important discovery was that the retinal capillaries could apparently actively absorb the fluorescein; since these capillaries have no glial sheaths, this means that the endothelial lining is capable of secretory activity.

GENERAL REFERENCES

ASCHER, K. W. (1961). *The Aqueous Veins.* Springfield: Charles C. Thomas.

DAVSON, H. (1956). *Physiology of the Ocular and Cerebrospinal Fluids.* London: Churchill.

DAVSON, H. (1969). The intra-ocular fluids. The intra-ocular pressure. In *The Eye,* Ed. Davson, pp. 67–186, 187–272. London: Academic Press.

DUKE-ELDER, S. & WYBAR, K. C. (1961). *System of Ophthalmology.* Vol. II: *The Anatomy of the Visual System.* London: Kimpton.

FRANÇOIS, J. & NEETENS, A. (1962). Comparative anatomy of the vascular supply of the eye in vertebrates. In *The Eye,* Ed. Davson, pp. 369–416. London: Academic Press.

LANGHAM, M. E. (1958). Aqueous humour and control of intra-ocular pressure. *Physiol Rev.* **38,** 215–242.

PAPPAS, G. D. & SMELSER, G. K. (1961). The fine structure of the ciliary epithelium in relation to aqueous humour secretion. In *The Structure of the Eye,* Ed. Smelser, pp. 453–467. New York: Academic Press.

ROHEN, J. (1961). Morphology and pathology of the trabecular meshwork. In *The Structure of the Eye,* Ed. Smelser, pp. 335–341. New York: Academic Press.

ROHEN, J. W. (1965). *Die Struktur des Auges.* Stuttgart: Schattauer Verlag.

THEOBALD, G. D. (1960). The limbal area. *Am. J. Ophthal.* **50,** 543–557.

WOLFF, E. (1968). *Anatomy of the Eye and Orbit* (revised by R. J. Last). London: H. K. Lewis.

SPECIAL REFERENCES

AMBACHE, N. (1957). Properties of irin, a physiological constituent of the rabbit's iris. *J. Physiol.* **135,** 114–132.

AMBACHE, N., BRUMMER, H. C., ROSE, J. G. & WHITING, J. (1966). Thin-layer chromatography of spasmogenic unsaturated hydroxy-acids from various tissues. *J. Physiol.* **185,** 77–78.

AMBACHE, N., KAVANAGH, L. & WHITING, J. (1965). Effect of mechanical stimulation on rabbits' eyes: release of active substance in anterior chamber perfusates. *J. Physiol.* **176**, 378–408.

AMSLER, M. & HUBER, A. (1946). Methodik und erste klinische Ergebnisse einer Funktionsprufung der Blut-Kammerwasser-Schranke. *Ophthalmologica, Basel,* **111**, 155–176.

ANDERSON, D. R. & GRANT, W. M. (1970). Re-evaluation of the Schiøtz tonometer calibration. *Invest. Ophthal.* **9**, 430–446.

ANSELMI, P., BRON, A. J. & MAURICE, D. M. (1968). Action of drugs on the aqueous flow in man measured by fluorophotometry. *Exp. Eye Res.* **7**, 487–496.

ARMALY, M. F. (1959). Studies on intraocular effect of orbital parasympathetic. *Arch. Ophthal., N.Y.* **62**, 117–124.

ARMALY, M. F. (1960). Schiøtz tonometer calibration and applanation tonometry. *Arch. Ophthal., N.Y.* **64**, 426–432.

ARMALY, M. F. (1964). Aqueous outflow facility in monkeys and the effect of topical corticoids. *Invest. Ophthal.* **3**, 534–538.

ASCHER, K. W. (1942). Aqueous veins. Preliminary note. *Am. J. Ophthal.* **25**, 31–38.

ASHTON, N. (1951). Anatomical study of Schlemm's canal and aqueous veins by means of Neoprene casts. *Br. J. Ophthal.* **35**, 291–303.

ASHTON, N., BRINI, A. & SMITH, R. (1956). Anatomical studies of the trabecular meshwork of the normal human eye. *Br. J. Ophthal.* **40**, 257–282.

AURICCHIO, G. & BÁRÁNY, E. H. (1959). On the role of osmotic water transport in the secretion of the aqueous humour. *Acta physiol. scand.* **45**, 190–210.

BAIRATI, A. & ORZALESI, N. (1966). The ultrastructure of the epithelium of the ciliary body. *Z. Zellforsch.* **69**, 635–658.

BÁRÁNY, E. H. (1947). The recovery of intraocular pressure, arterial blood pressure, and heat dissipation by the external ear after unilateral carotid ligation. *Acta Ophthal.* **25**, 81–94.

BÁRÁNY, E. H. (1967). The immediate effect on outflow resistance of intravenous pilocarpine in the vervet monkey, *Cercopithecus ethiops. Invest. Ophthal.* **6**, 373–380.

BÁRÁNY, E. H. (1968). Topical epinephrine effects on true outflow resistance and pseudofacility in vervet monkeys studied by a new anterior chamber perfusion technique. *Invest. Ophthal.* **7**, 88–104.

BÁRÁNY, E. H. & SCOTCHBROOK, S. (1954). Influence of testicular hyaluronidase on the resistance to flow through the angle of the anterior chamber. *Acta physiol. scand.* **30**, 240–248.

BAURMANN, M. (1930). Über die Ciliar-Fortsatz-gefässystem. *Ber. dtsch. ophth. Ges.* **48**, 364–371.

BECKER, B. (1967). Ascorbate transport in guinea pig eye. *Invest. Ophthal.* **6**, 410–415.

BECKER, B. & CONSTANT, M. A. (1955). The effect of the carbonic anhydrase inhibitor acetazolamide on aqueous flow. *Arch. Ophthal. N.Y.* **54**, 321–329.

BEITCH, B. R. & EAKINS, K. E. (1969). The effects of prostaglandins on the intraocular pressure of the rabbit. *Brit. J. Pharmacol.* **37**, 158–167.

BERGGREN, L. (1960). Intracellular potential measurements from the ciliary processes of the rabbit eye *in vivo* and *in vitro. Acta physiol. scand.* **48**, 461–470.

BERGGREN, L. (1964). Direct observation of secretory pumping *in vitro* of the rabbit eye ciliary processes. *Invest. Ophthal.* **3**, 266–272.

BERNSTEIN, H. N. & SCHWARTZ, B. (1962). Effects of long-term systemic steroids on ocular pressure and tonographic values. *Arch. Ophthal., N.Y.* **68**, 742–753.

BILL, A. (1962). The drainage of blood from the uvea and the elimination of aqueous humour in rabbits. *Exp. Eye Res.* **1**, 200–205.

BILL, A. (1964). The albumin exchange in the rabbit eye. *Acta physiol. scand.* **60**, 18–29.

BILL, A. (1967). The effects of atropine and pilocarpine on aqueous humour dynamics in cynomolgus monkeys (*Macaca irus*). *Exp. Eye. Res.* **6**, 120–125.

BILL, A. (1968). The effect of ocular hypertension caused by red cells on the rate of formation of aqueous humour. *Invest. Ophthal.* **7**, 162–168.

BILL, A. (1968). The elimination of red cells from the anterior chamber in vervet monkeys (*Cercopithecus ethiops*). *Invest. Ophthal.* **7**, 156–161.

BILL, A. (1970). Scanning electron microscopic studies of Schlemm's canal. *Exp. Eye Res.* **10**, 214–218.

BILL, A. (1970). Effects of norepinephrine, isoproterenol and sympathetic stimulation on aqueous humour dynamics in vervet monkeys. *Exp. Eye Res.* **10**, 31–46.

BILL, A. & BÁRÁNY, E. H. (1966). Gross facility, facility of conventional routes, and pseudofacility of aqueous humour outflow in the cynomolgus monkey. *Archs. Ophthal., N.Y.* **75**, 665–673.

BILL, A. & WÅLINDER, P. -E. (1966). The effects of pilocarpine on the dynamics of aqueous humour in a primate (*Macaca irus*). *Invest. Ophthal.* **5**, 170–175.

BITO, L. Z. (1970). Steady-state concentration gradients of magnesium, potassium and calcium in relation to the sites and mechanisms of ocular cation-transport processes. *Exp. Eye Res.* **10**, 102–116.

BITO, L. Z. & DAVSON, H. (1964). Steady-state concentrations of potassium in the ocular fluids. *Exp. Eye Res.* **3**, 283–297.

BITO, L. Z., DAVSON, H., LEVIN, E., MURRAY, M. & SNIDER, N. (1965). The relationship between the concentrations of amino acids in the ocular fluids and blood plasma of dogs. *Exp. Eye Res.* **4**, 374–380.

BITO, L. Z. & SALVADOR, E. V. (1970). Intraocular fluid dynamics. II. *Exp. Eye Res.* **10**, 273–287.

BONTING, S. L. & BECKER, B. (1964). Inhibition of enzyme activity and aqueous humor flow in the rabbit eye after intravitreal injection of ouabain. *Invest. Ophthal.* **3**, 523–533.

BRUBAKER, R. F. (1970). The measurement of pseudofacility and true facility by constant pressure perfusion in the normal rhesus monkey eye. *Invest. Ophthal.* **9**, 42–52.

BRUBAKER, R. F. & KUPFER, C. (1966). Determination of pseudofacility in the eye of the rhesus monkey. *Arch. Ophthal., N.Y.* **75**, 693–697.

CAMERON, E. & COLE, D. F. (1963). Succinic dehydrogenase in the rabbit ciliary epithelium. *Exp. Eye Res.* **2**, 25–27.

COLE, D. F. (1960). Rate of entrance of sodium into the aqueous humour of the rabbit, *Br. J. Ophthal.* 1960, **44**, 225–245.

COLE, D. F. (1961). Electrochemical changes associated with the formation of the aqueous humour. *Br. J. Ophthal.* **45**, 202–217.

COLE, D. F. (1961). Prevention of experimental ocular hypertension with polyphloretin phosphate. *Br. J. Ophthal.* **45**, 482–489.

COLE, D. F. (1961). Electrical potential across the ciliary body observed *in vitro*. *Br. J. Ophthal.* **45**, 641–653.

COLE, D. F. (1969). Evidence for active transport of chloride in ciliary epithelium of the rabbit. *Exp. Eye Res.* **8**, 5–15.

COPELAND, R. L. & KINSEY, V. E. (1950). Determination of the volume of the posterior chamber of the rabbit. *Arch. Ophthal., N.Y.* **44**, 515–516.

CUNHA-VAZ, J. G. & MAURICE, D. M. (1967). The active transport of fluorescein by the retinal vessels and the retina. *J. Physiol.* **191**, 467–486.

DAVSON, H. (1953). The penetration of large water-soluble molecules into the aqueous humour. *J. Physiol.* **122**, 10P.

DAVSON, H. & HUBER, A. (1950). Experimental hypertensive uveitis in the rabbit. *Opthalmologica, Basel*, **120**, 118–124.

DAVSON, H. & LUCK, C. P. (1956). A comparative study of the total carbon dioxide in the ocular fluids, cerebrospinal fluid, and plasma of some mammalian species. *J. Physiol.* **132**, 454–464.

DAVSON, H. & MATCHETT, P. A. (1951). The control of the intraocular pressure in the rabbit. *J. Physiol.* **113**, 387–397.

DAVSON, H., MATCHETT, P. A. & ROBERTS, J. R. E. (1952). Comparative studies of the distribution of chloride between plasma and aqueous humour. *J. Physiol.* **116**, 47P.

DAVSON, H. & QUILLIAM, J. P. (1947). The effect of nitrogen mustard on the permeability of the blood-aqueous humour barrier to Evans blue. *Br. J. Ophthal.* **31**, 717–721.

DAVSON, H. & SPAZIANI, E. (1960). The fate of substances injected into the anterior chamber of the eye. *J. Physiol.* **151**, 202–215.

DAVSON, H. & SPAZIANI, E. (1961). The effect of hypothermia on intraocular dynamics. *Exp. Eye Res.* **1**, 182–192.

DAVSON, H. & THOMASSEN, T. L. (1950). The effect of intravenous infusion of hypertonic saline on the intraocular pressure. *Br. J. Ophthal.* **34**, 355–359.

DRANCE, S. M. (1960). The coefficient of scleral rigidity in normal and glaucomatous eyes. *Arch. Ophthal. N.Y.* **63**, 668–674.

DUKE-ELDER, P. M. & DUKE-ELDER, W. S. (1931). The vascular responses of the eye. *Proc. Roy. Soc.* B, **109**, 19–28.

DURHAM, D. G., BIGLIANO, R. P. & MASINO, J. A. (1965). Pneumatic applanation tonometer. *Trans. Am. Acad. Ophthal. Oto-lar.* **69**, 1029–1047.

EAKINS, K. E. (1964). Effect of angiotensin on intraocular pressure. *Nature, Lond.* **202**, 813–814.

EDWARDS, J., HALLMAN, V. L. & PERKINS, E. S. (1967). Perfusion studies on the monkey. *Exp. Eye Res.* **6**, 316–326.

ERICSON, L. A. (1958). Twenty-four hourly variations of the aqueous flow. Examinations with perilimbal suction cup. *Acta ophthal.* Suppl. 50.

FORBES, M. & BECKER, B. (1960). *In vivo* transport of iodopyracet (diodrast). *Am. J. Ophthal.* **50**, Pt. 2, 862–867.

FORTIN, E. P. (1929). Action du muscle ciliare sur la circulation de l'oeil, etc. *C.r. Seanc. Soc. Biol.* **102**, 432–434.

FOWLKS, W. L. & HAVENER, V. R. (1964). Aqueous flow into the perivascular space of the rabbit ciliary body. *Invest. Ophthal.* **3**, 374–383.

FRANÇOIS, J. (1954). Cortisone et tension oculaire. *Annls Oculist.* **187**, 805–816.

FRIEDENWALD, J. S. (1957). Tonometer calibration. An attempt to remove discrepancies found in the 1954 calibration scale for Schiøtz tonometers. *Trans. Am. Acad. Ophthal. Otolaryngol.* **61**, 108–123.

FRIEDENWALD, J. S. & BECKER, B. (1956). Aqueous humour dynamics. Theoretical considerations. *Am. J. Ophthal.* **41**, 383–398.

GALIN, M. A., BARAS, I. & MANDELL, G. L. (1961). Measurements of aqueous flow utilizing the perilimbal suction cup. *Arch. Ophthal.* **66**, 65.

GALIN, M. A., DAVIDSON, R. & PASMANIK, S. (1963). An osmotic comparison of urea and mannitol. *Am. J. Ophthal.* **55**, 244–247.

GLOSTER, J. (1961). Influence of facial nerve on intraocular pressure. *Br. J. Ophthal.* **45**, 259–278.

GLOSTER, J. & GREAVES, D. P. (1957). Effect of diencephalic stimulation upon intraocular pressure. *Br. J. Ophthal.* **41**, 513–532.

GOLDMANN, H. (1946). Weitere Mitteilung uber den Abfluss des Kammerwassers beim Menschen. *Ophthalmologica, Basel,* **112**, 344–349.

GOLDMANN, H. (1955). Un nouveau tonomètre à aplanation. *Bull. Soc. franc. Ophtal.* **67**, 474.

GRANT, W. M. (1950). Tonographic method for measuring the facility and rate of aqueous outflow in human eyes. *Arch. Ophthal., N.Y.* **44**, 204–214.

GRANT, W. M. (1958). Further studies on facility of flow through trabecular meshwork. *Arch. Ophthal., N.Y.* **60**, 523–533.

GRANT, W. M. (1963). Experimental aqueous perfusion in enucleated human eyes. *Arch. Ophthal., N.Y.* **69**, 783–801.

GREAVES, D. P. & PERKINS, E. S. (1952). Influence of the sympathetic nervous system on the intraocular pressure and vascular circulation of the eye. *Br. J. Ophthal.* **36**, 258–264.

HAYREH, S. S. (1969). Blood supply of the optic nerve head and its role in optic atrophy, glaucoma, and oedema of the optic disc. *Br. J. Ophthal.* **53**, 721–748.

HEATH, H. & FIDDICK, R. (1966). The active transport of ascorbic acid by the rat retina. *Exp. Eye Res.* **5**, 156–163.

HETLAND-ERIKSEN, J. (1966). On Tonometry. I–IX. *Acta Ophthal.* **44**, 5–11, 12–19, 107–113, 114–120, 515–521, 522–538, 725–736, 881–892, 893–900.

HOLLAND, M. G. (1970). Chloride ion transport in the isolated ciliary body. II. *Invest. Ophthal.* **9**, 30–41.

HOLLAND, M. G. & GIPSON, C. C. (1970). Chloride ion transport in the isolated ciliary body. *Invest. Ophthal.* **9**, 20–29.

HOLLAND, M. G. & STOCKWELL, M. (1967). Sodium ion transport of the ciliary body in vitro. *Invest. Ophthal.* **6**, 401–409.

HOLMBERG, A. (1955). Studies on the ultrastructure of the non-pigmented epithelium in the ciliary body. *Acta ophthal. Kbh.* **33**, 377–381.

HOLMBERG, A. (1959). The fine structure of the inner wall of Schlemm's canal. *Archs. Ophthal. N.Y.* **62**, 956–958.

HOLMBERG, A. & BECKER, B. (1960). The effect of hypothermia on aqueous humour dynamics, *Am. J. Ophthal.* **49**, Pt. 2, 1134–1140.

HÖRVEN, I. (1961). Tonometry in newborn infants. *Acta ophthal. Kbh.* **39**, 911–918.

HUGGERT, A. (1958). Experimentally induced increase in intraocular pressure in the rabbit eye. *Acta ophthal. Kbh.* **36**, 750–760.

JACKSON, R. T. & WAITZMAN, M. B. (1965). Effect of some steroids on aqueous humor dynamics. *Exp. Eye Res.* **4**, 112–123.

JOCSON, V. L. & SEARS, M. L. (1968). Channels of aqueous outflow and related blood vessels. I. *Macaca mulatta* (Rhesus). *Arch. Ophthal.* **80**, 104–114.

KAYE, G. I. & PAPPAS, G. D. (1965). Studies on the ciliary epithelium and zonule. III. The fine structure of the rabbit ciliary epithelium in relation to the localization of ATPase activity. *J. Microscopie*, **4**, 497–508.

KAYES, J. (1967). Pore structure of the inner wall of Schlemm's canal. *Invest. Ophthal.* **6**, 381–394.

KINSEY, V. E. (1947). Transfer of ascorbic acid and related compounds across the blood-aqueous barrier *Am. J. Ophthal.* **30**, 1262–1266.

KINSEY, V. E., GRANT, M. & COGAN, D. G. (1942). Water movement and the eye. *Archs. Ophthal., N.Y.* **27**, 242–252.

KINSEY, V. E. & PALM, E. (1955). Posterior and anterior chamber aqueous humour formation. *Arch. Ophthal., N.Y.* **53**, 330–344.

KINSEY, V. E. & REDDY, D. V. N. (1959). An estimate of the ionic composition of the fluid secreted into the posterior chamber, inferred from a study of aqueous humour dynamics. *Doc. Ophthal.* **13**, 7–40.

KOZART, D. M. (1968). Light and electron microscopic study of regional morphologic differences in the processes of the ciliary body in the rabbit. *Invest. Ophthal.* **7**, 15–33.

KUPFER, C. (1961). Studies on intraocular pressure. I. *Arch. Ophthal., N.Y.*, **65**, 565–570.

LANGHAM, M. E. (1959). Influence of the intraocular pressure on the formation of the aqueous humour and the outflow resistance in the living eye. *Br. J. Ophthal.* **43**, 705–732.

LANGHAM, M. E. (1960). Steady-state pressure relationships in the living and dead cat. *Am. J. Ophthal.* **50**, Pt. 2, 950–958.

LANGHAM, M. E. & FRASER, L. K. (1966). The absence of supersensitivity to adrenergic amines in the eye of the conscious rabbit following preganglionic cervical sympathotomy. *Life Sciences*, **5**, 1699–1705.

LANGHAM, M. E. & ROSENTHAL, A. R. (1966). Role of cervical sympathetic nerve in regulating intraocular pressure and circulation. *Am. J. Physiol.* **210**, 786–794.

LANGHAM, M. E. & TAYLOR, C. B. (1960). The influence of pre- and post-ganglionic section of the cervical sympathetic on the intraocular pressure of rabbits and cats. *J. Physiol.* **152**, 437–446.

LANGHAM, M. E. & TAYLOR, C. B. (1960). The influence of superior cervical ganglionectomy on the intraocular pressure. *J. Physiol.* **152**, 447–458.

LANGHAM, M. E. & WYBAR, K. C. (1953). A fluorophotometer for the study of intraocular dynamics in the living animal. *J. Physiol.* **120**, 5P.

LEHMANN, G. & MEESMANN, A. (1924). Uber das Bestehen eines Donnangleichgewichtes zwischen Blut und Kammerwasser bzw. Liquor cerebrospinalis. *Pflüg. Arch.* **205**, 210–232.

LEYDHECKER, W., AKIYAMA, K. & NEUMANN, H. G. (1958). Der intraokulare Druck gesunder menschlicher Augen. *Klin. Mbl. Augenheilk.* **133**, 662–670.

LINNÉR, E. (1952). Effect of unilateral ligation of the common carotid artery on the blood flow through the uveal tract as measured directly in a vortex vein. *Acta physiol. scand.* **26**, 70–78.

McBain, E. H. (1958). Tonometer calibration. II: Ocular rigidity. *Arch. Ophthal., N.Y.* **60,** 1080–1091.

McBain, E. H. (1960). Tonometer calibration. III. *Arch. Ophthal., N.Y.* **63,** 936–942.

Macri, F. C. (1961). Interdependence of venous and eye pressure. *Arch. Ophthal., N.Y.* **65,** 442–449.

Macri, F. J. (1967). The pressure dependence of aqueous humor formation. *Arch. Ophthal., N.Y.* **78,** 629–633.

Macri, F. J., Dixon, R. L. & Rall, D. P. (1965). Aqueous humour turnover rates in the cat. I. Effect of acetazolamide. *Invest. Ophthal.* **4,** 927–934.

Macri, F. J., Wanko, T. & Grimes, P. A. (1958). The elastic properties of the human eye. *Arch. Ophthal., N.Y.* **60,** 1021–1026.

Mackay, R. S. & Marg. E. (1959). Fast automatic, electronic tonometers based on an exact theory. *Acta Ophthal.* **37,** 495–507.

Maurice, D. M. (1951). An applanation tonometer of new principle. *Br. J. Ophthal.* **35,** 178–182.

McEwen, W. K. (1958). Application of Poiseuille's law to aqueous outflow. *Arch. Ophthal., N.Y.* **60,** 290–294.

Maggiore, L. (1917). Struttura, comportamento e significato del canale di Schlemm nell'occhio umano, in condizioni normali e patologische. *Ann. Ottalm.* **40,** 317–462.

Melton, C. E. & De Ville, W. B. (1960). Perfusion studies on eyes of four species. *Am. J. Ophthal.* **50,** 302–308.

Miller, J. E. & Constant, M. A. (1960). The measurement of rabbit ciliary epithelial potentials *in vitro. Am. J. Ophthal.* **50,** Pt. 2, 855–862.

Moses, R. A. & Becker, B. (1958). Clinical tonography: the scleral rigidity correction. *Am. J. Ophthal.* **45,** 196–208.

Noell, W. K. (1963). Cellular physiology of the retina. *J. Opt. Soc. Am.* **53,** 36–48.

Oppelt, W. W. (1967). Measurement of aqueous humor formation rates by posterior-anterior chamber perfusion with inulin. *Invest. Ophthal.* **6,** 76–83.

Oppelt, W. W. & White, E. D. (1968). Effect of ouabain on aqueous humor formation rate in cats. *Invest. Ophthal.* **7,** 328–333.

Oppelt, W. W., White, E. D. & Halpert, E. S. (1969). The effect of corticosteroids on aqueous humor formation rate and outflow facility. *Invest. Ophthal.* **8,** 535–541.

Pandolfi, M. (1967). Fibrinolysis and outflow resistance in the eye. *Am. J. Ophthal.* **64,** 1141–1148.

Pandolfi, M. & Kwaan, H. C. (1967). Fibrinolysis in the anterior segment of the eye. *Arch. Ophthal., N.Y.* **77,** 99–104.

Perkins, E. S. (1955). Pressure in the canal of Schlemm. *Br. J. Ophthal.* **39,** 215–219.

Perkins, E. S. (1957). Influence of the fifth cranial nerve on the intraocular pressure in the rabbit eye. *Br. J. Ophthal.* **41,** 257–300.

Perkins, E. S. (1961). Sensory mechanisms and intraocular pressure. *Exp. Eye Res.* **1,** 160–167.

Perkins, E. S. & Gloster, J. (1957). Distensibility of the eye. *Br. J. Ophthal.* **41,** 93–102, 475–486.

Pohjola, S. (1966). The glucose content of the aqueous humour in man. *Acta Ophthal.* Suppl. 88.

Pollack, I. P., Becker, B. & Constant, M. A. (1960). The effect of hypothermia on aqueous humour dynamics. *Am. J. Ophthal.* **49,** Pt. 2, 1126–1132.

Reddy, D. V. N. & Kinsey, V. E. (1960). Composition of the vitreous humour in relation to that of the plasma and aqueous humour. *Arch. Ophthal., N.Y.* **63,** 715–720.

Reddy, D. V. N. & Kinsey, V. E. (1963). Transport of amino acids into intraocular fluids and lens in diabetic rabbits. *Invest. Ophthal.* **2,** 237–242.

Reddy, D. V. N., Rosenberg, C. & Kinsey, V. E. (1961). Steady-state distribution of free amino acids in the aqueous humors, vitreous body and plasma of the rabbit. *Exp. Eye Res.* **1,** 175–181.

REESE, T. S. & KARNOVSKY, M. J. (1967). Fine structural localization of a blood-brain barrier to exogenous peroxidase. *J. Cell Biol.* **34**, 207–217.

RILEY, M. V. (1966). The tricarboxylic acid cycle and glycolysis in relation to ion transport by the ciliary body. *Biochem. J.* **98**, 898–902.

RODRIGUEZ-PERALTA, L. (1962). Experiments on the site of the blood-ocular barrier. *Anat. Rec.* **142**, 273.

ROHEN, J. W. (1969). New studies on the functional morphology of the trabecular meshwork and the outflow channels. *Trans. Ophthal. Soc., U.K.* **89**, 431–447.

ROHEN, J. W. & LÜTJEN, E. (1968). Uber die Altersveränderungen des Trabekel-werke im menschlichen Auge. *v. Graefes' Arch. Ophthal.* **175**, 285–307.

ROHEN, J. W. & RENTSCH, F. J. (1969). Elektronenmikroskopische Untersuchungen über den Bau der Aussenwand des Schlemmschen Kanals unter besonderer Berücksichtigung der Abflusskanäle und Altersveränderungen. *v. Graefes' Arch. Ophthal.* **177**, 1–17.

ROHEN, J. W. & STRAUB, W. (1967). Elektronenmikroskopische Untersuchungen über die Hyalinisierung des Trabeculum corneosclerale beim Sekundär-glaukom. *v. Graefes' Arch. Ophthal.* **173**, 21–41.

RUSKELL, G. L. (1970). The orbital branches of the pterygopalatine ganglion and their relationship with internal carotid nerve branches in primates. *J. Anat.* **160**, 323–339.

RUSKELL, G. L. (1970). An ocular parasympathetic nerve pathway of facial nerve origin and its influence on intraocular pressure. *Exp. Eye Res.* **10**, 319–330.

SCHEIE, H. G., MOORE, E. & ADLER, F. H. (1943). Physiology of the aqueous in completely iridectomised eyes. *Arch. Ophthal., N.Y.* **30**, 70–74.

SEARS, M. L. (1960). Miosis and intraocular pressure changes during manometry. *Arch. Ophthal., N.Y.* **63**, 707–714.

SEARS, M. L. (1960). Outflow resistance of the rabbit eye: technique and effects of acetazolamide. *Arch. Ophthal., N.Y.* **64**, 823–838.

SEARS, M. L. (1966). The mechanism of action of adrenergic drugs in glaucoma. *Invest. Ophthal.* **5**, 115–119.

SEARS, M. L. & BÁRÁNY, E. H. (1960). Outflow resistance and adrenergic mechanisms. *Archs. Ophthal., N.Y.* **64**, 839–849.

SEARS, M. L. & GILLIS, C. N. (1967). Mydriasis and the increase in outflow of the aqueous humour of the rabbit eye after cervical ganglionectomy in relation to the release of norepinephrine from the iris. *Biochem. Pharmacol.* **16**, 777–782.

SEIDEL, E. (1918). Experimentelle Untersuchungen über die Quelle und den Verlauf der intraokularen Saftströmung. *v. Graefes' Arch. Ophthal.* **95**, 1–72.

SEIDEL, E. (1921). Uber den Abfluss des Kammerwassers aus der vorderen Augen-kammer. *v. Graefes' Arch. Ophthal.* **104**, 357–402.

SERR, H. (1937). Zur Analyse der spontanen Pulsercheinungen in den Netzhaut-gefässen. *v. Graefes' Arch. Ophthal.* **137**, 487–505.

SIMON, K. A., BONTING, S. L. & HAWKINS, N. M. (1962). Studies on sodium-potassium-activated adenosine triphosphatase. II. Formation of aqueous humour. *Exp. Eye Res.* **1**, 253–261.

SINGH, S. & DASS, R. (1960). The central artery of the retina. *Br. J. Ophthal.* **44**, 193–212, 280–299.

SLEZAK, H. (1969). Uber Fluorescein in der Hinterkammer des menschlichen Auges. *v. Graefes' Arch. Ophthal.* **178**, 260–267.

SMELSER, G. K. & PEI, Y. F. (1965). Cytological basis of protein leakage into the eye following paracentesis. *Investigative Ophthal.* **4**, 249–263.

SPEAKMAN, J. S. (1960). Drainage channels in the trabecular wall of Schlemm's canal. *Br. J. Ophthal.* **44**, 513–523.

v. SALLMANN, L., FUORTES, M. G. F., MACRI, F. J. & GRIMES, P. (1958). Study of afferent electric impulses induced by intraocular pressure changes. *Am. J. Ophthal.* **45**, Pt. 2, 211–220.

v. SALLMANN, L. & LOEWENSTEIN, O. (1955). Responses of intraocular pressure, blood pressure and cutaneous vessels to electrical stimulation of the dien-cephalon. *Am. J. Ophthal.* **39**, Pt. 2, 11–29.

TÖRNQVIST, G. (1970). Effect of oculomotor nerve stimulation on outflow facility and pupil diameter in a monkey (*Cercopithecus ethiops*). *Invest. Ophthal.* **9,** 220–225.

TREISTER, G. & BÁRÁNY, E. H. (1970). Mydriasis and intraocular pressure decrease in the conscious rabbit after unilateral superior cervical ganglionectomy *Invest. Ophthal.* **9,** 331–342.

TRIPATHI, R. C. (1968). Ultrastructure of Schlemm's canal in relation to aqueous outflow. *Exp. Eye Res.* **7,** 335–341.

TRIPATHI, R. C. (1969). Ultrastructure of the trabecular wall of Schlemm's canal. *Trans. Ophthal. Soc., U.K.* **89,** 449–465.

TRIPATHI, R. C. (1971). Mechanism of the aqueous outflow across the trabecular wall of Schlemm's canal. *Exp. Eye Res.* **11,** 111–116.

VEGGE, T. & RINGVOLD, A. (1969). Ultrastructure of the wall of the human iris vessels. *Z. Zellforsch.* **94,** 19–31.

VRÁBEC, F. (1954). L'innervation du système trabéculaire de l'angle irien. *Ophthalmologica, Basel,* **128,** 359–364.

WAITZMAN, M. B. (1970). Pupil size and ocular pressure after sympathectomy and prostaglandin-catecholamine treatment. *Exp. Eye Res.* **10,** 219–222.

WAITZMAN, M. B. & KING, C. D. (1967). Prostaglandin influences on intraocular pressure and pupil size. *Am. J. Physiol.* **212,** 329–334.

WÅLINDER, P.-E. & BILL, A. (1969). Influence of intraocular pressure and some drugs on aqueous flow and entry of cycloleucine into the aqueous humor of vervet monkeys (*Cercopithecus ethiops*) *Invest. Ophthal.* **8,** 446–458.

WEIGELIN, E. & LÖHLEIN, H. (1952). Blutdruckmessungen an den episkleral Gefässen des Auges bei kreislaufgesund Personen. *v. Graefes' Arch. Ophthal.* **153,** 202–213.

WESSELY, K. (1900). Uber die Wirkung des Suprarenins auf das Auge. *Ber. deutsch. ophth. Ges.* pp. 69–83.

WESSELY, K. (1908). Experimentelle Untersuchungen ü. d. Augendruck, sowie über qualitative und quantitative Beeinflussung des intraokularen Flussigkeitswechsels. *Arch. Augenheilk.* **60,** 1–48, 97–160.

YAMASHIDA, T. & ROSEN, D. A. (1965). Electron microscopic study of trabecular meshwork. *Am. J. Ophthal.* **60,** 427–434.

PART 2

THE VITREOUS BODY

THIS is a transparent jelly occupying, in man and most vertebrates, by far the largest part of the globe. As Fig. 1, p. 1, shows, it is bounded by the retina, ciliary body and the posterior capsule of the lens. If the sclera is cut through round the equator, the posterior half with the attached retina comes away from the vitreous body, leaving this suspended from its firm attachments to the ciliary body and pars plana of the retina. To isolate it completely from these attachments careful cutting and scraping are necessary.

CHEMISTRY

When discussing the dynamic relationships between aqueous humour and the vitreous body it was considered sufficient to treat the latter as an essentially aqueous medium; and this is justified by the observation that over 99 per cent of the material in this body is water and dissolved salts. The special features of the inorganic composition have been considered earlier, and here we are concerned with the composition in so far as it bears on the structure and stability of the vitreous body as a gel. If the vitreous body is held in a pair of forceps, or placed on a filter, it drips and after a time only a small skin of viscid material remains; this residue was called by Mörner the *residual protein*, and was considered to constitute the gel-forming material in essentially the same way that fibrin constitutes the gel-forming material of a plasma clot. The liquid dripping away, or passing through the filter, contains the dissolved crystalloids together with a mucopolysaccharide, *hyaluronic acid*, that imparts to the solution a high viscosity by virtue of its high molecular weight and highly asymmetrical molecules. In addition there is a small percentage of soluble proteins, some of which are peculiar to the vitreous whilst others are identical with plasma proteins (Laurent, Laurent & Howe, 1962). In particular, the glycoproteins, containing sialic acid, are in much higher concentration than in plasma. Figures given for the colloidal composition of bovine vitreous body are: Vitrosin, 150 mg/litre, Hyaluronic acid, 400 mg/litre and Soluble proteins 460 mg/litre. Thus the colloidal material constitutes less than 0·1 per cent of the total body.

Residual Protein or Vitrosin

Purified residual protein may be dissolved in alkali and subsequently precipitated by acid; when the precipitated material is examined in the electron microscope it appears fibrillar and very similar to collagen; this similarity is borne out by chemical analysis which reveals the characteristically high proportions of glycine and hydroxyproline; it differs from that in skin in having a high proportion of sugars, probably in the form of a polysaccharide firmly attached to the polypeptide skeleton (McEwan & Suran, 1960); in this respect it is similar to the collagens derived from Bowman's membrane and the lens capsule. The enzyme collagenase will liquefy the vitreous body.

STRUCTURE

Tentatively we may suppose that the structure of the vitreous gel is maintained by a framework of branching vitrosin fibrils, in the meshes of which the watery solution of crystalloids, hyaluronic acid and soluble proteins is imprisoned. The concentration of residual protein is so small (0·01 per cent) that the fibrils must be very fine, and the meshwork relatively coarse, if such a small amount of material is to form a coherent jelly.

Collagen Fibrils. This picture, which may be deduced on purely physico-chemical principles, is partly sustained by modern electron-microscopical studies, although there are still many points of uncertainty. The fibrillar basis is certainly well established by the examination of thin sections of frozen vitreous; this is illustrated by Fig. 36, Pl. III, from a paper by Fine & Tousimis which confirms earlier studies, particularly those of Schwarz, in this respect. As predicted, the fibrils are very thin being some 200 to 250 Å in diameter, and of indefinite length.* Olsen (1965) observed that some fibrils were made up of the parallel alignment of thin filaments, some 20–30 Å in diameter, and these could have represented chains of the fundamental unit of collagen, the tropocollagen molecule. The characteristic feature of collagen, prepared from most sources, is a cross-striation of period 640 Å; in general the periodicity observed along the vitrosin fibrils is much smaller than this, namely 110 Å.

Periodicity. The large period of 640 Å, usually observed in collagen, is due to the staggered side-by-side alignment of the individual tropocollagen molecules, so that at intervals certain polar groups come into register; these take up electron-stain and appear as dark bands on the large fibres made up of the aggregation of many of the fine fibrils. It is quite likely that the failure to observe the 640 Å banding is due to a less regular arrangement of filaments or, simply, that there are not sufficient in the very thin fibrils of vitrosin to render this periodicity visible; thus Smith & Serafini-Fracassini (1967) stated that, when these fibrils of diameter 140 Å were seen packed side-by-side, faint cross-bands of 640 Å periodicity were visible.

Hyaluronic Acid. By treating vitreous preparations with hyaluronidase, which breaks down the large hyaluronic acid molecules into their constituent sugar molecules and uronic acid, the vitrosin fibrils become more distinct, suggesting that hyaluronic acid is normally in close association with the vitrosin; when Smith & Serafini-Fracassini stained the vitreous gel with lead, which precipitates the hyaluronic acid, they observed dark aggregates at regular intervals along the vitrosin fibres suggesting that, in the normal state, the hyaluronic acid was attached at regular points.

Local Variations in Structure

The picture just presented, of a fibrous framework holding the fluid components of the vitreous in its meshes, would suggest that the vitreous body was essentially homogeneous throughout its bulk; a considerable amount of evidence indicates,

---

* Schwarz (1961) gives an even smaller value, namely 40–100 Å with an average of 67 Å, whilst Grignolo (1954) described three types with diameters ranging from 150 to 800 Å. Olsen's (1965) filaments had a diameter of 120–130 Å and those of Smith & Serafini-Fracassini 110 Å on average, the finest being 80 Å. The important point is the fineness compared with those in the cornea.

however, the existence of differentiation to give microscopically visible structures; these are most probably superimposed on the fundamental basis described above. To appreciate the significance of the differentiation we must review the embryonic history of the vitreous body.

Embryonic History. In the early stages, the optic cup is mainly occupied by the lens vesicle (p. 87); as the cup grows, the space formed is filled by a system of fibrillar material presumably secreted by the cells of the embryonic retina. Later, with the penetration of the hyaloid artery, more fibrillar material, apparently derived from the cells of the wall of the artery and other vessels, contributes to filling the space. The combined mass is known as *primary vitreous*. The *secondary vitreous* develops later and is associated with the increasing size of the vitreous cavity and the involution of the hyaloid system. The main hyaloid artery remains for some time but eventually disappears, and its place is left as a tube of primary vitreous surrounded by secondary vitreous; this tube is called *Cloquet's canal*, but is not a liquid-filled canal but simply a portion of differentiated gel. The *tertiary* vitreous is the name given to the grossly fibrillar material secreted by the inner walls of the optic cup in the region that will later become the ciliary body; this material eventually becomes the zonule of Zinn or suspensory ligament of the lens.

Phase-Contrast Microscopy. When examined in the ordinary light-microscope the vitreous body is said to be optically empty; in the phase-contrast microscope, however, Bembridge, Crawford & Pirie described three characteristic differentiations; in the anterior portion definite fibres of relatively large diameter were found, whilst in the bulk of the body there were much finer, but still optically resolvable, fibres. Chemically the large fibres seemed to be different from the fine fibres; thus the enzyme collagenase left these large fibres intact whilst it dissolved the fine fibres; again, the enzyme trypsin or chymotrypsin dissolved the large fibres but not the others. Finally, on the surface, there appeared to be a hyaline membrane (Fig. 37, Pl. III). Since phase-contrast microscopy is carried out on perfectly normal fresh tissue it is difficult to rule out these structures as artefacts; nevertheless the electron-microscopical observations, indicating a fibril of the order of 100–200 Å by contrast with these optically visible fibres of the order of 1–2 μ, i.e. 10,000–20,000 Å, indicate that the fundamental basis of structure is probably the fine fibril, whilst the larger fibres must be regarded as essentially localized densifications of this primary structure. This is especially true of the zonular fibres proper; electron-microscopical observations of the vitreous in this region indicate large fibres made up of closely aggregated fine filaments, whilst the neighbouring vitreous shows only the fine filaments loosely scattered.

Limiting Membrane. Again, in the electron-microscope there is little evidence for a structured limiting hyaline membrane belonging to the vitreous body; according to Tousimis & Fine the *internal limiting membrane* of the retina belongs to this tissue and cannot be regarded as a limiting membrane of the vitreous; when the vitreous is removed the membrane remains adherent to the cells of the retina and must be regarded as a basement membrane for these cells; as seen in the electron-microscope it is an acellular structure some 3000–5000 Å thick, and is essentially the basement membrane of the glial cells of the retina, being apparently continuous, embryologically, with the basement membrane of the ciliary epithelium, whose cell-bases face towards the vitreous body. It extends from the surface of the plasma membranes of the glial cells to the surface of the cortical layer of vitreous, which itself is covered by a delicate basal lamina made of the same basement-membrane material. The vitrosin fibrils run parallel with the surface, but make right-angle turns to connect with the basement membrane; the extent to which this happens varies with position, being greatest at the points of strong attachment, such as the ora serrata (Zimmerman & Straatsma, 1960).*

Cortical Zone. We must regard the hyaline layer, then, as a region in which the fibrils are more densely arranged, so as to give a sufficient difference of refractive index to create the appearance of a membrane in phase-contrast microscopy. This so-called *cortical zone* has been studied by Balazs who has emphasized the presence of numerous stellate cells buried in it; he has called these *hyalocytes* since he considers that they are responsible for synthesis of hyaluronic acid. Gärtner (1965) has made a careful study of the cells in the cortical layer in the rat embryo and young child; he characterizes the great majority of these cells as fibroblasts; in the region of the pars plana of the ciliary body they form a definite layer covering the epithelium; and the appearance of collagen-like fibrils in close association suggested that it was these cells that were responsible for secretion of vitrosin as well as hyaluronic acid.

Chemical Differences. There is chemical evidence for some differentiation of the vitreous body; for example Bembridge *et al.* found that the proportions of vitrosin varied, being highest in the anterior portion where also the electron-microscopists had differentiated the highest density of fibrils. Again Balazs found a very high concentration of hyaluronic acid and vitrosin in the cortical layer, corresponding with its more compact character; finally the concentrations of acid glycoproteins and water-soluble proteins vary with the region analysed.†

Hyaluronidase. This enzyme, which breaks hyaluronic acid down to its constituent amino sugar and hexuronic acid molecules, occurs naturally, for example in tears, where it is called *lysozyme* (Sect. III). It was considered to be present in the intraocular tissues where it might exert an influence on the fluidity of the vitreous body; recent studies have, however, ruled this out (Berman & Voaden, 1970).

Origin and Maintenance

It is customary to describe the vitreous as the product of secretion of ectodermal elements, namely the glial cells of the retina and the ciliary epithelium. The point has been discussed at some length by Gärtner (1965) in the light of his

---

* These regions of attachment have been examined with especial care by Hogan (1963); thus the vitreous fibrils attaching to the ciliary epithelium enter crypts and make contact with the basement membrane at the edges and depths of the crypts along with zonular fibres where they form a strong union. In eyes with vitreal detachment the cortex of the vitreous separates from the basement membrane of the glial cells; the separation extends as far forward as the peripheral third of the retina and seems to be limited by the radial attachments of the vitreous fibrils to the retina in this area.

† During development there are marked changes in composition; for example the concentration of hyaluronic acid steadily increases. Where different species are compared some very large differences are found; for example the vitreous of the bush-baby (*Galago*) and of the owl monkey (*Aotus*) is liquid except for a thin 50–100 μ thick cortical layer; it is this relatively solid cortex, which contains all the collagen of the vitreous, that acts as a barrier to flow between the posterior chamber and the fluid vitreous. Thus on removing the aqueous humour by paracentesis, vitreous is not withdrawn; the fluid material contains most of the hyaluronic acid (Österlin & Balazs, 1968). By contrast the vitreous bodies of the frog and hen are very solid, and have very little hyaluronic acid and large amounts of collagen (Balazs *et al.*, 1959a, b).

electron-microscopical studies; and he favours a mesenchymal origin for the primary vitreous and a connective tissue origin for the secondary vitreous. The tertiary vitreous, i.e. the zonule, he unequivocally attributes to fibroblasts, and his experimental evidence and arguments are convincing.

The eye reaches its full size within the first few years of life in man; after this there is no necessity for synthesis of new material for the vitreous body, unless there is an appreciable wear-and-tear. The problem as to whether vitreous may be replaced after loss is of considerable interest to the ophthalmologist, but experimental and pathological studies provide little evidence in favour of this. It seems that the hyalocytes, separated from the vitreous by centrifugation, can incorporate radioactive precursors of hyaluronic acid, e.g. glucosamine, into high-molecular weight compounds, and it is interesting that acellular parts of the vitreous also have activity, suggesting the presence of extracellular enzymes. Studies on the intact animal, e.g. those of Österlin in the owl monkey, indicate that, whatever turnover there is in the hyaluronic acid moiety of the vitreous, it is very slow (Österlin, 1968; Berman & Gombos, 1969).

Probable Structure

Modern studies on connective tissue indicate that both the collagen and mucopolysaccharide constituents contribute to the stability of the gel-structure that constitutes its basis. In the vitreous body the mucopolysaccharide is hyaluronic acid, a polysaccharide constituted by alternating acetyl glycosamine and glucuronic acid molecules, as indicated on p. 71. The molecular weights of such polysaccharides may be very large, of the order of a million, and in the case of the vitreous it would seem that the material exists in the form of a variety of different sized molecules, i.e. different degrees of polymerization. The molecule probably exists as a randomly kinked coil entraining relatively large quantities of water in

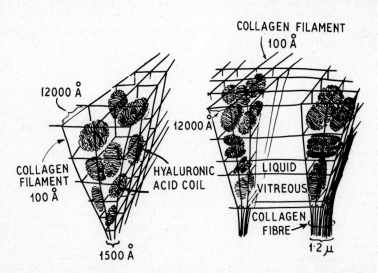

Fig. 38. Schematic picture of the fine structure of the vitreous gel showing network reinforced with hyaluronic acid molecules. *Left:* Random distribution of the structural elements. *Right:* Formation of liquid pool and partial collapse of the network. (Balazs, in *Structure of the Eye.*)

PLATE III

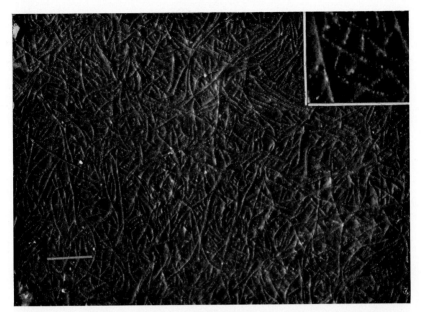

FIG. 36. Shadowcast section of posterior vitreous structure, approximately 0·5 mm anterior to surface of the equatorial retina of the human eye. The orientation of the filaments in the plane of the section suggests a tangential cut through a lamella. The filaments are uniform in width and show a distinct periodicity. × 12,000. The enlarged inset (× 38,000) demonstrates periodicity on filaments. (Fine & Tousimis, *Arch. Ophthal.*)

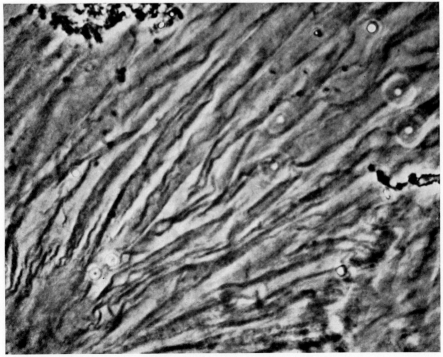

FIG. 37. Hyaline membrane of ox vitreous body. Phase-contrast.
(Courtesy Antoinette Pirie.)

To face p. 64.

PLATE IV

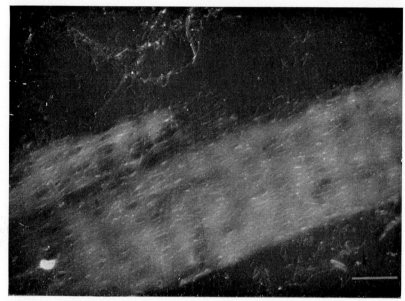

FIG. 39. Shadowcast section of adult human zonule and adjacent vitreous body framework. The widths of the filaments of both zonular and adjacent framework are similar; also the periodicities of the filaments (220 Å). × 13,000. (Fine & Tousimis, *Arch. Ophthal.*)

FIG. 41. Transverse section through the stroma of a human cornea showing several lamellae (the fibres in stained sections), in each of which the collagen fibrils are orientated parallel with each other and perpendicular to those in adjacent layers. Osmium fixation. × 10,230. (Jakus, in *The Structure of the Eye*, Academic Press.)

its framework so that in solution it occupies some 1000 times the volume it occupies in the dried state. The individual molecules probably entangle with each other to form a continuous three-dimensional network. In the vitreous body, then, such a network might act to stabilize the primary network formed by the vitrosin fibrils, as illustrated in Fig. 38. As early postulated by Davson & Duke-Elder (1935), and later by Pirie, Schmidt & Waters (1948), the effects of environmental changes on the state of the vitreous body are largely exerted through the hyaluronic acid moiety; and the same seems to be true of the cornea (p. 73).

Transport through Gel. The gel structure formed in this way is doubtless adequate to prevent significant bulk flow of fluid through it, so that any posterior flow of aqueous humour, if it occurs, would take place meridionally (Fowlks, 1963) and not through "Cloquet's canal". Diffusion of molecules would, however, be quite unrestricted, so that from the point of view of penetration of material across the blood-vitreal barrier the vitreous may be treated as an unstirred liquid; this applies to the movement of such large molecules as albumin too (Maurice, 1959); it has been suggested that the more tightly packed cortical region offers some restraint to entry of colloidal material, e.g. from the retina, as indicated above, but the diffusion of protein within the much denser cornea casts some doubt on this.

Transparency. Finally, we may note that the presence of solid fibrils in the vitreous in no way impairs its transparency; this is presumably because the fibrils are thin and their concentration is low (p. 61). Pathological aggregates of colloidal material would cause opacities.

THE ZONULE

Macroscopic Appearance. The zonule, or suspensory ligament of the lens, may be regarded as an extreme instance of differentiation of the vitreous body, its fibrous parts being composed presumably of the same material as that making up the large fibres of the vitreous body described by Bembridge *et al.*, since they are both dissolved by the enzyme trypsin; the difference between the two is essentially that the zonular fibres have found attachment to the lens capsule whilst the coarse fibres of the vitreous attach to the ciliary body. In the fresh specimen the zonule appears as a thick band-shaped structure attached internally to the equatorial region of the lens capsule; the anterior surface runs straight from the lens to the anterior region of the ciliary processes (Fig. 5, p. 4). The surfaces of the zonule show a marked striation running in a general sense from the lens to the ciliary body; these striations probably represent fibres of highly condensed residual protein; between them is a perfectly homogeneous transparent gel which gives the surface a membrane-like appearance. The space between the anterior and posterior surfaces is criss-crossed by strands of fibrous material, the spaces so formed being once again filled by transparent homogeneous jelly.

Chemistry. As indicated above, the zonular fibres are attacked by trypsin but not by collagenase, and this indicates a difference in chemical composition, which has been confirmed by the chemical analyses of Buddecke & Wollensak (1966), who found negligible amounts of hydroxyproline and only a small percentage of glycine—amino acids that are present in characteristically high concentration in collagen. There was a remarkably high proportion of cysteine (cystine) in the zonular protein, and this may have pathological interest for the Marfan syndrome associated with homocystinuria, i.e. the incomplete formation or tearing of the zonular fibres leading to ectopia lentis. The extracted material contained some 5 per cent of carbohydrate which was not attacked by hyaluronidase, and it may be that the polysaccharide acts as a cover to the fibrils, protecting them against the attack of proteases in the anterior chamber. Thus in old people there is less of the polysaccharide, and it may be for this reason that their zonules are relatively easily attacked by trypsin, a procedure that is employed in cataract surgery.

Electron-Microscopy. In the electron microscope the zonular fibres appear as dense aggregations of the fine fibrils that constitute the framework of the surrounding vitreous (Fig. 39, Pl. IV). Preparations of the ciliary epithelium made by Pappas & Smelser show that the fibres may be regarded as extensions of the basement membrane of the epithelial cells, which is itself fibrillar in nature. Thus the basement membrane invests the surface of the ciliary body with its processes very closely, so that the attachment of the zonular fibres to the ciliary body may be said to be similar to the "attachment of a tight-fitting glove on one's hand", the mere close approximation of the two membranes—epithelial cell membrane and basement membrane—on an irregular contour, giving a very efficient attachment. No fibrils of the zonule entered the epithelial cells.

GENERAL REFERENCES

BALAZS, E. A. (1961). Molecular morphology of the vitreous body. In *The Structure of the Eye*. Ed. Smelser, pp. 293–310. New York: Academic Press.

BALAZS, E. A. (1968). Die Mikrostruktur und Chemie des Glaskörpers. *Ber. deutsch. ophth. Ges.* **68**, 536–572.

GRIGNOLO, A. (1954). Studi sulla struttura submicroscopica dei tessuti oculari. *Boll. d'Oculist.* **33**, 1–144.

PIRIE, A. (1969). The vitreous body. In *The Eye*. Ed. Davson, pp. 273–297. London: Academic Press.

ROHEN, J. W., Ed. (1965). *Die Struktur des Auges*. Stuttgart: Schattauer Verlag.

SCHEPENS. C. L., Ed. (1960). *Importance of the Vitreous Body in Retina Surgery with Special Emphasis on Reoperations*. St. Louis: Mosby.

SCHWARZ, W. (1961). Electron microscopic observations on the human vitreous body. In *The Structure of the Eye*. Ed. Smelser, pp. 283–291. New York: Academic Press.

SPECIAL REFERENCES

BALAZS, E. A., LAURENT, T. C. & LAURENT, U. B. G. (1959a). Studies on the structure of the vitreous body. VI. *J. biol. Chem.* **234**, 422–430.

BALAZS, E. A., LAURENT, T. C., LAURENT, U. B. G., DE ROCHE, M. H. & DUNNEY, D. M. (1959b). Studies on the structure of the vitreous body. VIII. *Arch. Biochem. Biophys.* **81**, 464–479.

BEMBRIDGE, B. A., CRAWFORD, C. N. C. & PIRIE, A. (1952). Phase-contrast microscopy of the animal vitreous body. *Br. J. Ophthal.* **36**, 131–142.

BERMAN, E. R. & GOMBOS, G. M. (1969). Studies on the incorporation of U-^{14}C-glucose into vitreous polymers *in vitro* and *in vivo*. *Invest. Ophthal.* **8**, 521–534.

BERMAN, E. R. & VOADEN, M. (1970). The vitreous body. In *Biochemistry of the Eye*, Ed. C. Graymore, pp. 373–471. London: Academic Press.

BUDDECKE, E. & WOLLENSAK, J. (1966). Zur Biochemie der Zonulafaser des Rinderauges. *Z. Naturf.* **21b**, 337–341.

DAVSON, H. & DUKE-ELDER, W. S. (1935). Studies on the vitreous body. II. *Biochem. J.* **29**, 1121–1129.

FINE, B. S. & TOUSIMIS, A. J. (1961). The structure of the vitreous body and the suspensory ligaments of the lens. *Arch. Ophthal., Chicago*, **65**, 95–110.

FOWLKS, W. L. (1963). Meridional flow from the corona ciliaris through the pararetinal zone of the rabbit vitreous. *Invest. Ophthal.* **2**, 63–71.

GÄRTNER, J. (1965). Elektronemikroskopische Untersuchungen über Glaskörperrindenzellen und Zonulafasern. *Z. Zellforsch.* **66**, 737–764.

HOGAN, M. J. (1963). The vitreous, its structure, and relation to the ciliary body and retina. *Invest. Ophthal.* **2**, 418–445.

LAURENT, U. B. G., LAURENT, T. C. & HOWE, A. F. (1962). Chromatography of soluble proteins from the bovine vitreous body on DEAE-cellulose. *Exp. Eye Res.* **1**, 276–285.

McEWEN, W. K. & SURAN, A. A. (1960). Further studies on vitreous residual protein. *Am. J. Ophthal.* **50**, 228–231.

MAURICE, D. (1959). Protein dynamics in the eye studied with labelled proteins. *Am. J. Ophthal.* **47**, Pt. 11, 361–367.

MÖRNER, C. T. (1894). Untersuchungen der Proteinsubstanzen in den lichtbrechenden Medien des Auges. *Hoppe-Seyl. Z.* **18**, 61–106.

OLSEN, B. R. (1965). Electron microscope studies on collagen. IV. *J. Ultrastr. Res.* **13**, 172–191.

ÖSTERLIN, S. E. (1968). The synthesis of hyaluronic acid in vitreous. III. *Exp. Eye Res.* **7**, 524–533.

ÖSTERLIN, S. E. & BALAZS, E. A. (1968). Macromolecular composition and fine structure of the vitreous in the owl monkey. *Exp. Eye Res.* **7**, 534–545.

PAPPAS, G. D. & SMELSER, G. K. (1958). Studies of the ciliary epithelium and the zonule. I. *Am. J. Ophthal.* **46**, Pt. 2, 299–318.

PIRIE, A., SCHMIDT, G. & WATERS, J. W. (1948). Ox vitreous humour. 1. The residual protein. *Br. J. Ophthal.* **32**, 321–339.

SCHWARZ, W. (1956). Elektronenmikroskopische Untersuchungen an Glaskörperschnitten. *Anat. Anz.* **102**, 434–442.

SMITH, J. W. & SERAFINI-FRACASSINI, A. (1967). The relationship of hyaluronate and collagen in the bovine vitreous body. *J. Anat.* **101**, 99–112.

SZIRMAI, J. A. & BALAZS, E. A. (1958). Studies on the structure of the vitreous body. III. Cells in the cortical layer. *Arch. Ophthal., Chicago*, **59**, 34–48.

ZIMMERMAN, L. E. & STRAATSMA, B. R. (1960). In *Importance of the Vitreous Body in Retina Surgery with Special Emphasis on Reoperations*. Ed. C. L. Schepens. pp. 15–28. St. Louis: Mosby.

PART 3

THE CORNEA

STRUCTURE

THE cornea is illustrated by the meridional section in Fig. 40; the great bulk (up to 90 per cent of its thickness) is made up of the *stroma* which is bounded externally by Bowman's membrane and the epithelium, and internally by Descemet's membrane and the endothelium. The total thickness in man is just over 0·5 mm in the central region; towards the periphery it becomes some 50 per cent thicker.

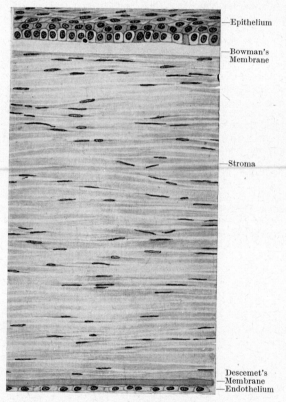

Epithelium

Bowman's Membrane

Stroma

Descemet's Membrane
Endothelium

FIG. 40. Meridional section through the human cornea.
(Wolff, *Anatomy of the Eye and Orbit*.)

Epithelium

This may be regarded as the forward continuation of the conjunctiva; it is a squamous epithelium consisting of some five or six layers of cells with a total thickness of 50–100 μ. The cells at the base are columnar, but as they are squeezed forward by new cells they become flatter so that three layers are recognized:

basal, wing-shaped, and *squamous cells*. The more superficial cells do not show keratinization, as with skin, the process of shedding being accompanied by the breaking up of the cell into fragments (Teng). Replacement of cells occurs by mitotic division of the basal layer, the average life of a cell being some 4–8 days (according to Hanna & O'Brien). Junctional complexes between cells in the various layers seem to be mainly of the desmosome type, not involving fusion of adjacent membranes. Nevertheless, because the permeability of the corneal epithelium as a whole is low, we must postulate tight junctions or zonulae occludentes between the cells of at least one layer, possibly the basal layer; according to Pedler (1962) these cells show extensive lateral interdigitation of plasma membranes. The surface layer of cells may be examined by the scanning electron microscope; viewed in this way the anterior surfaces of this outermost layer are covered by villous projections (Blümcke & Morgenroth, 1967). Amongst the cells a special, possibly secretory, type has been described by Teng; it may be that it secretes ths basement membrane upon which the epithelium rests; this latter is a layer some 100–300 Å thick resolved into an anterior apparently lipid layer and a reticular fibre meshwork in contact with Bowman's membrane.

Regeneration. When the epithelium is stripped off from the underlying Bowman's membrane there is a rapid regeneration, so that within twenty-four hours the surface may be completely covered by a new layer of regenerated epithelial cells derived from multiplication of the conjunctival cells at the periphery. This *single* layer subsequently develops to give, within several weeks, a complete epithelium 4–6 cells thick. Smaller lesions are repaired by migration of neighbouring epithelial cells from all layers, the migrated cells increasing their areas by flattening, and extending pseudopodia.

Stroma and Bowman's Membrane

The Lamellae and Fibrils. In meridional section the stroma appears as a set of lamellae, running parallel with the surface, and superimposed on each other like the leaves of a book; between the lamellae lie the *corneal corpuscles*, flattened in the plane of the lamellae; they are equivalent to fibrocytes of other connective tissue. The lamellae are made up of microscopically visible *fibres*, which run parallel to form sheets; in successive lamellae the fibres make a large angle with each other. In lower animals the lamellae extend over the whole surface of the cornea and the fibres in adjacent laminae are at right-angles to each other; in mammals and man the lamellae are said to be made up of bands some 90–260 μ wide; there is apparently no interweaving of these bands, however, so that essentially the structure consists of layers of fibres arranged parallel with themselves in any given lamella.

The lamellae in man are some 1·5 to 2·5 μ thick, so that there are at least 200 lamellae in the human cornea. The fibrous basis of the lamellae is made up of collagen which constitutes some 80 per cent of the solid matter of the cornea; by submitting the cornea to ultrasonic vibrations in order to break down its structure into its elementary components, Schwarz (1953) observed typical collagen fibrils of 250 to 330 Å diameter, and having the characteristic 640 Å periodicity (p. 61); the fibrils isolated in this way were covered by a dense layer of cement substance which was presumably the mucopolysaccharide. The fibrils were remarkably homogeneous in regard to diameter, in marked contrast to those of the sclera where the diameter varied from 280 Å to 2800 Å, and where the amount of ground substance was very much less. As we shall see, this uniformity in diameter is important for the transparency of the cornea by contrast with the opacity of the sclera. More recently Jakus (1961) has made thin sections of the cornea and examined these in the electron-microscope; these have revealed a remarkably uniform arrangement of fibrils, running parallel with each other and at right-angles with those in adjacent layers (Fig. 41, Pl. IV). The stromal cells in the human cornea appeared to lie within, rather than between, the laminae.

Bowman's Membrane. Immediately next to the epithelium, the classical histologists described a layer with no resolvable structure, some 8–14 μ thick in man—*Bowman's membrane** (Fig. 40). In the electron-microscope it became evident that Bowman's membrane is only a less ordered region of the stroma; thus, according to Jakus, and Kayes & Holmberg, the stroma immediately beneath the basement membrane of the epithelium consists of collagen fibrils closely but randomly packed into a felt-like layer which is not sharply differentiated from the remainder of the stroma beneath it. It should therefore be described as *Bowman's layer* rather than Bowman's membrane.

Descemet's Membrane

This is some 5–10 μ thick in man and is revealed as a characteristic differentiation of the stroma in the electron-microscope, although in man the organization is not so precise as in the ox cornea. The collagen is different, having an axial period of 1170 Å in place of the usual banding at intervals of 640 Å, and the carbohydrate is also different but similar chemically to that in the lens capsule (Dohlman & Balazs). Descemet's membrane is secreted by the endothelium.

Endothelium

This is a layer of flattened cells some 5 μ high and 20 μ wide. The intercellular spaces are sealed off by tight junctions so that when colloidal material is introduced into the anterior chamber its passage is held up in these regions. Some material does enter the endothelial cells, but this is probably by engulfment in vesicles (endocytosis) formed by the plasma membranes of the endothelial cells (Kaye & Pappas, 1962; Iwamoto & Smelser, 1965). Division of endothelial cells only occurs during growth and this is mitotic, in spite of claims to the contrary (v. Sallmann, Caravaggio & Grimes). During growth, multiplication of cells does not keep pace with the increase in area, so the cells become progressively flatter. By contrast with the epithelium, then, the endothelial cells are normally a non-renewable population. After damage, there is considerable mitotic division and neighbouring cells flatten and spread into the damaged area; Descemet's membrane is repaired after damage by the slow secretion of new material.

CHEMISTRY

The cornea contains some 22 per cent of solid matter, and is thus quite obviously built on a more rigid plan than the vitreous body; this solid matter is mainly collagen, but in addition there are mucopolysaccharides, other proteins than collagen, and crystalloids of which salts make up the bulk.

---

* In many lower mammals, such as the rabbit and mouse, Bowman's membrane is absent, so that only the basement membrane, which is common to all species, separates the epithelium from the stroma proper.

Figures given by Maurice (1969) are as follows:

	per cent
Water	78·0
Collagen	15·0
Other proteins	5·0
Keratan sulphate	0·7
Chondroitin sulphate	0·3
Salts	1·0

Mucopolysaccharides

This is the name commonly given to a group of polysaccharides containing an amino sugar and widely distributed throughout the animal body where they may act as lubricants in the joints and saliva, as the anticoagulant called heparin, as blood-group substances on the surface of red cells, as ground substance of connective tissue, as cement between cells, and so on. Strictly speaking the term mucopolysaccharide should be applied, however, only to the combination between this type of polysaccharide and a protein, the reaction of the complex being predominantly polysaccharidic, whilst the term *mucoprotein* is applied to a complex between protein and carbohydrate where the behaviour is predominantly protein in character.* Perhaps the best known of the polysaccharides is *hyaluronic acid*, already described in relation to the vitreous body; it consists of large asymmetrical molecules (weight of the order of a million or more) made up by the polymerization of N-acetylglucosamine and glucuronic acid, linked together as follows:

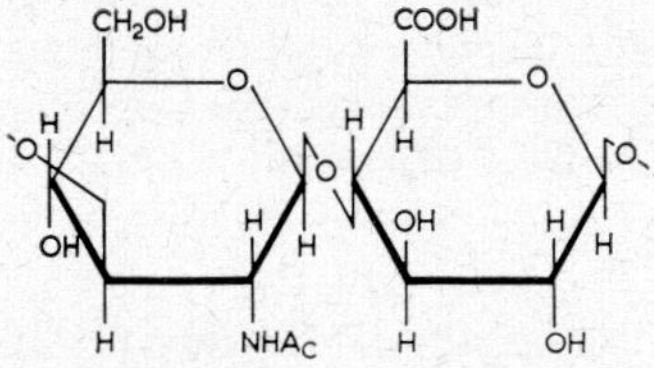

Hyaluronic Acid

Hyaluronic acid is specifically broken down by the enzyme, *hyaluronidase*, isolated from streptococci and pneumococci. Chondroitin sulphate was first found in the cornea but also occurs in cartilage; it is built up by the polymerization of N-acetylgalactosamine, glucuronic acid and sulphate, as below:

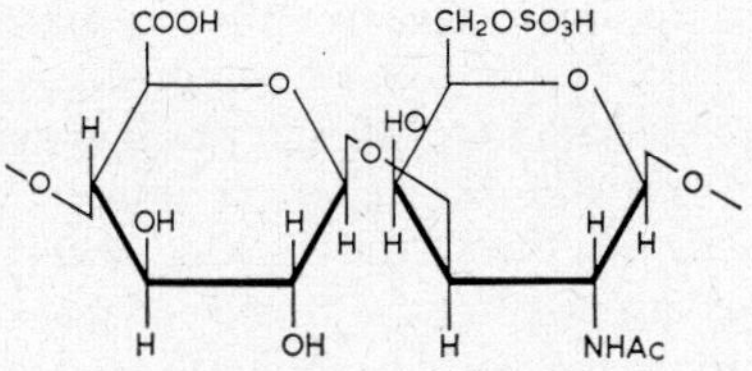

Chondroitin Sulphate

Chondroitin is the desulphated chondroitin sulphate, whilst the latter has been found in three different forms, called chondroitin sulphate A, B and C, the differences being not so much due to differences in the sugar moieties as in the nature of the linkages.

Histochemically these polysaccharides are identified by their so-called *metachromatic staining* with toluidine blue, whereby the stain acquires a different colour

* The term *glycosaminoglycan* is recommended for a polysaccharide containing a hexosamine; more specifically, *glucosaminoglycan* is a polysaccharide containing glucosamine; *galactosaminoglycan*, one containing galactosamine, and so on.

than it had in free solution. The *periodic acid-Schiff test* (P.A.S.) is also characteristic for the chondroitins although there is some doubt as to whether hyaluronic
acid reacts positively.

The polysaccharides react easily with proteins to form complexes, but it is by
no means certain as to what extent they are bound thus in the tissues. So far as the
cornea is concerned, at least a part of the polysaccharide is bound to non-colla-
genous protein so that extremely vigorous treatment is necessary to extract it
(Woodin, 1954).

HYDRATION AND TRANSPARENCY

Swelling

If an excised eye is kept at, say, 4°C in the refrigerator for 24 hours or more the
cornea will be seen to have lost some of its transparency, becoming "smoky"
in appearance; associated with this optical change there is found an increase in
thickness due to absorption of aqueous humour. Thus, with an ox eye, there is
an increase of water-content from 77·3 to 82 per cent. Expressed in this way
the increase in water-content is not impressive, but expressed as the amount
of water per gramme of solid matter the increase is from 3·4 g H_2O/g solids to
4·55 g/g, i.e. an increase of 33 per cent. The increase in water is associated with
an increase in the salt content and corresponds to an uptake of aqueous humour
rather than the selective absorption of water (Davson, 1949). If an excised cornea
is placed in saline the same increase in water-content, with absorption of salt,
is observed; if the epithelium is removed the uptake is just the same, if somewhat more rapid, so that it is clear that we are dealing with an increased hydration
of the stroma, which is behaving similarly to many of the inanimate gels studied
by the colloid chemists; like these gels the cornea shows a swelling pressure,
which may be measured by exerting a contrary pressure to prevent uptake of
fluid; it amounts to 75–85 g/cm² in the normal cornea; as the cornea takes up
water, the pressure it can exert decreases, whilst it increases as water is withdrawn
(see, for example, Fatt & Hedbys, 1970).

Effect of Temperature. This uptake of fluid by the dead cornea, and its
associated loss of transparency, was considered by Cogan & Kinsey to result
from the removal of the normal osmotic pressures of tears and aqueous humour
that were supposed, in the living eye, to remove water continuously from the
cornea as fast as it entered. In the first edition of this book the present author
showed the theroretical inadequacy of this theory and subsequent experimental
studies have confirmed this criticism. Thus, to quote a crucial experiment, on
the basis of the Cogan-Kinsey theory the swelling of the dead cornea should be
much more rapid and extensive if the eye were kept at body-temperature than in
the refrigerator, yet if the two eyes of a rabbit are excised and the one is kept at
4°C and the other in a moist chamber at the normal temperature of the aqueous
humour, namely 31°C, it is found that the eye kept in the warm is almost completely normal, with good transparency, and water-content only slightly different
from the average normal figure. Clearly, by cooling the eye we are robbing the
corneal tissue of its power of excluding fluid, and this is presumably due to
reducing the metabolism of the eye. In other words, the cornea is probably using
metabolic energy to resist the intrusion of fluid from the aqueous humour. To
understand the factors concerned, let us turn now to the fluid and salt relationships
between cornea, on the one hand, and aqueous humour and tears on the other.

Stroma-Aqueous Humour-Tears

Analysis of the salt content of the cornea and aqueous humour suggests the simple Gibbs-Donnan type of equilibrium between the stroma and the outside fluids, the stroma containing a high concentration of ionized colloid and being separated from the outside fluids, aqueous humour and tears, by membranes that allow movements of ions and water but not of the colloid. To quote Otori's figures for the rabbit, the concentrations of Na^+, K^+, and Cl^- in meq/kg H_2O are as follows:

	Aqueous Humour	Stroma
Na^+	144·0	171
K^+	4·9	22
Cl^-	103·0	108

The high concentration of K^+ in the stroma is doubtless due to the presence of keratocytes, whilst the high concentration of Na^+ is presumably due to the presence of ionized collagen and mucopolysaccharide; on this basis, it is to be remarked, however, that the concentration of Cl^- should be lower in the stroma than in the aqueous humour, and it may be that the bicarbonate ion is not at equilibrium (Hodson, 1971).

Swelling Tendency. The general picture of the relationships between the corneal stroma and the aqueous humour and tears is illustrated schematically in Fig. 42. The stroma, being a mixture of collagen and its associated mucopolysaccharides, may be treated as a gel, and largely by virtue of the polysaccharide this gel has a tendency to take up water.* It is separated from the

FIG. 42. Illustrating the tendency for the stroma of the cornea to hydrate by virtue of the inward movement of water and salts from aqueous humour, tears and limbal capillaries.

two fluids from which this water is obtainable by membranes, the epithelium and endothelium. These membranes are permeable to salts and water (Maurice, 1951) so that the swelling tendency of the gel must be resisted by some active process on the part of either the endothelium or epithelium or both. Removal of the epithelium or endothelium in the intact eye causes a rapid uptake of fluid (Maurice & Giardini). As Fig. 42 indicates, at the limbus the stroma may take up fluid and salts from the pericorneal capillary plexus, since there is no reason

---

* It seems well proved that it is the mucopolysaccharide that is responsible for the hydration tendency; thus the fibrils of a swollen cornea are no larger than normal when examined in the electron-microscope (François, Rabaey & Vandermeersche, 1954); again, extraction of the mucopolysaccharides from the cornea reduces its tendency to take up water (Leyns, Heringa & Weidinger, 1940).

to believe that there would not be filtration from these capillaries as in other parts of the body.

Reversal of Swelling. If the cornea of the excised eye is allowed to swell by keeping it for some time at 4°C, the process may be reversed by transferring the eye now to a moist chamber at 31°C (Davson, 1955) so that either the epithelium or endothelium has actively removed the salts and water that entered at the low temperature. Harris & Nordquist (1955) showed that this "temperature-reversal effect" would still occur when the epithelium was removed, so that the "pump" that removes fluid as it enters must be located in the endothelium. More recent studies have tended to confirm the importance of the endothelium; e.g. Green & Otori (1970) showed that the isolated cornea swelled up rapidly on removal of the endothelium, although when the endothelium was intact it could be maintained at normal thickness for hours provided glucose was in the incubation medium.

Active Process. As to the nature of the active process, it was early assumed that this could be achieved by the active transport of sodium from stroma to aqueous humour across the endothelial membrane, or from stroma to tears across the epithelium. The salt so transferred would carry with it its osmotic equivalent of water. Such an active transport, if it occurred, should be detectable by modern isotopic methods in which the fluxes of ^{24}Na are measured together with the potential across the cornea. This potential was found by Potts & Modrell, and Donn, Maurice & Mills to be of the order of 10–40 mV, the outside of the eye being negative to the inside; the epithelium was responsible.* This is just the opposite of what would be expected were the epithelium actively transporting sodium *from* the stroma, and measurement of the fluxes of this ion showed that it was being actively transported *into* the stroma, from the tears, thus increasing the tendency for this to hydrate. There was no evidence for an active transport of sodium out of the endothelium, and later workers, e.g. Freen (1967) have confirmed that the movement of ions across the endothelium appears to be passive whilst that across the epithelium is active. That the endothelium is closely concerned, nevertheless, and through some active transport process, is made very likely by the observation, by Trenberth & Mishima (1968) that ouabain caused a rapid swelling of the rabbit cornea when its posterior surface was perfused, in a chamber, with a medium containing 10^{-5} M; since these authors had shown that removal of the epithelium did not affect thickness under these conditions, this means that the ouabain was inhibiting pump activity on the part of the endothelium. Again, if a layer of plastic is implanted intralaminarly, it is the region *above* the implant that becomes oedematous whilst that below retains its normal thickness. Finally, Doane & Dohlman (1970) have actually maintained the cornea *in vivo* at normal thickness after total removal of the epithelium by sealing a plastic contact lens over the denuded surface. Maurice & Giardini had shown that removal of the epithelium caused a rapid increase in thickness of the cornea, and the fact that this does not occur when the surface is sealed means that an intact epithelium is important only because it restrains the passive influx of salts and water from the tears and the limbal blood vessels, i.e. the influx becomes too great for the endothelium to deal with.

* Maurice (1967) has shown that the potential occurs *in vivo*; it is very susceptible to damage so that drawing a fine thread over a localized area of the epithelium abolished the potential when measured across this area.

Metabolic Inhibition. Whatever the mechanism for maintaining normal corneal thickness, metabolic energy is necessary for it, so it is not surprising that when the excised eye is maintained in an atmosphere free of oxygen the hydration increases enormously (Schwartz, Danes & Leinfelder). Studies on the isolated cornea (Mishima *et al.*, 1969), which permitted the measurement of the effects of anoxia on either the epithelial or endothelial side, demonstrated that it is anoxia of the endothelium that is critical in causing swelling of the cornea; since the endothelium obtains its oxygen mainly from the aqueous humour, this explains the limited effects of contact glasses on the surface of the eye (Langham & Taylor, 1956). It must be appreciated, of course, that anoxia of the epithelium is damaging to this tissue, producing an epithelial oedema that results subjectively in halo formation (Smelser & Ozanics, 1953); as Heald & Langham had shown, the aqueous humour cannot supply the epithelium adequately with oxygen.

Amphibian Cornea. In the frog the ionic transport mechanism seems to be different, and it may well be that an epithelial ionic pump, transporting net amounts of salt from stroma to tears, is the mechanism that maintains hydration constant. Zadunaisky (1966, 1969) measured a potential between tears and aqueous humour of 10–60 mV, the aqueous humour being positive; this was due to the active transport of Cl^- from stroma to tears and could be abolished by removing this ion from the cornea. By splitting the cornea, the separate effects of epithelium and endothelium could be examined, and in this case it was the epithelium that reproduced the behaviour of the whole cornea. There was no evidence for active transport by the endothelium. When the transparency had been allowed to decrease by imbibition of fluid at a low temperature, the partial reversal of this by warming was accompanied by enhanced active transport of Cl^-; furthermore, arsenite caused a temporary acceleration of active transport, and this was accompanied by a definite increase of transparency presumably due to dehydration. Since Ploth & Hogben (1967) have also shown that the short-circuit current of the isolated frog cornea is equivalent to the net flux of Cl^-, we must conclude that in the amphibian cornea the predominant process is an active transport of this ion.

Transparency of the Cornea

Scattering of Light. By transparency we mean rather more than the ability of a material to transmit light, we are concerned with the manner in which it is transmitted; thus dark glasses may reduce the transmission to 10 per cent of the incident intensity yet the optical image remains perfect; on the other hand an opal screen may reduce the intensity much less, but because of the *scattering* of the light no image is possible. Thus diffraction, or scattering, of light is the important factor for transparency; and in this respect the cornea is very highly transparent, less than 1 per cent of the light being scattered. The question arises as to how this transparency is attained in an inhomogeneous medium consisting of fibres embedded in a matrix of different refractive index; and furthermore, why the transparency should be affected by increases in hydration of the cornea and changes in intraocular pressure, as when haloes are seen during an acute attack of glaucoma.

LATTICE THEORY

In general, when light strikes a small discontinuity in the medium the latter acts as origin for new wavelets of light travelling in all directions; hence a large number of discontinuities in a transparent medium will give rise to wavelets of light passing backwards as well as forwards and thus making the medium opaque. A solution of very small particles, such as NaCl, remains transparent in spite of these discontinuities in the medium; this is because the discontinuities are small compared with half the wavelength of light. When the particles are of comparable size or greater, the diffraction becomes important, and the question arises as to why the cornea, built up of fibrils with a diameter greater than the wavelength of visible light, should in fact be transparent. The problem has been examined in detail with considerable acumen by Maurice (1957) who has shown, from an analysis of the birefringence* of the cornea, that the size of the fibrils, and the difference of refractive index between them and the ground-substance in which they are embedded, are such as to lead one to expect a highly opaque structure, due to scattering of incident light. He has shown, nevertheless, that this scattering may be avoided if the fibres are of relatively uniform diameter and are arranged in parallel rows; in this event the diffracted rays passing forwards tend to cancel each other out by destructive interference, leaving the normal un-diffracted rays unaffected. In this way the opacity of the sclera is also explained, since there is no doubt from Schwarz' study that the prime distinction between the two tissues is the irregularity in arrangement and diameter of the scleral fibres.†

Swelling and Cloudiness. When the cornea swells (e.g. on being kept at low temperature), there is no appreciable change in the diameters of the collagen fibrils, not yet is there a significant swelling of the epithelial and endothelial layers. Thus the swelling is due to an increase in the amount of fluid in the interfibrillar spaces (see, for example, Langham, Hart & Cox, 1969); and we may assume that it is the irregularity in the arrangement of the fibrils that causes the cloudiness associated with such swelling. Heringa, Leyns & Weidinger (1940) showed that extraction of the mucopolysaccharides from cornea reduced its swelling pro-perties; thus the swelling results from the tendency of the mucopolysaccharide to hydrate. We may assume that the attachment of the mucopolysaccharide to the collagen fibrils ensures that they maintain a characteristic spacing, and this may be achieved by the mutual repulsion of ionized groupings on the mucopoly-saccharide chains. Excessive hydration of the mucopolysaccharide will lead to shielding of these ionized groups by water molecules, and thus allow the fibrils to aggregate; the disorganization so produced will prejudice transparency; at any rate, Langham *et al.* observed in the electron microscope islands devoid of fibrils. An increased intraocular pressure presumably leads to a disarrangement of the fibrils and causes cloudiness in this manner.

Elasmobranch Cornea. The elasmobranch cornea, for example that of the dogfish *Squalus acanthias*, does not swell even when both epithelium and endo-thelium are removed (Smelser, 1962); and this was attributed by Ranvier in 1881 to the presence of sutural fibres that ran at right-angles to the laminae, binding the basement membrane of the epithelium to Descemet's membrane. The electron

---

* The ordered arrangement of the collagen fibrils makes the velocity of light in one plane different from that in one at right-angles; this leads to double refraction or birefringence. From the magnitude and character of this birefringence important information regarding the orientation of the fibrils in the cornea may be obtained (Stanworth & Naylor, 1953; Maurice, 1957).

† Smith (1969) has pointed to wide variations in diameter and spacing of collagen fibrils in the cornea, and has suggested that it is rather the fineness of the fibrils, and the approxi-mate equality of refractive index of fibrils and matrix, that determine transparency. However his objections seem to have been largely answered by the experimental study and theoretical considerations of Cox, Farrell, Hart & Langham (1970). Schwarz & Keyserlingk (1966) have re-examined the fibrillar structure of the human cornea; the diameters are remarkably uniform, but the arrangement is not truly crystalline—the structure is "amorphous with a constant interfibrillar separation". They find nothing in this structure inconsistent with Maurice's hypothesis.

microscopical study of Goldman & Benedek (1967) has confirmed this picture of antero-posteriorly directed fibres constituting a "sutural complex", so that if these sutural fibres failed to stretch appreciably this would prevent swelling. Praus & Goldman (1970) have shown that the mucopolysaccharide of the shark cornea is more similar to that in cartilage than that found in the mammalian cornea, with very little keratan sulphate and a relatively high proportion of chondroitin sulphate.

PERMEABILITY OF THE CORNEA

In a homogeneous medium a dissolved substance will pass from a region of high to one of low concentration by the process of diffusion. When the diffusion from one region to another is restricted by the presence of a membrane, the process is described as one of "permeability", and the degree of restraint imposed on the migration of the solute molecules is measured inversely by the *permeability constant* of the membrane. In general, the ease with which a substance can penetrate cell membranes depends on its lipid-solubility; thus ethyl alcohol, which is highly lipid-soluble, penetrates cell membranes easily, whereas glycerol penetrates with difficulty. The membranes separating the stroma from its surrounding media are formed by layers of closely packed cells; lipid-soluble substances may be expected to pass easily into the cornea because they may pass easily into and out of the cells of these membranes; lipid-insoluble substances, including ions, may be expected to penetrate with difficulty; moreover, the endothelium, with its single layer of cells, might be expected to be the more permeable membrane.

Tight Junctions. However, the essential factor in permeability of such cellular layers is the extent to which the intercellular clefts are closed by zonulae occludentes or tight-junctions. If these clefts, or the majority of them, are not closed by the fusion of adjacent membranes, we may expect a high degree of permeability extending to such large molecules as inulin and serum albumin, and this permeability will be relatively unselective because the large diameter of the intercellular clefts will not permit any selective sieve action on the various solutes commonly studied. In general, the qualitative studies of Swan & White, and Cogan & Hirsch, confirm these general principles. Where the substance is an organic base or acid, such as atropine or salicyclic acid, the degree of dissociation is an important factor since the undissociated molecule penetrates more rapidly than the ion; hence the permeability of the cornea to these substances is influenced by pH.

Epithelium and Endothelium. More recent quantitative measurements, carried out mainly by Maurice and by Mishima & Trenberth, have allowed separate estimates of the permeabilities of the epithelium and endothelium. Thus *in vivo* Maurice (1951) showed that the endothelium was 100 times more permeable to the Na^+-ion than the epithelium; this does not mean that the endothelium offers no restraint to the passage of this ion, however; it means, rather, that the permeability of the epithelium is very low so that for many practical purposes the epithelium may be treated as a "semipermeable membrane", being permeable to water but effectively impermeable to the solutes of the tears. Maurice (1969) has calculated that, with Na^+, the endothelium offers some 1700 times the resistance to diffusion that would be offered by the same thickness of water; with sucrose it was 2000 times, with fluorescein, 4200, inulin 4400 and serum albumin greater than 100,000. It is interesting that glucose and methylglucose permeate the endothelium considerably more rapidly than the much smaller Na^+-ion; this illustrates the intervention of a facilitated transfer mechanism, which is presumably of importance in the transfer of glucose from the aqueous humour to the stroma and epithelium (Hale & Maurice, 1969). As with practically all cellular layers, the permeability to water is very high, and this is because the cell membranes have a high permeability to this solute.*

* The permeability coefficient of the epithelium for Na^+ is about $1 \cdot 10^{-7}$ cm/sec; the permeability of the capillary endothelium is about a thousand times bigger than this. The permeability of the epithelium to water is about $3 \cdot 10^{-5}$ cm/sec comparable with that found in a variety of cells (Donn, Miller & Mallett, 1963).

METABOLISM

Pathways. The metabolic energy of the cornea is derived from the break-down of glycogen and glucose; since the cornea utilizes oxygen, which it derives from the atmosphere, through the tears, and from the aqueous humour, at least some of the energy is derived from oxidative mechanisms; the high concentration of lactate in the cornea indicates that aerobic glycolysis is an important pathway for the breakdown of glucose. According to a recent study of Riley (1969) on the perfused cornea, 85 per cent of the glucose utilized is converted to lactate and only 15 per cent is oxidized. The lactate so formed is carried away into the aqueous humour and thus the cornea contributes to the high concentration of this anion in the fluid. A considerable fraction of the glucose apparently passes through the pentose phosphate shunt by which glucose-6-phosphate is converted to triose phosphate, a pentose phosphate being formed as an intermediate. The triose phosphate is metabolized to pyruvate, which can either be converted to lactate or enter the citric acid cycle to be oxidized by O_2. The pathways are indicated in Fig. 43.

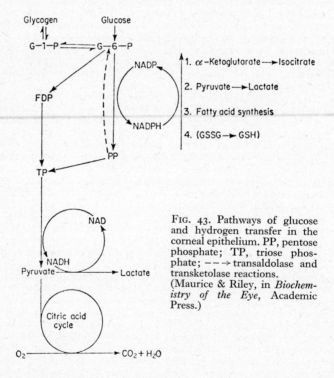

FIG. 43. Pathways of glucose and hydrogen transfer in the corneal epithelium. PP, pentose phosphate; TP, triose phosphate; $-- \rightarrow$ transaldolase and transketolase reactions. (Maurice & Riley, in *Biochemistry of the Eye*, Academic Press.)

Oxygen Tension. Under normal conditions, because of utilization, there is a gradient of oxygen-tension from tears, 155 mm Hg, to the aqueous humour, 55 mm Hg (Fatt & Bieber, 1968); thus the endothelium can apparently function normally at a relatively low oxygen tension; when access of oxygen to the outside of the cornea is prevented by a contact glass, the endothelium remains normal because it draws its O_2 from the aqueous humour; the epithelium, however, is in metabolic distress, and the concentration of lactate in the stroma increases. This metabolic distress is revealed in the swelling of the epithelial cells resulting in subjective haloes after wearing a contact glass for some time (Smelser & Ozanics, 1953). Langham showed that when the eyelids were closed the lids were able to supply the oxygen normally derived from the atmosphere.

Activities of Layers. A variety of studies have been concerned with the separate activities of the three layers of the cornea; as we might expect, the activities are determined by the relative proportions of cells so that on a cell-for-cell basis metabolic activity, however measured, is the same in epithelium, stroma and endothelium.*

Reserves. The excised cornea, or the cornea in the excised eye, can maintain normal metabolic function for many hours, and this is because of the reserves of glycogen in the epithelium as well as the lactate and glucose in the tissue (and aqueous humour of the intact eye).

Supply of Materials

Glucose. The supply to the cornea of materials required for metabolism, notably glucose, is theoretically possible by three routes; namely diffusion from the limbus, where there is a capillary network; diffusion from tears across the epithelium, and diffusion from aqueous humour. Maurice (1969) has examined the relative significance of these pathways and has shown that, because there is no significant flow of fluid through the stroma from limbus to central region, the transport by simple diffusion will be very inefficient. This is because, while the material diffuses slowly from limbus to central region, it will be exposed to concentration gradients from stroma to aqueous humour and tears (Fig. 44). Thus, with a small molecule or ion, if the concentration at the limbus is put equal to 100, the concentration at 6 mm from the limbus will be only 1 when a steady-state has been reached. On the other hand, where diffusion out into the aqueous humour and tears is restricted, as with plasma proteins, then a significant supply from the limbus is practicable, and would account for their presence in the cornea, as well as that of numerous enzymes. The supply of glucose, then, must be from aqueous humour and tears; however, the significance of the tears as a supply is probably negligible since, according to Giardini & Roberts, the concentration in this fluid is very small, whilst the permeability of the epithelium is low. A large number of experiments involving intralamellar implantation of plastic material in the cornea have shown that such a procedure leads to degeneration of the tissue above the layer of plastic, thereby emphasizing the importance of the diffusion of material from the aqueous humour; analysis of the epithelium showed that it had become deficient in glucose, glycogen and ATP (Turss, Friend & Dohlman, 1970).

Oxygen Supply. As indicated earlier, the oxygen supply for the epithelium is derived from the atmosphere; it would seem that normally the cornea is not

---

* DeRoetth (1951) claimed that the epithelium did not produce lactate but Langham showed that, in the rabbit's cornea, this tissue certainly produced lactate, its concentration in the epithelium being twice that in the stroma. Riley (1969) has confirmed Langham's work. Herrmann & Hickman (1948) argued that the metabolisms of stroma and epithelium were interdependent, the epithelium requiring lactate from the stroma and the stroma requiring a phosphate-acceptor from the epithelium. More modern work has shown that all layers have the correct metabolic apparatus (Maurice & Riley, 1970).

respiring, i.e. using oxygen, at its maximum rate since replacing the atmosphere with pure oxygen decreases the concentration of lactic acid in the cornea (Langham). Calculations suggest that the endothelium may obtain all the oxygen it requires from the aqueous humour. Fatt & Bieber (1968) have calculated the oxygen tensions at different regions of the cornea from the rates of utilization; the tension drops from 150 mm Hg at the epithelium to 55 mm Hg at the aqueous humour; it is possible that some O_2 is actually given up to the aqueous humour. When the eye is closed, the oxygen tension at the midpoint of the cornea falls from about 100 mm Hg, when the eye is open, to about 55 mm Hg.

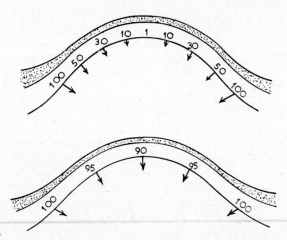

FIG. 44. Illustrating the effects of diffusion from cornea into aqueous humour on the transport of material from limbus to corneal centre. *Above:* Diffusion into aqueous humour is rapid, so that at the corneal centre the effective concentration is low. *Below:* Diffusion out is slow, so that a relatively high concentration can be built up at the corneal centre.

Corneal Vascularization. Under conditions of metabolic stress, for example after injury, when the tissue may be invaded by leucocytes and fibrocytes, the nutritive supply and metabolic reserves may become inadequate, with the result that vessels sprout from the limbal plexus and grow into the stroma—*corneal vascularization*. It is likely that the effects of vitamin deficiency fall into the same category of metabolic stress; thus riboflavine deficiency is associated with corneal vascularization, whilst deficiencies of essential aminoacids, such as tryptophan, lysine and methionine, will also produce the condition.

Stimulus to Vascularization. The nature of the stimulus to ingrowth of vessels is a matter of some interest; it has been argued that the prime factor is the loosening of the tissue associated with the injury, i.e. the corneal oedema, which permits the vessels to grow into the tissue. It is certain that corneal vascularization is very frequently, if not invariably, associated with oedema, but the fact that, after tissue repair, the vessels may regress well before the subsidence of the oedema suggests that an additional factor is necessary (Langham, 1953). Campbell & Michaelson found that, if the injury to the cornea occurred too far from the periphery, i.e. at the pole of the cornea, vascularization failed to take place, and it was argued that a vascularization factor had to diffuse from the point of injury to the limbus to stimulate growth; if the distance from the point of libera-tion to the limbus was too great, sufficient would be lost to the aqueous humour to reduce its effectiveness.

Tissue Factor

Maurice, Zauberman & Michaelson (1966) concluded from their own experi-ments on rabbits that, whereas oedema was a necessary precondition for vas-cularization, an additional factor was required; and their observations were consistent with the notion of a factor, liberated by the cells of the tissue, that had to diffuse to the limbus to provoke ingrowth of the vessels. Thus they implanted fine tubes, running radially in the stroma, and found that an injury to the centre of the cornea would provoke the growth of vessels inside the tube provided this was open at both ends; if the end nearest the pole of the cornea had been occluded, there was no ingrowth. Again, vascularization was more certain to occur if the aqueous humour was replaced by silicone oil, and it could be argued that this was because loss of the factor had been reduced. Tissue injury causes the liberation of a number of pharmacologically active substances such as histamine, 5-hydroxy-tryptamine and bradykinin; and Zauberman et al. (1969) showed that the growth of vessels into tubes implanted in the cornea was more likely to occur if these tubes contained acetylcholine, histamine and 5-hydroxytryptamine, although bradykinin had no effect.

NERVOUS SUPPLY

The innervation of the cornea is probably entirely sensory, mediated by the ophthalmic division of the trigeminal by way of the ciliary nerves. According to Zander & Weddell the fibres travel towards the limbus mainly in the perichoroidal space, and enter the substance of the cornea from three levels, scleral, episcleral and conjunctival. Most of the bundles entering the cornea are arranged radially and give rise, by subdivision, to a plexus in the stroma from which fibres supply the different regions. So great is the branching of any nerve fibre during its course in the cornea that a single axon may innervate an area equivalent to a quadrant. The endings are free, similar to those of pain fibres in other parts of the body; those reaching the epithelium are extremely fine, and send loops between the cells to within one cell-layer of the surface; similarly in the endothelium, the terminations are very fine.

Modalities of Sensation. Since v. Frey's time it has been categorically stated that the cornea is sensitive only to pain, in the sense that all forms of stimulus in man arouse only this sensation. According to Lele & Weddell (1956), however, the sensation of touch, as distinct from that of pain, may be evoked by gentle applica-tion of a nylon thread; again, a copper cylinder, maintained at $1 \cdot 5°C$ above or below the corneal temperature, evoked a sensation of warmth or cold. By recording from the ciliary nerves of the cat, Lele & Weddell (1959) found action potentials in the myelinated fibres in response to both thermal and mechanical stimuli, and it appeared that the same fibres responded to both types of stimulus. Thus, apparently in contradiction to Müller's Law of Specific Nerve Energies, the sensory fibres of the cornea were able to mediate both thermal and touch sense. According to Kenshalo, however, the sensations evoked by thermal stimulation of the human cornea are not those of warmth and cold but are of different types of "irritation"; thus the nerve endings are *sensitive* to thermal change as well as mechanical, but the sensations evoked with the different stimuli are essentially of the same modality.

Regeneration of Fibres. When the cornea is cut through, as in making a section for cataract extraction, there is a loss of sensibility of the affected region, followed by a return associated with the growth of new fibres from the cut bundles proximal to the scar; there is no sprouting of adjacent uninjured neurones, however, since Rexed & Rexed repeatedly renewed a cut in the periphery and found no recovery of sensibility in the area supplied by the cut fibres.*

RESPONSE OF THE CORNEA TO INJURY

Stromal Repair. The repair processes taking place in endothelium and epithelium have already been described; damage to the substantia propria is associated with the migration of new cells into the damaged area, either by chemotaxis from the limbal capillaries or, if the damage is severe, by the ingrowth of vessels into the stroma—the so-called corneal vascularization. The cells secrete new collagen fibres, but because the orientation of fibres is not so accurate as in normal tissue some degree of light-scattering remains in the scar. When the repair processes are over, the wandering cells that moved in settle down to become corneal corpuscles (Maumenee & Kornbluth).

Corneal Graft. The cornea is peculiar in that it is possible to transplant a portion from one individual to another of the same species, i.e. to carry out a *homograft*. This may be the result of the low antigenic powers of the collagen of the cornea, or of the absence of a vascular supply through which the antigen-antibody response has to be mediated. An important element in the "take" of the graft seems to be the sealing into place of the donor tissue by the migration of the host's corneal corpuscles to the cut edge of the trephine hole, and the extension of processes into the donor tissue (Hoffmann & Messier).

GENERAL REFERENCES

BINDER, R. F. & BINDER, H. F. (1957). Regenerative processes in the endothelium of the cornea. *Arch. Ophthal., N.Y.*, **57**, 11–13.

DAVSON, H. (1955). Anatomy and physiology of the cornea. In *Corneal Grafts*. Ed. Rycroft, pp. 11–35. London: Butterworth.

GRIGNOLO, A. (1954). Studi sulla struttura submicroscopica dei tessuti oculari. *Boll. d'Oculist.* **33**, 1–144.

HEYDENREICH, A. (1958). *Die Hornhautregeneration*. Halle: Marhold.

JAKUS, M. (1961). The fine structure of the human cornea. In *Structure of the Eye*. Ed. Smelser, pp. 343–366. New York: Academic Press.

KENT, P. W. & WHITEHOUSE, M. W. (1955). *Biochemistry of the Amino Sugars*. New York: Academic Press.

LANGHAM, M. E., Ed. (1969). *The Cornea*. Baltimore: Johns Hopkins Press.

MAURICE, D. M. (1969). The cornea and sclera. In *The Eye*. Ed. Davson, Vol. I, pp. 489–600. London: Academic Press.

MAURICE, D. M. & RILEY, M. V. (1970). The cornea. In *Biochemistry of the Eye*. Ed. C. Graymore, pp. 1–103. London: Academic Press.

PIRIE, A. & VAN HEYNINGEN, R. (1956). *Biochemistry of the Eye*. Oxford: Blackwell.

STACEY, M. & BARKER, S. A. (1962). *Carbohydrates of Living Tissues*. London: Van Nostrand.

THOMAS, C. I. (1955). *The Cornea*. Springfield: Thomas.

---

* A vegetative innervation of the cornea has been postulated and denied; the technique developed by Falck for identifying adrenergic fibres by virtue of their fluorescence has permitted the identification of this type of fibre in the anterior layers of the stroma of the embryonic cornea; in primates these fibres disappear after birth but in most other mammals they remain (Laties & Jacobowitz, 1964; Ehinger, 1966).

SPECIAL REFERENCES

BLÜMCKE, S. & MORGENROTH, K. (1967). The stereo ultrastructure of the external and internal surface of the cornea. *J. Ultrastr. Res.* **18**, 502–518.

CAMPBELL, F. W. & MICHAELSON, I. C. (1949). Blood-vessel formation in the cornea. *Br. J. Ophthal.* **33**, 248–255.

COGAN, D. G. & HIRSCH, E. D. (1944). The cornea. VII. Permeability to weak electrolytes. *Arch. Ophthal., N.Y.* **32**, 276–282.

COGAN, D. G. & KINSEY, V. E. (1942). Transfer of water and sodium by osmosis and diffusion through the excised cornea. *Arch. Ophthal., N.Y.* **27**, 466–476.

COGAN, D. G. & KINSEY, V. E. (1942). The cornea. V. Physiologic aspects. *Arch. Ophthal., N.Y.* **28**, 661–669.

COX, J. L., FARRELL, R. A., HART, R. W. & LANGHAM, M. E. (1970). *J. Physiol.* **210**, 601–616.

DAVSON, H. (1949). Some considerations on the salt content of fresh and old ox corneae. *Br. J. Ophthal.* **33**, 175–182.

DAVSON, H. (1955). The hydration of the cornea. *Biochem. J.* **59**, 24–28.

DOANE, M. G. & DOHLMAN, C. H. (1970). Physiological response of the cornea to an artificial epithelium. *Exp. Eye Res.* **9**, 158–164.

DOHLMAN, C. H. & BALAZS, E. A. (1955). Chemical studies on Descemet's membrane of the bovine cornea. *Arch. Biochem. Biophys.* **57**, 445–457.

DONN, A., MAURICE, D. M. & MILLS, N. L. (1959). The active transport of sodium across the epithelium. *Arch. Ophthal., N.Y.* **62**, 748–757.

DONN, A., MILLER, S. & MALLETT, N. (1963). Water permeability of the living cornea. *Arch. Ophthal., Chicago,* **70**, 515–521.

EHINGER, B. (1966). Adrenergic nerves to the eye and to related structures in man and in the cynomolgus monkey (*Macaca irus*). *Invest Ophthal.* **5**, 42–52.

FATT, I. & BIEBER, M. T. (1968). The steady-state distribution of oxygen and carbon dioxide in the *in vivo* cornea. I. *Exp. Eye Res.* **7**, 103–112.

FATT, I. & HEDBYS, B. O. (1970). Flow conductivity of human corneal stroma. *Exp. Eye Res.* **10**, 237–242.

FRANÇOIS, J., RABAEY, M. & VANDERMEERSCHE, G. (1954). Etude de la cornée et de la sclérotique. *Ophthalmologica, Basel,* **127**, 74–85.

GIARDINI, A. & ROBERTS, J. R. E. (1950). Concentration of glucose and total chloride in tears. *Br. J. Ophthal.* **34**, 737–743.

GOLDMAN, J. N. & BENEDEK, G. B. (1967). The relationship between morphology and transparency in the non-swelling corneal stroma of the shark. *Invest. Ophthal.* **6**, 574–600.

GREEN, K. & OTORI, T. (1970). Studies on corneal physiology *in vitro*. *Exp. Eye Res.* **9**, 268–280.

GRIGNOLO, A. (1952). Fibrous components of the vitreous body. *A.M.A. Arch. Ophthal.* **47**, 760–774.

HALE, P. N. & MAURICE, D. M. (1969). Sugar transport across the corneal endothelium. *Exp. Eye Res.* **8**, 205–215.

HANNA, C. & O'BRIEN, J. E. (1960). Cell production and migration in the epithelial layer of the cornea. *Arch. Ophthal., N.Y.* **64**, 536–539.

HARRIS, J. E. & NORDQUIST, L. T. (1955). The hydration of the cornea. I. Transport of water from the cornea. *Am. J. Ophthal.* **40**, 100–110.

HEALD, K. & LANGHAM, M. E. (1956). Permeability of the cornea and blood-aqueous barrier to oxygen. *Br. J. Ophthal.* **40**, 705–720.

HERINGA, G. C., LEYNS, W. F. & WEIDINGER, A. (1940). On the water adsorption of cornea. *Acta. neer. morph.* **3**, 196–201.

HERRMANN, H. & HICKMAN, F. H. (1948). *Bull. Johns Hopkins Hosp.* **82**, 182, 225, 260.

HODSON, S. (1971). Evidence for a bicarbonate dependent sodium pump in corneal endothelium. *Exp. Eye Res.* **11**, 20–29.

HOFFMANN, R. S. & MESSIER, P. E. (1949). *Arch. Ophthal., N.Y.* **42**, 140, 148.

IWAMOTO, T. & SMELSER, G. K. (1965). Electron microscopy of the human corneal endothelium with reference to transport mechanisms. *Invest. Ophthal.* **4**, 270–284.

JAKUS, M. (1946). The fine structure of Descemet's membrane. *J. Biophys. biochem. Cytol.* **2**, Suppl., 243–252.

KAYE, G. I. & PAPPAS, G. D. (1962). Studies on the cornea. I. The fine structure of the rabbit cornea and the uptake and transport of colloidal particles by the cornea *in vivo. J. Cell Biol.* **12**, 457–479.

KAYES, J. & HOLMBERG, A. (1960). The fine structure of Bowman's layer and the basement membrane of the corneal epithelium. *Am. J. Ophthal.* **50**, 1013–1021.

KENSHALO, D. R. (1960). Comparison of thermal sensitivity of the forehead, lip, conjunctiva and cornea. *J. app. Physiol.* **15**, 987–991.

LANGHAM, M. E. (1952). Utilization of oxygen by the component layers of the living cornea. *J. Physiol.* **117**, 461–470.

LANGHAM, M. E. (1953). Observations on the growth of blood vessels into the cornea. *Br. J. Ophthal.* **37**, 210–222.

LANGHAM, M. E., HART, R. W. & COX, J. (1969). The interaction of collagen and mucopolysaccharides. In *The Cornea*, Ed. M. E. Langham, pp. 157–184. Baltimore: Johns Hopkins Press.

LANGHAM, M. E. & TAYLOR, I. S. (1956). Factors affecting the hydration of the cornea in the excised eye and the living animal. *Br. J. Ophthal.* **40**, 321–340.

LATIES, A. & JACOBOWITZ, D. (1964). A histochemical study of the adrenergic and cholinergic innervation of the anterior segment of the rabbit eye. *Invest. Ophthal.* **3**, 592–600.

LELE, P. P. & WEDDELL, G. (1956). The relationship between neurohistology and corneal sensitivity. *Brain*, **79**, 119–154.

LELE, P. P. & WEDDELL, G. (1959). Sensory nerves of the cornea and cutaneous sensibility. *Exp. Neurol.* **1**, 334–359.

LEYNS, W. F., HERINGA, C. & WEIDINGER, A. (1940). Water binding capacity of the cornea. *Acta brev. Neerland. Physiol.* **10**, 25–26.

MAUMENEE, A. E. & KORNBLUTH, W. (1949). Regeneration of the corneal stromal cell. *Am. J. Ophthal.* **32**, 1051–1064.

MAURICE, D. M. (1951). The permeability to sodium ions of the living rabbit's cornea. *J. Physiol.* **122**, 367–391.

MAURICE, D. M. (1957). The structure and transparency of the cornea. *J. Physiol.* **136**, 263–286.

MAURICE, D. M. (1967). Epithelial potential of the cornea. *Exp. Eye Res.* **6**, 138–140.

MAURICE, D. M. & GIARDINI, A. A. (1951). Swelling of the cornea in vivo after destruction of its limiting layers. *Br. J. Ophthal.* **35**, 791–797.

MAURICE, D. M., ZAUBERMAN, H. & MICHAELSON, I. C. (1966). The stimulus to neovascularization in the cornea. *Exp. Eye Res.* **5**, 168–184.

MISHIMA, S., KAYE, G. I., TAKAHASHI, G. H., KUDO, T. & TRENBERTH, S. M. (1969). The function of the corneal endothelium in the regulation of corneal hydration. In *The Cornea*, Ed. M. Langham, pp. 207–235. Baltimore: Johns Hopkins Press.

MISHIMA, S. & TRENBERTH, S. M. (1968). Permeability of the corneal endothelium to nonelectrolytes. *Invest. Ophthal.* **7**, 34–43.

OTORI, T. (1967). Electrolyte content of the rabbit corneal stroma. *Exp. Eye Res.* **6**, 356–367.

PEDLER, C. (1962). The fine structure of the corneal epithelium. *Exp. Eye Res.* **1**, 286–289.

PLOTH, D. W. & HOGBEN, C. A. M. (1967). Ion transport by the isolated frog cornea. *Invest. Ophthal.* **6**, 340–347.

POTTS, A. M. & MODRELL, R. W. (1957). The transcorneal potential. *Am. J. Ophthal.* **44**, 284–290.

PRAUS, R. & GOLDMAN, J. N. (1970). Glycosaminoglycans in the nonswelling corneal stroma of dogfish shark. *Invest. Ophthal.* **9**, 131–136.

REXED, B. & REXED, V. (1951). Degeneration and regeneration of corneal nerves. *Br. J. Ophthal.* **35**, 38–49.

RILEY, M. V. (1969). Glucose and oxygen utilization by the rabbit cornea. *Exp. Eye Res.* **8**, 193–200.

RILEY, M. V. (1969). Aerobic glycolysis in the ox cornea *Exp. Eye Res.* **8**, 201–204.

DE ROETTH, A. (1951). Glycolytic activity of the cornea. *Arch. Ophthal., N.Y.* **45,** 139–148.

SCHWARTZ, B., DANES, B. & LEINFELDER, P. J. (1954). The role of metabolism in the hydration of the isolated lens and cornea. *Am. J. Ophthal.* **38,** Pt. 2, 182–194.

SCHWARZ, W. (1953). Elektronenmikroskopische Untersuchungen uber den Aufbau der Sklera und der Cornea des Menschen. *Z. Zellforsch.* **38,** 20–49.

SCHWARZ, W. & KEYSERLINGK, D. G. (1966). Uber die Feinstruktur der menschlichen Cornea, mit besonderer Berüchsichtigung des Problems der Transparenz. *Z. Zellforsch.* **73,** 540–548.

SMELSER, G. K. (1962). Corneal hydration. Comparative physiology of fish and mammals. *Invest. Ophthal.* **1,** 11–21.

SMELSER, G. K. & OZANICS, V. (1953). Structural changes in corneas of guinea pigs after wearing contact lenses. *Arch. Ophthal., N.Y.* **49,** 335–340.

SMITH, J. W. (1969). The transparency of the corneal stroma. *Vision Res.* **9,** 393–396.

STANWORTH, A. & NAYLOR, E. S. (1953). Polarized light studies of the cornea. *J. exp. Biol.* **30,** 164–169.

SWAN, K. C. & WHITE, N. G. (1942). Corneal permeability. *Am. J. Ophthal.* **25,** 1043–1058.

TENG, G. C. (1961). The fine structure of the corneal epithelium and basement membrane of the rabbit. *Am. J. Ophthal.* **51,** 278–297.

TRENBERTH, S. M. & MISHIMA, S. (1968). The effect of ouabain on the rabbit corneal endothelium. *Invest. Ophthal.* **7,** 44–52.

TURSS, R., FRIEND, J. & DOHLMAN, C. H. (1970). Effect of a corneal fluid barrier on the nutrition of the epithelium. *Exp. Eye Res.* **9,** 254–259.

V. SALLMANN, L., CARAVAGGIO, L. L. & GRIMES, P. (1961). Studies on the corneal endothelium of the rabbit. *Am. J. Ophthal.* **51,** Pt. 2, 955–969.

WOODIN, A. M. (1954). The properties of a complex of the mucopolysaccharide and proteins of cornea. *Biochem. J.* **58,** 50–57.

ZADUNAISKY, J. A. (1966). Active transport of chloride in frog cornea. *Am. J. Physiol.* **211,** 506–512.

ZADUNAISKY, J. A. (1969). The active chloride transport of the frog cornea. In *The Cornea,* Ed. M. E. Langham, pp. 3–34. Baltimore: Johns Hopkins Press.

ZANDER, E. & WEDDELL, G. (1951). Observations on the innervation of the cornea. *J. Anat. Lond.* **85,** 68–99.

ZAUBERMAN, H., MICHAELSON, I. C., BERGMAN, F. & MAURICE, D. M. (1969). Stimulation of neovascularization of the cornea by biogenic amines. *Exp. Eye Res.* **8,** 77–83.

PART 4

THE LENS

DEVELOPMENT AND STRUCTURE

THE lens is a transparent biconvex body (Fig. 45); the apices of its anterior and posterior surfaces are called the *anterior* and *posterior poles* respectively; the line passing through the lens and joining the poles is the *axis*, whilst lines on the surface passing from one pole to the other are called *meridians*. The *equator* is a circle on the surface at right-angles to the axis (Fig. 45). Structurally, the lens

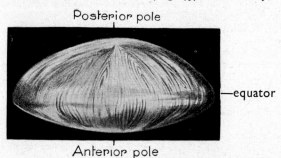

FIG. 45. The human lens. (Duke-Elder, *Textbook of Ophthalmology*.)

is divided into the *capsule*, an elastic envelope capable of moulding the lens substance during accommodative changes (Sect. III), and by which the attachment to the zonule is made; the *epithelium*, extending as a single layer of cells under the anterior portion of the capsule as far as the equator; and the *lens substance* made up of fibres and interstitial cement material. The significance of the intimate structure of the lens substance can only be appreciated with a knowledge of its embryonic and later development.

Embryology

The lens vesicle, on separation from the surface ectoderm, is a sphere whose wall consists of a single layer of epithelium. On their outer aspect the cells of this epithelium secrete a structureless hyaline membrane, the capsule; the cells of the posterior wall and the equatorial region, moreover, undergo marked differentiation and from these the adult lens is formed. The cells of the posterior wall differentiate first, becoming primitive fibres by a process of elongation (Fig. 46); their nuclei ultimately disappear and they remain as an optically clear *embryonic nucleus* in the centre of the mature lens. At the equator, meanwhile, epithelial cells likewise elongate to become fibres; the anterior end of each passes forward towards the anterior pole, insinuating itself under the epithelium, whilst the posterior end passes backwards towards the posterior pole. This process continues, and since each new fibre passes immediately under the epithelium it forces the earlier formed fibres deeper and deeper; successive layers thus encircle the central nucleus of primitive fibres.

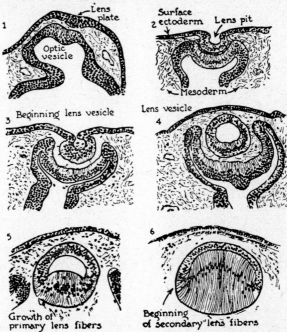

FIG. 46. Development of the human lens. (Bellows, *Cataract and Anomalies of the Lens.*)

Fibre Formation

The actual process whereby the developing lens fibres force themselves over less recently formed ones may now be described. The epithelial cells at the equator tend to become arranged in meridional rows (Fig. 47); they assume a pyramidal shape and finally their inner parts elongate and turn obliquely to run under the epithelium of the anterior capsule; the cell is now a young fibre and its inner end, which must now be referred to as its anterior end, grows forward whilst its outer (posterior) end is pushed backwards by a similarly developing cell immediately in front. As the cells grow into the lens substance, the nuclei gradually recede with them, becoming arranged in a wide bend which curves round anteriorly as the process of elongation proceeds; this configuration is termed the *nuclear zone.*

Sutures. A recently formed lens fibre thus has its centre approximately at the equator, whilst the two limbs pass forward and backward towards the anterior and posterior poles respectively to meet the limbs of another cell developing from an "antipodeal" position. If the system were completely symmetrical, the meeting point would be the same for all cells, as illustrated schematically in Fig. 48, A. In fact, no such complete symmetry exists and the fibres meet each other in quite complicated figures which are called *sutures*; the simplest is probably the line-suture of the dogfish (Fig. 48, B) and other lower vertebrates, whilst the human embryonic and adult nuclei have Y- and four-pointed stars respectively (Fig. 48, C and D).

The Lens Fibres. The superficial lens fibres are long, prismatic, ribbon-like cells some 8–10 mm long, 8–12 μ broad, and 2 μ thick. An equatorial section of

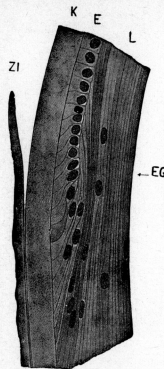

FIG. 47. Meridional section of adult human lens, illustrating the formation of meridional rows and the elongation of the epithelial cells. Zl, zonular lamella; K, lens capsule; E, lens epithelium; EG, level of epithelial border. (Salzmann, *Anatomie u. Histologie d. menschliche Augapfels.*)

the lens cuts these fibres tranversely, since they run in a generally antero-posterior direction; the appearance of such a section is shown in Fig. 49, Pl. V, the orderly development of layer after layer of fibres from the cells of the

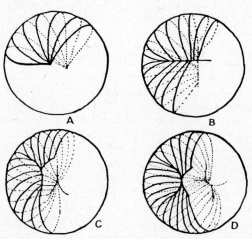

FIG. 48. Diagram showing the formation of lens sutures. A, simple arrangement with no sutures; B, linear sutures only; C, the Y-sutures of the human embryonic nucleus; D, four-pointed star of human adult nucleus. (Mann, *Development of the Human Eye.*)

PLATE V

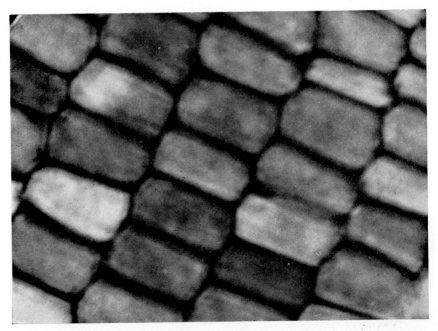

FIG. 49. Phase-contrast micrograph of equatorial section of guinea-pig lens. Haematoxylin-eosin. × 3,300. (Hueck & Kleifeld, *v. Graefes' Arch. Ophthal.*)

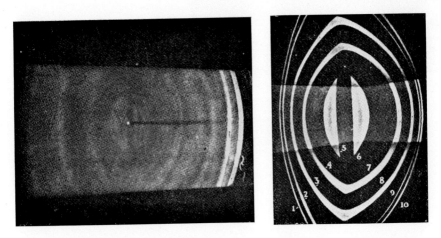

FIG. 50. *Left:* Optical section of normal human crystalline lens as seen in the slit-lamp microscope. *Right:* Schematic representation of the bands of discontinuity of normal lens as seen in optical section. 1, anterior capsule; 2, anterior disjunctive band; 3, anterior band of adult nucleus; 4, anterior external band of foetal nucleus; 5, anterior internal band of foetal nucleus, and so on. (Grignolo, *Boll. d'Oculist.*)

To face p. 88.

PLATE VI

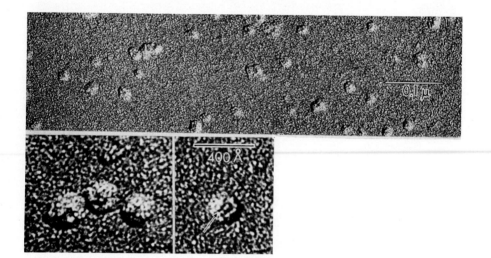

FIG. 54. α-Crystallin in solution at pH 7·8 spread on carbon film and shaded with carbon and platinum. *Above:* Large aggregates of rather uniform diameter (± 180 A) mixed with regularly shaped smaller particles. *Below:* Higher magnification. The particles seem to consist of globular or rod-like units. (Bloemendal, Bont & Benedetti, *Exp. Eye Res.*)

meridional rows leading to a lamellar appearance, the so-called *radial lamellae*. A meridional section, on the other hand, cuts the fibres longitudinally, giving the so-called "onion-scale" appearance (Fig. 47).

Cortex and Nucleus. Since the oldest fibres become more sclerosed and less translucent than the more recently formed ones, the lens is not optically homogeneous; it may be broadly differentiated into a central core—the *nucleus*—and a softer *cortex*. With the slit-lamp the nucleus can be further differentiated into zones, which correspond with the periods at which their fibres were laid down, as in Fig. 50, Pl. V. The lens grows throughout life, increasing from an average weight of 130 mg at 0–9 years to 255 mg at 80–89 years. The relative thickness of the cortex increases continuously until in old age it represents a third of the total. As a result of this growth, the curvature of the outer zone progressively decreases, i.e. the lens becomes flatter; the decrease in dioptric power that would otherwise ensue is prevented by the simultaneous increase in the refractive index of the nucleus. When the increase in refractive index fails to keep pace with the flattening of the surface, the eye becomes hypermetropic; in early cataract the nucleus becomes excessively sclerosed so that the eye becomes myopic.

Electron-Microscopy

The electron-microscope has added a few interesting facts, in particular Wanko & Gavin have described the relations between a given lens fibre and its neighbours. In many regions the cells seem to be regularly separated by a thin extracellular space, which in the dried specimen amounts to some 60 Å; in other regions, notably the equator and suture zones, there is considerable interdigitation of the surfaces. In the equator this is a lateral development to form apparent desmosomes, but in the suture zone there is interdigitation between the tips of the fibres from opposite zones (Fig. 51), an interdigitation that may be

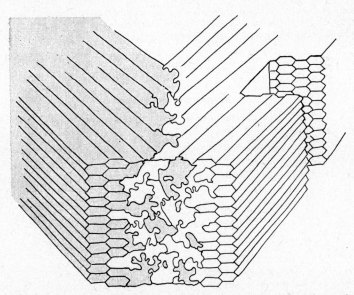

FIG. 51. Lens fibres in the suture area. The diagram illustrates how the cortical fibres converge from opposite sectors at the lens suture, where they interlace and at the same time interdigitate with one another and also with fibres located in the same lens sector. (Wanko & Gavin, in *Structure of the Eye*.)

responsible for the cohesion in this region, but sufficiently labile to permit the adjustments that must occur during accommodation (Sect. III). According to Tanaka & Iiono (1967), however, the interdigitations between processes from the sides of the fibres are so numerous and so intimate that a sliding of fibres over each other seems unlikely, so that changes in shape would have to be due to a redistribution of cellular content.

Tight Junctions. A few further details have been provided by Cohen (1965); of special interest are the tight junctions between epithelial cells that probably restrain passage of solutes along the intercellular clefts and thus give the epithelium, as a whole, the character of a highly selective membrane. The lens fibres of the cortex also show regions of occlusion, but the extent to which this occurs is not completely clear. The point is important since the posterior region of the lens lacks an epithelial covering, so that if the lens as a whole is to be limited by a selective membrane, we must postulate that the most superficial layer of lens fibres constitutes this barrier; this could be achieved were the fibres joined laterally by tight junctions or zonulae occludentes.

Capsule. The light-microscopists described the capsule as a "structureless membrane"; in the electron microscope, however, it appears as a regularly lamellated structure, with a cement substance between the layers; and there is no doubt that the laminae arise from successive depositions of basement membrane by the underlying cells. This was demonstrated autoradiographically by Young & Ocumpaugh (1966) who administered ^{3}H-glycine and ^{35}S-sulphate to young rats; labelling was at first exclusively in the epithelium and superficial fibres; an hour later there was some labelling of the capsule and after three to four hours practically all the label had been displaced to more anterior regions of the capsule, indicating that growth consisted of deposition from within. The capsule resembles Descemet's membrane and like this its basis is a collagen that shows characteristic differences from the "ordinary collagen" of connective tissue both electron-microscopically and chemically. This difference in ultrastructure is probably related to the special need of this elastic envelope which it shares with Descemet's membrane; this need is presumably an extra resistance to internal slip of the fibres under stretch. Its strong positive periodic acid-Schiff reaction is due to the presence of a polysaccharide which is difficult to separate from the collagen (Pirie). The capsule is obviously divided into two layers, the zonular fibres being attached to the outer one, which is therefore called the *zonular lamella*.

THE LENS AS A FUNCTIONAL UNIT

Regional Variations

The essentials of the chemical composition of the lens are shown in Table 5. Because of its history, however, we may expect to find differences in composition according to (a) the total age of the lens and (b), in a given lens, according to the region selected. Thus the average water-content of the calf lens, containing a high percentage of young fibres, was found by Amoore, Bartley & van Heyningen

TABLE 5

Chemical composition of the lens

	Young	Old
Water (per cent)	69.0	64.0
Protein (per cent)	30.0	35.0
Sodium (meq/kg H_2O)	17.0	21.0
Potassium	120.0	121.0
Chloride (meq/kg H_2O)	27.0	30.0
Ascorbic acid (mg/100 g)	35.0	36.0
Glutathione (mg/100 g)	2.2	1.5

to be 2·1 grammes of H_2O per gramme of dry weight (g/g), comparing with 1·8 g/g for ox lens. Again, by carefully peeling off concentric zones and analysing them, Amoore *et al*. found the following values from without inwards: 3·67, 2·86, 2·32, 1·79, 1·51 and 1·28 g/g respectively; thus the water-content of the innermost zone is less than half that of the outermost zone including the capsule.

Sodium and Potassium

The lens is essentially a cellular tissue, so that we may expect a high concentration of potassium and a low concentration of sodium; this is generally true since the figures given by Amoore *et al*., for example, indicate a value of approximately 20 m-molar for the concentration of sodium in the tissue-water and 120 m-molar for potassium. If this sodium is largely extracellular, whilst the potassium is largely intracellular, we may compute an extracellular space for the lens of about 13 per cent. Direct measurements, however, indicate that this is probably smaller, perhaps as low as 4–5 per cent (Thoft & Kinoshita, 1965; Paterson, 1970a) so that some of the sodium is intracellular.* In tissues such as nerve and

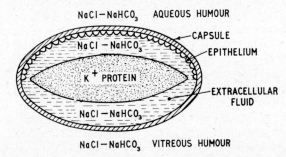

FIG. 52. Illustrating the salt and water relationships of the lens fibres, extracellular space, aqueous humour and vitreous body. For simplicity the lens fibres are represented as a single one containing predominantly potassium and protein. The actual volume of the extracellular fluid may be very small indeed, since the fibres are very tightly packed.

muscle, where we also have a mass of cells with high internal potassium and high external sodium, these gradients of concentration are maintained by active processes, in the sense that a part of the metabolic energy of the cells is directed towards "pumping out" any excess of sodium that penetrates into the cells, and towards pumping back any potassium that escapes from the cells. It may well be that the situation in the lens is similar, but it is complicated by the enclosure of the packed cells anteriorly by the lens epithelium and completely by the capsule. The situation may be represented schematically, to a first approximation, by Fig. 52, where a single fibre represents the total; it is surrounded by extracellular fluid of about the same composition as the vitreous and aqueous humours, these latter fluids being presumably in some sort of dynamic equilibrium since there is no doubt that the capsule is freely permeable to small crystalloids (Friedenwald).

Active Ion Transport. The importance of metabolism for the maintenance of

---

* When the lens is incubated in a medium containing ^{22}Na, exchange is apparently not complete, indicating the presence of a "bound fraction"; the subject has been studied most recently by Duncan (1970) and Paterson (1970b); the latter suggests that the bound material is associated with the albuminoid fraction of the proteins.

the normal sodium, potassium and chloride relationships has been demonstrated by several studies; thus Langham & Davson found that within a few hours *post mortem* the chloride concentration in the lens began to rise, indicating, presumably, an absorption of aqueous humour; again, Harris & Gehrsitz found that cooling the lens *in vitro*, thereby inhibiting its metabolism, caused the potassium concentration to fall and that of sodium to rise, processes that could be reversed by re-warming.

PUMP-LEAK SYSTEM

An obvious interpretation of these findings is that the individual lens fibres maintain their high internal K^+-concentration and low Na^+ by virtue of active pumps; cooling deprives the pumps of metabolic energy and allows K^+ to escape, and Na^+ and Cl^- to enter the fibres; the lost K^+ would diffuse from the extracellular space into the ambient medium. However, a number of studies have suggested an involvement of the epithelium in the active uptake of K^+ from the medium (e.g. Kinsey & Reddy, 1965; Becker & Cotlier, 1965; Harris & Becker, 1965) so that we may envisage a primary accumulation of K^+ into the extracellular space. Since the epithelium is absent from the posterior surface of the lens, the effects of this accumulation will be largely nullified by a passive escape of K^+, and such a "pump-leak" system has actually been envisaged, in which event the lens would effectively transport K^+ from the anterior aqueous humour to the posterior fluid and vitreous body; the distributions of K^+ between these fluids are not necessarily inconsistent with this view (Bito & Davson, 1964), whilst a kinetic study of the uptake of K^+, Rb^+ and Cs^+ by Kinsey & McLean (1970) was also consistent with the pump-leak hypothesis, in the sense that the measured uptakes could be made to fit the solutions to simple equations based on a primary accumulation process at the epithelium, showing competitive inhibition between the three cations, and a passive diffusion process at the posterior capsule.*

Lens Potential. A more realistic interpretation might, however, be provided by Fig. 53, in which the lens behaves, effectually, as a single cell, its outer membrane being made up of the epithelium plus a layer of fibres; the epithelium and layer of fibres are in free communication with the contents of the large mass of lens fibres that make up the body of the lens, and this communication is presumably achieved by a system of intercellular junctions. This view has been put forward by Duncan (1969) to account for the existence of a potential between inside and outside, and the influence of the external medium on this. Thus Brindley (1956) impaled the amphibian lens with a microelectrode; as soon as the capsule was penetrated a potential, inside negative, of 63–68 mV, was registered; this was decreased if the external concentration of K^+ was raised. Subsequent studies of Andrée (1958) and of Duncan have confirmed the existence of the potential, and have shown that it is determined by the most superficial layer beneath the capsule; thus the potential is independent of the depth of the electrode in the lens, so that it cannot be attributed exclusively to the epithelium. Even more significant is the observation of Andrée that removal of the capsule, with its adherent epithelium, only abolishes the potential temporarily; after some hours it is re-established to reach a value of 68 mV, compared with the original one of 74 mV. The potential may be

* The estimated transfer coefficients for the epithelium were 0·095, 0·056 and 0·024 hour^{-1} for K^+, Rb^+ and Cs^+ respectively; this means that passage is not through a simple water-filled pore, which would be unable to exert this kind of selectivity.

recorded with large 10–20 μ tipped electrodes, and so is definitely not a measure of the resting potential of individual lens fibres. In the eye of the trout, too, the lens potential is independent of the aspect penetrated by the electrode (Rae, Hoffert & Fromm, 1970).

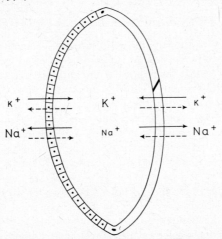

FIG. 53. Illustrating the equivalent anatomical basis for the behaviour of the amphibian lens. It behaves as though it were completely enclosed in a cellular membrane constituted anteriorly by the epithelium and posteriorly by the most superficial layer of lens fibres. Active transport is indicated by full lines and passive diffusion by broken lines.

AMPHIBIAN AND MAMMALIAN LENSES

On this basis, then, the amphibian lens as a whole acts like other epithelial systems, such as the skin, the outermost layer of cells apparently determining the electrical characteristics through active transport of solutes across them. Ouabain, by inhibiting Na^+- K^+-activated ATPase, would cause a loss of K^+ and gain of Na^+, as found; and this correlates with the presence of the enzyme in the epithelium (Bonting, Caravaggio & Hawkins, 1963). Measurements of the fluxes of Na^+ and K^+ into and out of the lens give values for the permeability coefficients for these ions that are consistent with the measured potential (Duncan, 1970). The main problems to be resolved, then, are the ultra-structural basis for this behaviour, i.e. in what way is this electrical continuity throughout the lens-mass achieved; and secondly, we must ascertain whether experiments on the amphibian and fish lens are applicable to the mammalian lens; according to Sperelakis & Potts (1959) the potential in the mammalian lens is more complex, the individual fibres, for example, exhibiting a potential of 23 mV, inside negative, whilst the capsule has a potential across it of 26 mV. It is unlikely, however, that the capsule *per se* generates any potential; all the evidence indicates that it is a freely permeable structure, allowing unrestricted passage of small solutes and even permitting the passage of protein (Francois & Rabaey, 1958).* It may be, as Riley (1970) has argued on the basis of his studies of ionic exchanges across partially damaged lenses, that the mammalian lens fibres exhibit active transport leading to extrusion

* The common observation that damage to, or removal of, the capsule, profoundly modifies its characteristics is misleading; in fact it is impossible to remove the capsule without also the epithelium and probably a thin layer of the most superficial lens fibres.

of Na$^+$ and accumulation of K$^+$, but that the "pumps" concerned with this are unable to compete with the "leaks" in the system without the cooperation of the lens epithelium which, by extruding Na$^+$ and accumulating K$^+$, assists in maintaining the appropriate intracellular concentrations in the fibres.

TRANSPARENCY AND CATARACT

The transparency of the lens must presumably depend on the avoidance of large transitions of refractive index; this can be achieved by a dense packing of the individual cells so that the regions of discontinuity are very small compared with the wavelength of light. Within a given lens fibre, the proteins, which are in a very high concentration, must be dispersed so as to give a non-opalescent solution.* Vacuolation—both intracellular and extracellular according to v. Sallmann—and local precipitation of protein are probably features of the loss of transparency that is described as *cataract*. Changes in transparency are certainly associated with changes in salt and water contents (Leinfelder & Schwartz), and the relationship here may be causal. As we shall see, metabolic disturbance invariably leads to cataract; metabolism is necessary for maintaining the normal water and ionic relationships so that we have here an obvious mechanism of cataract formation through metabolic upset; nevertheless metabolic disturbance may lead to quite gross histological alterations, including the death and break-up of lens fibres; and these must be associated with changes in the transmission of light by the lens as a whole.

The Proteins

An important factor in the maintenance of the transparency of the lens must be the physical state of the proteins, and it is for this reason that so much work has been, and is being, done on the isolation and characterization of the macro-molecules of the lens. The proteins were classically divided by Mörner into several fractions, as follows, where the percentages are those given by the more recent study of Krause:

Protein	per cent
Insoluble albuminoid	12·5
α-Crystallin	31·7
β-Crystallin	53·4
γ-Crystallin or albumin	1·5
Mucoprotein	0·8
Nucleoprotein	0·07

the "insoluble albuminoid" being the residue after extraction of the crystallins etc. The modern classification tends to retain this terminology, the segregation of the crystallins into α-, β- and γ-fractions being brought about experimentally by such procedures as migration in an electric field—electrophoresis—sedimentation in the ultracentrifuge, gel-filtration, and so on. Table 6 enumerates the main features of these fractions, as summarized by Waley (1969). Within any class, e.g. the α-crystallins, modern refinements of analysis, such as the technique of immuno-electrophoresis,† have often permitted the identification of sub-classes which Waley describes as αA, αB, and so on.

---

* Waley (1965) has pointed out that a macerated lens remains transparent; only when the *brei* has been diluted does it become cloudy.
† In the technique of immuno-electrophoresis, the protein sample is electrophoresed in the ordinary way, so that the proteins move in an electric field and separate from each

α-Crystallin. This protein has the largest molecular weight of the crystallins, values of 660,000 to 1,000,000 being given according to the mode of preparation. The preparation described by Bloemendal *et al.* (1962) had a molecular weight of some 810,000; on addition of hydrogen-bond breaking agents, such as 7 M urea, the molecular weight was reduced drastically to around 26,000, suggesting the breakdown into subunits. When the purified α-crystallin was examined in the

TABLE 6

Characteristics of the Main Classes of Soluble Proteins from Mammalian Lens
(Waley, The Eye.)

Property	Class		
	α	β	γ
Electrophoretic Mobility (at pH 8–9)	High	Medium	Low
Molecular Weight	High	Intermediate	Low
Thiol Content	Low	High	High
N-terminal Amino Group	Masked	Masked	Free

electron microscope spherical molecules of about 150 Å diameter were observed, and these appeared to be made up of smaller subunits (Fig. 54, Pl. VI). Björk (1964) examined the chemical character of the subunits and their reaggregation brought about by removing the urea and increasing the ionic strength of the medium. The subunits were definitely not identical chemically; moreover, complete reaggregation to the original α-crystallin was never achieved, so that it is likely that there are several types of subunit arranged in a definite manner; breakdown of this structure and reaggregation would lead to the production of a variety of different combinations which would show only partial immunochemical identity with the native material. Subsequent work on the subunit structure of α-crystallin has indicated that there are four of these, two acidic and two basic, differing in their amino acid compositions (Schoenmakers, Gerding & Bloemendal, 1969); each of these subunits is a single polypeptide chain; interestingly, the embryonic lens only contains three subunits, one of the acidic ones being missing (Schoenmakers & Bloemendal, 1968). The molecular weights of these subunits have been generally given as in the neighbourhood of 20,000 but more recently Leon, deGroot & Bloemendal (1970) have shown that they are in the region of 11–12,000.

Albuminoid. This is the material that fails to dissolve in aqueous salt solutions; as Rao, Mehta & Cooper (1965) have shown, however, it may be solubilized by hydrogen-bond breaking agents, such as 7 M urea; and the resulting product shows marked similarities with α-crystallin as revealed immunoelectrophoretically. Certainly there is a very close reciprocal relationship between the insoluble albuminoid and α-crystallin content of lenses; thus in cataractous conditions the

---

other by virtue of their different velocities. The electrophoresis is carried out in a gel, such as agar, and a channel is cut in this parallel to the direction of the electric field and antiserum is placed in it; the antibodies diffuse laterally from this channel to meet the various fractions of proteins where they precipitate them, to give characteristic "streaks" in different regions. The number of distinct streaks indicates the number of components in the protein sample that can react with the antiserum.

insoluble albuminoid fraction increases whilst the α-crystallin fraction decreases; similarly, in a given lens, the proportion of α-crystallin decreases on passing from cortex to nucleus whilst the proportion of insoluble albuminoid increases. The two proteins probably have the same primary structure, i.e. the same amino acid sequence (Mok & Waley, 1967), and a direct proof of the interconversion of the two has been given by the studies of Fulhorst & Young (1966), who injected ^{35}S-labelled methionine into day-old rats and examined the subsequent distribution of the label. If the lenses were removed one day after the injection, the label was uniformly distributed throughout the lens, as revealed autoradiographically; and chemical studies showed that practically all was in the soluble protein. Seven weeks later, however, the label was largely confined to the deeper layers, and analytically it was shown that 40 per cent of it was in the albuminoid fraction. We may conclude, then, that the molecules of α-crystallin aggregrate together by the formation of hydrogen-bonds between chains; the bond-formation presumably reduces the number of hydrophilic groups in the surface and thus induces insolubility. Although cataract is associated with increased formation of insoluble protein, it must be appreciated that the mere formation of albuminoid is not a cause of opacification since this material, although insoluble, probably exists, in the nuclear zone for example, as a transparent gel.

β-Crystallin. This protein is definitely not homogeneous, and relatively little work has been done on it; Björk (1964) has separated four main fractions with sedimentation coefficients of 13·6, 9·6, 4·9 and 4·2 S by a gel-filtration technique, but these were not individually pure; it may be that the larger molecules are built up during preparation by oxidation of SH–groups to form S–S links between chains, since the SH–groups are not shielded, reacting with the nitroprusside reagent.

γ-Crystallin. This is the slowest-migrating component of the lens proteins; it was first identified by François & Rabaey in 1957 and described by them as *embryonic lens protein;* its concentration is greatest in the nucleus, and this is because it is formed mainly before birth, further synthesis ceasing after birth; thus the absolute amount remains about the same during life but, since new α- and β-crystallins are formed, the percentage of γ-crystallin decreases. Björk (1964) has separated four distinct proteins from the γ-crystallin fraction; their molecular weights are very close, in the region of 19,000–20,000; three of the fractions were immunologically identical with each other whilst the fourth was only partially identical with these. Thus their differences reside in the amino acid composition, differences that presumably arose by mutations on specific parts of the gene during evolution, but since these mutations did not affect the transparency of the lens they were not eliminated by selection (Björk, 1964). As we shall see, the reversible "cold cataract" is due to precipitation of γ-crystallin. Within the same group is a fraction that inhibits mitosis of the lens epithelium (Voaden & Froomberg, 1967); this probably consists of several factors, one of which may be a chalone similar to that in epidermal tissue that acts with adrenaline to control mitosis (Bullough & Laurence, 1964); this could be responsible for the diurnal variation in mitotic activity of the lens epithelium (Voaden, 1968). Certainly, since the lens epithelium cells are dividing mitotically throughout life, some form of control is necessary; damage to the epithelium often leads to loss of this control with the resulting proliferation to give multilayered aggregates (see, for example, Kuwabara, Kinoshita & Cogan, 1969).

Glycoproteins. This is a class of sugar-containing proteins in which the sugar is linked to the protein covalently; they differ from the complexes between mucopolysaccharide and protein—*mucoproteins*—in, among other features, that they do not contain uronic acids and sulphate esters, although they do contain amino sugars; their characteristic constituents are N-acetylneuraminic acid, galactose, mannose glucosamine, galactosamine and L-fucose. It has been suggested that the glycoproteins isolated from the lens are largely derived from the cell membranes of the lens fibres, i.e. that they are part of the protein-lipid complex considered to represent the cell membrane; at any rate some two-thirds may be extracted by lipid solvents such as chloroform-methanol (Dische, 1965); quantitative estimates of the amount of this lipo-glycoprotein necessary to cover the surfaces of the lens fibres agreed reasonably well with the amount obtained.

Capsular Protein. As indicated earlier, the structural protein of the capsule is collagen; it is hydrolysed by collagenase and has the characteristic proportion of hydroxproline residues. The high proportion of carbohydrate differentiates it from the collagen of skin, and the fluorescent antibody-staining experiments of Roberts (1957) show that the material is analogous with the basement membrane material characteristically seen surrounding capillaries; embryologically it would seem to arise as a condensation of the basement membranes of lens cells (Cohen, 1965). The major sugar moiety is a sialohexosaminoglycan containing galactose, mannose, fucose and neuraminic acid; some sulphated esters are also present, an unusual feature for a connective-tissue glycan, possibly connected with the need for transparency of the capsule (Dische & Zelmenis, 1966).

Chicken Lens Proteins. Most work on lens proteins has been done on mammalian material; the study of the chicken is important experimentally since it is in this species that embryological studies are normally carried out. In general, the protein composition is very different, the only common protein being probably α-crystallin; the major component is what Rabaey (1962) called FISC or *first important soluble crystallin** and is the avian counterpart of the mammalian γ-crystallin, predominating before birth and becoming concentrated in the nucleus in the adult. As with mammalian α-crystallin, FISC probably is built on a subunit structure; for further details on the possible subunit basis of the chicken lens proteins the reader is referred to Clayton (1969).

Organ Specificity

As long ago as 1903 Uhlenhuth showed that the immunological characteristics of the lens were unusual. Thus if we inject an extract of, say, ox muscle into a rabbit, the rabbit produces antibodies in its serum that will react with this extract if injected again, to produce the so-called *immune reaction*; if the reaction is carried out *in vitro*, there will be a precipitation of the proteins by the antiserum. This reaction is on quite a strictly species basis, so that the rabbit's serum will not react to sheep muscle for example. If an ox antiserum is prepared to ox lens, by injecting a lens extract into a rabbit, it is found that this antiserum reacts to lens extracts of all other mammalian species, as well as to those of fishes, amphibians and birds (but not to that of the octopus). The lens is said, therefore, to be *organ*—rather than species—specific and we may conclude that the lens proteins that are responsible for this antigenicity are all built on a very similar plan in the lenses of different species. It was thought for some time that it was only the α-crystallin that behaved as an antigen, but by the technique of immuno-electrophoresis, for example, it has been shown that all three groups of crystallins are antigenic. Thus,

* Zwaan (1966) has called FISC δ-crystallin.

when a rabbit antiserum was prepared to human lens, this serum was shown to react with some nine components of the lens proteins, two in the alpha-, four in the beta- and three in the gamma-crystallin groups. In the fish there were only five, and in the squid none. According to Manski, Auerbach & Halbert, the origin of this organ-specificity lies in the retention, by the lens, of its evolutionary history; thus the mammalian lens can produce anti-bird, anti-amphibian, and anti-fish antibodies because it contains proteins common to these classes; the fish lens, on the other hand, cannot produce anti-bird or anti-mammal antibodies. Clayton, Campbell & Truman (1968) have shown, along with earlier workers, that many of the antigens in lens are not organ specific in so far as they react with antibodies to other tissues; and they conclude that the real finding is that the concentrations of certain antigens are very much higher in the lens than in other tissues. This is understandable, since the lens contains many enzymes, such as lactic dehydrogenase, that are common to the whole body; thus organ specificity is a tendency rather than an absolute fact, a tendency promoted by the isolated position of the lens.*

Phacoanaphylactic Endophthalmitis. Early workers were unable to detect *homologous* lens reactions, i.e. they were unable to produce anti-lens antibodies in a rabbit by injecting an extract of rabbit lens, but with the aid of Freund's adjuvants, Halbert *et al.* were able to do this, so the possibility naturally arises that an animal may produce antibodies to its own lens and thus cause, perhaps, a localized precipitin reaction in its lens leading to cataract. In fact, however, the production of very high titres of anti-lens antibodies in the rabbit failed to cause cataract, nor yet could congenital cataract be induced in the progeny of females with high anti-lens titres during pregnancy. Furthermore, human cataract patients had no anti-lens antibodies in their serum. Nevertheless, it is possible to induce a localized anaphylactic response in rabbits by producing anti-lens antibodies in the normal way and then to traumatize the lens; the resulting inflammatory reaction in the eye has been called by Burky *phacoanaphylactic endophthalmitis*.

Proteases. An excised lens, maintained under sterile conditions, eventually undergoes autolysis, i.e. non-bacterial decomposition. This is due to the activity of certain proteases that are normally suppressed. In general, because of the extremely slow turnover of the lens protein—not greater than 3 per cent per month according to Waley (1964)—proteolytic activity is not great; nevertheless the study of the lens proteinases, and the factors that normally hold them in check, is important from the point of view of the maintenance of the correct protein composition and the removal of precipitated material. The characteristics of a neutral† proteinase in lens have been described by Van Heyningen & Waley (1962); it is found in the fraction of α-crystallin, from which it may be separated. It converts protein, e.g. α-crystallin, to amino acids, and is to be distinguished from leucine aminopeptidase, which is also present in the α-crystallin fraction of lens; it seems likely that the protease breaks down the protein to smaller products, which are then attacked by leucine aminopeptidase.

Fluorescence

It is well known that in ultraviolet light the lens emits a blue fluorescence; the material responsible for this, extracted by Szent-Györgyi from an ox lens, appeared to be a protein; François, Rabaey & Recoules (1961) isolated a product from primate lenses that was fluorescent, and they considered it to be a polypeptide,

---

* Mehta, Cooper & Rao (1964) found that it was the β- and γ-crystallins that showed species-specificity, so that organ-specificity was peculiar to α-crystallins.
† The term neutral proteinase embraces a wide class of proteolytic enzymes that are most active at about pH 7·5; this property distinguishes them from *cathepsins* with a pH optimum of about 4.

whilst Cremer-Bartels (1962) isolated what he believed to be a pteridine from ox lenses; this material was strongly reactive to light and seemed to be involved in phosphate metabolism.*

METABOLISM

Lens Culture. The metabolism of the lens is most conveniently studied *in vitro*, and since Bakker in 1936 first maintained a lens in an artificial medium for several days with normal transparency, there have been frequent studies of the best medium for the purpose and the most suitable index of viability; outstanding in this respect has been the work of Kinsey and, more recently, Schwartz, and their collaborators. In general, the medium must contain a suitable ionic composition similar to that of the aqueous humour, whilst in order to allow for consumption of glucose and production of lactic and other acids, the medium must be changed regularly. As criterion of normal viability the best is probably the mitotic activity of the epithelium, the cells of which, it will be recalled, are continually dividing to produce new fibres (Kinsey *et al.*). Besides imitating the aqueous humour with regard to ionic make up, a variety of amino acids and vitamins, together with inositol, lactate and glutathione, were added. By using a constant flow of this solution, Schwartz kept a lens transparent for some 40 hours, but at the end of this time it was not completely normal, having a higher lactic acid content, for example, then when it began.

Metabolic Pathways

Glycolysis. The lens utilizes glucose as its source of metabolic energy; since the O_2-tension of the aqueous humour is low, most of this metabolism is anaerobic, the Q_{O_2} being only $0 \cdot 1$; thus the main pathway for the utilization of glucose is by way of the Embden-Meyerhof or glycolytic reactions leading to pyruvic acid, which is largely converted to lactic acid; the latter is carried away in the aqueous humour. The enzyme lactic dedydrogenase is involved in this process, together with the coenzyme $NADH_2$:

$$\text{Glucose} + 2Pi + 2ADP \rightarrow 2 \text{ Lactate} + 2ATP + 2H_2O.$$

Pentose Phosphate Shunt. As with the cornea, Kinoshita & Wachtl (1958) have shown that the pentose phosphate shunt contributes significantly to the oxidative breakdown pathway, in addition to the ordinary glycolytic pathway; this involves the coenzyme NADP.

Sorbitol. Finally Van Heyningen (1959) has shown that the lens contains sorbitol and fructose, together with enzymes capable of converting glucose to fructose by way of sorbitol. Thus the metabolic events in the lens may be summarized by the scheme of Fig. 55.

ATP. Glycolysis, the pentose phosphate shunt and the Krebs cycle all lead to the generation of ATP which acts as substrate, or phosphate donor, in a variety of endergonic reactions, e.g. the synthesis of protein and fatty acids, the active transport of ions, and so on. The pentose phosphate shunt is concerned with the production of a variety of intermediates in synthesis, e.g. nucleic acids; it also

---

* The yellow pigment that accumulates in the human lens with age, and reduces thereby the sensitivity of the eye to blue light, is probably a *urochrome*, similar to that found as a breakdown product of protein in urine (McEwan, 1959).

leads to the production of $NADPH_2$, the coenzyme concerned with many syntheses, especially of fatty acids.

Sorbitol Pathway. The significance of the sorbitol pathway, which converts glucose, via sorbitol, to fructose, is not clear; the enzymes concerned are aldol reductase, $NADPH_2$ and polyol reductase:

$$Glucose + NADPH_2 \rightarrow Sorbitol + NADP$$
Aldol reductase

$$Sorbitol + NAD \rightarrow Fructose + NADH_2$$
Polyol reductase

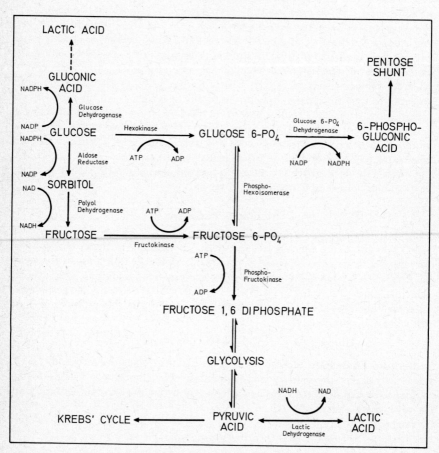

FIG. 55. Some of the possible metabolic routes of glucose in the lens. (After Spector. *Arch. Ophthal.*)

The same pathway allows xylose to be converted to xylitol, but dulcitol and arabitol, formed from galactose and arabinose by aldol reductase, are not metabolized further.

Oxidative Phosphorylation. Molecular oxygen is used in biological reactions by way of the cytochrome system, which is contained in the mitochrondria; this seems to be confined to the epithelium, according to Kinsey &

Frohman, so that it may be that respiration, as contrasted with glycolysis, is confined to this cellular layer. Because of the relative insignificance of O_2 in metabolism,* keeping a lens anaerobically has no effect on transparency, nor on its power to retain its high K^+ and low Na^+; if glucose is absent from the medium, however, anaerobiosis causes a drastic deterioration in the state of the lens, presumably because, in the absence of glucose, O_2 is utilized to oxidize other substrates than glucose (Kinoshita, 1965).

Control

Control over the rate at which a given metabolic process takes place is generally exerted through inhibition of one or more key enzymes; in the case of the lens these are hexokinase and phosphofructokinase, the former catalysing the conversion of glucose to glucose 6-phosphate and the latter the phosphorylation of fructose 6-phosphate to fructose 1,6-diphosphate:

$$Glucose + ATP \rightarrow Glucose\ 6\text{-phosphate} + ADP$$
$$Fructose\ 6\text{-phosphate} + ATP \rightarrow Fructose\ 1,6\text{-diphosphate} + ADP$$

Thus ATP inhibits phosphofructokinase activity whilst glucose-6-phosphate inhibits hexokinase activity; as a result, increasing the concentration of glucose in the lens does not increase lactate production or the level of glucose 6-phosphate. Increasing the level of glucose does, however, stimulate the sorbitol pathway, a finding of significance for the interpretation of diabetic cataract (p. 105).†

Protein Synthesis

As indicated earlier, there is a small turnover of protein in the adult lens; in addition, new fibres are being laid down throughout life, requiring further synthesis. The total amount of protein synthesized per day is, nevertheless, small, and this, together with an extremely efficient amino acid accumulating mechanism that maintains the concentrations of amino acids in the lens far above those in the aqueous humour and blood plasma, is presumably why the lens remains normal and continues to grow when the level of amino acids in the blood is so low as to inhibit body-growth entirely.

Amino Acid Accumulation. The active accumulation of amino acids by the lens was demonstrated by Kern (1962) and Reddy & Kinsey (1962), and is illustrated in Fig. 27 (p. 31); the process seems similar to that described in other cells (Brassil & Kern, 1967) and ensures that protein synthesis will not be limited by availability of amino acids. We may note that the lens, like the brain, may synthesize certain amino acids from glucose, namely glutamic acid and alanine (see, e.g. Van Heyningen, 1965).

Nucleic Acids. Protein synthesis occurs in association with nucleic acids by mechanisms that are now well understood and need not be repeated here. All the

* Because of the high efficiency with which ATP is produced through the oxidative phosphorylation mechanism, the amount produced by this pathway in the lens actually could be as high as 33 per cent; a definite Pasteur-effect can be demonstrated, moreover (Kern, 1962).
† The role of insulin in control of glucose utilization by the lens is a controversial topic; Farkas & Weese (1970) have shown that aqueous humour contains a stimulator to utilization which is enhanced by systemic administration of insulin in the presence of trace quantities of trivalent chromium.

evidence indicates that synthesis in the lens is in every way similar; the loss of the nucleus from the lens fibre, like that of the reticulocyte, presumably demands that the template messenger RNA be longer-lived than that in nucleated cells; in this case protein synthesis will not be inhibited completely by actinomycin D as is, indeed, found (Zigman & Lerman, 1968).

Glutathione

This tripeptide, glutamyl-cysteinyl-glycine, whatever its role, must be of importance for the lens since its concentration is some 1000 times that in the aqueous humour; studies on the incorporation of radioactive glycine into the lens show that glutathione is continually being synthesized; in all forms of cataract the fall in concentration of glutathione is striking. Glutathione constitutes a redox system:

$$2\,GSH + \tfrac{1}{2}O_2 \rightleftharpoons GSSG + H_2O_2$$

In the lens there is an enzyme, *glutathione reductase*, catalysing the reduction of oxidized glutathione. Reoxidation can take place by a complex series of reactions with the intermediate formation of H_2O_2 and the intervention of the enzyme *glutathione peroxidase*. Glutathione can behave as coenzyme in the glyoxalase reaction. Its function in the lens is obscure: it has been argued that it prevents the oxidation of SH–groups in lens proteins but this seems unlikely (Waley, 1969).*

Ascorbic Acid

Ascorbic acid (vitamin C) is present in the aqueous humour in high concentration in many species; when this occurs it is also concentrated in the lens, but its significance is not clear since in scurvy the lens shows no defect; furthermore, the *in vitro* preparation seems quite stable without it.

CATARACT

By cataract we mean any opacification of the lens, which may be a highly localized "spot" or a complete loss of transmission throughout the substance; it may occur as a senile change; as a result of trauma, which presumably injures the capsule and the underlying epithelium; as a result of metabolic or nutritional defects; or as a consequence of radiation damage.

Morphology. The histology of the cataractous lens has been described in some detail by v. Sallmann (1951, 1957), and more recently by Cogan (1962). The *nuclear cataract* in which the nucleus of the lens becomes yellow or brown, is probably an exaggeration of the normally occurring senile changes. The cortical cataracts, on the other hand, represent opacities developing in the more recently laid-down lenticular fibres, and thus reflect the response to some abnormality in metabolism. The gross morphology of these cortical cataracts consists in the appearance of aberrant forms of epithelial cells, which become large and round and are called "balloon cells"; again, lens fibres may become fused into discrete masses, called morgagnian globules; if this process goes far enough the whole

---

* Waley (1969) has described a number of other S-containing compounds identified in the lens, e.g. *ophthalmic acid* or glutamyl-α-aminobutyryl-glycine, which inhibits the glyoxalase reaction; *norophthalmic acid*, γ-glutamylalanyl-glycine; *sulphoglutathione*, and so on.

cortex becomes a milky liquid mass, and material may escape from the lens into the aqueous humour and vitreous body.

Of more interest, of course, are the early changes which, in the nature of things, must be examined in experimental forms of cataract in animals. In galactose cataract, Kuwabara, Kinoshita & Cogan (1969) have shown, in the electron microscope, that the first change is a vacuolization of the epithelial and superficial cortical lens cells; also the intercellular spaces between cortical fibres are expanded to form cysts. All these changes were reversible by returning the animals to a normal diet. After 7 days the lenses were opaque, and now the intercellular cysts had enlarged and become vacuoles; the membranes of individual lens fibres had broken and the cells had liquefied. After 10 to 15 days the lens was densely opaque, and the whole cortex had liquefied whilst the epithelium was irregular and showed regions of cell proliferation. Thus the early feature seems to be the appearance of vacuoles in the cells and enlargements of the extracellular space to form cysts; if this process is not held in check there is irreversible rupture of lens fibres.

TABLE 7

Effects of Age on Protein Content of Human Lenses.
(*After* Mach. Klin. Mbl. Augenheilk.)

Age (years)	Weight (mg)	Total	Protein Content (mg/100 g lens)		%-Soluble
			Soluble	Insoluble	
0–10	122	64·0	61·0	2·3	96·3
10–20	122	63·0	59·0	3·8	94·1
21–30	140	62·8	57·3	5·5	91·1
30–40	153	57·6	51·7	5·9	90·2
40–50	170	55·6	49·7	5·6	89·2
50–60	212	50·7	43·3	7·6	84·9
60–70	230	49·7	39·5	10·3	79·4
70–80	215	47·0	37·0	10·0	78·6
>80	269	47·6	36·0	11·5	75·9
Cataract					
Normal (40–86)	297	50·2	41·4	9·1	81·7
Cataract	190	44·0	23·4	19·9	51·4

Metabolic Defects. The way in which a metabolic defect can lead to a loss of transparency is not clear, but an important factor must unquestionably be the amount and physical condition of the proteins in the lens fibres; the lens continually produces new proteins and the existing ones are in a continuous state of "turnover", in the sense that they are broken down and resynthesized. These processes require energy which they derive from the metabolic breakdown of glucose, and it may be surmised that defects in the protein synthetic mechanism could cause cataract. Other factors may be an alteration in pH causing a precipitation of protein in the fibres; a failure of the ionic pumps that maintain the normal salt and water contents of the fibres; finally, there is the death of the lens fibres, producing gaps in the ordered arrangement, or the partial death leading to vacuolization of the cells.

Senile Cataract. The most characteristic feature of the ageing of the human lens is the decrease in the percentage of the soluble proteins; this is shown in Table 7 from Mach (1963); it will be seen that the total protein actually falls too, so that the loss of soluble material is accompanied by an increase in the insoluble albuminoid. In cataractous lenses there is a similar decrease in the total protein and in the percentage of soluble proteins, but this now falls to about 50 per cent compared with an average of 82 per cent in normal lenses from subjects aged 40 to 86. François, Rabaey & Stockmans (1965) have emphasized that the most significant feature of these changes in composition is the decrease in the low-molecular weight material, probably γ-crystallin, whilst the concentration of α-crystallin remains remarkably constant.

LENTICULAR SCLEROSIS

The gradual hardening of the central zone of the lens with age, called lenticular sclerosis, presumably is related to the process of senile cataract formation. Zigman, Schultz & Yulo (1969) have examined the insoluble protein in lenses of rats at increasing ages. As Fig. 56 shows, the most characteristic feature is the fall in

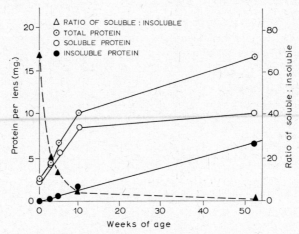

FIG. 56. Changes in rat lens protein levels with age. (Zigman, Schultz & Yulo, *Biochem. Biophys. Res. Comm.*)

the ratio of soluble to insoluble protein with age, especially during the first ten weeks. They found that the insoluble material could be divided into two fractions, namely that which could be redissolved with 7 M urea, which breaks loose bonding between side-chains, and the urea-insoluble material presumably made insoluble by much firmer links between side-chains, e.g. disulphide links. As Fig. 57 shows, at about 9 weeks there is a precipitate fall in the urea-soluble at the expense of the urea-insoluble material, and this is coincidental with a fall in synthesis of γ-crystallin. The urea-soluble material contained about equal parts of α- and γ-crystallins whilst the urea-insoluble material was almost completely γ-crystallin, so that it seems that the hardening of the lens, which is accompanied by, or due to, formation of urea-insoluble protein, is due ultimately to the conversion of soluble γ-crystallin to an insoluble form, presumably by the formation of S–S linkages and oxidized phenolic groups, both of which are abundant in γ-crystallin. It is interesting that the chicken lens contains no γ-crystallin and does not develop a nucleus, even in old age, whilst only traces of urea-insoluble material can be extracted from it. Thus, as Zigman *et al.* picture the process, the soluble proteins are first converted into insoluble material by a loose binding that can be broken down by urea; with further ageing, the cortex is compressed into the nucleus and the loose gel becomes a tightly bound one by the aggregation of γ-crystallin. As the authors point out, it may be that the same process applies to the ageing human lens, although the percentage of α-crystallin in the urea-insoluble material is much higher.

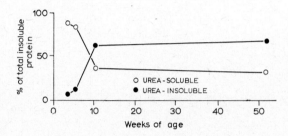

Fig. 57. Changes in rat lens insoluble proteins with age. (Zigman, Schultz & Yulo, *Biochem. Biophys. Res. Comm.*)

Sugar Cataracts. The essential factor in this form of cataract seems to be the high concentration of sugar in the aqueous humour, be it glucose, as in diabetes, or galactose and xylose in the experimental cataracts obtained by feeding these sugars to certain animals such as the rat and hamster. The only metabolic pathway common to all the cataractogenic sugars is the production of a polyhydric alcohol, e.g. sorbitol in the case of a high blood-glucose, dulcitol with high blood galactose, etc. Van Heyningen (1959) observed that feeding these sugars led to accumulation of the alcohols in the lens; and it was suggested by Kinoshita, Merola & Dikmak (1962) that the primary event was, indeed, the intracellular accumulation of the alcohol, which, not being able to escape from the lens fibres, caused an increased osmolality leading to osmotic swelling.

SWELLING OF LENS FIBRES

Fig. 58 shows changes in water- and dulcitol-content of rat lenses when the animals were maintained on a galactose diet. *In vitro*, also, exposure of the lens to galactose caused corresponding increases in hydration and alcohol concentration. When an aldose reductase inhibitor was added to the galactose medium, the formation of dulcitol was prevented, and likewise the increase in hydration (Kinoshita *et al.*, 1968). As to the ultimate steps leading to the mature cataract, i.e. the opacity of the lens, the position is not clear; Patterson & Bunting (1966) consider that this does not result from the progressive swelling and disruption of the lens fibres, since this happens suddenly over a period of 24 hours, and up to this point the changes are reversible by removing the sugar from the diet; according to Kinoshita, the swelling induces secondary changes in the lens fibres that reduce the efficacy of their active transport mechanisms, e.g. the extrusion of Na^+, accumulation of K^+ and amino acids. Thus Kinoshita *et al.* (1969) were able to inhibit the loss of amino acids and myoinositol, found in galactose-exposed lenses, by preventing the swelling through addition of osmotic material to the medium. This last finding seems to be conclusive proof that the primary insult is the osmotic swelling; this presumably causes an increase in passive permeability to ions and metabolic substrates such as amino acids, thereby making the active transport mechanisms ineffective; the failure of the Na^+-pump would eventually lead to bursting of the cells, and failure to accumulate amino acids might reduce protein synthesis and turnover.*

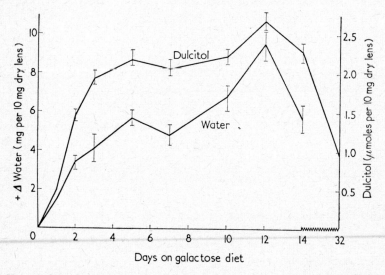

FIG. 58. The accumulation of dulcitol and water in the lenses during development of galactose cataract in rats. (Kinoshita, Merola & Dikmak, *Exp. Eye Res.*)

Radiation Cataract. The lens is subject to damage by ionizing radiations because it is essentially only those cells that undergo division that are susceptible. If the radiation is confined to the central area of the lens, by screening the equatorial region, no cataract develops (Goldmann & Leichti). If only a sector of the lens is exposed, the damage in this region is much less severe than if the whole lens had been exposed, and it would seem that the undamaged region can exert a protective action on the remainder (Pirie & Flanders). As with other forms of radiation damage, injections of cysteine and certain other protective substances, e.g. cysteamine, reduce the severity of the lesion.

Glassblower's Cataract. Exposure of the eyes to strong infra-red radiation causes cataract and this was shown by Goldmann in 1933 to be due to the high temperature to which the lens was raised by the absorption of heat; absorption of light by the ciliary body and iris, if sufficiently intense, will also raise the temperature of the lens and cause cataract (Langley, Mortimer & McCulloch).

Toxic Cataracts. A number of drugs will cause opacities in the lens; one that

* Lerman & Zigman (1967) found that the onset of galactose cataract was associated with a fall in the level of soluble γ-crystallin; they also found that ³H-sorbitol was bound to protein, in this condition, especially to γ-crystallin. Holt & Kinoshita (1968) have described alterations in the β- and γ-crystallin fractions in this form of cataract.

has been studied experimentally in some detail is naphthalene, which is converted in the body to, among other substances, naphthalene diol; according to Van Heyningen & Pirie (1967), this can be converted in the eye to 1-2-dihydroxy-naphthalene, which is probably responsible for the blue fluorescence in the eye of the naphthalene-fed rabbit. This is oxidized to β-napthaquinone (Rees & Pirie, 1967) and is highly reactive; for example, it reacts with ascorbic acid to form H_2O_2 which is toxic. This reaction converts ascorbic acid into its oxidized product, dehydroascorbic acid; and it is interesting that the concentration of ascorbic acid in the lens actually increases in the napthalene-fed rabbit; it has been suggested by Van Heyningen (1970) that the dehydroascorbic acid, which readily penetrates the lens, is reduced there to ascorbic acid from which it cannot escape so easily, being less lipid-soluble than dehydroascorbic acid. The actual means whereby β-naphthaquinone causes cataract has not so far been clarified. In addition to naphthalene, many drugs, often used therapeutically at one time, such as dinitrophenol, myeleran, dimethyl sulphoxide, will cause cataracts but not necessarily in man in therapeutic doses; for example dimethyl sulphoxide has not reportedly caused eye damage in man although in experimental animals, such as the dog, doses of 5 g/kg will cause cataract, albeit of an unusual type (Rubin & Barnett, 1967). Such knowledge as we have of the mode of action of a variety of drugs has been summarized by Van Heyningen (1969), whilst Pirie (1968) has described the early pathology of napthalene lesions in lens and retina.

Reversible Cataracts. Finally, we may mention two interesting forms of cataract that are reversible; when a young animal, such as a rat, is cooled, cataracts may appear, which disappear on rewarming, and they are due to a reversible precipitation of protein in the lens fibres. Zigman & Lerman (1965) isolated the precipitated material and showed that its sedimentation constant of 4 S was characteristic of a single low-molecular weight protein, although immunochemically it had characteristics of α-, β-, and γ-crystallins. According to Balazs & Testa (1965) the material is γ-crystallin. Lerman & Zigman (1967) made the interesting observation that the α- and β-crystallins, in high enough concentration, will inhibit cold-precipitation of γ-crystallin. The other form of opacity is hardly a true cataract; it follows the administration of morphia-type alkaloids to rodents and seems to be the consequence of a precipitation of the altered drug on the anterior surface of the lens; it develops in a few minutes and disappears within a few hours (Weinstock & Stewart).

The Calcium Content. Great interest has attached to the calcium content of the lens since it is known that interference in the general calcium metabolism of the organism, such as parathyroidectomy, or in rickets, is associated with the development of cataract. Moreover, in aged, sclerotic or cataractous lenses the calcium content increases. It is unlikely, however, that the deposition of insoluble calcium salts is the primary change in any form of cataract; most probably the breakdown of the proteins and possibly of the lipids, as a result of metabolic disturbances, liberates sulphates, phosphates, and carbonates which are precipitated by calcium already present in the lens. Thus Bellows has shown that in galactose cataract the calcium deposition always follows the primary cataractous changes. Finally we may note that an adequate concentration of calcium in the medium surrounding a cultured lens is important if it is to maintain its normal hydration and cation content (Harris & Gehrsitz, 1951).

GENERAL REFERENCES

BELLOWS, J. G. (1944). *Cataract and Anomalies of the Lens*. London: Kimpton.
GRIGNOLO, A. (1954). Studi sulla struttura submicroscopica dei tessuti oculari. *Boll. d'Oculist.* **33**, 513–652.
MANN, I. (1949). *The Development of the Human Eye*. London: B.M.A. (2nd ed.)
PIRIE, A. & VAN HEYNINGEN, R. (1956). *Biochemistry of the Eye*. Oxford: Blackwell.
VAN HEYNINGEN, R. (1969). The Lens Metabolism and Cataract. *In The Eye*, Ed. DAVSON. Vol. I, pp. 381–488. London: Academic Press.
WALEY, S. G. (1969). The lens: function and macromolecular composition. In *The Eye*. Ed. H. Davson, pp. 299–379. New York: Academic Press.

SPECIAL REFERENCES

AMOORE, J. E., BARTLEY, W. & VAN HEYNINGEN, R. (1959). Distribution of sodium and potassium within cattle lens. *Biochem. J.* **72**, 126–133.
ANDRÉE, G. (1958). Uber die Natur des transkapsularen Potentials der Linse *Pflüg. Arch. ges. Physiol.* **267**, 109–116.
BAKKER, A. (1936). Eine Methode, die Linsen erwachsener Kaninchen ausserhalb des Körpers am Leben zu erhalten. *v. Graefes' Arch. Ophthal.* **135**, 581–592.
BALAZS, E. A. & TESTA, M. (1965). The cryoproteins of the lens. *Invest. Ophthal.* **4**, 944.
BECKER, B. & COTLIER, E. (1965). The efflux of [86]rubidium from the rabbit lens. *Invest. Ophthal.* **4**, 117–121.
BITO, L. & DAVSON, H. (1964). Steady-state concentrations of potassium in the ocular fluids. *Exp. Eye Res.* **3**, 283–297.
BJÖRK, I. (1964). Studies on the subunits of α-crystallin and their recombination. *Exp. Eye Res.* **3**, 239–247.
BJÖRK, I. (1964). Fractionation of β-crystallin from calf lens by gel filtration. *Exp. Eye Res.* **3**, 248–253.
BJÖRK, I. (1964). Studies on γ-crystallin from calf lens. II. *Exp. Eye Res.* **3**, 254–261.
BLOEMENDAL, H., BONT, W. S. & BENEDETTI, E. L. (1965). On the sub-units of α-crystallin. *Exp. Eye Res.* **4**, 319–326.
BLOEMENDAL, H., BONT, W. S., JONGKIND, J. F. & WISSE, J. H. (1962). Splitting and recombination of α-crystallin. *Exp. Eye Res.* **1**, 300–305.
BONTING, S. L., CARAVAGGIO, L. L. & HAWKINS, N. M. (1963). Studies on sodium-potassium-activated adenosinetriphosphate. VI. *Arch. Biochem. Biophys.* **101**, 47–55.
BRASSIL, D. & KERN, H. L. (1967). Characterization of the transport of neutral amino acids by the calf lens. *Invest. Ophthal.* **7**, 441–451.
BRINDLEY, G. S. (1956). Resting potential of the lens. *Br. J. Ophthal.* **40**, 385–391.
BULLOUGH, W. S. & LAURENCE, E. B. (1964). Mitotic control by internal secretion: the role of the chalone-adrenalin complex. *Exp. Cell Res.* **33**, 176–194.
BURKY, E. L. (1934). Experimental endopthalmitis phacoanaphylactica. *Arch. Ophthal.* **12**, 536–546.
CLAYTON, R. M. (1969). Properties of the crystallins of the chick in terms of their subunit composition. *Exp. Eye Res.* **8**, 326–339.
CLAYTON, R. M., CAMPBELL, J. C. & TRUMAN, D. E. S. (1968). A re-examination of the organ specificity of lens antigens. *Exp. Eye Res.* **7**, 11–29.
COGAN, D. G. (1962). Anatomy of lens and pathology of cataracts. *Exp. Eye Res.* **1**, 291–295.
COHEN, A. I. (1965). The electron microscopy of the normal human lens. *Invest. Ophthal.* **4**, 433–446.
CREMER-BARTELS, G. (1962). A light-sensitive, fluorescent substance in bovine and rabbit lenses. *Exp. Eye Res.* **1**, 443–448.
DISCHE, Z. (1965). The glycoproteins and glycolipoproteins of the bovine lens and their relation to albuminoid. *Invest. Ophthal.* **4**, 759–778.
DISCHE, Z. & ZELMENIS, G. (1966). The sialohexosaminoglycan of the bovine lens capsule. *Doc. Ophthal.* **20**, 54–72.

DUNCAN, G. (1969). Relative permeabilities of the lens membranes to sodium and potassium. *Exp. Eye Res.* **8**, 315–325.

DUNCAN, G. (1969). The site of the ion restricting membranes in the toad lens. *Exp. Eye Res.* **8**, 406–412.

DUNCAN, G. (1970). Movement of sodium and chloride across amphibian lens membranes. *Exp. Eye Res.* **10**, 117–128.

FARKAS, T. G. & WEESE, W. C. (1970). The role of aqueous humor, insulin and trivalent chromium in glucose utilization of rat lenses. *Exp. Eye Res.* **9**, 132–136.

FRANÇOIS, J. & RABAEY, M. (1957). The protein composition of the human lens. *Am. J. Ophthal.* **44**, Pt. 2, 347–357.

FRANÇOIS, J. & RABAEY, M. (1958). Permeability of the capsule for the lens proteins. *Acta ophthal., Kbh.* **36**, 837–844.

FRANÇOIS, J., RABAEY, M. & RECOULÈS, N. (1961). A fluorescent substance of low molecular weight in the lens of primates. *Arch. Ophthal., N.Y.* **65**, 118–126.

FRANÇOIS, J., RABAEY, M. & STOCKMANS, L. (1965). Gel filtration of the soluble proteins from normal and cataractous human lenses. *Exp. Eye Res.* **4**, 312–318.

FRIEDENWALD, J. S. (1930). Permeability of the lens capsule. *Arch. Ophthal.* **3**, 182–193.

FUCHS, R. & KLEIFELD, O. (1956). Uber das Verhalten des wasserlöslichen Linseneiweiss junger und alter Tiere bei papierelektrophoretische Untersuchung. *v. Graefes' Arch. Ophthal.* **158**, 29–33.

FULHORST, H. W. & YOUNG, R. W. (1966). Conversion of soluble lens protein to albuminoid. *Invest. Ophthal.* **5**, 298–303.

GOLDMANN, H. & LEICHTI, A. (1938). Experimentelle Untersuchungen über die Genese des Röntgenstars. *v. Graefes' Arch. Ophthal.* **138**, 722–736.

HALBERT, S. P., LOCATCHER-KHORAZO, D., SWICK, L., WITMER, R., SEEGAL, B. & FITZGERALD, P. (1957). Homologous immunological studies of ocular lens. I and II. *J. exp. Med.* **105**, 439–452, 453–462.

HARRIS, J. E. & BECKER, B. (1965). Cation transport of the lens. *Invest. Ophthal.* **4**, 709–722.

HARRIS, J. E. & GEHRSITZ, L. B. (1951). Significance of changes in potassium and sodium content of the lens. *Am. J. Ophthal.* **34**, Pt. 2, 131–138.

HARRIS, J. E., GEHRSITZ, L. B. & NORDQUIST, L. (1953). The *in vitro* reversal of the lenticular cation shift induced by cold or calcium deficiency. *Am. J. Ophthal.* **36**, 39–49.

HOLT, W. S. & KINOSHITA, J. H. (1968). Starch-gel electrophoresis of the soluble lens proteins from normal and galactosemic animals. *Invest. Ophthal.* **7**, 169–178.

KERN, H. L. (1962). Accumulation of amino acids by calf lens. *Invest. Ophthal.* **1**, 368–376.

KINOSHITA, J. H. (1965). Pathways of glucose metabolism in the lens. *Invest. Opthal.* **4**, 619–628.

KINOSHITA, J. H., BARBER, G. W., MEROLA, L. O. & TUNG, B. (1969). Changes in the level of free amino acids and myo-inositol in the galactose exposed lens. *Invest. Ophthal.* **8**, 625–632.

KINOSHITA, J. H., DVORNIK, D., KRAML, M. & GABBAY, K. H. (1968). The effect of an aldose reductase inhibitor on the galactose-exposed rabbit lens. *Biochim. biophys. Acta.* **158**, 472–475.

KINOSHITA, J. H., MEROLA, L. O. & DIKMAK, E. (1962). Osmotic changes in experimental galactose cataracts. *Exp. Eye. Res.* **1**, 405–410.

KINSEY, V. E. & FROHMAN, C. E. (1951). Distribution of cytochrome, total riboflavin, lactate and pyruvate and its metabolic significance. *Arch. Ophthal., N.Y.* **46**, 536–541.

KINSEY, V. E. & McLEAN, I. W. (1970). Characterization of active transport and diffusion of potassium, rubidium, and cesium. *Invest. Ophthal.* **9**, 769–784.

KINSEY, V. E. & REDDY, D. V. N. (1965). Studies on the crystalline lens. XI. *Invest. Ophthal.* **4**, 104–116.

KINSEY, V. E., WACHTL, C., CONSTANT, M. A. & CAMACHO, E. (1955). Mitotic activity in the epithelia of lenses cultured in various media. *Am. J. Ophthal.* **40**, Pt. 2, 216–223.

KINOSHITA, J. H. & WACHTL, C. (1958). A study of the ^{14}C-glucose metabolism of the rabbit lens. *J. biol. Chem.* **233**, 5–7.

KUWABARA, T., KINOSHITA, J. H. & COGAN, D. G. (1969). Electron microscopic study of galactose-induced cataract. *Invest. Ophthal.* **8**, 133–149.

LANGHAM, M. & DAVSON, H. (1949). Studies on the lens. *Biochem. J.* **44**, 467–470.

LANGLEY, R. K., MORTIMER, C. B. & McCULLOCH, C. (1960). The experimental production of cataracts by exposure to heat and light. *Arch. Ophthal., N.Y.* **63**, 473–488.

LEINFELDER, P. J., & SCHWARTZ, B. (1954). Respiration and glycolysis in lens transparency. *Acta XVII Conc. Ophth.* pp. 984–991.

LEON, A. E., DE GROOT, K. & BLOEMENDAL, H. (1970). The molecular weight of the subunits of α-crystallin. *Exp. Eye Res.* **10**, 75–79.

LERMAN, S. & ZIGMAN, S. (1967). Metabolic studies on the cold precipitable protein of the lens. *Acta Ophthal.* **45**, 193–199.

MACH, H. (1963). Untersuchungen von Linseneiweiss und Mikroelektrophorese von Wasserlöslichem Eiweiss im Altersstar. *Klin. Mbl. Augenheilk,* **143**, 689–710.

MANSKI, W., AUERACH, T. P. & HALBERT, S. P. (1960). The evolutionary significance of lens organ specificity. *Am. J. Ophthal.* **50**, Pt. 2, 985–991.

MEHTA, P. D., COOPER, S. N. & RAO, S. S. (1964). Identification of species-specific and organ-specific antigens in lens proteins. *Exp. Eye Res.* **3**, 192–197.

McEWAN, W. K. (1959). The yellow pigment of human lenses. *Am. J. Ophthal.* **47**, Pt. 2, 144–146.

MOK, C. -C. & WALEY, S. G. (1967). Structural studies on lens proteins. *Biochem. J.* **104**, 128–134.

MÖRNER, C. T. (1894). Untersuchung der Proteinsubstanzen in der lichtbrechenden Medien des Auges. III. *Z. physiol. Chem.* **18**, 233–256.

PATERSON, C. A. (1970a). Extracellular space of the crystalline lens. *Am. J. Physiol.* **218**, 797–802.

PATERSON, C. A. (1970b). Sodium exchange in the crystalline lens. I. *Exp. Eye Res.* **10**, 151–155.

PATTERSON, J. W. & BUNTING, K. W. (1966). Sugar cataracts, polyol levels and lens swelling. *Doc. Ophthal.* **20**, 64–72.

PIRIE, A. (1951). Composition of the lens capsule. *Biochem. J.* **48**, 368–371.

PIRIE, A. (1968). Pathology of the eye of the naphthalene-fed rabbit. *Exp. Eye Res.* **7**, 354–357.

PIRIE, A. & FLANDERS, H. P. (1957). Effect of X-rays on partially shielded lens of the rabbit. *Arch. Ophthal., N.Y.* **57**, 849–854.

RABAEY, M. (1962). Electrophoretic and immuno-electrophoretic studies on the soluble proteins in the developing lens of birds. *Exp. Eye Res.* **1**, 310–316.

RAE, J. L., HOFFERT, J. R. & FROMM, P. O. (1970). Studies on the normal lens potential of the rainbow trout (*Salmo gairdneri*). *Exp. Eye Res.* **10**, 93–101.

RAO, S. S., MEHTA, P. D. & COOPER, S. N. (1965). Antigenic relationship between insoluble and soluble lens proteins. *Exp. Eye Res.* **4**, 36–41.

REDDY, D. V. N. & KINSEY, V. E. (1962). Studies on the crystalline lens. Quantitative analysis of free amino acids and related compounds. *Invest. Ophthal.* **1**, 635–641.

REES, J. R. & PIRIE, A. (1967). Possible reactions of 1,2-naphthaquinone in the eye. *Biochem. J.* **102**, 853–863.

RILEY, M. V. (1970). Ion transport in damaged lenses and by isolated lens epithelium. *Exp. Eye Res.* **9**, 28–37.

ROBERTS, D. St. C. (1957). Studies on the antigenic structure of the eye using the fluorescent antibody technique. *Br. J. Ophthal.* **41**, 338–347.

RUBIN, L. F. & BARNETT, K. C. (1967). Ocular effects of oral and dermal application of dimethyl sulfoxide in animals. *Ann. N.Y. Acad. Sci.* **141**, 333–345.

SCHOENMAKERS, J. G. G. & BLOEMENDAL, H. (1968). Subunits of alpha-crystallin from adult and embryonic cattle lens. *Nature,* **220**, 790–791.

SCHOENMAKERS, J. G. G., GERDING, J. J. T. & BLOEMENDAL, H. (1969). The subunit structure of α-crystallin. *Eur. J. Biochem.* **11**, 472–481.

SCHWARTZ, B. (1960). Initial studies of the use of an open system for the culture of the rabbit lens. *Arch. Ophthal., N.Y.* **63**, 643–659.

SPECTOR, A. (1969). Physiological chemistry of the eye. *Arch. Ophthal., N.Y.* **81**, 127–143.

SPERELAKIS, N. & POTTS, A. M. (1959). Additional observations on the bioelectric potentials of the lens. *Am. J. Ophthal.* **47**, 395–409.

SZENT-GYÖRGYI, A. (1955). Fluorescent globulin of the lens. *Biochim. biophys. acta.* **16**, 167.

TANAKA, K. & IINO, A. (1967). Zur Frage der Verbindung der Linsenfasern im Rinderauge. *Z. Zellforsch.* **82**, 604–612.

THOFT, R. A. & KINOSHITA, J. H. (1965). The effect of calcium on rat lens permeability. *Invest. Ophthal.* **4**, 122–128.

UHLENHUTH, P. T. (1903). Zur Lehre von der Unter scheidungverschiedener Eiweissarten mit Hilfe spezifischer Sera. In 'Koch Festschrift, p. 49. Jena: G. Fischer.

VAN HEYNINGEN, R. (1959). Formation of polyols by the lens of the rat with "sugar" cataract. *Nature, Lond.* **184**, 194–195.

VAN HEYNINGEN, R. (1965). The metabolism of glucose by the rabbit lens in the presence and absence of oxygen. *Biochem. J.* **96**, 419–431.

VAN HEYNINGEN, R. (1970). Ascorbic acid in the lens of the naphthalene-fed rabbit. *Exp. Eye Res.* **9**, 38–48.

VAN HEYNINGEN, R. & PIRIE, A. (1967). The metabolism of naphthalene and its toxic effect on the eye. *Biochem. J.* **102**, 842–852.

VAN HEYNINGEN, R. & WALEY, S. G. (1962). Search for a neutral proteinase in bovine lens. *Exp. Eye Res.* **1**, 336–342.

V. SALLMANN, L. (1951). Experimental studies on early lens changes after roentgen irradiation. *Arch. Ophthal., N.Y.* **45**, 149–164.

V. SALLMANN, L. (1957). The lens epithelium in the genesis of cataract. *Am. J. Ophthal.* **44**, 159–170.

VOADEN, M. J. (1968). A chalone in the rabbit lens? *Exp. Eye Res.* **7**, 326–331.

VOADEN, M. J. & FROOMBERG, D. (1967). Topography of a mitotic inhibitor in the rabbit lens. *Exp. Eye Res.* **6**, 213–218.

WALEY, S. G. (1964). Metabolism of amino acids in the lens. *Biochem. J.* **91**, 576–583.

WALEY, S. G. (1965). The problem of albuminoid. *Exp. Eye Res.* **4**, 293–297.

WANKO, T. & GAVIN, M. A. (1961). Cell surfaces in the crystalline lens. In *The Structure of the Eye*, Ed. Smelser, pp. 221–233. New York: Academic Press.

WEINSTOCK, M. & STEWART, H. C. (1961). Occurrence in rodents of reversible drug-induced opacities of the lens. *Br. J. Ophthal.* **45**, 408–414.

YOUNG, R. W. & OCUMPAUGH, D. E. (1966). Autoradiographic studies on the growth and development of the lens capsule in the rat. *Invest. Ophthal.* **5**, 583–593.

ZIGMAN, S. & LERMAN, S. (1965). Properties of a cold-precipitable protein fraction in the lens. *Exp. Eye Res.* **4**, 24–30.

ZIGMAN, S. & LERMAN, S. (1968). Effect of actinomycin D on rat lens protein synthesis. *Exp. Eye Res.* **7**, 556–560.

ZIGMAN, S., SCHULTZ, J. & YULO, T. (1969). Chemistry of lens nuclear sclerosis. *Biochem. Biophys. Res. Comm.* **35**, 931–938.

ZWAAN, J. (1966). Sulfhydryl groups of the lens proteins of the chicken in embryonic and adult stages. *Exp. Eye Res.* **5**, 267–275.

SECTION II

The Mechanism of Vision

PART I

RETINAL STRUCTURE AND ORGANIZATION

IN this section we shall be concerned with the peripheral mechanisms in the visual process. When light strikes the retina, physical and chemical changes are induced that lead eventually to a discharge of electrical impulses in the optic nerve fibres. The "messages" in these nerve fibres are a record of the events taking place in the retina; because of the variety of sensations evoked by visual stimuli, these records must be of almost fantastic complexity when even a simple image is formed on the retina. The aim of investigators in this field of eye physiology is to disentangle the skein; their approaches are various but may be classed generally as psychophysical, chemical, and electrophysiological. Remarkable progress has been made, especially in recent years, but there is still a long way to go before anything approaching an exact picture of the retinal processes can be elaborated. The following description is designed more to show the modes of approach to the various problems rather than to present a theoretical picture of the "mechanism of vision".

THE STRUCTURE OF THE RETINA

Sensory Pathways. The retina is a complex nervous structure made up of a number of layers (Fig. 59) whose significance is best understood by recalling that in other parts of the body, for example the skin, the nervous elements involved in the transmission of sensory impulses to the cerebral cortex consist of (a) a specialized arrangement of cells called a receptor-organ, e.g. a Pacinian corpuscle sensitive to pressure; (b) a neurone connecting the receptor to the spinal cord and medulla (neurone I) with its cell body in the posterior root ganglion; (c) a connector neurone to the thalamus (neurone II) with its cell body in the nucleus cuneatus or gracilis, and finally (d) a neurone to transmit the impulses to the cortex (neurone III) with its cell body in the thalamus.

With the vertebrate eye, not only are the specialized receptor cells, the *rods* and *cones*, in the retina, but also the cell bodies and connecting fibres of neurones I and II so that the nerve fibres leading out of the eye, which constitute the optic nerve, are the axons of neurone II; the majority of the fibres run to the lateral geniculate body where the impulses, initiated in the rods and cones, are relayed to the occipital lobe of the cortex. The analogy between the two systems is demonstrated in Fig. 60.

The Retinal Layers. We may thus expect to differentiate in the retina a layer of receptors (*layer of rods and cones*), a layer of cell bodies of neurone I, the *bipolar cells* (*inner nuclear layer*) and a layer of cell bodies of neurone II, the *ganglion cells* (*ganglion layer*) (Fig. 59). The rods and cones are highly differentiated cells, and their orderly arrangement gives rise to a *bacillary layer*, consisting of their outer segments; an *outer nuclear layer* containing their cell bodies and nuclei; and an *outer plexiform layer* made up of their fibres and synapses with the bipolar cells. The region of synapse between bipolar and ganglion cells is the *inner plexiform layer*. Two other types of nerve cell are present in the retina,

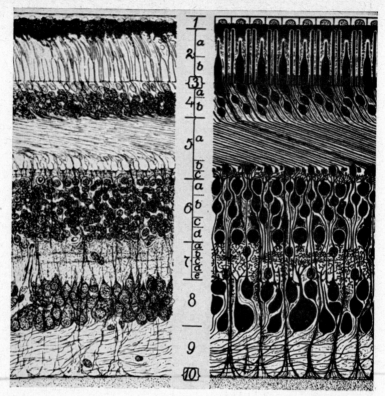

FIG. 59. Cross-section through the adult human retina in the periphery of the
central area, showing details of stratification at a moderate magnification. The left-
hand figure shows the structures as they appear when stained with a non-
selective method (e.g. haematoxylin and eosin). The right-hand figure reproduces
schematically the same view as it would appear from a study with an analytical
method such as Golgi's or Ehrlich's, using many sections to get all details.
(1) Pigment epithelium; (2) bacillary layer; (2-*a*) outer segments and (2-*b*) inner
segments of the thinner rods and the thicker cones; (3) outer limiting membrane;
(4) outer nuclear layer; (4-*a*) outer zone of the outer nuclear layer chiefly composed of the cone nuclei; (4-*b*) inner zone of the outer nuclear layer composed
exclusively of the rod nuclei; (5) outer plexiform layer; (5-*a*) outer zone of the
outer plexiform layer composed of the inner rod and cone fibres and of the
corresponding portions of Müller's "radial fibres" enveloping the first (here
only one shown, beginning with the outer limiting membrane encasing the body
and the fibre of a cone and continuing through ventral layers down to the inner
limiting membrane); (5-*b*) middle zone of the outer plexiform layer composed of
the rod spherules and cone pedicles; (5-*c*) inner zone of the outer plexiform
layer made up of the expansions of the nerve cells and of Müller's supporting
"radial fibres" whose bodies and nuclei reside in the sixth layer; (6) inner
nuclear layer with its four zones: 6-*a*, 6-*b*, 6-*c*, and 6-*d*; (7) inner plexiform
layer with its five zones: 7-*a*, 7-*b*, 7-*c*, 7-*d*, and 7-*e*; (8) layer of the ganglion
cells; (9) layer of the optic nerve fibres; (10) inner limiting membrane, next to
the vitreous. (Polyak, *The Retina*.)

namely the *horizontal* and *amacrine* cells, with their bodies in the inner nuclear
layer; the ramifications of their dendritic and axonal processes contribute to the
outer and inner plexiform layers respectively; their largely horizontal organizations permits them to mediate connections between receptors, bipolars and

ganglion cells. Besides the nerve cells there are numerous *neuroglial cells*, e.g. those giving rise to the radial fibres of Müller, which act as supporting and insulating structures.

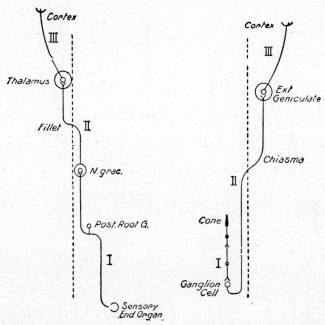

FIG. 60. Illustrating the analogy between a typical cutaneous afferent pathway and the visual pathway. Note the cone has not been treated as a neurone.

The Rods and Cones. The photosensitive cells are, in the primate and most vertebrate retinae, of two kinds, called *rods* and *cones*, the rods being usually much thinner than the cones but both being built on the same general plan as illustrated in Fig. 61. The light-sensitive pigment is contained in the *outer segment, o*, which rests on the pigment epithelium; the other end is called the *synaptic body, s*, and it is through this that the effects of light on the receptor are transmitted to the bipolar or horizontal cell. In the rods this is called the *rod spherule*, and in the cone the *cone pedicle*. The human rod is some 2 μ thick and 60 μ long; the cones vary greatly in size and shape with their position in the retina, becoming long and thin and not easily distinguished from rods in the most central portion (Fig. 62).

Synaptic Relationships. If it is the rod or cone that absorbs the light-energy of the stimulus, it is the bipolar cell that must transmit any effect of this absorption to higher regions of the central nervous system; consequently the relationship between the two types of cell is of great interest. They have been examined in the electron microscope in some detail.* In general, the relation

* Since Golgi's and Polyak's classical studies, developments on the primate retina have been due largely to Missotten (1965), Sjöstrand (1969), Dowling & Boycott (1966, 1969), Kolb (1970); of special interest is the study of Dowling & Werblin (1969) on the retina of the mudpuppy *Necturus*, since this has been accompanied by a parallel study of the electrophysiology of its connections (p. 169).

FIG. 61. Diagrammatic representation of retinal photo-receptor cells. A is a schematic drawing of a vertebrate photoreceptor, showing the compartmentation of organelles as revealed by the electron microscope. The following subdivisions of the cell may be distinguished: outer segment (*o*); connecting structure (*c*); ellipsoid (*e*) and myoid (*m*), which comprise the inner segment; fibre (*f*); nucleus (*n*); and synaptic body (*s*). The scleral (apical) end of the cell is at the top in this diagram. B depicts a photoreceptor cell (rod) from the rat. (Young. *J. Cell Biol.*)

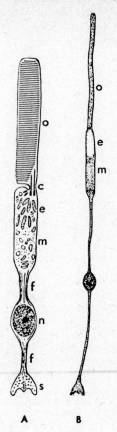

FIG. 62. Human cones. 1, Near ora serrata; 2, at equator; 3, at macula. *a*, Outer segment; *b*, inner segment; *c*, cone fibre; *d*, cell body and nucleus; *e*, cone foot; *f*, ellipsoid; *g*, myoid. (Duke-Elder, *Textbook of Ophthalmology*.)

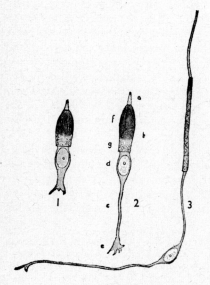

between receptor and neurone is of the synaptic type, in the sense that the dendrite of the bipolar cell, for example, comes into very close contact with the rod spherule or cone pedicle, and fusion of membranes does not occur, a definite synaptic cleft being maintained. The contacts are of two main types, namely

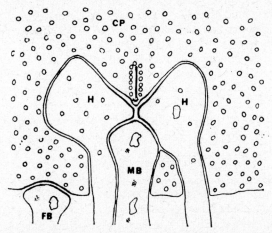

FIG. 63. Synaptic junction of the cone pedicle, CP, called a triad. Two processes of horizontal cells, H, and a midget bipolar dendrite invaginate into the cytoplasm of the cone pedicle. (Dowling & Boycott, 1966.)

when the dendritic process penetrates deep into an invagination of the pedicle or spherule, and when the process makes a simple contact with the surface. Fig. 63 illustrates a characteristic form of the invaginating synapse called by Missotten (1962) the *triad*; the figure shows a portion of the cone pedicle, CP, containing many vesicles reminiscent of synaptic vesicles as seen elsewhere in the central

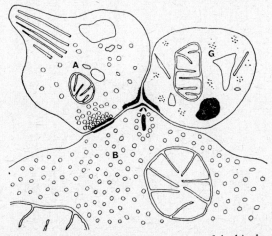

FIG. 64. Illustrating the dyad type of synaptic contact of the bipolar terminal, *B*, with ganglion cell dendrite, *G*, and amacrine process, *A*. Frequently the amacrine cell process makes a synaptic contact backwards on to the bipolar terminal, forming a reciprocal contact between bipolar terminal and amacrine process. (Dowling & Boycott, *Proc. Roy. Soc.*)

nervous system. The triad is formed by the penetration into the invagination of a central dendritic process from a bipolar cell, flanked by two processes from horizontal cells. In the same figure, a simpler *contact synapse* can be seen formed by a dendritic process from another bipolar cell on the cone pedicle. A *dyad* form of synapse is found in the inner plexiform layer (Fig. 64); here a bipolar axon and an amacrine process make adjacent synaptic contacts with a ganglion cell dendrite, and it seems that the amacrine cell is being both postsynaptic and pre-synaptic in its relations with the bipolar cell. Thus the bipolar cell involved in the dyad, like the cone pedicle involved in the triad, contains a linear portion of heavily staining material surrounded by vesicles, the so-called *ribbon synapse*; the conventional type of synapse is illustrated by the relation of the amacrine cell to the bipolar cell in Fig. 64; here there are dense lines on the apposing membranes, and opposite them in the presynaptic cell—amacrine—is a collection of synaptic vesicles.

Organization of the Retina

The main organization of the retina is on a vertical basis, in the sense that activity in the rods and cones is transmitted to the bipolar cells and thence to the ganglion cells and out of the eye into higher neural centres—the lateral geniculate body, and so on. However, a given receptor may well activate many bipolar cells, and a single bipolar cell may activate many ganglion cells, so that the effect of a light-stimulus may spread horizontally as it moves vertically. The connections brought about by the horizontal and amacrine cells further increase the possi-bilities of horizontal spread and interaction, so that the problem of the analysis of the pathways of activity in the retina is highly complex, and only a beginning has so far been made in its solution. The basic picture of a vertical retinal organiza-

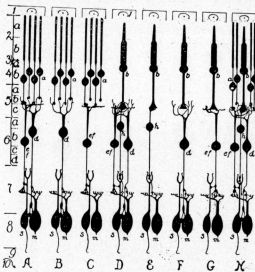

FIG. 65. Some of the types of receptor-bipolar synapses in the retina. *a*, rod; *b*, cone; *d, e, f*, bipolars that make several synaptic relationships with receptors; *h*, midget bipolar; *s*, midget ganglion cell; *m*, parasol ganglion cell (used here to typify the type that makes several synaptic relationships with bipolars). (Polyak, *The Retina.*)

tion, as conceived by Polyak, is illustrated by Fig. 65; according to this, the bipolar cells are of two main classes, namely *midget bipolars, h*, which connect only to a cone, and the diffuse type, which makes synaptic relations with several receptors, *d, e* and *f*. Similarly, the ganglion cells fall into two classes, the *midget ganglion cell*, connecting only with a single midget bipolar cell, and the diffuse type connecting with groups of bipolar cells. On this basis, then, convergence of receptors on ganglion cells would be largely achieved by the diffuse types of bipolar and ganglion cells, whilst a virtual one-to-one relationship would exist between those cones that made a connection with a midget bipolar cell. As the figure indicates, however, this relationship is not strictly one-for-one, since a cone, connected to a midget bipolar, also connects with a diffuse type of bipolar, and so its effects may spread laterally. The more recent studies of Boycott &

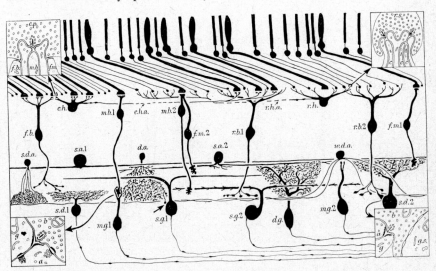

FIG. 66. The main kinds of nerve cell identified in the primate retina. *Inset, top left* shows how the cones (c.p.) contact their nerve cells. The *inset top right* shows how nerve cell processes invaginate into a rod spherule (r.s.). The *inset, lower left* represents a dyad in which a bipolar cell *b* is presynaptic to a ganglion cell dendrite *g* and an amacrine cell process *a*. The amacrine cell process has a reciprocal contact back on to the bipolar terminal and an amacrine to ganglion cell dendrite synapse. The inset *lower right* shows that the rod bipolar cell axosomatic contacts may consist of 'close contacts' on to the ganglion cell perikaryon (g.s.) (though which types of ganglion cell is unknown). This inset also shows that these same bipolar terminals have dyad contacts. *c.h.* and *r.g.* are horizontal cells considered to connect cones and rods to each other respectively; more recent work shows that horizontal cells only connect cones to rods, however; *f.b.*, flat bipolar making contact with several cones; *m.b.*1 and *m.b.*2 are midget bipolars; *f.m.*1 and *f.m.*2 are flat midget bipolars, making contact with only the base of the cone pedicle by contrast with the *m.b.'s* which invaginate into the triad. *r.b.*1 and *r.b.*2 are rod bipolars whose dendrites make the central elements of the rod spherule (*Upper right inset*). *s.a.*1 and *s.a.*2 are unistratified amacrine cells whose processes ramify immediately under the inner nuclear layer, *d.a.* and *s.d.a.* are amacrine cells with branches extending much deeper; *d.a.* may be seen making an axo-somatic contact with the ganglion cell *s.g.*1. *w.d.a.* indicates a wide-field diffuse amacrine cell, with little or no ramification of the main branches as they run to the ganglion cell perikarya. Midget ganglion cells, *m.g.*1 and *m.g.*2 have a single dendrite branching at its tip to embrace the terminal of a midget bipolar cell. *s.g.*1 and *s.g.*2 are unistratified ganglion cells, the dendrites branching in a single plane, whilst the diffuse ganglion cell, *d.g.* has dendrites branching in planes throughout the inner plexiform layer. *s.d.*1 and *s.d.*2 are stratified diffuse ganglion cells corresponding to Polyak's "parasol ganglion cell". (Boycott & Dowling, *Phil. Trans.*)

Dowling (1969) have largely confirmed Polyak's view but have corrected some errors; their general picture is indicated in Fig. 66.

Bipolar Cells. According to this study, the bipolar cells related to cones are of two types; the *midget*, making a unique type of synapse with the cone, and the *flat* cone bipolar, making connections with some seven cones; the synapses are characteristically different, the dendritic process from the midget bipolar cell penetrating into an invagination of the cone pedicle, whereas the process from the flat cone bipolar makes a more simple type of contact (Fig. 66, *top left inset*). The third type of bipolar is also diffuse, and serves to collect messages from rods, being connected to as many as fifty of these; the synaptic arrangement is one of invagination, as seen in Fig. 66, *top right inset*. Bipolars that synapse with both rods and cones, as indicated in Polyak's figure, almost certainly do not occur.

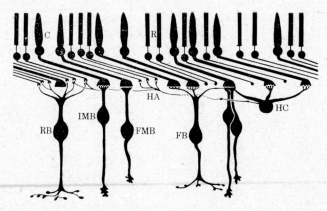

Fig. 67. Illustrating organization in the outer plexiform layer of the primate retina. The cells are drawn diagrammatically and do not, except for the midget bipolar cells, show the full extent of their dendritic spead. In general the drawing does not show the overlap of cell types onto their respective receptors, the exception being the third cone from the right, where all the cell types connecting with a single cone pedicle are shown. The rod bipolar (RB) makes contact with rod spherules only. Two or three rod bipolars contact an individual rod, but a single rod bipolar only makes one dendritic contact per rod. Horizontal cell axon terminals (HA) end in rod spherules only. Probably only two axons contribute a terminal to an individual rod. The axon is presented as a dotted line because the direct evidence for axon terminals being connected to the horizontal cell is not yet available.

The invaginating midget bipolar (IMB) is connected to a single cone. The flat midget bipolar (FMB) is also exclusive to a single cone pedicle. The flat bipolar (FB) is a diffuse cone bipolar and contacts six to seven cones. Probably six or seven flat bipolars overlap onto a single cone pedicle. The horizontal cell (HC) has dendrites ending in cone pedicles only. (Kolb, *Phil. Trans.*)

MIDGET BIPOLAR

The midget bipolar has been examined in serial sections by Kolb (1970), and she has shown that a single bipolar may make some twenty-five connections with a single cone pedicle, providing single central dendritic processes for the twenty-five invaginations. Thus, since these twenty-five invaginations constituted the total for the cone pedicle, this proved that the bipolar cell made synapses with only a single cone. In the peripheral retina there were some branched midgets, probably making contact with two cones; whilst in some cases it seemed likely that the pedicle accommodated processes from two midget bipolars. In addition to this connection with a midget, each cone also connected with one or more flat bipolars (Boycott & Dowling, 1969). Kolb also showed that the midget bipolar occurred in two types, the *invaginating midget* as described above, and also a *flat midget*, which at first was mistaken for the invaginating type; in fact, however, its dendritic "tree-top" consists in a large number of very fine processes that do not enter the invaginations of the pedicle but only make a superficial contact with the base. Thus a cone pedicle contained twenty-five invaginating contacts and some forty-eight superficial contacts. Like the invaginating midget, the flat midget makes contact with only a single cone pedicle.

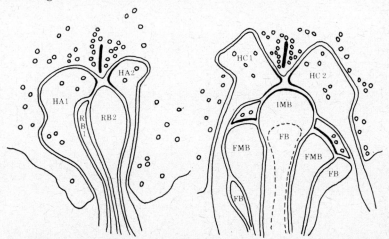

Fig. 68. Illustrating synaptic arrangements in rod spherule (left) and cone pedicle (right) as deduced from electron-microscopy.
    Left: The lobed lateral elements are horizontal cell axon terminals, each from a different horizontal cell (HA1 and HA2). The central elements are rod bipolar dendrites from different rod bipolars (RB1 and RB2).
    Right: The horizontal cell dendrites, usually from different horizontal cells (HC1 and HC2) form the lateral elements of the invagination. The invaginating midget bipolar (IMB) pushes up into the invagination to lie between the two lateral elements. The synaptic ribbon and synaptic vesicles point into the junction of the three processes. The flat midget bipolar (FMB) dendrites lie alongside the invaginating midget bipolar dendrite and push part of the way up into the triad, but are in contact with the cone pedicle base. The flat bipolar (FB) terminals are also clustered around the invaginating bipolar dendrite and make superficial contact on the cone pedicle base. (Kolb, *Phil. Trans.*)

Horizontal Cells. The dendritic processes of these cells make synaptic contacts with the cone pedicles as lateral processes of the triads (Fig. 63); they were considered by Boycott & Dowling (1969) to be of two types, Type A making dendritic connection with cones and Type B with rods (Fig. 66); however Kolb (1970) has shown that both types actually make dendritic connection only with cones; and the only differentiation consists in the numbers of cones with which a single horizontal cell makes contact, the "small-field" contacting some seven cones and the "large-field" some twelve. Of greatest interest, of course, is the manner in which the axons of the horizontal cells terminate; thus their arrangement with the cones is post-synaptic, i.e. they receive messages from these; according to Kolb the axon processes terminate exclusively in rod spherules, constituting lateral elements in triads; each rod spherule thereby receives input from two horizontal cells. Thus the horizontal cells in the primate retina apparently mix the rod and cone messages.

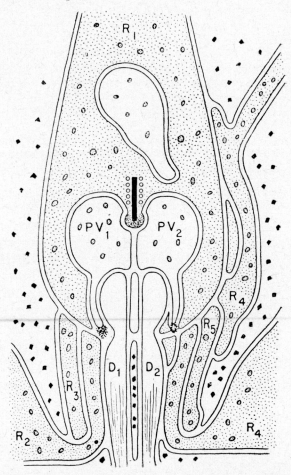

FIG. 69. Schematic drawing showing the relationship between the ends of the lateral processes from β-type receptor cells (R_2, R_3, R_4, R_5) and the synaptic body of one α-type cell (R_1), as well as between these processes and the end branches of bipolar cells (D_1, D_2) in the guinea pig retina. (Sjöstrand. *The Retina*. University of California Press.)

Outer Plexiform Layer. In the light of this work, the outer plexiform layer may be illustrated by Fig. 67, whilst the triad structures of rod and cone are shown in Fig. 68; it will be noted that the structure of the cone triad is much more complex than at first thought, containing, in addition to the invaginating midget bipolar process (IMB) and the lateral horizontal processes (HC 1 and HC 2), more superficially placed processes from flat midget bipolars (FMB) and from flat cone bipolars FB. The possibilities for spread of information are thus quite large.

Inner Plexiform Layer. Essentially this process of vertical and horizontal connections is repeated in the inner plexiform layer, which consists of axons and terminals of bipolar cells, processes of amacrine cells, and dendrites of ganglion cells. The ganglion cells fall into two main types, namely the midget ganglion cell which connects with a midget bipolar cell and thus has a very close relationship with a single cone, providing the latter with an exclusive pathway as far as the inner plexiform layer; the other type is diffuse, being connected with many bipolar cells; this type falls into several categories according to the manner in which their dendrites branch. The extent to which a ganglion cell may relay only from rod bipolars or cone bipolars is not clear from the anatomical evidence; physiological evidence indicates that a ganglion cell may respond to both rod and cone stimulation, whilst another may respond exclusively to cone stimulation; the latter is probably of the midget type.

Amacrine Cells. The amacrine cells are of several types with respect to the nature and extent of their ramifications; they may, apparently, receive influences from bipolar cells and from other amacrines, whilst they may transmit influences to other bipolar terminals and to ganglion cells. Certain ganglion cells have extremely wide receptive fields (p. 158) in the sense that they may respond to light falling on retinal points that are far remote from each other; this means that the ganglion cell is collecting information from a wide area of rods and cones, and the manner in which their effects are funnelled into the ganglion cell is probably through the amacrine cells. This *convergence* of the retinal message is also assisted by the diffuse type of bipolar and ganglion cells, and it is not surprising that the total number of optic nerve fibres emerging from the human retina, namely about 1 million, is far less than the total number of receptors (7 million cones and 70–150 million rods). It is interesting, nevertheless, that in the central retina, there are three times as many bipolar cells as cones.

Inter-Receptor Contacts. In the rabbit and guinea pig Sjöstrand described two types of rod, differentiated by the nature of their synaptic bodies; in the α-type cells this is ovoid whilst in the β-type it is conical with the base of the cone facing the neuropil of the outer plexiform layer; the β-type rods make lateral contacts with the α-type cells on the lateral aspects of their triad synaptic bodies as illustrated by Fig. 69; in addition, the processes from these β-type cells make synaptic contacts with the vitread pole of several surrounding α-type cells. Sometimes the processes are long so that contacts are made with cells some distance away. In addition, the β-type cells are interconnected with each other. The lateral processes of the β-type cells always contain vesicles. Sjöstrand (1969) has suggested that these contacts mediate some form of lateral inhibition analogous with that described by Hartline (p. 154).

Centrifugal Pathways. Polyak described centrifugal bipolar cells that transmitted messages backwards to the receptors, but the more recent work of Dowling & Boycott on the primate retina, and of Brindley & Hamasaki (1966) on the cat retina, has failed to confirm their existence in these higher vertebrates; the same may well apply for centrifugal pathways from the higher centres of the brain back to the retina, at any rate so far as primates are concerned. In the pigeon, Dowling & Cowan (1966) confirmed Cajal's description of centrifugal fibres terminating in the outer edge of the inner plexiform layer; destruction of the isthmo-optic nucleus caused characteristic terminations on amacrine cells to lose their synaptic vesicles and finally to disappear, so that it would seem that the avian brain exerts control over transmission in the retina through effects on amacrine cells, which seem to be presynaptic to the terminals of bipolar cells.*

The Retinal Zones

The retina is divided morphologically into two main zones, the *central area* (containing Regions I–III) and the *extra-areal periphery* (containing Regions IV–VII).

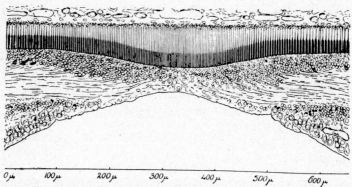

FIG. 70. Illustrating the outer fovea, filled with the long thin cones. Note the thinning of the outer nuclear layer and the practical disappearance of the remaining inner layers in the foveal centre. (Polyak, *The Retina*.)

Fovea. Region I, the *central fovea* (Fig. 70), is a small depression, shaped like a shallow bowl with a concave floor, on the vitreal face of the retina, caused by the bending away of layers 5–9 and a thickening of the bacillary and outer nuclear layers. From edge to edge it is some 1500 μ across in man and thus subtends about 5° at the nodal point of the eye; the floor of the fovea is about 400 μ across. The fovea, described in this way, is generally referred to as the *inner fovea*, the word "inner" being used to denote the aspect from which the depression is regarded; the *outer fovea*, some 400 μ across, represents a depression in the opposite sense, i.e. a cupping of the outer limiting membrane from outwards inwards (Fig. 70), and corresponds roughly in extent to the floor of the inner fovea; it is caused by the lengthening of the cones in the bacillary layer, the cone proper increasing in length more than the remaining portion. The outer fovea is thus the central portion of the inner fovea; here the receptor elements consist entirely of cones which, as we have seen, are longer and thinner than elsewhere in the retina, being some 70 μ long and 1·5 μ thick at the base and 1 μ thick at the tip. The "*rod-free territory*" is actually rather greater in extent than the outer fovea, being some 500–600 μ across and containing some 34,000 cones. Beyond this limit rods appear and their proportion to cones increases progressively. The total number of cones in the territory of the inner fovea amounts to 115,000. As we shall see, the most central, rod-free, portion of the retina is concerned with the highest degree of visual acuity since it is here that the obstruction to light caused by the nerve fibres and other layers is reduced to a minimum and the density of receptors is highest; vision mediated by this portion of the retina is thus spoken of as "foveal vision"; it must be emphasized, however, that the fovea proper embraces both this zone and the walls of the inner fovea where obstructions to the passage of light are not removed, where the cone density is considerably less, and where rods are present, so that it is inaccurate to speak of foveal vision when what is meant is vision mediated by the rod-free

---

* Cowan & Powell (1963) have shown that the avian optic nerve fibres relay in the optic tectum; they run in the tractus isthmo-tectalis to the isthmo-optic nucleus; here they relay and centrifugal fibres run through the lateral geniculate nucleus to the bipolar cells of the retina.

area or perhaps even a more restricted central zone (the "central bouquet of cones", for example, some 50–75 μ across).

Macula. Mention may be made here of the *yellow spot* or *macula lutea*; this is an ill-defined area of the central retina characterized by the presence of a yellow pigment in the nervous layers; in actuality the region extends over the whole of the central area (Regions I–III) but the intensely pigmented part is limited to the fovea (Region I); the central portion of the fovea (outer fovea) is only slightly pigmented in man, since the pigment is associated with layers beyond number 4 which, as we have seen, are very much reduced or absent here.

Parafovea and Perifovea. Regions II and III are the *parafoveal* and *perifoveal* regions respectively, the outer edge of the former being 1250 μ from the centre of the fovea and that of the latter 2750 μ from the centre, the diameter of the entire central area being thus about 5000–6000 μ. The cone density is considerably diminished in these regions whilst that of the rods is increased, so that in Region III there are only some twelve cones per 100 μ (compared with some fifty per 100 μ in the outer fovea) with two rods between each pair of cones. The *extra-areal periphery* (Regions IV to VII) is associated with a further decrease in the density of cones and an increase in the proportion of rods. In the whole human retina there are said to be about 7 million cones and 75–150 million rods.

GENERAL REFERENCES

BOYCOTT, B. B. & DOWLING, J. E. (1969). Organization of the primate retina: light microscopy. *Phil. Trans.* **255,** 109–184.

BRINDLEY, G. S. (1970). *Physiology of the Retina and Visual Pathway.* 2nd. Edn. London: Arnold.

DOWLING, J. E. & BOYCOTT, B. B. (1966). Organization of the primate retina: electron microscopy. *Proc. Roy. Soc. B,* **166,** 80–111.

MISSOTTEN, L. (1965). *The Ultrastructure of the Retina.* Brussels: Arscia. Uitgaven N.V.

POLYAK, S. L. (1941). *The Retina.* Chicago: University Press.

POLYAK, S. L. (1957). *The Vetrebrate Visual System.* Chicago: University Press.

ØSTERBERG, G. (1935). Topography of the layer of rods and cones in the human retina. *Acta ophthal. Kbh.,* Suppl. 6.

PIRENNE, M. H. (1967). *Vision and the Eye.* London: Chapman & Hall.

SJÖSTRAND, F. S. (1969). The outer plexiform layer and the neural organization of the retina. In *The Retina,* Ed. B. R. Straatsma *et al.,* pp. 63–100. Los Angeles: University of California Press.

SPECIAL REFERENCES

BRINDLEY, G. S. & HAMASAKI, D. I. (1966). Histological evidence against the view that the cat's optic nerve contains centrifugal fibres. *J. Physiol.* **184,** 444–449.

COWAN, W. M. & POWELL, T. P. S. (1963). Centrifugal fibres in the avian visual system. *Proc. Roy. Soc. B.,* **158,** 232–252.

DOWLING, J. E. & COWAN, W. M. (1966). An electron microscope study of normal and degenerating centrifugal fiber terminals in the pigeon retina. *Z. Zellforsch.* **71,** 14–28.

DOWLING, J. E. & WERBLIN, F. S. (1969). Organization of retina of the mudpuppy *Necturus maculosus.* I. Synaptic structure. *J. Neurophysiol.* **32,** 315–335.

KOLB, H. (1970). Organization of the outer plexiform layer of the primate retina: electron microscopy of Golgi-impregnated cells. *Phil. Trans.* **258,** 261–283.

PART 2

MEASUREMENT OF THE STIMULUS

PHOTOMETRY

Visible Light. Visible light consists of electromagnetic vibrations limited to a certain band of wavelengths—the *visible spectrum*. The limits are given classically as 4000 Å and 7600 Å, wavelengths shorter than 4000 Å being described as ultraviolet, and longer than 7600 Å as infrared. As we shall see, however, the retina is sensitive to wavelengths shorter and longer than these limits. Light is a form of energy and, as such, is amenable to quantitative measurement in absolute units, namely ergs. For physiological purposes, however, that subjective quality of light we call "brightness" or "luminosity" is the one that is of greatest interest, and since this is not uniquely determined by the energy content of the light, but varies widely with the wavelength (p. 256), it has been considered desirable to base measurements of light on a somewhat arbitrary "luminosity basis." Thus, on a luminosity basis, two sources of light are said to be of equal intensity when they produce equal sensations of brightness when viewed by the same observer under identical conditions.

Luminous Intensity. The luminous intensity (I) of a source of light is measured in terms of an arbitrarily chosen standard source of light-energy. Classically, this was a sperm-wax candle made to burn at a certain rate, and it gave rise to the term *candle-power*, a source of light appearing x times as bright as the standard candle having x candlepower. The modern *candela* (abbreviated cd) is fixed in terms of a standard filament lamp. It is such that one square centimetre of the surface of a full radiator at the temperature of solidification of platinum has a luminous intensity of 60 cd in the direction normal to the surface. In effect it has about the same value as the classical candle.

If two sources of light illuminate a screen under identical conditions, to produce identical degrees of illumination, they are said to be of equal intensity. An observer may easily determine whether two identical screens, side by side, are equally illuminated; if they are not equally illuminated (i.e. if one appears brighter than the other) he is unable to state exactly how much more intensely illuminated one is than the other, so that in comparing intensities of illumination recourse is had to the photometer, one type of which utilizes as its basis the law that the intensity of illumination of a screen, placed normal to the rays from a point source of light, varies inversely as the square of its distance from the source. Thus, if we have two sources of light, of intensities I_1 and I_2 and at distances r_1 and r_2 from a screen such that they illuminate it equally, we can say:

$$I_1/I_2 = r_2{}^2/r_1{}^2.$$

Hence, if one of the light sources (I_2) is the standard candle, the intensity of the other is given by:

$$I_1 = r_2{}^2/r_1{}^2 \text{ candelas.}$$

Intensity of Illumination. In introducing the measurement of the intensity of a light source, we have mentioned the *degree of illumination*; this is, of course, determined entirely by the amount of light falling on a surface and hence by the intensity of the light source, the distance of this source from the object, and the angle the surface makes with the direction of the incident light. Thus, if we have a screen perpendicular to the direction of the light incident from a point source of intensity I, we may define the *intensity of illumination*, E, by the equation:

$$E = I/r^2, \text{ where } r \text{ is the distance from the source.}$$

When I equals unity, i.e. one candela, and r equals 1 foot, E is unity and is called 1 *foot-candela*. If r is expressed in metres, the unit of illumination is the *metre-candela* or *lux*; expressed in words, it is the illumination of a surface, normal to the incident rays from a source of 1 candela one metre away. The *centimetre-candela* is called a *phot*. If a surface, normal to the incident rays from a source, has an illumination E, its illumination when the surface is inclined at an angle θ to the rays is given by $E \cos \theta$.

Luminance. The illumination of the surface does not necessarily define its intensity as a stimulus, since the latter depends on the light entering the eye and so is determined by the light reflected from the surface and the direction of view. If, however, we imagine a white surface that reflects all the light it receives and is, moreover, a "perfect diffuser" in the sense that it appears equally bright from all directions, then the *luminance* of a surface may be defined in terms of its illumination from a standard source. Thus, on the so-called *Lambert system*, a perfectly diffusing surface, normal to the incident light from a source of 1 candela 1 centimetre away, has a luminance of 1 *lambert*. This unit is too large for most purposes and the *millilambert* (mL) represents a thousandth of a lambert; for studies of light thresholds (p. 134) the log micro-microlambert scale is often used. A luminance of 3 log micro-microlamberts means a luminance of $10^3 \times 10^{-12} = 10^{-9}$ lamberts or a millionth of a millilambert; 2 log micro-microlamberts are equivalent to $10^2 \times 10^{-12} = 10^{-10}$ lamberts, and so on. The *equivalent foot-candela*, or *foot-lambert*, is the luminance of a similar screen distant 1 foot from the source.

Surface Intensity. This mode of defining the luminance of a surface in terms of the light incident on it is limited by the ideal nature of a perfectly diffusing surface; a definition that does not involve this would be superior from many points of view. Such a definition is given by describing the luminance in terms of the light emitted by the surface, i.e. we can treat the surface as a luminous source, whether it is self-luminous as in the case of a hot body, or whether its luminosity depends on reflected light. Thus if 1 square foot of a surface emits the same amount of light in a given direction as a source of 10 candelas, for instance, we may define the *luminous intensity of the surface* as 10 candelas; the subjective brightness of the surface and hence its intensity as a stimulus will clearly depend on the area from which this light emanates; if it all comes from a small area the brightness will be greater than if it came from a large area. The luminance is thus defined in terms of the candlepower of the surface per unit area, e.g. candelas per square foot or candelas per square centimetre. We may then calculate the luminance, in candelas per square foot, of a perfectly diffusing surface illuminated by a source of 1 candela at a distance of 1 foot; this turns out to be $1/\pi$ candelas per square foot, i.e. $1/\pi$ *candelas per square foot* = 1 *equivalent foot-candela*. Similarly $1/\pi$ candelas per square centimetre = 1 lambert.

Effectively Point Source. The definition of the luminance of a surface in terms of its candlepower *per unit area* has an obvious physiological basis. By the luminance of a surface we wish to express in a quantitative manner the intensity of the stimulus when its image falls on the retina. An illuminated surface of given area forms an image of a definite area on the retina and the sensation of brightness, or rather the stimulus, depends on the amount of light falling on this area of the retina. If the surface is moved closer to the eye, the amount of light entering the eye is increased in accordance with the inverse square law, but the surface does not appear brighter because the size of the image has increased, also in accordance with an inverse square law, and consequently the actual *illumination* of the retina has not altered. Thus the subjective brightness of an extended source is independent of its distance from the observer. It may be argued that the brightness of a lamp decreases as it is moved away from the eye and eventually becomes so faint that it disappears if it is moved far enough away. This, however, is a consequence of the fact that at a certain distance from the eye the lamp becomes an "effectively point source", i.e. its image on the retina is of the same order of size as that of a receptor element; consequently, when it is moved further away still, although its image on the retina becomes smaller, the amount of light per receptor element now begins to fall in accordance with the inverse square law. Thus the definition of luminance in terms of candles per unit area of necessity implies that the surface has a finite area; when this surface becomes effectively a point from the aspect of the size of its image on the retina, its power of stimulating the retina will be determined only by its total luminous intensity, i.e. its candlepower, and its distance from the observer. The limiting angle, subtended by an object at the eye, below which it behaves as an effectively point source depends on the particular visual effect being studied and the conditions of observation; for the study of absolute thresholds in the dark it is as big as 30–60 minutes of arc.

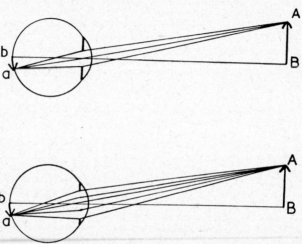

FIG. 71. Illustrating the importance of the size of the pupil in defining the stimulus. Above: the pupil is small and only a thin pencil of light from A enters the eye. Below: the pupil is large, permitting a large pencil to enter the eye.

Retinal Illumination. We have indicated that the characterization of the luminance of an extended surface by its candlepower per unit area is a sound one physiologically in so far as it indicates the illumination of the retina; the retinal illumination is not, however, completely defined by the candlepower per unit area of the surface being viewed, since the amount of light falling on unit area of the retina depends also on the pupil size; thus from Fig. 71 it is clear that the pencils of light that form the image of AB on the retina are larger in the case of the larger pupil. Hence the stimulus required to produce a definite sensation in the eye must be defined in terms of both the luminance, in candles per unit area, and the pupillary area. The unit of retinal illumination suggested by Troland* and now given his name is defined by the statement: A troland is that intensity

---

* This unit of retinal illumination was originally called the *photon*, a term that had already been applied to the quantum of light energy. We may note that, in order to convert the luminance of an extended surface, expressed in millilamberts, to trolands of "retinal illumination", it must be multiplied by $5d^2/2$, where d is the pupillary diameter in milli-metres. Because of the different sensitivities of the eye to the different wavelengths, according as it is light- or dark-adapted (p. 256), the equivalent stimuli for white light are not the same under photopic and scotopic conditions, so that we must speak of a *photopic* and *scotopic* troland; the theory behind the estimation of the conversion factor (1 scotopic lumen = 1·4 photopic lumens at a colour-temperature of 2750°C) has been discussed by Schneider & Baumgardt (1966). In terms of energy flux, a scotopic troland is equivalent to 4·46 quanta at 5070 Å per square degree of visual field per second, incident on the eye.

of stimulation that accompanies the use of a pupillary area of one square millimetre and an external stimulus surface luminance of one candela per square
metre. It is to be noted that although the troland is referred to as a unit of
"retinal illumination," it is not truly so, but is rather an indication of this quantity
that is independent of the size of the observer's pupil. To define a stimulus in
terms of true retinal illumination we must know the magnification and transmission of the optical media of the eye. On the basis of a value of 16·7 mm for the
distance of the nodal point from the retina, the true retinal illumination is
given by:

$$D = 0{\cdot}36\,\tau_\lambda\,s\,B$$

where B is the luminance of the extended source of light, s is the area of the pupil
in cm² and τ_λ is the *transmission factor*, being the ratio of the intensities of light
falling on the retina over the light falling on the cornea. This depends on wavelength, varying from 0·1 in the extreme violet where absorption by the yellow
pigment of the lens is important, to 0·7 in the deep red; on average with white
light it is about 0·5 indicating 50 per cent loss of light (Ludvigh & McCarthy).

When the stimulus in question is due to an effectively point source the pupillary area is of course important, and the stimulus is defined by the candlepower
of the source, its distance from the eye and the pupillary area, i.e. by the total
light flux entering the eye, which is given by the illumination of the pupil multiplied by its area.*

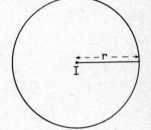

FIG. 72. A source of I candlepower at the centre of a sphere
of radius r. $4\pi I$ lumens are emitted per sec. and spread over
a surface of $4\pi r^2$; the flux density is I/r^2.

In view of the importance of the adequate definition of the stimuli used in
experimental studies of vision, it is worth recapitulating the substance of the
last few paragraphs. The luminance of an extended source is defined as its
candlepower per unit area; its value as a stimulus is independent of its area and
of its distance from the observer. For a given pupillary area it is an indication
of the retinal illumination; the troland is a measure of retinal illumination which
is independent of the observer's pupillary area; the true retinal illumination may
be derived from the troland if the magnification and transmission characteristics
of the eye are known. If the stimulus is due to an effectively point source, it is
sufficient to characterize it by the total light flux entering the eye.

Light Flux. For the sake of simplicity, we have so far made our definitions
without introducing the concept of *light flux*; this is defined as the amount of

* Stiles and Crawford (p. 289) have shown that pencils of light entering the eye obliquely
are less effective as stimuli than those entering the pupil centrally; this effect is not due
to aberrations in the optical system but is most likely related to the orientation of the
receptors in the retina. A correct definition of the stimulus must therefore take into
account the directions of the pencils of light entering the eye.

visible light energy passing any plane in unit time; its relationship to the luminous intensity of a point source will be clear from the following.

Suppose we have a point source of light of I candlepower; let us imagine the point to be the centre of a sphere of radius r (Fig. 72). The light energy from a point source travels in all directions uniformly through space, and we may think of the sphere being occupied by energy passing through it at a definite rate. The flux, considered thus, is defined in relation to the unit of luminous intensity by saying that 4π times the candlepower is the flux emitted by the source, expressed in *lumens*:

$$4\pi \times I \text{ (candelas)} = \text{flux (lumens)}$$

(We may note in passing that the lumen, which is an arbitrary unit of power, depending as it does on the candela, may be converted into absolute units by measurement of the energy emitted by the lamp defining the candela. The "mechanical equivalent of light" so obtained is $\dfrac{1}{679 \cdot 6}$ watts/lumen.

TABLE 8

Some recommended illuminations

	foot-candela
General stores	2–4
Classroom	5–10
Billiard table	20
Type-setting	35–50
Operating table	100

TABLE 9

Some commonly encountered luminances

	candela/sq ft
Starlit sky	0·00005
Moonlit sky	0·002
Full moonlight on snow	0·007
Ploughed land at noon in average daylight	130
Clear sky at noon	1,000
Sun's disk	150,000,000

E, the intensity of illumination of the inner surface of the sphere, as defined before, is:

$$E = I/r^2$$
$$= 4\pi I / 4\pi r^2.$$

Now $4\pi I$ represents the number of lumens of flux falling on the surface of the sphere, and $4\pi r^2$ is the area of its surface. Hence E may be expressed as the ratio of the flux emitted by the source of light over the area on which it falls: flux per unit area, or *flux density*. The units of illumination may therefore be, besides the foot- or metre-candela, a lumen per square foot or a lumen per square metre. Thus:

$$1 \text{ foot-candela} = 1 \text{ lumen/sq ft}$$
$$1 \text{ metre-candela} = 1 \text{ lumen/sq m}$$

In Table 8 the illuminations recommended for different tasks are shown; in Table 9 some luminance levels commonly encountered are presented, and in Tables 10 and 11 some conversion factors for different units of illumination and luminance.

TABLE 10

Conversion Factors. Illumination. (Value in unit in left-hand column times conversion factor = value in unit at top of column; e.g. 3 foot-candela = 3 × 1·08 × 10 = 32·4 metre-candela.)

	Foot-candela	Centimetre-candela	Metre-candela
Foot-candela	1	$1·08 \times 10^{-3}$	$1·08 \times 10$
Cm-candela (Phot)	$9·29 \times 10^{2}$	1	1×10^{4}
Metre-candela (Lux)	$9·29 \times 10^{-2}$	10^{-4}	1

TABLE 11

Conversion Factors. Luminance. (Value in unit in left-hand column times conversion factor = value in unit at top of column)

	Candela/ft^2	Equiv. Fc.	Candela/cm^2	Millilambert
Candela/ft^2	1	$3·14$	$1·08 \times 10^{-3}$	$3·38$
Equiv. Fc.	$3·18 \times 10^{-1}$	1	$3·43 \times 10^{-4}$	$1·076$
Candela/cm^2	$9·3 \times 10^{2}$	$2·92 \times 10^{3}$	1	$3·14 \times 10^{3}$
Millilambert	$2·96 \times 10^{-1}$	$9·29 \times 10^{-1}$	$3·18 \times 10^{-4}$	1

GENERAL REFERENCES

WALSH, J. W. T. (1953). *Photometry*. London: Constable.

SPECIAL REFERENCE

LUDVIGH, E. & McCARTHY, E. F. (1938). Absorption of visible light by the refractive media of the human eye. *Arch. Ophthal., Chicago* **20**, 37–51.

SCHNEIDER, H. & BAUMGARDT, E. (1966). Sur l'emploi en optique physiologique des grandeurs scotopiques. *Vision Res.* **7**, 59–63.

PART 3

SOME GENERAL ASPECTS OF VISION

Photochemical Reaction. The mechanism by which the energy contained in visible light is transmitted to the nervous tissue of the retina in such a way as to produce that change in its structure which we call a stimulus is a fascinating problem and one that lends itself to physicochemical study. There is a group of chemical reactions that derive a part, or all, of their energy from absorbed light— the *photochemical reactions*—a biological instance of this type of reaction is the synthesis of carbohydrate by the green leaves of plants, the pigment chlorophyll absorbing light and transmitting the energy to the chemical reactants. It is thus reasonable to inquire whether there is an analogous chemical reaction at the basis of the response of the retina to light, i.e. whether or not there are substances in the retina that can absorb visible light and, as a consequence of this absorption of energy, can modify the fundamental structure of the rod or cone to initiate a propagated impulse. One such substance, visual purple, was discovered in the retina as long ago as 1851 by H. Muller, but we owe the early knowledge of its function as an absorber of light to the researches of Kühne, described in his classical monograph of 1878. He was the first to obtain this pigment in solution by extracting the retina and he called it "*Sehpurpur*", which has been translated as *visual purple*. As we shall see, thanks to the recent studies of Dartnall and Wald, the term visual purple must now be used to describe a class of visual pigments rather than a single entity.

Categories of Vision. Before the detailed study of the photochemical and electrical changes in the retina associated with the visual process is entered into, it would be profitable to discuss some of the more elementary phenomena of vision itself and to indicate the experimental methods available for its study. The visual perception of an object is a complicated physiological process, and attempts have been made to resolve it into simpler categories. Thus, it is common to speak of the *light sense*, defined as the appreciation or awareness of light, as such, and of modifications in its intensity. Secondly, the *colour sense*, which enables us to distinguish between the qualities of two or more lights in terms of their wavelengths, and finally the *form sense* that permits the discrimination of the different parts of a visual image. For the present we shall preserve this simple classification for its convenience, but modern psychophysicists would hesitate to use the term "sense" in relation to the discrimination of form.

THE LIGHT SENSE

The Absolute Threshold. The simplest way to study the light sense is to measure the *threshold* to white light, i.e. the minimum light-stimulus necessary to evoke the sensation of light. There are many ways of measuring this, and the general technique will be discussed later. In general, the subject is placed in a completely dark room and the illumination of a test-patch is increased until he is aware of its presence. The patch may be illuminated all the time and its luminance steadily increased, or it may be presented in flashes of varying luminance.

Dark-Adaptation. Early attempts to measure the threshold soon showed that, in order to obtain other than very irregular results, the observer must be allowed to remain in the dark for at least half an hour (i.e. he must be *dark-adapted*) before any attempt is made to obtain a definite value. If this is not done, it is found that the threshold luminance of the patch is very high if the test is made immediately after the subject has entered the dark room, and that the threshold falls progressively until it reaches a fairly constant value which may be as small as one ten-thousandth of its initial figure. This effect is demonstrated by Fig. 73, in which the illumination of the test patch (in microlux) required to produce the sensation of light in a subject, who had gazed at a bright light for some minutes before entering the dark room, is plotted against time in the dark; the curve is referred to as a *dark-adaptation curve.*

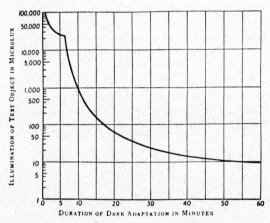

FIG. 73. Typical dark-adaptation curve. Ordinate: illumination of the test-patch in microlux. Abscissa: duration of dark-adaptation in minutes. (Kohlrausch, in *Handbuch der Normalen und Pathologischen Physiologie.*)

This extraordinary increase in sensitivity of a sense organ on removing it from its normal background of stimuli has not been met in the study of the other senses, and an inference to be drawn from the dark-adaptation curve is that a second mechanism of vision is being brought into play requiring half an hour, or more, to be established at its full level of efficiency. This inference represents the basis of the *Duplicity Theory of Vision*, a theory which assumes that at luminance levels below a certain value (about 0·01 mL) vision is mediated primarily by one mechanism—*scotopic vision*, involving the rods—and above this value by another mechanism—*photopic vision*, involving the cones; the transition is not abrupt so that there is a range in which the two mechanisms operate together, the *mesopic range.*

Duplicity Theory

Rod Vision. The duplex nature of vision is readily appreciable under conditions of very low illumination. Thus, whereas in daylight, or photopic, vision the subject sees best by direct fixation, so that the image of the object or luminous point he wishes to see falls on the fovea, under night, or scotopic, conditions his best vision is *peripheral*; he must look away from the object he is trying to distinguish so that its image falls on the peripheral retina. Thus, in looking at a dim star one must, as the astronomer Arago said, not look at it; if one fixates it, it disappears. This gives the clue to the anatomical basis of the duplicity theory; the fovea, in its central region specially, contains only cones. Since the fovea is apparently blind at low intensities of illumination, we may conclude that it is the rods that are functional under these conditions. That the rods are indeed responsible for night vision is now well recognized; it was Schultze who put forward this hypothesis on the basis of his comparative studies of animal retinae; thus he found that the nocturnal owl had a preponderance of rods whilst a diurnal bird, such as the falcon, had a preponderance of cones in its retina.

ACHROMATICITY

Another feature of vision at low luminance is its *achromaticity*, i.e. its failure to distinguish colours. Suppose, for example, that the threshold for vision is measured using different wavelengths of light; it is found that these thresholds are different, in the sense that more yellow energy, for instance, is required to give the sensation of light than blue energy, but the "yellow" and blue" lights are both colourless. Only when the intensities are raised well above threshold do colours appear, i.e. when the threshold for stimulation of the *cones* is reached. Thus, on a moonless night colours are not distinguished, whereas in the *mesopic range*, corresponding to around full moonlight, colours may be appreciated but not very distinctly.

LOW VISUAL ACUITY

A very obvious difference between scotopic and photopic vision is the low visual acuity at low luminances; in other words, in spite of the high sensitivity of the eye to light under scotopic conditions, the ability to discriminate detail (e.g. to read print) is very low.

Other features of the duplicity theory will be considered later; for the moment we may say that it is now a well established *principle* rather than a theory. In essence it states that at high levels of luminance the relatively insensitive cones are able to operate, whilst the highly sensitive rods are out of action (perhaps through overstimulation). Vision mediated by the cones is coloured and exhibits high powers of discrimination of detail. When the light available is restricted, the insensitive cones no longer operate whilst the rods, after a period of dark-adaptation as shown by Fig. 73, become operative, mediating an achromatic, peripheral, low-acuity type of vision. Over an intermediate range, which may extend up to 1000 times the threshold for cone vision, both rods and cones are operative—the *mesopic range*.

The Differential Threshold or L.B.I.

Instead of using the minimum luminance of a test patch as a means of studying the light sense, we may measure the *differential threshold*—the amount an existing stimulus, which itself produces a sensation of light, must be increased or decreased to produce a change in the sensation.

Measurement. Thus the subject may be presented with a luminous field

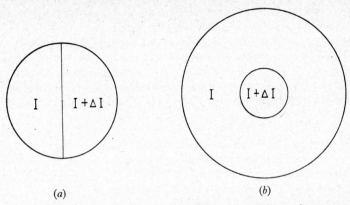

FIG. 74. Illustrating the measurement of the liminal brightness increment.

divided into two, as in Fig. 74a, the luminances of either being varied independently. Beginning with them both equal at I, the right-hand one may be increased to a value $I+\Delta I$, such that the observer can just detect a difference of subjective brightness. The increment, ΔI, is called the *liminal brightness increment* (l.b.i.) or *differential threshold*. Alternatively, as in Fig. 74b, the subject may look at a large circular field of luminance I, and a smaller concentric field may be thrown on to this with a variable luminance greater than I. With this experimental technique, the measurements need not be confined to those intensities of illumination in which only scotopic vision is operative, but rather the whole range of intensities may be used.

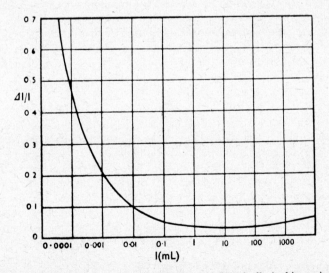

FIG. 75. Intensity discrimination curve. The ratio $\Delta I/I$ is the liminal intensity (ΔI) increment divided by the intensity (I). If the Weber-Fechner Law held, this ratio would be independent of the value of I. The curve shows that, over a certain range (0·1 to 1000 mL), this is approximately true. In the scotopic region, however, the ratio increases greatly, i.e. the power to discriminate changes in luminance falls off at low levels.

Weber-Fechner Law. The differential threshold, or l.b.i., varies with the initial value of the prevailing luminance, I; for example if I is 100 mL then ΔI may be in the region of 2 mL; if I is 1000 mL then ΔI is about 20 mL. Over a certain range, however, it may be expressed as a constant fraction of the prevailing luminance, so that the ratio: $\Delta I/I$ is constant. This is the basis of *Weber's Law of Sensation* which applies approximately to all sensory modalities, namely that the difference-limen is proportional to the prevailing stimulus. For example, in the case of the sense of touch or pressure, Weber showed that when two weights resting on the skin differed by about 1 in 30 they could be discriminated as different weights. Thus if an individual could discriminate between a 30 and a 31 gm weight he could also discriminate between a 3 and 3.1 gm weight; in other words, the *Weber fraction*, $\Delta S/S$, is constant, where S is the prevailing stimulus and ΔS its liminal difference for perception.

Fechner Fraction. Fechner extended Weber's work to the study of visual discrimination, and the fraction $\Delta I/I$ is frequently called the Fechner-fraction. Fig. 75 represents a more modern study, where the Fechner-fraction, $\Delta I/I$, has been plotted against the intensity level, I. At high intensities, between about 0·1 and 1000 mL, this fraction is fairly constant at about 0·02 to 0·03, indicating that a change of two to three per cent in luminance can be detected; at low luminances the power to discriminate falls off sharply so that at 0·0001 mL two patches must differ by about 50 per cent in luminance if they are to be discriminated.

THE COLOUR SENSE

By means of the colour sense we become aware of those qualitative characteristics of light that depend on its spectral composition. On passing white light through a prism it is resolved spatially into its component wavelengths, and the result is the well-known spectrum; this resolution shows that the sensation of white is evoked by the summated effects of stimuli which, when operating alone, cause entirely different sensations, namely, those of red, orange, yellow, green, etc.

Hue

The spectrum shows a number of characteristic regions of colour: red, orange, yellow, green, blue, indigo and violet; these regions represent large numbers of individual wavelengths; thus the red extends roughly from 7500 to 6500 Å; the yellow from 6300 to 5600 Å; green from 5400 to 5000 Å; blue from 5000 to 4200 Å, and violet from 4200 to 4000 Å. Within these bands of colour there are thus a variety of subtle variations, and a spectral colour or *hue*, has been defined somewhat idealistically as the sensation corresponding to a single wavelength. This is too rigid a definition, both because of the limitations to discrimination of the human eye, and the experimental difficulty in obtaining light confined to a single wavelength.

Thus coloured lights may be obtained by the use of filters, which cut out many wavelengths and selectively transmit others; these filters are never sufficiently selective to confine the transmitted light to a single wavelength and at best they transmit a *narrow band* about a mean position. Hence by a green filter of 5300 Å is meant a filter that transmits mainly green light with a peak of transmission at

5300 Å; transmission at nearby wavelengths is also considerable, however, whilst even at remote wavelengths there may be a little transmission.

Monochromator. A more precise method is by the use of the monochromator, the essential principle of this technique being to select, from a spectrum, just that narrow band of wavelengths that one requires. In practice the band is selected by a movable slit that passes over the spectrum; and its width, in terms of the range of wavelengths, can be reduced by making the slit narrower and narrower. Unfortunately, this decreases the amount of light-energy transmitted, so that once again the experimenter must be content with a band of wavelengths rather than a single one.

Hue-Discrimination. The physiological limit to the differentiation of hues is given by the *hue-discrimination curve*, shown in Fig. 76. The subject is presented with a split field, the two halves of which may be illuminated independently by lights of equal subjective intensity but differing wavelength. Starting with, say, a red of 7000 Å in the left-hand field the wavelength of the light of the right-hand field is shortened till the subject can just discriminate a change in hue, say at 6800 Å; the hue-limen is thus 200 Å. With both fields at 6800 Å, the right-hand field wavelength is again shortened and a new hue-limen obtained. Fig. 76 shows the results for a normal human observer; discrimination is best at about 5000 Å in the blue-green and at 6000 in the yellow, but even here the smallest detectable change in wavelength is about 10 Å.*

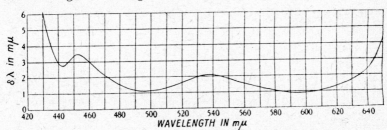

FIG. 76. Hue-discrimination curve. At any given wavelength, indicated on the abscissa, there is a certain change in wavelength, δλ, that must be made before the change in hue can be discriminated. (Wright, *Researches on Normal and Defective Colour Vision*.)

Saturation. In practice, then, coloured lights are not monochromatic, being generally a mixture of different wavelengths; the hue is then said to be *impure*, and in the case when the impurity can be accounted for by an admixture of white light, it is said more specifically to be *unsaturated*, the actual colour being called a *tint* of the hue that has been thus diluted. The percentage saturation of a hue is the percentage contributed by the monochromatic light, estimated on a luminosity basis. By mixing white light with a spectral hue, the latter actually changes; thus red becomes pink, orange becomes yellow, yellow becomes green, green becomes yellow, and violet becomes salmon-pink; the yellow-green does, in effect, remain constant and the blue changes very little.

Limits of the Visual Spectrum. The limits are usually given as 4000 Å and 7600 Å, wavelengths shorter than 4000 Å and longer than 7600 Å being characterized as *ultraviolet* and *infrared* respectively. In fact, however, the retina is

* Cornu & Harlay (1969) have examined the effects of luminance on hue discrimination, and have summarized the recent literature on this point.

sensitive to wavelengths well beyond these limits; for example, Wald showed that the insensitivity of the eye to ultraviolet up to 3500 Å is entirely due to absorption by the lens; aphakic subjects are able to see easily with these wavelengths, calling the sensation blue or violet. Again, the failure of early workers to evoke sensation with wavelengths greater than 7600 Å was due entirely to their inability to obtain intense enough sources, the retina being so insensitive to red light. Griffin, Hubbard & Wald were able to evoke sensation with wavelengths as long as 10,000–10,500 Å, and Brindley has found that, when compared on an equi-energy basis (p. 256), the far reds are much more orange than the near reds; for example, a red of 8870 Å could be matched by an orange of 6400 Å.

Just as with white light, we may refer to the luminosity of a given hue and measure it in the same units, and although it may seem somewhat unreal to speak of comparative luminosities of widely different hues, by making use of a suitable colour photometer it is possible to use a scale of luminosity which applies to all the spectral colours in such a way that if all the luminosities of the hues of a spectrum are determined separately, the luminosity of the white light obtained by re-combining the spectrum is equal to the sum of these individual spectral luminosities.

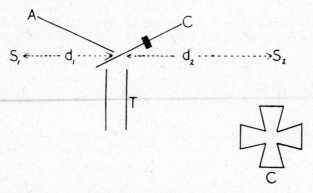

FIG. 77. The flicker-photometer. Light from S_1 is reflected from A down the viewing tube, T. Light from S_2 is reflected from the rotating cross, C. When A and C are equally illuminated flicker ceases. (After Barton.)

The Flicker Photometer. If it is desired to measure the relative luminosities of two differently coloured sources, the most satisfactory instrument to use is the *flicker photometer*. This is represented diagrammatically in Fig. 77. It consists of a white screen, A, and a Maltese Cross, C, which is placed at such an angle that light from one of the sources (S_2) is reflected down the viewing tube, T. The screen, A, is likewise placed so that light from the other source (S_1) is reflected down the tube. When the cross is rotated, the eye receives light alternately from S_2 and S_1 by reflection from C and A respectively. If the luminances of the two surfaces are different, a sensation of flicker is experienced which can be made to disappear by adjusting the distances of the two sources from the surfaces which they illuminate. When flicker disappears, the luminances of the two surfaces are equal. If A and C make the same angle with the direction of the incident light, and if their reflection factors are equal, the luminous intensities of S_1 and S_2 will be in the ratio of $d_2{}^2 : d_1{}^2$.

Variation of Hue with Intensity. We have seen that hue and saturation are not absolutely independent, since altering the saturation of a yellow, for example, can actually make it into a green. In a similar way, hue is not independent of luminosity. Thus, as the intensity of a coloured light is increased its subjective hue makes a characteristic shift, and ultimately all wavelengths, if made intense enough, appear the same, evoking a yellowish-white sensation. These shifts are described as the *Bezold-Brücke phenomenon*, after their discoverers. On lowering the intensity there are also characteristic changes which will be discussed later in connection with the Purkinje phenomenon, and ultimately the whole spectrum becomes achromatic, in the sense that at sufficiently low intensity levels all wavelengths appear white and indistinguishable from each other qualitatively. It is therefore important to define the level of luminance at which studies of colour vision are made; the value generally chosen is of the order of 2–3 mL.

Colour Mixtures. If two colours, not too far apart in the spectrum—say a red and a yellow—are mixed, the resulting sensation is that produced by a single spectral hue of intermediate wavelength. The actual position of the resultant hue in the spectrum is determined by the relative luminosities of the two hues that are mixed; if they are equal, the resultant lies midway between; if the red predominates, the resultant lies towards the red end. As the distance apart of the two hues is increased, the resultant hue is found to become progressively unsaturated, i.e. it can only be matched with an intermediate spectral hue provided that white light is added to the latter. Eventually a point is reached when the mixture results in white; any two hues that, when combined, give white light are called *complementary colours*. As a result of the classical researches of Young and Helmholtz and more recently those of Wright, the laws of colour mixture have been put on a quantitative basis. In general, it may be stated that any colour sensation, within an average range of intensities, can be produced by a mixture of not more than three *primary spectral wavelengths*, with the reservations that (1) in many cases white light must be added to the comparison colour to obtain a match in saturation as well as in hue. Thus, if it is required to match a given standard colour, say a bluish green, then by varying the intensities of the three primaries it will usually be found that it is impossible to make a perfect match with this standard, the mixture appearing unsaturated by comparison. We must therefore add white to the standard until the two match or, as it is often put, we must subtract white from the mixture to obtain a perfect match in hue and saturation. The three primaries are chosen so that any one cannot be produced by mixing the other two; thus red, green and blue are primaries.*

A given hue may thus be expressed in terms of three coefficients, namely:

$$\text{Colour} = \alpha R + \beta G + \gamma B$$

where the latter determine the relative quantities of these arbitarily chosen primaries. Light is measured on a luminosity basis in most work, so that it would be natural to measure the contributions of the primaries to any given hue in terms of luminosity units, remembering that the luminosity of the hue will be equal to the sum of the separate luminosities of the primaries required to make a match in hue and luminosity. Since the resultant mixture of the primaries will

---

* There are several ways of expressing the "Laws of Colour Mixture". A common one is as follows: given any four lights, it is possible to place three of them in one half of a photometric field and one in the other half, or else two in one half and two in the other, and by adjusting the intensities of three of the four lights to make the two halves of the field indistinguishable to the eye. This appears different from the statement in the text, but is actually the same in so far as it is another statement of the *trivariance* of colour matching, the three variables being the intensities of the three colours that are varied, the fourth remaining constant. The laws governing colour-matching are called *Grassmann's Laws.*

not necessarily correspond in saturation with the given hue to be matched, it is necessary to subtract a certain quantity of white—that is, certain quantities of the three primaries—and this may lead to negative coefficients. There is no particular theoretical objection to negative coefficients, but they can be avoided by choosing certain hypothetical primaries which do not correspond to any spectral hues. Thus we have seen that a mixture of three spectral primaries gives an unsaturated spectral hue; if the amounts of the primaries which must be subtracted from this unsaturated spectral hue, in order to saturate it, are subtracted from the primaries themselves, we obtain a set of hypothetical primaries which give positive coefficients. Thus Wright used hypothetical primaries, R', G', B', defined by the relationship:

$$R' = 1 \cdot 015\,R - 0 \cdot 015\,G$$
$$G' = -0 \cdot 90\,R + 2 \cdot 00\,G - 0 \cdot 10\,B$$
$$B' = -0 \cdot 05\,G + 1 \cdot 05\,B$$

where R, G, and B were spectral hues of 6500, 5300, and 4600 Å respectively.

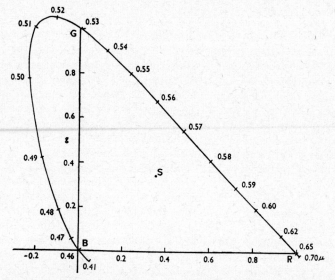

FIG. 78. Chromaticity chart. Units of green are indicated on the vertical line; units of red on the horizontal line. The amount of blue in a given mixture is given by the unit colour equation, R+G+B = 1. The line joining R to G and continuing as a curved loop to B is called the *spectral locus* and shows how the individual spectral hues can be matched by the three primaries. S is standard white. (Wright, *Researches on Normal and Defective Colour Vision*.)

For ease in manipulation, the quantities of the primaries are often expressed in terms of their contributions to colour quality rather than to luminosity; thus the coefficients are defined by the statement that equal quantities of the primaries make white:

White = 0·333 R+0·333 G+0·333 B, the units defined in this way being referred to as *trichromatic units*, and since the sum of the coefficients is equal to unity, the equation defining a colour is called a *unit colour equation*. It is a relatively simple matter to transpose from one set of coefficients to another.

The Chromaticity Chart. The unit colour equation is useful because it enables us to express graphically the composition of any hue in terms of the primaries; thus if we construct rectangular co-ordinates so that the ordinates represent green coefficients and the abscissae red coefficients, then any hue will be given by a point representing its red and green coefficients; since $\alpha+\beta+\gamma = 1$, the blue coefficient will be given by subtracting the red and green coefficients from unity. The locus of spectral hues will be a line (Fig. 78) and white, if defined by the equation above, will be represented by a point whose ordinate and abscissa are 0.333.

FORM

The form or structure of the visual field is determined by the power to discriminate between two or more separate stimuli in regard to intensity and spatial extension; we perceive the shape of an object by virtue of our ability to integrate our responses to a large number of separate stimuli of varying intensity arising from different positions on its surface.

Resolving Power and Visual Acuity

The power of the eye to discriminate on a spatial basis is studied experimentally by measurements of the *resolving power*, the smallest angle subtended at the nodal point of the eye by two points or lines such that they are appreciated as separate, i.e. the so-called *minimum separable*. The visual acuity is defined as the reciprocal of this angle when it is measured in minutes of arc. Thus, under good experimental conditions, resolving powers of the order of 30 seconds of arc, or half a minute, may be measured, corresponding to a visual acuity of $1/0.5 = 2$.

FIG. 79. Snellen letter, grating and Landolt-C. In each case the angle subtended at the eye by the distance, *s*, apparently determines whether the gap will be recognized.

Measurement. The most commonly used test-objects for measuring resolving power are the Landolt-C and the grating (Fig. 79). The Landolt-C can be rotated into eight separate positions which the subject is asked to identify; C's of diminishing size are presented and the errors made in indicating the positions are recorded. Experimentally it is found, as with all psychophysical studies of thresholds, that there is no sharp division between the size of C that can be discriminated and that which cannot; near the "end-point" we will have a range of sizes where the average number of errors in any set of trials increases with decreasing size of C until eventually the answers are purely random. We must therefore somewhat arbitrarily choose an end-point; Lythgoe chose that size of C with which 4.5 out of eight answers were correct, and the actual end-point was computed from the errors obtained with several sizes of C by the application of statistical theory. So far as the resolving power of the eye is concerned, it would appear that the ability to detect the position of the break in the C, which distinguishes it from a complete circle, depends on the size of the break, *s*; in fact, the angle subtended at the eye by this distance, *s*, at the end-point, is indeed the measure of the resolving power of the eye. Thus at a distance of 6 metres, under given conditions, the size of the break, s, was 3·6 mm, at the end-point; this distance subtended 2 minutes at the eye and so the visual acuity was 0·5.

GRATING ACUITY

If the grating of Fig. 79, consisting of equally spaced black and white lines, is removed sufficiently far from the observer, it appears uniformly grey; at a certain point the lines can be just discriminated and the angle subtended at the eye by the distance s once again is said to indicate the resolving power of the eye. According to Shlaer (1937) the limiting size for resolution subtends some 28 seconds of arc at the eye under optimal conditions, corresponding to an acuity of 2·12.

SNELLEN TEST

The clinical *Snellen test* is essentially a resolution test; it is assumed that the average individual can resolve the details of a letter if they subtend one minute of arc at the eye, and the test letters are constructed so that their details (e.g. the gap in a C) subtend this angle, whilst the whole letter subtends five minutes of arc (Fig. 79), when they are at a definite distance from the eye. The chart has several rows of these letters, the sizes in the different rows being graded so that their details subtend one minute at different distances, e.g. the smallest letters subtend this angle at 4 metres whilst the largest do so at 60 metres. The subject is placed 6 metres from the chart and the smallest row of letters that can be read correctly is determined. The visual acuity is expressed as the ratio of the distance of the individual from the chart to the distance at which this row of minimum legible letters should be read, assuming a minimum visual angle of one minute. Thus at 6 metres if the individual can read down to the letters whose details subtend one minute when viewed from a distance of 18 metres his visual acuity is 6/18; if he can read letters so small that they subtend the same angle at only 5 metres, his visual acuity is 6/5, and so on.

Experimental Results. Factors influencing visual acuity will be dealt with in detail later; for the present we need only note that it is markedly influenced by the illumination of the test object, the state of adaptation of the eye, and the pupil size. Under optimal conditions it has a value in the region between 2 and 3 (i.e. a minimum separable of about 30–20 seconds of arc) when a grating or Landolt-C is used as test-object.

VISUAL ACUITY IN ANIMALS

Visual acuity can be measured in animals by training them to choose between alternate pathways marked with a grating or Landolt-C of variable visual angle; selection of the required pathway is rewarded with food. When the choice becomes random the animal is unable to resolve the grating or Landolt-C. Farrer & Graham (1967) found a limiting resolution of 0·5–1·0 minutes of arc in a monkey, indicating a visual acuity as good as that of man; in the rabbit, Van Hof (1967) found values between 10 and 20 minutes, and this corresponds with the absence of a fovea in this animal.

INDUCTION

The effect of light falling on a given portion of the retina is generally not confined to the retinal elements stimulated; moreover, the characteristics of those elements that have been stimulated are altered for an appreciable time, so that their response to a second stimulus is modified. The influence of one part of the retina on another is described as *spatial induction*; by *temporal induction* is meant the effect of a primary stimulus on the response of the retina to a succeeding stimulus.

Spatial Induction

Spatial induction, or *simultaneous contrast*, can be demonstrated by comparing the appearance of a grey square, when seen against a black background, with its appearance when seen against a white background; against the black background the grey appears almost white, whereas it appears quite dark when seen against the white background. The white surround apparently depresses the sensitivity of the whole retina, thereby making the grey appear darker; the net effect of this depression in sensitivity is to accentuate the contrast between the grey and the white surfaces. Where coloured lights are concerned the same principle of a reduction in sensitivity of the retina applies, but now this reduction is confined to certain wavebands, and the subjective result is to enhance the appearance of the complementary colour. Thus, if a white and blue light patch are placed side-by-side, the white patch appears yellow; apparently the blue light falling on one half of the retina has depressed the sensitivity of the whole retina to blue light; the white light falling on the retina is now losing some of its "blueness" because of this loss of blue-sensitivity; removing blue from white leaves a yellow colour, since blue and yellow are complementaries.

Temporal Induction

In a similar way, if the whole retina is stimulated with white light the sensitivity to a given stimulus following this is reduced, so that the second stimulus appears less bright than it would have done had there been no "conditioning stimulus", as this preliminary, adapting, stimulus is called. This form of induction, which has been called *light-adaptation*, has been studied in detail by Wright in England, and Schouten in Holland.

Binocular Matching. The technique employed was that of binocular matching, based on the principle that the states of adaptation of the two eyes are independent. Thus the observer looks through a binocular eye-piece, each eye seeing the half of a circular patch of light, and these halves have to be matched for equal subjective brightness. If, say, the right eye is exposed to a bright light, we may determine the changes in sensitivity of this eye during its exposure to the bright light by matching its half-field with the left half-field, as seen by the dark-adapted eye. This means varying the intensity of the left half-field until the match is made. It is found that the intensity of the left half-field has to be continuously reduced during exposure of the right eye, until it levels off at a constant value which may be about one-sixteenth of the value at first. This means that the right eye becomes rapidly less sensitive to light from the moment it is exposed to it; this is the phenomenon of light-adaptation and is illustrated by Fig. 80. The phenomenon is, of course reversible, so that if the two eyes are now allowed to match fields of moderate luminance, the left field will, at first, have to be made much weaker than the right because the left eye has remained dark-adapted and has remained the more sensitive; very rapidly, however, this matching field will be made stronger until they finally become equal when both eyes are completely dark-adapted.*

Practical Demonstration. This adaptation is very easy to demonstrate qualitatively; thus if an observer is instructed to stare at a brilliantly illuminated screen for a few minutes with one eye closed, and then this is replaced quickly by one of moderate intensity, he finds, on comparing the subjective brightness of the screen as seen through either eye separately, that the eye that had been kept closed sees the screen as very much brighter. We may note here that the phenomenon of *dark-adaptation*, as revealed by the progressive decrease in the absolute threshold, is a manifestation of temporal induction, in so far as the phenomenon reveals a change in the retina following the cessation of a stimulus, namely of the original bright light to which the observer was exposed before being placed in the dark.

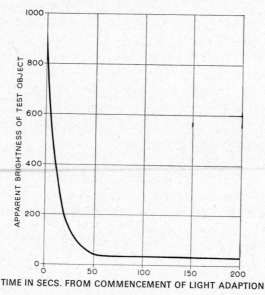

FIG. 80. Curve showing the fall in sensitivity of the fovea during exposure to a bright light. The right eye views the bright light whilst the dark-adapted left eye views a patch that must be matched in brightness with the sensation experienced through the right eye. The matching intensity decreases, indicating that the apparent brightness of the light, as seen through the right eye, becomes less and less. (Wright, *Trans. Illum. Eng. Soc.*)

Adaptation to Coloured Lights. The same process of light-adaptation, or successive contrast, is seen with respect to colours; thus with the same binocular matching technique, except that the two fields must be matched for colour rather than subjective brightness, the effects of adapting to a given wavelength of light can be found. The same principle emerges, namely that adaptation of the one eye to, say, green, reduces the sensitivity of this eye to this wavelength, and in

* These studies of light-adaptation and their reversal are made with foveal vision and are essentially studies of cone vision, i.e. although comparisons are made with a dark-adapted eye, the intensities of the half-fields to be compared are always maintained well above cone-thresholds.

order that the other eye may make a match it has to employ much less green than before. Thus two half-fields may be made to match, the right eye viewing a yellow and the left a mixture of red and green, their intensities being adjusted to give the match with the right eye. The right eye is now exposed to an adapting green light for a few minutes and it is found that, on viewing the two fields, the match no longer holds, the right field appearing reddish by comparison with the yellow of the left field. The left field must now be made less green to make the match, and the extent to which the green must be reduced in intensity in this field is a measure of the reduction of sensitivity of the right eye to green.

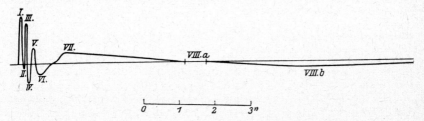

FIG. 81. The successive after-images said to be experienced after viewing a bright flash. Time-scale in seconds. (Tschermak, in *Handbuch der Normalen und Pathologischen Physiologie*.)

After-Images. When the eye is exposed to a brief flash of light the subjective sensations last for much longer than the stimulus, and are described as a succession of *after-images*. The phenomena may be classed as inductive, in so far as they reveal changes in the retina or its higher pathways following a primary

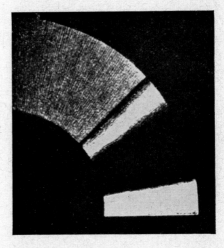

FIG. 82. Bidwell's ghost. A black disc with a sector cut out rotates in front of a brightly illuminated opal screen. The after-images of the sector are projected to the left to give the appearance shown. (McDougall, *Brit. J. Psychol.*)

stimulus; although they have attracted a lot of interest, little of fundamental value has emerged from their study and it would be a pity to spend much space on them. As classically described, the responses to a single flash are illustrated schematically in Fig. 81 where upward movements indicate sensations of brightness (positive after-images) and downward movements sensations of blackness (negative after-images). If a coloured primary stimulus is used, some of the after-images are coloured, often with the complementary. The events depicted by Fig. 81 occur rapidly and are exceedingly difficult to resolve, except by such devices as Bidwell's disc, illustrated by Fig, 82. Here a disc, with a narrow slit in it, is rotating in front of a bright opal screen, and the after-images are projected on to the black disc to give the appearances of *Bidwell's ghost*. A more slowly developing and longer-lasting after-image is that observed after staring for a few minutes at a bright source and then transferring the gaze to a dimly lit screen. In a little while a "negative" image of the bright source appears; if the source was coloured this has the complementary colour, e.g. magenta after looking at green; yellow after looking at blue, and so on.

Intermediate Exposures. The after-effects of exposures intermediate in duration between the brief flash and the lengthy gaze have been studied under well controlled conditions by Trezona. Thus, with an exposure of the order of 8 seconds, the after-effects may be classified as follows:

(a) *Persistence of vision*, this being the apparent continuation of the stimulus after it has been cut off, and lasting about 1 second at rapidly diminishing subjective brightness.

(b) *After-colours*; these may last for seconds or minutes, changing colour frequently. The threshold for these after-colours was very high, of the order of 100,000 trolands of retinal illumination.

(c) *After-blueness*; this was a characteristically different effect from the after-colours mentioned above; it lasts for some seconds and remains the same colour. The threshold was much lower, of the order of 100–1000 trolands and the effect could only be obtained with peripheral vision, suggesting mediation by the rods.

(d) *After-blackness*. This was favoured by long exposures, and consisted of several appearances of the black image of the source.

Liminal Brightness Increment. If a subject was exposed to pairs of bright flashes in the two halves of a test-field, the after-images could be distinguished if the intensities of the flashes differed, the Fechner-fraction, $\Delta I/I$, comparing with that found for direct vision; this was in spite of the fact that with direct vision the two halves of the field could not be distinguished because of the extreme brightness and shortness of the flashes. When the intensities of the flashes became too high, however, there was a fall in discrimination, i.e. a rise in the Fechner-fraction, and this was probably due to excessive bleaching of the cones (Brindley, 1959).

Peripheral Origin. This finding, together with the study of the applicability of Bloch's Law (p. 298), strongly suggests that after-images have a peripheral origin, i.e. are determined by photochemical events in the receptors. Support for this is given by Craik's experiment, in which the eye was temporarily blinded by exerting pressure on the globe. On exposure of the eye to a bright light nothing was seen, but on restoring the circulation an after-image appeared. The possibility that some of the phases have a central origin must not be ruled out, however; thus Brindley was unable to find any regularity in behaviour of his after-images unless he ignored those appearing within the first 15 seconds; after this time they conformed to rules predictable on photochemical grounds (p. 298).

Value of Contrast

Simultaneous and successive contrast are essentially manifestations of an inductive process that tends to reduce the sensitivity of the retina to an existing stimulus and to enhance its sensitivity to a stimulus of a different kind, e.g. the sensitivity to the complementary of the existing, coloured, stimulus. As a result of simultaneous contrast the contours of an image become better defined and, as a result of successive contrast, the confusion that would result from the mixing of successively presented images is reduced.

GENERAL REFERENCES

BRINDLEY, G. S. (1970). *Physiology of the Retina and Visual Pathway*, 2nd ed. London: Arnold.

LEGRAND, Y. (1957). *Light, Colour and Vision* (translation by R. W. G. Hunt, J. W. T. Walsh & F. R. W. Hunt). London: Chapman & Hall.

PIRENNE, M. H. (1967). *Vison and the Eye*, 2nd ed. London: Chapman & Hall.

WRIGHT, W. D. (1946). *Researches on Normal and Defective Colour Vision*. London: Kimpton.

SPECIAL REFERENCES

BRINDLEY, G. S. (1955). The colour of light of very long wavelength. *J. Physiol.* **130**, 35–44.

BRINDLEY, G. S. (1959). The discrimination of after-images. *J. Physiol.* **147**, 194–203.

CORNU, L. & HARLAY, F. (1969). Modifications de la discrimination chromatique en fonction de l'éclairement. *Vision Res.* **9**, 1273–1287.

FARRER, D. N. & GRAHAM, E. S. (1967). Visual acuity in monkeys. *Vision Res.* **7**, 743–747.

GRIFFIN, D. R., HUBBARD, R. & WALD, G. (1947). The sensitivity of the human eye to infra-red radiation. *J. opt. Soc. Amer.* **37**, 546–554.

KOHLRAUSCH, A. (1931). Tagessehen, Dammersehen, Adaptation. *Handbuch der normalen und pathologischen Physiologie*. Bd. XII/2, pp. 1499–1594. Berlin: Springer.

LYTHGOE, R. J. (1932). The measurement of visual acuity. *M.R.C. Sp. Rep. Ser.*, No. 173.

McDOUGALL, W. (1904). The sensations excited by a single momentary stimulation of the eye. *Brit. J. Psychol.* **1**, 78–113.

SCHOUTEN, J. F. & ORNSTEIN, L. S. (1939). Measurement on direct and indirect adaptation by means of a binocular method. *J. opt. Soc. Amer.* **29**, 168–182.

SHLAER, S. (1937). The relation between visual acuity and illumination. *J. gen. Physiol.* **21**, 165–188.

TREZONA, P. (1960). The after-effects of a white light stimulus. *J. Physiol.* **150**, 67–78.

TSCHERMAK, A. (1929). Licht- und Farbensinn. *Handbuch der normalen und pathologischen Physiologie*. Bd. XII/1, pp. 295–499. Berlin: Springer.

VAN HOF, M. W. (1967). Visual acuity in the rabbit. *Vision Res.* **7**, 749–751.

WRIGHT, W. D. (1939). The response of the eye to light in relation to measurement of subjective brightness and contrast. *Trans. Illum. Eng. Soc.* **4**, 1–8.

PART 4

ELECTROPHYSIOLOGY OF THE RETINA

THE SENSORY MESSAGE

WHEN a receptor is stimulated, a succession of electrical changes takes place in its conducting nerve fibre. These changes are measured by placing electrodes on the nerve fibre and connecting them to a device for recording rapidly occurring potential differences, an *oscillograph*. A series of electrical variations, called *action potentials* or *"spikes"*, is obtained when the receptor is stimulated. The duration of an individual spike is very short (it is measured in milliseconds) and, as the strength of the stimulus is increased, the *frequency* of the discharge (the number of spikes per second) increases, but the size of the individual spikes remains unaltered (*"all-or-none" effect*).

Spike Potentials. Thus the message sent to the central nervous system is in a *code* consisting of a series of spike potentials apparently all with the same magnitude, the frequency being the variable that indicates how strongly the receptor has been stimulated. When the nerve fibres from different receptors are compared, e.g. touch, temperature, vision, hearing, the action-potentials are very similar in magnitude and time-course; and it has not been possible to correlate any given type of action potential with the type of sensation that it is mediating. In other words, it appears that the elements of the message, like the dots and dashes of the Morse code, are the same for all, whether they indicate a change in temperature or the flash of a light, and we must conclude that it is the places where the messages are carried in the brain—the *central terminations*—that determine the character of the sensation. The intensity of the sensation, as indicated above, is determined by the frequency of the spikes, i.e. the number that pass under our recording electrode in unit time.

Recording the Message. The accurate study of the events taking place when a receptor is stimulated demands that the recording electrode should be on or within the receptor or, if this cannot be achieved, on the sensory nerve that makes direct connection with this receptor. Moreover, if a recording from a nerve is to be made, the leads should be taken from a single fibre, otherwise the record of events taking place in the many thousands of fibres of a whole nerve would be a very confused and attenuated picture of the events taking place in the single fibres. With the vertebrate eye, these requirements are very difficult to meet, because the receptors—rods and cones—and the first sensory neurones, the bipolar cells, are buried in the depth of the retina; hence if we wish to record from a reasonably accessible nerve, namely the optic nerve, we will be recording from a second-order neurone, namely the ganglion cell. In its passage through the ganglion cell, the original message from the rod or cone and the bipolar cell has been submitted to a process of what the neurophysiologist calls *re-coding*; unnecessary parts of the message being dropped and other parts being accentuated at their expense. Thus, to understand what happens in the receptor cell, it would be a help if we could place our electrode on or within a rod or cone, or failing that, on a bipolar cell, which makes such intimate relationships with the rods and cones. For this reason the invertebrate eye, built on a simpler plan neurologically, was studied by Hartline; and it is to his pioneering studies that we owe so much of our knowledge of sensory mechanisms. Hartline studied the compound eye of the horse-shoe crab *Limulus*, the fibres of its optic nerve being, apparently, processes leading away from the retinula cells on which the light impinges. Thus the study of the electrical changes in these fibres brings us close to the electrical changes occurring directly in response to the impinging of light on a light-sensitive cell.

THE INVERTEBRATE EYE

Ommatidium

Essentially the compound eye of insects and crustaceans is an aggregation of many unit-eyes or *ommatidia*, each of these units having its own dioptric apparatus that concentrates light on to its individual light-sensitive surface, made up of specialized *retinula cells*. The facetted appearance of the compound eye is thus due to the regular arrangement of the transparent surfaces of these unit-eyes (Fig. 83*a*). The *ommatidium* is illustrated schematically in Fig. 83*b*; light is focused by the

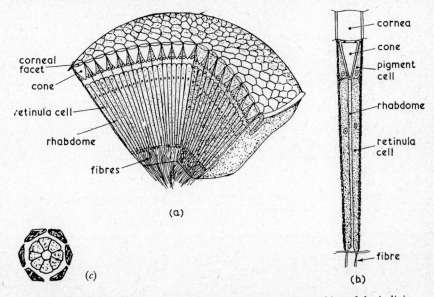

FIG. 83. The facetted eye. (*a*) Cut-away view showing the packing of the individual ommatidia; (*b*) longitudinal section of an ommatidium; (*c*) cross-section of ommatidium.

crystalline cone on to the highly refractile *rhabdom*, which constitutes an aggregation of specialized regions of the *retinula cells*. These retinula cells are wedge-shaped and are symmetrically arranged around the axial canal like the segments of an orange (Fig. 83*c*); the rhabdom is made up of the apical portions of these wedge-shaped cells, the separate portions being called *rhabdomeres*. As illustrated in Fig. 84, which shows a cross-section of the *Limulus* (horse-shoe crab) ommatidium, the regions of junction of the retinula cells are characterized by dense staining, so that a cross-section appears like the spokes of a wheel, the axle being the central canal. This canal contains the process of the *eccentric cell*; this is apparently a neurone, with its cell body squeezed between the retinula cells at the base of the ommatidium; its long spear-like process fills the central canal and may be regarded as a dendrite; the central end of the cell consists of a long process that passes out of the ommatidium in the optic nerve.

Eccentric Cell. The eccentric cell thus constitutes the nerve fibre of the ommatidium, whilst we may regard the retinula cells, with their specialized rhabdomeres, as the cells responsible for absorbing the incident light. Provisionally, then, we may say that the light-stimulus falls on the rhabdom; as a result of chemical changes taking place in this region, electrical changes are induced in the eccentric cell that culminate in the passage of impulses along the axon-like process of this cell.

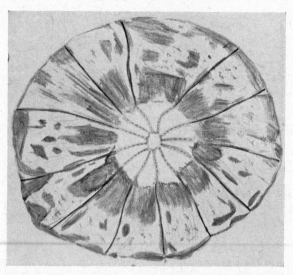

Fig. 84. Cross-section of ommatidium of *Limulus*, showing individual retinula cells with the rhabdomeres represented in white. (After Miller, *Ann. N.Y. Acad. Sci.*)

Retinula Processes. Actually each retinula cell also has a long proximal process leading out of the ommatidium in the optic nerve; consequently the optic nerve is made up of some 10–20 retinula processes to every one eccentric cell process. The significance of these retinula processes is not clear; they do not, apparently, carry action potentials, so that it would appear that the great majority of the "optic nerve fibres" of the compound eye are not nerve fibres at all.

Fig. 85. Oscillogram of the amplified potential changes in a single *Limulus* optic nerve fibre due to steady illumination of the eye. Intensity of illumination with upper record 10 times greater than for lower record. Full length of each record corresponds to 1 second. (Hartline, *J. Opt. Soc. Amer.*)

The electron-microscope has contributed a lot to the understanding of the structure of the rhabdomere; the spokes illustrated in Fig. 84 are apparently made up by innumerable interdigitating villi from the surfaces of contiguous retinula cells; they thus constitute a vast increase in the surface area of the retinula cells and it is here that the photosensitive pigment, responsible for absorbing the energy of the light-stimulus, is apparently concentrated.

Single Fibre Responses

In Fig. 85 are shown Hartline's oscillograms of the amplified potential changes in a single optic nerve fibre of *Limulus* when the eye is exposed to a steady illumination of two separate intensities; that for the upper record is ten times greater than that for the lower. The figure shows that, as with other receptors, the response to a continuous stimulus is intermittent, and that the effect of increasing the intensity of the stimulus is to increase the frequency of the action potentials, but not to modify the magnitude of the individual spikes.

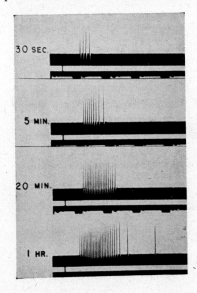

FIG. 86. The effect of dark-adaptation on the response of the eye of *Limulus* to a flash of constant intensity. As the eye dark-adapts, the frequency and duration of the response increase. (Hartline, *J. Opt. Soc. Amer.*)

FIG. 87. Responses to flashes of light of decreasing duration as indicated by the size of the break in the upper white line. Note that in the lower records the response occurs entirely after the stimulus; also the long latency in the lowest record. (Hartline, Wagner & MacNichol, *Cold Spr. Harb. Symp. quant. Biol.*)

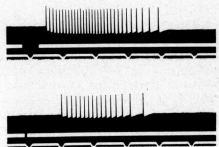

Dark-Adaptation. The phenomenon of dark-adaptation has already been described; in Fig. 86 are shown the electrical discharges of a *Limulus* single fibre in response to light stimuli of the same duration and intensity at intervals of 30 seconds, 5, 20 and 60 minutes after a light-adapting exposure of the eye to a bright light. It is quite clear the frequency of the response increases with the period of dark-adaptation. This suggests that the increase in sensitivity that occurs during a stay in the dark is a feature of the receptors, i.e. the rods and cones of the vertebrate eye, rather than of events in the central nervous system. As we shall see, however, there are many features of adaptation that indicate the intervention of higher nervous activity.

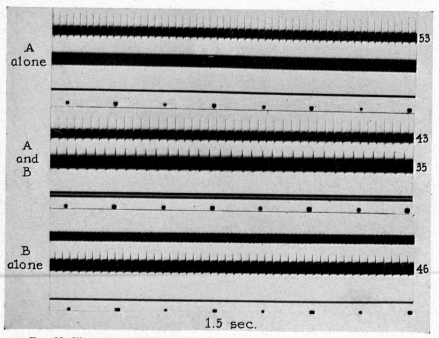

FIG. 88. Illustrating inhibition of discharge in one ommatidium by excitation of a neighbouring ommatidium. Ommatidium A, when illuminated alone, gives a frequency of 53 spikes/sec; ommatidium B gives a frequency of 46/sec; when A and B are illuminated together, A has a frequency of 43, and B one of 35/sec, i.e. they have mutually inhibited each other. (Hartline, Wagner & Ratliff, *J. Gen. Physiol.*)

Latency. When the stimulus is a brief flash of light, as in Fig. 87, the discharge occurs entirely after the stimulus is over. The weaker the flash the longer this latency, as Fig. 87 shows; the latency represents the time elapsing between the absorption of light by the rhabdom and the induction of an electrical change in the nearby eccentric cell, i.e. it is the time required for some photochemical change to take place in the rhabdom. The stronger the flash of light the more rapid will be this photochemical reaction and thus the shorter the latency.

Inhibition. It was considered in the early work on ommatidia that they were entirely independent of each other so far as the discharges in the optic nerve fibres were concerned; recently, however, Hartline, Wagner & Ratliff have shown that adjacent ommatidia may inhibit each other's response. An example is shown by Fig. 88. This inhibition must have resulted from an interaction between the responses in the nerve fibres from individual ommatidia, taking place presumably in the plexus of fibres just outside the ommatidia.

This plexus, or *neuropil*, has been examined in the electron microscope by Whitehead & Purple (1970); regions of close contact between processes were common, and these were associated with the presence of electron-dense material and vesicles in the cytoplasm suggestive of a synaptic arrangement. In the more complex eyes of vertebrates we may expect inhibitory effects to be more marked because of the greater possibilities of interaction between individual pathways resulting from the well defined system of interneuronal connections at the level of the bipolar and ganglion cells.

SIGNIFICANCE OF INHIBITION

The importance of inhibition in the sensory pathway may not be immediately evident but will become so when we discuss such phenomena as flicker and visual acuity; for the moment we need only note that it is just as important to prevent a discharge—inhibit—as to initiate one—excite. Thus there is some evidence that a single cone can be excited separately by a spot of light, although we know on optical principles that the image of this spot must cover several cones. In fact several cones are stimulated, in the sense that light falls on them, but only one may actually be excited because the activities of adjacent ones are inhibited by the one that receives the most light, i.e. the cone in the centre of the image of the spot of light where the image is most intense. It is through inhibition then, that unnecessary responses fail to reach the higher levels of the central nervous system. The example shown in Fig. 88 reveals inhibition at the lowest level; in the vertebrate retina it may occur at the level of the bipolar cells and ganglion cells and higher still in the geniculate body and cortex.

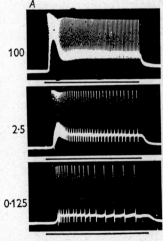

FIG. 89. Generator and spike potentials from impaled eccentric cell of *Limulus* ommatidium. Duration of illumination indicated by black lines below records. Intensity of illumination decreases from above downwards. Note that the frequency of the spikes increases with the size of the generator potential. (Fuortes, *J. Physiol.*)

Mechanism of Activation of the Sensory Neurones

Generator Potential. The electrical events following stimulation have been described as a series of spikes, brief changes in potential taking place across the nerve fibre's membrane. Those concerned with the way in which light, or some other stimulus, can provoke this electrical change have studied the events leading up to this spike discharge; and in general it appears that the common feature is the establishment in a sensory cell of what is called a *generator potential;* this is a more slowly developing and longer lasting electrical negativity of the cell than the brief spike, and is essentially a "preliminary change" that, if carried far enough, sets off the discharge. The development of this generator potential in the *Limulus* eye was followed by Fuortes, who placed an electrode inside the eccentric cell of an ommatidium; when light fell on this ommatidium the energy was presumably taken up by the rhabdom which, as we have seen (p. 152) surrounds the process of the eccentric cell. As a result of the chemical changes taking place in the rhabdom, the eccentric cell developed a generator potential, becoming negative internally; superimposed on this negativity were the much more rapid and abrupt changes that we call the spikes. This is illustrated by Fig. 89 where the ommatidium has been stimulated with three intensities (100, 2·5 and 0·125 units); the generator potential is indicated by the rise in the baseline, and the spikes by the strokes taking off from this; with increasing intensity the generator potential increases, and the spike frequency too.

RETINULA AND ECCENTRIC CELLS

Behrens & Wulff (1965) were able to insert microelectrodes into two cells of the same ommatidium and, after recording their responses to illumination, they injected dye so as to be able to identify the types of cell from which records were made. In general, as others had found before, the responses fell into two classes: (*a*) a slow wave of depolarization with superimposed small spikes, and these could be identified with retinula cells; (*b*) a slow wave with large spikes identified with the eccentric cell; they showed that the retinula responses always preceded, by some 7 to 30 msec, the eccentric cell response; furthermore, the retinula responses, including the small spikes, were always synchronous with each other when two retinula cells were recorded from. Thus the retinula cell's slow potential wave seems to be the initial reaction to light; since the small spikes in this cell are not propagated along its nerve fibre, we may look on these as transmitted effects from the large spikes developed in the eccentric cell. It is possible, then, that the effect of light on the retinula cells is, after some photochemical change, to develop a slow depolarization which acts as a generator potential for the development of a depolarization of the eccentric cell, which finally leads to spike activity. The fact that the electrical changes recorded from the retinula cells of the same ommatidium were always synchronous, indicates an electrical coupling between them, as suggested by Tomita, Kukuchi & Tanaka (1960) and proved by Smith, Baumann & Fuortes (1965).

SUMMATION

With a subthreshold stimulus there is only a small generator potential, too small to initiate spikes. However, this subthreshold generator potential will add with a second potential, so that one factor in the summation of subthreshold stimuli, observed in psychophysical studies, must be the ability of the generator potentials to add, provided, of course, that the interval between stimuli is not too long.

Chemical Transmitter. According to Fuortes and Rushton, the electrical events can be best described on the assumption that the retinula cells (rhabdom) liberate a substance (transmitter) after absorbing light, and it is this chemical transmitter that activates the sensory neurone, in this case the eccentric cell.*

VERTEBRATE RESPONSES

It is now time to turn our attention to the messages developed in the vertebrate retina, a system that poses more difficulties to the experimenter because the easily accessible fibres are those from the ganglion cells, whose responses are only distantly related to the initial effects of light on the receptors. It will be recalled that the receptors synapse in the outer plexiform layer with the bipolar and horizontal cells; the bipolar cells, in their turn, synapse in the inner plexiform layer with the ganglion cells; and it is the axons of these, emerging from the eye, that constitute the optic nerve. Logically it would be reasonable to begin this account with the responses in the receptors, but since it is only very recently that such responses have been unequivocally identified, the more historical approach is preferable.

Frog Optic Nerve

Single fibres of the frog's optic nerve were first studied by Hartline. In general, three types of fibre were found, differing in their responses to a flash of light. In the first place there is a group of fibres (representing on the average some 20 per cent of the total) which behave in essentially the same way as *Limulus* fibres, showing a response which starts almost immediately after the light has been switched on and ceasing when the light has been switched off (Fig. 90, A); these are called ON-fibres; secondly, there is a group (some 50 per cent of the total) which respond with an initial outburst of impulses when the light is switched on and again when the light is switched off (Fig. 90, B); these are called ON-OFF-fibres; during the intermediate period no impulses are recorded. Finally, there is a group of fibres (some 30 per cent of the total) which respond only when the light is switched off (Fig. 90, C); these are called OFF-fibres. The group B, responding to the switching on and off of the light, shows a remarkable sensitivity to changes in the intensity of the stimulus and to the slightest movement of the image on the retina, the greater the change in intensity and the more rapid and extensive the movement of the image, the greater the frequency of the resultant discharge. These OFF-discharges probably represent a release from *inhibition*, the ganglion cells giving rise to them presumably being subjected to an inhibitory discharge by bipolar and other retinal neurones (horizontal and amacrine cells?) during the period of exposure to light; when the light is switched off this inhibition ceases and the ganglion cells discharge, the phenomenon being called more generally "inhibitory rebound".

Mammalian Eye. Hartline's work was extended by Granit, and more recently by Kuffler and Barlow, to the mammalian eye where the same pattern of ON-OFF-, and ON-OFF-responses was revealed, although the phenomena were even

---

* In the thoroughly dark-adapted eye of *Limulus*, Werblin (1968) has examined discrete potentials in the retinula cells that occur in response to a weak light stimulus; these are of two types, namely small discrete "quantum bumps", perhaps the result of absorption of a quantum of light, and larger potentials that have many of the features of regenerative potentials described in nerve, although they are not abolished by tetrodotoxin. Werblin suggests that this regenerative type has its origin in the retinula cell and represents an amplification of the "quantal bump" permitting the eventual activation of the eccentric cell.

more complex. The mammalian retina shows a background of activity in the dark, so that ON- and OFF-effects were manifest as accentuations or diminutions of this normal discharge. In general, ON-elements showed an increased discharge when the light was switched on, and an inhibition of the background discharge when the light was switched off. Similarly, an OFF-element showed an inhibited discharge when the light was switched on but gave a powerful discharge when the light was switched off. The OFF-effect is thus quite definitely a *release from inhibition*.*

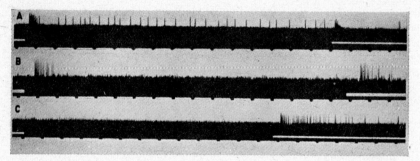

FIG. 90. Discharges in a single optic nerve fibre of the vertebrate (frog) eye. Type A: Pure ON-element responding at ON with a discharge maintained during illumination. Type B: ON-OFF-element responding with discharges at ON and OFF but silent during the main period of illumination. Type C: Pure OFF-element responding only at OFF. Signal of the exposure to light blackens out white line above time-marker. Time marked in 0·2 second. (Hartline, *J. Opt. Soc. Amer.*)

Receptive Fields

The number of fibres in an optic nerve is of the order of one million, whilst the number of receptors is of the order of 150 million. This means, in neurophysiological terms, that there is a high degree of *convergence* of the receptors on the ganglion cells, so that a single optic nerve fibre must have connections with many bipolar cells which themselves have connections with many rods and cones. In consequence, the area of retina over which a light-stimulus may evoke a response in a single optic nerve fibre may be quite large; the *receptive field* of a single fibre may be plotted by isolating a single fibre and, with a small spot of light, exploring the retina for those parts that will cause a discharge in this fibre. As Fig. 91 shows, the field increases with the strength of stimulus, so that in order that a light-stimulus falling at some distance from the centre of the field may affect this particular fibre it must be much more intense than one falling on the centre of the field. This shows that some synaptic pathways are more favoured than others.

Summation. As we shall see, this convergence is the anatomical basis for the phenomena of summation; thus a single spot of light falling on the periphery of

* In psychophysical terms the beginning of the ON-discharge may well correspond to the primary sensation of light when the eye is exposed to a flash; the so-called Hering after-image, that follows rapidly on the primary sensation, would then correspond to the activation of the ON-OFF-elements, whilst the dark-interval following this would represent activation of the OFF-elements. Finally, the appearance of the Purkinje after-image, rather later, could be due to a secondary activation of the ON-elements described by Grüsser & Rabelo (1958).

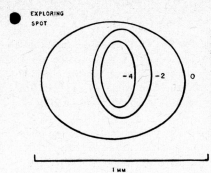

FIG. 91. Showing the retinal region with which a single optic nerve fibre makes connections. Each curve encloses the retinal region within which an exploring spot (relative size shown at upper left) produced responses in the fibre at an intensity of illumination whose logarithm is given in the respective curve. Note that the greater the stimulus the greater is the receptive field of the fibre. (Hartline, *J. Opt. Soc. Amer.*)

the receptive field may be too weak to provoke a discharge in the ganglion cell; several spots each too weak to excite will, if presented together in different regions of the field, provoke a discharge. In neurophysiological terms we may say that the ganglion cell requires a certain bombardment by discharges from the bipolar cells, with which it makes synaptic contacts, before it discharges; this bombardment can be provided by a few bipolar cells discharging strongly or by many bipolar cells discharging more weakly. The electrical basis for this summation of effects is essentially the same as that for the summation in the receptors; the ganglion cells develop generator potentials which are this time called *synaptic potentials*; they are not all-or-none, and so can add their effects and ultimately build up to a sufficient height to initiate a spike discharge.

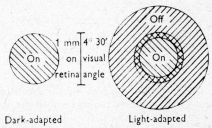

FIG. 92. Receptive field of ON-centre unit.

Dark-adapted   Light-adapted

Centre and Periphery. A careful study of the receptive fields in the cat's retina by Barlow, Fitzhugh & Kuffler showed that they were not so simple as had been thought, the more peripheral part of the field giving the opposite type of response to that given by the centre. Thus, if at the centre of a receptive field the response was at ON, the response farther away, in the same fibre, was at OFF, and in an intermediate zone it was often mixed to give an ON-OFF-element as illustrated in Fig. 92. In order to characterize an element, therefore, it was necessary to describe it as an *ON-centre* or *OFF-centre* element, meaning thereby that at the centre of its receptive field its response was at ON or OFF respectively. By studying the effects of small spot-stimuli on centre and periphery separately and together, Kuffler demonstrated a mutual inhibition between the two.

EFFECT OF ADAPTATION

The striking feature of this organization of the receptive field is that it was altered on the dark-adapted state, as illustrated in Fig. 92, the surrounding area of opposite activity becoming ineffective in the dark-adapted state. In this sense, therefore, the receptive field shrinks during dark-adaptation, but as it is a reduction in inhibitory effect between centre and periphery it means, in effect, that the effective field actually increases, i.e. the regions over which summation can occur.

Non-Response to Uniformity. This opponence between centre and periphery of the receptive field seems to be a nearly universal character of the organization of the retinal response, and is continued to further stages in the visual pathway; thus the geniculate neurones exhibit the opponence to the point that centre and surround usually balance each other's effects exactly, so that uniform illumination of the retina usually fails to excite geniculate cells.* The physiological significance of this opponence becomes clear if we appreciate that under ordinary conditions of illumination all receptors might be expected to be responding to light, and if these responses were all transmitted to the central nervous system along the million odd optic nerve fibres, this would mean a tremendous bombardment of the higher centres with information; the centre-surround organization means that for the great majority of ganglion cells uniform illumination of their receptive field will produce no, or only a small, effect, the influence of light on the receptors that lie in the centre of the field being balanced by the opposite influence on the receptors that lie in the surround. On the other hand, a spatial discontinuity in the field, given by a bar on a white field, for instance, would activate centre and surround to different extents and so give rise to a message that would enable the perception of form. To generalize father, we may say that the receptive system is organized so that the response to uniformity is small, whilst that to change is large; the uniformity applies to both temporal and spatial features of the stimulus; adaptation reduces the response to a light—this is the temporal feature—whilst the failure to respond to uniform illumination, mentioned above, represents the spatial feature.

Quantitative Aspects of Centre-Surround Interaction. Rodieck & Stone (1965) measured the responses of centre and periphery of cat ganglion cell receptive fields to small spots of light, using, as a measure of the response, the summated spikes during the interval of summation; the summation was carried out over small intervals of time and the result presented as a response histogram (Fig. 93). Special attention was paid to the sizes of receptive fields, summation within the field, and interaction between centre and surround. In general, the fields were radially symmetrical, and the maximal response of the central region was usually at the geometrical centre, but not always. The sizes of the central areas varied from 0.5 to 4 degrees. Beyond the centre, the strength of the surround-response increased quickly to a maximum and then gradually decreased until no response

---

* The anatomical basis for the centre-surround organization of the receptive field may well be that the centre corresponds with the dendritic field of the ganglion cell whilst the surround represents the influence of neighbouring horizontal cells. Brown & Major (1966) found that the dendritic fields of cat ganglion cells fell within two groups with diameters of 70–200 μ and 400–700 μ, and suggested that these corresponded with the centres of the receptive fields. Leicester & Stone (1967) found that the commonest centre-surround fields had centres with diameters of 110–880 μ, whilst dendritic fields of the deep multidendritic type of ganglion cell were in the range 70–710 μ, sufficiently close to make the suggestion of Brown & Major reasonable.

could be obtained, the limiting size of the peripheral field being determined by the intensity of the spot stimulus. When the retina was stimulated by a large uniform field, the response of a given unit was that of its centre, i.e. surround and centre did not neutralize each other completely; as Barlow *et al.* had found, dark-adaptation abolished the surround effect. The ON-OFF region of a field was usually confined to a narrow ring; Fig. 93 shows the response histograms to a small

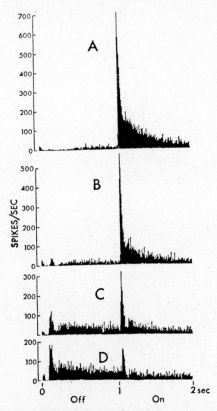

FIG. 93. Illustrating the transition from a pure ON-response in A to an ON-OFF response in D. A small spot of light was flashed for 1 sec off and 1 sec on. A shows the prominent ON-response of the centre with suppression of background during the OFF-phase. At B to D the flashes were on more and more peripheral parts of the receptive field. (Rodieck & Stone, *J. Neurophysiol.*)

spot of light flashed 1 second off and 1 second on; it is an ON-centre field and the spot is projected first on the centre (A) and then more and more peripherally, and the transition from a pure ON-response to the ON-OFF response in D is clearly seen. A moving spot was a most effective stimulus, and the response to a small movement varied with the initial position in the receptive field and the direction of the movement; for an ON-centre unit, centripetal movement of the spot anywhere within the border of maximum OFF-surround caused excitation and outside it, inhibition. The converse was true for OFF-centre fields. Only to this extent was the response to movement direction-sensitive; unlike the situation in the rabbit, units specifically sensitive to a given direction in the visual field were not found. Summation of responses, both spatial and temporal, was found to be virtually complete.

CALCULATED RESPONSES

Rodieck & Stone, on the basis of studies of this sort, computed the effects of stimulating a receptive field by a moving pattern; to do this it was necessary to make assumptions as to the way in which the different regions of a given field interacted, i.e. the manner in which similar portions summated and opposite portions subtracted. They accepted the suggestion of Wagner, MacNichol & Wohlbarsht (1963) that the surround field, did, in fact, extend into the centre, so that the response to a spot on the centre was really the algebraic sum of centre and "surround" responses; the situation is illustrated in Fig. 94, the curves

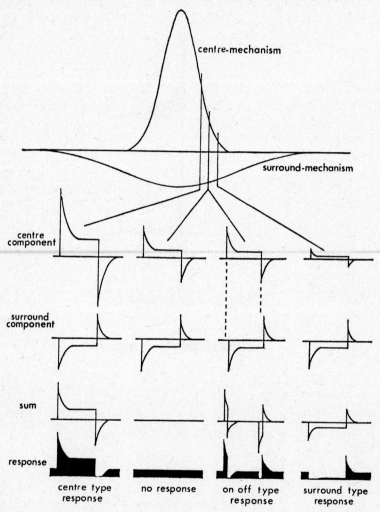

FIG. 94. At the top the theoretical responses of the centre and surround mechanisms are depicted, in this case an ON-centre unit with positive response in the centre and a smaller negative response over the whole field. Near the centre of the field, as the left column of hypothetical responses shows, the centre mechanism predominates; a little farther out, centre and surround cancel completely; still farther out the response becomes ON-OFF, and in the extreme periphery of the field the response is that of the surround only. (Rodieck & Stone, *J. Neurophysiol.*)

indicating the relative effectiveness of stimuli falling on different portions of the centre and surround. The centre mechanism has a steep outline whilst that of the surround is altogether broader; the diagrams below indicate the hypothetical results of interaction when stimuli fall in more and more peripheral parts of the field, leading to typical centre-type response, no response, ON-OFF response, and surround-type response. The response of a unit to the movement of a contoured body passing across it will be given by the responses to movement of its leading and trailing edges, and Rodieck (1965) was able to predict these, on the basis of his simple theoretical treatment. Thus Fig. 95 illustrates the calculated responses to moving bars of increasing width across the receptive field for two types of unit, those giving a centre-activated (CA) and centre-suppressed (CS) response; these agree remarkably well with the experimental responses (Rodieck & Stone, 1965a).*

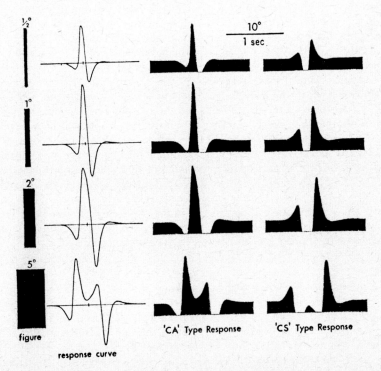

Fig. 95. Calculated response curves for moving bars of increasing width across the receptive field of a cat's ganglion cell. (Rodieck, *Vision Res.*)

Rod and Cone Interaction. Psychophysical studies suggest a great deal of independence of rod and cone activities when thresholds are measured; when the sensation of brightness is studied, however, the Purkinje phenomenon (p. 256) indicates that the rods may modify colour sensation, and this is presumably through ganglion cells that make neural connections with both rods and cones, possibly through the horizontal cells. Many studies of ganglion cells in lower mammals have indicated a convergence of the two types of receptor; in the monkey, too, Gouras & Link (1966) demonstrated this, and the interaction of rod and cone

---

* Varela & Maturana (1970) have shown that the latency of the centre is always less than that of its surround; moreover, it is possible to abolish the effect of stimulating one region by a stimulus applied to the other if the timing is chosen correctly.

influences at this level; thus the shorter latency of the cone response permits its differentiation from that of the rod, and by using red flashes to excite cones, and violet flashes to excite rods, intervals between the two may be found such that the rod response is occluded by the cones or *vice versa*. In general, with both threshold and suprathreshold stimuli, rods and cones do not contribute simultaneously to a ganglion cell's response; because the cone system is so much faster it manages to control ganglion cell function whenever stimuli are sufficient to excite it.

Directional Sensitivity. As we shall see, neurones within the cerebral cortex were early shown to be sensitive to the direction in which a moving spot passed over the retina; that a single neurone may respond so selectively to the retinal stimulus requires a considerable degree of integration and selection during the transmission of information; thus it has to ignore messages coming from receptors not on this line of movement, and has to take account of the fact that the stimuli in receptors along this preferred direction are sequential, and so on. Barlow, Hill & Levick (1964) showed that, already at the ganglion cell level in the rabbit, this discriminative power is present; they observed many cells with the typical centre-surround arrangement with either ON- or OFF-responses in the centre and OFF- or ON-responses in the surrounds. Others, however, resembled the ON-OFF units of the frog over the whole field, and these* showed directional sensitivity in the sense that their maximal response occurred when a moving spot traversed a certain "preferred" direction, whilst there was no response at all to movement in the opposite, "null", direction. That inhibition was concerned was suggested by the observation that movement in the null direction would inhibit spontaneous activity; again, very slow motion in the null direction would evoke a response. Thus we may imagine that the bipolar cells related to a series of receptors, located along a certain direction in the retina, all feed their responses into a ganglion cell; these responses are excitatory and inhibitory, the inhibitory effect being transmitted, after a small latency, to the ganglion cell.

INHIBITION

The picture that fitted best with the experimental facts is illustrated by Fig. 96. Activity roused at either ON or OFF in the receptors C, B, etc., is passed laterally to inhibit the bipolar cell B . $\sim$C', A . $\sim$B' and, so on. This inhibition prevents the excitatory effect on B and A from getting through when the motion of the spot is in the null direction, but arrives too late if motion is in the preferred direction. If the movement is very slow we can imagine easily that the inhibition may not last sufficiently long to inhibit the responses of adjacent elements. On this basis, two stationary stimuli, falling on regions of the retina separated along the preferred direction in sequence, will give a large response when the sequence corresponds with motion along the preferred direction and a smaller response for the null direction. In fact this was found, provided the separation was small; in the null direction sequence, the second stimulus could give no discharge at all (see also, Michael, 1968). Barlow & Levick (1965) postulated that the inhibitory action was mediated by the horizontal cells, and a study of Werblin (1970) on the responses of the different retinal cells—receptors, bipolars, etc.—to moving spots of light are consistent with this inhibitory role of the horizontal cells.

---

* In the visual streak area of the retina, characterized by densely packed ganglion cells, Levick (1967) found that the ON-OFF-units were not all direction-sensitive; many were local-edge detectors (p. 165).

THE PREFERRED DIRECTIONS

Oyster & Barlow (1967) determined the preferred directions in many units of the rabbit retina and found that they fell into four groups, and these corresponded accurately with the directions of movement of the fixation axis when the four rectus muscles were allowed to act separately; these were not purely horizontal and vertical, but inclined to these by a few degrees. Thus if the retina is to act as a servo-system correcting movements of the eyes, the direction-sensitive units fulfil an obvious role here.*

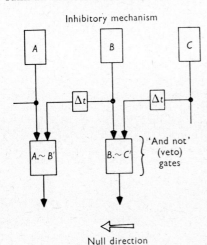

Inhibitory mechanism

Null direction

FIG. 96. Hypothetical mechanism for direction-sensitive ganglion cell. The receptor elements A, B, C have a direct excitatory influence on bipolar cells and an inhibitory action on adjacent bipolar cells in the null direction, the inhibitory action occurring after a latency of Δt. These are said to act as "And not" veto gates. A spot of light passing in the null direction, i.e. from right to left, would tend to inhibit its influence on successive receptors. (Barlow & Levick, *J. Physiol.*)

Other Units. Levick (1967) has examined the nature of the responses of several types of ganglion cell that are different from the typical concentric ON-OFF centre-surround type. One of these he described as *orientation-selective*; these would respond to a rectangle of light or dark, only when this was oriented in a definite manner; the response was usually evoked by movement of the target but not necessarily so, a stationary rectangular patch of light causing a discharge only when correctly oriented. *Local edge detectors* had a small receptive field; although no manifest effects could be obtained by stimulating the surround, there was no doubt that it was, in fact, inhibitory. The receptive field was of the ON-OFF type, and excited by contrasting borders within the receptive field, and inhibited by similar stimulation of the surrounding area. Finally *uniformity detectors* gave a uniform discharge only when the stimulus conditions were constant, e.g. constant illumination; any change in these caused a reduction in the steady discharge; thus moving targets caused sustained suppression; an increase in illumination caused a suppression that disappeared when the new level was maintained long enough.

More Peripheral Responses

Attempts to record from more peripheral parts of the retina by puncturing receptors and bipolar cells, or by placing external electrodes very close to them or

* These direction-sensitive ganglion cells are presumably analogous in some respects to Grüsser *et al.*'s (1967) Class 2 movement-sensitive units in the frog's retina; a number of quantitative aspects of their responses to moving stimuli, e.g. area of the stimulus, angular velocity, and so on, have been examined, and a model, based on convergence of excitatory and inhibitory influences from bipolar cells on to the ganglion cell, has been constructed to account for the experimental findings (Grüsser, Finkelstein & Grüsser-Cornehls, 1968). An interesting attempt to relate the electrophysiological findings to the frog's fly-catching activities is contained in the paper by Butenandt & Grüsser (1968).

their synapses in the plexiform layers, have been numerous, but until recently the identification of the source of the recorded electrical changes has been equivocal. Some of these studies, e.g. the S-potentials,* will be referred to later under the heading of colour vision, and others in relation to the interpretation of the electro-retinogram; and we may profitably concentrate on a recent study of Werblin & Dowling (1969) on the retina of the mudpuppy, *Necturus*, the *neuronal* organization of this retina having been exhaustively analysed by Dowling & Werblin (1969).

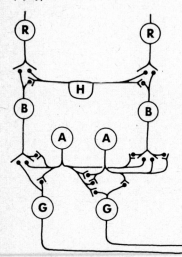

Fig. 97. Schematic "wiring diagram" of the mud-puppy retina. R, receptor; H, horizontal cell; B, bipolar cell; A, amacrine cell; G, ganglion cell. Note that in this retina the horizontal cell influences both receptor and bipolar cell, and the amacrine cell influences both bipolar cell and ganglion cell. (Dowling & Werblin, *J. Neurophysiol.*)

"Wiring Diagram". The synaptic structures and relations between retinal cells in *Necturus* are similar to those of the primate retina, but the relations between horizontal cells and receptors and bipolar cells would appear to be different. The "wiring diagram" of the *Necturus* retina is illustrated schematically in Fig. 97, and it will be seen that the horizontal cell is not a mere mixer of rods and cones, since it forms presynaptic relations with bipolar cells, i.e. it transmits from receptor to bipolar cell and thus contributes to vertical transmission. In a similar way, the amacrine cell not only links bipolars horizontally but also transmits from bipolars to ganglion cells. Thus there are two pathways from rod to bipolar, and two from bipolar to ganglion, either direct or through a horizontal or amacrine cell.

Recording. Microelectrodes were inserted into different regions of the retina and intracellular recordings of the responses to light stimuli were made; the stimuli consisted of a small spot or of an annulus, either 250 μ or 500 μ wide, either singly or in combination with the spot. In this way the existence, or otherwise, of opponent responses typical of those decribed for ganglion cells, could be ascertained. After the recordings had been made, the position of the electrode was identified by ejecting Niagara blue from the electrode and subsequent histological examination. The results are summarized in Fig. 98.

* Interest in the S-potentials transcends the phenomena of colour vision because the horizontal cells that give rise to it collect responses from all classes of cones; Naka & Rushton (1966a, b, c) and Naka (1969) have carried out an exhaustive analysis of the S-potentials with a view to identifying the nature of the inputs from cones and the "cross-talk" between them.

Receptor Response. The striking feature of the receptor response is that it consists of a hyperpolarization of the cell, rather than a depolarization as found in the invertebrate receptor. The response is graded, like the invertebrate generator potential, but there is no all-or-none spike activity. Annular stimulation did not affect the element, nor did it affect the response to a spot stimulus; thus there was no opponent organization of centre and surround, as found in the ganglion cell.

Horizontal Cell. This cell's response, too, is hyperpolarizing, and comes on with a longer latency than that of the receptor; it is graded and does not lead to spikes, and the effects of light are summated over quite a wide field since a $500\ \mu$ annulus of light gave the same response as a $250\ \mu$ annulus containing the same number of quanta of energy (Fig. 98).

Bipolar Cell. Once again, this responds with a slow potential change, graded according to the strength of the stimulus and giving no spikes.* The receptive field, however, is concentrically organized in an opponent fashion so that if the centre hyperpolarizes, the periphery causes depolarization, and *vice versa*. In the middle column of Fig. 98 the small annulus stimulated both centre and periphery of the receptive field so that both hyper- and de-polarizing responses were observed; at the right the large annulus was combined with the central spot; now the central spot has been suppressed and the response is purely depolarizing. The central response always preceded the antagonizing surround response; hence when both centre and surround are illuminated, and then the intensity is changed, there is always a transient central response even if the contrast between centre and surround is unchanged.

Amacrine Cell. Responses in these are recorded from the inner plexiform layer; they consist typically of one or two spikes superimposed on a transient depolarizing potential (Fig. 98). The transient responses occurred at both ON and OFF. In these cells the effect of variation of intensity of the stimulus was essentially a change in the latency, the number of spikes being independent of the intensity. In some units the receptive field was organized in an opponent fashion, but in others the responses were uniform over the whole area.

Ganglion Cells. These responded with a spike discharge taking off from an initial depolarization; the frequency of discharge was proportional to the degree of membrane depolarization. ON-, OFF- and ON-OFF-responses were obtained in different cells. Centre-surround antagonism was also a feature of some cells and was similar to that of the bipolar cells; moreover, the extents of the fields were very similar.

Retinal Organization. The central part of the receptive field of the bipolar cell is about $100\ \mu$ wide and is surrounded by an annulus of about $250\ \mu$. The dendritic field of a bipolar cell in *Necturus* retina is about $100\ \mu$ in diameter, so that the large annular area, stimulation of which gives responses in a single bipolar cell, must transmit its effects through another type of cell that synapses

---

* Kaneko and Hashimoto (1969) recorded spike activity from cells in the inner nuclear layer of the carp's retina, which they attributed to bipolar cells; Murakami & Shigematsu (1970) have also observed these, in the frog retina. Tetrodotoxin, which abolishes spike activity but not slow potential changes, allowed some form of transmission to ganglion cells in response to light, so that it was argued that the bipolar cells could transmit their slow potential changes to the ganglion cell as well as their spike activity. It may be, however, that the spike activity occurring normally was, in fact, recorded from amacrine cells, as indicated by Werblin & Dowling. Certainly Kaneko (1970) found no spike activity in bipolar cells of the goldfish retina.

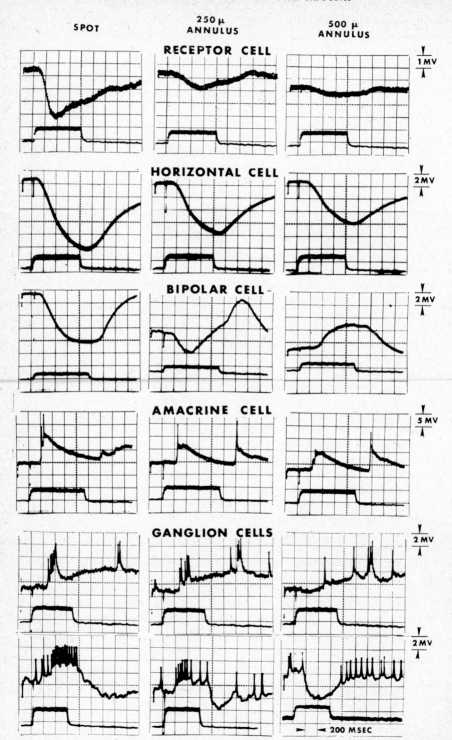

FIG. 98. *See caption opposite.*

with the bipolar. This other type is presumably the horizontal cell; this receives influences from the receptors in the surrounding annulus and transmits them to bipolar cells connected with the central receptors, exerting an opposite type of action; thus, if the central receptors hyperpolarize the bipolar, the horizontal cell depolarizes it. Hence it is the horizontal cells that organize the contrast-detection, as exhibited in the centre-surround antagonism. The further transformation of the message, which gives it its transient character, is probably done through the amacrine cells; these respond transiently, by contrast with the more sustained responses of bipolar cells, and they thus impose a transient type of response on the ganglion cells; the ganglion cells, driven in this way, presumably are able to carry information about *changes* in the visual field, whilst other ganglion cells, driven by bipolar cells directly, would be better able to carry information about the static character, i.e. contrast. These principles are illustrated by Fig. 99.

THE ELECTRORETINOGRAM

The electroretinogram* (ERG) is a record of comparatively slow changes of potential that take place across the retina in a radial direction when light falls on it. As classically studied in the intact eye, one electrode is placed on the cornea whilst the other, indifferent, electrode may be placed in the mouth or, in experimental animals, on the optic nerve. A better record may be obtained in experimental animals by placing the active electrode in the vitreous body. It is essentially because the retina is built up of large numbers of extremely well orientated units, arranged in layers, that the successive states of depolarization or hyperpolarization in these units lead to definite changes of polarity across the whole structure.

* The "retinal currents" now called the ERG were first described by Holmgren in 1865 but were independently discovered by Dewar and McKendrick. Kühne & Steiner (1880, 1881) showed that the same electrical changes could be obtained from the retina removed from the eye, whilst the remaining sclera, choroid and pigment epithelium were unresponsive.

FIG. 98. Responses of different retinal cells to a spot-stimulus (extreme left) and to annular stimuli of 250 μ and 500 μ radius. Receptors have relatively narrow receptive fields, so that annular stimulation evokes very little response. The horizontal cell responds over a broader region of the retina, so that annular illumination with the same total energy as the spot (left column) does not reduce the response significantly (right columns). The bipolar cell responds by hyperpolarization when the centre of its receptive field is illuminated (left column). With central illumination maintained (right trace; note lowered base line of the recording and the elevated base line of the stimulus trace in the records) annular illumination antagonizes the sustained polarization elicited by central illumination, and a response of opposite polarity is observed. In the middle column the annulus was so small that it stimulated the centre and periphery of the field simultaneously. The amacrine cell was stimulated under the same conditions as the bipolar cell, and gave transient responses at both the onset and cessation of illumination. Its receptive field was somewhat concentrically organized, giving a larger ON-response to spot illumination, and a larger OFF-response to annular illumination of 500μ radius. With an annulus of 250μ radius, the cell responded with large responses at both ON and OFF. The ganglion cell shown in the upper row was of the transient type and gave bursts of impulses at both ON and OFF. Its receptive-field organization was similar to the amacrine cell illustrated above. The ganglion cell shown in the lower row was of the sustained type. It gave a maintained discharge of impulses with spot illumination. With central illumination maintained (right column) inhibited impulse firing for the duration of the stimulus. The smaller annulus (middle column) elicited a brief depolarization and discharge of impulses at ON, and a brief hyperpolarization and inhibition of impulses at OFF. (Werblin & Dowling, *J. Neurophysiol.*)

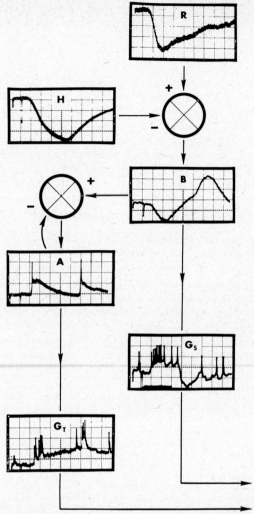

FIG. 99. Summary diagram of synaptic organization of the retina. Transforma-
tions taking place in each plexiform layer are represented here by summing
junctions. At the outer plexiform layer the direct input to the bipolar cell from the
receptor is modified by input from the horizontal cells. At the inner plexiform
layer the bipolar cell drives some ganglion cells directly. These ganglion cells
generate a sustained response to central illumination, which is inhibited by
additional annular illumination: and thus these cells follow the slow, sustained
changes in the bipolar cell's response. Bipolar cells also drive amacrine cells, and
the diagram suggests that it is the amacrine-to-bipolar feedback synapse that converts the sustained bipolar response to a transient polarization in the amacrine
cell. These amacrine cells then drive ganglion cells which, following the amacrine
cell input, respond transiently. (Werblin & Dowling, *J. Neurophysiol.*)

Typical Record. In Fig. 100 is shown a typical example; the first sign of activity
is shown as an *a*-wave corresponding to a decrease in the normal positivity of the
corneal electrode; this is conventionally recorded as a down-stroke on the record,
and is called a *negative* wave; it lasts less than 0·1 second. The *a*-wave is followed

by a reversal in polarity so that the cornea now becomes strongly positive, and this gives the sharp upward *b*-wave conventionally described as a *positive* wave; the *c*-wave is a much slower rise in corneal positivity and is terminated, when the light is switched off, by the *d*-wave, or OFF-effect.

Adaptation. According to the type of retina, and the state of adaptation of the eye, the records are qualitatively different; and this is because of the predominance of cones or rods in producing the net record. Thus, when the cones dominate the record, e.g. in the light-adapted mixed retina of the frog or the pure cone retina of the squirrel, the slow rise of the *c*-wave disappears, and the *a*-wave and OFF-effect become prominent. In the dark-adapted state, or in the virtually pure rod retinae of many nocturnal animals, the *a*-wave is not so pronounced whilst the *c*-wave is prominent.

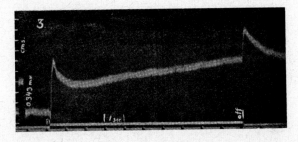

Fig. 100. Typical electroretinogram (ERG). Lowest record, time scale; middle record, stimulus; top record, galvanometer deflection. (Hartline, *Am. J. Physiol.*)

Origin of the ERG

It is well established that the ERG is the record of changes taking place in the retina, and we may ask what is the cause of these fluctuations in potential. Essentially they are changes in the potential difference across the retina since they may be best recorded by a pair of electrodes one on the inner face and the other on the scleral face of the retina. We may recall that a neurone may develop a rapid spike which is reflected in a very brief phase of negativity of an electrode placed exactly at the site of this activity. This change is so rapid, however, that it is unlikely that spike activity on the part of the retinal neurones will have much to do with the ERG.

Slow Potentials. Neurones will also, however, develop a much more slowly occurring change of potential; an example of this is the negativity which has been called the generator potential. In other cases we may record an increase in the normal membrane potential—a hyperpolarization—so that the neurone is said to become *positive*; this change is associated with inhibition of activity. We may assume, therefore, that it is this type of change, rather than spike activity, that is at the basis of the ERG and the questions that present themselves are (*a*) What are the actual changes taking place in the retina? and (*b*) Where are they occurring, i.e. in what layer or layers?

Partial or complete answers to these questions have been provided by the studies of Tomita and Brown and their colleagues, based largely on the insertion of microelectrodes to different depths of the retina and noting the changes in the records when the retina was exposed to light. Before describing these, we may consider the classical analysis of Granit.

Granit's Analysis. Granit considered that the ERG was the record of several events taking place very nearly simultaneously, so that the measured potential changes were the algebraic sum, at any instant, of these. On the basis of experiments designed to block some processes, and thus allowing others to appear in a less obscure form, Granit deduced the presence of three "components", namely PI, PII and PIII. PI is a slowly developing positive component; with strong stimuli, or after light-adaptation, it tends to disappear; it accounts for the secondary rise in positivity, the *c*-wave. PII is likewise positive but is much more rapid in development and is chiefly responsible for the *b*-wave. PIII is the negative component; it comes on rapidly and is responsible for the intial *a*-wave. The *d*-wave, or "off-effect," is considered to represent an interference phenomenon resulting from the rapid swing of the negative PIII component back to the baseline; if this occurs more rapidly than the fall of the PI and PII components, it will clearly be reflected in a wave of positivity, the *d*-wave. Fig. 101 illustrates the analysis for the dark-adapted (above) and light-adapted eye of the mixed frog retina. The dotted lines indicate the probable courses of the pure components acting in isolation, and the full lines the algebraic sums.

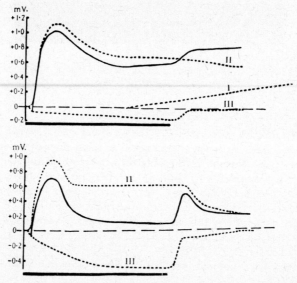

FIG. 101. Analysis of the ERG. Above: Dark-adapted frog's eye; PIII component and OFF-effect small. Below: Light-adapted frog's eye; PIII component and OFF-effect large; PI is missing so that the *c*-wave is absent. The black line indicates the duration of the stimulus (0·2 second). (Granit and Riddell, *J. Physiol.*)

E- and I-Retinae. The results of numerous experiments by different workers have shown that there is a strong correlation between the magnitude of the negative *a*-wave and the magnitude of the "off-effect"; thus in the light-adapted frog's eye the *a*-wave and "off-effect" are both marked (Fig. 101); in the dark-adapted frog's eye, or the cat's eye, the *a*-wave and "off-effect" are small, in fact the latter is just shown as a slight check in the downward trend of positivity. Granit has divided the retinae studied into two groups, *E-* and *I-retinae*, according principally to the predominance of the negative PIII component; thus the retinae of the cat, rat, and guinea pig are all E-retinae with small PIII components in their electroretinograms; the pigeon and light-adapted frog retinae belong to the I-type, with pronounced PIII characteristics. Since those retinae that give the E-type of response contain predominantly rods, whilst the I-type of response is shown either by retinae with predominantly cones, as with the pigeon or tortoise, or by retinae with both types of receptor in which rod function is suppressed (light-adapted frog), it is reasonable to assume that the negative PIII component is mainly associated with cone activity. The available evidence suggests that it is connected with *inhibition* of impulses, as opposed to their initiation.

Excitation. The PII component, on the other hand, is closely associated with the passage of impulses along the optic nerve, i.e. with excitation; thus when the PII component is blocked by ether anaesthesia, simultaneous records of the electrical changes in the optic nerve show that the action potentials here are likewise blocked. So strong, indeed, is the correlation that the height of the *b*-wave has been used as an index to the intensity of the optic nerve response. Thus in Fig. 102 we have the ERG analogue of a dark-adaptation curve, obtained by plotting the intensity of the light stimulus necessary to evoke a *b*-wave of fixed height against the time of dark-adaptation of a frog's retina. Dark-adaptation causes the sensitivity of the retina to increase; this is reflected in a decrease in the stimulus necessary to evoke the *b*-wave.

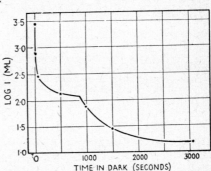

FIG. 102. The ERG analogue of the dark-adaptation curve. The logarithm of the stimulus necessary to evoke a *b*-wave of fixed magnitude is plotted against time in the dark. As the eye adapts, the stimulus necessary becomes smaller (cf. Fig. 73). (Re-drawn from Riggs.)

Pre-Excitatory Inhibition. Granit refers to the PIII component as a process of *"pre-excitatory inhibition"*; the correlations between the ERG and sensory phenomena of vision (p. 292) indicate that the first effect of stimulation of the light-adapted retina is one of inhibition; it is as though the slate were being wiped clean in preparation for a new picture. The component is probably associated with those single fibres of the optic nerve that show a discharge at "Off" (Fig. 90, C). The comparative rarity of this type of fibre in the mammalian retina is probably due to the fact that mainly E-type retinae have been studied (cat, rat, guinea pig).

Further Developments

In general, Granit's analysis remains essentially valid, in the sense that the record is one of independently generated potential changes, although modifications have been introduced in the light of microelectrode studies.

Records from a Distance. Thus the conventional ERG electrode on the cornea is recording events at a considerable distance from where they take place, and the actual record must be considerably attenuated so that it is, at first thought, surprising that it is not completely extinguished. It is only because the cells of the retina are all so well orientated that a record is possible. Thus, we may suppose the situation to be that illustrated by Fig. 103, where the rectangles indi-

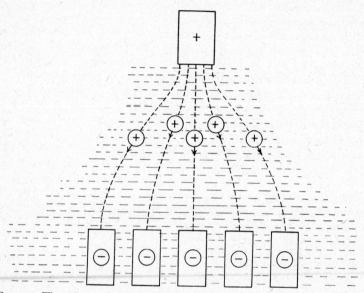

FIG. 103. The orientated cells, indicated by the rectangles, have become negative in relation to their surroundings and thus become what is known as a *sink* for positive current; the electrode at the top of the figure, at some distance from the sink, thus becomes positive, i.e. a *source* of positive current which flows into the sink.

cate radially orientated retinal cells; they have become active in the sense that a part of them has become negative in relation to other parts and the rest of the retina. This part would be described as a *sink of current* into which positive current would flow, and it may be shown that a distant electrode, e.g. one on the cornea, or closer still, in the vitreous, would record a phase of *positivity*. If the position were reversed, i.e. if later this cell, or another, became positive, we should expect to record negativity. It different cells showed opposite effects simultaneously we should expect cancellation of effects, so that the magnitude of the changes recorded at a distance might be much smaller than if the electrode were close to the origins of activity. Thus, on causing our electrode to penetrate into the retina we may expect to find larger potential changes; furthermore, when the electrode comes very close to the cells the record changes polarity; the electrode becomes negative when the cell does, and positive when it becomes positive. At the site of electrical activity, then, we may expect to find a *reversal of the ERG*, or perhaps, if the events take place in different layers, reversal of some components only.

The PIII Component. We may begin with the PIII component, which is the first to manifest itself in the record as a negative downstroke. This was early

shown to be localized in the region of the outer segments of the receptors (Brown
& Wiesel, 1961); a microelectrode was inserted to successive depths, which
were later identified; the *a*-wave reached a maximum when the active electrode
was some 120 μ from the scleral surface of the retina and this corresponds to the
layer of receptors. Again Murakami & Kaneko (1966), using a rather more elaborate recording procedure, identified the origin of the *a*-wave at some 120 μ from

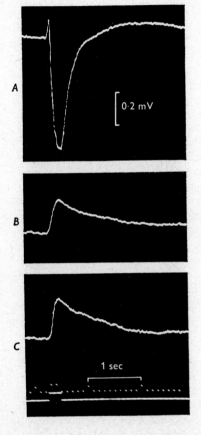

FIG. 104. Effect of sodium aspartate on the
ERG, recorded from the isolated, inverted
frog retina. A, control, before treatment with
aspartate; B, after washing the retina with 10
mM aspartate-Ringer; and C, with 110 mM
aspartate-Ringer. (Sillman, Ito & Tomita,
Vision Res.)

the surface, but they noted that at this level there was a considerable *a*-wave,
recorded from the receptor layer and the inner surface of the retina, i.e. they
identified what they called a proximal component due to activity in the outer
plexiform layer. Sillman, Ito & Tomita (1969) showed that this component, as
well as the *b*-wave, was blocked by aspartate, so that, in the presence of this
inhibitor, they were able to record what was probably the pure receptor potential,
i.e. the potential change taking place in the outer segments of the receptors. This
is illustrated by Fig. 104, where the top record shows the normal record, inverted
because the retina had been inverted. The lower records show the *a*-wave in
isolation after the *b*-wave had been blocked by two concentrations of aspartate.

Brown & Watanabe (1962) were able to isolate the *a*-wave by making the retina
anoxic by compressing the retinal vessels; in this case activity in all cells proximal

to the receptors was abolished and thus the *b*-wave was suppressed; in this way they were able to compare the *a*-wave derived from cones—over the fovea—with that from both rods and cones—periphery—and they observed that the pure cone response from the fovea came on and off rapidly, whilst the mixed response consisted of two steps, indicating the presence of a much slower rod component.

INTRACELLULAR RECORDS

The unequivocal localization of electrical activity is given by insertion of the recording electrode into the cells responsible for this. We have seen that the response of the receptor of the mudpuppy, *Necturus*, is a wave of hyperpolarization; a similar type of response was found by Tomita (1965) in the large cones of the carp, and by Baylor & Fuortes (1970) in the turtle. Such a change in potential is consistent with the polarity of the *a*-wave, which indicates that the outermost layer of the retina becomes positive in relation to the indifferent electrode outside the eye. The transmission of such an effect to the bipolar cell might well inhibit spontaneous activity so that the first response to light could well be what Granit had described as *pre-excitatory inhibition*.

PII Component. The study of Brown & Wiesel (1961) located the maximum amplitude of the *b*-wave at a more proximal site, about half-way through the retina near the inner nuclear layer; by reducing the strength of the stimulus, the interfering effects of the *a*- and *c*-waves were removed. Since it could be abolished by local anaesthetics and anoxia, it was thought to represent synchronized neural activity. Rather surprisingly Miller & Dowling (1970) have shown that the time-course of the depolarizing responses of Muller glial cells to light is similar to that of the *b*-wave; glial cells, in general, do not generate action potentials but they do become depolarized as a result of the escape of K^+ from adjacent neurones when these are excited, so that it has been postulated that the responses of the distal retinal neurones to light induce depolarizing responses in the glial cells. Thus Granit's finding that the *b*-wave is associated with nervous activity is consistent with this hypothesis.

PI Component. This is defined as that responsible for the *c*-wave, and is prominent in the dark-adapted state; microelectrode studies have shown that this, as well as the *a*-wave, is generated close to the scleral surface of the retina; Noell (1954) showed that iodate selectively reduced the *c*-wave and destroyed the pigment epithelium, and he concluded that the *c*-wave was generated by these cells. Brown & Wiesel (1961) concluded from their intracellular recording that the *c*-wave was, indeed, derived from pigment epithelium cells. Brindley (1960) pointed out that the pigment of these cells is melanin so that we might expect the action spectrum for the *c*-wave to be that of the absorption spectrum of this pigment, whereas it is, in fact, similar to that of rhodopsin (Granit & Munsterhjelm, 1937). Presumably, then, the *c*-wave is a secondary event to electrical changes taking place in the receptors.*

* Noell (1958) has correlated changes in the ERG with development of the retina in the newborn. The appearance of the ERG, on the 8th day, coincides with the appearance of miniature receptors with distinguishable inner and outer segments; since ganglion cell discharges in response to light-stimulation are also obtained at this time, the neural pathway is complete. The ERG appears as an *a*-wave. Subsequent development during the 11th–18th days is associated with lengthening of the miniature receptors and the emergence of the *b*-wave; by the 18th day the ERG is adult in appearance, although the *b*-wave is not so high.

D.C. Component. Brown & Wiesel attribute the OFF-effect, as seen in the cat's ERG, essentially to the cessation of what they called the d.c. component, a phase of negativity that comes on immediately with the light-stimulus and remains at a constant level during this, to fall abruptly when the light is switched off. Thus the d.c. component is a part of what Granit included in his PII, which gives rise not only to the *b*-wave but the sustained negativity during maintained stimulation. Steinberg (1969) has recently analysed both *b*-wave and d.c. component, and his results tend to confirm the distinction between the two.*

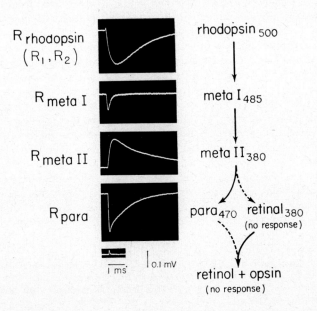

FIG. 105. Showing changes in the early receptor potential (ERP) at successive intervals following a bleaching exposure, correlated with the changes in rhodopsin. $R_{rhodopsin}$ is the response of the dark-adapted eye; $R_{meta\,I}$ was obtained by presenting a white stimulus-flash 0·1 sec after a flash-bleach; $R_{meta\,II}$ with a violet stimulus-flash after the start of a 10 sec bleaching exposure; R_{para} with a blue-green stimulus-flash 3 min after the bleaching exposure. (Cone & Cobbs, *Nature*.)

Early and Late Receptor Potentials. Brown & Murakami (1964), when making their ERG records with a microelectrode in the retina, found that the *a*-wave was preceded by a diphasic potential change with no measurable latency; this was obtained with the tip of the electrode in the region of the junction of the outer and inner segments of the receptors. We may therefore speak of an *early* and *late* receptor potential, (ERP and RP), the latter being the PIII component of

* Brindley (1956) noticed a profound change in the characteristics of the local record of the ERG obtained when his electrode passed to a certain depth in the retina; it was as though the electrode had passed through a membrane of very high resistance; he provisionally identified this *R-membrane* with the external limiting membrane. Tomita, Murakami & Hashimoto (1960) and Brown & Wiesel (1961) considered that it was Bruch's membrane, but this is unlikely in view of its connective-tissue nature; according to Byzov (1968), however, the electrode, when it has crossed the R-membrane has really entered a complex of proximal processes of pigment epithelium cells, receptors and the external limiting membrane, i.e. the record has become *intracellular*.

the electroretinogram, and the former being over before this is normally recorded. A comparison of the effectiveness of different wavelengths of light showed that, in the virtually all-rod retina of the albino rat, the pigment responsible was rhodopsin (Cone, 1964)*; however, the early receptor potential is only obtained with intense flashes, the relative amount of energy for late and early potentials being in the ratio of one to a million. The potential is diphasic, the initial positive phase occurring at sub-zero temperatures that abolished the later phase. All the evidence indicates that both phases have nothing in common with bioelectric potentials as usually recorded; thus they can be recorded from fixed tissue, and, as indicated, at sub-zero temperatures. Some organization of retinal structure is necessary, however, since the potential vanishes if the eye is heated above 58°C, and this is accompanied by a measurable disorientation of the rhodopsin (Cone & Brown, 1967). The currently accepted hypothesis is that the potential arises through transfer of electrical charge from one part of the rhodopsin molecule to another as a result of the absorption of light-energy (Brindley & Gardner-Medwin, 1966), possibly with the formation of free radicles (Grady & Borg, 1968); thus the first effect of light is to isomerize the 11-cis retinal on the rhodopsin molecule (p. 197); in order that rotation of a double-bond may occur, this must be broken, and this will result in the transfer of electrons. Cone & Cobbs (1969) have argued that the nature of the potential should change with the state of the photopigment; thus a single intense flash results in the presence of metarhodopsins I and II (p. 195) and a compound that Wald calls pararhodopsin, absorbing maximally at 4700 Å; they have shown that, corresponding with the appearance of these pigments, the early potential changes its form (Fig. 105).†

Intracellular Record

Recently Murakami & Pak (1970) have compared intra- with extra-cellular recording of the ERP in the gekko, axolotl and mud-puppy retinae; the amplitude of the intracellular record was 1.7 mV on average compared with only 0.09 mV extracellularly; the rise and decay of the second component, R_2, were comparable in their time-relations with synaptic potentials, and from the time-constant a reasonable estimate of the membrane capacitance was deduced, provided the whole area of the cone discs was involved.

Pigment Epithelial Potentials. Both early and late receptor potentials may be recorded from the eye-cup with retina removed and pigment epithelium intact; all three components have been shown by Brown & Crawford (1967) to be generated by pigment epithelium cells and to have action-spectra corresponding with melanin rather than rhodopsin.

Fovea and Periphery. Early attempts at showing a difference between the

---

* In the mixed retina, it is the cones that determine the early receptor potential in spite of the predominance of the rod pigment; thus the action-spectrum for the monkey retina has a maximum between 5350–5700 Å (Carr & Siegel, 1970), whilst the recovery of the potential after a flash is at a rate commensurate with cone pigment regeneration rather than that of rhodopsin (Goldstein & Berson, 1969).
† Arden et al. (1968) have questioned this interpretation, in the light of their experiments on the effects of fixatives and other chemical reagents on the potential; especially cogent is the absence of significant effects of pH whilst the Meta I to Meta II reaction is strongly affected by this. Hydroxylamine has no effect, yet it accelerates the breakdown of meta-rhodopsin at room temperature. Villermet & Weale (1969) have shown that insults, such as high temperatures, that destroy the birefringence of rods also prevent the recording of the negative wave of ERP. Thus intactness of the normal orientation of the molecules in the outer limb seems to be essential for eliciting the negative wave.

electroretinograms recorded from fovea and periphery failed; and this was due to the fact that, even though light is concentrated on a small area of retina, the ERG is that corresponding to stimulation of the whole retina, because of stray light due to scatter; thus a spot of light confined to the blind-spot had the same effect as one on an adjacent region (Asher, 1951). Brindley & Westheimer (1965) suppressed the effects of scattered light by using a background illumination and imposing the flashing spot on this: if the background luminance was not less than one-sixth of that of the flash, the scattered light had a negligible effect. Fig. 106 shows the foveal record, *a*, and records obtained by illuminating larger and large retinal areas; the most obvious difference is the prominent *b*-wave that emerges when the peripheral retina is stimulated.

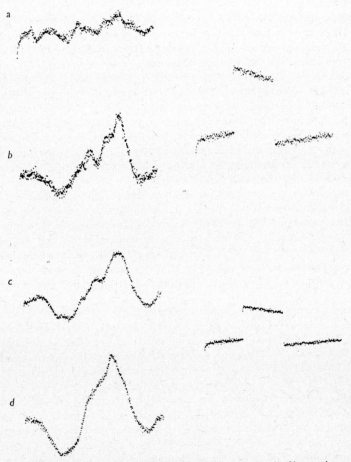

FIG. 106. The human ERG obtained for centrally fixated stimuli of increasing area (a) 2°, (b) 4°30′, (c) 7°22′, (d) 11°34′. Records *b*, *c* and *d* are sums of 1428 responses, record *a* the sum of 2856 responses. Record *a* is displayed at half the vertical gain used for record *b* to compensate for its being the sum of twice as many responses. The upper calibration (5·5 µV and 20 msec) is thus applicable to both of them. Records *c* and *d* are displayed at the same gain as *a*, and the lower calibration (also 5·5 µV and 20 msec) is applicable to them. (Brindley & Westheimer. *J. Physiol.*)

The Mechanism of the Hyperpolarizing Response

The depolarizing responses seen typically in the generator potentials of invertebrate visual cells, or in mechanoreceptors such as the Pacinian corpuscle, are attributable to an increase in membrane permeability to certain ions, which is reflected in a decrease in membrane resistance; thus the stimulus causes a "leakage" of current as the potential falls to a lower value. The hyperpolarization is thus a potential change of opposite character; furthermore, the external recording of this change from a distance, as in the electroretinogram, shows as a cornea-positive deflection, which is the opposite to what might be expected from such a change, were it analogous with the commonly occurring depolarizing changes, since it is these that would give a positive deflection when recorded from a distance.

Outer Segment Potential. Bortoff & Norton (1967) suggested that, in the dark, the membrane potential of the outer segment was less than that of the inner segment, leading to a flow of positive current from inner to outer segment. This would make the vitreous side of the retina positive in relation to the sclera, and account for the normal d.c. potential measured across the retina. If the effect of light were to increase the resistance of the outer segment, the flow of current would be reduced or reversed, and this would make the outer segment more positive in relation to the inner segment, thus accounting for the positive deflection of the externally recorded electroretinogram; the effect would also be to increase the negativity inside the receptor, i.e. to hyperpolarize it. Penn & Hagins showed, by an ingenious placement of electrodes in the rat's retina, that in the dark the rods normally leaked current, the outer segments acting as a sink to current from more proximal regions. On illumination, an oppositely directed current was generated so that the sink tended to disappear; as a result, the potential at the tips of the outer segments became relatively positive. The cause of the photocurrent seems to be an increase in membrane resistance; this was measured directly by Toyoda, Nosaki & Tomita (1969), whilst experimental passage of current across the receptor membrane during illumination showed, once again, that the resistance had increased (Bortoff & Norton, 1967; Baylor & Fuortes, 1970).

Sodium Pump. The functional activity of the retina is acutely dependent on the active transport of sodium, since Frank & Goldsmith (1967) found that treatment of the frog retina with ouabain, which, as we have seen, inhibits the $Na^{+}+K^{+}$-activated ATPase concerned with pumping sodium across membranes, rapidly blocked the ERG, the b-wave disappearing first. The effect was irreversible unless the retina was maintained, during treatment, in a Na^{+}-free medium, in which event return to a normal environment permitted the return of the ERG. It seems, therefore, that the normal functioning of the retina depends on a suitable gradient of Na^{+} between outside medium and the cellular elements of the retina, probably the receptors themselves, since Sillman, Ito & Tomita (1969) found a strong dependence of the receptor-potential on the concentration of Na^{+} in the medium; the potential was measured with aspartate in the medium, which isolates the distal PI from the proximal PI and PII components (p. 175).*

---

* During their studies on the effects of intravitreal injections of ouabain on the intraocular pressure, Langham, Ryan & Kostelnik (1967) noticed that the injections caused loss of vision in the rabbit; this was paralleled by inhibition of retinal ATPase. For a more elaborate experimental analysis of the ionic basis of the currents flowing through

Electrical Stimulation of the Retina. When a current-pulse is passed radially across the retina, the scleral side being positive, while recording from a horizontal cell, the result is a depolarization of the cell; i.e. the opposite effect to that of illumination. It was suggested that the receptors influenced the horizontal cell by liberating a depolarizing transmitter substance, and that this normally took place continuously in the dark, thus maintaining a low level of polarization of the horizontal cell in the dark; illumination would cause cessation of the liberation, by hyperpolarizing the receptor and thus the degree of polarization of the horizontal cell would increase, i.e. the response to light would be a hyperpolarizing one (Byzov & Trifonov, 1968). The effects of electrical stimulation fit into this pattern, as well as the observation that anoxia actually causes a repolarization of the horizontal cell (Fatechand *et al.*, 1966).

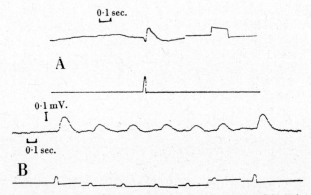

FIG. 107. Human ERG's. A. Response to flash of white light in light-adapted eye. Note initial negative deflection, the *a*-wave. B. Scotopic type of response with no *a*-wave. The lower record of each pair indicates time of stimulation. (Adrian, *J. Physiol.*)

The Human ERG

In man the ERG may be measured with a simple wick-electrode in the inner canthus, or by an electrode maintained in position on the cornea with a contact-lens (Karpe). It was first studied independently by Motokawa & Mita in Japan and Adrian in England. In the light-adapted subject the response to a single bright flash was found by Adrian to be diphasic, as in Fig. 107, a negative *a*-wave being followed by a large positive *b*-wave. Presumably this was mainly a cone response. In the dark-adapted state, with a low-intensity flash, the response was monophasic and much slower—the rod response. With different coloured lights the rod and cone types of response could also be separated, a red flash giving a diphasic cone-response, a blue light a rod-response, and an orange-red light a mixed response in which the *b*-wave contained two components, presumably due to the different time-courses of the rod and cone PII components (Fig. 108). Thus, with a mixed retina, and with stimulation such as to excite both rods and cones, we may expect to be able to resolve the *b*-wave into a rod- and cone-component, unless one type of receptor completely dominates the record.

the receptors of the pigeon cone, and the effects of light on these, the reader is referred to a recent paper by Arden & Ernst (1970), whilst Bowmaker (1970) describes the effect of ouabain and bleaching on the wash-out of sodium from rod outer segments.

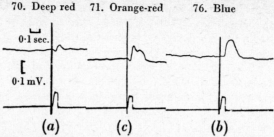

FIG. 108. Human ERG's. Responses to red, orange and blue light stimuli showing transition from photopic to scotopic types of response. (The lower record indicates time of stimulation.) (Adrian, *J. Physiol.*)

Cone ERG. It might be thought that, by confining the stimulus-area to the fovea, a pure cone ERG could be obtained, but this is to ignore the scattering of light within the eye that results in the production of an easily measurable ERG even when the light-stimulus is confined to the blind-spot (p. 179). Aiba, Alpern

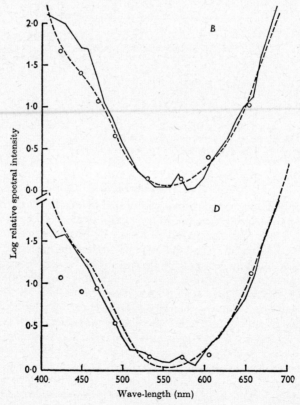

FIG. 109. Spectral sensitivity curves for fovea of two observers using as index to the response the foveal electroretinogram. Open circles are the measured points whilst the continuous lines are the psychophysically determined spectral sensitivity curves for the same observers. Interrupted lines show the C.I.E. photopic luminosity curve. (Aiba, Alpern & Maaseidvaag. *J. Physiol.*)

& Maaseidvaag (1967) employed a technique developed by Brindley & West-heimer (1965) for minimizing the effects of scattered light on the rods, which consisted in presenting the foveal stimulus against a blue background of sufficient intensity to reduce the sensitivity of the rods to the scattered light and so to pre-vent their contributing significantly to the ERG. With this technique, and by enhancing the signal-to-noise ratio by feeding the amplified responses of some 200–2000 flashes into a computer of average transients (CAT), they obtained a spectral sensitivity curve quite similar to the C.I.E. photopic sensitivity curve (Fig. 109).

GENERAL REFERENCES

ADRIAN, E. D. (1932). *Mechanism of Nervous Action*. Philadelphia.
BRAZIER, M. (1960). *The Electrical Activity of the Nervous System*. 2nd. Ed. London: Pitman.
BROWN, K. T., WATANABE, K. & MURAKAMI, M. (1965). The early and late receptor potentials of monkey cones and rods. *Cold Spr. Harb. Symp. quant. Biol.* **30**, 457–482.
BROWN, K. T. (1968). The electroretinogram: its components and their origins. *Vision Res.* **8**, 633–677.
BRINDLEY, G. S. (1970). *Physiology of the Retina and Visual Pathway*. 2nd. Ed. London: Arnold.
GRANIT, R. (1947). *Sensory Mechanisms of the Retina*. London: Oxford University Press.
GRANIT, R. (1955). *Receptors and Sensory Perception*. Newhaven: Yale University Press.
HARTLINE, H. (1941–2). The neural mechanisms of vision. Harvey Lectures, Series 37.
TOMITA, T. (1970). Electrical activity of vertebrate photoreceptors. *Quart Rev. Biophys.* **3**, 179–222.

SPECIAL REFERENCES

ADRIAN, E. D. (1945). The electric response of the human eye. *J. Physiol.* **104**, 84–104.
AIBA, T. S., ALPERN, M. & MAASEIDVAAG, F. (1967). The electroretinogram evoked by the excitation of human foveal cones. *J. Physiol.* **189**, 43–62.
ARDEN, G. B., BRIDGES, C. D. B., IKEDA, H. & SIEGEL, I. M. (1968). Mode of generation of the early receptor potential. *Vision Res.* **8**, 3–24.
ARDEN, G. B. & ERNST, W. (1970). The effect of ions on the photoresponses of pigeon cones. *J. Physiol.* **211**, 311–339.
ASHER, H. (1951). The electroretinogram of the blind spot. *J. Physiol.* **112**, 40 P.
BARLOW, H. B., FITZHUGH, R. & KUFFLER, S. W. (1957). Change of organization in the receptive fields of the cat's retina during dark adaptation. *J. Physiol.* **137**, 338–354.
BARLOW, H. B., HILL, R. M. & LEVICK, W. R. (1964). Retinal ganglion cells responding selectively to direction and speed of image motion in the rabbit. *J. Physiol.* **173**, 377–407.
BARLOW, H. B. & LEVICK, W. R. (1965). The mechanism of directionally selective units in rabbit's retina. *J. Physiol.* **178**, 477–504.
BAYLOR, D. A. & FUORTES, M. G. F. (1970). Electrical responses of single cones in the retina of the turtle. *J. Physiol.* **207**, 77–92.
BEHRENS, M. E. & WULFF, V. J. (1965). Light-initiated responses of retinula and eccentric cells in the *Limulus* lateral eye. *J. gen. Physiol.* **48**, 1081–1093.
BORNSCHEIN, H. (1958). Spontan- und Belichtungsaktivität in Einzelfasern der Katze. *Z. Biol.* **110**, 210–222, 223–231.
BORTOFF, A. & NORTON, A. L. (1967). An electrical model of the vertebrate photo-receptor cell. *Vision Res.* **7**, 253–263.

BOWMAKER, J. K. (1970). Sodium movements from frog outer segments. *Vision Res.* **10**, 601–604.

BRINDLEY, G. S. (1956). The passive electrical properties of the frog's retina, choroid and sclera for radial fields and currents. *J. Physiol.* **134**, 339–352.

BRINDLEY, G. S. (1960). *Physiology of the Retina and Visual Pathway* (1st. ed.). London: Arnold.

BRINDLEY, G. S. & GARDNER-MEDWIN, A. R. (1966). The origin of the early receptor potential of the retina. *J. Physiol.* **182**, 185–194.

BRINDLEY, G. S. & WESTHEIMER, G. (1965). The spatial properties of the human electroretinogram. *J. Physiol.* **179**, 518–537.

BROWN, J. E. & MAJOR, D. (1966). Cat retinal ganglion cell dendritic fields. *Exp. Neurol.* **15**, 70–78.

BROWN, K. T. & CRAWFORD, J. M. (1967). Intracellular recording of rapid light-evoked responses from pigment epithelium cells of the frog eye. *Vision Res.* **7**, 149–163.

BROWN, K. T. & CRAWFORD, J. M. (1967). Melanin and the rapid light-evoked responses from pigment epithelium cells of the frog eye. *Vision Res.* **7**, 165–178.

BROWN, K. T. & MURAKAMI, M. (1964). A new receptor potential of the monkey retina with no detectable latency. *Nature*, **201**, 626–628.

BROWN, K. T. & WATANABE, K. (1962). Isolation and identification of a receptor potential from the pure cone fovea of the monkey retina. *Nature*, **193**, 958–960.

BROWN, K. T. & WIESEL, T. N. (1958). Intraretinal recording in the unopened cat eye. *Am. J. Ophthal.* **46**, Pt. 2, 91–98.

BROWN, K. T. & WIESEL, T. N. (1959). Intraretinal recording with micropipette electrodes in the intact cat eye. *J. Physiol.* **149**, 537–562.

BROWN, K. T. & WIESEL, T. N. (1961). Analysis of the intraretinal electroretinogram in the intact cat eye. *J. Physiol.* **158**, 229–256.

BROWN, K. T. & WIESEL, T. N. (1961). Localization of origins of electroretinogram components by intraretinal recording in the intact cat eye. *J. Physiol.* **158**, 257–280.

BUTENANDT, E. & GRÜSSER, O. -J. (1968). The effect of stimulus area on the response of movement detecting neurons in the frog's retina. *Pflüg. Arch. ges. Physiol.* **298**, 283–293.

BYZOV, A. L. (1968). Localization of the R-membrane in the frog eye by means of an electrode marking method. *Vision Res.* **8**, 697–700.

BYZOV, A. L. & TRIFONOV, J. A. (1968). The response to electric stimulation of horizontal cells in the carp retina. *Vision Res.* **8**, 817–822.

CARR, R. E. & SIEGEL, I. M. (1970). Action spectrum of the human early receptor potential. *Nature*, **225**, 89–90.

CONE, R. A. (1964). Early receptor potential of the vertebrate retina. *Nature*, **204**, 736–739.

CONE, R. A. & BROWN, P. K. (1967). Dependence of the early receptor potential on the orientation of rhodopsin. *Science*, **156**, 536.

CONE, R. A. & COBBS, W. H. (1969). Rhodopsin cycle in the living eye of the rat. *Nature*, **221**, 820–822.

DOWLING, J. E. (1960). Night blindness, dark adaptation and the electroretinogram. *Am. J. Ophthal.* **50**, 875–889.

DOWLING, J. E. & WERBLIN, F. S. (1969). Organization of retina of the mudpuppy *Necturus maculosus.* I. Synaptic structure. *J. Neurophysiol.* **32**, 315–335.

FATECHAND, R., SVAETICHIN, G., NEGISHI, K. & DRUHAN, B. (1966). Effects of anoxia and metabolic inhibitors on the S-potential of isolated fish retinas. *Vision Res.* **6**, 271–283.

FRANK, R. N. & GOLDSMITH, T. H. (1967). Effects of cardiac glycosides on electrical activity in the isolated retina of the frog. *J. gen. Physiol.* **50**, 1585–1606.

FUORTES, M. G. F. (1959). Initiation of impulses in visual cells of *Limulus. J. Physiol.* **148**, 14–28.

GOLDSTEIN, E. B. & BERSON, E. L. (1969). Cone dominance of the human early receptor potential. *Nature*, **222**, 1272–1273.

GOURAS, P. (1967). The effect of light-adaptation on rod and cone receptive field organization of monkey ganglion cells. *J. Physiol.* **192**, 747–760.

GOURAS, P. & LINK, K. (1966). Rod and cone interaction in dark-adapted monkey ganglion cells. *J. Physiol.* **184**, 499–510.

GRADY, F. J. & BORG, D. C. (1968). Light-induced free radicals of retinal, retinol, and rhodopsin. *Biochem.* **7**, 675–682.

GRANIT, R. (1933). The components of the retinal action potential in mammals and their relation to the discharge in the optic nerve. *J. Physiol.* **77**, 207–240.

GRANIT, R. & MUNSTERHJELM, A. (1937). The electrical responses of dark-adapted frogs' eyes to monochromatic stimuli. *J. Physiol.* **88**, 436–458.

GRANIT, R. & RIDDELL, H. A. (1934). The electrical responses of light- and dark-adapted frog's eyes to rhythmic and continuous stimuli. *J. Physiol.* **81**, 1–28.

GRÜSSER, O. -J. & RABELO, C. (1958). Reaktioner retinaler Neurone nach Lichtblitzen. *Pflüg. Arch.* **265**, 501–525.

GRÜSSER, O. -J. *et al.* (1967). A quantitative analysis of movement detecting neurons in the frog's retina. *Pflüg. Arch. ges. Physiol.* **293**, 100–106.

GRÜSSER, O. -J., FINKELSTEIN, D. & GRÜSSER-CORNEHLS, U. (1968). The effect of stimulus velocity on the response of movement sensitive neurons of the frog retina. *Pflüg. Arch. ges. Physiol.* **300**, 49–66.

HARTLINE, H. K. (1940). The nerve messages in the fibres of the visual pathway. *J. opt. Soc. Amer.* **30**, 239–247.

HARTLINE, H. K., WAGNER, H. G. & MACNICHOL, E. F. (1952). The peripheral origin of nervous activity in the visual system. *Cold Spr. Harb. Symp. Quant. Biol.* **17**, 125–141.

HARTLINE, H. K., WAGNER, H. G. & RATLIFF, F. (1956). Inhibition in the eye of *Limulus*. *J. gen. Physiol.* **39**, 651–673.

KANEKO, A. (1970). Physiological and morphological identification of horizontal bipolar and amacrine cells in goldfish retina. *J. Physiol.* **207**, 623–633.

KANEKO, A. & HASHIMOTO, H. (1969). Electrophysiological study of single neurons in the inner nuclear layer of the carp retina. *Vision Res.* **9**, 37–55.

KARPE, G. (1945). The basis of clinical electroretinography. *Acta ophthal. Kbh.*, Suppl. **24**, pp. 118.

KUFFLER, S. W. (1953). Discharge patterns and functional organization of mammalian retina. *J. Neurophysiol.* **16**, 37–68.

KÜHNE, W. & STEINER, J. (1880). Uber das electromotorische Verhalten der Netzhaut. *Unt. physiol. Inst. Univ., Heidelberg* **3**, 327–377.

KÜHNE, W. & STEINER, J. (1881). Uber electrische Vorgänge im Sehorgane, *loc. cit.* **4**, 64–168.

LANGHAM, M. E., RYAN, S. J. & KOSTELNIK, M. (1967). The Na, K ion dependent adenosine triphosphatase of the retina and the mechanism of visual loss caused by cardiac glycosides. *Life Sci.* **6**, 2037–2047.

LEICESTER, J. & STONE, J. (1967). Ganglion, amacrine and horizontal cells of the cat's retina. *Vision Res.* **7**, 695–705.

LEVICK, W. R. (1967). Receptive fields and trigger features of ganglion cells in the visual streak of the rabbit's retina. *J. Physiol.* **188**, 285–307.

MICHAEL, C. R. (1968). Receptive fields of single optic nerve fibers in a mammal with an all-cone retina. I. Contrast-sensitive units. II. Directionally sensitive units. *J. Neurophysiol.* **31**, 249–256, 257–267.

MILLER, R. F. & DOWLING, J. E. (1970). Intracellular responses of the Müller (glial) cells of mudpuppy retina: their relation to *b*-wave of the electroretinogram. *J. Neurophysiol.* **33**, 323–341.

MILLER, W. H. (1957). Morphology of the ommatidia of the compound eye of *Limulus*. *J. biophys. biochem. Cytol.* **3**, 421–428.

MOTOKAWA, J. & MITA, T. Uber eine einfachere Untersuchungsmethode und Eigenschaften der Aktionsströme der Netzhaut des Menschen. *Tohoku J. Exp. Med.* 1942, **42**, 114–133.

MURKAMI, M. & KANEKO, A. (1966). Differentiation of PIII subcomponents in cold blooded vertebrate retinas. *Vision Res.* **6**, 627–636.

MURAKAMI, M. & PAK, W. L. (1970). Intracellularly recorded early receptor potential of the vertebrate photoreceptors. *Vision Res.* **10**, 965–975.

MURAKAMI, M. & SHIGEMATSU, Y. (1970). Duality of conduction mechanism in bipolar cells of the frog retina. *Vision Res.* **10**, 1–10.

NOELL, W. K. (1954). The origin of the electroretinogram. *Am. J. Ophthal.* **38**, 78–90.

NOELL, W. K. (1958). Differentiation, metabolic organization and viability of the visual cell. *Arch. Ophthal.* **60**, 702–733.

OYSTER, C. W. & BARLOW, H. B. (1967). Direction-selective units in rabbit retina: distribution of preferred directions. *Science*, **155**, 841–842.

PENN, R. D. & HAGINS, W. A. (1969). Signal transmission along retinal rods and the origin of the electroretinographic *a*-wave. *Nature*, **233**, 201–205.

RODIECK, R. W. (1965). Quantitative analysis of cat retinal ganglion cell response to visual stimuli. *Vision Res.* **5**, 583–601.

RODIECK, R. W. & STONE, J. (1965a). Response of cat retinal ganglion cells to moving visual patterns. *J. Neurophysiol.* **28**, 819–832.

RODIECK, R. W. & STONE, J. (1965b). Analysis of receptive fields of cat retinal ganglion cells. *J. Neurophysiol.* **28**, 833–849.

SILLMAN, A. J., ITO, H. & TOMITA, T. (1969). Studies on the mass receptor potential of the isolated frog retina. I. *Vision Res.* **9**, 1435–1442.

SMITH, T. G., BAUMANN, F. & FUORTES, M. G. F. (1965). Electrical connexions between visual cells in the ommatidium of *Limulus*. *Science* **147**, 1446–1447.

STEINBERG, R. H. (1969). Comparison of the intraretinal b-wave and d.c. component in the area centralis of cat retina. *Vision Res.* **9**, 317–331.

RUSHTON, W. A. H. (1959). A theoretical treatment of Fuortes's observations upon eccentric cell activity in *Limulus*. *J. Physiol.* **148**, 29–38.

TOMITA, T., KIKUCHI, R. & TANAKA, I. (1960). Excitation and inhibition in lateral eye of horseshoe crab. In *Electrical Activity of Single Cells*, pp. 11–23. Ed. Y. Katsuki. Tokyo: I. Shoin.

TOMITA, T., MURAKAMI, M. & HASHIMOTO, Y. (1960). On the R membrane in the frog's eye: Its localization and relation to the retinal action potential. *J. gen. Physiol.*, **43**, Suppl. 81–94.

TOYODA, J., NOSAKI, H. & TOMITA, T. (1969). Light-induced resistance changes in single photoreceptors of *Necturus* and *Gekko*. *Vision Res.* **9**, 453–463.

VARELA, F. G. & MATURANA, H. R. (1970). Time courses of excitation and inhibition in retinal ganglion cells. *Exp. Neurol.* **26**, 53–59.

VILLERMET, G. M. & WEALE, R. A. (1969). The optical activity of bleached retinal receptors. *J. Physiol.* **201**, 425–435.

WAGNER, H. G., MacNICHOL, E. F. & WOLBARSHT, M. L. (1963). Functional basis for 'on'-centre and 'off'-centre receptive fields in the retina. *J. opt. Soc. Am.* **53**, 66–70.

WERBLIN, F. S. (1970). Response of retinal cells to moving spots: intracellular recording in *Necturus maculosus*. *J. Neurophysiol.* **33**, 342–350.

WERBLIN, F. S. & DOWLING, J. E. (1969). Organization of the retina of the mudpuppy, *Necturus maculosus*. II. Intracellular recording. *J. Neurophysiol.* **32**, 339–354.

WHITEHEAD, R. & PURPLE, R. L. (1970). Synaptic organization in the neuropile of the lateral eye of *Limulus*. *Vision Res.* **10**, 129–133.

PART 5

PHOTOCHEMICAL ASPECTS OF VISION

THE receptors that mediate vision at night are the rods which, as mentioned earlier, contain a pigment—*visual purple* or *rhodopsin*—responsible for the absorption of light in the primary photochemical event leading to sensation. According to the photochemist, then, the visual process consists of the absorption of light by a specialized molecule, visual purple or rhodopsin. The absorption of light provides the rhodopsin molecule with a supply of extra energy and it is said, in this state, to be "activated". In this activated state it is highly unstable and so it will change to a new form, i.e. the molecule will undergo some kind of chemical change by virtue of this absorption of energy. The effects of this change will be to cause an "excited" condition of the rod as a whole and it will be this excited condition that will ultimately lead to the sensation of light.

Photochemical and Thermal Changes

As we shall see, the effects of light-absorption on the rhodopsin molecule are complex because the new molecule formed is not stable but immediately undergoes a succession of changes, which are called *thermal reactions* to distinguish them from the primary *photochemical* change. From an economic point of view we may expect the photochemical and other changes taking place in the retina to be at least partly reversible, otherwise enormous quantities of rhodopsin would have to be synthesized continuously to keep pace with the material destroyed by light.

Regeneration. In fact, the early investigators who observed the bleaching of visual purple in the retina also observed that it would regenerate if the intact retina was allowed to remain in the dark. Subsequent studies on solutions of rhodopsin have indeed shown that, provided the necessary enzymes are present, the photochemical and subsequent chemical changes are largely reversible. Tentatively, then, the visual process may be represented schematically as follows:

Rhodopsin ⇌ Activated State ⇌ Bleached Products

As we shall see, the "bleached products" are the results of a highly complex series of changes, so that the rhodopsin molecule alters its "colour", i.e. its light-absorbing properties, from the original magenta through orange to yellow and ultimately to white, when it has become vitamin A and a protein, *opsin*. In general, when the retina has been kept away from light for some time—i.e. when it has been completely dark-adapted—we may expect to find only rhodopsin in the rods, whilst when the retina is being exposed to light we may expect to find a mixture of rhodopsin with many of its "bleached products", the relative amounts of each product being determined by the intensity of light and the duration of exposure.

Ultra-Structure of the Receptors

Rods. Let us consider the structure of the rod and cone in more detail, as revealed by the electron-microscope. As Fig. 62, p. 118, showed, the rod may be divided into outer and inner segments; since rhodopsin has been shown to be confined to the *outer segment*, the structure of this region is of special interest. Fig. 110, Pl. VII, shows that this part consists essentially of a pile of flattened sacs which are probably to be regarded as infoldings of the cell membrane. Moody & Robertson (1960) showed that the sacs did, indeed, arise through the deep invagination of the cell membrane and later became separated from this by a process of nipping off.

The structure of the inner segment is quite different (Fig. 111, Pl. VIII), and it contains large mitochondria tightly packed together to give what the light-microscopists called the *ellipsoid*. This part of the rod is thus presumably concerned with the energy-requiring metabolic events, whilst we may look upon the outer segments as specialized regions for the absorption of light. Thus, there is optical evidence, based on the behaviour of polarized light (Denton 1954; Weale, 1970) suggesting that the molecules of rhodopsin are spread on to the surfaces of the sacs, the chromophore-groups of the molecule (p. 193) lying in the plane of the discs, i.e. at right-angles to the incident light, an arrangement that provides for much more efficient absorption than if the molecules were scattered at random. The outer segment is connected to the inner segment by a cilium (Fig. 111) which suggests that the rod has developed from a primitive ciliated photosensitive cell. The cilium has been examined in some detail by Richardson (1969) who has shown that, in the mammalian rod, it is not the sole connecting link between inner and outer segments; in addition there is a bridge of cytoplasm and through this ribosomes, mitochrondria and small saccules have direct access to the disc membranes.

TURNOVER OF THE DISCS

The process of disc formation proceeds from the base, so that new discs are added successively from below, displacing the first-formed discs towards the apex of the segment (Nilsson, 1964). According to Young's (1967) radioautographic study of the incorporation of labelled protein into the outer segment, the process of disc formation continues in the adult retina; at any rate, the newly incorporated material appears first in the inner segment and proceeds as a "reaction band" through the outer segment to reach, finally, the apex from which it disappears, presumably into the adjacent pigment epithelium.* Fig. 112, from a more recent study by Young & Droz (1968), illustrates schematically the progressive shift of labelled protein, as recognized autoradiographically, from its region of synthesis in the endoplasmic reticulum of the myoid portion of the inner segment of the rod to the Golgi apparatus, and thence anteriorly to the outer segment. From the most proximal part of the outer segment it passed forward, as before, indicating a regular turnover of the structure by repeated additions of new membranous discs at the base of the outer segment and their ultimate removal at the extremity. With the cones, however, the protein became diffusely distributed over the whole cell, and there was no evidence of a comparable turnover. Young & Droz estimated that the discs were replaced at a rate of some 25 to 36 per day, according to the type of rod.

The Cones. These are built on the same plan, the outer segments consisting of

---

* Absorption of light occurs in the outer segment; its effects must presumably be transmitted to the inner segment, and it is suggested that the primitive cilium and ciliated rootlet constitute a conducting pathway; the rootlet makes close contact with the mitochondria and endoplasmic reticulum of the inner segment, and may touch its plasma membrane (Matsusaka, 1967).

PLATE VII

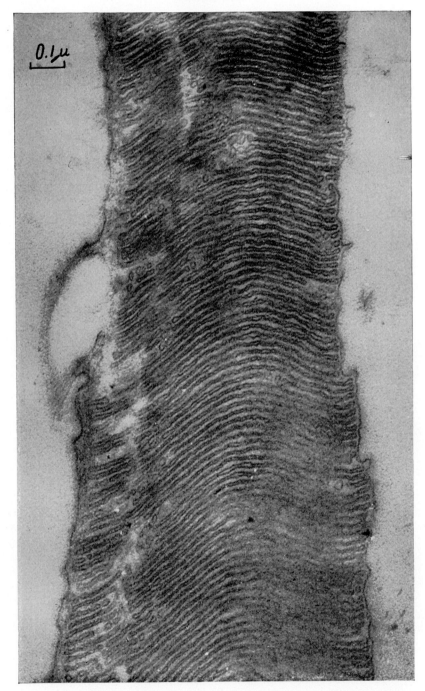

FIG. 110. Electron-micrograph of thin section of outer segment of retinal rod. Osmium fixation. × 90,000. (Courtesy F. S. Sjöstrand.)

To face p. 188.

PLATE VIII

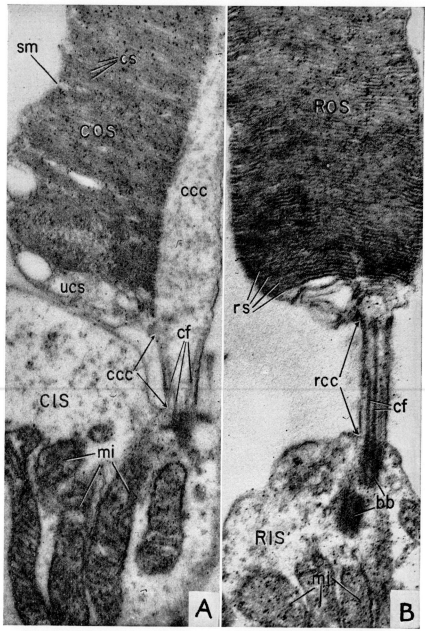

Fig. 111. A. Cone showing part of the outer (COS) and inner (CIS) segments and the connecting cilium (*ccc*). Compare the difference in size and shape of this connecting cilium with that of the rod (B). Some filaments of the cilium are indicated (*cf*) and the continuation of the cilium into the outer segment is shown. The outer segment shows a proximal portion in which the structure is less orientated (*ucs*). × 32,000. B. Rod, showing part of the outer (ROS) and inner (RIS) segment and connecting cilium (*rcc*) with two basal bodies (*bb*). *cs*, cone sac; *sm*, surface membrane; *mi*, mitochondria; *rs*, rod sac; *rcc*, rod connecting cilium. × 40,000. (de Robertis & Lasansky, *J. biophys. biochem. Cytol.*)

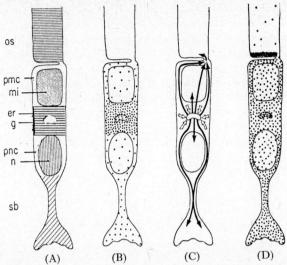

os
pmc
mi
er
g
pnc
n
sb

(A) (B) (C) (D)

FIG. 112. Illustrating the spread of labelled protein in the rod from its region of synthesis in the myoid portion of the inner segment. (A) indicates the cell components that were analysed. *os*, outer segment; *pmc*, perimitochondrial cytoplasm; *mi*, mitochondria; *er*, ergastoplasm; *g*, Golgi complex; *pnc*, perinuclear cytoplasm; *n*, nucleus; *sb*, synaptic body. (B) indicates distribution of label after 10 min. (C) illustrates the probable pathways by which the label is distributed within the cell. (D) indicates distribution after 8 hr. At this time the concentration of labelling in the basal outer segment discs is nearly 20 times greater than that in the ergastoplasm. (Young & Droz, *J. Cell Biol.*)

a pile of discs or saccules. In the rod, the saccules are clearly separate from the enveloping plasma membrane, and this is true of higher vertebrates such as the monkey (Dowling, 1965) but in many lower vertebrates, such as the gekko and axolotl, continuity between the saccules and the plasma membrane is retained, so that the intersaccular space is apparently continuous with the extracellular space (Cohen, 1970). Development of the outer segment follows the same pattern as that described for the rod, but each invaginated sac moves outwards by a shift of the point of invagination rather than a movement of the isolated sac (Nilsson, 1964). Cohen (1968) showed that when lanthanum was precipitated in the glutaraldehyde-fixed retina, the dense precipitate could be seen within the cone-, but not within the rod-, sacs.* Cohen observed in the cones of the pigeon retina that the diameter of the sac was not necessarily the same as that of the outer segment, so that the outer segment could consist of two or even three piles of sacs all in accurate register, with no evidence of lateral fusion between sacs. The outer segments could therefore be called mono-, di- and tri-columnar.

Extraction and Characterization of Photopigment

As mentioned earlier, Kühne observed that visual purple in the intact retina went first yellow and then white when exposed to light; on allowing the eye to

---

* Preliminary fixation with glutaraldehyde is generally necessary for the success of the experiment, since exposure of the unfixed retina to lanthanum resulted in disintegration of the outer segments; when some survived, the same difference between rods and cones was observed. Subsequent experiments using ferritin have shown that this appears in the body of the outer segment of the cone but not of the rod (Cohen, 1970).

remain in the dark the changes were reversed. The problem of the photochemist is to extract from the retina the visual purple, and to elucidate its structural chemistry and the changes it undergoes during exposure to light. Extraction of the pigment is achieved by the use of a "solubilizing agent" such as digitonin or bile salts; these form complexes with the visual purple that permit it to form a homogeneous solution. The amount of pigments obtainable from a single retina is so small that special methods for characterizing them, and the changes they undergo, must be employed; of these the study of absorption spectra has been easily the most profitable.

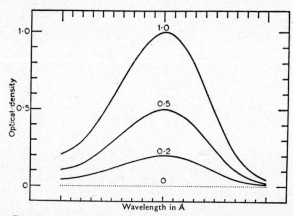

FIG. 113. Density spectra of different concentrations of visual purple. Note that as the concentration is reduced the curve becomes flatter although the wavelength of maximum absorption remains the same. (Dartnall, *The Visual Pigments*.)

Density Spectra. A solution of a pigment is coloured because the pigment molecules absorb certain wavelengths more strongly than others, so that white light, after transmission through the solution, has had some of its wavelengths selectively removed. Thus chlorophyll solutions are green because its molecules selectively remove red and blue wavelengths, and hence allow the green and yellow parts of the spectrum to pass through. Visual purple is magenta coloured because its molecules selectively absorb the middle region of the spectrum— blues and greens—allowing a mixture of red and violet lights to pass through with relatively little absorption. The selective absorption of visual purple solutions can be measured by passing a beam of light, whose wavelength can be varied throughout the spectrum at will, through the solution and measuring, with a suitable photocell, the amount of light-energy transmitted. This may be compared with that transmitted through water alone, and the difference gives the light-energy absorbed. The results may be plotted as a density spectrum, as in Fig. 113, where the optical density of the solution is plotted against wavelength. It will be seen that the spectrum has a maximum at about 5000 Å in the blue-green. It is this *wavelength of maximal absorption*, indicated by λ_{max}, that we may use to characterize a given pigment extracted from the retina.

Optical Density. If the absorption of light obeys Beer's Law, it is determined by the length of the path through which the light travels, and the concentration of pigment, in accordance with the equation:

$$\log_e I_{inc}/I_{trans} = d_\lambda \times c \times x$$

where I_{inc} and I_{trans} are the intensities of the incident and transmitted beams respectively, c is the concentration, x the path-length, and d_λ is the *extinction coefficient* for this particular wavelength. This extinction coefficient is the measure of the tendency for the molecules to absorb light of a given wavelength, so that by plotting d_λ against wavelength the *extinction spectrum* so obtained is a measure of the absorption of the different wavelengths. If logarithms to base 10 are used in the above equation, and the concentration is expressed in moles per litre, the *optical density, D_λ,* of a solution is defined as:

$$D_\lambda = \log_{10} I_{\text{inc}}/I_{\text{trans}} = \epsilon_\lambda \times c \times x$$

where ϵ_λ is the *molar extinction coefficient*. For a given solution the optical density may be used as the measure of absorption of a given wavelength, and the curve obtained by plotting optical density against wavelength is called a *density spectrum** (Fig. 113). It will be clear from the above equations that the extinction coefficients will be independent of concentration of pigments so that extinction spectra will also be independent of this variable; the optical density, on the other hand, increases with increasing concentration, and the shape of the density spectrum will, indeed, vary according to the concentration of the pigment in the solution being studied. Where the concentration of absorbing material is not directly determined, as in studies on the intact retina, then the shape of the density spectrum gives a clue to the actual concentration present. We may note, finally, that if the density for any given wavelength is expressed as a percentage of the maximum density (in the case of rhodopsin 5000 Å) then the density spectrum does become independent of concentration, and it may be used to characterize the pigment.

Difference Spectra. It usually happens that visual pigment solutions are contaminated by light-absorbing impurities; if these are unaffected by light, in the sense that they retain their light-absorbing qualities unchanged after exposure, then the true absorption spectrum of the visual pigment may usually be deduced from the *difference spectrum*, given by measuring the absorption spectrum of the solution before and after bleaching with white light; thus, before the bleaching, the absorption will be due to visual pigment plus impurities; after bleaching it will be due to impurities alone if the visual pigment has become colourless. In many cases, of course, the visual pigment does not become colourless, but allowance can usually be made for this residual absorption.

Purity. Retinal extracts may contain mixtures of visual pigments, in which case it becomes important to decide when any extract is homogeneous or not. Dartnall has developed a technique of partial bleaching with selectively chosen wavelengths that permits of an unequivocal demonstration whether or not the preparation contains more than one visual pigment.

DENSITY SPECTRUM AND SCOTOPIC SENSITIVITY

Human Scotopic Sensitivity. The density spectrum of the pigment extracted from the rods of the dark-adapted retina may be used to determine whether this pigment is, indeed, responsible for night vision, since the efficiency of any wavelength in evoking a sensation of light must be closely related to the efficiency with which it is absorbed by the rods. Clearly, if only one per cent of orange light falling on the retina were absorbed, whilst 90 per cent of the blue light were absorbed, we should have to use ninety times as much orange light-energy to evoke the sensation of light. In Fig. 114 two things have been plotted; the smooth curve shows the absorption spectrum of visual purple extracted from a dark-adapted human eye, whilst the plotted points are the sensitivities of the dark-adapted human eye to the different wavelengths, the so-called action spectrum

---

* The *absorption spectrum* is given by plotting the percentage of light absorbed against wavelength. Terminology is loose in this respect, however, so that absorption spectrum and density spectrum are terms used interchangeably.

of the human eye. As indicated earlier, the dark-adapted eye is achromatic at light intensities in the region of the threshold; the different wavelengths do not evoke the sensation of colour but simply of light, and the experiment consisted of measuring the thresholds for the different wavelengths; the reciprocals of these, measured in quanta, represent the visual efficiencies of the wavelengths. The agreement between the two measurements is striking.*

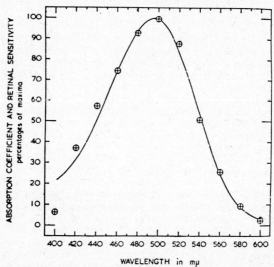

FIG. 114. Comparison of human retinal scotopic sensitivity at different wavelengths, as determined by Crawford, with the absorption spectrum of human visual purple. The curve represents the absorption spectrum whilst the plotted points represent human scotopic sensitivity. (Crescitelli & Dartnall, *Nature*.)

Sensitivity in Animals. In animals, also, the scotopic sensitivity of the eye can be measured; for example, by determining the intensities of different wavelengths required to evoke a *b*-wave of fixed magnitude in the electroretinogram, or to evoke a discharge of fixed frequency in a single optic nerve fibre. The reciprocals of these values give the sensitivities of the retina at the different wavelengths and, when plotted against wavelength, give, usually, the characteristic "rhodopsin-type" of curve with a maximum in the region of 5000 Å. If the light-adapted retina is employed, on the other hand, there is a characteristic shift in the maximum towards, the longer wavelengths to give a maximum in the yellow-green at 5600 Å. This is the *Purkinje shift* and represents the change in spectral sensitivity of the eye due to passing from rod to cone vision. Under these conditions, of course, the eye appreciates the colours of the stimuli.

The Nature of Rhodopsin

The rhodopsin extracted from the retina is a chromoprotein; early estimates of Hubbard (1954) put the molecular weight at about 40,000 but the more recent

---

* Because we are concerned with the effectivity of the light as it actually reaches the retina we must obviously make allowance for the absorption by the transparent media; thus the lens absorbs blue light preferentially and so reduces its apparent effectivity. Crawford's figures for spectral sensitivity have been corrected for the effects of absorption, scattering, etc.

work of Heller (1968) and Shici (1969) indicates a value of 27,000 to 30,000. Each molecule has one light-absorbing chromophore group on it. The nature of this chromophore group, responsible for the colour of the molecule and thus of its characteristic absorption, and the nature of its linkage to the protein moiety, are the keys to the photochemistry of vision.

Changes on Exposure to Light

The early workers on visual pigments observed that, after exposure to light, rhodopsin produced a "visual yellow" which was shown by Lythgoe to result from two steps, a substance *transient orange* being first produced which was then converted in the dark to *indicator yellow*, a substance that changed colour with pH. Wald later showed that the final products of bleaching were *retinene*, which was yellow, and *vitamin A* into which retinene was converted if the retina was exposed to light for long enough to prevent the retinene from being converted back to rhodopsin. Vitamin A is an alcohol and it was shown by Morton that retinene was vitamin A aldehyde, and it is now more correctly called *retinal* (Fig. 115). Thus retinal must be regarded as the chromophore group of rhodopsin; when attached to the protein moiety—*opsin*—the whole has a magenta colour, due to some interaction between the two molecules. It was originally considered that the action of light was to split off retinene from opsin, but more recent studies from several laboratories have left us with the picture of events illustrated schematically by Fig. 116, based on experiments in which chemical changes, subsequent to the initial effect of light, could be arrested by cooling the rhodopsin

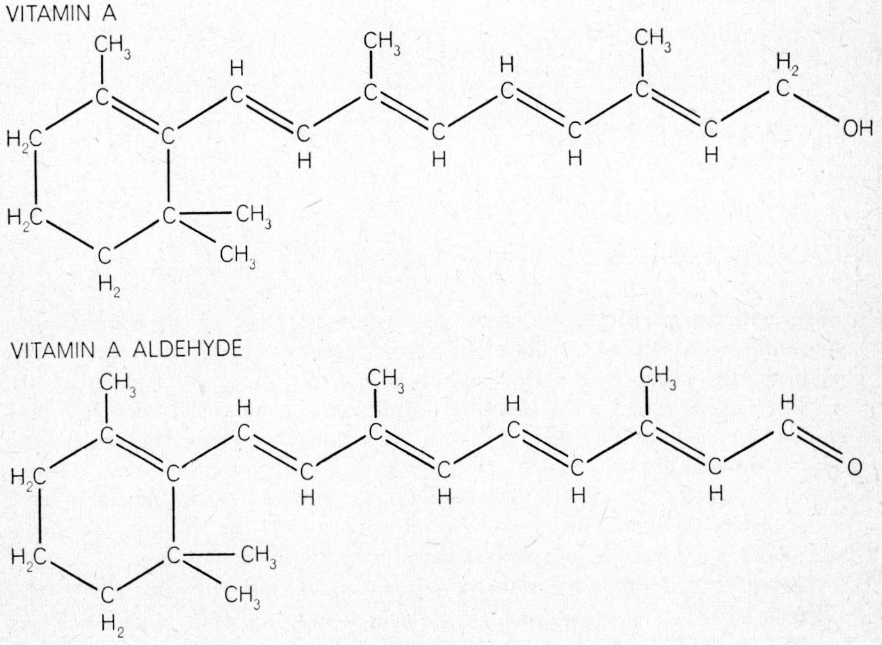

FIG. 115. Structural formulae of vitamin A, or retinol, and its aldehyde, retinal (retinene).

solution to a solid; by controlled warming, the subsequent steps could be identified by the changes in absorption spectrum of the solution. The scheme proposed by Abrahamson & Ostroy (1967), based mainly on the studies in Wald's laboratory, is shown in Fig. 116; the successive steps are recognized by the change in wavelength of maximum absorption (λ_{max}), and the new molecules so formed are indicated by their λ_{max} deduced from the difference spectra.*

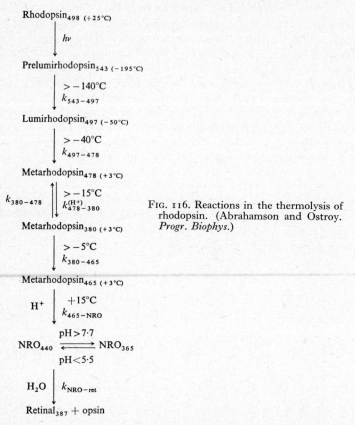

FIG. 116. Reactions in the thermolysis of rhodopsin. (Abrahamson and Ostroy. *Progr. Biophys.*)

Prelumirhodopsin. According to this scheme, the first change, and the only one requiring absorption of light, is the conversion of rhodopsin to prelumirhodopsin; the thermal conversion of this to the next stage, lumirhodopsin, is prevented by carrying out the exposure to light in a glycerol-water mixture at the temperature of liquid nitrogen, $-195°C$. Since light can reverse this transition, there is an equilibrium mixture of rhodopsin and prelumirhodopsin during continued exposure to light, and the equilibrium position is determined by the wavelength of the light, since a short wavelength favours the production of prelumirhodopsin and a long wavelength the production of rhodopsin.

Subsequent Changes. By warming to $-140°C$ or greater, prelumirhodopsin

* As Williams (1970) has emphasized, intermediates in the breakdown of rhodopsin are characterized by their wavelength of maximal absorption of light, so that if an intermediate is formed isochromically it escapes detection; he has provided chemical evidence favouring an intermediate step between metarhodopsins I and II.

is converted to lumirhodopsin, which remains stable at temperatures up to $-40°C$, above which it is converted to a series of *metarhodopsins*, differing in their λ_{max}, then to N-retinylidene opsin (NRO) previously called *indicator yellow* because of its change of colour with pH, and finally NRO is hydrolysed to retinal and opsin. In all but the last of these successive changes, the rhodopsin molecule has retained its attached chromophore group, retinal, so that the alterations have been essentially changes in the configuration of the whole molecule. The final splitting of the chromophore group from the protein moiety represents the breaking of what is known as a Schiff-base link.

The Retinene-Opsin Link. According to the studies of Collins & Morton retinene forms a Schiff base with amines; thus, methylamine reacts with retinene to give retinene methylimine:

$$R . CHO + CH_3 . NH_2 \rightarrow \underset{\underset{H}{|}}{R . C} = N . CH_3$$

a substance with indicator properties analogous to those of indicator yellow; in acid solution the molecule becomes a cation:

$$\underset{\underset{H}{|}}{R . C} = \overset{+}{\underset{\underset{H}{|}}{N}} . CH_3$$

a transformation that could well cause a shift in the absorption spectrum (Pitt, Collins, Morton & Stok, 1955). By analogy, then, the structure of indicator yellow would be:

$$\underset{\underset{C_{19}H_{27}C}{|}}{\overset{H}{}} = N - Opsin$$

i.e. N- retinylidene opsin.*

In neutral solution there will be an equilibrium between the protonated and neutral forms:

$$\underset{\text{Alkaline}}{C_{19}H_{27}CH = N - R} \rightleftharpoons \underset{\text{Acid}}{C_{19}H_{27}CH = {}^{+}NHR}$$

Finally, the conversion of N-retinylidene opsin to retinal plus opsin is a process of hydrolysis requiring water:

$$\underset{\text{N-Retinylidene opsin}}{C_{19}H_{27}CH = NR} + H_2O \rightarrow \underset{\text{Retinal}}{C_{19}H_{27}CH = O} + \underset{\text{Opsin}}{RNH_2}$$

Regeneration of Rhodopsin

In the retina the products of bleaching—retinene and vitamin A—are reconverted to rhodopsin in the dark, and it has been generally considered that this regeneration is the basis for dark-adaptation (p. 207). In isolated systems the

* The scheme of decay of rhodopsin given in Fig. 116 includes the formation of N-retinylidene opsin before the final splitting off of retinal from the opsin moiety; this substance, the indicator yellow of Lythgoe, has always been treated as an artefact by Wald, being formed, according to him, from the retinal liberated from rhodopsin. Abrahamson & Ostroy (1967) point out that the failure to identify NRO during bleaching is accounted for by its ready hydrolysis at physiological pH.

regeneration was not nearly so efficient, and the main cause for this turned out to be the necessity for a special isomer of retinene for the combination with opsin (Hubbard & Wald).

$$
\begin{array}{ccc}
\text{H} & & \text{H} \\
\diagdown & & \diagup \\
& \text{C} = \text{C} & \\
\diagup & & \diagdown \\
\text{R} & & \text{R}'
\end{array}
\qquad
\begin{array}{ccc}
\text{H} & & \text{R}' \\
\diagdown & & \diagup \\
& \text{C} = \text{C} & \\
\diagup & & \diagdown \\
\text{R} & & \text{H}
\end{array}
$$

Cis   *Trans*

(a)   FIG. 117.   (b)

Retinene Isomers. Retinene and vitamin A contain five double-bonds joining C-atoms, and the presence of these gives the possibility of numerous *cis-trans* isomers, due to the fact that rotation about a double bond is not possible, so that a molecule illustrated by Fig. 117a is different from that illustrated by Fig. 117b. Of the various possible isomers obtained by rotating the molecule about different double bonds Wald found two that were capable of condensing with opsin to give photosensitive pigments. One, neo-b retine or 11-*cis* retinene

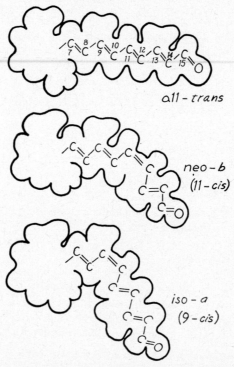

FIG. 118. Shapes of retinene isomers. The *all-trans* isomer is straight; a *cis*-linkage introduces a bend into the molecule. In addition, the 11-*cis* linkage encounters steric hindrance and therefore twists the molecule out of the plane in a manner not shown in the diagram. (Hubbard & Kropf, *Ann. N.Y. Acad. Sci.*)

(Fig. 118), when incubated with opsin gave rhodopsin, whilst the other, iso-retinene-a or 9-*cis* retinene (Fig. 118) gave a photosensitive pigment, *iso-rhodopsin*, with λ_{max} of 4870 Å. The reason why reversal of the changes undergone by bleaching is not always possible is because the retinene that splits off from rhodopsin has the *all-trans* form, which is not active for resynthesis; thus retinene in the rhodopsin molecule is in the 11-*cis* of neo-b form, but when split off is in the all-*trans* form. In order to be of further use the all-*trans* must be isomerized, a process that can take place in the retina in the presence of an enzyme, *retinene isomerase*, although mere exposure to light also causes iso-merization.

While the retina is being exposed to light, therefore, and for some time after, the rods will contain a mixture of substances which will presumably include isorhodopsin, a stable photosensitive pigment obtained by condensing 9-*cis* retinene with opsin, the 9-*cis* retinene becoming available by the isomerization of all-*trans* retinene liberated from rhodopsin. Examination of the products obtained by exposing the retina to flashes of very bright light—*flash-photolysis*—has shown that isorhodopsin may, indeed, be formed under these conditions as well as other products intermediate between rhodopsin and retinal+opsin (Bridges, 1961; Frank, 1969; Ripps & Weale, 1969).

The Primary Change. The fact that only two isomers, both with a similar configuration, will condense with opsin, suggests that there are certain groupings in the opsin molecule into which the retinene must fit (Dartnall, 1957) and this view has been developed by Hubbard & Kropf to account for some of the events taking place on exposing rhodopsin to light. Their scheme is illustrated by Fig. 119. The retinal moiety is held in place on the opsin molecule by virtue of the special shape of the 11-*cis* configuration; photoisomerization to the all-*trans* form would make a neat fit impossible, and this might then allow the opsin to alter its configuration, as illustrated by Fig. 119. The change in configuration would expose previously shielded reactive groups, such as the SH-group as demonstrated chemically by Wald & Brown; and in this way there would be a *molecular amplification* of the primary photochemical event. On this basis, the changes taking place constitute varying degrees of reversible denaturation of the opsin molecule, and the fact that the retinal moiety can be photoisomerized to the 11-*cis* condition while still *in situ* on the opsin molecule means that rhodopsin may be regenerated from the intermediate steps. Thus Yoshizawa & Wald (1963) demonstrated the photo-conversion of lumirhodopsin to prelumirhodopsin at $-195°C$; the conversion of metarhodopsin to prelumirhodopsin required a higher temperature, so that the low temperature prevents the change in the opsin molecule in both directions.

The Metarhodopsins

According to the scheme illustrated by Fig. 116, lumirhodopsin decays through three intermediates, metarhodopsins I, II and III, to N-retinylidene opsin, or indicator yellow. The chemical identities of the metarhodopsins, in which the retinal moiety is attached to opsin in the all-*trans* form, are by no means clear; essentially the changes are in the opsin molecule, involving alterations in shape that expose reactive groupings formerly shielded. Thus treatment of rhodopsin with borohydride is without effect, but after exposure to light a dihydro-retinylidene rhodopsin is formed that is stable to light; it is at the Meta II stage that this reduction by borohydride becomes possible, presumably because the $CN{=}N$ group becomes exposed at this stage:

$$C_{19}H_{27}CH{=}N \ . \ opsin \xrightarrow{NaBH_4} C_{19}H_{27}CH_2NH \ . \ opsin$$

Bownds (1967) showed that the retinal was bound to the opsin molecule through the ϵ-amino-group of lysine. A very significant finding was that, in rhodopsin, the retinal was apparently bound to phosphatidyl ethanolamine (Akhtar & Hirtenstein, 1969), i.e. to the lipid component of rhodopsin, so that, as Poincelot *et al.* (1969) demonstrated, the conversion of metarhodopsin I to metarhodopsin II consisted in the transfer of the chromophore from lipid to the amino-acid skeleton of opsin. Since this change occurs within 100 μsec in the retina, it may very well be the critical one that produces sufficient separation of electronic charges to cause the early receptor potential (p. 177).

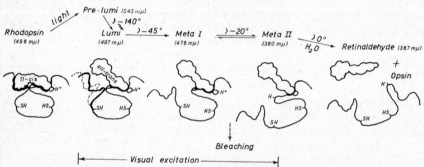

FIG. 119. Pictorial representation of stages in the bleaching of rhodopsin. Rhodopsin has, as a chromophore, 11-cis retinal, which fits closely a section of the opsin structure. The only action of light is to isomerize retinal to the all-*trans* configuration (prelumirhodopsin). Then the structure of opsin opens in stages (lumirhodopsin, metarhodopsin I and II), until finally the retinaldehyde is hydrolysed away from opsin. Bleaching occurs in the transition from metarhodopsin I and II; and by this stage visual excitation must also have occurred. The opening of the opsin structure exposes new groups, including two —SH groups and one H^+-binding group. The absorption maxima shown are for prelumirhodopsin at $-190°C$, lumirhodopsin at $-65°$, and other pigments at room temperature. (Wald, *Proc. Int. Congr. Physiol.*)

Pararhodopsin. Metarhodopsin III, or meta$_{465}$, was identified in the bleached products by Matthews *et al.* (1963), and Wald has suggested that it be called *pararhodopsin*; it is probably equivalent to the *transient orange* identified by Lythgoe & Quilliam (1938) as an early product of bleaching; there is some doubt as to whether it is necessary for metarhodopsin II to pass through this stage in the bleaching process.

In Vivo Changes

The various compounds formed as intermediates in the bleaching of rhodopsin have been identified by optical means in solutions of the pigment; it is by no means certain that significant concentrations of all these compounds build up in the retina. There have been a variety of studies on the intact retina, and it would seem that it is metarhodopsin II and pararhodopsin that are the main long-lived photoproducts (see, for example, Frank, 1969; Ripps & Weale, 1969). It has been argued that these, and other photoproducts, are capable of influencing the excitatory state of the rods, thereby raising ganglion cell thresholds (Donner & Reuter, 1968) but the studies of Frank & Dowling (1968) make this very unlikely, the loss of sensitivity being due to loss of rhodopsin, *per se*, rather than the presence of metarhodopsin. Thus it is essentially the *process* of change, rather than the change itself, that acts as the trigger for the visual process.*

The Pigment Epithelium. It was early discovered that regeneration of rhodopsin after exposure to light would not occur in the retina if it was detached from the pigment epithelium; and this is due to the close dependence of the receptors on exchanges with this tissue. Thus Jancsó & Jancsó profited by the fluorescence of vitamin A in ultra-violet light to demonstrate that after maximal light-adaptation there is a high concentration of the vitamin in the pigment epithelium; during full dark-adaptation none could be demonstrated. Again, the chemical studies of Hubbard & Colman and of Dowling have confirmed this migration of vitamin A during light- and dark-adaptation, and have shown that the total amount of vitamin A in the eye remains constant, if we include under this title the retinene bound to rhodopsin.†

COMPARATIVE PHYSIOLOGY OF THE PIGMENTS

Rhodopsin and Porphyropsin

Köttgen & Abelsdorff (1896) observed that the visual purples extracted from mammals, birds and amphibians were different from those extracted from various fresh-water fishes; the former had the usual maximal absorption near 5000 Å whilst the fish pigments had a maximal absorption near 5400, and thus appeared more violet. Wald showed that the retinae of freshwater fishes gave retinene and vitamin A on bleaching, but it became evident that these were different and have now been called retinene$_2$ and vitamin A$_2$, differing by the presence of a double bond in the beta-ionone ring from what have now been called retinene$_1$ and vitamin A$_1$ of mammals and amphibians. Wald called the pigments obtained from the fish retinae *porphyropsins* to distinguish them from the rhodopsins of λ_{max} 5000 approx. Subsequent studies showed that marine fish had the rhodopsin type of pigment based on retinene$_1$ whilst euryhaline fishes, spending part of their time in freshwater and part in sea water, gave extracts of intermediate absorption characteristics which were assumed to contain mixtures of rhodopsin and porphyropsin.

Multiplicity of Pigments

Wald's statement that there were only two scotopic types of pigment, rhodopsin and porphyropsin, has been shown, mainly by Dartnall's work, to be incorrect. It is true that all the known pigments are based on only two retinenes, 1 and 2, but presumably because of the widely differing opsins, a very wide range of pigments with characteristic λ_{max} ranging from 5630 Å in the yellow to 4300 Å in the violet has now been demonstrated; in general those with λ_{max} in the long wavelength region—so-called porphyropsins—are based on retinene$_2$, but there is considerable overlap, so that the carp pigment based on retinene$_2$ has a λ_{max} of 5230 Å, whilst the gecko pigment, based on retinene$_1$ has a λ_{max} of 5240 Å. Furthermore, some

---

* Weale (1967) has brought forward some evidence that metarhodopsin III can stimulate the retina on absorption of light. We have seen that Cone & Cobbs (p. 177) have attributed changes in the early receptor potential to the presence of different photoproducts.

† Moyer (1969) has described the structure of the pigment epithelium in some detail; at the apical (vitread) ends of the cells the plasma membrane interdigitates with the outer segments of rods and cones, and this may promote exchanges between the two types of cell. The intercellular spaces between pigment epithelium cells are sealed by tight junctions.

retinae gave extracts containing at least two photosensitive pigments of the scotopic type. Even the typical rhodopsin, moreover, differs significantly according to the species of retina from which it is extracted; thus cattle rhodopsin has a λ_{max} of 4990 Å, human 4970 Å, squirrel and frog 5020 Å.

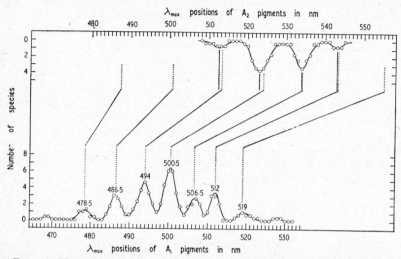

FIG. 120. The relationship between the spectral distributions of A_1 and A_2 pigments in teleost fishes. The data for the A_1 pigments are derived from those fishes that possess *only* A_1 pigments; the data for the A_2 pigments are not restricted: *i.e.* they include all fishes that possess A_2 pigments, some of which have A_1 pigments as well. (Dartnall and Lythgoe, *Vision Res.*)

Adaptation to Environment. The significance of the wide variations in absorption characteristics of the scotopic visual pigments, found in fish especially, is probably to be sought in the requirements of the animal's habitat. Thus Denton & Warren and Munz independently showed that deep-sea fishes, such as the conger eel, had golden coloured pigments that were apparently adapted for the reception of the predominantly blue light of their habitat, and subsequent study has indeed shown that the λ_{max} of bathypelagic fishes range between 4780 Å and 4900 Å, and are based on retinene₁. As Munz has suggested, since some of these deep-sea fish have light-emitting organs, it is probable that their pigments are adapted to correspond with the λ_{max} for *emission* of these organs.* In addition to modification of the pigment to meet the environmental need, some deep-sea fishes show adaptations of retinal structure. Locket (1970) has described the long outer segments in many of these that permit a longer optical pathway through the pigment; again, in some species many rods are packed tightly together, and the presence of tight-junctions between laterally apposing cell membranes suggests that these groups might act as one in response to light.

* Munz has established some very clear correlations between the spectral quality of the light available in different habitats and the λ_{max} of the predominant visual pigment. Thus in clear oceanic waters the λ_{max} of transmitted light at 200 metres is 4750 Å and this corresponds to that of the absorption maxima of bathypelagic fishes; closer to the surface the λ_{max} for transmission is about 4850–4950 Å corresponding to the λ_{max} of the visual pigments of surface pelagic fishes. With rocky coasts the transmitted light is different from that in the turbid waters near sandy beaches, and there is a corresponding difference in the visual pigments (Munz, 1958). J. N. Lythgoe (1968) has pointed out, however, that for detection of a target, sensitivity to light is not the entire story since it is rather the contrast that is the determining factor, and better contrast may, in some circumstances, be obtained with a wavelength of light that is not the same as the wavelength of maximal absorption of the visual pigment. This is because the spectral composition of the light reflected from an object is different from that reaching the eye from the background.

Temporal Variations. There seems no doubt, moreover, that the pigment composition of a retina is not an invariable characteristic; for example, immature (yellow) eels have a mixture of pigments based on retinene$_1$ and retinene$_2$, whereas the adult (silver) eel has only the pigment based on retinene$_1$. Again Dartnall, Lander & Munz found that, in a given population of rudd, the proportions of the two retinal pigments based on retinenes 1 and 2 varied with time of the year, and this was due to the variations in the length of the days; in winter the dominant pigment was that with λ_{max} 5430 based on retinene$_2$, whilst in summer it was that with λ_{max} 5100 based on retinene$_1$. By artificially varying the environmental illumination in an aquarium the proportions of the pigments could be varied.

Retinene and Opsin Variations. In general, we can expect two ways of varying the character of visual pigment; first with a given retinene we may vary the opsin part of the molecule; with retinene$_1$ this gives quite a wide range of possibilities permitting pigments that enable the honey-bee, for example, to make full use of ultra-violet light (λ_{max} 4400–4500); the deep-sea fish to have golden retinae with λ_{max} in the region of 4800 Å; mammals to have pigments with λ_{max} around 5000 and so on up to the chicken's pigment, iodopsin, with λ_{max} on the long-wavelength side in the region of 5600 Å. Another series of pigments may be obtained by varying the opsins attached to retinene$_2$. Fig. 120 shows the results of studies by Dartnall & Lythgoe (1965) and of Bridges (1964) on the λ_{max} of pigments extracted from a variety of fishes based on retinene$_2$ (above) and retinene$_1$ (below). The figure illustrates, first, that there is considerable overlap between the two types, and it also shows that there is a tendency for the λ_{max} to cluster around certain wavelengths. Dartnall & Lythgoe established an empirical relationship between the λ_{max} of pigments with the same opsin, and retinene$_1$ or retinene$_2$ attached; with the aid of this, they predicted the shifts in λ_{max} of pigments containing retinene$_1$ that would be obtained if the same opsin were attached to retinene$_2$. It will be seen that the predictions, indicated by the positions of the upper vertical dotted lines, agree with actual pigments very well.

Phylogenetic Aspects. Wald considered that the retinene$_2$-pigments, his porphyropsins, were exclusively found in fresh-water fishes; and this was the basis for some interesting phylogenetic generalizations. However, there is no doubt from Schwanzara's (1967) study that many freshwater fishes have a retinene$_1$ pigment, in fact 35 per cent of those examined had this, unmixed with a retinene$_2$ pigment; moreover, mixtures commonly were found in fishes that were not euryhaline. Again, marine fishes did not necessarily have exclusively retinene$_1$ pigments. Schwanzara found a dominance of retinene$_2$ pigments in freshwater fishes of temperate waters and of retinene$_1$ pigments in tropical freshwater fishes; and he attributed this to the different spectral qualities of the light in tropical and temperate fresh-water.

Photopic Pigments

By means of reflexion densitometry Rushton and Weale have shown that both foveal and peripheral cones undergo a bleaching process on exposure to light; and the action spectrum for this bleaching corresponds with the photopic luminosity curve, with a maximum at longer wavelengths than for scotopic vision (*cf.* Ripps & Weale, 1970). The cones serve not only to respond to light at high intensities but also to discriminate its spectral quality; they are the receptors for colour vision. We shall see that this discrimination in man and many other species requires that the cones be of at least three kinds, containing different pigments; hence we need not expect to extract a single pigment with the absorption spectrum corresponding with the photopic spectral sensitivity curve. In fact, as we shall see, the presence of three different pigments has been described, although they have not been prepared in the pure state.

Iodopsin. When the relatively pure cone retina of the chicken was extracted Wald (1937) did, indeed, obtain a pigment with 5620 which he called *iodopsin*.

It is based on retinene$_1$ and undergoes the same cycle of changes undergone by rhodopsin and thus differs from the latter in the opsin part of the molecule (Yoshizawa & Wald, 1967); by condensing retinene$_2$ with the chicken opsin Wald Brown & Smith obtained a new pigment which they called *cyanopsin* with λ_{max} of 6200 Å; this has so far not been extracted from any retina, but it is interesting that the tench and tortoise have maximal photopic sensitivities at 6200 Å.

These pigments could, indeed, subserve the function of responding to light and indicating its relative luminosity, in the same way that rhodopsin does at low intensities, but in the primate and fish cones examined spectrophotometrically (p. 275) there was no evidence for such pigments. Moreover, Sillman (1969) in his extensive study of bird photopigments found as major pigment, and the only pigment in some 14 species, one with a λ_{max} in the range 5000 to 5060 Å, whilst in five species of passerines a second photopigment with λ_{max} 4800 to 4900 Å occurred; there is every reason to believe that both these are scotopic pigments.

GENERAL REFERENCES

BRIDGES, C. D. B. (1965). Absorption properties, interconversions, and environmental adaptation of pigments from fish photoreceptors. *Cold Spr. Harb. Symp. quant. Biol.* **30**, 317–334.

COHEN, A. I. (1969). Rods and cones and the problem of visual excitation. In *The Retina*, Ed. B. R. Straatsma *et al.*, pp. 31–62. Los Angeles: University of California Press.

DARTNALL, H. J. A. (1957). *The Visual Pigments.* London: Methuen.

DARTNALL, H. J. A. (1962). The photobiology of visual processes. In *The Eye*, Ed. Davson, chapters 17–20. London: Academic Press.

DE ROBERTIS, E. & LASANSKY, A. (1961). Ultrastructure and chemical organization of photoreceptors. In *The Structure of the Eye*, Ed. Smelser, pp. 29–49. New York: Academic Press.

GRANIT, R. (1947). *Sensory Mechanisms of the Retina.* London: Oxford University Press.

MOMMAERTS, W. H. F. M. (1969). Circular dichroism and the conformational properties of visual pigments. In *The Retina.* Ed. B. R. Straatsma *et al.*, pp. 225–234. Los Angeles: University of California Press.

WALD, G. (1956). The biochemistry of visual excitation. In *Enzymes*, pp. 355–367. New York: Academic Press.

WALD, G. (1960). The distribution and evolution of visual systems. In *Comparative Biochemistry*, vol. I. New York: Academic Press.

WOLKEN, J. J. (1961). A structural model for a retinal rod. In *The Structure of the Eye*, Ed. Smelser, pp. 173–192. New York: Academic Press.

SPECIAL REFERENCES

ABRAHAMSON, E. W. & OSTROY, S. E. (1967). The photochemical and macromolecular aspects of vision. *Progr. Biophys.* **17**, 181–215.

AKHTAR, M. & HIRTENSTEIN, M. D. (1969). Chemistry of the active site of rhodopsin. *Biochem. J.* **115**, 607–608.

BOWNDS, D. (1967). Site of attachment of retinal in rhodopsin. *Nature*, **216**, 1178–1181.

BRIDGES, C. D. B. (1961). Studies on the flash photolysis of visual pigments. *Biochem. J.* **79**, 128–134.

BRIDGES, C. D. B. (1964). The distribution of visual pigments in freshwater fishes. *Abstr. Fourth Internat. Congr. Photobiol.*, Oxford, p. 53. Bucks: Beacon Press.

COHEN, A. I. (1968). New evidence supporting the linkage to extracellular space of outer segment saccules of frog cones but not rods. *J. Cell Biol.* **37**, 424–444.

COHEN, A. I. (1970). Further studies on the question of the patency of saccules in outer segments of vertebrate photoreceptors. *Vision Res.* **10**, 445–453.

COLLINS, F. D. (1953). Rhodopsin and indicator yellow. *Nature* **171**, 469–471.

COLLINS, F. D. & MORTON, R. A. (1950). Studies on rhodopsin. I–III. *Biochem. J.* **47**, 3–9, 10–17, 18–24.

CRAWFORD, B. H. (1949). The scotopic visibility function. *Proc. Phys. Soc. Lond.*, B, **62**, 321–334.

CRESCITELLI, F. & DARTNALL, H. J. A. (1953). Human visual purple. *Nature* **172**, 195–196.

DARTNALL, H. J. A. LANDER, M. R. & MUNZ, F. W. (1961). Periodic changes in the visual pigment of a fish. In *Progress in Photobiology*, pp. 203–213. Amsterdam, Elsevier.

DARTNALL, H. J. A. & LYTHGOE, J. N. (1965). The spectral clustering of visual pigments. *Vision Res.* **5**, 81–100.

DENTON, E. J. (1954). On the orientation of molecules in the visual rods of *Salamandra maculosa*. *J. Physiol.* **124**, 17 P.

DENTON, E. J. & WARREN, F. J. (1956). Visual pigments of deep sea fish. *Nature* **178**, 1059.

DE ROBERTIS, E. & LASANSKY, A. (1958). Submicroscopic organization of retinal cones of the rabbit. *J. biophys. biochem. Cytol.*, **4**, 743–746.

DONNER, K. O. & REUTER, T. (1968). Visual adaptation of the rhodopsin rods in the frog's retina. *J. Physiol.* **199**, 58–87.

DOWLING, J. E. (1960). Chemistry of visual adaptation in the rat. *Nature* **188**, 114–118.

DOWLING, J. E. (1965). Foveal receptors of the monkey retina: fine structure. *Science* **147**, 57–59.

FRANK, R. N. (1969). Photoproducts of rhodopsin bleaching in the isolated, perfused frog retina. *Vison Res.* **9**, 1415–1433.

FRANK, R. N. & DOWLING, J. E. (1968). Rhodopsin photoproducts: effects on electroretinogram sensitivity in isolated perfused rat retina. *Science* **161**, 487–489.

HELLER, J. (1968). Purification, molecular weight, and composition of bovine visual pigment. *Biochem.* **7**, 2906–2913.

HUBBARD, R. (1954). The molecular weight of rhodopsin and the nature of the rhodopsin–digitonin complex. *J. gen. Physiol.* **37**, 373–379.

HUBBARD, R. & COLMAN, A. D. (1959). Vitamin A content of the frog eye during light and dark adaptation. *Science* **130**, 977–978.

HUBBARD, R. & KROPF, A. (1959). Molecular aspects of visual excitation. *Ann. N.Y. Acad. Sci.* **81**, 388–398.

HUBBARD, R. & WALD, G. (1952). Cis-trans isomers of vitamin A and retinene in the rhodopsin system. *J. gen. Physiol.* **36**, 269–315.

JANCSÓ, N. v. & JANCSÓ, H. v. (1936). Fluoreszenmikroskopische Beobachtung der reversiblen Vitamin-A Bildung in der Netzhaut wahrend des Sehaktes. *Biochem. Z.* **287**, 289–290.

KÖTTGEN, E. & ABELSDORFF, G. (1896). Absorption und Zersetzung des Sehpurpurs bei den Wirbeltieren. *Z. Psychol. Physiol. Sinnesorg.* **12**, 161–184.

LOCKET, N. A. (1970). Deep-sea fish retinas. *Brit. med. Bull.* **26**, 107–111.

LYTHGOE, J. N. (1968). Visual pigments and visual range. *Vision Res.* **8**, 997–1011.

LYTHGOE, R. J. (1937). Absorption spectra of visual purple and visual yellow. *J. Physiol.* **89**, 331–358.

LYTHGOE, R. J. & QUILLIAM, J. P. (1938). The relation of transient orange to visual purple and indicator yellow. *J. Physiol.* **94**, 399–410.

MATSUSAKA, T. (1967). Lamellar bodies in the synaptic cytoplasm of the accessory cone from the chick retina as revealed by electron microscopy. *J. Ultrastr. Res.* **18**, 55–70.

MATTHEWS, R. G., HUBBARD, R., BROWN, P. K. & WALD, G. (1963). Tautomeric forms of metarhodopsin. *J. gen. Physiol.* **47**, 215–240.

MOODY, M. F. & ROBERTSON, J. D. (1960). The fine structure of some retinal photo-receptors. *J. Biophys. biochem. Cytol.* **7**, 87–92.

MOYER, F. H. (1969). Development, structure and function of the retinal pigmented epithelium. In *The Retina*, Ed. B. R. Straatsma *et al.*, pp. 1–30. Los Angeles: University of California Press.

MUNZ, F. W. (1958). Photosensitive pigments from the retinae of certain deep-sea fishes. *J. Physiol.* **140**, 220–235.

MUNZ, F. W. (1958). The photosensitive retinal pigments of fishes from relatively turbid coastal waters. *J. gen. Physiol.* **42**, 445–459.

NAKA, K. I. & RUSHTON, W. A. H. (1966a). S-potentials from colour units in the retina of fish (*Cypridinae*). *J. Physiol.* **185**, 536–555.

NAKA, K. I. & RUSHTON, W. A. H. (1966b). An attempt to analyse colour reception by electrophysiology. *J. Physiol.* **185**, 556–586.

NAKA, K. I. & RUSHTON, W. A. H. (1966c). S-potentials from luminosity units in the retina of fish (*Cypridinae*). *J. Physiol.* **185**, 587–599.

NILSSON, S. E. G. (1964). Receptor cell outer segment development and ultra-structure of the disk membranes in the retina of the tadpole (*Rana pipiens*). *J. Ultrastr. Res.* **11**, 581–620.

PITT, G. A. J., COLLINS, F. D., MORTON, R. A. & STOK, P. (1955). Retinylidene-methylamine an indicator yellow analogue. *Biochem. J.* **59**, 122–128.

POINCELOT, R. P., MILLAR, P. G., KIMBEL, R. L. & ABRAHAMSON, E. W. (1969). Lipid to protein chromophore transfer in the photolysis of visual pigments. *Nature* **221**, 256–257.

RICHARDSON, T. M. (1969). Cytoplasmic and ciliary connections between the inner and outer segments of mammalian visual receptors. *Vision Res.* **9**, 727–731.

RIPPS, H. & WEALE, R. A. (1969). Flash bleaching of rhodopsin in the human retina. *J. Physiol.* **200**, 151–159.

SCHICHI, H., LEWIS, M. S., IRREVIERE, F. & STONE, A. L. (1969). Purification and properties of bovine rhodopsin. *J. biol. Chem.* **244**, 529–536.

SCHWANZARA, S. A. (1967). The visual pigments of freshwater fishes. *Vision Res.* **7**, 121–148.

SILLMAN, A. J. (1969). The visual pigments of birds. *Vision Res.* **9**, 1063–1077.

SJÖSTRAND, F. S. (1953). The ultrastructure of the outer segments of rods and cones of the eye as revealed by electron microscopy. *J. cell. comp. Physiol.* **42**, 45–70.

WALD, G. (1935). Carotenoids and the visual cycle. *J. gen. Physiol.* **19**, 351–371.

WALD, G. (1937). Photo-labile pigments of the chicken retina. *Nature* **140**, 545–546.

WALD, G. (1939). The porphyropsin visual system. *J. gen. Physiol.* **22**, 775–794.

WALD, G. & BROWN, P. K. (1952). The role of sulphydryl groups in the bleaching and synthesis of rhodopsin. *J. gen. Physiol.* **35**, 797–821.

WALD, G. BROWN, P. K. & SMITH, P. H. (1955). Iodopsin. *J. gen. Physiol.* **38**, 623–681.

WEALE, R. A. (1967). On an early stage of rhodopsin regeneration in man. *Vision Res.* **7**, 819–827.

WEALE, R. A. (1970). Optical properties of photoreceptors. *Brit. med. Bull.* **26**, 134–137.

WILLIAMS, T. P. (1970). An isochromic change in the bleaching of rhodopsin. *Vision Res.* **10**, 525–533.

YOSHIZAWA, T. & WALD, G. (1963). Pre-lumirhodopsin and the bleaching of visual pigment. *Nature* **197**, 1279–1286.

YOSHIZAWA, T. & WALD, G. (1967). Photochemistry of iodopsin. *Nature*, **214**, 566–571.

YOUNG, R. W. (1967). The renewal of receptor cell outer segments. *J. Cell Biol.* **33**, 61–72.

YOUNG, R. W. & DROZ, B. (1968). The renewal of protein in retinal rods and cones. *J. Cell Biol.* **39**, 169–184.

PART 6

DARK-ADAPTATION
AND THE MINIMUM STIMULUS FOR VISION

THE DARK-ADAPTATION CURVE

WE have seen that, in the dark, the sensitivity of the human eye to light increases until a maximum is reached after about 30 minutes; this gives the dark-adaptation curve showing the progressive fall in threshold. We have already concluded that the curve represents the course of recovery of function of the rods. This curve, however, is not smooth, but shows a "kink" so that adaptation consists in an initial rapid phase of recovery of sensitivity, which is finished in about 5–10 minutes, and a slower phase.

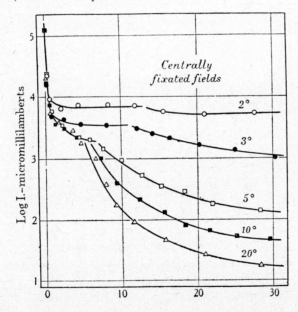

FIG. 121. Dark-adaptation curves with centrally fixated fields subtending larger and larger angles at the eye. With the larger fields peripheral vision comes into play giving the characteristic break in the curve. (Hecht, Haig & Wald, *J. gen. Physiol.*)

Cone and Rod Portions. The first rapid phase has been shown, by many experiments, to represent the increase in sensitivity of the cones in the dark, so that the threshold during the first few minutes in the dark is determined by the cones, the thresholds of the rods still being higher than those of the cones. Numerous lines of evidence support this conclusion. For example, if the area of the light-stimulus used for measuring the threshold is small, and is concentrated on the fovea, the curve of dark-adaptation consists only of the initial part. As the size of the stimulus is increased, so that overlap on to the parafovea, containing rods, occurs, the dark-adaptation curve shows the typical kink with cone and rod portions (Fig. 121). Again, if coloured stimuli are used to test the threshold, during the early portion of the curve the subject appreciates the colour of the threshold stimulus, whilst at the later stages the light is said to be white. Yet again, by varying the state of light-adaptation of the subject before he is placed in the dark, the type of curve may be changed. After a period of weak light-adaptation the cones are nearly at their maximum sensitivity, so that the threshold on going into the dark is very soon determined by the rods; hence the kink comes early or not at all (Fig. 122). If the wavelength of the adapting light is varied, the relative extents to which the thresholds of the rods and cones are raised will be different, so that on placing the subject in the dark after adapting him to red light, for instance, his rods will be relatively active—because red light barely influences these—whilst the cones will have a high threshold; the curve for dark-adaptation will therefore have no kink, the rods determining the threshold from the beginning.

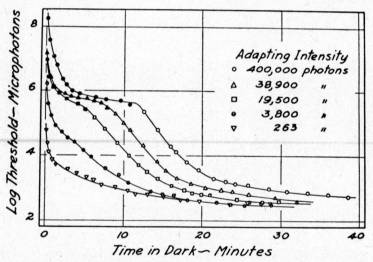

FIG. 122. The course of dark-adaptation as measured with violet light following different degrees of light-adaptation. The filled-in symbols indicate that a violet colour was apparent at the threshold, while the empty symbols indicate that the threshold was colourless. (Hecht, Haig & Chase, *J. gen. Physiol.*)

Rod Monochromat. Finally, a human subject defective in cone vision, the so-called *rod monochromat*, has a dark-adaptation curve with no kink (Fig. 123). This type of subject is useful since he permits the study of the rod thresholds without interference from the cones; thus, immediately on placing the normal light-adapted subject in the dark his threshold is determined by the cones; in other words the rods are less sensitive than the cones and so we cannot tell what their threshold is. In the rod-monochromat, however, we can measure the rod threshold as soon as he is placed in the dark and this is found to be very high; nevertheless the rods are not incapable of being stimulated, so that in ordinary daylight, although their thresholds are higher than those of the cones, they may be responding to light. However, there seems no doubt from Aguilar & Stiles' study, that the rods become "saturated" at high levels of illumination (2000–

2500 trolands or 100–300 cd/m²) in the sense that they become incapable of responding to changes of illumination (the l.b.i. is infinitely high); the same was shown by Fuortes, Gunkel & Rushton on their rod-monochromat who was virtually blind at high levels of illumination; the rods were responding to light, but not to changes, so that there was no awareness of contrast.

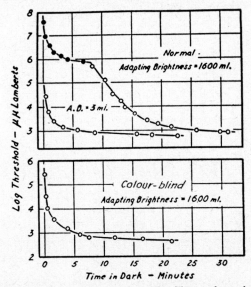

FIG. 123. Dark-adaptation curves for a normal subject and a rod monochromat. After adaptation to 1600 mL, the normal subject shows a rapid cone adaptation (solid circles) followed by a slow rod adaptation (open circles). The colour-blind shows only a rod curve of the rapid type. (Hecht, Shlaer, Smith, Haig & Peskin, *J. gen. Physiol.*)

Regeneration of Rhodopsin

Light-adaptation is accompanied by the bleaching of rhodopsin as shown by experiments involving extraction of pigment from eyes before and after exposure to light. It seems reasonable to suppose, therefore, that the process of dark-adaptation has as its basis the regeneration of rhodopsin in the rods; and the early studies of Tansley and Peskin certainly showed that the time-course of regeneration, as measured by extraction of frog's retinae after different periods in the dark, corresponded with the time-course of dark-adaptation. By the technique of *reflexion densitometry* the changes in concentration of pigment in the retina have been followed in the intact eye in man and lower animals.*

Reflexion Densitometry. The technique is based on the principle of measuring the light reflected from the fundus of the eye; if the retina contains rhodopsin it will absorb blue-green light of 5000 Å in preference to red and violet light, hence the light reflected from the fundus will contain less light of 5000 Å and more of these other wavelengths. The loss of intensity of light, moreover, will give a direct measure of the degree of absorption, and thus of the optical

---

* Brindley & Willmer (1952) initiated this ingenious technique, and it has been applied by both Rushton (Rushton & Cohen, 1954; Campbell & Rushton, 1955) and Weale (1953, 1962) to human and animal eyes respectively.

density of the rhodopsin in the retina. Careful analysis of the changes in amount of light after reflexion showed that they were indeed due, in the dark-adapted eye, to the presence of rhodopsin (Campbell & Rushton), the degree of absorption of the different wavelengths corresponding approximately with the extinction spectrum of rhodopsin. As Fig. 124 shows, exposure to lights of increasing intensity caused decreases in the amount of rhodopsin.

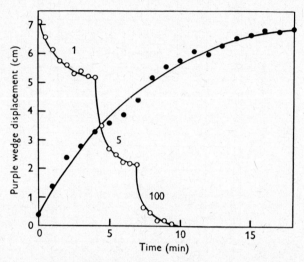

FIG. 124. ○—○, effect of bleaching the dark-adapted retina with lights of intensities 1, 5 and 100 units (1 unit = 20,000 trolands). ●—●, course of regeneration of rhodopsin. The ordinate indicates the degree of absorption of light in terms of displacement of an optical wedge. (Campbell & Rushton, *J. Physiol.*)

Rates of Regeneration. After complete bleaching the curve of regeneration is slow, the total regeneration requiring 30 minutes, and thus being equivalent to the time for complete dark-adaptation in man. In the cat, where dark-adaptation is considerably slower, the regeneration, measured in a similar way, takes much longer (Weale). Again, in the alligator, which dark-adapts relatively rapidly when the *b*-wave of the ERG is used as a measure of retinal sensitivity, the course of regeneration of pigment is correspondingly rapid.

Although there is thus a correlation between pigment regeneration and increase in scotopic sensitivity, the quantitative relationship between the two is not the simple linear one that might be predicted on the basis of photochemistry. Thus, exposure of the dark-adapted eye to a flash that only bleached 1 per cent of its rhodopsin caused an increase of tenfold in its threshold. Clearly, absorption of light by rhodopsin has reduced its sensitivity out of all proportion to the amount of bleaching; again, the final 1 per cent of regeneration during the last fifteen minutes of dark-adaptation brings about a disproportionate increase in sensitivity of the eye.

Rhodopsin Concentration and Retinal Sensitivity

Nevertheless, of course, there is a relationship between the amount of rhodopsin in the retina and visual threshold; thus Rushton (1961) varied the amount of bleaching by different exposures of humans to light and found that the threshold for rod vision always occurred when 92 per cent of the rhodopsin had regenerated; before this had happened the measured threshold in the dark was always the cone threshold; the remainder of the dark-adaptation curve, i.e. beyond the "kink" (Fig. 73, p. 135) corresponded to the regeneration of the remaining 8 per cent. Rushton was led to the belief that the relationship between threshold and rhodopsin concentration would be logarithmic, so that on plotting the log threshold against rhodopsin concentration a straight line should be obtained. Since, with a normal human subject in the dark, the threshold measured is that of the cones during the initial portion of dark-adaptation, it is not possible to find out what the condition of the rods is in the early stages, i.e. we cannot measure their threshold by simply exposing the eye to light, because it is higher that that of the cones. Thus we cannot pursue the relationship between rod threshold and rhodopsin concentration below concentrations corresponding to less than 92 per cent of the value in the completely dark-adapted eye.

Logarithmic Relationship. In a study on a rod-monochromat, a subject with defective cones so that the threshold was determined at all states of adaptation apparently by the rods, Rushton showed that the logarithmic relationship applied over a large range of rhodopsin concentration. Independently Dowling, using the time required for the b-wave of the ERG to appear after bleaching the rhodopsin in the rat's eye, showed that the logarithm of the threshold was indeed linearly related to the rhodopsin concentration measured chemically by extraction; this is shown in Fig. 125, where the amount of rhodopsin in the retina was varied, either by light-adaptation or by vitamin-A deficiency.*

Finally, Weinstein, Hobson & Dowling (1967) measured the changes in concentration of rhodopsin in the isolated retina and correlated these with visual sensitivity, as measured by the b-wave of the electroretinogram. The isolated retina remains functional in an artificial medium, but it does not regenerate rhodopsin after bleaching since it has lost its attachement to the pigment epithelium; thus the preparation is ideal for making controlled changes in the amount of unbleached rhodopsin. A perfect straight-line relationship between percentage of rhodopsin in the retina and the logarithm of the sensitivity was obtained.

Nervous Events. The fact that the sensitivity of the eye to light is not related in a simple linear way to the amount of visual pigment in its receptors suggests that the sensitivity is determined by other factors than this; and a great deal of evidence supports this. Thus Arden & Weale found that the rate of adaptation depended on the size of the test-field, the larger the field the greater the rate, and they concluded that this was because, during dark-adaptation, the size of the receptive field was actually increasing, i.e. the power of the retina to summate was improving. The importance of the neural aspect of adaptation was emphasized by Pirenne's and Rushton's experiments, showing that it was apparently unnecessary to expose a given portion of the retina to light in order to reduce this portion's sensitivity to light. Rushton exposed the eye to an adapting light made up of a

---

* Granit, Munsterhjelm & Zewi (1939) were unable to establish any relationship between height of b-wave and rhodopsin concentration, but Dowling considers that this is because the *height* of the wave is not the parameter that should be measured. Wald (1954) has suggested an ingenious explanation for the experimental fact, namely that the absorption of a little light has an inordinate effect on threshold in the completely dark-adapted eye. According to this the rhodopsin is compartmented in such a way that the absorption of a quantum of light by any molecule in a compartment prevents all the other molecules from absorbing.

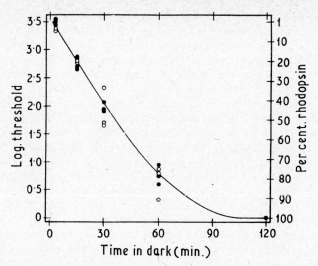

FIG. 125 (*a*). Dark-adaptation in the rat. As ordinates are plotted the threshold for producing an ERG (left) and the amount of rhodopsin extractable from the retina as a percentage of the dark-adapted value (right). Following light-adaptation the log threshold falls whilst the percentage of rhodopsin rises. ● Log threshold; ○, rhodopsin.

black and white grating of uniform stripes subtending 0·25° at the eye; the test flash, used to measure the threshold, was another identical grating, and the threshold was measured first when the two gratings were in phase, i.e. when white fell on white, and then when they were out of phase, white falling on black. It was

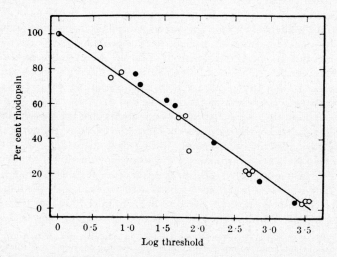

FIG. 125 (*b*). The relation between rhodopsin content of the retina and the visual threshold in animals dark-adapted after exposure to bright light, and in animals night-blind due to vitamin A deficiency. In both cases the log threshold varies linearly with the per cent of rhodopsin in the retina. ●, Night-blindness; ○, dark-adaptation. (Dowling, *Nature*.)

found that there was no difference between the thresholds, the striped background raising the threshold to the same extent in both cases.*

"Dark Light"

Thus we may speak of an effect of "dark light", the sensitivity of part of the retina having been reduced in exactly the same way as if light had actually fallen on it. This concept, originally developed by Crawford in 1937, showed that the threshold for vision in the dark at any period of dark-adaptation after an initial bleach of the retina could be expressed in terms of the background illumination required to give the same threshold. Thus, if, after say 5 minutes in the dark, the threshold had fallen to a luminance of I cd/m², then a background of $I-\Delta I$ could be found that would give an increment threshold, ΔI, equal to the measured threshold. In this way the eye, at any given state of adaptation, could be considered to be looking at its test-spot through a "veiling luminance" or through dark light.†

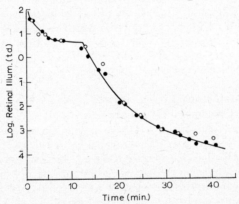

FIG. 126. Dark-adaptation curves. Solid circles are the equivalent luminance of a positive after-image obtained from the stabilized image that matched it. Open circles represent the equivalent background luminance that would be required to give the actual threshold, measured, now, as a liminal brightness increment, as postulated by Crawford. (Barlow & Sparrock, *Science*.)

"Retinal Noise"

Barlow (1964) has developed this view still further and has suggested that the dark light is essentially an expression of "retinal noise", i.e. of unnecessary retinal signals on to which a stimulus-light has to be imposed in order to be appreciated. Thus, when a subject has been exposed to light and his threshold is being determined in the dark, we may ask whether the high threshold we observe can be considered to be the high value he would have were he looking through a dark filter, or alternatively were he looking through dark light. The situations are different intrinsically; in the former case the signals from the stimulated rods are

---

* Barlow & Andrews (1967) have been unable to repeat this finding exactly; they found the threshold on those parts of the retina exposed to the dark bars of the grating lower than on those exposed to the bright bars, but higher than if there had been no illumination at all.
† The validity of this concept of equivalent background has been confirmed and extended by Blakemore & Rushton (1965) in a rod monochromat whose scotopic visual function could be examined over a wide range of luminance without contamination by cones.

attenuated, whilst in the latter they are enhanced. Barlow provided convincing arguments against the filter hypothesis; he asked, however, why, if the rods in this light-adapted state are highly active, the subject is not aware of light when placed in the dark? The situation should be of someone looking at the test spot through an after-image of the adapting light. However, after-images fade rapidly, and this is presumably because they are, in effect, stabilized on the retina; thus if we wish to prove that the after-image of the adapting light is really reducing the sensitivity of the retina to a test spot, we must stabilize the test-spot on the retina and compare its subjective brightness with that of the after-image. Barlow & Sparrock (1964) carried out this experiment; the eye was exposed to a bright annulus which caused bleaching of the retina; then, in the dark, a stabilized image of a spot, seen in the centre of the annulus, was compared with the after-image, and its luminance adjusted until a match was obtained. As time progressed this fell, and the typical dark-adaptation curve was obtained (Fig. 126, solid circles). In separate experiments the increment-threshold relationship for the test lights used was determined, against a background provided by real light, and the equivalent background for each of the matching spots was obtained and plotted against time on the same curve (Fig. 126, open circles). The coincidence between the two demonstrates that when one measures the threshold in a partially adapted eye, one is measuring, in effect, the incremental threshold when the test light is seen against a background corresponding to a certain after-image luminance.

Adaptation Pool. Intrinsically, Barlow's concept accords with modern thinking on signal detection; most sensory neurones give a background discharge, and this is certainly true of the retina when completely dark-adapted and not exposed to light; thus in order that the signal created by the impinging of light on the retina may be detected it must achieve a certain "signal-to-noise" ratio. In detail, however, this may be only partly the story since Rushton (1965a) has devised a critical experiment, the result of which is not consistent with the idea that bleached rods are sending out signals that are similar to those in a luminous background. Rushton (1965b) has developed the concept of an "adaptation pool" which, at any given state of adaptation, measures the extent to which spatial summation of rod responses is possible; during complete dark-adaptation this is maximal and after complete rod bleaching it is minimal; in some way, then, the presence of bleached rhodopsin in the rod causes it to signal to the pool and control the degree of summation and hence the threshold.

Rate of Regeneration. It could be argued that it is not the concentration of rhodopsin that determines the state of dark-adaptation but the rate of regeneration at any given moment. It is difficult to disprove or prove this hypothesis because in both situations a logarithmic relation between threshold and time in the dark would be expected; however, with foveal vision Rushton & Henry (1968) found that the rate of regeneration after a long bleach was twice as fast as after a short bleach, and yet under these conditions a subject's threshold was determined entirely by the amount of visual pigment bleached, in spite of differences in rate at any moment (Rushton, Fulton & Baker, 1969).

THE ABSOLUTE THRESHOLD

Frequency-of-Seeing. In the fully dark-adapted state, the minimum stimulus necessary to evoke the sensation of light is called the absolute threshold. It must be appreciated, however, that this threshold is not a fixed quantity but changes from moment to moment, so that if a subject is presented with a luminous screen in the dark, we may find a range of luminance over which there will be some probability, but not certainty, of his seeing it. Experimentally, therefore, the screen must be presented frequently at different luminances and the experimenter must plot the "frequency-of-seeing" against the luminance, to give the typical "frequency-of-seeing" curve shown in Fig. 127. From this curve it is a simple matter to compute a threshold on the basis of an arbitrarily chosen frequency, e.g. 55 per cent.

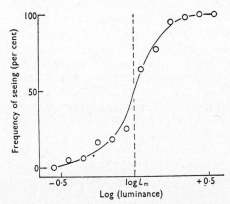

FIG. 127. Typical frequency-of-seeing curve for dark-adapted human observer. (Pirenne, Marriott & O'Doherty, *M.R.C. Sp. Rep.*)

Minimum Retinal Illumination. The threshold *stimulus* may be defined in several ways according to the experimental method of measurement. Thus we may express it as the luminance of a large test-screen, exposed for a relatively long time, e.g. one second. Since the size of the pupil is important, this must be measured, and the threshold expressed in trolands; furthermore the spectral composition of the light must be defined so that, if necessary, the number of quanta falling on unit area of the retina in unit time may be derived. This definition of the threshold stimulus is thus a statement of the *retinal illumination*. Experimentally this type of threshold has been found to be of the order of $0.75 . 10^{-6}$ cd/m² varying betwen 0.4 and $2.0 . 10^{-6}$ (Pirenne, Marriott & O'Doherty). This is equivalent to a retinal illumination of $4.4 . 10^{-5}$ scotopic trolands, and corresponds to the absorption of some 2300 quanta per second by some 13.4 million rods over which the large test-patch fell. Thus, in each second of time, only one rod in 5800 absorbs a single quantum when the eye is being stimulated at threshold intensity.

Minimum Flux of Energy. If an effectively point source of light is used, the image is concentrated on to a point on the retina so that the concept of retinal illumination loses its application; instead, then, we define the threshold as the *minimum flux* of light-energy necessary for vision, i.e. the number of lumens per second or, better, the number of quanta per second entering the eye. Experimentally Marriott, Morris & Pirenne obtained a value of 90 to 144 quanta/second.

Minimum Amount of Energy. Finally, if a very brief flash of light is used as the stimulus (less than 0.1 second) we may express the threshold as simply the total number of quanta that must enter the eye to produce a sensation. Thus there are three thresholds, the *minimum retinal illumination*, the *minimum light flux* and the *minimum amount of energy*. Of these, it is the last that interests us most since from this we are able to compute just how many quanta of light a single receptor must absorb to be excited.

Minimum Stimulus

This was determined in 1942 by Hecht, Shlaer & Pirenne by the method of presenting short flashes and determining the frequency-of-seeing curves. With a 60 per cent frequency as the end-point for threshold, the amount of energy striking the cornea was 54 to 148 quanta, depending on the observer; after allowing for scattering and absorption by the ocular media the range of useful light-energy was some 5–14 quanta. The image of the flash fell on about 500 rods, so that it was unlikely that one rod received more than a single quantum. This might suggest, therefore, that the minimum stimulus consisted of the simultaneous stimulation of some 11 rods, each absorbing a single quantum, their effects being transmitted to bipolar and ganglion cells by convergence. On this basis, then, the excitation of a single rod at threshold levels is not sufficient for vision, the excitatory effect being insufficient to permit conduction through the various synapses on the visual pathway. Only if several rods are excited will their effects be sufficient to break down these synaptic barriers and permit the passage of the excitation along the visual pathway to higher centres in the brain.

Quantum Fluctuations

The threshold varies from moment to moment and, according to Hecht, Shlaer & Pirenne, this must be due partly, if not wholly, to fluctuations in the number of quanta that a given flash contains when it is repeated. Thus, when we are dealing with very small amounts of light-energy, the actual number of quanta in any given flash will vary. Hence, according to the *uncertainty principle*, when we speak of a flash having an energy, a, we mean that this is the *average energy* the flash would have if it were repeated a large number of times, and the total number of quanta actually measured were divided by the number of flashes. For any given flash, with a mean energy-content of a quanta, we can only speak of the *probability* that it will contain a given number n, or more, quanta; and this probability may be simply calculated. On raising the stimulus-strength, measured by a, over a range of, say, tenfold, the probability that a flash will have a given number of quanta will increase, and finally become unity. The curve obtained by plotting this probability against the stimulus-strength has a characteristic shape, the steepness depending on the value of n as Fig. 128 shows.

Frequency-of-Seeing. Now, if the seeing of a flash depends only on the flash's containing the right number of quanta, say n, clearly the frequency-of-seeing will increase as the strength of the stimulus, measured by a, increases, because the probability that there will be n quanta in a flash increases with increasing value of a. Moreover, the shape of the frequency-of-seeing curve should be similar to the shape of the probability curve. As Fig. 129 shows, there is remarkably good agreement between the two, the range of stimulus-strength over which a flash is never seen and always seen being about a factor of tenfold, corresponding with the range over which the probability that a flash has n quanta in it varies from 0 to unity. We may note that the slope of the probability curve depends on the number of quanta, n, required, becoming steeper the greater the number, so that in order that our frequency-of-seeing curve shall match the probability curve we must choose a suitable value of n, the number of quanta required for vision. Hecht *et al.* found that, with different observers, n varied between 5 and 7, a fact suggesting that the absorption of these numbers of quanta corresponded with the stimulus for threshold vision. Direct measurements of the mean amount of quanta in a threshold stimulus gave values ranging from 6 to 17 quanta actually absorbed; since there is reason to believe that only one in two of absorbed quanta are effective —the *quantum efficiency* of the process is said to be 0·5—this would correspond then to 3 to 8 effective quanta.

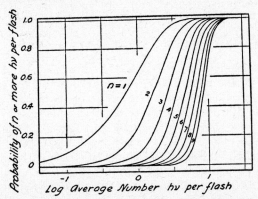

FIG. 128. These are probability curves based on quantum theory, showing the chance that a given flash, with average energy shown on the abscissa, will have *n* quanta in it or more. For example, if we employ curve belonging to *n* = 2, the chance of having 2 or more quanta in a flash is about 0·2 (i.e. 1 in 5) if the average energy of the flashes is 1 quantum (log 1 = 0). (Hecht, Shlaer & Pirenne, *J. gen. Physiol.*)

Two-Quanta Theories. Since the publication of this study of Hecht *et al.* there have been a number of discussions on the actual minimum number of quanta required to excite vision. There is no doubt that the absorption of one quantum *can* excite a rod, but the phenomena of summation, both temporal and spatial, prove that the excitation of more than one rod is necessary to evoke *sensation*. Thus, it has been found that, within certain limits of area, a given number of quanta in a flash are equally effective if they are concentrated on a small area of retina or spread over a larger one. If only one absorbed quantum was necessary to act as a stimulus, spreading these quanta over a large area should reduce the chance of any one rod

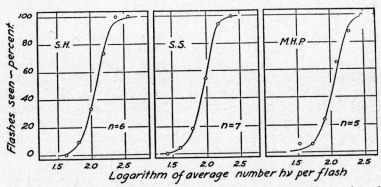

FIG. 129. Frequency-of-seeing curves for three human subjects. The number of times the flash was seen is plotted against the logarithm of the average energy per flash. The curves are theoretical ones, based on the theory of quantum fluctuations in the stimulus, and the fact that the experimental points fall on the curves suggests that the failure to see a given flash was largely due to its not having the required number of quanta in it. (Hecht, Shlaer & Pirenne, *J. gen. Physiol.*)

absorbing this quantum, so that the threshold should rise. The fact that, within an area subtending some 10 minutes of arc, the stimulus is equally effective, however spread out, means that at threshold more than one rod must be stimulated, their combined effects, transmitted to a bipolar cell, constituting the minimum stimulus for vision. Pirenne inclines to the belief that n is either equal to or greater than 4 although Bouman considers that his own and van der Velden's observations are better fitted by a value of 2.

SPATIAL AND TEMPORAL SUMMATION

The minimum stimulus has been defined in three ways, as we have seen, but the significance of the differences in definition may not be clear unless the importance of spatial and temporal summation is appreciated. It is because of spatial summation that the size of the test-stimulus must usually be stated, and because of temporal summation that the duration of the stimulus acquires significance.

Receptive Fields. Studies on single optic nerve fibres (p. 158) have shown that they have quite extensive "receptive fields" in the sense that a small stimulus applied over a large number of receptor cells can evoke a response in this single fibre. The anatomical basis for this is the convergence of receptors on bipolar cells and the bipolar cells on ganglion cells, so that a single ganglion cell may have synaptic relationships ultimately with many thousands of rods and cones. Because of this convergence, two spots of light falling within the receptive field may be expected to be more effective than a single one of the same light-energy, so that if we are measuring thresholds, the luminance of a single test-patch must be greater than that of two test-patches of the same area falling within the receptive field of a single ganglion cell. Our threshold luminance, expressed in terms of retinal illumination, will be smaller, then, the larger the area. If summation is perfect, moreover, we may expect the following relationship to hold:

$$Intensity \times Area = Constant$$

the greater the area of the stimulus the smaller the luminous intensity required to excite.

Ricco's Law. This is *Ricco's Law*. Expressed in terms of the actual number of quanta of energy reaching the retina this means that the threshold is independent of the area of the stimulus; thus if we have two stimuli forming images of 1 mm² and 2 mm² on the retina, according to Ricco's Law the luminous intensity of the smaller image will be four times that of the larger one at threshold. The *amounts* of energy falling on the two areas of retina in unit time will be equal, however, since the greater area of the one stimulus exactly compensates for its smaller retinal illumination. The area over which Ricco's Law holds varies with position on the retina, being about 30 minutes of arc in the parafoveal region with eccentricity of 4–7°, and increasing to as much as 2° at an eccentricity of 35° (Hallett, Marriott & Rodger, 1962). Over these areas, then, there is complete summation, in the sense that there is no loss of efficiency on spreading the available light-energy over a certain area instead of concentrating it on a much smaller one. When we are determining the minimum amount of energy necessary to excite the sensation of light, then, we must clearly ensure that the area of stimulus is within the limit over which Ricco's Law applies, since beyond this area the summation is only *partial*, in the sense that the larger stimulus will require more quanta than the smaller one, and will therefore have to have a higher luminous intensity. The existence of this partial summation will be indicated by the fact that the *luminous intensity* of the larger stimulus will be less than that of the smaller one, whilst its partial nature will be indicated by the fact that the *number of quanta* required to excite vision will be greater the greater the area over which they are spread.

Piper's Law. By Piper's Law is meant the statement that over a certain range of areas, where summation is partial, the relationship:

$$\sqrt{Area} \times Intensity = Constant$$

holds.* In practice partial summation extends over large areas, up to about 24°. Beyond this area there is no further summation, so that increasing the area of the stimulus requires no diminution in its intensity for threshold excitation; beyond this area then, the *luminous intensity* of the test-patch is independent of the area. The *amount of energy* entering the eye, of course, increases as the square of the size of the retinal image so that the threshold, expressed as the number of quanta per second required to excite, will increase as the area of test-patch increases.†

Probability Summation. Since partial summation extends over large areas of retina, it seems unlikely that convergence can be the sole basis for it, and much more likely that it results from an increased probability that the stimulus will have the necessary number of quanta to stimulate the retina. Thus, we may measure the threshold first with a single patch of light and next with two patches such that their images fall widely apart. If there is any summation at all, the two patches will have a lower luminance than the single one. If we suppose that the two patches are too far apart to permit any convergence of receptors on the same optic nerve fibres, we may nevertheless explain the greater effectiveness of the two patches on a probability basis. Thus, with a single patch the probability of seeing it has a certain value, p, so that the chance that it will not be seen is $(1-p)$. The chance of *not seeing* two patches is $(1-p)(1-p)$ which is smaller, so that a lower luminance is necessary for these two patches to have a given "frequency-of-seeing" than for the single one whose probability of not seeing is only $(1-p)$. We may thus speak of *physiological summation*, when this is due to convergence of retinal elements on optic nerve fibres, and *probability summation*, when the parts of the retina on which the summation occurs may be treated as independent of each other.‡

* According to Baumgardt (1959) this relationship follows from probability theory, but there seems little doubt from Barlow's (1958) studies that no simple Law will cover all the phenomena; thus the degree of spatial summation depends on the duration of the stimulus whilst the degree of temporal summation depends on the size of the stimulus. Some of the limitations of studies on area are considered by Hallett *et al.* (1962).

† At the fovea threshold measurements are those for stimulation of the cones and, as we have seen, in the completely dark-adapted condition the thresholds are much higher. This may well be due, not so much to a lower intrinsic sensitivity of the cone as compared with the rod, but to the much smaller degree of convergence exhibited by cones, so that the area over which complete summation is possible is very small. Hence, as ordinarily measured, the threshold will be in the region of partial or no summation. By reducing the size of the test-stimulus to a subtense of 2·7 minutes of arc Arden & Weale (1954) found the threshold intensities to be the same for fovea and parafovea, but their contention that the sensitivities of foveal cones and peripheral rods are equal is probably unsound (Pirenne, 1962, p. 106).

‡ An extreme example of probability summation is given by measuring the frequency of seeing a flash, when two observers are employed, with the frequency of seeing when only a single observer is used; from these we may compute thresholds in the usual way and it is found that the threshold for the "double observer" is less than that for the single

Ganglion Cell Responses. An interesting example of physiological summation that was "more than complete" was provided by Easter (1968) using the goldfish retinal ganglion cells stimulated by red light to ensure that records were obtained from single cones. Easter measured the intensity, $k \times I$, of a single spot of light falling in the receptive field of a ganglion cell, such that it gave the same response as two spots of intensity, I, falling on separate parts of the receptive field simultaneously. On the basis of complete summation we would expect k to be 2. In fact it was usually 3 or 4, indicating that the energy in the two spots acting together was much more efficient than the same amount of energy on a single spot. Essentially this means that the intensity of the response is not a linear function of the intensity of the stimulus, and analysis of the two-spot measurements indicated that a power function was more applicable, i.e. Response $=$ Constant $\times I^n$ with n equal to about 0·5. The logarithmic relation postulated by "Fechner's Law" was found to be inapplicable.

Temporal Summation. By this we mean that a stimulus lasting a short time must be stronger in luminance than a stimulus lasting a long time; thus the stimulus with the longer duration can be considered to be made up of a lot of successively repeated stimuli, and it is these successive stimuli that apparently add their effects. Over a period up to 0·1 second the Bunsen-Roscoe Law holds:

$$I \times t = Constant$$

indicating that up to this period the effective stimulus is only the number of quanta entering the eye, and we may say that the summation is *complete*. On a photochemical basis this complete summation is what one might expect, since the limiting factor in the threshold is presumably the absorption of a definite number of quanta, and the chances that this will happen will increase proportionately with the number of quanta entering the eye.*

PARTIAL SUMMATION

Beyond a certain time, which has been called the *action time*, the absorption of a second quantum of light will be of no use to help the effects of the absorption of the first, because these effects, e.g. production of an electrical change, will subside. We can therefore understand the limited period over which the Bunsen-Roscoe Law applies. Beyond this period, however, we may expect summation on purely probability grounds, since the longer the stimulus the greater the probability it will contain a certain minimal number of quanta, and according to Baumgardt (1959) there should be a square-root relationship between threshold energy and duration of the stimulus; over a certain range the experimental results agree with those predicted on this basis.

---

one. In a similar way the difference between the binocular and uniocular thresholds of a given subject could be accounted for by the increased probability of seeing a flash when using two eyes than one, i.e. the two eyes of a subject may be considered as independent from this point of view (Pirenne, 1943). There is no doubt that there are cortical neurones activated by impulses from both eyes (p. 507) so that a physiological summation must undoubtedly occur; Lythgoe & Phillips (1938) found that the binocular threshold, Th_B, was given by: $1 \cdot 4\ Th_B = \frac{1}{2}(Th_R + Th_L)$, a relationship to be expected were the threshold determined by areal summation of the two fields and if Piper's Law applied.
* Levick & Zacks (1970) have examined the application of the Bunsen-Roscoe Law to the spike response of cat ganglion cells; the *critical duration*, or *period of complete summation*, lasts for 64 msec, when the threshold, defined as the stimulus required to produce an extra spike over that contributed by the background activity, is employed. They suggest a redefinition of the critical duration as "the time interval within which an arbitrary manipulation of the wave-form of stimuli leads to responses indistinguishable from the response to a very brief flash of the same energy".

Signal Theory. Barlow & Levick (1969) have examined the responses of single ganglion cells of the cat to incremental light-stimuli, with a view to defining a threshold in terms of alteration in the background spike-discharge, the idea being to obtain, from the recorded spike activity before, during, and after a flash, a measurable quantity that would enable the cat to distinguish between a change in retinal illumination and a casual fluctuation in the background spike activity. The prime factor was undoubtedly the quantum/spike ratio, i.e. the extra quanta absorbed required to produce an extra spike in the sampling time, τ. Additional factors are the time-course of the response, so that alterations in this may affect the threshold, as in Bloch's Law; and finally the statistical distribution of the impulses that occur in the time-interval τ in the absence of a stimulus.

GENERAL REFERENCES

BRINDLEY, G. S. (1970). *Physiology of the Retina and Visual Pathway*, 2nd. ed. London: Arnold.

PIRENNE, M. H. (1956). Physiological mechanisms of vision and the quantum nature of light. *Biol. Rev.* **31**, 194–241.

PIRENNE, M. H. (1962). Visual function in man. In *The Eye*, vol. II, chapters I–II. Ed. Davson. London: Academic Press.

RUSHTON, W. A. H. (1959). Visual pigments in man and animals and their relation to seeing. *Progr. Biophys.* **9**, 239–283.

RUSHTON, W. A. H. (1965b). Visual adaptation. *Proc. Roy. Soc. B* **162**, 20–46.

SPECIAL REFERENCES

AGUILAR, M. & STILES, W. S. (1954). Saturation of the rod mechanism in the retina at high levels of illumination. *Optica Acta* **1**, 59–65.

ARDEN, G. B. & WEALE, R. A. (1954). Nervous mechanisms and dark-adaptation. *J. Physiol.* **125**, 417–426.

BARLOW, H. B. (1958). Temporal and spatial summation in human vision at different background intensities. *J. Physiol.* **141**, 337–350.

BARLOW, H. B. (1964). Dark-adaptation: a new hypothesis. *Vision Res.* **4**, 47–58.

BARLOW, H. B. & ANDREWS, D. P. (1967). Sensitivity of receptors and receptor 'pools'. *J. opt. Soc. Am.* **57**, 837–838.

BARLOW, H. B. & LEVICK, W. R. (1969). Three factors limiting the reliable detection of light by retinal ganglion cells of the cat. *J. Physiol.* **200**, 1–24.

BARLOW, H. B. & SPARROCK, J. M. B. (1964). The role of afterimages in dark adaptation. *Science* **144**, 1309–1314.

BAUMGARDT, E. (1959). Visual spatial and temporal summation. *Nature* **184**, 1951–1952.

BLAKEMORE, C. B. & RUSHTON, W. A. H. (1965). Dark adaptation and increment threshold in a rod monochromat. *J. Physiol.* **181**, 612–628.

BOUMAN, M. A. (1955). Absolute threshold conditions for visual perception. *J. opt. Soc. Am.* **45**, 36–43.

BRINDLEY, G. S. & WILLMER, E. N. (1952). The reflexion of light from the macular and peripheral fundus oculi in man. *J. Physiol.* **116**, 350–356.

CAMPBELL, F. W. & RUSHTON, W. A. H. (1955). Measurement of the scotopic pigment in the living human eye. *J. Physiol.* **130**, 131–147.

CRAWFORD, B. H. (1937). The change of visual sensitivity with time. *Proc. Roy. Soc. B* **123**, 69–89.

DOWLING, J. E. (1960). Night blindness, dark adaptation and the electroretinogram. *Am. J. Ophthal.* **50**, 875–889.

DOWLING, J. E. (1960). Chemistry of visual adaptation in the rat. *Nature* **188**, 114–118.

EASTER, S. S. (1968). Excitation in the goldfish retina: evidence for a non-linear intensity code. *J. Physiol.* **195**, 253–271.

FUORTES, M. G. F., GUNKEL, R. D. & RUSHTON, W. A. H. (1961). Increment thresholds in a subject deficient in cone vision. *J. Physiol.* **156,** 179–192.

GRANIT, R., MUNSTERHJELM, A. & ZEWI, M. (1939). The relation between concentration of visual purple and retinal sensitivity to light during dark adaptation. *J. Physiol.* **96,** 31–44.

HALLETT, P. E., MARRIOTT, F. H. C. & RODGER, F. C. (1962). The relationship of visual threshold to retinal position and area. *J. Physiol.* **160,** 364–373.

HECHT, S., HAIG, C. & CHASE, A. M. (1937). The influence of light adaptation on subsequent dark adaptation of the eye. *J. gen. Physiol.* **20,** 831–850.

HECHT, S., HAIG, C. & WALD, G. (1935). The dark adaptation of retinal fields of different size and location. *J. gen. Physiol.* **19,** 321–337.

HECHT, S., SHLAER, S. & PIRENNE, M. H. (1942). Energy, quanta and vision. *J. gen. Physiol.* **25,** 819–840.

HECHT, S., SHLAER, S., SMITH, E. L., HAIG, C. & PESKIN, J. C. (1948). The visual functions of the complete colorblind. *J. gen. Physiol.* **31,** 459–472.

LEVICK, W. R. & ZACKS, J. L. (1970). Responses of cat retinal ganglion cells to brief flashes of light. *J. Physiol.* **206,** 677–700.

LYTHGOE, R. J. & PHILLIPS, L. R. (1938). Binocular summation during dark adaptation. *J. Physiol.* **91,** 427–436.

MARRIOTT, F. H. C., MORRIS, V. B. & PIRENNE, M. H. (1959). The minimum flux of energy detectable by the human eye. *J. Physiol.* **145,** 369–373.

PIRENNE, M. H. (1943). Binocular and uniocular thresholds of vision. *Nature* **152,** 698–699.

PIRENNE, M. H., MARRIOTT, F. H. C. & O'DOHERTY, E. F. (1957). Individual differences in night-vision efficiency. *Med. Res. Council Sp. Rep. Ser.*, No. 294.

RUSHTON, W. A. H. (1965a). Bleached rhodopsin and visual adaptation. *J. Physiol.* **181,** 645–655.

RUSHTON, W. A. H. & COHEN, R. D. (1954). Visual purple level and the course of dark adaptation. *Nature* **173,** 301–304.

RUSHTON, W.A. H. (1961). Dark-adaptation and the regeneration of rhodopsin. *J. Physiol.* **156,** 166–178.

RUSHTON, W. A. H. (1961). Rhodopsin measurement and dark-adaptation in a subject deficient in cone vision. *J. Physiol.* **156,** 193–205.

RUSHTON, W. A. H., FULTON, A. B. & BAKER, H. D. (1969). Dark adaptation and the rate of pigment regeneration. *Vision Res.* **9,** 1473–1479.

RUSHTON, W. A. H. & HENRY, G. H. (1968). Bleaching and regeneration of cone pigments in man. *Vision Res.* **8,** 617–631.

RUSHTON, W. A. H. & WESTHEIMER, G. (1962). The effect upon the rod threshold of bleaching neighbouring rods. *J. Physiol.* **164,** 318–329.

WALD, G. (1954). On the mechanism of the visual threshold and visual adaptation. *Science* **119,** 887–892.

WEALE, R. A. (1953). Photochemical reactions in the living cat's retina. *J. Physiol.* **122,** 322–331.

WEALE, R. A. (1962). Further studies of photo-chemical reactions in living human eyes. *Vision Res.* **1,** 354–378.

WEINSTEIN, G. W., HOBSON, R. W. & DOWLING, J. E. (1967). Light and dark adaptation in the isolated rat retina. *Nature* **215,** 134–138.

PART 7

FLICKER

CRITICAL FUSION FREQUENCY

THE sensation of "flicker" is evoked when intermittent light stimuli are presented to the eye; as the frequency of presentation is increased a point is reached—the *critical fusion frequency*—at which the flicker sensation disappears to be replaced by the sensation of continuous stimulation. The study of flicker has turned out to be a valuable method of approach to the fundamental problems of visual phenomena—it is amenable to fairly accurate measurement and it represents a perceptual process intermediate in complexity between intensity discrimination and form perception.

The Talbot-Plateau Law. This generalization states that, when the critical fusion frequency has been reached, the intensity of the resultant sensation, i.e. the brightness of the intermittently illuminated, but non-flickering, patch, is the mean of the brightness during a cycle. Thus if the subject's view of an illuminated patch, of luminance 5 mL, is interrupted by rotating an opaque disc in front of it from which a sector has been removed, the resultant sensation will be equal to that given by a continuous stimulus of luminance equal to 5 × Area of Sector/ Total Area of Disc. This law finds a useful application in many experimental studies in which it is desired to cut down the illumination by accurately determined amounts; by varying the size of the sector in the rotating disc, different average intensities of illumination per cycle may be achieved and, so long as the critical fusion frequency is exceeded, these will appear as different steady luminances.

EFFECT OF LUMINANCE ON FUSION FREQUENCY

In general, the greater the luminous intensity of the flickering light, the higher must be its frequency to attain fusion. This is illustrated by Fig. 130 where the critical fusion frequency has been plotted against the logarithm of the retinal illumination in trolands. With foveal observation, the relationship is linear over a wide range, between 0·5 and 10,000 trolands, and this is the basis of the so-called *Ferry-Porter Law* which states that the critical fusion frequency is proportional to the logarithm of the luminance of the flickering patch. At very high luminances the fusion frequency passes through a maximum in the region of 50 to 60 cycles/sec. At very low luminances, in the scotopic range, the fusion frequency is remarkably low, of the order of 5/sec, so that under these conditions the temporal resolution of the fovea is extremely small, a separation of 200 msec between successive stimuli not being discriminated.

Peripheral Vision. With peripheral vision, where use of both rods and cones is possible, there is an obvious discontinuity in the graph, indicating the limited application of the Ferry-Porter Law; by plotting the logarithm of the fusion frequency against the logarithm of the luminance, as in Fig. 131, the break in continuity becomes more obvious and may be seen to occur in the region of a

221

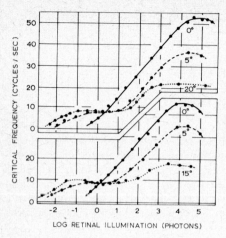

FIG. 130. Dependence of critical fusion frequency on the intensity of the stimulus. With foveal fixation (0°), the Ferry-Porter Law holds over a wide range. With eccentric fixation, the curves show discontinuities indicating the activity of the rods. (Hecht, Shlaer & Smith, *Cold Spr. Harb. Symp. Quant. Biol.*)

retinal illumination of 10 trolands (log 10 = unity), and corresponds to the transition from predominantly cone to predominantly rod vision. In Fig. 131 the different curves correspond to the use of different wavelengths of light, and it is seen that no break occurs with a red light of 6700 Å, as one would expect since this wavelength tends to stimulate only cones.

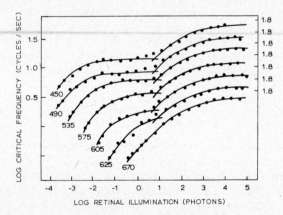

FIG. 131. The log of the fusion frequency has been plotted against the log of the stimulus intensity. With red light, no break occurs in the curve. The ordinates to the left apply to the topmost curve; the others have been moved down in steps of 0·2 log units, and their exact positions are indicated to the right. (Hecht, Shlaer & Smith, *Cold Spr. Harb. Symp. Quant. Biol.*)

Rod Vision. The low fusion frequencies obtainable with rod vision might suggest that these receptors would not give as high fusion frequencies as the cones at high luminances, if they could be studied in isolation. This view of the rods as less discriminating from the point of view of temporal resolution has been amply confirmed by many experiments. Thus Hecht *et al.* (1948) studied a rod monochromat, a subject with apparently non-functional cones. The curve obtained by plotting fusion frequency against log luminance did, indeed, show a break, suggesting the operation of two mechanisms, but it was argued that these corresponded to two types of rod, since the break occurred with red light;* the highest fusion frequency measured was only 21/sec so that the rods do, apparently, have less powers of mediating temporal resolution than the cones.

Significance of Ferry-Porter Law. The experimental finding—that to produce fusion the frequency of stimulation must be increased when the intensity of the light stimulus is increased—is, at first thought, surprising. The fact that fusion is possible is clearly dependent on the existence of an "after-effect" resulting from a single stimulus, so that a succeeding stimulus can fall on the retina while it is still responding positively to the first; on purely photochemical grounds one might expect the after-effect to last longer the more powerful the primary stimulus, since a greater quantity of light-sensitive substance would be decomposed; if this were the case, clearly the minimum time elapsing between this stimulus and the next, such that the latter arrives while the eye is still responding strongly to the first stimulus, will be greater with the more powerful stimulus and therefore the critical fusion frequency should be lowered. The response of the eye to illumination, however, is far too complex to be described simply in terms of a primary photochemical reaction—the latter is the first stage only in a process that is elaborated in the nervous structures of the retina and higher visual centres. The study of the ERG has thrown valuable light on the mechanism of flicker and has emphasized the importance of inhibition, besides excitation, in determining this form of resolution.

THE ERG AND THE INTERPRETATION OF FLICKER

The ERG in response to a single flash of light is a characteristic sequence of potential changes considerably outlasting the duration of the flash; a second flash, falling on the eye during these changes, starts off a new series of events, but the extent to which it will modify the existing state of the retinal potential will depend on how soon it follows; and we may envisage a condition, with rapidly repeated stimuli, such that the potential never falls to its baseline and, moreover, if the frequency is high enough, such that the record appears unbroken and smooth. From the point of view of the ERG we may call this the critical fusion frequency but this need not mean, at any rate without further proof, that the subjective sensation is likewise smooth. Fig. 132 from Granit & Riddell illustrates the phenomenon of fusion in the ERG of the light-adapted frog; as the frequency of the intermittent light-stimuli increases, the record becomes smoother, whilst the amplitude of the potential change decreases.

Importance of Inhibition. In the dark-adapted frog's eye, or the rod-dominated retinae of the cat and guinea pig, the fusion frequency was low, and below fusion successive stimuli seemed merely to result in successive *b*-waves imposed on the declining limbs of the previous *b*-wave. In the light-adapted frog eye, or the *I*-type of retina (pigeon, squirrel, etc.), a much higher fusion frequency was obtained, and an analysis of the ERG at frequencies below fusion indicated that the fluctuations of potential with successive stimuli were now characteristically different, the results of successive stimuli being to impose a negative *a*-wave on the preceding *b*-wave, i.e. the succession of events was *a-b-a-b-a* as opposed to *b-b-b-b-b*. If the *a*-wave really is a measure of pre-excitatory inhibition,

* The features of vision by the rod monochromat are discussed further on p. 269.

then the high frequency of flicker is possible with the cone-dominated *I*-retinae because, before the excitatory effects of each flash, there is a preliminary inhibition of the excited state remaining from the previous flash. Since the size of the *a*-wave, i.e. of the inhibitory P III component, increases with increasing stimulus-strength, we have a reasonable explanation of the effects of luminance on fusion frequency.*

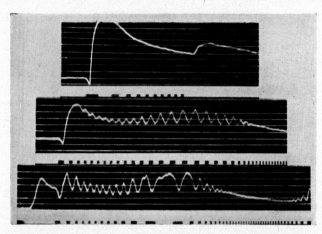

FIG. 132. Flicker as shown by the frog's ERG. The light stimuli are shown at the bottom of each record as white rectangles. Top: Dark-adapted eye with no appreciable flicker. Middle and bottom: Light-adapted eye showing flicker with fusion when the stimuli are sufficiently frequent. (Granit & Riddell, *J. Physiol.*)

ROD-TYPE ERG

Furthermore, the great difference between rod and cone vision might be explained on the basis that the rods are less susceptible to inhibition than the cones. Thus in the rod-dominated retina of the cat at ordinary stimulus intensities the flicker is determined, according to Granit, by a different process; here the negative, inhibitory, component is relatively unimportant, and the sensation of flicker is dependent on the primary excitation (the *b*-wave in the ERG) dying down sufficiently to allow the next stimulus to evoke a new *b*-wave which will appear as a notch on the descending limb. In the absence of a pronounced P III component, this will necessitate a considerable pause between individual light-stimuli if they are to be appreciated as distinct, i.e. the fusion frequency will be low.†

Subjective and ERG-Flicker. Subsequent studies of the flicker-ERG have tended to confirm Granit & Riddell's view, but we must remember that the

---

* The falling off in fusion frequency with very high luminances seems to be due to the intensification of the *a*-wave under these conditions, the flicker-ERG consisting solely of *a*-waves (Heck, 1957).
† We may note that in the typical rod-ERG the response to the first flash of a series of intermittent flashes is always very much larger than the succeeding responses; this is apparently not due to the light-adaptation that takes place to some extent during repeated exposure of the retina to flash stimuli; the relative amplitudes of first and successive responses depend critically on the duration of the dark phase between the two, as though the retina had to recover from an inhibitory condition imposed by the first flash (Arden, Granit & Ponte, 1960).

explanation begs the question as to the justification of considering the *a*-wave as peculiarly inhibitory. Moreover, we cannot be certain that unevenness on the record necessarily means a flicker sensation, nor can we be certain that a smooth record corresponds to subjective fusion. This is well exemplified by the study of the human ERG when the subjective sensation can be compared with the electrical record. Subjectively, as we have seen, we may obtain fusion frequencies of the order of 60/sec or more by suitably increasing the luminance of the flickering patch; in the early studies of the human ERG, however, the fusion frequency rose to a maximum at about 25/sec, further increases in luminance causing no increase in fusion frequency. The human ERG, under photopic conditions, was apparently similar to that given by a rod-dominated retina, so that a smooth record did not correlate with subjective sensation. This could be because the ERG is, in general, the pooled response of all the receptors in the retina, and since rods are in such a huge majority in the human retina the ERG-response to flickering stimuli should be a rod-response. The *subjective sensation*, however, might be determined by the cones, in spite of their numerical inferiority, presumably because, under photopic conditions, their pathways to higher centres are more favoured, or because of their greater cortical representation.

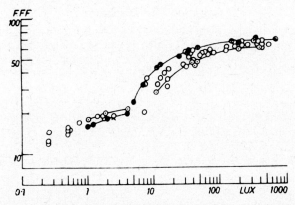

FIG. 133. The critical fusion frequency (FFF) as a function of stimulus-intensity as indicated by the human ERG. Ordinate: flashes/second; Abscissa: luminance in lux. Note logarithmic scales. (Dodt & Wadensten, *Acta ophthal. Kbh.*)

Cone Activity. If this were true it ought to be possible, by adequate amplification of the ERG, and by increasing the intensity of the light stimuli, to bring out the cone activity in the ERG when the rod activity was suppressed. How can we suppress the rod activity, however? The flicker-ERG gives us just this opportunity since, as we have seen, the ERG record becomes smooth at a luminance of about 10 lux with a fusion frequency of about 25/sec. Thus the rod-response can no longer follow the stimuli and any irregularity in the record obtained by increasing the luminance must now be due to cone activity. Dodt in 1951 was able to demonstrate just this; he increased the luminance of the flashes to far greater levels than earlier workers had used, and by employing an opal contact lens that enabled large areas of retina to be stimulated he found that, beyond 10 lux, flicker again appeared on the record, and the curve of fusion frequency against luminance corresponded with the subjective sensation. Fusion frequencies as high as 70/sec were obtained as Fig. 133 from Dodt & Wadensten shows. Even in the cat, moreover, Dodt & Walther were able to demonstrate a cone-type of ERG at sufficiently high luminance.

The Cone-ERG. The capacity of the rods to have their contribution to the ERG suppressed by use of an intermittent stimulus has permitted the study of the contribution of cones in some detail, and from thence it has been a simple step to the analysis of cone function in the intact retina. For example, by studying the effectiveness of different wavelengths in producing a cone-ERG under intermittent stimulation, Dodt & Walther found a typical photopic sensitivity curve with maximum in the yellow-green at 5500 Å compared with the scotopic maximum of about 5000 Å determined by measuring sensitivity of the retina in the dark-adapted state. This is the characteristic Purkinje shift (p. 256). Again, the characteristics of a single cycle of potential change, i.e. a single ERG, can also be studied by choosing frequencies that are well below fusion of the cone-ERG but of course above those for rod fusion; in general the *a*-wave is well defined and may be resolved into *two*, whilst the *b*-wave has *three*, components, suggesting the activities of different types of cone (Heck). In the completely colour-blind (rod-monochromat) there was no cone-ERG at all.

Ganglion Cell Discharges

We may ask what takes place in the optic nerve during flicker and fusion; The point was examined in detail by Enroth.* It will be recalled that single cells, or optic nerve fibres, respond to a flash of light in different manners and are described as Pure ON-, Pure OFF- and ON-OFF-elements according as they give a burst of spikes at ON, OFF or both at ON and OFF. In the cat the ON-OFF ganglion cells are in the majority so that these were studied most, but the ON-response of such a cell was essentially the same as a pure ON-response from the point of view of flicker, so that the same general principles of behaviour were valid; the same applied to the OFF-responses.

Fusion. To consider the pure ON-element, the first few flashes usually caused a simple spike discharge with no silent period between; this corresponds with the non-flickering period of the ERG. Soon each flash caused a burst of spikes in step with the stimulus (Fig. 134); as the frequency of flashing increased, the bursts became shorter but remained in step. At a certain frequency of flashing fusion occurs, in the sense that now the spike discharges, if they occur at all, bear no relationship with the stimuli. The unit shown in Fig. 134 gave a discharge corresponding to what would be seen with steady illumination, but others (usually OFF- and ON-OFF-elements) remained completely silent during fusion. Decreasing the flash-rate brought about an immediate flicker in the record. With ON-OFF elements, at low flickering rates the ON- and OFF-discharges behaved independently, each "flickering" with the ON- and OFF-phases of the light cycle. Usually fusion of the one response occurred before that of the other, so that if, for example, the ON-element fused first, the record was the same as that for a flickering OFF-element.

Ferry-Porter Law. The remarkable feature emerging from this study was the relationship between the fusion frequency for a given ganglion cell and the frequency of its spike discharge in response to a single flash; if the spike discharge

---

* The main results were described by Enroth in her thesis (1952); they have been extended by Dodt & Enroth (1954) and Grüsser & Rabelo (1958).

had a high frequency the fusion frequency was high, and *vice versa*. Since, usually, spike discharge frequency increases with increasing intensity of light, we have an electrophysiological basis for the Ferry-Porter Law relating fusion frequency with light intensity.

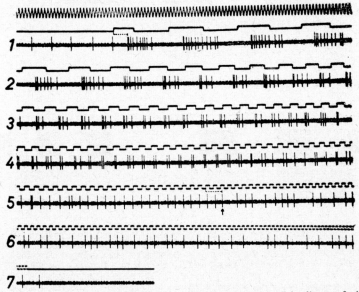

FIG. 134. Response of ON-element to repetitive flash stimuli leading to fusion, in the sense that when the flash-frequency reaches a certain value, marked by the arrow, the response is a discharge that bears no relationship to the frequency of flash-stimulation. Before this phase is reached the spike bursts are synchronous with the light flashes, beginning after a latency indicated by the dotted lines on the records marked 1 and 5. Upward movement of signal indicates onset of flash. (Enroth, *Acta physiol. scand.*)

Pre-Excitatory Inhibition. The mechanism of fusion is doubtless related to the pre-excitatory inhibition of the succeeding stimulus on the excitatory effect of the preceding one, and we may assume that the actual period between flashes that permits fusion is determined by the latency of this pre-excitatory inhibition and the latency of the excitatory response. Thus, to consider an OFF-element; at the end of the first flash it discharges, after a latency of a few msec. The second flash will inhibit this discharge provided the latency of its inhibitory effect is not too long, and as Fig. 135 shows, it will be when the dark period is equal to the difference between the two latencies. These latencies are variable, so that simple relationships between them and flicker-rate are not to be expected, although in general there is, indeed, a strong correlation between the length of the dark period and the difference of latencies (Grüsser & Rabelo).

Post-Excitatory Inhibition. With ON-elements, the significant factor is probably the post-excitatory inhibition that brings to an end the discharge in response to the onset of the flash. This inhibition is seen when the element is stimulated with a single flash of light; the response is a *primary activation* lasting about 20–70 msec and consisting of a burst of spikes; this is followed by a *discharge pause* of 80–250 msec and then a *secondary activation*. The discharge pause represents post-excitatory inhibition, and fusion will presumably occur when the flashes are so timed that the primary activation of the one flash is due to fall in the discharge pause of the preceding one. According to Grüsser & Rabelo, the latency of onset of the primary activation is the determining factor.

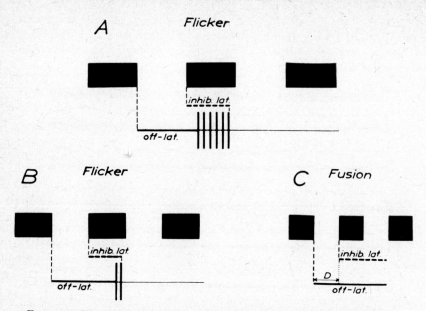

FIG. 135. Diagrammatic representation of OFF-latency and pre-excitatory inhibition latency during flicker of an OFF-element. Flashes are indicated by black rectangles. *A*, the dark interval is too long for the inhibitory effect of the light flash to affect the OFF-response; *B*, the interval is shorter, but the OFF-response has not been completely inhibited; *C*, there is fusion because the dark period is equal to the difference between the OFF-latency and the inhibitory latency. (Enroth, *Acta physiol. scand.*)

Spike Frequency Analysis. By using modern methods of analysis of the responses in single ganglion cells of the cat, Ogawa, Bishop & Levick (1966) have made a more precise analysis of critical fusion frequency and the factors determining it. Fig. 136 shows the response of a single unit to a flashing spot of light at 8 cycles/sec, i.e. well below fusion frequency; after an initial inhibition of the spontaneous discharge, because the unit examined is an OFF-centre type, spikes occur in response to the flashes, but in the early stages, e.g. in C, there is a virtually

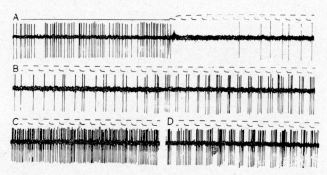

FIG. 136. Development of steady-state discharge pattern by a dark-adapted OFF-centre unit in response to an 8 cycles/sec flashing spot. Upper trace: phases of flash-cycle; downward, light off. A: immediately before and after the onset of the flashing light; B: continued from A; C: 25 sec after B; D: 2·5 min after C. (Ogawa, Bishop & Levick, *J. Neurophysiol.*)

continuous discharge with little relation to the flash-rate; after a time, D, the discharges become grouped in accordance with the flash-rate. Subjectively this doubtless corresponds with the observation of Granit & Hammond (1931) that when a flickering light is first presented it may appear continuous, the sensation of flicker developing later. Ogawa et al. used as a criterion of fusion the point when the record, presented as an "interval histogram", became identical with that of the spontaneous discharge during constant illumination. They found that the Ferry-Porter Law applied, the typical rod-cone break occurring at luminance levels some 100-fold those at which it occurs in man. The fusion frequencies at high luminance were high, of the order of 70 cycles/sec which is much higher than that obtained in man, but may well correspond with that in the intact cat, since Kappauf (1937) obtained frequencies as high as these in some cats using a behavioural method of study. When the steady discharges during fusion were measured at different intensities of illumination, these showed no obvious relation although the Talbot-Plateau Law would seem to demand this; however, the optic nerve message indicating luminance need not necessarily be the frequency of discharge in a given ganglion cell, in fact in view of the ON-OFF characteristics of so many, this is unlikely; it may be that it is the small movements of the eyes, provoking ON-OFF discharges, that are important, and these were absent in the experimental preparation.

The Visually Evoked Response

When the sensation of fusion is attained, the organism is unable to utilize, or else it ignores, the information reaching it at the receptor level; an obvious reason for this could be the failure of the receptors, or elements in the conducting pathway, to follow the repetitive stimuli with discrete responses, so that it is of some interest to find where the failure to transmit information occurs. The work with the ERG and ganglion cells suggests that fusion occurs at the retinal level, but a number of studies in which the cortical responses to visual stimuli have been compared with subjective sensation in man, or behavioural responses in animals, indicate that the sensation of fusion may intervene long before the cortical response becomes flat.* Thus a repetitive response in the retina is transmitted to the cortex, and may be recorded from its surface as a series of potential changes, of the same frequency as the flash-rate, but this does not necessarily evoke a continuous sensation. The cortical response is referred to as the *visually evoked response (VER)*, and will be described in more detail in Section IV (p. 519). In the rabbit, Schneider (1968) recorded the integrated VER at different frequencies; he found that frequencies that gave fusion, so far as the animal's behaviour was concerned, still gave VER's in phase with the stimuli; only when the flash-rate was some 20 cycles/sec greater than the behavioural critical fusion frequency did the record become smooth.

The amplitude of the VER diminished as the flash-rate increased, and it is interesting that behavioural fusion always occurred when the VER had attenuated to 10–20 per cent of its maximal value. The attenuation of the VER is due to failure of a proportion of the cortical cells to follow, so that their responses become equivalent to those induced by steady illumination; and examination of individual cortical cells shows that these vary greatly in their power to follow intermittent stimuli (Ogawa et al., 1966). Thus it seems to require a definite number of cortical cells to respond in phase with the stimuli, for flicker to be perceived, or more correctly, a definite amount of cortical activity.

* Walker et al. (1943) compared the responses to repetitive photic stimuli in the optic nerve, lateral geniculate body and cortex of the monkey; the optic nerve and lateral geniculate body could be "driven" up to rates of 62 and 59 cycles/sec respectively, whereas the cortex gave fusion at 34 cycles/sec. The critical fusion frequency for the monkey is probably about 35 cycles/sec, so that here fusion is determined by the cortex rather than the retina. Brindley (1962) has shown that, when the eye is intermittently stimulated both electrically and photically, it is easy to obtain beats by adjusting the phase-relations appropriately; these beats consist of a repetitive emphasis of some feature of the visual field. The fact that beats could be obtained when the photic stimulation rate (111–125 cycles/sec) was far above the critical fusion frequency (76–87 cycles/sec) indicated that the photoreceptors were responding repetitively at these high frequencies.

Effect of Dark-Adaptation

On the basis of Granit's analysis, we may expect the subjective phenomenon of flicker to show marked variations according as the human eye is light- or dark-adapted, i.e. according as the cone or rod mechanism predominates. This is, indeed, the case but the effects are complex for reasons that will become evident.

Let us suppose that a subject is placed in a box and that he looks at a flickering test-patch through a small window; the walls of the box may be illuminated to any desired intensity and thus the subject's state of light- or dark-adaptation may be controlled accurately. The flickering patch may likewise be given any desired luminance, so that we may make two kinds of measurement of the effects of dark-adaptation, i.e. we can make the walls of the box darker and darker and keep the luminance of the test-patch at a level such that rods only will be stimulated; alternatively we may make the walls of the box darker and darker as before, thereby increasing the dark-adaptation, but we can keep the luminance of the flickering test-patch at a level above the threshold for cone vision. It is very important to keep in mind this distinction in experimental approach as the effects of dark-adaptation are greatly different in the two cases. A third method, extensively adopted in the earlier work on this theme, consists in placing the subject in a dark room and varying the luminance of the flickering test-patch from the very highest levels to the lowest, threshold ones; in this case, however, it is very difficult to control the adaptation of the eyes—at the low levels of illumination of the test-patch the eyes will be more or less completely dark-adapted, but at the high levels they will be only partially so, and to different degrees, owing to the effects of the test-patch.

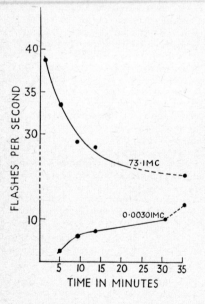

Fig. 137. The effect of dark-adaptation on critical fusion frequency. When the patch is of photopic intensity, dark-adaptation decreases c.f.f. When the patch is of scotopic intensity, dark-adaptation increases c.f.f. (10° eccentric fixation). (After Lythgoe & Tansley, *M.R.C. Sp. Rep.*)

Rod Vision. If the subject is placed in the box and is allowed to view a flickering patch of luminance below the cone threshold, then the effects of dark-adaptation are precisely those we should expect on the basis of the Ferry-Porter

Law; as the dark-adaptation proceeds, the test-patch appears brighter and brighter and consequently we may expect the fusion frequency to increase with dark-adaptation—this is the case as Fig. 137 (lower curve) shows; the general level of fusion frequency is low (3 to 12 cycles/sec).

Photopic Vision. If, on the other hand, the flickering patch has a high luminance, so that both rods and cones are stimulated, the effect of dark-adaptation (decreasing the illumination of the walls of the box) is in the opposite sense, the critical fusion frequency falling from a value of nearly 40 cycles/sec in the light-adapted eye to about 25 cycles/sec in the completely dark-adapted condition (Fig. 137, upper curve). The curves shown in Fig. 137 were obtained with 10° eccentric view, i.e. both rods and cones were being stimulated by the test patch; if central vision alone is used, the effects of dark-adaptation are said by Granit to be very small, hence the peripheral cones are apparently behaving differently from foveal cones. As we shall see in more detail in the next section, the peripheral cones differ in their organization from the foveal ones, in the sense that many share the same bipolar cell with rods, and it is just possible that these effects of dark-adaptation are due to the increased activity of rods that takes place.

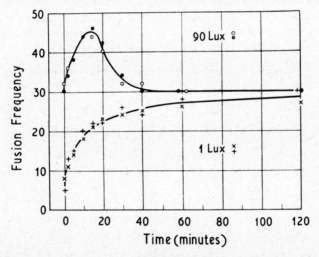

FIG. 138. Fusion frequency of cat's ERG. Lower curve shows that with a weak stimulus (1 lux) the fusion frequency increases with dark-adaptation to reach a maximum, characteristic of the rods. Upper curve shows that, with a moderately high intensity of light (90 lux), dark-adaptation causes an increase of fusion frequency, due to cone activity, but after a time the fusion frequency falls to that characteristic of rod vision, presumably because the rods are inhibiting the cones. (Dodt & Heck, *Pflüg. Arch.*)

Rod-Cone Interaction. In other words, the rods may well inhibit the cones. A particularly good example of this rod-cone interaction was described by Dodt & Heck in the cat, using the ERG as the criterion of flicker and fusion. Fig. 138 shows the effects of dark-adaptation on fusion frequency; with a scotopic stimulus, below the cone threshold (1 lux), the fusion frequency increases to a limiting rate of about 30/sec, the highest value for rod vision in the cat. This is the effect we have already seen in man. When a moderately high stimulus-intensity is employed (90 lux), the fusion frequency corresponds at first to that of the rods; during dark-adaptation, however, the cones apparently take over, since the fusion frequency rises above the rod limit. After 15 minutes of dark-adaptation, nevertheless, the fusion frequency falls and eventually becomes equal to that obtained with pure rod stimuli; presumably the rods have taken over again and suppressed cone activity.

Size of Patch and the Light-Dark Ratio

Granit-Harper Law. In general, if the size of the flickering patch is increased, its luminance being held constant, the fusion frequency increases; the effects can be large, so that a retinal area of 10 mm² may have a fusion frequency of say 60 cycles/sec whilst an area of 0·001 mm² of the same luminance may have just about half this. According to the so-called Granit-Harper Law, the fusion frequency is proportional to the logarithm of area. Varying the stimulus area, however, causes alterations not only in the total number of retinal elements stimulated, but in the relative extents to which peripheral and foveal elements enter into the response, and it would only be accidental if any simple relationship were found to hold over the whole retina. As Landis states: "Area as a determinant of fusion frequency is a complicated affair. Not only is the size of the retinal area a determinant, but its position on the retina, its shape, whether it is discrete or composed of several parts, all enter into the areal effect."

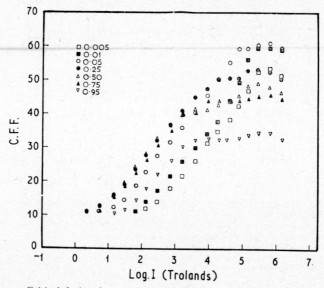

FIG. 139. Critical fusion frequency (C.F.F.) plotted against logarithm of light intensity (log I) for different light-dark ratios. (Lloyd & Landis, *J. opt. Soc. Amer.*)

Light-Dark Ratio. The effects of varying the light-dark ratio are even more complex and the results described are so frequently contradictory that one hesitates to make any generalization. Some results of Lloyd & Landis are illustrated in Fig. 139 where fusion frequency has been plotted against logarithm of luminance, the different symbols representing different light-dark ratios. With very small and very large ratios fusion frequency is lower than with intermediate ratios.*

Subjective Aspects

Brücke-Effect. v. Brücke in 1864 observed that the subjective brightness of a flickering patch of light could be higher than that experienced when the patch was presented continuously; the best effect was obtained with a flicker-rate of some 10 cycles/sec. It will be recalled that, according to the Talbot-Plateau Law, the sensation at fusion is the mean of the sensation during the light and dark intervals, hence the subjective brightness of the flickering light is very considerably greater than what would be experienced at the critical fusion frequency, since with a light-dark ratio of 0·5 this would be one half that experienced with steady presentation of the bright patch. According to Bartley the optimum condition for eliciting this Brücke-effect is when the bright phase lasts one third the time of the dark phase. Under these conditions, then, the after-effect of the first stimulus accentuates the effect of the second, a finding that probably explains the observation of Strughold that visual acuity may be improved by use of a flickering light of some 5–6 cycles/sec. It must be appreciated that when a very short set of flashes is presented, at a rate high enough to give fusion, the Talbot-Plateau Law need not apply, in fact there is reason to believe that it will not, and Nelson & Bartley (1964) have shown that, when the duration of the presentation of the set of flashes gets very short, the sensation of brightness increases inversely with the duration, and may go above that for steady illumination. This seems to be an expression of the Broca-Sulzer effect, i.e. the initial rapid increase in visual sensation when a light is turned on, followed by a decline corresponding with the α-adaptation of Schouten (p. 290). As with the Broca-Sulzer effect, the "Brücke-Bartley effect" is dependent on the wavelength of the light; moreover there is a very definite shift in subjective hue and of saturation (see, for example, Van der Horst & Muis, 1969).

Subjective Flicker-Rate. Another interesting subjective observation, recorded by Bartley, is that the actual flicker sensation, at the instant when flicker is about to disappear, does not correspond at all with the objective flash-rate. Thus at low levels of luminance it is possible to obtain fusion at 4 flashes/sec, nevertheless the flicker sensation at a flash-frequency just below this is that corresponding with a flashing light in the region of 10–20/sec. Conversely, at a high luminance, such that fusion is attained with 40 flashes/sec, the flicker *sensation*, as it gives place to fusion, has about the same subjective frequency as that observed with an objective rate of 4 flashes/sec, i.e. in the region of 10–20/sec. Thus the final flickering sensation, obtained just before flicker disappears, seems to be independent of the actual flicker-rate.

Electrical Excitation of the Eye

If a current is passed through an electrode on the surface of the eye and another, indifferent electrode, a sensation of light is evoked at make and break—the

---

* The effects of light-dark ratio have been exhaustively described by Landis (1954). Collins (1956) has made an elaborate analysis of the effects of varying not only the light-dark ratio but also the *wave-form*, i.e. the manner in which the flash comes on and goes off; he has combined these characteristics of the intermittent stimulus into a single parameter, which he calls the *flicker index*.

electrical phosphene—and Brindley (1955) has shown that this is most probably due to excitation of radially arranged retinal elements, either receptors or bipolar cells but almost certainly not of optic nerve fibres. When the current is alternated at, say, 50 cycles/sec, a flickering sensation is evoked. Wolff *et al.* (1968) have described in some detail the patterned sensations produced by much higher frequencies; these consisted of the appearance of blue patches of light with yellow lines radiating from them; as the frequency of alternation increased beyond 190 cycles/sec the blue disappeared and grey lines appeared looking like smoke-rings. At greater than 210 cycles/sec all patterns disappeared. When a green and red light were presented alternately at around the flicker fusion frequency, the predominance of either component in the fused sensation could be altered by electrical stimulation at the same frequency, the actual effect being determined by the phase-relations of the electrical and photic stimuli. This was presumably due to either a weakening or strengthening of the visual response by the electric currents and, depending on the phase relations, this could preferentially affect the response to red or green (Brindley, 1964).

GENERAL REFERENCES

BARTLEY, S. H. (1941). *Vision: A Study of its Basis.* New York: Nostrand.
GRANIT, R. (1947). *Sensory Mechanisms of the Retina.* London: O.U.P.
LANDIS, C. (1954). Determinants of the critical flicker-fusion threshold. *Physiol. Rev.* **34,** 259–286.

SPECIAL REFERENCES

ARDEN, G., GRANIT, R. & PONTE, F. (1960). Phase of suppression following each retinal *b*-wave in flicker. *J. Neurophysiol.* **23,** 305–314.
BRINDLEY, G. S. (1955). The site of electrical excitation of the human eye. *J. Physiol.* **127,** 189–200.
BRINDLEY, G. S. (1962). Beats produced by simultaneous stimulation of the human eye with intermittent light and intermittent or alternating electric current. *J. Physiol.* **164,** 157–167.
BRINDLEY, G. S. (1964). A new interaction of light and electricity in stimulating the human retina. *J. Physiol.* **171,** 514–520.
BROCA, D. & SULZER, A. (1903). Sensation lumineuse en fonction du temps pour les lumières colorées. *C.r. hebd. Séanc. Acad. Sci. Paris,* **137,** 944–946, 977–979, 1046–1049.
COLLINS, J. B. (1956). The influence of characteristics of a fluctuating visual stimulus on flicker sensation. *Ophthalmologica, Basel* **131,** 83–104.
DODT, E. (1951). Cone electroretinography by flicker, *Nature* **168,** 738.
DODT, E. & ENROTH, C. (1954). Retinal flicker response in cat. *Acta physiol. scand.* **30,** 375–390.
DODT, E. & HECK, J. (1954). Einflüsse des Adaptationszustandes auf die Rezeption intermittierender Lichtreize. *Pflüg. Arch.* **259,** 212–225.
DODT, E. & WADENSTEN, L. (1954). The use of flicker electroretinography in the human eye. *Acta ophthal. Kbh.* **32,** 165–180.
DODT, E. & WALTHER, J. B. (1958). Der photopischer Dominator im Flimmer-ERG der Katze. *Pflüg. Arch.* **266,** 175–186.
ENROTH, C. (1952). The mechanism of flicker and fusion studied on single retinal elements in the dark-adapted cat. *Acta physiol. scand.* **27,** Suppl. 100, p. 67.
GRANIT, R. & HAMMOND, E. L. (1931). The sensation time curve and the time-course of the fusion frequency of intermittent stimulation. *Am. J. Physiol.* **98,** 654–663.

GRANIT, R. & RIDDELL, H. A. (1934). The electrical responses of the light- and dark-adapted frog's eyes to rhythmic and continuous stimuli. *J. Physiol.* **81**, 1–28.

GRÜSSER, O. J. & RABELO, C. (1958). Reaktionen retinaler Neurone nach Lichtblitzen. *Pflüg. Arch.* **265**, 501–525.

HECHT, S., SHLAER, S. & SMITH, E. L. (1935). Intermittent light stimuli and the duplicity theory of vision. *Cold Spr. Harb. Symp. Quant. Biol.* **3**, 237–244.

HECHT, S., SHLAER, S., SMITH, E. L., HAIG, C. & PESKIN, J. C. (1948). The visual functions of the complete colorblind. *J. gen. Physiol.* **31**, 459–472.

HECK, J. (1957). The flicker electroretinogram of the human eye. *Acta physiol. scand.* **39**, 158–166.

KAPPAUF, W. E. (1937). The relation between brightness and critical frequency for flicker discrimination in the Cat. Ph.D. Thesis, Rochester. N.Y. University of Rochester. (Quoted by Ogawa *et al.*, 1966.)

LLOYD, V. V. & LANDIS, C. (1960). Role of light-dark ratio as a determinant of the flicker fusion threshold. *J. opt. Soc. Am.* **50**, 332–336.

LYTHGOE, R. J. & TANSLEY, K. (1929). The adaptation of the eye: its relation to the critical frequency of flicker. *Med. Res. Council Sp. Rep. Ser.* No. 134.

NELSON, T. M. & BARTLEY, S. H. (1964). The Talbot-Plateau law and the brightness of restricted numbers of photic repetitions at CFF. *Vision Res.* **4**, 403–411.

OGAWA, T., BISHOP, P. O. & LEVICK, W. R. (1966). Temporal characteristics of responses to photic stimulation by single ganglion cells in the unopened eye of the cat. *J. Neurophysiol.* **29**, 1–30.

SCHNEIDER, C. W. (1968). Electrophysiological analysis of the mechanisms underlying critical flicker frequency. *Vision Res.* **8**, 1233–1244.

WALKER, A. E., WOOLF, J. I., HALSTEAD, W. C. & CASE, T. J. (1943). Mechanism of temporal fusion effect of photic stimulation on electrical activity of visual structures. *J. Neurophysiol.* **6**, 213–219.

VAN DER HORST, G. J. C. & MUIS, W. (1969). Hue shift and brightness enhancement of flickering light. *Vision Res.* **9**, 953–963.

WOLFF, J. G., DELACOUR, J., CARPENTER, R. H. S. & BRINDLEY, G. S. (1968). The patterns seen when alternating electric current is passed through the eye. *Quart. J. exp. Psychol.* **20**, 1–10.

PART 8

VISUAL ACUITY

THE perception of flicker demands the resolution of two stimuli separated in time; by visual acuity is generally meant the power of the eye to resolve two stimuli separated in space, i.e. it is fundamentally related to the spatial relationships between receptor elements as opposed to the temporal characteristics of the response of a single element.

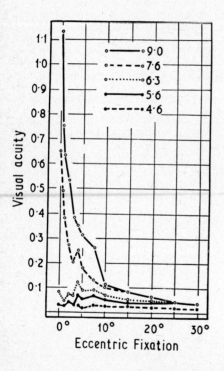

FIG. 140. The variation of visual acuity with eccentricity of vision as measured along the horizontal meridian of the temporal retina. Five intensity levels, varying from 9·0 to 4·6 log micromicrolamberts were employed (9·0 log micromicrolamberts = 10^9 micromicrolamberts = 1 mL). At 25° and 30° all the curves, except for 4·6 log micromicrolamberts, practically run together and are designated by single points. (Mandelbaum & Sloan, *Am. J. Ophthal.*)

Central and Peripheral Viewing

As with the other aspects of vision that we have so far considered in detail, the resolving power of the human eye differs markedly according as the conditions of viewing favour rod or cone vision. Thus, visual acuity in the dark-adapted eye with parafoveal vision and scotopic luminous intensities is only a small fraction of that obtained with foveal vision and high luminous intensities. Nevertheless it would be wrong to consider this as a difference in rod and cone functions *per se*, since peripheral vision under photopic conditions gives a low order of visual acuity. This is illustrated by Fig. 140, which shows visual acuity as a function of angle of view, foveal acuity being put equal to unity. Thus at 5° the acuity is only about a quarter of that found with foveal vision, and at 15° only about one-seventh.

Cone Density. The fall in visual acuity correlates fairly well with the fall in density of distribution of cones, which become more and more rare as we progress to the periphery of the retina. Density of receptors *per se*, however, is also not the determining factor since the rod density is very high throughout the retina yet visual acuity under scotopic conditions is always less than that under photopic conditions as Fig. 140 shows except, perhaps, in the extreme periphery. As we shall see, it is the synaptic organization of the receptors that is the important factor.

PHOTOPIC VISUAL ACUITY

Effects of Luminance of Test-Object and Adaptation

If the luminance of the test-object is such as to stimulate the cone mechanism, Lythgoe has shown that the curve showing the variation of visual acuity with the luminance of the test-object varies according to the experimental conditions; if the subject is placed in a dark box and is allowed to see the test-object through a small window, he can be considered to be in a state of dark-adaptation even though the luminance of the test-object is high enough to stimulate cone vision. In this case the acuity increases up to a maximum at approximately 10 e.f.c. and then begins to fall (Fig. 141, A); if the walls of the box are given a luminance

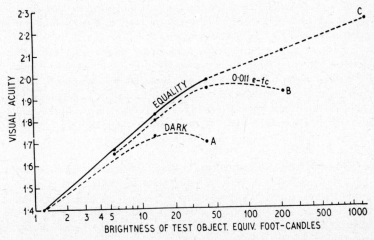

FIG. 141. Illustrating the effects of the luminance of the surround (i.e. state of adaptation of the eye) on visual acuity. A, black surround: visual acuity increases with increasing luminance of test-object only up to a certain point. B, surround luminance 0·011 e.f.c. C, surround luminance varied continuously so as to be equal to that of test-object. (Lythgoe, *Med. Res. Council Sp. Rep.* By permission of the Controller of H.M. Stationery Office.)

of 0·011 e.f.c., i.e. if the subject is only partly dark-adapted, the acuity continues to increase to a maximum at approximately 50 e.f.c. but then falls off (Fig. 141, B). If the luminance of the box is continuously adjusted so that it is equal to that of the test-object, it is found that visual acuity increases progressively with the luminance of the test-object (Curve C). If the luminance of the box is made greater than that of the test-object, generally the acuity is lower than when the luminances are equal. Thus with visual acuity we are confronted again with some form of retinal interaction in so far as the acuity of the central retina is influenced by the general illumination of the eye, i.e. by light-adaptation.

Adaptation. We have seen that dark-adaptation increases visual acuity under scotopic conditions; this effect is doubtless due to the increased sensitivity of the rods; the curves in Fig. 141 indicate that light-adaptation, under photopic conditions, increases visual acuity, and consequently dark-adaptation decreases it; since dark-adaptation increases the sensitivity of the cones (although not to the same degree as that of the rods) it is clear that the effect of dark-adaptation on photopic visual acuity is not closely connected with changes in sensitivity of the cones *per se*.

Normal Visual Acuity

We have so far considered variations in the visual acuity, but have said little about its absolute magnitude. It is customary to think of the visual acuity being determined by the "grain" of the retinal mosaic of cones—thus the Snellen letters are constructed on the assumption that the average person can resolve points separated by 1 minute of arc, and this limit has been considered to be roughly determined by the diameter of a single cone. Thus if we consider two white points on a black field, if these are so close together that their images fall on a pair of adjacent cones it is clear that they would not be appreciated as separate since a single, larger, white point would produce the same effect; only if one unstimulated cone is between the two images should the points appear as discrete. If we take Polyak's figure of 1.5μ for the diameter of the base of a foveal cone, this would permit a resolving power of approximately 20 seconds of arc; the best figure for visual acuity indicates a resolving power of this order. When, however, visual acuity is measured by the power to detect a break in a contour, e.g. with a vernier scale adjustment, or to detect the presence of a single line on a uniform background, a far greater resolving power is found (of the order of 4 seconds of arc for the contour break and 0.5 second for the line), and numerous explanation have been put forward to explain the apparent anomaly.

The "One-to-One" Relationship. If the individual retinal cones are to be considered as the units on the basis of which resolving power is to be determined, as opposed to groups of cones, there must be a so-called "one-to-one" relationship between cones and optic nerve fibres; in other words, two adjacent cones must be able to convey their impulses along independent paths; if two cones, close together, were connected to the same bipolar cell, the impulses relayed from this cell would be characteristic of neither cone separately, so that it would be difficult to conceive of a "local sign" being attached to either cone, and consequently resolving power could not be so fine as to permit the differentiation of stimuli separated by one unstimulated cone.

Midget Cells

The organization of the retina has been discussed earlier, and we have seen that the midget bipolar, whose significance was first emphasized by Polyak (1941), makes a special type of contact with the cone pedicle; furthermore, a given midget bipolar only receives messages from one cone. The same exclusive relationship is found with the midget ganglion cells, so that in this sense a given cone may have an exclusive pathway to the lateral geniculate body, a pathway not shared by another cone. Strictly speaking, however, there is no absolute "one-to-one" relationship between cones, bipolars, and ganglion cells owing to the fact that cones, even in the central area, make connections with diffuse bipolars besides with a midget; similarly, the same midget bipolar that makes a "private" synapse with a midget ganglion also makes contact with the diffuse varieties of ganglion cells. The "one-to-one" relationship thus exists only by virtue of a peculiar synaptical arrangement between a cone and its related midget bipolar and ganglion cells; although a share of the impulses excited in a cone is directed into collateral channels (diffuse bipolars, horizontal and amacrine cells) the main or specific cone influence seems to remain restricted to this special channel.

SPATIAL SUMMATION

Thus spatial summation of stimuli, such as is observed with rod vision, could only be mediated in the central (cone) area by means of the subsidiary synaptic connections and it is therefore not surprising that very little evidence of this type of summation by the cones has been obtained. Lythgoe suggested that the increased visual acuity obtained by light-adaptation (see, e.g. Fig. 141) is due to an inhibition of these subsidiary nervous paths so that the cones, during light-adaptation, tend to react more and more as single units.

Diffraction and Chromatic Aberration. Having established the existence of a "one-to-one" relationship between cone and optic nerve fibre for the central region of the retina, albeit restricted in scope, we can postulate a certain local sign attached to any foveal cone, so that we have a basis for the differentiation by the higher centres between the stimulation of two adjacent cones and two separated by one unstimulated cone. In the absence of diffraction and aberration effects, then, a figure of 20 seconds of arc could be taken as the anatomical limit of resolution of two points by the most central portion of the fovea. Diffraction and chromatic aberration, however, must certainly be taken into consideration; it is all very well to assume that the image of a grating or Landolt-C will be a clear-cut replica of the original, but let us suppose that the white break in the Landolt-C at the limit of resolution produces an image two, three, or four times its "geometrical size"; clearly it will stimulate two, three, or four cones so that it will be impossible to localize the break exactly; moreover, the black line of the Landolt-C has the same thickness as its break; if the white points on the inside and outside of the black contour all produce images extending over one or more cones there will be no unstimulated cone on the retina and, on the simple view so far taken, a Landolt-C whose black line subtends only 20 seconds of arc should be invisible.

DIFFRACTION RINGS

As a result of diffraction, a point of light produces an image of finite size on the retina; the image consists of a central bright spot surrounded by successive dark and bright rings; with a pupillary diameter of 3 mm and with light of wavelength 5550 Å, the central spot will have a diameter of rather more than $3.8\ \mu$, subtending some 47 seconds of arc at the nodal point of the eye; that is, there should be a definite overlapping of cone stimulation even when the stimulus arises from a theoretical point source. In general, the light falling in the bright rings concentric with the central bright disc only amounts to 16 per cent of the whole, so that the much greater overlapping due to these rings can probably be ignored.

CHROMATIC ABERRATION

The effects of chromatic aberration are even more serious; they will be discussed in more detail in a later section (p. 617); for the moment we must note that the effect is to make the image of a point source into a large disc covering some four cones.

RETINAL IMAGE

As a result of these combined effects the image of a grating, for example, will not consist of a series of bright and black patches on the retina; long before the resolution limit of such a grating has been reached, the image will be such that *no cone will be unstimulated*; this is shown by Fig. 142 from Hartridge where the relative luminances of the different parts of the image of a grating on the retina are plotted; in the middle of the geometrical shadow of the black lines, the intensity is some 40 per cent of what it would be if there were no black lines at all; beneath the image of the white bar the intensity is only 60 per cent. In the same figure the distribution of light in the region of the geometrical edge of an extended object is shown; it is seen that there is no clear-cut shadow, light falling on the retina several cones' width within the geometrical shadow.

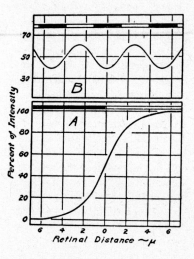

FIG. 142. The distribution of light intensity in the image of an edge (A) and a grating (B) on the retina as calculated by Hartridge. (Shlaer, *J. Gen. Physiol.*)

Criterion for Resolution. Clearly the classical conception of the basis of visual acuity—which demands that two resolvable points should be such that their images have an unstimulated cone between them—must be abandoned, since it fails to fit the observed facts of visual acuity. If we adopt Hartridge's criterion of the limit of resolution of two points, namely that their images must be sufficiently far apart so that there is a cone between that is *less stimulated than the rest*, we are not only provided with a way round our difficulty, but we can also explain the extremely high order of visual acuity exhibited in the detection of the break in a contour. Thus although a cone, falling under the geometrical image of a black bar, is stimulated, it receives less light than adjoining cones falling under the geometrical images of the white bars, and there is thus a peripheral basis for the resolution of points closer together than might otherwise have been expected.

INDUCTION EFFECTS

According to Hartridge's analysis, the limit to resolution is determined almost exactly by the "grain" of the retinal mosaic of cones, i.e. by the diameter of a central retinal cone. It may be mentioned here that local inductive effects, whereby the sensitivity of cones adjacent to the most strongly stimulated one is depressed (pp. 145, 293), assist in the resolution process. i.e. the inductive effects *amplify* the differences in response of adjoining cones on which the image of a grating falls.

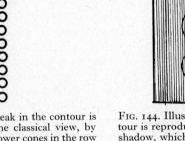

FIG. 143. The break in the contour is recognized, on the classical view, by the fact that the lower cones in the row are stimulated whilst the upper ones are not.

FIG. 144. Illustrating how the break in the contour is reproduced accurately in the diffraction shadow, which extends to some distance to the right of the geometrical shadow. Above the break, the shadow is slightly deeper than below, hence the cones above are slightly less strongly stimulated than those below.

Detection of Breaks in Contours. The high degree of resolution exhibited in the recognition of breaks in contours, where the limit is of the order of 4 seconds of arc—"vernier acuity"—appears, at first sight, to present greater difficulties. On the classical view, the limit to detection of a break in a contour would be determined by the break of such a size that the image of the line above the break fell on a row of cones whilst the image of the line below the break fell on another row (Fig. 143); the break in the contour would be recognized by the existence of some stimulated and some unstimulated cones in a row. Such a view demands, however, a cone size much too small for the observed resolution.

ABERRATIONS

When diffraction and chromatic aberration are taken into account, the image of the contour extends to a considerable distance beyond the geometrical shadow, but the *break* in the contour is accurately reproduced, even though at a considerable distance (Fig. 144). Let us imagine that the edge of the shadow falls along a row of cones (Fig. 144); at the break in the contour the cones below the break receive less illumination than those above it, but it is impossible, of course, that any number of cones in a row could be unstimulated whilst others were stimulated. The basis for the recognition of the break in the contour must be that in the region of the break some cones will be slightly less strongly stimulated than others. Hartridge has calculated that the variation of luminance in the shadow is sufficiently great to permit a significant difference in stimulus-intensity for the cones in the region of the break, even when the latter only subtends an angle of a few seconds at the eye; i.e., by moving along the retina a distance corresponding to 4 seconds of arc, a sufficient change in illumination is obtained to have an effect on the response of the cones. On this view, the edge of the shadow does not have to fall on any line of cones; the shadow varies continuously in luminance from left to right; at any point above the break, the luminance will be always rather less than at any point in the same vertical line below the break. So long as that difference in luminance is adequate to cause a differential stimulation of cones at the two points, we have the basis for the recognition of the break in the contour. Diffraction and aberration, by causing a progressive decrease in luminance of the image of the line to left and right, thus provide the basis for discrimination which would be impossible with a geometrical image.

Resolution of a Single Line. Hecht & Mintz have shown that a line may be detected against a uniformly bright background when it subtends only 0.5 seconds of arc at the eye; the image of such a line on the retina consists of nothing more than a slight variation in luminance to left and right of the geometrical image; this is so slight that even along the line corresponding to the geometical shadow the luminance is only 1 per cent less than that in regions unaffected by the line; nevertheless this difference of luminance is considered to be adequate to cause a differential cone stimulation that could be interpreted as a localized change in luminance, i.e. the increment corresponds to the value of $\Delta I/I$ found for the prevailing level of luminance. On this view, it would appear that the limiting factor in the detection of a line is the liminal brightness increment.

Effect of Luminance of Test-Object

We have seen that visual acuity increases progressively with illumination of the target; the explanation for this is probably not simple, since this improvement is only achieved up to a point unless the surround illumination is increased too (Fig. 141, p. 237). This influence of the surround illumination suggests that neuronal interaction in the retina is also affected by luminance. Hecht considered that increasing the luminance brought into operation more and more cones, and so the effective mosaic became finer and finer and thereby permitted better and better resolution. In the simple form in which it was presented the theory was inadequate, but as modified recently by Pirenne and his colleagues, it may well account for many of the phenomena.

Multiple Unit Hypothesis. Their point of view is best illustrated by a diagram (Fig. 145). We have seen that the requirement for maximal visual acuity is the effective one-to-one relationship between cones and optic nerve fibres. Any deviations from this relationship must reduce resolving power. Thus, as Fig. 145 shows, a perfect three-to-one relationship would give a visual acuity of one third the maximum, since the effective size of the receptor is three times as large. The visual acuity, other things being equal, must therefore depend on the size of the effective retinal units than can be brought into operation, and Pirenne, Marriott & O'Doherty have suggested that it is mainly because increasing the luminance of the test-object brings into operation smaller units, i.e. smaller groups of receptors converging on a single optic nerve fibre, that the basic relationship between visual acuity and luminance pertains.

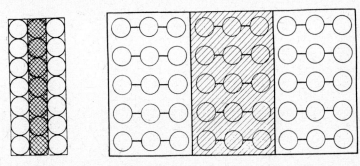

FIG. 145. Illustrating the concept of retinal units in defining resolving power. *Left:* The cones behave in a perfect one-to-one manner, so that the effective retinal mosaic for discriminating a grating is constituted by rows of single cones. *Right:* Because of convergence in the visual pathway, three cones behave as units, hence the grating at limiting resolution must be three times as wide.

LIMITING LIGHT-FLUX

Thus the large units are able to summate light-stimuli falling over large areas and therefore their threshold will be lower than that of the smaller units in which summation will be more limited. On this basis we might expect that the flux of light-energy reaching the eye from the white break in the Landolt-C, corresponding to maximal visual acuity, would be the same for the various luminances. Thus under scotopic conditions we might expect it to be equal to the minimum flux necessary for vision, of the order of 150 quanta/sec, and according to Pirenne *et al.* this is roughly true. Similarly, with photopic vision, the minimum flux to excite the cones is of the order of $2 \cdot 5 \cdot 10^{-12}$ lumens/sec, and if the fluxes emitted from the breaks of the Landolt-C's are calculated for different luminances these correspond approximately up to visual acuities of about 2. Beyond this, however, the amount of light entering the eye from the break is well above the threshold flux. As Pirenne has indeed pointed out, it is highly unlikely that this can be a complete explanation since threshold measurements are made under vastly different conditions from measurements of visual acuity, where large parts of the retina are stimulated by the surroundings of the Landolt-C.

Contrast. In considering the luminance of the test object there is an unfortunate tendency to ignore the real factor, namely the contrast between the black details on the white background, or *vice versa*; there is a tendency to treat the black as absolutely black, and not reflecting light from its surface, thereby implying that increasing the luminance will increase the contrast in a parallel fashion.

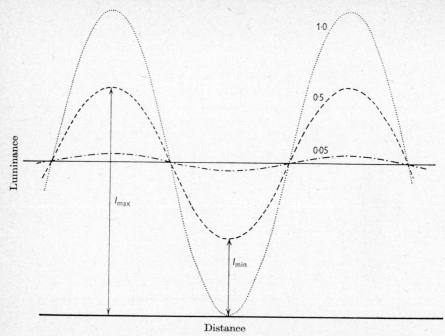

FIG. 146. Illustrating the contrast-ratio for a grating generated by a sinusoidal variation in luminance on a cathode-ray screen. Contrast is defined as $(I_{max} - I_{min})/(I_{max} + I_{min})$. Three ratios of 1·0, 0·5 and 0·05 are shown. Note that the mean luminance level remains constant for different contrast-ratios. Spatial frequency is defined as the reciprocal of the angular distance between successive maxima in the sinusoidal distribution. (Campbell & Green, *J. Physiol.*)

This is not necessarily true, so that it is more logical to express the lighting conditions of the target in terms of its *contrast*. A useful experimental method of establishing gratings with accurately defined contrast is to generate the grating on a cathode-ray oscilloscope screen by means of an appropriate oscillator; in the set-up described by Campbell & Green (1965) the intensity of emitted light from the screen varied in a sinusoidal manner with distance, as illustrated in Fig. 146; the contrast is given by: $(I_{max} - I_{min})/(I_{max} + I_{min})$, and three different contrast-ratios are illustrated for the same grating-frequency. The advantage of this system, besides that of accurately defining the important experimental parameter, namely contrast, is that the average luminance of the screen during a cycle remains the same independently of altered contrast-ratios. We may define the *grating-*, or *spatial, frequency* as the angle subtended at the eye by the distance between two peaks of intensity; when the subject is just able to discriminate the grating, this grating-frequency, in cycles per degree, gives a measure of the resolving power, as usually defined, and its reciprocal is the visual acuity.*

* The resolving power of the eye is measured by the angle subtended at the eye by the minimum separable, i.e. by the line of a grating; the grating-frequency of the system illustrated by Fig. 146 is given in cycles per degree subtended at the eye, the angular measure of the cycle being that between peaks of intensity, i.e. twice the angular width of the lines of the grating. Thus with a grating or spatial frequency of 60 cycles/degree, the distance between peaks is one minute of arc, and the grating-width is 30 seconds. Thus a resolving power of 60 cycles/degree is equivalent to one of 30 seconds of arc, or a visual acuity of two.

CONTRAST-SENSITIVITY

The subject is presented with a given grating-frequency, and the contrast below which resolution is impossible indicates the threshold; the reciprocal of this contrast is called the *contrast-sensitivity*. The curve through the circles in Fig. 147 shows the relation between contrast-sensitivity and grating-frequency; at low grating-frequencies, i.e. with wide gratings, the contrast-sensitivity is high, i.e. small degrees of contrast are necessary.

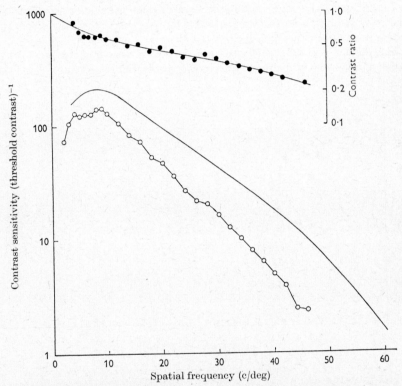

FIG. 147. The line through the open circles represents the "contrast-sensitivity function" of a human observer. As ordinates are plotted the reciprocals of the contrast required to make gratings, of frequencies indicated as the abscissae, just resolvable. The continuous curve is a corresponding function for gratings that have been formed by interference on the retina and thus have suffered no optical defects. The filled circles indicate the ratios of the contrast-sensitivities for the two modes of presentation of the gratings. (Campbell & Green, *J. Physiol.*)

OPTICAL DEFECTS IN IMAGE

Campbell & Green have applied this technique to assess the extent to which optical defects in the image contribute to a loss of visual acuity; to do this it was necessary to project on to the retina a grating of known contrast on the retina rather than of known contrast on the screen, since the effect of diffraction and aberrations is, by spreading light over the geometrical image of a black line, to reduce contrast. This was achieved by a technique developed by LeGrand in 1937; monochromatic light from two coherent sources (p. 622) was projected into the eye in such a way as to form interference fringes with calculable contrast-ratios, and the width and contrast of these fringes were varied until the subject failed to resolve them. The results have been drawn in Fig. 147 as the continuous line, whilst the ratios of the contrast-sensitivities are plotted above in solid circles. Essentially these contrast-ratios are a measure of how the optics of the system reduce visual acuity.

Effect of Orientation. The resolution of gratings is better if the lines are arranged vertically or horizontally (Taylor, 1963); Campbell, Kulikowski & Levinson (1966) have shown that the effects are similar with high and low spatial-frequency gratings, when they are assessed in terms of contrast-sensitivity (p. 245), so that it is unlikely that optical effects are the cause; and this was confirmed by finding similar orientational effects when the optical system of the eye was by-passed through the formation of interference fringes on the retina.

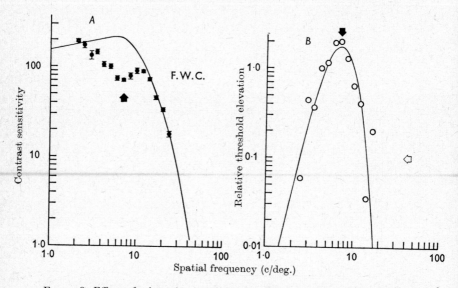

Fig. 148. Effect of adaptation to a given grating-frequency on the contrast-sensitivity function. A. The curve shows the contrast-sensitivity function for the observer F.W.C. The plotted points show the effects of adapting to a grating-frequency of 7·1 cycles/degree. B. The relative elevation of the threshold has been plotted against grating-frequency, giving a peak at about 7 cycles/degree. (Blakemore & Campbell, *J. Physiol.*)

Adaptational Effects

Gilinsky (1968) observed that the grating-acuity is lowered by adaptation to an identical pattern and bar-width but of higher mean luminance, an effect that was decreased if the orientation of the two was altered in relation to each other. Blakemore & Campbell (1969) showed that the contrast-threshold for a grating increased by 0·5 logunits after viewing for one minute a high-contrast grating; within a further minute the threshold returned to normal; the greater the contrast of the adapting grating, the greater the rise in threshold. When the effect of a given adapting grating on the contrast-sensitivity function (Fig. 147, p. 245) was studied, it was found that the adaptive effect was limited to a certain range of spatial frequencies, as illustrated in Fig. 148 where the effect of a grating of 7·1 cycles/degree is seen to depress contrast sensitivity over only a part of the range of sensitivity, so that when the relative elevation of the threshold is plotted against spatial frequency, a narrow curve, peaking at about 7 cycles/degree, is obtained suggesting the preferential depression of some spatial frequency-sensitive channels. When different adapting grating-frequencies were used, different peaks were obtained, except at low spatial frequencies, i.e. wide gratings, when there was only a depression of sensitivity without any change of the peak frequency; thus at 2·5, 1·8, 1·5 and 1·3 cycles/degree the peaks all occurred at 3 cycles/degree. Again, Campbell & Kulikowski (1966) showed that simultaneous viewing of one grating reduced the sensitivity to another, the effect depending on the angles made by the two with each other, falling off to nearly zero over a range of 15–20 degrees. An interesting observation was that, not only does adaptation alter the contrast-sensitivity, but it also changes the perceived spatial frequency, so that, after adaptation to a given grating, another grating, of about the same spatial frequency appears different, one with bars narrower than those of the adaptation grating appearing even narrower and one with bars wider appearing even wider. Such an effect is predictable if the cortical neurones showing the adaptation effect actually encode the information regarding spatial frequency of the stimulus (Blakemore, Nachmias & Sutton, 1970).

In general, then, it appears that the human visual system is organized in such a way as to give a number of channels tuned to a peak spatial-frequency, and presumably they would operate as size-detectors.*

Some Electrophysiological Aspects

Although the resolving power of the retina depends in the last analysis on the size and density of packing of the receptors in the retina, we must remember that it is the neural organization of these receptors, to give receptive fields of higher-order neurones such as the ganglion cell, that determines whether or not the maximal theoretical resolving power is attainable. It is of some interest, therefore, to determine how a ganglion cell responds to the presence of a grating form of stimulus in its receptive field. The subject has been examined in some detail by Enroth-Cugell & Campbell (1966) in the cat. When a stationary series of black and white lines is projected on to the retina, the responses of ganglion cells fall into two classes, X- and Y-cells. The X-cells behaved as though the responses to individual units of area in the field were linearly additive, the units in the centre adding to give a combined Centre-response, and those of the periphery to give a combined Surround-response, the final response being determined by subtraction of the combined Centre and Surround responses. With the Y-cells, no such linearity was present. To confine attention to the X-cells, it is interesting that two positions of the grating could be found that gave no response, the positions depending on the phase of the grating in relation to the centre of the field (Fig. 149); the existence of these null-points means that, when the illuminations over both halves of the total receptive field are equal, their responses just balance. When the grating was caused to drift over the retina in a cyclical fashion, then the ganglion cell responded in a cyclical fashion too, in phase with the drift, and it was found that this was the best way of stimulating a ganglion cell in order to examine the effects of contrast and grating-width, or, its reciprocal, the *grating-frequency* in cycles/degree. In this way the *contrast-sensitivity* of a given unit may be determined as the change in contrast of the drifting grating that can produce a measurable effect on the discharge. When contrast-sensitivity was plotted against grating-width, characteristic curves—called contrast-sensitivity functions—were obtained with an exponential falling off at high grating-frequencies. This falling off at high frequencies corresponds to the limit of resolving power of the ganglion cell as the

* Studies were usually carried out with gratings designed so that the luminous intensity along the grating varied in a sinusoidal fashion; when a square-wave form was employed, Campbell & Robson (1968) showed that the eye behaved as though it responded to the first higher harmonic, as well as to the fundamental spatial frequency; this is the third, with three times the spatial frequency of the fundamental and one third of the amplitude; it is interesting that Blakemore & Campbell (1969) found that the third harmonic of a square wave did, in fact, produce a substantial elevation of the threshold for a sine-wave grating, in addition to that produced by the fundamental, when the eye was adapted to this square-wave grating.

grating becomes fine; and a theoretical treatment, based on summation* of effects in the linear fashion indicated above, permitted the derivation of curves that could be made to match these "contrast sensitivity functions" (Fig. 150). The contrast-sensitivity functions for different ganglion cells varied, indicating selective sensitivity to different ranges of grating-frequency, and it followed from theory that these were determined by the size of the central region of the receptive field which, in the cat, was estimated to range from 0·5 to 4·4 degrees.

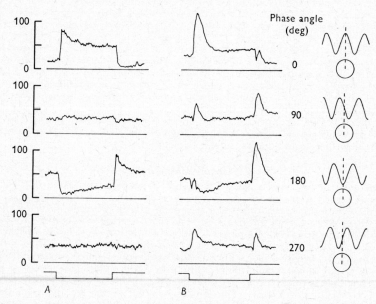

FIG. 149. Illustrating the responses of two types of ganglion cell to the presence of a grating on their receptive fields. A is the response of an OFF-centre X-cell and B that of an OFF-centre Y-cell. The phases of the grating in relation to the centre of the field are indicated at the right. In order to evoke a response, the contrast of the grating was turned on and off, as indicated in the lowest traces where a downward deflection indicates turning off. Note that with the X-cell there are two positions of the grating that give a null response. Ordinates at the left give pulses per sec of the ganglion cell discharge. (Enroth-Cugell & Robson, *J. Physiol.*)

The interesting feature of this variability in the contrast-sensitivity function is that different ganglion cells would presumably mediate the discrimination of different gratings according to the position of the optimum in their contrast-sensitivity function.

The Visually Evoked Response. Campbell & Maffei (1970) have used the potential recorded from an electrode on the occipital cortex (VER) p. 519 as a measure of the human visual response to gratings; in general their findings paralleled remarkably well the psychophysical studies, so that a characteristic contrast-sensitivity function was obtained when the contrast required to give threshold response was plotted against different spatial frequencies. Again, adaptation to a grating parallel with the test-grating caused a decrease in the evoked potential; as the angle between the two increased, the adaptation-effect decreased, until it was zero at 15–20 degrees. This adaptation-effect required that the spatial frequencies of test and adaptational gratings were close to each other,

* The assumed basis of this summation is through a "weighting function" by which points more distant from the centre contribute less than points closer to it; the actual function chosen was a Gaussian one (Rodieck, 1965).

so that the neurones depressed by the exposure were both frequency- and orientation-selective. Thus the studies with the VER seem to confirm the suggestion of Campbell & Robson (1968) that there exists an array of functionally separate mechanisms in the visual system each responding maximally at some particular grating-frequency and hardly at all to frequencies differing by a factor of two.

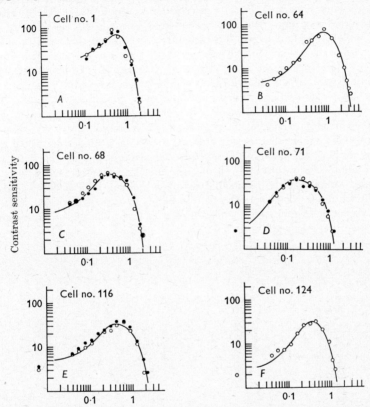

FIG. 150. Contrast-sensitivity functions for five ON-centre X-cells (A-E) and one OFF-centre X-cell (F). Gratings of different frequencies drifted over the receptive field at a frequency of 1 cycle/sec, and the contrast was altered until the observer could detect a modulation of the discharge frequency at 1 cycle/sec (i.e. synchronously with the temporal luminance modulation). (Enroth-Cugell & Robson, *J. Physiol.*)

Scotopic Visual Acuity

The essential features of scotopic visual acuity have already been described, and are illustrated by Fig. 140 (p. 236); the generally low order is presumably due to the synaptic organization of the rods such that the functional units are large. A study of the completely colour-blind subject with defective cone vision confirms this limitation of the rods; however great the luminance employed the visual acuity could not be improved beyond 0·17, i.e. a value only about one-tenth that found for cone vision in the normal eye. Under mesopic conditions, and with peripheral vision, both rods and cones are active, yet Mandelbaum & Sloan found that usually the rods dominated the picture, presumably because they were operating far above their thresholds by comparison with the cones.

Pupil-size, Wavelength and Eye Movements

The Effect of Pupil-Size. Decreasing the size of the pupil will decrease the effects of aberrations on the optical image, and thus should improve visual acuity; on the other hand, when the pupil becomes sufficiently small, diffraction effects become important. Finally, the smaller the pupil the smaller the amount of light that may enter the eye; under scotopic conditions, at any rate, this must be an important factor. Experimentally it is found that with pupil-sizes below 3 mm the visual acuity is independent of pupil-size because the improvement in the retinal image due to reducing aberrations is compensated by the deleterious effects of diffraction. Increasing the pupil-size beyond 3 mm decreases visual acuity at high intensities of illumination, presumably because of the aberrations in the optical image. If a human subject is placed in a room that is darkened steadily the size of the pupil naturally increases, and it was shown by Campbell & Gregory that the size attained was, in fact, optimal for visual acuity at this particular luminance. Thus, by using artificial pupils they determined the optimum sizes for given levels of luminance, and they compared these with the actual values that the pupils took up under the same luminances. The results are shown in Fig. 151 where the plotted points indicate the optimal pupil size for different luminance levels, whilst the broken lines indicate the natural pupil-size under the same conditions.

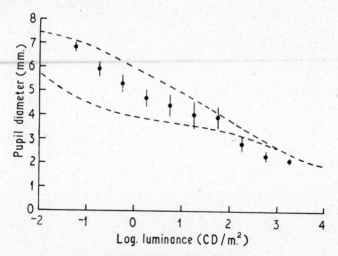

FIG. 151. The broken lines indicate the natural pupil-sizes corresponding with the luminances shown on the abscissa; the points indicate the optimum pupil-sizes appropriate to these luminance levels. (Campbell & Gregory, *Nature.*)

DIFFRACTION

The spread of light from a point source, due to diffraction, is given by $\sin \theta = 1 \cdot 22 \, \lambda a$, a being the diameter of the entrance-pupil, λ the wavelength of light and θ the angle subtended at the nodal point of the eye by the distance from the centre of the diffraction pattern to the first zero, i.e. the edge of Airy's disc. Westheimer (1964) has discussed the effects of diffraction on the theoretical visual acuity to be expected with varying pupil diameters; his theoretical values are given in Table 12, and they represent the limits of a grating that would provide a contrast-ratio between the images of the black and white lines greater than zero. This is obviously a theoretical upper limit, since the ability of the retina to detect this difference must be considered, whilst aberrations in the optical system have been ignored.

TABLE 12

Theoretical Maximal Visual Acuity and Pupil Diameter (Westheimer, 1964).

Entrance-Pupil Diameter (mm)	Visual Acuity
0·5	0·5
1·0	1·0
1·5	1·5
2·0	2·0
2·5	2·5
3·0	3·4

Influence of Wavelength. In general, the use of different coloured lights has little effect on visual acuity provided their luminance be adequately maintained; the use of monochromatic, as opposed to white, light should abolish chromatic aberration so the fact that monochromatic light is not preferable indicates that physiological mechanisms, leading to suppression of the coloured fringes, come into play; in fact Hartridge found that artificially doubling the chromatic aberration of the human eye failed to influence visual acuity. It has been argued that, because colour vision is mediated by at least three types of cone, the use of monochromatic light would impair visual acuity since, with green light, say, only the green-sensitive cones would be operative and so the retinal mosaic would be much grosser, the intervening red-sensitive and blue-sensitive cones being functionless in this respect. Whilst under certain critical conditions it is possible to isolate one cone-mechanism from the rest, e.g. a blue-sensitive one, and when this is done to find that the visual acuity is indeed reduced, nevertheless under the ordinary conditions of measuring visual acuity it seems very likely that all types of cone would be operating, and so effects of wavelength would be negligibly small.

VERNIER ACUITY

Foley-Fisher (1968) has examined the effects of varying the colour of the illumination on vernier acuity and compared his results with earlier work. The acuity of different subjects varied between 4 and 10 seconds of arc, and the subjects fell into two groups, the one with a peak performance in the red and the other in the yellow-green. The best performance was never obtained with white light.

The Effect of Eye Movements. The question as to whether involuntary movements of the eyes play a role in determining visual acuity has been agitated repeatedly. As we shall see, the eyes are never absolutely still, so that the contours of the retinal image are repeatedly falling on new sets of receptors; and we may expect the OFF-effects to be just as important as the ON-effects, if not more so, in the interpretation of the retinal image by the brain. Troxler noted that when fixation was maintained by an effort, peripheral images tended to disappear, to reappear immediately on moving the eyes; this is the *Troxler phenomenon*, and suggests that, for peripheral vision at any rate, the eyes must move if the perception of contours is to be maintained.

STABILIZED RETINAL IMAGE

With foveal fixation it requires a special optical device to demonstrate the same effect; Ditchburn & Ginsborg and Riggs *et al.*, working independently, showed that with a "stabilized retinal image", such that however the eyes moved the image of a fixated object remained on the same part of the retina, the image would disappear within a few seconds and, if it was of a fine line, would fail to reappear. The actual involuntary movements made by the eye are small, so that it is presumably the renewal of the stimuli at the contour that determines the retention of the visual percept.*

Significance of Eye Movements. Experimental studies on visual acuity, however, indicate that these movements have little influence on visual acuity. Mandelbaum & Sloan, in the study illustrated by Fig. 140, used flash-exposures of their Landolt-C's lasting 0·2 second, so that it is very unlikely that eye movements could have contributed appreciably, yet their recorded visual acuities are not unusually low. Furthermore, if visual acuity were better with long exposures than with short this could be attributed rather to the extra time given the subject to make several tries at resolution rather than to the eye movements; these extra tries would improve resolution on a purely statistical basis. Direct experiments designed to test this point have been carried out by Riggs *et al.* (1953) and Keesey (1960), and both sets of experiments have shown that the involuntary movements are unimportant. Thus Keesey measured visual acuity when the targets were exposed from 0·02 to 1·28 seconds; in one series of experiments the target was viewed normally and in the other as stabilized images; as Fig. 152 shows, the curves relating visual acuity to exposure-time are identical.

OFF-EFFECTS

According to Pirenne (1958) visual acuity under scotopic conditions is, indeed, improved if the subject is allowed to make quite large movements of the eyes, so that it can be argued that it is the OFF-effects, resulting from these movements, that are significant, rather than the continuous ON-effects from the white break in the C. This view is illustrated by Fig. 153 where the Landolt-C is large enough for resolution at the given level of illumination. The C has been moved and the receptors that have been covered by the movement have indicated this by their OFF-discharges. At the break, however, there has been no OFF-effect, and it is the absence of OFF-effect in the region of the break that determines the perception of its position. If the break were too small, or the illumination too low, then the C would behave as an O, in the sense that the receptors under the break would not be stimulated, and movement of the C would cause OFF-effects to take place over the whole contour. For scotopic vision the movements would have to be large and greater than the involuntary movements whose maximum extent is only of the order of 10

---

* The stabilized retinal image has acquired a large literature; the *relative visibility factor* is defined as $t/T \times 100$ per cent, where t is the time during which the image is visible and T is the total viewing time. If the fading of the image is due to failure to stimulate ON-OFF elements, then a flickering image should increase the visibility factor (Yarbus, 1960) or moving the target in a controlled manner (Ditchburn, Fender & Mayne, 1959). West (1968) has studied the effects of target-size, and of position on the retina, on the optimal flicker frequency necessary to give maximal relative visibility factor. Sparrock (1969) has shown that the logarithmic relation between increment threshold and background (Weber's Law) is unaffected by stabilization, so that although the subjective sensation of brightness is reduced the power to discriminate changes has not been affected.

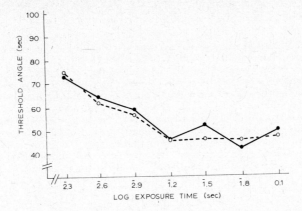

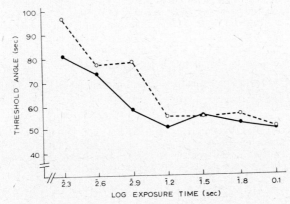

FIG. 152. Showing unimportance of involuntary eye movements for visual acuity. Threshold visual angle has been plotted as a function of viewing time, the longer the time the lower the angle subtended by the grating elements at the eye. Dashed lines refer to normal viewing, the full lines to viewing with a stabilized retinal image. Curves for two subjects are shown. (Keesey, *J. Opt. Soc. Am.*)

FIG. 153. Illustrating the effects of an eye movement on the retina on which is projected the image of a Landolt-C. The initial position is in black; cones with crosses on them experience an OFF-effect whilst the remainder remain unaffected; it may be that it is the absence of OFF-effect at the break that signals the presence of the break.

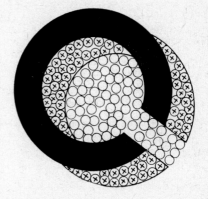

minutes of arc; with photopic vision, on the supposition that OFF-effects were also of importance, the involuntary movements might well be adequate.*

Binocular Visual Acuity. If the two eyes had identical visual acuities we might, at first thought, expect to find no difference in visual acuity according as monocular or binocular vision were employed. Measurements of acuity are carried out, however, at the very limit of the subject's resolving power, the observer investigating how frequently a correct answer is made in the discrimination of breaks in successively presented Landolt-C's; there is thus no definite size of Landolt-C that can be invariably resolved whilst a slightly smaller one invariably cannot—the investigator chooses a certain "end-point," e.g. five correct answers out of eight, so that he really estimates a probability that the subject will correctly resolve a Landolt-C of a given size. Bárány (1946) has shown that merely on a probability basis the apparent visual acuity should be increased in binocular vision, and that the experimentally determined increase conforms to the predictions derived from probability theory. Campbell & Green (1965) have considered the matter from the point of view of a genuine summation of the responses from the two eyes, these being fed into a central integrating circuit. If these outputs are "noisy", the standard error of the sum of n independent measurements of a random noisy process decreases as $\sqrt{n}$; an observer using two eyes would make $\sqrt{2n}$ measurements, so the ratio of visual acuities in binocular and monocular vision should be $1/\sqrt{2}$; they did, indeed, find experimentally that the ratio was 1·414.

GENERAL REFERENCES

HARTRIDGE, H. (1947). The visual perception of fine detail. *Phil. Trans.* **232**, 519–671.

HECHT, S. (1937). Rods, cones and the chemical basis of vision. *Physiol. Rev.* **17**, 239–290.

PIRENNE, M. H. (1962). Visual acuity. In *The Eye*, vol. II, Chapter 9. Ed. Davson. London: Academic Press.

RIGGS, L. A. (1965). Visual acuity. In *Vision and Visual Perception*. Ed. C. H. Graham, pp. 321–349. New York: Wiley.

SPECIAL REFERENCES

BÁRÁNY, E. H. (1946). A theory of visual acuity and an analysis of the variability of visual acuity. *Acta ophthal. Kbh.* **24**, 63–92.

BLAKEMORE, C. & CAMPBELL, F. W. (1969). On the existence of neurones in the human visual system selectively sensitive to the orientation and size of retinal images. *J. Physiol.* **203**, 237–260.

BLAKEMORE, C., NACHMIAS, J. & SUTTON, P. (1970). The perceived spatial frequency shift—evidence for frequency-selective neurones in the human brain. *J. Physiol.* **210**, 727–750.

CAMPBELL, F. W. & GREEN, D. G. (1965). Optical and retinal factors affecting visual resolution. *J. Physiol.* **181**, 576–593.

CAMPBELL, F. W. & GREEN, D. G. (1965). Monocular versus binocular visual acuity. *Nature* **208**, 191–192.

CAMPBELL, F. W. & GREGORY, A. H. (1960). Effect of size of pupil on visual acuity. *Nature* **187**, 1121–1123.

* When the test-chart is caused to move across the field of vision the visual acuity falls, and it is customary to speak of the *dynamic* visual acuity under these conditions, as opposed to the normally measured *static* acuity. With a movement of 40 degrees/sec, the acuity is about half the static value, and at 80 degrees/sec it is about one third (Methling, 1970). At these angular velocities the pursuit mechanism (p. 341) is insufficient to maintain a steady picture on the retina, and it is presumably the blurring of the retinal image that reduces acuity.

CAMPBELL, F. W. & KULIKOWSKI, J. J. (1966). Orientational selectivity of the human visual system. *J. Physiol.* **187**, 437–445.

CAMPBELL, F. W., KULIKOWSKI, J. J. & LEVINSON, J. (1966). The effect of orientation on the visual resolution of gratings. *J. Physiol.* **187**, 427–436.

CAMPBELL, F. W. & MAFFEI, L. (1970). Electrophysiological evidence for the existence of orientation and size detection in the human visual system. *J. Physiol.* **207**, 635–652.

CAMPBELL, F. W. & ROBSON, J. G. (1968). Application of Fourier analysis to the visibility of gratings. *J. Physiol.* **197**, 551–566.

DITCHBURN, R. W., FENDER, D. H. & MAYNE, S. (1959). Vision with controlled movements of the retinal image. *J. Physiol.* **145**, 98–107.

DITCHBURN, R. W. & GINSBORG, B. L. (1952). Vision with a stabilised retinal image. *Nature* **170**, 36–37.

ENROTH-CUGELL, C. & ROBSON, J. G. (1966). The contrast sensitivity of retinal ganglion cells of the cat. *J. Physiol.* **187**, 517–552.

FOLEY-FISHER, J. A. (1968). Measurements of vernier acuity in white and coloured light. *Vision Res.* **8**, 1055–1065.

GILINSKY, A. S. (1968). Orientation-specific effects of patterns of adapting light on visual acuity. *J. opt. Soc. Am.* **58**, 13–18.

HECHT, S., & MINTZ, E. U. (1939). The visibility of single lines at various illuminations and the retinal basis of visual resolution. *J. gen. Physiol.* **22**, 593–612.

KEESEY, U. T. (1960). Effects of involuntary movements on visual acuity. *J. opt. Soc. Am.* **50**, 769–774.

LYTHGOE, R. J. (1932). The measurement of visual acuity. *Med. Res. Council Sp. Rep. Ser.* No. 173.

MANDELBAUM, J. & SLOAN, L. L. (1947). Peripheral visual acuity. *Am. J. Ophthal.* **30**, 581–588.

METHLING, D. (1970). Schschärfe bei Augenfolgebewegungen in Abhängigkeit von der Gesichtsfeldleuchtdicke. *Vision Res.* **10**, 535–541.

PIRENNE, M. H. (1957). 'Off' mechanisms and human visual acuity. *J. Physiol.* **137**, 48–49 P.

PIRENNE, M. H. (1958). Some aspects of the sensitivity of the eye. *Ann. N.Y. Acad. Sci.* **74**, 377–384.

PIRENNE, M. H., MARRIOTT, F. H. C. & O'DOHERTY, E. F. (1957). Individual differences in night-vision efficiency. *Med. Res. Council Sp. Rep. Ser.* No. 294.

POLYAK, S. L. (1941). *The Retina.* Chicago: Chicago University Press.

POLYAK, S. L. (1957). *The Vertebrate Visual System.* Chicago: Chicago University Press.

RIGGS, L. A., RATLIFF, F., CORNSWEET, J. C. & CORNSWEET, E. F. (1953). The disappearance of steadily fixated visual test objects. *J. opt. Soc. Am.* **43**, 495–501.

RODIECK, R. W. (1965). Quantitative analysis of cat retinal ganglion cell response to visual stimuli. *Vision Res.* **5**, 583–601.

SHLAER, S. (1937). The relation between visual acuity and illumination. *J. gen. Physiol.* **21**, 165–188.

SPARROCK, J. M. B. (1969). Stabilized images: increment thresholds and subjective brightness. *J. opt. Soc. Am.* **59**, 872–874.

TAYLOR, M. M. (1963). Visual discrimination and orientation. *J. opt. Soc. Am.* **53**, 763–765.

WEST, D. C. (1968). Flicker and the stabilized retinal image. *Vision Res.* **8**, 719–745.

WESTHEIMER, G. (1964). Pupil size and visual resolution. *Vision Res.* **4**, 39–45.

YARBUS, A. L. (1960). Perception of images of variable brightness fixed with respect to the retina of the eye. *Biophysics* **5**, 183–187.

PART 9

WAVELENGTH DISCRIMINATION AND THE THEORY OF COLOUR VISION

THE problem of colour vision has attracted the interest and excited the speculation not only of physiologists but of physicists, psychologists, anatomists, philosophers, and poets with the inevitable result that the subject has abounded with theories; in the present treatment we shall confine ourselves to the main experimental findings and show how they are related to the Young-Helmholtz trichromatic theory.

THE PURKINJE PHENOMENON

We may introduce this account by studying the alterations in visual characteristics that take place on passing from scotopic to photopic vision.

Scotopic and Photopic Spectral Sensitivity Curves

Scotopic Sensitivity Curve. As we have seen, the scotopic sensitivity curve is obtained by measuring the absolute threshold of the dark-adapted eye for different wavelengths, and plotting the reciprocals of these thresholds, which measure sensitivity, against the wavelength. In a sense, then, this curve represents the relative luminosities of the different wavelengths so that, if we had a spectrum containing equal energies in all the wavelengths, the blue-green at about 5000 Å would appear brightest, and the red and violet darkest. The spectral sensitivity curve is thus sometimes described as the *luminosity curve for an equal-energy spectrum*.

Photopic Sensitivity Curve. The photopic sensitivity curve is the measure of the relative effectiveness of the different wavelengths in exciting the light sense at levels well above threshold; at these levels the lights excite the sensation of colour as well and, since this is mediated by the cones, it would not be surprising if the curve of spectral sensitivities really indicated the relative sensitivities of the cones to different wavelengths. Thus we may take two patches, A and B, of light of different wavelengths and adjust their intensities until they appear equally bright by the method of flicker-photometry (p. 140). We may then compare B with a new patch of light, C, of another wavelength, and so on through the spectrum. The reciprocals of these amounts of energies will be measures of the sensitivity of the eye to the different wavelengths and they give the curve of Fig. 154. The maximum occurs at 5550 Å in the yellow-green instead of at 5000 in the blue-green for scotopic vision. On comparing the two sensitivity curves, as in Fig. 155, it will be seen that there are characteristic changes in sensitivity of the eye on passing from scotopic to photopic vision, which are given the general name of *Purkinje shift*.

Shift in Sensitivity. Purkinje in 1823 observed that the relative luminosities of hues changed as the luminance was reduced (the *Purkinje shift*), and the nature of the change can be predicted from the curves of Fig. 155. Thus let us consider an orange of 6500 Å and a yellow-green of 5500 Å; under photopic conditions

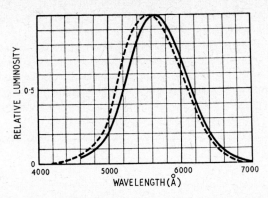

FIG. 154. Photopic luminosity curves. The curve in full line is that shown by Jainski whilst that in broken lines is the C.I.E. curve. (Le Grand, *Optique Physiologique*.)

their luminosities are roughly in the ratio of 13:100, the yellow-green being some 8 times brighter than the orange; on referring to the scotopic curve it is seen that the same colours have luminosities in the ratio: 2:56, i.e. the yellow-green is now some 28 times brighter than the orange. In general, the short wave-lengths will become brighter in comparison with the long wavelengths as the luminance level is reduced. It should be noted that the Purkinje phenomenon becomes apparent at levels where colour is still perceived, in the mesopic range where both rods and cones are operative; the shift in the apparent luminances of coloured lights is thus due to the superimposition of a rod effect—stimulating only the light sense—on the cone effect. The photopic and scotopic sensitivity curves are thus curves for cones and rods acting essentially alone; at intermediate luminances, where both are operative, the sensitivity curve lies between these limits.

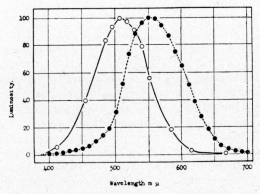

FIG. 155. The scotopic luminosity curve (open circles) compared with the photopic curve (full circles). (After Hecht.)

Red Light and Dark-Adaptation. We may note from the luminosity curves that red light is much less luminous to the dark-adapted than to the light-adapted eye in comparison with other wavelengths; in fact it is now generally considered that the rods are barely sensitive to the extreme red end of the spectrum, so that the dark-adapted eye may be exposed to fairly high luminance levels of deep red light—bright enough to permit reading with ease and hence to allow the photopic mechanism to function—without serious loss of dark-adaptation. This principle was applied extensively during the war; chartrooms were illuminated with red light and night-fighter pilots wore special red dark-adaptation goggles which permitted them to read in a well-illuminated room and yet maintain a considerable degree of dark-adaptation.

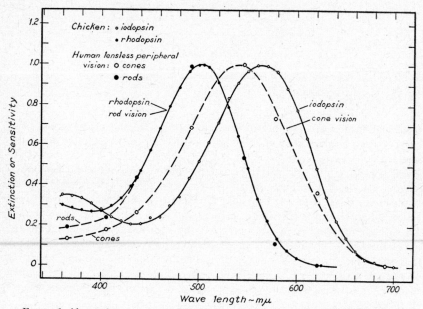

FIG. 156. Absorption spectra of chicken rhodopsin and iodopsin compared with the spectral sensitivity of human rod and cone vision. The scotopic (rod) sensitivity agrees with the absorption spectrum of rhodopsin over most of its course. The photopic (cone) sensitivity curve does not coincide exactly with the absorption spectrum of iodopsin, being displaced some 20 mμ toward the blue from this absorption spectrum. (Wald, Brown & Smith, *J. gen. Physiol.*)

Iodopsin. Photochemical studies have demonstrated the remarkable correspondence between the absorption spectrum of rhodopsin and the scotopic sensitivity curve, from which it has been concluded that scotopic vision is mediated by this single pigment, rhodopsin. May we conclude that the photopic sensitivity curve corresponds to the presence of a cone-pigment with a maximum in the yellow-green at about 5500 Å? This question is not easy to answer because cone vision is obviously more complex than rod vision, changes in wavelength being accompanied by changes in qualitative sensation that we call colour; and, as we shall see, wavelength discrimination requires the presence of several pigments, contained in separate types of receptor. The chicken retina contains practically only cones and Wald was able to extract from this a pigment, which he called *iodopsin*, with an absorption spectrum not greatly different from the photopic sensitivity curve. This is illustrated by Fig. 156, where both scotopic and photopic sensitivity curves have been drawn, whilst the plotted points represent absorption spectra of rhodopsin and iodopsin respectively. The agreement between rhodopsin absorption and scotopic sensitivity is good, as we have seen before; the maximum absorption of iodopsin, at 5620 Å, is some 200 Å closer to the red end of the spectrum than the photopic sensitivity maximum measured, in this case, on a subject with a lens-less eye to reduce effects of absorption by the ocular media.

Electrophysiological Aspects

Frog ERG. We may expect the Purkinje effect to be manifest in all mixed retinae, whilst in pure rod (guinea pig) or pure cone (squirrel) retinae it should be absent. The electroretinogram has been a useful tool in the study of this phenomenon; thus we may use the reciprocals of the relative amounts of light-energy to evoke a b-wave of fixed magnitude as the index to sensitivity of the retina to different wavelengths; if the light- and dark-adapted eyes are compared in the frog, for example, there is the same characteristic shift in sensitivity to longer wavelengths on passing from dark- to light-adaptation (Fig. 157).

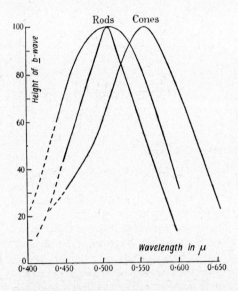

Fig. 157. The Purkinje phenomenon in the frog's retina. The curves are luminosity curves and it is seen that the wavelength of maximum luminosity changes from 5600 Å to 5000 Å on passing from photopic to scotopic conditions. Double contour of rod graph indicates limits of variability. (Granit & Wrede, *J. Physiol.*)

Flicker ERG. Even in the cat, whose retina is very much dominated by rods, a shift in spectral sensitivity can be obtained by employing the flicker method of analysis of the ERG. It will be recalled that the rods are unable to follow a light-stimulus flickering at more than about 25 cycles/sec. With a very strong light, flickering at, say, 50 cycles/sec the responses on the record are thus cone responses, and we may measure the effectiveness of the different wavelengths in producing an electrical effect of fixed magnitude. Under these conditions the maximum sensitivity shifts from 5000 Å to 5500 Å (Dodt & Walther). In the pure cone eye of the squirrel there is no Purkinje shift so that only high critical fusion frequencies of flicker are found and the spectral sensitivity shows a maximum in the yellow-green that is barely affected by dark-adaptation (Arden & Tansley).

Single Fibres. Studies with single fibres of the optic nerve, or of large ganglion cells in the retina, show essentially the same sort of Purkinje shift but, as we shall see, this change belongs to a certain type of element which Granit has called the *dominator*.

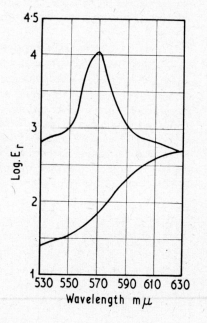

Fig. 158. Illustrating the photochromatic interval. The lower curve indicates the thresholds for the sensation of light (absolute thresholds) whilst the upper curve indicates the thresholds for the sensation of colour. Note that the thresholds coincide in the orange-red. (After Dagher *et al.*, *Visual Problems of Colour*.)

Photochromatic Interval

The threshold stimulus for the different wavelengths gives us the scotopic sensitivity curve which, as we have seen, refers only to the light sense since no colour sensation is evoked (except possibly in the deep red). If the threshold stimulus for a given wavelength is increased, at a certain value the sensation of colour is reached (the *specific threshold*). The interval in luminance between the absolute light threshold and the specific threshold is called the *photochromatic interval*. In Fig. 158 some results obtained by Dagher, Cruz & Plaza on a human subject are shown for the interval between 5300 and 6300 Å, i.e. over the green-orange-red part of the spectrum. The lower curve represents the absolute thresholds whilst the upper curve represents the specific thresholds. It will be seen that the interval becomes negligibly small at the red end of the spectrum, which means that the threshold for red light corresponds to the threshold for cone vision. At the fovea, too, the photochromatic interval is very small or non-existent.

Cone Thresholds

Dark-Adaptation Curves. As stated above, if the photopic sensitivity curve represents the sensitivities of the cones, then if we can determine the thresholds for cone vision at different wavelengths we may obtain a spectral sensitivity curve which should be similar to that obtained by simple matching techniques. The problem is, of course, to be sure of stimulating only cones when measuring a threshold; in the completely dark-adapted eye this is impossible, except at the extreme red end of the spectrum, because the rods are the more sensitive. Wald made use of the fact that immediately after light-adaptation the first portion of the dark-adaptation curve (Fig. 73, p. 135) represents cone sensitivity i.e., under these conditions we may actually measure cone thresholds. By using different wavelengths for the light-stimulus he was able to plot a cone sensitivity curve which had a maximum at 5550 Å and was similar in essentials to the C.I.E. curve of Fig. 154.

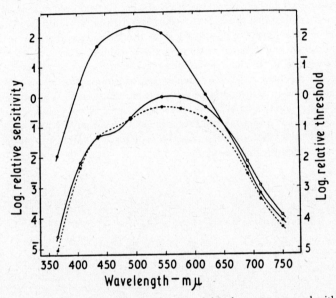

FIG. 159. Sensitivity curves for foveal and peripheral cones compared with that for rods. Top curve: rods. Bottom curves: cones. The broken line refers to peripheral cones. (After Wald, *Science*.)

Foveal Thresholds. Again, if vision is confined to the fovea by making use of flash stimuli of small extent, the thresholds should be determined by the cones. Under these conditions the subject usually states that he perceived the colour of the stimulus. By comparing results obtained in this way with those obtained by the dark-adaptation technique, we may compare the sensitivities of foveal and peripheral cones. As Fig. 159 shows, the curves are very similar, the foveal cones being less sensitive, except in the red part of the spectrum where they are about the same.

Increment Thresholds. Finally, we may note that measurement of the liminal brightness increment is a measure of the sensitivity of the retina under particular conditions of exposure to light; and according as to whether the rods or cones are the more sensitive under the particular conditions, the one or other type of receptor will determine this increment, or differential threshold as we have called it. By choosing suitable conditions of light-adaptation it is possible to ensure that the cones will be the more sensitive (e.g. by adapting with blue light to which the rods are so much more sensitive than the cones) and therefore it is possible to measure their thresholds with different wavelengths of light and obtain spectral sensitivity curves. This technique was developed by Stiles & Crawford and has been exploited widely in the measurement of spectral sensitivity curves. As we shall see, the measurements may be so refined that we may determine not only the thresholds of the cones in general, but thresholds of different types of cone (p. 266).

WAVELENGTH DISCRIMINATION

As emphasized earlier, the difference between the scotopic and photopic sensitivity curves is not just the shift in maximal sensitivity to a different wavelength described as the Purkinje effect; there is the difference that vision becomes chromatic under photopic conditions, and the subject is able to state not only that light is present but that it belongs to a certain part of the spectrum. Before studying the mechanism of this wavelength discrimination, we may ask why it is that the retina is achromatic under scotopic conditions. The answer is that under scotopic conditions only one type of receptor is operating, the rods, and that these rods are physiologically identical in that they all have the pigment rhodopsin as their photopigment. Such a functionally homogeneous retina cannot act as wavelength discriminator.

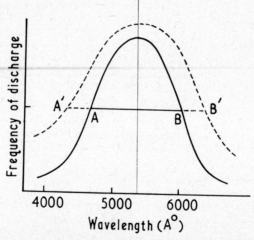

FIG. 160. Theoretical wavelength-response curve for a single receptor of *Limulus* retina. The maximum response will occur at 5200 Å, the absorption maximum of the retinal pigment. With wavelengths of light corresponding to the points A and B, the responses will be identical, so that no discrimination between these two wavelengths is possible.

Homogeneous Retina. To make this clear we may consider the responses in the optic nerve fibres of *Limulus* which, we may recall, are derived directly from the photoreceptors. A study of their responses to different wavelengths gives a typical sensitivity curve, with maximal sensitivity at 5200 Å corresponding to the wavelength of maximum absorption of its photopigment. If we were to plot the intensity of response in a single fibre against wavelength we would obtain a curve roughly like Fig. 160. In a sense, then, this receptor, and thus the eye, is showing some wavelength discrimination, in that it is responding with different frequencies of spike discharge to different wavelengths. A little consideration will show, however, that this receptor could not discriminate between changes of wavelength and changes of intensity; and even if we could keep intensity constant its discrimination is limited. Thus, at the points A and B on the curve, the responses are the same but the wavelengths are widely different; by drawing other lines parallel to the abscissa, cutting the curve in two points, many other pairs of spectral hues could be obtained producing identical responses. Moreover, this curve was obtained by keeping the energy content of the stimulus the same; new curves would be obtained by varying the energy content, as indicated by dotted lines, so that in fact with a single wavelength we would be able to mimic the responses of all other wavelengths of the spectrum simply by varying the intensity of this single wavelength. Hence a retina containing a single type of receptor is unable to discriminate changes in both wavelength and intensity.

Multiple Receptor Hypothesis. The position becomes different if we have several types of receptor, with different photosensitive pigments; these would have different response curves and it is not difficult to show that the effects of varying wavelength on the pattern of discharges in the several types might be unique and not matchable by a change in intensity. Thus, for example, with six receptors a green light of wavelength 5300 Å might evoke spike responses in these having say, 15, 34, 65, 20, 4, 2 spikes/sec. A yellow light of 6000 Å might give a different set of responses, say 35, 65, 45, 20, 15, 7 spikes/sec; and it would be these *patterns* of discharges that would be characteristic of the respective wavelengths for this particular energy level of the stimuli. Increasing the energy would increase the frequencies of discharge, but it could well be that they would be able to maintain a unique pattern that could not be mimicked by other lights however much we varied the intensity. Clearly, the more types of receptor we have, the more certain we will be of having a unique type of response for a given combination of wavelength and intensity. As we shall see, Nature has economized in this respect, since it appears that only three types of receptor suffice for human wavelength discrimination.

THE YOUNG-HELMHOLTZ THEORY

Trivariance of Colour Vision

We have already seen that colour-matching experiments have led to the generalization that we may match any colour by a mixture of three primary spectral wavelengths provided we admit that the match is not perfect so far as saturation is concerned. Thus if we wish to match a given colour X, we may place it in one half of a photometric matching field and in the other half we may project a mixture of red, green and blue lights; by varying the relative intensities of these we may match our colour from the point of view of chromaticity, but the mixture that we have made will be unsaturated, so we must add some white to X to make the match perfect in respect to both chromaticity and saturation. Thus our mixing equation would take the form:

$$\alpha R + \beta G + \gamma B = X + x \ White$$

or,

$$\alpha R + \beta G + \gamma B - x \ White = X$$

But, since white may be represented by a mixture of red, green and blue, say: $\alpha'R + \beta'G + \gamma'B$, we may subtract these from the left-hand side of the equation, and we are left with an equation that defines our colour, X, in terms of only three variables:

$$(\alpha - \alpha')R + (\beta - \beta')G + (\gamma - \gamma')B = X$$

It is in this sense, then, that we are able to say that colour vision is a *trivariant phenomenon*; we can, in essence, describe any colour-stimulus in terms of three variables; normal human colour vision is thus *trichromatic*.

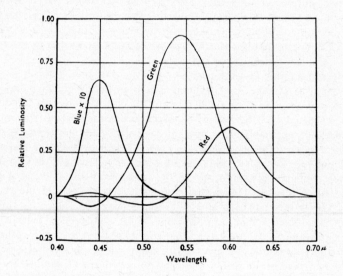

FIG. 161. Spectral mixtures curve indicating the relative amounts of the three primaries that are required for matching any spectral wavelength. For example, with a blue of 0·50 μ (5000 Å), the relative contributions are about 0·33 Green, 0·1 (÷ 10) Blue, and 0·05 Red. (Wright, *Researches on Normal and Defective Colour Vision*.)

Spectral Mixture Curves. The trichromaticity of colour vision tells us that *any* colour can be matched in terms of three primary stimuli. If we confine ourselves to matching the spectral colours we may represent this process by the chromaticity chart of Fig. 78, p. 142; alternatively we may plot the individual contributions of our red, green and blue primaries to the different spectral colours as in Fig. 161. As abscissae we have the wavelengths of the spectral hues we wish to match, and as ordinates are plotted the amounts of each individual primary required for each match. These *spectral mixture curves* show us at a glance how much of a given primary is required to match a spectral hue; thus a yellow of 6000 Å requires some 0·25 units of green and some 0·4 units of red with no blue. A blue-green of 5000 Å requires about 0·4 units of green, 0·01 unit of blue* and some "negative red".

---

* The contribution of blue to any mixture is very small, in terms of luminosity, so that the values have been multiplied by ten in the Figure.

Three-Receptor Hypothesis

These curves lead naturally to the hypothesis that any colour sensation may be evoked by stimulating three fundamental receptors to varying extents. Thus we see that a yellow sensation may be created either by passing yellow light of, say, wavelength 6000 Å into the eye, or we may create exactly the same sensation by passing into the eye a mixture of quite different lights, lights that by themselves would have excited the sensations of green and red respectively. Thus the colour mixing phenomena are essentially phenomena of confusion or failure to distinguish; the light of wavelength 6000 Å is physically different from the lights of 5300 Å and 6500 Å that are used for the match, but the eye cannot distinguish between the light of 6000 Å and the mixture. The phenomena illustrate too the economy of Nature; if an eye is to discriminate single wavelengths from mixtures then a large number of receptors must be necessary, but it is very questionable whether this discriminatory power would have much biological survival value, and instead the trivariance of colour sensation suggests that Nature "makes do" with a very limited number of receptors.

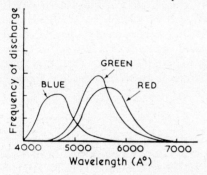

FIG. 162. Theoretical wavelength-response curves for three hypothetical receptors maximally sensitive to red, green and blue respectively.

The Fundamental Mechanisms

The three-receptor hypothesis is the essential basis of the Young-Helmholtz trichromatic theory of colour vision; in essence it has come to mean today that the retina contains three types of cone that respond differently to the same wavelengths; thus a "blue" cone would respond most prominently to the short end of the spectrum, and only a little to lights of the middle region and perhaps not at all to very long wavelengths; a "green cone" would respond preferentially to the middle regions, and so on. Each mechanism, or cone-type, may therefore be regarded as having its own spectral sensitivity curve, and we may look on the spectral mixture curves of Fig. 161 as being related to the spectral sensitivity curves of the separate mechanisms. Thus, to refer to the figure, we might say that the yellow light of wavelength 6000 Å had excited the green mechanism by 0·25 units and the red mechanism by 0·4 units and had failed to excite the blue mechanism at all. Thus the match occurred because the yellow light did exactly what the green and red lights would have done if they had fallen simultaneously on the retina. On this basis, then, we could plot three sensitivity curves for the three mechanisms, using as ordinates the frequency of discharge, say, in the nerve fibres leading from these fundamental mechanisms as in Fig. 162.

Spectral Sensitivities. The next problem, of course, is to decide on the true spectral characteristics of these mechanisms; the colour-mixing phenomena tell us we can use any three primaries, so we can have an infinite number of sets of spectral mixture curves. Which, if any of these, is the correct set? Various methods were used to derive the true fundamental response curves, as they were called.

ADAPTATION

For example, Wright used as his criterion the effects of adaptation. He has shown that if the eye is adapted to white light the sensitivity to all wavelengths is depressed in proportion; however, if the adapting light is coloured, there is a certain preferential depression of sensitivity to this colour, so that a subject who has been adapted to red, for instance, requires more red to make a match with white than he did before the adaptation. If the adapting light is yellow, both the red and green sensations are depressed, and so on. Thus if there are, indeed, three fundamental sensations, it should be possible to adapt the eye to at least one or two hues such that sensitivity to these alone is depressed, since these alone should be stimulated. Wright found that a blue of 4600 Å and a red of 7600 Å behaved in this way, and he suggested that these were two of the primary sensations; violet he concluded was definitely not a fundamental sensation since it was considerably modified by adaptation of the eye to red light. The wavelength corresponding to the third sensation, which from its position cannot be stimulated alone, was deduced to have the value of 5300 Å, i.e. in the green. On the basis of this reasoning Wright drew a set of spectral sensitivity curves.

COLOUR-BLIND

Other workers studied the so-called colour-blind, and on the assumption that their colour defect was the consequence of one of the mechanisms' being absent (p. 268), they were able to deduce, from the colour-blind person's matching phenomena, the characteristics of the spectral sensitivity curve of the missing mechanism, and thence that of the remainder.

Two-Colour Threshold Technique

Liminal Brightness Increment. Stiles developed an extremely ingenious technique for the identification of separate cone mechanisms in the human eye. The method is based on the measurement of increment thresholds, or liminal brightness increments (l.b.i.). It will be recalled that the l.b.i. for white light is measured by the extra luminance of a test patch, ΔI, necessary for it to be distinguished against its white background of luminance I. By successively increasing the background luminance, I, we get a series of steps, ΔI, and by plotting $\log \Delta I$ against $\log I$ we find that the graph has a characteristic shape. Thus, over the range of scotopic vision it may be represented by Fig. 163; at very low luminance, the increment threshold, ΔI, is approximately constant and equal to the absolute threshold; at a certain point the graph becomes a straight line of slope unity,* and finally there is a point where the rod mechanism is said to be saturated.

* It is over this range that the Fechner fraction, $\Delta I/I$, is constant; so that we have the relationship:

$$\log \Delta I = \log I + Constant$$

so that with a logarithmic plot we should have a straight line with slope equal to unity.

Rod and Cone Mechanisms. This type of graph is only obtained, however, if a single mechanism is operating over the whole range of luminance, i.e. if the rods only are active; and the type of curve shown in Fig. 163 could only be obtained by using special techniques to keep the cones from coming into play. In general, then, when the luminance comes up to a certain point the cones become the more sensitive and they now determine the l.b.i., and the graph of log ΔI against I shows a characteristic break, or kink, indicating the taking over of the new "mechanism", in this case the taking over of the threshold by cones in place of rods. This is illustrated by Fig. 164. In this way we have identified, then, the operation of a new mechanism. By varying the wavelength of the background, and of the test-stimulus, we may create conditions that favour one or the other mechanism and so study it over a large range of luminance; e.g. by using a red background and measuring the liminal brightness increment with a blue test-stimulus we may suppress cone activity and study rod activity. In this way, then, we may not only identify the presence of rod and cone mechanisms but we may also measure their spectral sensitivities, using the l.b.i. as a measure of threshold for any given wavelength, care being taken that the background and test-stimulus are such that only one of the two mechanisms is operative.

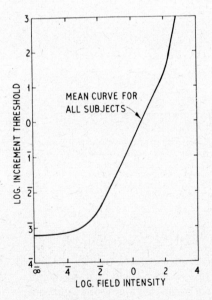

FIG. 163. Showing the variation of the liminal brightness increment (l.b.i. or increment threshold) with field intensity. For details see text. (After Aguilar & Stiles, *Optica Acta*.)

Cone Mechanisms. The same technique can be applied to identify different cone mechanisms, and Stiles found, indeed, that under appropriate conditions of background and test-stimulus the presence of three mechanisms could be established with maximum sensitivities in the red, green and blue. Subsequent studies indicated that the blue mechanism is complex, and can be described as being made up of three, π_1, π_2, and π_3, whilst the green and red mechanisms are called π_4 and π_5 respectively. The discovery of five, as opposed to three, cone mechanisms may demand some modification of the Young-Helmholtz trichromatic theory, but since this really is a statement of fact, namely the practical trivariance of colour vision, in the sense that the phenomena of mixing can be explained in terms of only three variables, this suggests that the extra information that is sent to the higher centres by virtue of the three types of blue mechanism is somehow lost.*

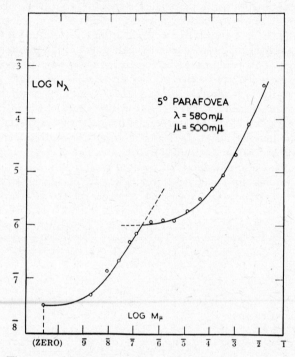

Fig. 164. The curve obtained by plotting the logarithm of the liminal brightness increment ($N\lambda$) against logarithm of the background intensity ($M\mu$). In this case the background was blue-green and the central patch a yellow of 5800 Å. The left part of the curve represents increment thresholds for rod vision; at the break and beyond, the increment thresholds are determined by the cones. $N\lambda$ and $M\mu$ are expressed as erg sec^{-1} (degree of arc)$^{-2}$. (Stiles, *Proc. Nat. Acad. Sci. Wash.*)

COLOUR-BLINDNESS

Since the study of the colour-defective human subject (usually described somewhat inaccurately as the colour-blind) has thrown some light on the more general aspects of colour discrimination, it is right that we devote some space to the main experimental findings at this juncture. In general, a colour-defective subject is one whose powers of discrimination between lights of different wavelengths are more limited than normal; the extent of this defect may range from complete failure of wavelength discrimination—*monochromatism*—through

* Stiles' increment-threshold technique is one of several that have been employed to isolate the individual mechanisms; Brindley's studies on adaptation to very strong monochromatic stimuli are described later (p. 296); the blue mechanism may be identified by virtue of its much lower critical fusion frequency for flicker (Brindley, Du Croz & Rushton, 1966), and the greater area over which Ricco's Law applies, (12–18′ compared with 4′ for red and green), suggesting that the blue-sensitive cones are connected to a more diffuse type of bipolar (Brindley, 1954).

dichromatism, where the discrimination is so limited that we must postulate the absence of one of the three fundamental mechanisms, to the *colour-anomalous*, where the defect is a difference from average behaviour, in the sense that the protanomalous uses more red than the average in matching a yellow with a mixture of red and green.

Rod Monochromatism

The more frequently occurring type of completely colour-blind is one whose cones are apparently pathologically defective; the subject is therefore called a *rod monochromat*, and we have already seen that his visual functions are largely what one might expect on the basis of the duplicity theory; thus he shows no Purkinje phenomenon, having a spectral sensitivity curve corresponding only to the scotopic type. Such monochromats occur with a frequency of about one in thirty thousand.

Visual Characteristics. There is by no means complete agreement as to the essential feature, if there is one, of the so-called rod monochromat; being more common than the cone monochromat, the rod monochromat is called "typical", but various studies indicate that all the features of this condition cannot be forced into a single type. Walls & Heath have emphasized the complexities in some of the visual characteristics of the "typical" achromat; thus we have already seen that the plot of critical fusion frequency against luminance shows a break, as in normals, suggesting the operation of two types of receptor, and sometimes the dark-adaptation curve also shows a "kink". These findings led v. Koenig and later Hecht to postulate the presence of "photopic rods" in the retina of the monochromat. In general, we may regard "typical" achromatopsia as a retinal dystrophy with the following signs, not all of which need be present in any subject: (*a*) achromatic vision; (*b*) low visual acuity, of the order of 0·1 to 0·3, better at moderate luminance than at high; (*c*) no Purkinje shift; (*d*) avoidance of very bright lights; (*e*) nystagmus; (*f*) some signs of macular dystrophy. The post-mortem study of Larsen showed that cones were indeed present in the retina, with most of them apparently normal.

Three Types. Walls & Heath suggested that some cases could be accounted for on the assumption of the absence of functional cones together with the presence of rods in the fovea, whilst others could be described as containing, in addition to rods, the "blue-type" cones postulated by trichromatic theory, the "red" and "green" cones being non-functional or absent. More recently Blackwell & Blackwell have reported on a study of nine subjects whose psychophysical findings allowed them to be placed in three categories: (*a*) typical rod monochromats with all visual functions determined by rods; (*b*) a group with a scotopic luminosity curve very similar to normal except in the blue where sensitivity was low, possibly due to the presence of the yellow xanthophyll macular pigment (they were using central fixation); this suggested that these achromats possessed rods in the fovea. At photopic luminances the sensitivity curve shifted towards the short-wave end of the spectrum, to give a maximum at about 4400 Å, suggesting the coming into play of "blue cones". The third group were intermediate between (*a*) and (*b*).

Cone Monochromatism

The cone, or atypical, monochromat is extremely rare, occurring with a frequency of only one in a hundred million; in this condition visual acuity is perfectly normal, but there is no discrimination of coloured lights if their luminosities are made equal. It might be thought that the retina, in this condition, contained only one type of cone, but if this were so we should expect the photopic spectral sensitivity curve to be abnormal. Thus, under photopic conditions the sensitivity of the eye to the different wavelengths must be the average sensitivity of the three or more cone mechanisms in the retina; removal of two of the mechanisms would leave the retina with only one type, and now the spectral sensitivity curve should coincide with the sensitivity curve of a single mechanism. In fact, Weale found the photopic sensitivity curve of the cone monochromat to lie within the normal range. When Weale (1959) applied his foveal bleaching techniques to a cone monochromat (p. 274), he found that the difference-spectrum for light reflected from the fundus indicated the presence of at least two cone mechanisms, and Gibson (1962), using Stiles' incremental threshold technique, found that three mechanisms were operative. Finally Fincham (1953) showed that a cone monochromat was able to make use of coloured aberration fringes in the accommodation reflex (p. 405), so that all the evidence indicates that the defect is not in foveal pigments but in the central integrating mechanism.

Dichromatism

Colour-Matching. The dichromat is one who can match all colours with suitable mixtures of only two, instead of three, primaries. The colour mixing equation will therefore be of the form:

$$C = \alpha A + \beta B$$

where A and B are the primaries; instead of a pair of rectangular co-ordinates being required to represent their colour mixing data (p. 142), the latter may be exhibited in the form of a straight line where any point on this line corresponds to a definite mixture of the two primaries. It will be clear, without any mathematical analysis, that the range of colours that a dichromat can appreciate as distinct will be very restricted compared with the range of the normal individual, since it is obviously possible to obtain more variations by mixing three things than two.

Missing Mechanisms. According to the classical interpretation, the *protanope* is missing the fundamental red sensation—he is red-blind; the *deuteranope* lacks the green sensation and the *tritanope* the blue. If this viewpoint is correct, we may anticipate the appearance of the spectrum to the protanope, for example, from a consideration of the curves in Fig. 161, and imagining that the curve corresponding to the red sensation has been removed. In the region between 7600 and about 6500 Å no colour sensation should be experienced, i.e. the protanope's spectrum should be shortened; between 6500 and about 5000 Å the sensation should be that of green only (although it may be called yellow); with the introduction of the blue sensation at about 5000 Å, changes in hue should be appreciated and colour discrimination should be approximately normal.

Hue-Discrimination Curves. Thus the curve, obtained by plotting the difference in wavelength necessary for a dichromat to appreciate a change in hue, as ordinates, against the wavelength, as abscissae, (the hue discrimination curve) should be much simpler than that of a trichromat; in Fig. 165 are shown typical curves for the protanope and deuteranope (that of the normal trichomat is shown in Fig. 76, p. 139); it is seen that the maximum discrimination occurs in the region 4800–5000 Å where the blue mechanism is operative. Discrimination falls away rapidly on both sides of this point as we should expect.

Neutral Point. The dichromat can match all hues (spectral and non-spectral and including white) by appropriate mixtures of two primaries; since white can be formed by a certain mixture of these primaries it follows that there will be a certain point in the spectrum which will appear to the dichromat as white— this point is called the *neutral point* and occurs at 4955 Å with the protanope and at 5000 Å with the deuteranope.

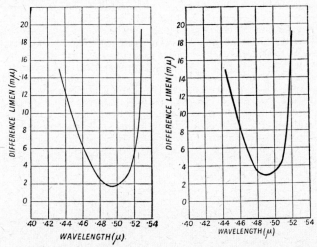

FIG. 165. Hue-discrimination curves for deuteranope (left) and protanope (right). The smallest change in wavelength, necessary to produce a change in the sensation of hue, has been plotted against the wavelength. Greatest sensitivity is in the blue-green region. (Wright & Pitt.)

An important characteristic of the dichromat, emphasized by Pitt, is the fact that those colour matches holding good for the normal trichromat are appreciated as colour matches by the dichromat, although the inverse proposition is of course not true. Thus trichromatic white is appreciated as white by the dichromat; this white is made up of the three fundamental primaries, red, green and blue, in certain proportions; the protanope does not see the red in this white and we may thus expect him to match a mixture of green and blue with white; similarly a deuteranope would match a purple (red plus blue) with normal white.

Iso-Colour Chart. We have seen that for the normal trichromat the locus of spectral colours is given by a curved line in the colour chart, where each point represents by its co-ordinates the coefficients in the unit colour equation; by joining the two ends of this curve, to make a closed figure, we get the locus of certain non-spectral hues obtained by mixing blue and red, the purples (Fig. 78). All points within this closed figure represent non-spectral hues appreciated by the trichromat. The locus of spectral colours for the dichromat is a straight line only, and since this cannot enclose space, all colours appreciated by the dichromat are spectral hues. This means that for all those numerous hues lying within the normal trichromatic locus, which are appreciated by the trichromat as different, there will correspond only a few spectral colours in the dichromat's colour range; in other words, to the dichromat, for any given spectral hue there will be a large number of non-spectral hues that are indistinguishable from it.

A determination of the colour-mixing data for dichromats permits the construction of iso-colour charts; in Fig. 166 that for a protanope is given. The normal trichromatic spectral locus is drawn in the usual manner, and trichromatic white is indicated by a star. The neutral point of the protanope, i.e. the point on his spectral locus where the mixture of his two primaries gives white, lies at 4955 Å; and an iso-colour line can be drawn through this point A, and the star representing trichromatic white, since the latter is appreciated as white by the protanope too. The line cuts the red axis at C, which corresponds to a very reddish purple. All colours represented by this line, ranging from blue to reddish purple, appear to the protanope as identical and white. From the data on the spectral hue discrimination of the protanope (Fig. 165), it is possible to mark off points on the spectral locus corresponding to just discriminable differences in hue, and the directions of the iso-colour lines passing through these points may be calculated from colour-mixing data (it will be remembered that between 5300 and 7600 Å all spectral colours appear the same to the dichromat). The iso-colour lines are drawn in the figure, and they divide the trichromatic colour area into a series of zones; all trichromatic hues represented by points in any one zone are indistinguishable by the dichromat, provided the luminosity is kept constant. For the protanope there are seventeen zones, whilst the deuteranope has some twenty-seven. The iso-colour charts permit one to determine whether any pair of hues can be discriminated by a dichromat, and provide a rational interpretation of the numerous and hitherto uncorrelated facts of dichromatic matching and errors in discrimination.

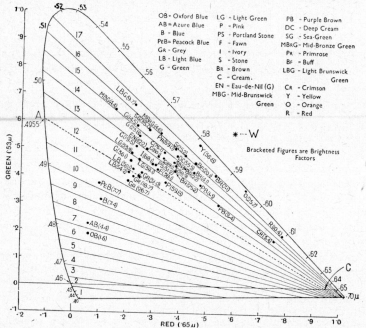

FIG. 166. Iso-colour chart for protanope. (Pitt, *Med. Res. Council Sp. Rep.* By permission of the Controller of H.M. Stationery Office.)

Luminosity Curves. If dichromatism does, indeed, represent the loss of one of the cone mechanisms we may expect the photopic luminosity curve to be different from normal in dichromatism. The regions of maximum luminosity are, in fact, shifted; that for the protanope lying at about 5400 Å and for the deuteranope at about 5700 Å. Furthermore we should expect the loss of the red mechanism to cause a shortening of the visible spectrum, and in the protanope this is indeed found, the reds at the end of the spectrum appearing grey. Again the loss of one of the "colour mechanisms" presumably corresponds to the dysfunction of one group of cones; in this case we may expect a loss of absolute luminosity, i.e. the thresholds for vision in the dichromat should be generally higher than for the trichromat. With the protanope this is very well established in the red end of the spectrum and, according to Hecht, both protanopes and deuteranopes have a general loss of sensitivity over most of the spectrum, the average rise in threshold of the protanope being 49 per cent and that of the deuteranope 39 per cent. Subsequent studies on a uniocular deuteranope, i.e. on a subject with one eye normal and the other deuteranopic, showed that this difference in threshold was quite definite (Graham & Hsia, 1958).

Deuteranope. Pitt's analysis of deuteranopia suggested, however, that this could not be simply accounted for by the loss of the green mechanism; rather the deuteranope behaved as though he had, in place of the red and green mechanisms, a single one corresponding to a weighted mean of the normal red and green mechanisms; the deuteranope, on this basis, is indeed a dichromat, in the sense that his colour-mixing data can be expressed as a function of only two variables, but he behaves as though his red and green mechanisms are in some way linked so that they cannot act independently. More recent work, however, has shown that the deuteranope does, indeed, lack the green mechanism (p. 275).

Tritanope. The tritanope is rare, constituting, according to Wright (1952), only one in 13-65,000 of the population by comparison with one in a hundred males for protanopia and deuteranopia. The tritanope seems to lack the blue mechanism, so that his discrimination is good in the red-green part of the spectrum but bad in the blue-green region; the neutral point is in the yellow at 5750 Å. By employing Stiles' increment-threshold technique, Cole & Watkins (1967) showed that the tritanope lacked all three of the "blue mechanisms", namely π_1, π_2 and π_3.

Anomalous Trichromats

Here the subject requires three primaries to match all colours, but uses different proportions from those of the average; frequently this deviation from the average or "normal" is associated with defective discrimination, but not always, according to Pickford. The abnormalities may be roughly classified as *protans*, *deuterans* and *tritans*; thus a protan is a trichromat who requires more red to match a yellow than does a normal trichromat. The anomalous trichromat differs from the dichromat in that he will not accept matches that the normal makes; thus the protanope matches a blue-green with white, but he will also accept as white a mixture of this blue-green with red. The protan will match a spectral yellow with an orange, i.e. he requires more red for his match than normal, but he will not accept the normal's red-green mixture as a yellow. Anomalous trichromasy is common, amounting to nearly 6 per cent of the male population.

Cone Photopigments

We have seen that, by the method of reflexion-densitometry, Campbell & Rushton and Weale were able to demonstrate the presence of rhodopsin in the human and animal eye, and to measure the changes that occurred during exposure to light. In the fovea, too, these authors have demonstrated the presence of two pigments absorbing preferentially in the red (*erythrolabe*) and green (*chloro-labe*) parts of the spectrum. The existence of a cone pigment was first established by exposing the eye to a strong orange light, which might be expected to bleach this pigment in preference to rhodopsin; by subsequently measuring the absorption of yellow light after reflexion from the fovea, a curve for regeneration of cone pigment was obtained. This regeneration was much faster than that of rhodopsin, being complete in 5–6 minutes.

Erythrolabe and Chlorolabe. A study of the effects of bleaching with different wavelengths indicated the presence of two pigments, and these were isolated by studying a protanope who, according to classical theory, is missing the red mechanism and thus might be expected to have one fewer pigment. The curve of Fig. 167 shows the absorption spectrum of what is apparently a single photopigment in the protanope's fovea, and corresponds remarkably well with the photopic spectral sensitivity curve of the protanope, as determined by Pitt.

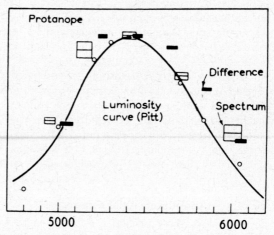

FIG. 167. The black and white rectangles represent the difference spectrum of the foveal pigments in the protanope, and presumably therefore correspond to the absorption spectrum of the green-sensitive pigment, chlorolabe. The curve represents the luminosity curve of the protanope as determined by Pitt, whilst the circles represent the actual luminosity curve of one of the subjects studied by reflexion densitometry. (Rushton, *Progress in Biophysics*.)

The peak absorption of this green-sensitive pigment was near 5400 A, and was called *chlorolabe*. In effect the retina of the protanope, under the conditions studied, was behaving as though only one pigment were present, and it was found that lights of different wavelengths, matched by the protanope as equally bright by means of flicker photometry, were, indeed, equally effective in bleaching the foveal photosensitive pigment. The failure to detect blue-sensitive pigment was presumably due to the relatively small contribution the blue mechanism makes to luminosity.

The deuteranope, like the protanope, fails to discriminate hues in the red-green part of the spectrum, and his colour-mixing data suggest that he lacks a green mechanism, so that, like the protanope, he has only a single mechanism, or pigment, operating in the red-green part of the spectrum. Rushton (1965) showed that, over this range, only a single pigment absorbed light significantly, and its absorption maximum corresponded reasonably well with the maximum of the deuteranope's luminosity curve. Rushton showed that the effects of bleaching the retina with different coloured lights were not consistent with the presence of a mixture of red- and green-absorbing pigments, as had been at first thought to be the case, so that we can conclude that the deuteranope is red-green blind because he lacks a green-absorbing pigment, *chlorolabe*, and not because, in some way, his red and green cones have become "mixed". In the normal eye we may expect to demonstrate the presence of both erythrolabe and chlorolabe; and this was in fact done by Baker & Rushton (Fig. 168) with the use of selective bleaching techniques.*

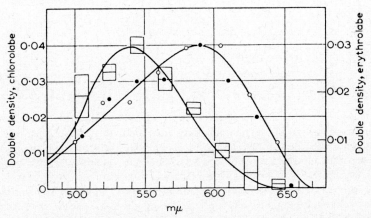

FIG. 168. Difference spectra of two pigments in the normal fovea. The left curve represents chlorolabe and the other erythrolabe. (Rushton, *Progress in Biophysics*.)

Absorption Spectra of Cones

A more precise characterization of the types of pigment in cones is given by measuring the absorption spectra of individual cones exposed to a minute pencil of light on a microscope slide. The pioneering study was made by Hanoaka & Fujimoto in 1957, using the large fish cones, and it was shown that different cones had different absorption spectra, with maxima in the regions of 5000, 5350, 5750 and 6200; more recent studies of Marks (1965), also on fish cones, gave maxima at 4550, 5300 and 6250 Å, i.e. in the blue, green and red parts of the spectrum. Subsequent refinements of technique have permitted the study of the smaller human and monkey cones; Fig. 169 shows the presence of three types of cone with λ_{max} at 4500, 5250 Å and 5550 Å (Brown & Wald, 1964); essentially similar results were found by Marks, Dobelle & MacNichol (1964), their maxima occurring at 4450, 5350 and 5700 Å. These studies thus confirm in a striking manner the predictions of the trichromatic theory derived from colour mixing.

---

* It has been argued that certain deuteranopes, with a normal luminosity curve, have both red- and green-absorbing pigments, but that their mechanisms are in some way "fused"; Alpern, Mindel & Torii (1968) have shown that certain deuteranopes do, indeed, have a normal luminosity curve, but this is because of a deficiency in macular pigment. Consequently the view that all deuteranopes lack chlorolabe is sustained.

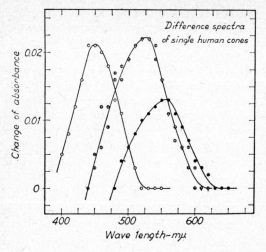

FIG. 169. Difference spectra of the visual pigments in single cones of the human parafovea. In each case the absorption spectrum was recorded in the dark, then again after bleaching with a flash of yellow light. They involve one blue sensitive cone, two green-sensitive cones, and one red-sensitive cone. (Brown and Wald, *Science*.)

ELECTROPHYSIOLOGY OF WAVELENGTH DISCRIMINATION

It is not, of course, sufficient to demonstrate the presence of "mechanisms" by either subjective or by photochemical methods; the physiologist will only be satisfied when he can tap the messages whereby these mechanisms transmit their separate information. Thus, if there are red-, green- and blue-sensitive cones, these must respond in specific fashions to light of any given wavelength falling on the retina, and somehow these responses must be transmitted to bipolar cells and ganglion cells and pass along the optic nerves to the lateral geniculate bodies. The messages will, of course, be altered, but it is essential to appreciate that the neurones through which they are carried will be unable to *add* anything; the wavelength discrimination must have been carried out at the moment of reception of the light-stimulus, and subsequent alterations in the messages can only be described as a *re-coding* of the information, with perhaps the scrapping of some details and alteration of emphasis of the different parts. Thus, to consider the "blue mechanisms," there might be three types of "blue cone", but if their separate messages were all pooled in a single type of bipolar cell, the extra information would be lost. The fact that these mechanisms are shown to exist by psychophysical methods, however, proves that their special messages have indeed survived the re-coding processes so that it is difficult to escape the conclusion that there are several types of cone that are specially sensitive to blue light.

Intracellular Records from Single Receptors

S-Potentials. In discussing the electrophysiology of the retina earlier, it was indicated that several workers had claimed to have recorded from individual cones by inserting a microelectrode into the retina; some of these showed some interesting chromatic responses, namely the so-called *S-potentials* of Svaetichin in the fish retina*; subsequent work has shown that these are derived from horizontal

---

* The fish retina has been studied with microelectrodes because the cones are very large so that intracellular recording becomes practicable; moreover, the colour vision of fishes

cells, i.e. one step later than the receptors themselves. In general these responses to light were not spike discharges but graded potential responses consisting of either an increase of the normal resting potential, a *hyperpolarization*, or a decrease, *depolarization*. The responses were graded in the sense that the change of potential increased with increasing intensity, whereas a spike potential is all-or-none in character. According to the depth of the electrode in the retina the responses were of two main types; an L-type which responded in the same way with all wavelengths and therefore gave a "luminosity message" (Fig. 170) and a C-type which gave opposite electrical effects according as the retina was stimulated with red and green or blue and yellow lights (Fig. 171, *a*, *b*). The

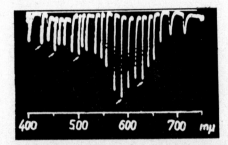

FIG. 170. L-type of response of unit in fish retina; the polarity of the electrical change does not alter with varying wavelength of stimulus. (Svaetichin & MacNichol, *Ann. N.Y. Acad. Sci.*)

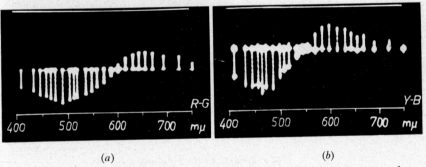

(*a*)   (*b*)

FIG. 171. (*a*) C-type of unit, of the red-green variety, giving responses of opposite polarity according as the wavelength of the stimulus lies in the red or green regions of the spectrum. (*b*) C-type, of yellow-blue variety. (Svaetichin & MacNichol, *Ann. N.Y. Acad. Sci.*)

messages carried by this second type thus indicated the chromatic nature of the stimulus, and the units were called *chromatic*, or C-units, being subdivided into *R-G* (red-green) and *Y-B* (yellow-blue) categories. Thus the C-potentials exhibit an "opponent colour" basis so that a given element may be hyperpolarized by one band of wavelengths and depolarized by another band; if hyperpolarization indicates inhibition, and depolarization activation, then these opposite responses may indeed be described as opponent. An interesting characteristic of the S-responses was the very large receptive fields extending across the whole

has been studied very extensively from a behavioural point of view, and there is no doubt that their colour discrimination is good and similar to that of man (see Svaetichin & MacNichol, 1958, for references).

retina; in these fields summation was complete so that Ricco's Law applied perfectly (Norton *et. al.*, 1968).*

Cone Responses. Of more interest are the responses of individual cones; and these have been described by Tomita *et al.* (1967) in the carp retina. The response to light was always a hyperpolarization, and when units were examined with light of different wavelengths, regions of maximum sensitivity were obtained in three broad regions in the blue, green and red. Fig. 172 illustrates the averaged results; the peaks occurred in the following regions:

	Blue	Green	Red	
	4620 ± 150	5290 ± 140	6110 ± 230	Carp Retina
	4550 ± 150	5300 ± 50	6250 ± 50	Goldfish

these are compared with the absorption maxima of goldfish cones described by Marks (p. 275).† The authors point out that their results provide no evidence for cones with maximal responses in the region of 6800 Å, as deduced by Naka & Rushton from their analysis of S-potentials.

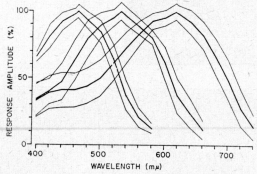

FIG. 172. Responses of three types of cone to light of different wavelengths indicating maximum sensitivity in the blue, green and red parts of the spectrum. Thin lines indicate standard deviation curves. (Tomita, Kaneka, Murakami & Pautler, *Vision Res.*)

Optic Nerve Fibre Response

Dominators and Modulators. Granit was the first to study the spectral sensitivity of optic nerve fibres. With the dark-adapted frog eye, and scotopic stimuli, the responses were the same whatever unit was picked up, the maximal response corresponding to the wavelength of maximal absorption of rhodopsin,

---

* The L-type of response is more complex than originally thought; thus it may be modified by adapting to a given colour, e.g. red; and it would seem that the response to any given wavelength is an integrated response from two or more types of cone (Naka & Rushton, 1966), in fact, according to Svaetichin (1967), three different types of response-curves, with amplitude maxima occurring in the red, green and blue regions of the spectrum, may be recorded from units in the three layers of horizontal cells. In the cat, Steinberg (1969) demonstrated both rod and cone influences, and this presumably corresponds with the anatomical connections of the horizontal cells (p. 121). Witkovsky (1967) has compared ganglion cell and S-potential response properties in the carp retina, and concluded that their origin was in the horizontal cells.

† Tomita *et al.* found the red responses predominant in 74 per cent of the total; blue came next at 16 per cent and green at 10 per cent; this is in sharp contrast with Marks' results, which indicated that there were 5–10 times as many red and green as blue units; with microelectrode work there is always the danger of selective sampling since a larger or more accessible cell will necessarily be penetrated more frequently.

about 5000 Å. In the light-adapted frog eye the wavelength-response curves varied with the unit picked up; they fell into two categories described as *dominators* and *modulators*. The dominators were the most common and gave broad sensitivity curves with a maximum at the wavelength for maximal photopic sensitivity as found by the ERG, namely at about 5550 Å (Fig. 173). The modulators had relatively narrow spectral sensitivity curves confined to one portion of the spectrum, so that it was possible to speak of a red-sensitive modulator whose peak of maximal sensitivity occurred at, say, 6000 Å; a green-sensitive modulator with a peak at 5300 Å, and so on. Fig. 174 illustrates the modulators isolated by this electrophysiological method from various eyes.

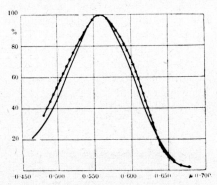

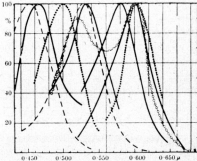

FIG. 173. Wavelength-response curves in single fibres of the frog (continuous line) and snake (interrupted by dots) optic nerve. The greatest response has been put equal to 100. The dominator. (Granit, *Nature.*)

FIG. 174. Wavelength-response curves in single fibres of the optic nerve of the rat (dots), guinea pig (broken lines), frog (line in full) and snake (line interrupted by circles). The modulators. (Granit, *Nature.*)

The Optic Nerve Message. Thus, so far as the *message* carried by the ganglion cells to the lateral geniculate bodies is concerned, we may say that it is organized apparently on a dual basis; the dominators indicate, by their frequency of discharge, the intensity of the prevailing light; under scotopic conditions their sensitivities correspond to the absorption of the scotopic pigment; rhodopsin in the case of the frog and mammals, and porphyropsin in the case of many fresh-water fishes. In the light-adapted mixed retina the dominator has the sensitivity corresponding to the photopic sensitivity curve. The second half of the message conveys information about the spectral composition; thus the red modulators would respond preferentially to red light falling on the retina, whilst the green and blue modulators would be much less affected, hence the message corresponding to the illumination of the retina with red light at photopic intensity might be a small discharge in the dominator, high discharge in the red modulator and small or negligible discharges in the green and blue modulators.

Opponent-Type Responses. In the goldfish Wagner, MacNichol & Wolbarsht found a rather more complex set of messages. In this species the ON-OFF type of response is most frequent and it was these elements that were studied from the point of view of wavelength discrimination. It at once became apparent that an element that, with white light, behaved as an ON-OFF element, with monochromatic light would behave either as a pure ON- or a pure OFF-element depending on the wavelength; this is illustrated by Fig. 175 where it is seen that with wavelengths of 4000 to 5500 Å the element behaves as a pure ON-element, whilst from 6000 to 7500 Å it is behaving as a pure OFF-element. It will be recalled that the essential feature of the OFF-element is really the silent period during the illumination of the retina, which is due to an active inhibition of the retina, the OFF-discharge being a release from this inhibition. The type of response illustrated by Fig. 175 thus demonstrates that the messages carried by a ganglion cell may be inhibitory over the red end of the spectrum and excitatory over the blue-green region. In other units the opposite relationship was found the blue-green region being inhibitory. Furthermore, when the receptive field of any given unit was studied it was found that its wavelength-dependence altered according as the centre of the field was studied or the periphery.* In general, two types were found; units that were ON to red in the centre and OFF to green over a wider area, and those that were OFF to red in the centre and ON to green over a wider area. As the authors emphasized, the messages tended to be opponent in character so that if red light caused inhibition its complementary, green, tended to excite, whilst diffuse white light would tend to have only a small effect because of cancellation of opposite influences.

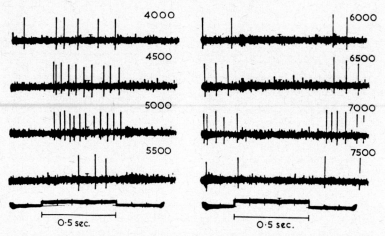

FIG. 175. Variation of type of response from a single ganglion cell with change of wavelength of stimulus. Duration of stimulus is indicated by step in the signal trace at the base of each series of records. With the shorter wavelengths (4000–5500 Å) the element behaves as a pure ON-element, whilst at the longer wavelengths (6000–7500 Å) it behaves as a pure OFF-element. (Wagner, MacNichol & Wolbarsht, *J. gen. Physiol.*)

TYPES -P AND -O

A more careful examination of the gold-fish retinal units by Daw (1968) showed that only some 5 per cent responded in this fashion, which were designated as Type-P; the majority, designated as Type-O, gave a more complex response in which both centre and periphery gave opponent responses to red and green.

* In the monkey's ganglion cells Gouras (1968) described opponent colour organization similar to that which De Valois had found higher up in the visual pathway (p. 282); the cells consisted of ON-centre-OFF-surround units, one type being sensitive to both red and green whilst the other was sensitive to only one wavelength band; the former gave only transient responses whilst those in the latter were sustained and were more common in the fovea.

Thus the centre could be ON-red and OFF-green whilst the periphery would be OFF-red and ON-green. The peripheral response required that a large part of it be excited at once, in the form of an annular stimulus, since a spot-stimulus was usually ineffective, however intense. It is because Wagner *et al.* examined the periphery with spot stimuli that they failed to observe its red-green opponence. When the centre was stimulated with one colour and the surround with another, the effects could be predicted; thus a red-OFF green-ON centre unit was stimulated by a red spot in a green surround; the response was a strong OFF-response; a green spot in a red surround gave a strong ON-response, whilst uniform illumination with red or green gave a weak ON-OFF response. Daw pointed out that the Type-O units, by contrast with the Type-P, are suited for mediating simultaneous contrast; e.g. a green spot on a red background gives a maximal OFF-response, whilst a red spot on a green background gives a maximal ON-response, or *vice versa*.

Dominator. So far as the photopic dominator type of response is concerned, it seems most likely that this really represents activity in ganglion cells that make connections with all three types of cone, red- green- and blue-sensitive; it is interesting that Witkovsky (1965) was unable to find this type of response in the light-adapted carp retina, although in the dark-adapted state dominators with maximal sensitivity at 5230 Å are found, i.e. corresponding to the λ_{max} of the carp's scotopic pigment. This was perhaps because he used very small stimulus-fields of 30–50 μ diameter, by contrast with the much larger fields employed by Granit.

Ground Squirrel. An interesting study is that of Michael on the pure-cone retina of the ground squirrel since here there are probably only two types of cone, a blue- and a green-sensitive one. The responses of single ganglion cells fell into two groups which either responded with excitation to green and inhibition (OFF-response) to blue or *vice versa*. Because of the antagonistic effects of blue and green, white light had very little effect. When the receptive fields were studied, using spot stimuli, these two groups fell into a number of classes; thus Class I showed no evidence for centre-surround antagonism, so that the receptive fields for ON- and OFF-responses were identical. Class II showed an obvious centre-surround antagonism, which could be demonstrated with white spots or annuli; once again diffuse white light had little effect because of the antagonistic effects of centre and surround. Analysis of the effects of large and small coloured spots indicated that the centre was sensitive to one wavelength band, e.g. green, whilst the surround was sensitive to both blue and green. The third class was similar in having centre-surround antagonism and differed from Class II in that the centre was sensitive to one band (green) and the periphery to the other (blue); this is the most common organization for the geniculate units of the monkey (p. 282).

As Michael points out, any of these unit-types could be responsible for successive colour contrast (p. 145), a flash of one colour inducing an after-image of the complementary in the same area of retina.

So far as the colour vision of the ground squirrel is concerned, it seems generally agreed that it is similar to that of the red-blind human (protanope), able to distinguish blues from greens and reds but not between reds and greens (Crescitelli & Pollack, 1965).*

* Colour vision, or its absence, in the cat has been well studied perhaps because of its status as a domestic pet and certainly because of the use of its ganglion cells for study of unit activity. Behaviourally, Daw & Perlman (1970) have been able to establish discrimination between red and cyan, and orange and cyan; this was at luminance levels that saturated the rods so that one must postulate two types of cone at least. Some lateral geniculate units were found showing opponent properties suggesting inputs from green- and blue-absorbing photoreceptors, so that the evidence suggests that the cat is a dichromat, and more specifically a tritanomalous protanope.

Responses in the Lateral Geniculate Body

The optic nerve fibres relay in an ordered fashion to the cells of the lateral geniculate body (p. 484); De Valois was the first to record from these third-order neurones, showing that the messages sent to the cortex in the monkey, an animal with colour-vision comparable with that of man, were colour-coded. As with the ganglion cells and optic nerve fibres, there were pure ON-elements that responded by increasing frequency of discharge on illuminating the retina; the responses were sensitive to changes of wavelength so that their wavelength-response curves seemed to be typical modulators; these modulator type elements were especially numerous in the region to which the foveal ganglion cells projected.

ON- or OFF-Elements. A second type of element, the most common, was the ON or OFF, responding with increased discharge at ON or else responding with inhibition during ON and a discharge at OFF. As to whether the cell behaved as an ON- or OFF-element depended on the wavelength of the stimulating light. Thus cells could be found to respond at ON with red stimulation and at OFF with green stimulation, so that if the response, measured by the number of spikes, is plotted against wavelength for equal energy stimuli, we get the two curves of Fig. 176. From this we see that with a green of 5300 Å there is an inhibition of the existing spontaneous discharge followed by a large discharge at OFF; with a red of 6400 Å there is a discharge at ON whilst at OFF there is some inhibition of spontaneous activity. With other elements the opposing wavelength bands were in the blue and yellow. It is clear that units of this type should be hardly affected by white light since this may be regarded as the combination of two of the complementary opponent stimuli, red-green, blue-yellow, etc. In fact, up to very high intensities, white light produced no response in these elements.

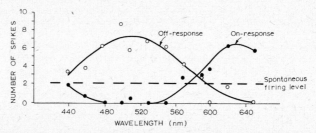

Fig. 176. The responses of a Red-ON, Green-OFF element to different wavelengths of light, the energy-contents of the stimuli being the same. (De Valois, *J. gen. Physiol.*)

Pure OFF-Elements. Finally pure OFF-elements were of the dominator type with maxima at 5100 and 5500 Å, presumably corresponding to scotopic and photopic pigments. At any rate Jacobs (1964), studying the squirrel monkey in which behavioural experiments had given the scotopic and photopic sensitivity curves, found reasonable concordance with the corresponding wavelength-sensitivity curves of dominator type units of the lateral geniculate body.

The Complex Message. Once again, then, the study of the messages in the optic pathway indicates the importance of inhibition and the OFF-discharge, besides simple excitation, in determining the code in which the nature of the stimulus is transmitted to the brain. Thus when a green light falls on the retina some fibres will carry bursts of discharge at ON whilst others will be inhibited, so that their spontaneous discharge is reduced; still others will perhaps show no response at all, neither excitation nor inhibition. The message sent to the brain is therefore complex, and it is by virtue of its complexity that it tends to be unique for the given band of wavelengths and flux of energy, so that a red light cannot mimic the responses evoked by a green light.

Hue-Discrimination Curve. De Valois, Abramov & Mead (1967) examined the responses of their four classes of lateral geniculate unit to different pairs of wavelength bands in order to determine the power of this system to discriminate small differences, i.e. they prepared a hue-discrimination curve analogous with the psychophysical one of Fig. 76. They predicted that greatest discrimination would be in the regions of cross-over from excitation to inhibition, and this proved to be true. Fig. 177 illustrates the hue-discrimination curves determined electrophysiologically.

Spatial Effects. De Valois used large light-stimuli and so he examined the overall effects of stimulating all the receptors in the receptive field of a given geniculate cell. Wiesel & Hubel (1966) showed that, like the ganglion cell units, the geniculate units were also spatially organized, having opposing centre and periphery effects with white light; when spectral sensitivity was studied a number of types were identified. The Type I unit gave an ON-response at its centre with a limited range of wavelengths whilst the surround gave an OFF-response to the complementary band, e.g. the commonest was a Red ON-centre, Green OFF-surround. These corresponded with De Valois' ON-OFF elements since he illuminated the whole receptive field. At a certain wavelength, the *neutral point*, the opposing effects of centre and surround just balanced; thus at 5400 Å a large spot of light had no effect. There were relatively few Blue ON-centre-Green-OFF-surround, and no Blue-Red units. With this Type-I, then, it appears as though the geniculate cell is connected to two types of cone, one of which inhibits and the other excites, or *vice versa*. When the intensity of the light-stimulus was reduced, a Purkinje-shift in sensitivity could be obtained with some Type I units, indicating connections with rods as well as cones.

The Type II units showed no centre-surround organization and consisted of a receptive field of some $\frac{1}{2}$ degree; the responses to different wavelength-bands were opposite in character, e.g. ON to blue and OFF to green, with a neutral point of 5000 Å, ON to red and OFF to green with a neutral point of 6000 Å, and so on. This type obviously receives input from two populations of cones, one excitatory and the other inhibitory, and the two types are distributed uniformly throughout the $\frac{1}{2}$ degree field. No Purkinje-effect was observed with these.

Type III units revealed no opponent colour effects; they had the centre-surround organization, the periphery usually suppressing the centre, but the responses were similar at all wavelengths. In these the representation of cone types was the same for inhibitory and excitatory areas of the field.

All these types were found in the dorsal layer of the lateral geniculate body; in the ventral layer an additional type was found, Type IV; they were organized on a centre-surround basis and were characterized by the very effective inhibition exerted by the surround.

In general these results show an ingenious arrangement whereby receptors subserving colour discrimination can be effectively neutralized by white light extending over the whole receptive field of a geniculate cell, the opponent responses to complementary colours of centre and surround being neutralized. With coloured lights, on the other hand, the colour-sensitive units will respond in a characteristic fashion, whilst the centre-surround organization will permit analysis of form as well as colour.

Cortical Responses

Motokawa, Taira & Okuda (1962) recorded from single neurones in the striate cortex of the monkey; many of these gave simple ON-responses to

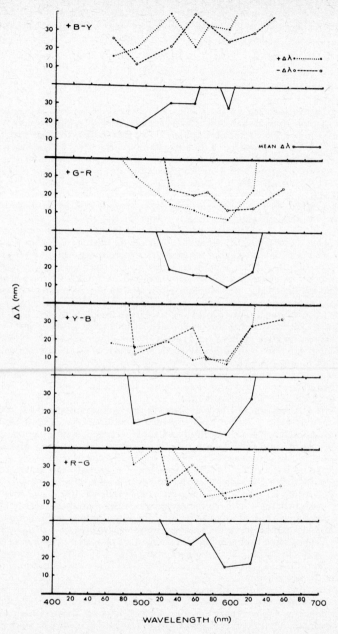

FIG. 177. Wavelength discrimination functions for units of the monkey lateral geniculate nucleus. As ordinate is plotted the shift in wavelength required to produce a measurable change in unit discharge when the eye is exposed to light of the wavelength indicated by the abscissa. The dotted and dashed lines represent shifts to wavelengths longer and shorter than the standard, respectively. Solid line: mean of + and − Δλ functions. (De Valois, Abramov & Mead, *J. Neurophysiol.*)

illumination with maximal responses to one part of the spectrum; the regions of maximum response were 6000–6400, 5200–5400 and 4600, i.e. in the red, green and blue; others gave opponent type responses, similar to those described by De Valois for the lateral geniculate body, the wavelengths for maximum ON- and OFF-responses being in different regions, e.g. in the red and blue. Some units gave ON-OFF responses to light over the whole range of wavelengths with maximum effect at 5000 Å; these were presumably related to rods, but their sensitivity to light was low suggesting an inhibitory activity in the pathway. With other units there was no clear-cut peak over the whole spectrum, whilst with still others there were peaks at the two ends, suggesting connections with two types of cone; as the intensity of the stimulus was reduced the peaks diminished and a new peak at 5000 Å emerged, indicating that rods and cones were converging on the same cortical neurone; in the light of the studies on geniculate cells this was to be expected.

Simple Cells. More recently Hubel & Wiesel (1968) have re-examined the cortical cells; as we shall see, these have receptive fields organized on a linear, as opposed to a circular, basis; in the striate cortex some 6 out of 25 simple cells showed chromatic sensitivity; they responded to long narrow fields with greatest sensitivity to long wavelengths; these fields were flanked by a more inhibitory area with blue-green sensitivity. Thus the cortical cell behaved as though it were connected with a set of Type I red-ON-centre green-OFF-surround geniculate cells organized in a linear fashion. Twelve out of 177 complex cells showed chromatic preference; these could be stimulated by a blue slit orientated in a specific direction; they did not respond to white; other units were similar but favoured longer wavelengths.

THE HERING THEORY

Although this theory was considered as antagonistic to the Young-Helmholtz theory, it is now appreciated that in essence it is simply a modification of this, since it accepts the essential trichromasy of human vision and merely puts forward hypotheses as to the manner in which this trichromasy is brought into being.

Opponent Process Theory. As restated by Hurvich & Jameson in their so-called *opponent process theory*, the essence of the theory is the postulation of opposite and mutually exclusive responses, somewhere in the neural pathway, to certain bands of wavelengths and to black and white. The paired stimuli are Red-Green, Yellow-Blue and White-Black, and they are considered to oppose each other since mixtures of the pairs do, indeed, tend to cancel each other out from a chromatic viewpoint, red and green and blue and yellow becoming eventually white, whilst black and white give the neutral grey sensation that one experiences in the dark. Space will not permit of a detailed description of the theory, which has been developed in quantitative form by Hurvich & Jameson; as indicated above it is really a trichromatic theory in spite of the postulation of four fundamental sensations, red, green, yellow and blue. The concept of opponent processes in the nervous pathways, whereby certain bands of wavelengths cause inhibition of ganglion cells whilst others have excitatory effects, is obviously supported by electrophysiological studies. We may assume that some receptors, when excited by light, inhibit bipolar cells and through them ganglion cells by virtue of the liberation of a special type of chemical mediator, similar to the inhibitory action of Renshaw cells on motor neurones. Other receptors might excite by virtue of having a different chemical mediator. If, now, the two types of receptor contained different photopigments, it is easy to see how different wavelengths could exert opposite nervous influences, the one inhibitory the other excitatory.

GENERAL REFERENCES

BRINDLEY, G. S. (1970). *Physiology of the Retina and Visual Pathway.* 2nd. Ed. London: Arnold.

GRANIT, R. (1947). *Sensory Mechanisms of the Retina.* London: Oxford University Press.

LE GRAND, Y. (1948). *Optique Physiologique.* Vol. 2. Paris: Editions de la 'Revue D'Optique'.

MARRIOTT, F. H. C. (1962). Colour vision. In *The Eye*, vol. II, Chapters 12–16. Ed. Davson. London: Academic Press.

RIPPS, H. & WEALE, R. A. (1970). The photophysiology of vertebrate color vision. *Photophysiol.* **5**, 127–168.

RUSHTON, W. A. H. (1959). Visual pigments in man and animals and their relation to seeing. *Progr. Biophys.* **9**, 239–283.

STILES, W. S. (1959). Colour vision: the approach through increment-threshold sensitivity. *Proc. Natl Acad. Sci. Wash.* **45**, 100–114.

WALD, G. (1969). The molecular basis of human vision. In *The Retina*, Ed. B. R. Straatsma *et al.*, pp. 281–295. Los Angeles: University of California Press.

WEALE, R. A. (1965). Vision and fundus reflectometry. *Documenta Ophthalmologica* **19**, 252–286.

WRIGHT, W. D. (1946). *Researches on Normal and Defective Colour Vision.* London: Kimpton.

SPECIAL REFERENCES

AGUILAR, M. & STILES, W. S. (1954). Saturation of the rod mechanism at high levels of illumination. *Optica Acta* **1**, 59–65.

ALPERN, M., MINDEL, J. & TORII, S. (1968). Are there two types of deuteranopes? *J. Physiol.* **199**, 443–456.

ARDEN, G. B. & TANSLEY, K. (1955). The spectral sensitivity of the pure cone retina of the squirrel (*Sciurus carolinensis leucotis*). *J. Physiol.* **127**, 592–602.

BAKER, H. D. & RUSHTON, W. A. H. (1965). The red-sensitive pigment in normal cones. *J. Physiol.* **176**, 56–72.

BLACKWELL, H. R. & BLACKWELL, O. M. (1961). Rod and cone mechanisms in typical and atypical congenital achromatopsia. *Vision Res.* **1**, 62–107.

BRINDLEY, G. S. (1954). The summation areas of human colour-receptive mechanisms at increment threshold. *J. Physiol.* **124**, 400–408.

BRINDLEY, G. S., DU CROZ, J. J. & RUSHTON, W. A. H. (1966). The flicker fusion frequency of the blue-sensitive mechanism of colour vision. *J. Physiol.* **183**, 497–500.

BROWN, P. K. & WALD, G. (1964). Visual pigments in single rods and cones of the human retina. *Science* **144**, 45–52.

CAMPBELL, F. W. & RUSHTON, W. A. H. (1955). Measurement of the scotopic pigment in the living human eye. *J. Physiol.* **130**, 131–147.

CRESCITELLI, F. & POLLACK, J. D. (1965). Color vision in the antelope ground squirrel. *Science* **150**, 1316–1317.

COLE, B. L. & WATKINS, R. D. (1967). Increment thresholds in tritanopia. *Vision Res.* **7**, 939–947.

DAGHER, M., CRUZ, A. & PLAZA, L. (1958). Colour thresholds with monochromatic stimuli in the spectral region 530–630 μ. In Visual Problems of Colour. *N.P.L. Symp.* No. 8, pp. 389–398.

DAW, N. W. (1968). Colour-coded ganglion cells in the goldfish retina: extension of their receptive fields by means of new stimuli. *J. Physiol.* **197**, 567–592.

DAW, N. W. & PERLMAN, A. L. (1970). Cat colour vision: evidence for more than one cone process. *J. Physiol.* **211**, 125–137.

DE VALOIS, R. L. (1960). Colour vision mechanisms in the monkey. *J. gen. Physiol.*, **43**, Suppl. 115–128.

DE VALOIS, R. L., ABRAMOV, I. & MEAD, W. R. (1967). Single cell analysis of wavelength discrimination at the lateral geniculate nucleus in the macaque. *J. Neurophysiol.* **30**, 415–433.

DODT, E. & WALTHER, J. B. (1958). Der photopischer Dominator im Flimmer-ERG der Katze. *Pflüg. Arch.* **266**, 175–186.

FINCHAM, E. F. (1953). Defects of the colour-sense mechanism as indicated by the accommodation reflex. *J. Physiol.* **121**, 570–580.

GIBSON, I. M. (1962). Visual mechanisms in a cone-monochromat. *J. Physiol.* **161**, 10–11 P.

GOURAS, P. (1968). Identification of cone mechanisms in monkey ganglion cells. *J. Physiol.* **199**, 533–547.

GRAHAM, C. H. & HSIA, Y. (1958). Colour defect and colour theory. *Science* **127**, 675–682.

GRANIT, R. (1943). A physiological theory of colour perception. *Nature, Lond.* **151**, 11–14.

GRANIT, R. & WREDE, C. M. (1937). The electrical responses of light-adapted frog's eyes to monochromatic stimuli. *J. Physiol.* **89**, 239–256.

HANOAKA, T. & FUJIMOTO, K. (1957). Absorption spectrum of a single cone in carp retina. *Jap. J. Physiol.* **7**, 276–285.

HECHT, S. (1947). Colourblind vision. I. Luminosity losses in the spectrum from dichromats. *J. gen. Physiol.* **31**, 141–152.

HUBEL, D. H. & WIESEL, T. N. (1968). Receptive fields and functional architecture of monkey striate cortex. *J. Physiol.* **195**, 215–243.

HURVICH, L. M. & JAMESON, D. (1957). An opponent-process theory of colour vision. *J. Psychol. Rev.* **64**, 384–404.

JACOBS, G. H. (1964). Single cells in squirrel monkey lateral geniculate nucleus with broad spectral sensitivity. *Vision Res.* **4**, 221–232.

MacNICHOL, E. J. & SVAETICHIN, G. (1958). Electric responses from the isolated retinas of fishes. *Am. J. Ophthal.*, Pt. 2 **46**, 26–46.

MARKS, W. B. (1965). Visual pigments of single goldfish cones. *J. Physiol.* **178**, 14–32.

MARKS, W. B., DOBELLE, W. H. & MacNICHOL, E. F. (1964). Visual pigments of single primate cones. *Science* **143**, 1181–1183.

MICHAEL, C. R. (1968). Receptive fields of single optic nerve fibers in a mammal with an all-cone retina. III. Opponent color units. *J. Neurophysiol.* **31**, 268–282.

MOTOKAWA, K., TAIRA, N. & OKUDA, J. (1962). Spectral responses of single units in the primate visual cortex. *Tohoku J. exp. Med.* **78**, 320–337.

NAKA, K. I. & RUSHTON, W. A. H. (1966a). S-potentials from colour units in the retina of fish (*Cypridinae*). *J. Physiol.* **185**, 536–555.

NAKA, K. I. & RUSHTON, W. A. H. (1966b). An attempt to analyse colour reception by electrophysiology. *J. Physiol.* **185**, 556–586.

NAKA, K. I. & RUSHTON, W. A. H. (1966c). S-potentials from luminosity units in the retina of fish (*Cypridinae*). *J. Physiol.* **185**, 587–599.

NORTON, A. L., SPEKREIJSE, H., WOHLBARSHT, M. L. & WAGNER, H. G. (1968). Receptive field organization of the S-potential. *Science* **160**, 1021–1022.

PITT, F. H. G. (1935). Characteristics of dichromatic vision. *Med. Res. Council Sp. Rep. Ser.*, No. 200.

RUSHTON, W. A. H. (1963). A cone pigment in the protanope. *J. Physiol.* **168**, 345–359.

RUSHTON, W. A. H. (1965). A foveal pigment in the deuteranope. *J. Physiol.* **176**, 24–37.

STEINBERG, R. H. (1969). Rod and cone contributions to S-potentials from the cat retina. *Vision Res.* **9**, 1318–1329.

SVAETICHIN, G. (1956). Spectral response curves from single cones. *Acta physiol. scand.*, **39**, Suppl. 134, 17–46.

SVAETICHIN, G. & MacNICHOL, E. F. (1958). Retinal mechanisms for chromatic and achromatic vision. *Ann. N.Y. Acad. Sci.* **74**, 385–404.

TOMITA, T., KANEKA, A., MURAKAMI, M. & PAUTLER, E. L. (1967). Spectral response curves of single cones in the carp. *Vision Res.* **7**, 519–531.

WAGNER, H. G., MacNICHOL, E. F. & WOLBARSHT, M. L. (1960). The response properties of single ganglion cells in the goldfish retina. *J. gen. Physiol.*, **43**, Suppl. 45–62.

WALD, G. (1945). Human vision and the spectrum. *Science* **101**, 653–658.

WALD, G., BROWN, P. K. & SMITH, P. H. (1955). Iodopsin. *J. gen. Physiol.* **38**, 623–681.

WALLS, G. L. & HEATH, G. G. (1954). Typical total colour blindness reinterpreted. *Acta ophthal. Kbh.* **32**, 253–297.

WEALE, R. A. (1953). Cone monochromatism. *J. Physiol.* **121**, 548–569.

WEALE, R. A. (1959). Photosensitive reactions in fovea of normal and cone-monochromatic observers. *Optica Acta* **6**, 158–174.

WEALE, R. A. (1962). Further studies of photo-chemical reactions in living human eyes. *Vision Res.* **1**, 354–378.

WIESEL, T. N. & HUBEL, D. H. (1966). Spatial and chromatic interactions in the lateral geniculate body of the rhesus monkey. *J. Neurophysiol.* **24**, 1115–1156.

WITKOVSKY, P. (1965). The spectral sensitivity of retinal ganglion cells in the carp. *Vision Res.* **5**, 603–614.

WITKOVSKY, P. (1967). A comparison of ganglion cell and S-potential response properties in carp retina. *J. Neurophysiol.* **30**, 546–561.

WRIGHT, W. D. (1932). The characteristics of tritanopia. *J. opt. Soc. Am.* **42**, 509–521.

PART 10

STILES-CRAWFORD EFFECT; ADAPTATION; PHOTOPIC SENSITIVITY CURVES

WE may conclude this account of modern research devoted to the elucidation of the mechanism of vision with a few special aspects of phenomena that have already been discussed in a more elementary fashion.

DIRECTIONAL SENSITIVITY OF THE RETINA

Pupillary Diameter. Stiles & Crawford decided to develop an indirect method of measuring pupillary diameter. They argued that the larger the pupil the greater the amount of light that would enter the eye and hence the greater the apparent brightness of a luminous field; the amount of light would increase in proportion to the area of the pupil and this in proportion to the square of its diameter. They found that the calculated diameter never increased beyond 5·5 mm although the maximum diameter actually measured was about 8 mm. This discrepancy was shown to be due to the circumstance that, as the pupil became larger, the retina received relatively more rays of light that struck it obliquely; if these oblique rays were less efficient than those entering through the centre of the pupil, then the discrepancy could be explained. The effects are large, so that if the subjective brightness of a pencil entering at the very periphery of the pupil is compared with one entering centrally, it may be only 15 per cent of this.

Cone Phenomenon. It was shown that the phenomenon was peculiar to the cones, both foveal and peripheral, and this has provided a valuable method for determining whether rods or cones are concerned in a given visual event. Thus Donner & Rushton, studying the responses of a single ganglion cell in the frog, showed that when it behaved as a scotopic dominator its responses were independent of the direction of incidence of the pencil of light; when it behaved as a photopic dominator, after light-adaptation, it showed directional sensitivity; moreover, in the mesopic range when a hump appears on the sensitivity curve in the blue, this "blue hump" was shown to be insensitive to direction, and thus to be due to activity of rods, probably of a special type called "green rods" (Denton & Wyllie).

Variation with Wavelength. The magnitude of the directional effect varies with wavelength (Stiles, 1937); moreover, if monochromatic light is used there is a change in its subjective hue as the angle of incidence of the rays varies; this is called the Stiles-Crawford effect of the second kind. Brindley showed that colour-matches could be upset by changing from direct to oblique incidence, so that it would seem that the forms of the fundamental response curves of the cone mechanisms responsible for wavelength discrimination change with angle of incidence of the radiation. Since the shape of an absorption spectrum depends on the density of the pigment in solution (p. 190), and since the shape of the sensitivity curve presumably depends on that of the absorption spectrum of the cone-pigment, we can conclude that the effective optical density of pigment in the cone varies with direction of incidence.

Wave-Guide Theory. The most plausible explanation for the phenomenon is that put forward by Wright & Nelson to the effect that light entering directly along the axis of a cone is more likely to be retained by a series of total internal reflexions than light entering obliquely; this *wave-guide theory* has been developed quantitatively by Enoch.*

ADAPTATION

Rod Adaptation

The main phenomena of adaptation have been already described throughout this Section, and here we shall discuss only some special aspects. By dark-adaptation we mean the recovery of sensitivity of the retina following exposure to light; so far as rod vision in man is concerned, there is little doubt that this process is largely determined by the regeneration of rhodopsin, although the relationship between retinal sensitivity and concentration of rhodopsin in the retina is not so simple as had been at first thought (p. 208).

Foveal Adaptation

With human foveal, i.e. cone, vision, Wright has established the existence of an essentially similar process that takes place far more rapidly however, presumably because cone pigments regenerate more rapidly (p. 208). The reverse process, which we may call foveal light-adaptation, constitutes a decrease in sensitivity of the retina during continued exposure to light; it was studied by Wright with his binocular matching technique, the change in sensitivity of one eye, during exposure to light, being measured by comparing the subjective sensations in the two eyes. The curve of decrease in sensitivity falls over a period of some 50 seconds to reach a plateau with a sensitivity about one sixteenth of the original value (Fig. 80, p. 146).

Alpha-Adaptation. A much more rapid adaptive effect was described by Schouten & Ornstein, who measured the effect of a small glare-source of light, stimulating a peripheral part of the retina, on the brightness-sensation experienced by the fovea. Thus the two eyes would view independently the two halves of an illuminated field, and these would be adjusted in luminosity to give a match. The peripheral glare-source, entering one eye only, would now be switched on, and immediately the match would be disturbed, the half of the field seen by the eye with the glare-source appearing much darker. By raising its luminance till a match was obtained the decrease in sensitivity could be determined. This α-*adaptation*, as Schouten & Ornstein called it, is complete within a fraction of a second and was thought to differ from the β-adaptation of Wright in that its effect extended to the whole retina, even though only a

---

* Enoch has attempted to implicate this wave-guide phenomenon in the mechanism of wavelength discrimination; thus the various wavelengths would be brought to modes in different parts of the cone, and if pigments were distributed unevenly we could envisage different responses in the same cone to the various wavelengths. The fact that the appearance of monochromatic light is practically unchanged when it reaches the retina through the sclera (Brindley & Rushton) rather discounts this theory. It also shows that wavelength discrimination in man is not mediated by colour filters in front of the cones. In birds, however, the coloured oil-droplets seem to fulfil this role (see, for example, Donner).

small part was stimulated. Provided exposure to the glare-source was short, recovery of sensitivity was rapid. Thus long before Wright, in his binocular matching studies, had made his first measurement of retinal sensitivity, this had presumably decreased by a factor of about five because of this rapid α-adaptive process.

VEILING LUMINANCE

Subsequent work, notably that of Fry & Alpern has shown, however, that the phenomenon is not one of retinal interaction but is due to the scattered light from the glare-source. The effect of this scattered light is to behave as a "veiling luminance", decreasing the sensitivity of the whole retina. We may regard the phenomenon as one of very rapid adaptation, but not, as originally thought, of inhibition of the whole retina by a stimulus falling on only a part.

Changes of Sensation. According to Wright, the change in sensitivity of the fovea, following exposure to a bright light may be indicated graphically as in Fig. 178. The dotted line represents the almost instantaneous α-adaptive process. On switching off the light, there is a rapid increase in foveal sensitivity, continued as a slower β-adaptive recovery.

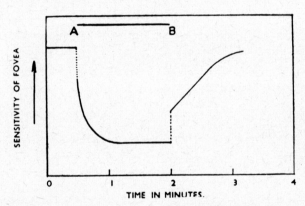

FIG. 178. Schematic representation of α- and β-adaptation. The eye is exposed to light during the interval AB. The dotted lines represent the virtually instantaneous α-adaptive decrease and increase in sensitivity. (Wright, *Trans. Illum. Eng. Soc.*)

Development of the Visual Sensation

If we take into account the general phenomena of adaptation, we may predict, in a crude way, the course of the visual sensation when the eye is exposed to a bright light. At first, because of the relatively high accumulation of photo-pigments in the cones, the sensation will be intense, but because of the decom-position of these we may presume that the sensation, on this photochemical ground, will fall off, until a steady state is reached with regeneration of pigment keeping pace with decomposition. On these photochemical grounds then we may expect sensation to grow to a peak, decline and level off to a steady state. Because of the rapid α-adaptive process, however, these slower changes will be modified, in the sense that the rise in sensation will be slowed by the reduced sensitivity of the retina, whilst the fall will be more rapid, and to a lower level than would have prevailed if there had been no α-adaptation. The actual response, as measured, is shown in Fig. 179 where it is seen how rapidly the sensation of brightness declines; in Fig. 180 this curve of sensation is shown to be made up of the algebraic sum of a "primary change" depending on the photochemistry of the receptors, and the inhibitory α-adaptation. In the same figure there is the analogy with the electroretinogram, the α-adaptive process being likened to the inhibitory PIII component.

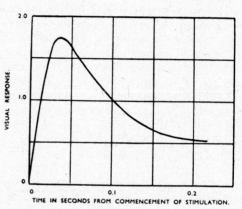

FIG. 179. Actual sensation curve during the first 200 msec of stimulation by a bright light. (Wright, from Schouten.)

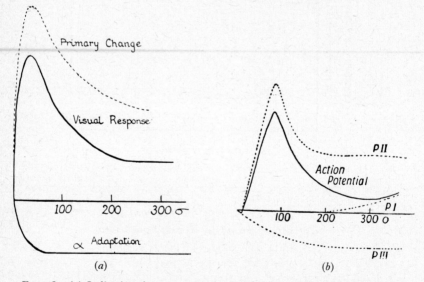

FIG. 180. (a) Indicating the composite nature of the sensation curve resulting from stimulation with a bright light. (b) Illustrating the analogy between α-adaptation and the PIII component of the ERG. (Wright & Granit, *Brit. J. Ophthal.*)

Simultaneous Contrast

At the time when the effects of the glare-source described by Schouten & Ornstein were interpreted as being due to an adaptive effect of one part of the retina on the rest, it was natural to assume that this α-adaptation lay at the basis of simultaneous contrast, whereby the apparent brightness of a grey patch is diminished by surrounding it with a white background. However, with the demonstration that the glare-source effect was due to scattered light within the eye, this interpretation becomes suspect. However, the fact of simultaneous contrast remains, in the sense that a patch of light appears much darker if surrounded by a white background than if surrounded by a black one; so it may well be that the stimulation of the surrounding retina does, indeed, reduce the sensitivity of the central retina to light.

Retinal Interaction. The studies of Westheimer (1965, 1967) on the effects of surround-brightness on the threshold have certainly demonstrated an inhibitory action. Westheimer caused a subject to view a test spot through an illuminated annulus and to adjust the luminance of the spot so that it could be discriminated from its surround; i.e. the value of ΔI, or the *incremental threshold*, was obtained. When the size of this surround was increased, the threshold first increased, but beyond a certain point, namely 0.75 degrees, in scotopic vision it decreased by as much as one logunit up to a diameter of 2.5 degrees. The initial threshold-raising effect with increasing area is simply an expression of Ricco's Law; the annular surround is increasing its subjective brightness by spatial summation; at the point where the threshold falls a new factor is becoming evident, and this Westheimer attributes to an inhibitory action of the peripheral part of the annular surround on the more central part, so that the subjective sensation from the annulus is reduced and thus requires a smaller value of luminance of the central test spot for a match. In the fovea, too, there is a similar effect of increasing surround-area, the limit beyond with Ricco's Law fails is smaller, namely 5 minutes of arc.*

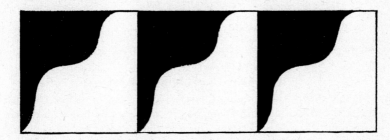

FIG. 181. Illustrating the formation of Mach bands. If this figure is rotated on a drum, at a rate above the critical fusion frequency, the subject sees a grey surface with alternate light and dark stripes, although there are no actual sharp transitions of luminance on the figure. (Graham, *Vision and Visual Perception.* John Wiley Inc.)

Mach Bands. A manifestation of simultaneous contrast or spatial inductive effects is given by the so-called *Mach bands*. If a picture, such as Fig. 181, is rotated above the critical fusion frequency in front of the eye, the subject sees a

* Sakmann, Creutzfeld & Schleich (1969) have shown that the area for summation of surround adaptive effects is not the same as that for excitation of a test spot in the cat's retina, in fact the receptive field of a ganglion cell for adaptive effects shows no centre-surround antagonism. Easter (1968) has measured adaptive effects of light stimuli on goldfish retinal ganglion cells, using cone-stimuli. The receptive field for the adaptive raising of threshold was certainly much larger than that for excitation of the ganglion cell, so that we may speak of an "adaptation pool's field"; this may well be mediated by horizontal and amacrine cells.

grey surface whose objective luminance varies in accordance with the Talbot-Plateau Law. However, although Fig. 181 exhibits no sharp transitions of luminance, the subject sees alternate light and dark stripes on the height axis, adjacent stripes bordering each other at fairly clear lines. The light stripes correspond to the parts of the light-distribution curve that are concave to the height-axis (Fig. 182) and the dark to those that are convex. We may presume that in the regions of fairly rapid change of luminosity, inhibition and facilitation occur, leading to the sharpening of the transition from one level of subjective sensation of greyness to another.

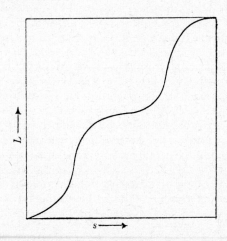

F*ig*. 182. Illustrating the variation of luminance of the picture presented to the eye by Fig. 181, when rotated so as to cause fusion of successive images. The luminance is determined, under these conditions, by the Talbot-Plateau Law. (Graham, *Vision and Visual Perception*. John Wiley Inc.)

Successive and Meta-Contrast

It is thought that the phenomena of successive contrast, or temporal induction, are associated with the slower β-adaptation; here we are dealing with the effects of one stimulus on a succeeding one, both of them falling on the same part of the retina; the phenomena are essentially of an adaptive nature in the sense that the first stimulus depresses the sensitivity of the retina for a succeeding, like, stimulus.

Meta-Contrast. By *meta-contrast* is meant the inductive effect of a primary light stimulus on the sensitivity of the eye to a previously presented light stimulus on an adjoining part of the retina; it is a combination of temporal and spatial induction, or successive and simultaneous contrast. The effect is produced by illuminating the two halves of a circular patch consecutively for a brief duration. If the left half only, for example, is illuminated for 10 msec, it produces a definite sensation of brightness. If, now, both are illuminated for the same period, but the right half some 20–50 msec later, the left half of the field appears much darker than before, and near the centre may be completely extinguished. The left field has thus been inhibited by the succeeding, nearby, stimulus. The right field, moreover, appears darker than when exposed alone—it has been inhibited by the first stimulus (*para-contrast*). The phenomenon may be described as α-adaptation—a nervous inhibitory action of one stimulus on the response to another, but it emphasizes what the conception of α-adaptation fails to do, namely the mutual effects of two stimuli *on each other*; the second stimulus inhibits the first whilst the first inhibits the second. Baumgardt & Segal have shown that, by varying the interval between the presentation of the two halves of the illuminated field, a variety of intermediate degrees of meta-contrast may be achieved; moreover, these may be reproduced as after-images, but in the latter process the time relationships are altered so that if, for example, the left half of the field was dark and the right half bright in the primary sensation, the reverse relationship could be found in the after-image.

Adaptation and the Colour Sense

So far we have considered the effects of adaptation on the light and form senses; we know from the phenomena of spatial and temporal induction that the colour sense can be appreciably affected by adaptation, but it is only comparatively recently that serious quantitative measurements of the specific effects of any given wavelength have been made.

Colour Matches. Wright has applied his binocular matching technique to the problem. For example, the right eye was made to view a yellow patch; the left eye was presented with a patch which could be illuminated by the subject with three primary colours, red, green, and blue, and their proportions and intensities could be varied so that a match, both in hue and intensity, could be made with the yellow viewed by the right eye. The right eye was then exposed to an adapting light for three minutes and both eyes once again viewed their respective test and comparison patches. As a result of the adaptation it was found that the previous match no longer held in respect to either brightness or hue; and the change was measured by adjusting the intensities of the primaries as quickly as possible until a correct match was made. By repeating the matches every thirty seconds the course of recovery from adaptation could be plotted as three separate curves corresponding to the recoveries of the red, green, and blue sensations. In general, it was found that adaptation to any given wavelength had a general and specific effect; there was a general reduction in sensitivity of the eye to all wavelengths, together with a specific reduction in sensitivity to the adapting wavelength. Consequently, a match that holds for one state of adaptation of the eye does not necessarily hold good for another. Thus with a yellow test-colour, a match was made with equal amounts of red and green before adaptation of the right eye to a white light; after adaptation, a match was made with red and green in the proportions of 38 : 60, i.e. far more green was required by the dark-adapted left eye to match the sensation experienced by the light-adapted right eye, a fact which indicates that an adapting white light depresses the sensitivity to red more than to green.

Colour-Mixing Equations. This observation does not, however, indicate that the colour-mixing equations, applicable for one level of luminance, are inapplicable for another; in fact, there is some reason to expect that, in the same eye, matches will hold independently of the condition of this eye, at any rate over a fairly wide range of light-adaptation. Thus if we take a mixture of red and green in one half of a field and match it with a mixture of yellow and white in the other, then the fact that there is a match means that the retina is responding in essentially the same way in the two halves of the field; any

change in the retina, caused by adaptation to any wavelength, will influence
the receptors receiving both halves of the field in the same way, so that although
the subjective sensation will be altered the two halves may be expected to appear
the same, and the match will be said to hold in spite of the adaptation of the
retina. Thus, presentation of the two halves of the field to the two eyes separately,
as in the binocular matching technique, would indicate a failure of the match,
but that is because we are comparing two eyes in different states of adaptation.
In practice, colour-matches do, indeed, hold under different conditions of
adaptation.

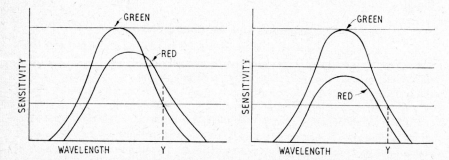

Fig. 183. Showing that, if the forms of the fundamental response curves are
altered by an adapting light, colour matching will be upset. *Left:* the yellow
stimulus, Y, produces responses in the green and red mechanisms in the ratio
of 2 : 3. *Right:* the same yellow stimulus produces responses in the ratio 1 : 2
because selective adaptation has altered the shape of the red mechanism's response curve.

Artificial Colour-Defectiveness. Nevertheless, if very high adapting
luminances are employed, Wright showed that uniocular colour matches may
be upset, and the phenomenon was investigated in some detail by Brindley.
On the basis of the trichromatic theory, the maintenance of colour matches in
spite of adaptation implies that the forms of the three spectral response curves
remain unaltered, since it is essentially because of their characteristics that the
laws of colour matching are what they are. Thus, to consider the Red and Green
response curves and their contribution to Yellow, the matching of Red+
Green to equal yellow is represented by Fig. 183; here the yellow stimulus
matches the mixed Red and Green stimuli because it stimulates the Red by
3 units and the Green by 2 units. If we now adapt to red light so that the shape
of the Red curve is altered, then now our yellow radiation is stimulating the Red
and Green mechanisms in different proportions and the match may no longer
hold. By using very extreme conditions of adaptation such as might be expected
to bleach the photopigment specifically from one type of cone completely, we
might expect to achieve a state of artificial dichromatism, or, if we put two
types of receptor out of action, one of monochromatism. In fact, all these disturbances of colour-matching may be produced (Brindley, 1953). For example,
by adapting to a very bright violet light, an artificial tritanopia may be produced
so that all spectral wavelengths may be matched by suitable mixtures of only
two primaries, a red and a blue. With a very bright red or blue-green light this
artificial tritanopia may be converted into an artificial monochromatism, in

which large parts of the spectrum can be matched with a single wavelength; thus a "red monochromatism" was obtained by strong violet and blue-green adaptation, presumably because only the red mechanism was operative. Under these conditions all spectral colours from 5000 Å to 7000 Å could be matched by suitably varying the intensity of a yellow stimulus of 5780 Å. The sensation was that of a very unsaturated red or pink. After adaptation with violet and red, a "green monochromatism" was obtained, so that all wavelengths between 4800 Å and 6200–6300 Å could be matched with a yellow, both fields appearing this time as an unsaturated blue-green.* Finally, a "violet monochromatism" was obtained by adaptation to a bright yellow light; over the range of 4000 Å to 5000 Å all wavelengths were matched with violet of 4470 Å, both fields appearing as a very saturated violet. Presumably, then, Brindley was putting the fundamental mechanisms out of action selectively, but of course the remaining mechanisms could hardly be called normal since they would have been subjected to quite considerable bleaching, so that the density of pigment in them would be less than that pertaining with normal trichromatic vision.

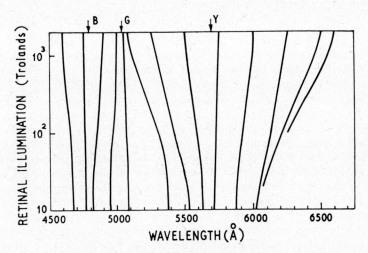

FIG. 184. Wavelengths having the same hue at different levels of retinal illumination. For example, an orange of about 6000 Å shifts towards the red when the retinal illuminance is increased from 10 to 1000 trolands. The arrows indicate the invariant hues. (Walraven, *J. opt. Soc. Amer.*)

Bezold-Brücke Phenomenon. When the intensity of a monochromatic light is increased considerably, there is a change of its subjective hue so that it can be matched with a different monochromatic light of lower intensity; this is the Bezold-Brücke phenomenon. The shifts are illustrated in Fig. 184, where the lines join wavelengths of the same apparent hue when viewed at ten, and over a thousand, trolands; the arrows indicate the invariant hues, namely a blue of 4760 Å, a blue-green of 5080 Å and a yellow of 5700 Å; thus oranges

* In a paper published in the *Philosophical Transactions* in 1898, Burch described these "red" and "green" monochromatisms after adapting to strong violet+green or violet+red lights. It should be noted that these states last only for a short time, of the order of 10 seconds.

and reds at high intensities are equivalent to a yellow at moderate intensity, and so on. Peirce in 1877 suggested that these shifts could be explained on the basis of the Young-Helmholtz theory, with the assumption that the three sensitivity curves altered relatively to each other when the intensity-level of stimulation increased. Thus, if an orange at moderate intensity stimulated the red mechanism very strongly and the green mechanism only weakly, then increasing the intensity of the orange might be expected to increase the green's response more strongly than that of the red, and thus to shift the subjective sensation towards the green end of the spectrum, i.e. to make an orange approach a yellow. On this basis, the invariant hues would correspond to the crossing-over points of the three sensitivity curves. This is true of the invariant yellow at 5070 Å if we accept Pitt's sensitivity curves, but as Walraven has shown, further assumptions are necessary to account for the other two invariant colours. As presented, the theory of Peirce, and its modification by Walraven, are obviously too naive, but the general principle that the relative shapes of the response curves of the cone mechanisms will alter with intensity is in accordance with what would be expected on electrophysiological grounds.

Bloch's Law

If the sensation of brightness is due primarily to the absorption of quanta of light, we may expect that the intensity of this sensation, induced by a flash, other things being equal, will be determined by the product: $I \times t$, a large intensity, I, and short time, t, being equivalent to a smaller intensity and longer time. For threshold studies this relationship has already been described, and is the basis of the so-called Bunsen-Roscoe Law familiar to the photochemist. For flashes above threshold this relationship was found to hold by Bloch in 1885 and is called *Bloch's Law*. According to Brindley (1952) the relationship holds down to flash-durations of extreme brevity, at least as short as $4 \cdot 1 \cdot 10^{-7}$ sec when the retinal illumination was 3.10^8 trolands, and the greatest duration over which it extends is probably of the order of 0·03 second (Brindley, 1959).

After-Images. According to Brindley, Bloch's Law applies with some precision to the subjective brightness of after-images, so that if a subject is exposed to two half-fields illuminated independently, the product of $I \times t$ being the same for each half-field but the values of I and t being different, then the after-images are indistinguishable.* This emphasizes that the long-lasting effects of a light stimulus are largely determined by the photochemical changes in the receptors. Especially strong support for this view is the finding that Bloch's Law breaks down at extremely high intensities and short exposures; thus if two flashes are exposed in the top half-field at an interval of 4 msec and the same two flashes in the bottom half-field at an interval of 0·28 msec, the total quantities of light are the same but, if the intensities are very high, the after-images of the upper and lower fields are distinguishable. The breakdown is probably related to Hagins' discovery that with a single bright flash-exposure of the rabbit's retina, the amount of rhodopsin bleached, as determined by reflexion densitometry, was never greater than 50 per cent of the total, however intense the stimulus, although a second flash, separated from the first by a few hundredths of a second, would bleach a further 50 per cent.

* That is, provided the after-images appearing during the first 15 seconds were ignored.

PHOTOPIC SENSITIVITY CURVES

The curves shown in Fig. 154, p. 257, give the relative sensitivities of the light-adapted eye to different wavelengths, i.e. under conditions where the cones apparently determined the sensation. As we have seen, this sensation may well be mediated by three types of cone, so that one might expect to find discontinuities in the curve in regions where one type of cone tends to "take over" from another. In fact, as we have seen, Stiles was able to create conditions where the "taking over" by the separate mechanisms could be demonstrated with some precision, but this required conditions that ensured that he was measuring the threshold for the new mechanism, i.e. by light-adaptation with carefully chosen wavelengths, the mechanisms with which he was not concerned could be depressed in sensitivity sufficiently to allow the single mechanism to manifest itself. Under the ordinary conditions of measurement of photopic sensitivity, where essentially what we are doing is to match different wavelengths for brightness sensation, thereby finding the relative energies required for equal sensation, we may expect all mechanisms to be operative to some extent, and thus to obtain a smooth curve of spectral sensitivity.

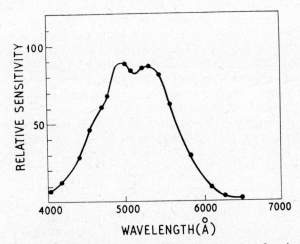

FIG. 185. Mean spectral sensitivity curve for several species of squirrel, determined electroretinographically. Ordinate: percentage sensitivity. Abscissa: wavelength in Å of stimulating light. Intensity was adjusted so that at each wavelength the same number of quanta/sec fell on to the retina. (Tansley, Copenhaver & Gunkel, *Vision Res.*)

"Humps" on the Curves

By reducing the size of the test-patch, or by using peripheral parts of the retina, however, some evidence for discontinuities, in the form of "humps" on the sensitivity curves, can be obtained; at these humps the retina is presumably specially sensitive to a selected region of wavelengths. For example, Thomson found evidence for extra sensitivity in three regions of the spectrum, one at 4600 Å, another at 5200 Å and a third between 5800 and 6100 Å. Again, by studying the periphery, Weale found two humps indicating extra sensitivity in the blue at 4600 Å and in the green at 5400 Å. Using a rather different technique, Auerbach & Wald identified what they called a violet receptor with maximum sensitivity at 4350 Å.

Squirrel. Humps on the sensitivity curve are not peculiar to man; thus Fig. 185 shows the mean results for a number of different species of squirrel, animals that are interesting because their retina is a pure cone one. There are two obvious peaks at 5350 and 4900 Å respectively, and here it is interesting that neither of these coincides with the absorption peaks for the pigments that can be extracted from the squirrel retina, which is actually a rhodopsin with λ_{max} of 5020 Å (Dartnall). By selectively adapting the retina with blue or green light the two humps could be selectively depressed, and this suggests that there are different pigments mediating blue and green sensitivity.

Colour Matching

Not only may the sensitivity curves be "abnormal" under certain conditions but also the phenomena of colour-matching.

Foveal Tritanopia. König, in 1894, discovered that the central fovea is tritanopic, blues and greens, and whites and yellows being confused if the matching fields were kept very small and central fixation was accurately maintained. Later Willmer rediscovered the phenomenon and suggested that the blue receptor—apparently lacking in the central fovea—was the rod, which behaved as a colour receptor under photopic conditions. However, this tritanopia is probably due, as Hartridge showed, to the use of very small fields *per se*, since a similar dichromacy could be observed in the periphery, provided the field was made small enough. This loss of blue sensitivity may be due to a relative rarity of the "blue" cones, so that unless the field is large enough an insufficient number are stimulated to permit their effects to be manifested.

Fovea and Periphery. Frequently, however, an increased sensitivity to blue may be demonstrated under suitable conditions; in general, according to Weale, the visual functions of the different regions of the retina may be summarized as follows:

Retinal Area	Extent from centre nasally	Type of Vision
Foveal centre	50′	Tritanopia
Fovea	2° 30′	Trichromatic
Parafovea	4° 10′	Trichromatic, with enhanced blue sensitivity
Perifovea	9° 10′	
Inner periphery		Anomalous trichromacy with enhanced blue sensitivity
Intermediate periphery	14° 10′ to 24° 10′	
External periphery	77° 35′	Dichromacy with enhanced blue sensitivity

Thus, in the periphery, the sensitivity to blue seems to be *enhanced*.

MESOPIC SENSITIVITY CURVES

The Purkinje shift has been described as a change in the luminosity curve, or spectral sensitivity function, as one passes from scotopic to photopic levels of luminance; at the intermediate ranges the sensitivity function becomes intermediate between that for rods and that for cones, and it is as though both were contributing to the sense of luminosity at the different wavelengths. When the quantitative aspects of these changes are investigated, however, it may be shown that a given mesopic luminosity curve cannot be constructed by linear addition of the pure scotopic and photopic curves, so that we must envisage interaction between rods and cones at some level in the visual pathway (Hough, 1968). It is interesting that when one studies the successive changes in the luminosity curve as the luminance is increased, it is as though Stiles' blue cone mechanism were coming into play with greater and greater effect, rather than all three cone-types. This was strikingly confirmed by Hough & Ruddock (1969), who found that in the tritanope, whose blue cone mechanism is defective, the Purkinje shift is very defective, the luminosity curve at a mesopic level of 10 photopic trolands being identical with the scotopic luminosity curve, whereas in a normal trichromat the maximum sensitivity had changed from the scotopic 5050 Å to some 5400 Å. In view of the mixing of rod and cone responses that may occur as a result of the connections of horizontal cells (p. 123), it may well be that the mesopic luminosity curve reveals this interaction.*

SPECIAL REFERENCES

AUERBACH, E. & WALD, G. (1954). Identification of a violet receptor in human colour vision. *Science* **120**, 401–405.

BAUMGARDT, E. & SÉGAL, J. (1946). La localisation du mécanisme inhibiteur dans la perception visuelle. *C. R. Soc. Biol., Paris* **140**, 431–432.

BRINDLEY, G. S. (1952). The Bunsen-Roscoe law for the human eye at very short durations. *J. Physiol.* **118**, 135–139.

BRINDLEY, G. S. (1953). The effects on colour vision of adaptation to very bright lights. *J. Physiol.* **122**, 332–350.

BRINDLEY, G. S. (1959). The discrimination of after-images. *J. Physiol.* **147**, 194–203.

BRINDLEY, G. S. & RUSHTON, W. A. H. (1959). The colour of monochromatic light when passed into the human retina from behind. *J. Physiol.* **147**, 204–208.

DARTNALL, H. J. A. (1960). Visual pigment from an all-cone retina. *Nature* **188**, 475–479.

DENTON, E. J. & WYLLIE, J. H. (1955). Study of the photosensitive pigments in the pink and green rods of the frog. *J. Physiol.* **127**, 81–89.

DODT, E. (1967). Purkinje-shift in the rod eye of the bush-baby, *Galago crassicaudatus*. *Vision Res.* **7**, 509–517.

DONNER, K. O. (1953). The spectral sensitivity of the pigeon's retinal elements. *J. Physiol.* **122**, 524–537.

DONNER, K. O. & RUSHTON, W. A. H. (1959). Rod-cone interaction in the frog's retina analysed by the Stiles-Crawford effect and by dark-adaptation. *J. Physiol.* **149**, 303–317.

EASTER, S. S. (1968). Adaptation in the goldfish retina. *J. Physiol.* **195**, 273–281.

ENOCH, J. M. (1960). Waveguide modes: are they present, and what is their possible role in the visual system. *J. opt. Soc. Am.* **50**, 1025–1026.

ENOCH, J. M. (1961). Visualization of wave-guide modes in retinal receptors. *Amer. J. Ophthal.*, Pt. 2 **51**, 1107–1118.

FRY, G. A. & ALPERN, M. (1953). The effect of a peripheral glare source upon the apparent brightness of an object. *J. opt. Soc. Am.*, **43**, 189–195.

HAGINS, W. A. (1955). The quantum efficiency of bleaching of rhodopsin *in situ*. *J. Physiol.* **129**, 22–23 P.

HOUGH, E. A. (1968). The spectral sensitivity functions for parafoveal vision. *Vision Res.* **8**, 1423–1430.

* Dodt (1967) has described a Purkinje shift in the pure rod eye of the bush-baby, *Galago crassicaudatus*, when the visual response was measured by the height of the *b*-wave of the ERG.

HOUGH, E. A. & RUDDOCK, K. H. (1969). The parafoveal visual response of a tritanope and an interpretation of the Vλ sensitivity functions of mesopic vision. *Vision Res.* **9,** 935–946.

GRAHAM, C. H. (1965). Visual form perception. In *Vision and Visual Perception,* Ed. C. H. Graham *et al.,* pp. 548–574. New York: Wiley.

SAKMANN, B., CREUTZFELDT, O. & SCHLEICH, H. (1969). An experimental comparison between the ganglion cell receptive field and the receptive field of the adaptation pool in the cat retina. *Pflüg. Arch. ges. Physiol.* **307,** 133–137.

SCHOUTEN, J. F. & ORNSTEIN, L. S. (1939). Measurements on direct and indirect adaptation by means of a binocular method. *J. opt. Soc. Am.* **29,** 168–182.

STILES, W. S. (1937). The luminous efficiency of monochromatic rays entering the eye pupil at different points and a new colour effect. *Proc. Roy. Soc.* B **123,** 90–118.

STILES, W. S. & CRAWFORD, B. H. (1933). The luminous efficiency of rays entering the eye pupil at different points. *Proc. Roy. Soc.* B **112,** 428–450.

TANSLEY, K., COPENHAVER, R. M. & GUNKEL, R. D. (1961). Spectral sensitivity curves of diurnal squirrels. *Vision Res.* **1,** 154–165.

THOMSON, L. C. (1951). The spectral sensitivity of the central fovea. *J. Physiol.* **112,** 114–132.

WALRAVEN, P. L. (1961). On the Bezold-Brücke phenomenon. *J. opt. Soc. Am.* **51,** 1113–1116.

WEALE, R. A. (1953). Spectral sensitivity and wavelength discrimination of the peripheral retina. *J. Physiol.* **119,** 170–190.

WESTHEIMER, G. (1965). Spatial interaction in the human retina during scotopic vision. *J. Physiol.* **181,** 881–894.

WESTHEIMER, G. (1967). Spatial interaction in human cone vision. *J. Physiol.* **190,** 139–154.

WILLMER, E. N. (1946). *Retinal Structure and Colour Vision.* Cambridge: Cambridge University Press.

WRIGHT, W. D. (1939). The response of the eye to light in relation to the measurement of subjective brightness and contrast. *Trans. Illum. Eng. Soc.* **4,** 1–8.

WRIGHT, W. D. (1946). *Researches on Normal and Defective Colour Vision.* London: Kimpton.

WRIGHT, W. D. & GRANIT, R. (1938). On the correlation of some sensory and physiological phenomena of vision. *Brit. J. Ophthal.* Mon., Suppl. No. 9.

WRIGHT, W. D. & NELSON, J. H. (1936). The relation between the apparent intensity of a beam of light and the angle at which the beam strikes the retina. *Proc. Phys. Soc.* **48,** 401–405.

SECTION III

The Muscular Mechanisms

PART I

THE EXTRAOCULAR MUSCLES AND THEIR ACTIONS

APART from its importance in the treatment of squint, an adequate knowledge of the movements of the eyes, and their neuromuscular control, is of considerable value in understanding the physiological mechanisms concerned with the maintenance of posture; moreover, it would seem that our perceptions of objects, and particularly of their spatial relationships, are determined in part by the laws governing the movements of the eyes.

THE MUSCLES

The Recti. The muscles concerned with the eye movements are six in number: the *medial* and *lateral recti*, which determine sideways movements; the *superior* and *inferior recti* causing primarily upward and downward movements, and the *superior* and *inferior obliques*, which, when acting alone, primarily cause the eye to twist around its antero-posterior axis.

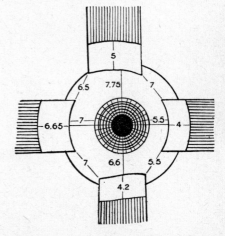

FIG. 186. Insertions of the rectus muscles. The medial rectus is to the right. (Duke-Elder, *Text-book of Ophthalmology*.)

The insertions of the rectus muscles are indicated in Fig. 186, and since the muscles arise from a common origin at the apex of the orbit, their main actions of elevation, depression, adduction (i.e. pulling towards the nose) and abduction, are not difficult to appreciate.

The Obliques. The superior oblique (Fig. 187) likewise arises from the apex of the orbit but after running above the medial rectus almost to the orbital margin, it becomes tendinous as it passes through a cartilaginous pulley, the trochlea, and turning sharply backward, downwards, and laterally over the eye it is inserted into the postero-lateral aspect of the sclera. Since the muscle pulls from above on an attachment behind the equator of the eye, it acts as a depressor.

The inferior oblique (Fig. 188) arises from the anterior part of the orbit, in the antero medial corner of its floor. It passes laterally and backwards beneath

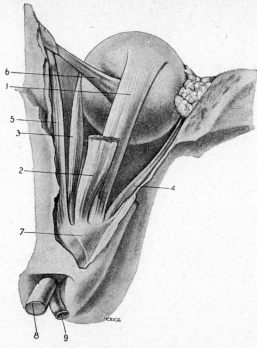

FIG. 187. Extrinsic muscles of the eye from above. 1, superior rectus; 2, levator palpebrae superioris; 3, medial rectus; 4, lateral rectus; 5, superior oblique; 6, reflected tendon of superior oblique; 7, annulus of Zinn; 8, optic nerve; 9, ophthalmic artery. (Duke-Elder, *Text-book of Ophthalmology*.)

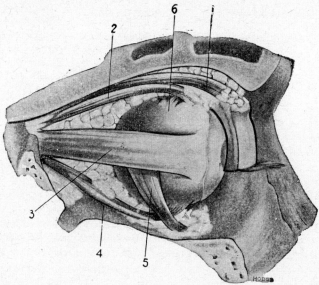

FIG. 188. Extrinsic muscles of the eye. Lateral view. 1, levator palpebrae superioris; 2, superior rectus; 3, lateral rectus; 4, inferior rectus; 5, inferior oblique; 6, superior oblique. (Duke-Elder, *Text-book of Ophthalmology*.)

the inferior rectus and is inserted into the postero-lateral aspect of the globe. Pulling from below on an attachment behind the equator the inferior oblique thus acts as an elevator.

The Muscle Fibres and their Innervation

The mammalian extraocular muscles have several structural and physiological features that differentiate them from most other striated muscles, features that they share with some submammalian skeletal muscles and with mammalian intrafusal fibres.

Twitch and Slow Fibres. Thus, the muscles of the frog, for example, fall into two classes, characterized by the structure and innervation of their fibres; the *twitch*, or *fast*, fibres are usually large and innervated by large, rapidly conducting, nerve fibres that make typically a single *en plaque* or *sole-plate* ending on each muscle fibre; this type of muscle is concerned with rapid phasic movements and, on electrical stimulation of its nerve, gives a twitch response. The other type of muscle has thinner fibres with multiple *en grappe* nerve terminals distributed over its whole length; this type is associated with the tonic type of sustained contraction; electrical stimulation of its motor nerve leads to graded mechanical responses without the spread of action potentials. Pharmacologically, the fibres are distinguished by their responses to acetylcholine, the slow fibres giving a substantial contraction whilst the twitch fibres give a single twitch. Structurally there are striking differences between the fibres; Krüger pointed out that the slow fibres lacked the M-band whilst their fibrils were irregularly arranged, being clustered together in clumps to give what he called the *Felderstruktur*, which contrasted with *Fibrillenstruktur* of the twitch fibre, in which the myofibrils were regularly arranged in a more abundant sarcoplasm. Subsequent electron-microscopical studies of Peachey & Huxley (1962) and Page (1965) have emphasized these differences; in the fibrillar, or twitch, type of fibre the myofibrils are surrounded by a well defined system of sarcoplasmic reticulum made up of longitudinal and transverse systems giving rise to the *triads*, which are considered to be developments permitting the rapid transmission of the electrical changes occurring on the surface into the depths of the fibre. In the slow fibres the myofibrils lack regular arrangement, the sarcoplasmic reticulum is scarcer and lacks the transverse system, so that the triads are absent.

Mammalian Skeletal Muscles

In the skeletal muscles of mammals this sharp differentiation into twitch and slow muscles does not occur; nevertheless, it is possible to describe muscles as either fast or slow on the basis of their contraction-times, i.e. the time to reach peak tension after a nerve stimulus; the fast muscles are usually white, capable of rapid action but soon fatiguing, whilst the slow, concerned with sustained postural contractions, are red, having large amounts of myoglobin and usually plentiful mitochondria.* Fibres with intermediate contraction-times, that would otherwise be classed with the slow fibres, are also described. Histologically and histochemically the fibres may be distinguished, but not in such a discrete manner as

---

* It must be appreciated that many muscles are mixed, containing both twitch and slow fibres; in the mammal, for example, Close (1967) described slow and intermediate types of unit in the soleus, whereas the extensor digitorum longus contained only fast fibres.

with the submammalian system. Thus, at one extreme we have fibres rich in oxidative enzymes, such as succinate dehydrogenase, which probably depend on the citric acid cycle for their energy; these correspond with the slow type. At the other extreme there are fibres rich in the enzymes for glycolysis, such as α-glycerophosphate dehydrogenase, and these are of the twitch type. Histologically, these extreme forms would be classed as *Felderstruktur* and *Fibrillenstruktur*, but between these extremes there are many intermediate forms, whilst all these mammalian muscle fibres respond to electrical stimulation by an action potential and an all-or-none type of contraction.

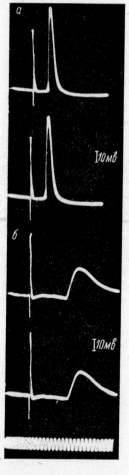

Fig. 189. Intracellular recordings of superior oblique muscle fibres in response to trochlear nerve stimulation. a. Responses in phasic fibre; b. Responses in tonic fibre. (Matyushkin, *Sechenov Physiol. J.*)

Extraocular Muscles. The mammalian extraocular muscles exhibit the presence of both *Fibrillen-* and *Felderstruktur-*fibres, as seen in the submammalian muscle; the large twitch fibres have typical *en plaque* endings whilst the smaller fibres have the typical *en grappe* terminations of the submammalian slow fibres; both types of ending are associated with cholinesterase activity, so that they are both motor.

The Soviet physiologist, Matyushkin (1961), was the first to demonstrate that the mammalian extraocular muscles contained two types of fibre that responded differently to electrical stimulation of the nerve nucleus; Fig. 189 shows intra-cellularly recorded responses of muscle fibres of the superior oblique to a single stimulus in the trochlear nucleus; the records in (a) are responses in a phasic fibre, being spikes of 60–90 mV amplitude, and duration 1·3–1·5 msec, whilst the records in (b) are from a tonic fibre of low amplitude (12–28 mV and duration greater than 20 msec). At rest, the phasic fibre was silent whilst the tonic fibre gave small potentials at 20/sec. The longer latent period for the tonic fibre indicated that it was supplied by finer fibres. In a later study (1964), the effects of repetitive stimulation of the nerve supplying the superior oblique indicated the phasic and tonic natures of the two types of fibre; the spikes were associated with the rapid development of tension whilst the small potential changes were only associated with significant tension when repetitive stimulation was maintained.

In a comparative anatomical and physiological study, Hess & Pilar (1963) showed that the two types of muscle fibre could, indeed, be classed as twitch and slow, as revealed by their responses to repetitive stimulation. Thus, if the nerve to the superior oblique, for example, is stimulated repetitively at 250–300 impulses/sec, maximal tension develops, and this soon falls off due to Wedensky block, but the tension does not fall to zero, remaining at about 30 per cent of the maximum value, whilst records of activity in the muscle fibres now show no evidence of spikes but only the slow monophasic potentials characteristic of the non-propagated type of change occurring in slow muscles (Pilar, 1967).*

POSTURAL AND PHASIC REQUIREMENTS

In submammalian species the slow mucles subserve the sustained contractions required for the maintenance of postural states; this function is fulfilled in mammals by fibres that are of the twitch-type, in that they respond to a nerve impulse in an all-or-none manner. In the eye, however, the requirement for two very different types of contraction arises, namely the very fast contraction in a saccade (p. 339) and the much slower contractions required for a following movement or convergence; and it would seem that the submammalian solution of this problem, namely a muscle with both twitch and slow fibres, as in the iliofibularis of the frog, has been retained in the mammal.

ACETYLCHOLINE

Duke-Elder & Duke-Elder (1930) observed that mammalian extraocular muscles resembled the slow skeletal muscles of lower vertebrates in responding by a contracture to applied acetylcholine.† In the superior rectus of the rabbit, where the slow and fast types of fibre are separated in two layers within the muscle plate,

* Bach y Rita & Ito (1966) disputed this difference between slow and fast fibres and considered that both were of the twitch type, but Pilar (1967) has confirmed, with additional experiments, the qualitative difference between the responses in the two types of fibre. It may be, however, that there are three types of fibre: fast and slow twitch-types, and slow graded fibres (Peachey, 1968). In some extraocular muscles, e.g. the superior oblique studied by Hess & Pilar, the *Felderstruktur* fibres are large and plentiful by comparison with the inferior oblique. Hess (1967) has summarized the similarities and differences between submammalian skeletal and mammalian extraocular muscle fibres.
† The effects of several cholinomimetics on the cat's extraocular muscles have been examined by Sanghvi & Smith (1969); spike activity certainly occurs, perhaps due to the twitch fibres only. Muscarine has a smaller effect than that of acetylcholine, which is blocked by atropine, so that the receptors are mainly "nicotinic".

Kern (1965) demonstrated contracture-type responses in the slow fibres. The tension developed during acetylcholine contracture of the superior rectus muscle is about 30 per cent of the maximal tetanic tension, and it is interesting that the tension developed during repetitive stimulation of this muscle settles down to 30 per cent of the maximal tension tension developed (Pilar, 1967); this corresponds with the situation when the fast fibres have ceased to respond because of Wedensky inhibition, so that the tension finally reached during repetitive stimulation is due to the slow fibres.

SYMPATHETIC

Paralysis of the sympathetic system produces Horner's syndrome consisting of ptosis, due to paralysis of the adrenergic Muller's muscle, miosis due to paralysis of the dilator pupillae, and enophthalmos. The last effect suggests that the extra-ocular musculature may be influenced by the sympathetic system. Kern (1968) has found that isoproterenol causes relaxation of tension in the monkey's extra-ocular muscles, an effect that is blocked by DCI, indicating a β-action. Since the amine caused relaxation of tension developed during treatment with acetylcholine, he argued that it was the slow fibres that were sensitive to the sympathomimetic agent. On the same basis, then, the exophthalmos accompanying extreme fear could be due to the β-action of the adrenaline secreted by the adrenal gland under these conditions.

PRIMATE MUSCLES

The early studies on the anatomy and physiology of the extraocular muscles were carried out on the rabbit, guinea pig and cat; more recently these studies have been extended to primates such as the monkey (Mayr, Stockinger & Zenker, 1966; Miller, 1967) and man (Cheng & Breinin, 1965; Dietert, 1965; Brandt & Lansing, 1966). In general, the distinction between twitch-type fibres, with single *en plaque* innervation, and slow-type with multiple *en grappe* innervation, was sustained, although the levator palpebrae had only the twitch type. Miller (1967) has emphasized the many gradations in fibre-type between the two extremes, so that with many fibres, so far as histochemistry and ultrastructure are concerned, it is impossible to classify them as either fast or slow, a situation similar to that in other mammalian muscles; he pointed out a definite topographical arrangement of the main types, however, the fast singly innervated fibres being closest to the globe and the slow, *en grappe* innervated, being on the surface, whilst fibres of intermediate types were in between.

Motor Unit. The motor unit, i.e. the number of muscle fibres supplied by a single motor axon, is remarkably small being probably of the order of 6, although the difficulty in establishing whether a given fibre is sensory or motor may result in some error in the estimate. This compares with some 100 to 150 for limb muscles (Torre, 1953). Because of its short twitch-time the fusion frequency for an extraocular muscle is very high indeed, of the order of 350 impulses per sec, and this permits a large variation in tension—over a range of tenfold—on passing from a single twitch to a fused tetanus. These two features, namely small size of motor unit and small twitch/tetanus tension ratio, permit a high degree of gradation of the strength of contraction.

SENSORY INNERVATION

In 1946 Daniel described spiral nerve endings surrounding fibres of the human extraocular muscle; he surmised that they were sensory in function, responding perhaps to thickening during contraction. Later Cooper & Daniel identified definite muscle spindles in human muscles, a single rectus containing as many as 50, a number comparable with those in a lumbrical muscle of the hand; the spindles were concentrated near the origin of the muscle, and that was why they had not been discovered earlier. In general the structure of the spindle was similar to that classically described in other muscles; 2–10 fine muscle fibres (intrafusal) are enclosed in a capsule whilst an adjacent nerve trunk sends fibres into the capsule to end on the intrafusal fibres. The innervation is both motor and sensory; thus the gamma efferents—fine motor fibres—by causing the muscle fibres in the capsule to contract are able to increase the sensitivity of the spindle to stretch; whilst the sensory endings encircle the nuclear bag. Besides man, the higher apes and artiodactyls—goat and sheep—possessed well defined spindles in their extraocular muscles. In certain species, notably the cat and monkey, no spindles were found but the richness of the innervation of the muscles certainly suggests the presence of simpler organs, perhaps just spirally wound endings encircling single muscle fibres (Cooper & Fillenz).* Subsequent work, notably that of Wolter, suggests that the sensory apparatus in the human extraocular muscles is even more varied; in all, he has described some six different types of ending, varying in complexity from the spindle to relatively simple terminations, which may be distinguished from motor endings by the circumstance that they end in connective tissue between fibres. A simplified scheme of the innervation of a muscle fibre is shown in Fig. 190.

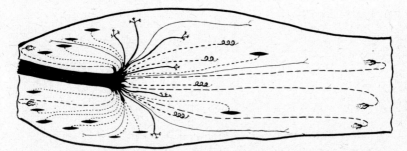

FIG. 190. Much simplified diagram of a human rectus oculi muscle, full length. The origin is on the left and the insertion on the right. The sensorimotor nerve (black trunk) is seen entering the muscle at the junction of the proximal and middle thirds. The muscle spindles are most numerous peripherally and in the proximal third of the muscle. Their nerves (composed of sensory fibres and gamma efferent fibres) are indicated by dotted lines. In the middle third of the muscle are the motor nerve endings whose nerve fibres are shown as solid black lines. Here also, mainly just distal to the band of motor nerve endings, are the simple spiral endings. At both the origin and insertion of the muscle tendon endings are seen (with their afferent nerve fibres shown as broken lines). The lines ending in small Ys represent nerve fibres supplying various of the other endings that are to be seen in eye muscles. (Cooper, Daniel & Whitteridge, *Brain*.)

On purely anatomical grounds, therefore, we may conclude that the muscles are able to send messages to the brain indicating their state of tension at any moment, a *sine qua non* for the efficient performance of any muscular act requiring precision. The role of proprioception in muscular activity and visual perception will be discussed later.

* The responses recorded from N III when the cat's inferior oblique was stretched have been examined in great detail by Bach-y-Rita & Ito (1966), and compared with the responses from true muscle spindles in skeletal muscle. In general, it seems that the receptor-organ of the cat extraocular muscle is simpler, without any gamma-efferent control.

Defining the Movements

Axes of Rotation. In describing the actions of the muscles, we are concerned with changes in the direction in which the eye is "pointing" i.e. the orientation of the *fixation axis* (defined as a line joining the point of fixation with the centre of rotation of the eye; Fig. 191). As a *primary position* or starting point, we may consider that position in which the eye is looking straight ahead when the head is erect. If the fixation axis swings horizontally, the eye is said to *adduct* or *abduct*, according as it swings towards or away from the nose; if the fixation axis swings vertically, the eye is *elevated* or *depressed*. After any purely horizontal or vertical movement of the fixation axis from the primary position, the eye is said to have reached a *secondary position*. If the eye executes a movement which causes both a horizontal and vertical displacement of the fixation axis (e.g. if the eye looks up and to the right) it is said to adopt a *tertiary position*.

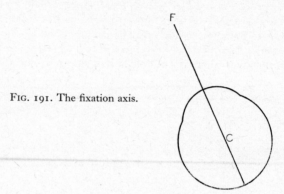

Fig. 191. The fixation axis.

Experiment has shown that the movements of the eye in general can be approximately described as rotations about a fixed point,* the *centre of rotation*, situated some 13·4 mm behind the anterior surface of the cornea. To define a movement it is, of course, not sufficient to indicate the centre of rotation, but an *axis*, passing through the centre, about which the eye turns, must be given; thus adduction or abduction is achieved by rotation about a vertical axis, the *z-axis* (Fig. 192); elevation or depression by rotation about a transverse horizontal axis, the *x-axis*. The fixation axis can be made to point in any direction by rotating the eye first about one axis and then about the other—alternatively the necessary position of the eye can be attained by a single rotation about an axis intermediate between the *x-* and *z-*axes. It is worth noting that when the eye is in the primary position the *x-* and *z-*axes are in a vertical frontal plane— called *Listing's plane* (Fig. 192)—and that *any* direction of the fixation axis can be obtained by rotating the eye about an axis lying in this plane.

Torsion or Rolling. The physiologist is concerned not only with the direction of the fixation axis—which tells him where the eye is pointing—but also with the orientation of a certain fixed plane in the eye—the *retinal horizon* (Fig. 193).

* This is, indeed, an approximation; it would seem from the most recent work on this subject that the "centre of rotation" varies within a range of about 2 mm according to the nature of the eye movement; in Van der Hoeve's view, this variability permits a more accurate adjustment of the eyes than would be possible with a fixed centre of rotation.

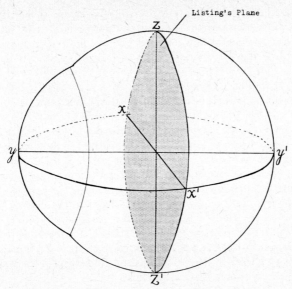

FIG. 192. The primary axes of rotation and Listing's plane. *xx'*, transverse horizontal giving elevation and depression; *yy'*, antero-posterior giving rolling or torsion; *zz'*, vertical giving adduction or abduction.

Imagine the eye in the primary position looking straight ahead; a horizontal plane through the pole of the cornea is the retinal horizon and the image of any point in space lying on this plane falls on a horizontal line on the retina—the *horizontal meridian*. The retinal horizon, being fixed in relation to the eye, moves with it. Now we shall see that the appreciation of direction—of a line for example—depends in large measure on the directions the images of external objects make on the retina, and if the orientation of the retina is upset, estimates of direction can become erroneous. Thus a horizontal line is appreciated as such because its image falls on, or is parallel to, the horizontal meridian; if the retina is twisted round, the horizontal line appears to be inclined unless the twist is compensated psychologically (p. 436).

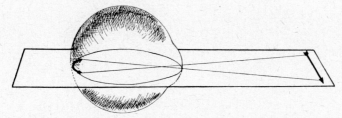

FIG. 193. The retinal horizon. The image of the arrow, fixated in the primary position of the eye, falls on the horizontal meridian.

We have thus to take into account the influence of movements of the eye on the orientation of the retina, or what amounts to the same thing, the orientation of the retinal horizon. So long as this plane keeps a definite orientation in space the images of horizontal lines in a frontal plane fall on the horizontal meridian or are parallel to it; the problem is how to define a position of the retinal horizon which we may consider normal—i.e. where there is no "twist" of the retina; for the primary position of the eye this is easy, the retinal horizon must be horizontal and, if the eye rotates about its antero-posterior axis (the y-axis) from this position, the retinal horizon becomes inclined to the horizontal and the eye is said to have *rolled* or to have attained an *angle of rolling* or *torsion*. For other positions of the eye we cannot use the horizontal as our reference plane, e.g. if the eye is directed vertically upwards the retinal horizon is not twisted round but it is not horizontal. The normal position of the retinal horizon can be defined in relation to a fixed plane in space—the *median plane*, a vertical plane through the nose (Fig. 194); so long as the retinal horizon remains perpendicular to this, its position is normal and there is no torsion. Thus in Fig. 195 the eye is represented as looking up and to the right; the retinal horizon is inclined upwards, but it remains at right-angles to the median plane; there is no torsion and a horizontal line on the wall is parallel with the retinal horizon. In Fig. 196 the retinal horizon has been twisted, i.e. there has been torsion; it is now no longer at right-angles to the median plane and a horizontal line on the wall is not parallel with the retinal horizon; its image on the retina is therefore not parallel with the horizontal meridian.*

FIG. 194. The median plane.

Rotations about Intermediate Axes. Rolling is brought about by rotation of the eye around its antero-posterior axis; it is not difficult to see that a combined elevation and rolling can be achieved by rotating the eye first about the x-axis

---

* The definition of the primary position can now be modified to take into account the orientation of the retina; it is the position when the eye is looking straight ahead with head erect and the retinal horizon horizontal.

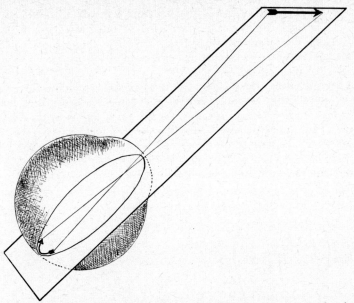

FIG. 195. The eye is inclined up and to the right. No rolling has occurred and the arrow produces an image coinciding with the horizontal meridian.

and then about the *y*-axis, or alternatively by rotating about a horizontal axis intermediate between the two. A motion consisting of all three types, e.g. elevation, adduction, and rolling, can be obtained by choosing an axis with components along each of the three primary *x*-, *y*-, and *z*-axes.

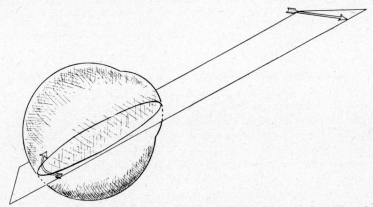

FIG. 196. The eye is inclined up and to the right. Torsion has occurred so that the horizontal arrow produces an image inclined across the horizontal meridian.

Actions of the Individual Muscles

We are now in a position to analyse the actions of the individual muscles.

The Medial and Lateral Recti. In Fig. 197 is shown a horizontal meridional section through the eye; the medial and lateral recti are indicated by the heavy lines; they run over the surface of the globe but the pull takes place where the muscle is tangential to this surface, at T, the *tangential point*, and the direction of pull is long the *tangential line of force*, TT'. The *muscle plane* is defined as the plane containing the tangential line of force, TT', and the centre of rotation C, i.e. it is the plane of the paper in Fig. 197. If C, the centre of rotation, is to remain fixed in space, a pull along TT' must result in a rotation about an axis through C projecting vertically through the paper; in general we may state that *any muscle will rotate the eye about an axis perpendicular to the muscle plane concerned.* The muscle planes of the medial and lateral recti may be taken as horizontal and they thus rotate the eye about a vertical axis through C which has already been defined as the *z*-axis. Such a rotation leads, as we have seen, to a simple movement of adduction or abduction.

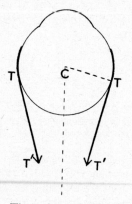

FIG. 197. The actions of the medial and lateral recti. T, tangential point; TT', tangential line of force; C, centre of rotation. The muscle plane is the plane of the paper.

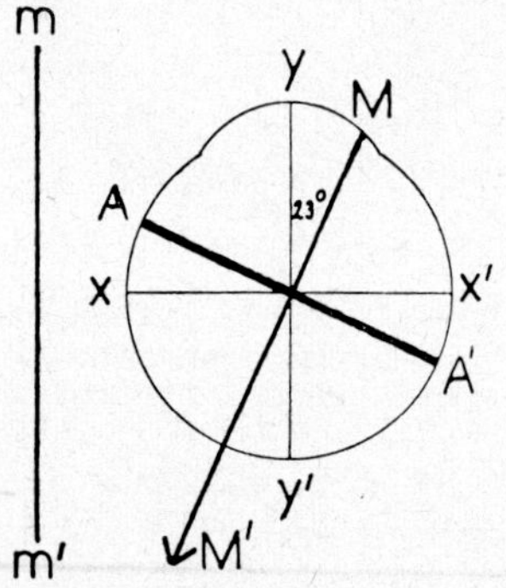

FIG. 198. Actions of the superior and inferior recti (right eye). The muscle plane intersects the horizontal section of the eye in MM'. AA' is the axis of rotation. *mm'* is the intersection of the median plane with the plane of the paper.

The Superior and Inferior Recti. The muscle planes of the superior and inferior recti are vertical and approximately identical; however they do not pass through the antero-posterior axis of the eye but are inclined at about 23° to it. The axis of rotation for these muscles is perpendicular to this plane and therefore lies horizontally but, as a result of this inclination, it does not correspond with the transverse horizontal *x*-axis but is inclined 23° away from it. The position is indicated in Fig. 198, which shows a horizontal section through a right eye. The vertical muscle plane intersects this section in the line MM' inclined at an angle of 23° to the antero-posterior axis, *yy'*. The axis of rotation is the line AA' inclined at an angle of 23° to the transverse horizontal axis, *xx'*.

Rotation of the eye about *xx'* causes pure elevation or depression; rotation about *yy'* causes pure rolling; hence rotation about the intermediate axis AA' causes a movement involving both these effects; thus the superior rectus causes a rotation such that M rises out of the paper and its action is therefore *elevation and inward rolling*, the movement of the retinal horizon being clockwise; the inferior rectus causes *depression and outward* rolling.

ADDUCTION

Besides causing elevation (or depression) and rolling, the superior and inferior recti act as adductors, causing the fixation axis to move towards the median plane. This is not immediately evident since it is clear that there can be no rotation about the vertical z-axis, the axis of rotation AA' being perpendicular to it; this third action results from referring the orientation of a fixed direction in the eye—the fixation axis—to a plane outside the eye—e.g. the median plane through the head.

Thus in Fig. 198 let us imagine that the superior rectus acts; yy' corresponds to the fixation axis, which in the primary position remains parallel to the median plane, mm'; when the eye is elevated, y moves out of the plane of the paper and moves through a circle which is clearly not parallel to mm'; hency y becomes closer to mm' whilst y' moves farther away, and this can only result in the line yy', the fixation axis, being directed towards the median plane, mm', i.e. the superior rectus causes adduction. The inferior rectus causes y to move downwards, but it still moves towards mm' whilst y' moves away; hence both superior and inferior recti cause adduction.

The difficulty arises from the physiological necessity of referring a direction in the eye, the fixation axis for example, to a plane outside it, the median plane; if the axis of rotation is parallel with, or perpendicular to, this plane of reference the inclination of this direction in the eye is unchanged during rotation; thus rotation about xx' gives pure elevation because the axis of rotation is at right angles to the reference plane, mm'; similarly rotation about yy' causes pure rolling because yy' is parallel to mm'; AA', the axis of rotation for the superior and inferior recti, is neither parallel with, nor perpendicular to mm', hence the inclination of the fixation axis to mm' must alter during rotation, i.e. adduction must occur.

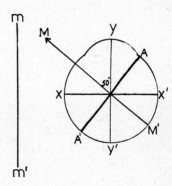

FIG. 199. Actions of the obliques. (Right eye.) AA' is the axis of rotation.

MODEL EXPERIMENT

The problem becomes very much clearer with a model; a tennis ball may be used; a point is marked to represent the pole of the cornea and a knitting needle is inserted through the centre to represent any desired axis of rotation. If the needle is passed vertically through the ball, rotation causes adduction or abduction; a horizontal needle through the equator permits elevation and depression; a horizontal one through the pole of the cornea causes pure rolling. A horizontal needle passed in a direction intermediate between these two positions, which gives us the axis of rotation for the superior and inferior rectus muscles, causes both rolling and elevation or depression, and it will be seen also that the direction in which the pole of the cornea points, referred to a vertical plane parallel to the original direction of the fixation axis, changes during the rotation. It becomes quite clear that this change of direction is not brought about by rotation about a vertical axis since the knitting needle is held horizontally all the time.

The Superior and Inferior Obliques. The muscle planes of the superior and inferior obliques may be considered to be vertical, as with the recti, but to run at an angle of about 50° to the antero-posterior axis, and on the opposite side, as in Fig. 199. The axis of rotation, AA′, is likewise inclined at 50° to the transverse horizontal axis. Contraction of the inferior oblique causes rotation about AA′ such that y moves up out of the paper; the eye is thus elevated and rolled outwards; moreover, since the axis of rotation is inclined towards the median plane, the direction of the fixation axis, $yy′$, does not remain constant in respect to this median plane; as y comes up out of the paper it clearly moves away from $mm′$ and thus the inferior oblique causes *abduction*; the superior oblique causes depression, inward rolling and abduction.

Summary. The individual actions of the muscles, thus deduced, may be summarized as follows:

Muscle	Main action	Secondary action
Lateral Rectus	Abduction	None
Medial Rectus	Adduction	None
Superior Rectus	Elevation	Inward Rolling and Adduction
Inferior Rectus	Depression	Outward Rolling and Adduction
Inferior Oblique	Outward Rolling	Elevation and Abduction
Superior Oblique	Inward Rolling	Depression and Abduction

Combined Actions of Muscles

As we shall see, a given movement involves activity in, or inhibition of, several if not all of the six extraocular muscles. It is therefore important to consider

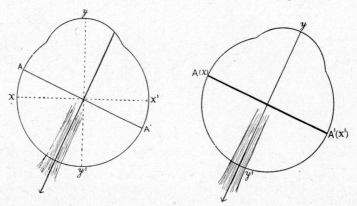

FIG. 200. Illustrating the effect of abduction on the action of the superior rectus. Abduction causes the x-axis of the eye to approach the axis of rotation, AA′, so that rolling and adduction are decreased.

their combined actions. Furthermore, we have so far only discussed the direction of pull of a muscle from the primary position of the eye; since this must change frequently as the eye moves out of the primary position, the theoretical axis of rotation must also alter.

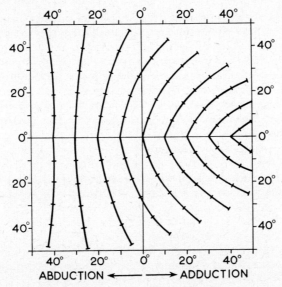

FIG. 201. Trace of line of fixation on a sphere concentric with the eye when the superior and inferior recti act individually from positions of adduction or abduction. If we imagine we are viewing inside the sphere the diagram corresponds to a left eye. Each complete curve represents rotation of 50° about the axis of rotation. Small cross-lines indicate cyclotorsion with respect to the objective vertical at every 10° of rotation. (After Boeder, *Amer. J. Ophthal.*)

Effect of Changed Position of Eye. The simplest example of the effects of moving out of the primary position on the action of a muscle is given by Fig. 200 which illustrates the effects of abduction of the right eye on the direction of pull of the superior rectus. Obviously the rolling and adduction will decrease as the antero-posterior axis of the eye approaches the plane of the muscle, whilst the elevating action will increase. An exactly reverse relationship must exist with the obliques. The variations in the degree of elevation or depression exerted by the superior and inferior rectus, according as the eye is caused to adduct or abduct before the muscle acts, are shown in Fig. 201 after Boeder. Here we imagine that we are looking into a sphere concentric with the left eye, and the lines are those that would be traced out by the fixation axis when the superior or inferior rectus acted alone, and caused a rotation of 50° about the axis of rotation. It is seen that, as the initial position becomes more and more one of adduction, the elevation caused by a given degree of rotation about the axis of rotation becomes smaller and smaller, whilst it increases with abduction.*

* This assumes of course that the insertion of the muscle remains the same during abduction; in fact the effective insertion of an extraocular muscle varies very little because it is wide with its tendinous fibres fanning out; hence the axis of rotation tends to remain in the same *position in space* and therefore different in respect to the antero-posterior axis of the eye.

Diagnostic Directions of Gaze. These essentially theoretical considerations on the change in the direction of pull with adduction and abduction, according to which the superior and inferior recti become more efficient in abduction whilst the inferior and superior obliques become more efficient in adduction, have led to the clinical concept of the *diagnostic directions of gaze.* In most movements of elevation and depression the obliques and the recti cooperate; thus a pathologically reduced ability to elevate could be due to paresis of either the superior rectus or the inferior oblique. If, however, the defect is accentuated when the eye is caused to abduct to the extent that the elevating action of the inferior oblique is very small, then the superior rectus is probably the defective muscle. As Boeder has shown, however, because the axes of rotation of the eyes, due to single muscle-pairs, vary so considerably as elevation or depression takes place, this simple view may well be illusory, so that in fact the superior and inferior recti probably always contribute more to elevation and depression than their synergistic obliques. According to Boeder's calculation, their proportionate contractions remain always in the ratio of about three to two. This is borne out by the observations of Breinin (1957) and of Tamler, Jampolsky & Marg (1959) that when the eye is looking upwards in strong adduction and then moves horizontally to abduction, e.g. if the left eye starts by looking up and to the right and then looks up and to the left, the superior rectus actually shows increased activity in the electromyogram although, on the classical view, the reverse would be expected since the eye is passing into a position where the action of the superior rectus becomes more and more efficient and therefore, presumably, requires less powerful contraction of its fibres.

Evidence from the Electromyogram (EMG). Until the application of the electromyograph to the extraocular muscles the decision as to what muscles were concerned in a given act was largely based on speculation, although the general principle of reciprocal innervation, whereby during a given movement the antagonist was inhibited during the activation of the agonist, was established by the classical experiment of Sherrington. This investigator cut the nerve supply to an eye so that only the lateral rectus could respond to a cortical stimulus (p. 355); the eye was deviated outwards owing to the unopposed tonic action of the lateral rectus. On stimulating the motor cortex in such a way as to produce adduction in the normal eye, the eye moved inwards to the mid-line as a result of an active inhibition of the tonic impulses to the lateral rectus.

ELECTROMYOGRAM

The activity in a muscle may be measured by the action potentials of its muscle fibres recorded from a needle-electrode inserted into the muscle; the more intense the contraction the greater will be the frequency of the spike discharges in individual units and the larger will be the number of units active. Because of the danger of various artefacts, the results of any study must always be interpreted with caution, but in the hands of experienced workers the EMG has been a valuable tool in assessing the contributions of individual muscles to given eye movements.*

* The technique and many of the pitfalls in interpretation of the EMG have been described in some detail by Marg, Jampolsky & Tamler (1959).

RECIPROCAL INNERVATION

Björk & Kugelberg showed that, when the eye looked straight ahead, a given extraocular muscle showed a steady tonic discharge, single units firing with frequencies as high as 50/sec.* This is in marked contrast with skeletal muscles which are usually almost or completely silent during relaxation. With needles in the medial and lateral recti, reciprocal innervation was very well demonstrated; thus on abduction, there was an increased discharge in the lateral rectus associated with a simultaneous inhibition of the tonic discharge in the medial rectus. This is illustrated by Fig. 202; the records marked A are obtained with

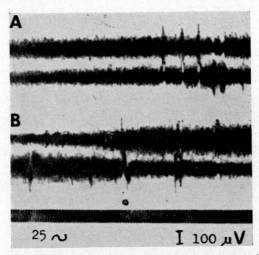

FIG. 202. Electromyograms of human medial and lateral recti during steady fixation (A) and a lateral movement (B). The upper record of each pair is from the lateral rectus, the lower is from the medial rectus. A. The gaze is maintained in the straightforward direction; note intense activity in both muscles. B. Slow following movement from 40° inward to 40° outward showing continuous diminution of activity in medial rectus and corresponding increase in the lateral rectus. The dot below the traces in record B indicates the moment at which the gaze passes the midline. Records read from right to left. (Björk & Kugelberg, *EEG clin. Neurophysiol.*)

steady fixation, whilst B shows the discharges in the medial (top) and lateral (bottom) recti during a slow following movement from 40° inward to 40° outward, reading from right to left. The gradual diminution in activity in the medial rectus (top record), accompanied by the increase in activity of the lateral rectus, is well brought out. With this slow type of movement there was never a complete inhibition of the antagonist, by contrast with the rapid "saccade" (p. 339) where the antagonist becomes quite silent (Fig. 203).

CO-CONTRACTION

The EMG has also permitted a solution of the problem of co-contraction, the simultaneous contraction of agonist and antagonist that is a feature of certain

* This is true of all the extraocular muscles, so that all three antagonistic pairs are constantly pulling against each other at rest; the oblique muscles are less active than the recti. According to Björk & Wåhlin (1960) the resting activity is absent in sleep and general anaesthesia.

acts involving the skeletal musculature. The phenomenon of reciprocal inhibition shows that in the execution of any movement of the eyes agonist and antagonist do not work against each other, and this applies even to the point at which the movement terminates; thus the movement of abduction, say, is not arrested by activity in the medial rectus, the reciprocal inhibition observed at the beginning and during the course of the movement being maintained to the end, at which time the muscle exhibits a discharge that is characteristic for the particular position of the eye. Thus the movement is not *ballistic*, as a movement arrested by active contraction of antagonists would be called (Miller, 1958). Nevertheless it might be argued that the movement would be steadied by co-contraction of the elevator and depressor pairs (Boeder), but Breinin and later workers have shown that there is no increase over the normal tonic resting activity in these muscle-pairs during a lateral movement; similarly, during a pure elevation or depression, the medial and lateral recti fail to show augmented activity.

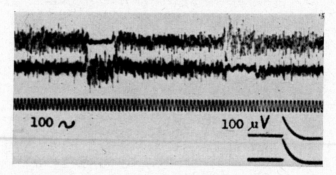

FIG. 203. Electromyograms from lateral rectus (upper tracing) and medial rectus (lower tracing) during quick changes of fixation from 10° outward to straight ahead and back to 10° outward, showing brief acceleration of activity in the agonist accompanied by abrupt cessation of activity in the antagonist. (Björk & Kugelberg, *EEG clin. Neurophysiol.*)

Single Motor Units. The EMG gives only a gross, averaged, and not highly quantitative measure of muscular activity; a more precise indication of the manner in which muscle fibres are activated during eye movements is given by a study of the spike potentials in the motor neurones innervating the muscle. Modern techniques have permitted this in the conscious primate, so that spike-frequency in single motor units may be recorded continuously during any required movement of the eyes (Fuchs & Luschei, 1970; Robinson, 1970). When an electrode is inserted into the nucleus of a motor nerve supplying an eye muscle, e.g. abducens, individual neurones give rise to a strong regular tonic discharge which varies in intensity according to the position of the eye in relation to the field of action of the muscle supplied by the neurone; thus in the abducens nucleus, supplying the lateral rectus, the discharge is maximal when the eye is abducted, and it falls linearly as the eye adopts positions of adduction until a point is reached where it becomes completely silent. This is called the *threshold position*, and in Fuchs & Luschei's study of the lateral rectus this occurred at deviations of 15 to 45° in the medial direction. Thus, as the eye is moved from extreme adduction, where all or most of the neurones are silent, more and more neurones are recruited; at about

15–20° all of these have come into action so that further activity is due to an increased frequency of discharge in all units. It is customary to speak of the *On-direction* as that in which activity increases—abduction for a neurone activating the lateral rectus; elevation for the superior rectus, and so on; the *Off-direction* is that which causes reduction of electrical activity.

Thus, so far as the static positions of the eyes are concerned, these studies emphasize that the eye muscles are active even when the eyes are in the zero position; for example, when the eye was looking straight ahead the abducens units gave steady discharges at an average rate of 100/sec. In order to counteract the tensions developed when the eye rotates from the zero position greater muscular activity, as revealed by the higher frequency of discharge in the motor units, is required. In the antagonistic muscle, however, the steady discharge is reduced.

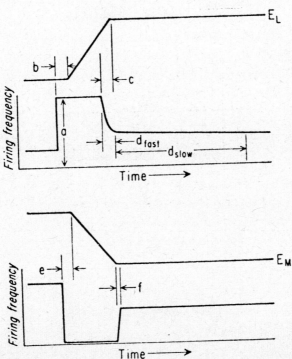

FIG. 204. Schematic representation of changes in firing frequency of abducens units during a lateral saccade (E_L) and a medial saccade (E_M). Lower case letters designate measurements used to characterize the unit firing patterns. Upper trace represents EOG; lower trace represents firing frequency. (Fuchs & Luschei, *J. Neurophysiol.*)

SACCADE

When the eyes make a saccade in fixating a new object, there is a sudden burst of activity in a single unit, which may reach a discharge-frequency of 700/sec, far greater than that required to cause maximal tension.* If the movement is in the

---

* These very high rates of discharge seem to be unnecessary, since maximal tension is reached at about 250–300 stim/sec; however, the maximal *rate of development* of tension continues to increase above the fusion-frequency.

Off-direction, there is an equally sudden decrease in spike-frequency. Fig. 204 illustrates schematically the changes in firing frequency for a single unit of the abducens nucleus, whilst Fig. 205 shows some typical records, during fixation and saccadic movements, of a motor neurone of what was probably the inferior rectus muscle. In general, the discharge began some 2–11 msec before any sign

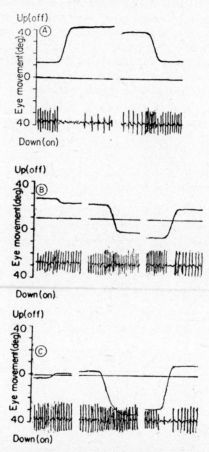

Fig. 205. Discharges of single units in the oculomotor nucleus of the intact, alert monkey during fixation and saccadic eye movements. The upper record of each section indicates eye movement, and the lower, unit discharge. The sections were selected from a continuous recording of the activity of a unit associated with downward movements and thought to be a motor neurone of the inferior rectus muscle. (Robinson, *J. Neurophysiol.*)

of muscular activity, and it typically started to decay before the end of the movement, rapidly at first, and then slowly, to reach the tonic discharge-rate appropriate to the new eye position. The discharge-rate was measured by Robinson for different velocities of movement, and a linear relation was found, so that the actual discharge-rate at any moment was compounded of two factors:

$$D = k\theta + r \, d\theta/dt$$

Robinson's mechanical studies (p. 339) had shown that, in a saccade, the force developed in a muscle was given by a similar equation:

$$F = K\theta + R \, d\theta/dt$$

where K is an elastic and R is a viscous coefficient whose ratio, R/K, is the time-constant of the mechanical load. Thus r/k might also be expected to indicate the time-constant, if spike discharge frequency were directly related to tension; and this is likely since in the eye the mechanical load is essentially constant. Actually, the ratio for the electrical parameters was 198 msec and that for the mechanical parameters 176 msec.

SACCADE AND PURSUIT MOVEMENT

Both Robinson & Fuchs and Luschei found that all units studied became involved during saccades and slow pursuit movements, so that the study of unit activity does not lend support to the operation of different types of motor unit in the two types of activity (p. 342).

Proprioception and Position Sense

An important element in the mechanism of reciprocal innervation is the sensory message received from the muscles involved, messages that are initiated by spindles and tendon-organs. Thus, the contraction of the agonist would stretch the antagonist, and the resulting sensory discharge from the spindles and tendon-organs might be the basis for the reflex inhibition of this muscle. The presence of spindles and tendon-organs, as well as several other types of apparently sensory terminations in extraocular muscles, has already been described, but the exact function of these sensory elements has been a matter of dispute that even now is not resolved. This is because of the confusion between two possible roles of muscle sensory organs; as classically regarded, the main role of such an organ as a muscle spindle was to act as the sensory pathway for position sense, whereby the individual is aware of the position of a limb in relation to the rest of the body; the second, and much better understood role, is that of being the sensory element in the myotatic reflex—whereby a stretched muscle contracts actively—or in the clasp-knife response, whereby a muscle will, if stretched beyond a certain point, be caused to relax. Modern studies, moreover, have shown that the muscle spindle, by virtue of its "gamma efferent" system, represents an extremely valuable feedback system that permits an accurate control of the length of a muscle in any given act (Merton, 1951). Thus position sense is not served by spindles or Golgi tendon-organs, but rather by other sensory elements in the joints.

Position Sense. The absence of proprioceptive sensory information from the muscles of the eye was argued because there is no "position sense" of the eyes (Irvine & Ludvigh); if a subject is placed in the dark and sensation from the conjunctiva is abolished by application of a local anaesthetic, then the subject in unaware of his involuntary eye movements as recorded by the EOG (Reinecke & Poser), whilst artificial movements, caused by experimentally pulling on a muscle, fail to be appreciated (Brindley & Merton). As we shall see, it is the absence of this "position sense" that is the basis for illusions of movement and false projection (p. 431). This absence of position sense, however, simply limits the function that we must attribute to the sensory end-organs in the extraocular muscles; we must regard them as feedback receivers, indicating the degree of contraction or stretch of the muscles and thereby permitting rapid adjustments during the course of a given eye movement. That sensory information is conveyed from the muscles, during stretch, was shown without doubt by Cooper, Daniel & Whitteridge for the goat where spindles were identified, and also for the cat and monkey in which spindles were not found (Fillenz); action potentials being recorded from the brainstem, or from the nerve supplying the muscle, e.g. the oculomotor nerve (N III), on stretching the inferior oblique (thus proving that this nerve is not purely motor, incidentally).

Gamma Efferents. Again, it will be recalled that the sensitivity of a muscle spindle to stretch can be augmented greatly by stimulating the small motor fibres (gamma-efferents) innervating the intrafusal fibres. In the goat and sheep Whitteridge (1959) isolated fibres in the fourth nerve that, when stimulated, increased the discharge of a muscle spindle. Thus the spindles in the extraocular muscles are exactly similar in electrophysiological behaviour to those found in the limb muscles, and it is therefore impossible to ignore their role in adjusting the force of contraction through a feedback mechanism that indicates the length of a muscle at any moment.

Sensory Pathway. The evidence relating to the pathway for this sensory information has been summarized by Rogers and Whitteridge. There seems little doubt that there are connections between the muscle-nerves and the trigeminal (N V), often within the orbit; recording from these branches, in the goat, Whitteridge (1955) obtained spindle-responses on stretching the eye-muscle. In the sheep, Manni, Bortolami & Desole (1967) found a group of neurones in the Gasserian ganglion with a steady discharge of 10–20/sec; on stretching an extraocular muscle, the discharge-rate suddenly increased to up to 300/sec, finally settling down to 150–90/sec when the stretch was maintained. On release of the muscle, the discharge ceased abruptly, and later the steady discharge of 10–20/sec was reassumed. This is typical of the behaviour of spindle afferent neurones. Subsequent studies in both sheep and pig leave no doubt that in these species the sensory neurones innervating the spindles are located in the Gasserian ganglion, and not in the mesencephalic root of the trigeminal (Manni, Bortolami & Deriu, 1970), nor yet in ganglion cells located along the oculomotor nerve (Manni, Desole & Palmieri, 1970). In the cat, destruction of the eye-muscle nuclei completely denervated the muscles, according to Corbin & Oliver, so that it may well be that in this species the sensory neurones have migrated from N V to the motor nuclei. Certainly Manni, Bortolami & Desole (1968) were unable to find neurones in the Gasserian ganglion responding to stretch of eye muscles, as in the sheep and pig.

Electromyographic Evidence. According to Breinin (1957) the resting activity of a muscle falls to practically nothing on cutting its tendon, suggesting that the normal resting activity is a reflex response to stretch; however, the failure to augment the discharge on pulling the muscle rather discountenances this view of the origin of *tonic* activity, which is therefore probably central in origin. The participation of the spindles in a feed-back mechanism is indicated, however, by Breinin's failure to observe the finely graded reciprocal innervation of agonists and antagonists in the muscles of an enucleated eye; thus, under these conditions, the antagonist is abruptly inhibited as soon as the agonist acts instead of showing the gradual decrease illustrated by Fig. 202. More recently Sears, Teasdall & Stone (1959) have shown that pulling on the agonist or on the antagonist during a movement causes a decrease in the spike discharge measured electromyographically, whilst mere cutting the insertion of the muscle had no effect on spontaneous activity. On balance, then, the electromyographic evidence suggests that the motor discharge to the muscles is influenced by sensory messages from the muscles themselves, but more work is obviously required.

Torsion and Listing's Law

We have indicated (p. 313) that a rolling movement of the eye—e.g. by rotation about its antero-posterior y-axis, would upset the individual's estimate of direction in space unless psychologically compensated. It has been seen that the associated actions of the superior oblique with the inferior rectus tend to cancel out any rolling motion, and in actual fact a study of the movements of the eyes shows that they are executed in such a way as, in general, if not to exclude rotations about the antero-posterior axis, to reduce them to a minimum. This is the basis of *Listing's Law* which will be enunciated more exactly later; for the moment we need only note that since rotations about this third, antero-posterior, axis are apparently excluded, there are virtually only two degrees of freedom in the motions of the eyes, i.e. the motions can be resolved into rotations about the horizontal *x*- and vertical *y*-axes. Can we conclude, therefore, that there is no torsion during all movements of the eyes, so that the image of a horizontal line in the frontal plane always falls on the horizontal meridian or parallel with it? We have seen, in the analysis of the actions of the superior and inferior recti, that the latter adduct the eye even though they rotate it about an axis perpendicular to the vertical, i.e. in spite of the fact that there is no rotation about the vertical *z*-axis, and it can be shown that rolling or torsion will occur in spite of the exclusion of rotations about the antero-posterior axis.

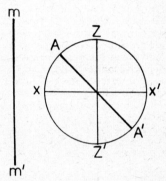

FIG. 206. Illustrating torsion. A vertical frontal section through the right eye. Rotation about AA' brings *z* nearer to *mm'* and takes *z'* farther away.

Torsion in the Tertiary Position. In Fig. 206 a vertical section through the equator of the eye in the primary position is shown, so that the plane of the paper corresponds to Listing's plane; *xx'* is the transverse horizontal axis, rotation of the eye about which causes pure elevation or depression; *zz'* is the vertical axis, rotation about which causes adduction or abduction. *xx'* coincides in direction with the horizontal meridian and may be considered as the intersection of the retinal horizon with the plane of the paper; it is at right-angles to the median plane, *mm'*, coming out of the plane of the paper; in this, the primary position, there is no torsion and, so long as the retinal horizon, passing through *xx'*, remains at right-angles to this plane, torsion will be absent.

Rotation about *xx'* obviously does not alter its inclination, and therefore that of the retinal horizon, to *mm'*, so that with pure elevation from the primary position there is no torsion; similarly, rotation about *zz'*, causing adduction or abduction, does not change the inclination of the retinal horizon to the plane *mm'*.

Rotation about the intermediate axis, AA', lying in the plane of the paper (Listing's plane), causes a combined movement of adduction and depression or abduction and elevation; during this rotation the point *x'*, for example, moves in a circle which is obviously not parallel with the plane *mm'* nor at right angles to it, consequently the retinal horizon passing through *xx'* becomes inclined to the plane *mm'* and the eye experiences torsion in spite of the fact that the axis of rotation, AA', is in Listing's plane, at right-angles to the antero-posterior axis. If we imagine that we are looking into the eye from behind, a rotation that brings *x'* up out of the paper corresponds to looking up and to the right; in this event the inclination of the retinal horizon towards *mm'* follows an anti-clockwise direction and if the eye is a right eye it may be said to have undergone inward rolling. (It is perhaps easier to visualize the events by considering a median plane through *zz'*, corresponding to the vertical meridian of the retina; on rotation about AA' *z* clearly moves closer to *mm'*.)

TORSION AND ROLLING

Thus rotation of the eye about an axis intermediate between the horizontal and vertical, whereby the eye passes from the primary position to a tertiary one, results in an angle of torsion; it is customary, but confusing, to draw a distinction between the two general ways of producing this deviation of the retinal horizon from perpendicularity to the median plane; if this is caused by rotation about the antero-posterior axis, or an axis with a component along this direction, it is said to be due to *rolling*; if it is a consequence of rotation about an axis in Listing's plane, it is called *torsion* or *false torsion*. No one has suggested a similar differentiation between adduction or abduction caused by rotation about a vertical axis and that caused by rotation about an axis perpendicular to this (p. 317), and the differentiation between the two processes is entirely unnecessary; the physiologist is chiefly concerned with the position of the retinal horizon in respect to the median plane, and the degree to which it deviates from perpendicularity may be called the angle of torsion or angle of rolling indifferently. Rather than increase the already existing confusion, however, it is probably advisable to retain the distinction; thus, with Quereau (1955) we may define rolling as a rotation about an antero-posterior axis and torsion (or false torsion as it is often called) as the consequence of oblique rotations about an axis in Listing's Plane.

So long as the head is vertical and no torsion has occurred, the retinal horizon will intersect a vertical frontal plane, e.g. the wall of a room, in a horizontal straight line (Fig. 195); if the eye fixates any horizontal line in this plane, therefore, its image will fall on the horizontal meridian of the retina so long as no torsion has been experienced during the movement of fixation.

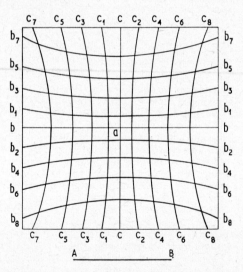

FIG. 207. The inclination of the projection of the after-image of a horizontal line, fixated in the primary position of the eye, on moving the latter into various other positions. *a* is the original fixation point and AB represents the relative distance of the eye from this point. (Helmholtz, *Physiological Optics*.)

Donders' Law. The position of the horizontal meridian of the eye in relation to the median plane, i.e. the degree of torsion present, can be ascertained by after-image experiments. If the eye, in the primary position, is directed at a horizontal straight line, the image of the latter falls on the horizontal meridian; if the stimulus is strong, an after-image may be evoked and the sensation is projected outwards, so that wherever the gaze is directed the line is perceived in the position it must have if its image is to fall on the horizontal meridian. If this projected after-image coincides with a true horizontal line, it can be said that the truly horizontal line produces an image on the horizontal meridian, i.e. that no torsion is present. If, on the other hand, torsion has occurred, the projection of the horizontal meridian appears inclined to a truly horizontal line, and the angle between the after-image and the horizontal gives the angle of torsion.

With the eye in the primary position, a black horizontal line is fixated for a time and the eye is then moved over a screen with horizontal lines on it. If the eye moves vertically upwards or downwards, or horizontally to the left or right, the after-image is found to coincide with the lines on the screen, indicating that there is no torsion. If the eye is moved to a tertiary position, the after-image no longer coincides with the horizontal lines; e.g. if the eye looks up and to the right, or down to the left, the after-image is inclined over to the left, as we should expect if rotation took place about an intermediate axis in Listing's plane. The actual directions, in which the projected after-image lies for different movements are indicated in Fig. 207. It is found that no matter how the eye is moved to any position—whether in several steps or as a result of a single rotation—the torsion is the same; this is the basis of *Donders' Law* which states: *"When the position of the line of fixation is given with respect to the head, the angle of torsion will have a perfectly definite value for that particular adjustment; which is independent not only of the volition of the observer but of the way in which the line of fixation arrived in the position in question."*

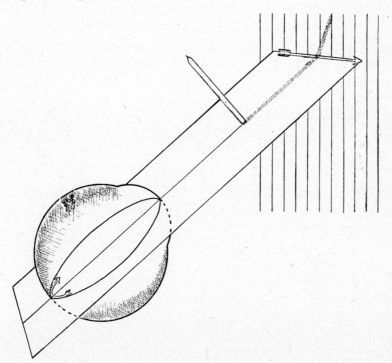

Fig. 208. Illustrating the projection of the vertical meridian of the retina. The eye is looking up and to the right, and it is supposed that torsion has been corrected. The image of the pencil thus falls on the vertical meridian of the retina; its projection against the frontal plane is nevertheless inclined over to the right.

Projection of Vertical Lines. With vertical lines, the position is more complicated; a vertical after-image is found to be inclined as though the eye had twisted in the opposite sense, i.e. on looking up and to the right the vertical line after-image is tilted over to the right instead of to the left; since the eye cannot exhibit torsion in two ways at once, and since the twist of the horizontal line agrees with theory, this apparent torsion in the opposite sense must be due to some other cause. Suppose now the eye is moved from the primary position upwards and to the right, and the torsion is corrected by an appropriate rolling movement. The after-image of a horizontal line, fixated in the primary position, now corresponds in direction with the truly horizontal line through the new point of fixation; the vertical meridian of the retina is at right angles to the retinal horizon so that we might expect that its projection outwards would be at right angles to a horizontal line on the wall, coinciding with a vertical line. This, however, is not true; the retinal horizon is inclined upwards and to the right, and a line perpendicular to it—which represents the projection of the vertical meridian in the absence of torsion—is inclined over to the right when projected on the screen (Fig. 208).

We are now in a position to explain the opposite appearances of horizontal and vertical after-images; due to torsion the horizontal and vertical meridians are tilted over to the left in the case examined. Owing to the characteristics of the projection of the vertical meridian on a vertical frontal plane in space, the after-image of a vertical line, fixated in the primary position, appears tilted over to the right, and more so than the tilt to the left due to torsion.

As a result of these combined effects, therefore, the after-image of a cross, fixated in the primary position, will have the appearances shown in Fig. 209 when the eyes are moved into various tertiary positions.

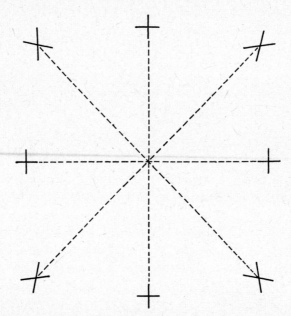

FIG. 209. Approximate appearances of the projections of the after-image of a cross, fixated in the primary position of the eye, for various directions of the fixation axis.

Listing's Law. It is clear that if, during a rotation of the eyes from the primary position, there is to be no component of rolling about the antero-posterior axis, then the axis of rotation must be perpendicular to this axis, i.e. it must lie in Listing's plane (Fig. 192). According to Listing this is, indeed, true so that movements of the eye are confined to two, instead of three, degrees of freedom. Given this limitation—which is the basis of Listing's Law—the experimental finding of Donders that the torsion of the eye is characteristic for its position however this was achieved, follows at once, since the torsion due to rotation about an axis in Listing's plane is predictable on geometrical grounds. More recent experimental studies on the subject have shown that Listing's Law is only an approximation; thus, when the eye is made to look up and outwards, Quereau found that some rotation about an antero-posterior axis occurred so that the resulting torsion was not that predictable on geometrical grounds on the assumption of a rotation about an axis in Listing's plane.*

BINOCULAR MOVEMENTS

Hering's Law. The movements of the two eyes are so co-ordinated as to permit them to operate as a single entity; this unity in behaviour is the basis for *Hering's Law of the Ocular Movements* which states: "The movements of the two eyes are equal and symmetrical."

A movement is said to be *conjugate* when the fixation axes are parallel and the movements of the separate eyes are equal in all respects (Fig. 210*a*). When the fixation axes are not parallel, as in convergence or divergence, the movement is said to be *disjunctive* (Fig. 210*b*); in this case the movements of the two eyes are symmetrical, and, as we shall see, they may well be equal in extent even though it can appear that one eye does not move at all (e.g. when converging on a point directly in front of one eye).

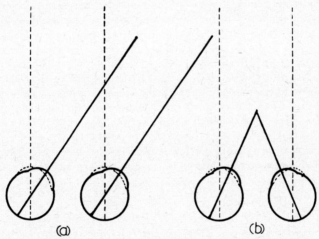

FIG. 210. Conjugate (*a*) and disjunctive (*b*) movements of the eyes.

The Position of Rest. We have so far defined a primary position of the eyes as a reference point from which movements may be described; when the eyes are free from any stimulus determining their mutual orientation (accommodation, fusion of images, etc.) they adopt the so called *fusion-free position* or *physiological position of rest* and it generally happens that in this position the fixation axes diverge slightly (physiological heterophoria) and are thus not in their primary positions; hence, even for a distant point, the act of fixation requires a tonic action of the extraocular muscles whereby the fixation axes are maintained parallel. As we shall see, this "tonic" action is the result of the

---

* This account of torsion is a slightly abbreviated version of that in the first edition of this book; in writing it I relied almost exclusively on Helmholtz's account. Since then several writers have contributed papers designed to clear up the confusion that enters once one departs from Helmholtz (Quereau 1954, 1955; Marquez, 1949; Moses, 1950). Since I find it very difficult to believe that Helmholtz was wrong in his geometry I have not been tempted to alter my first version.

operation of the fixation reflex that ensures that both eyes will fixate a given point accurately and thereby prevent the diplopia that would occur were this accurate binocular fixation not maintained (p. 449). To determine this fusion-free position various optical devices are employed to remove the tendency to binocular fixation; thus if a red spot on a green background is viewed by one eye through a red glass and by the other through a green one, the two eyes see quite different pictures and there is no tendency to fuse these, so that the fixation reflex is not operative and the eyes adopt their "natural positions". The muscle balance of the eyes would be said to be perfect if the two eyes maintained accurate alignment with the same point; if they tended to converge they would be said to be *esophoric*; if they diverged *exophoric*, and so on.

Conjugate Movements

Conjugate Innervation. In conjugate binocular movements both eyes must move to the same extent if diplopia is not to result from any shift of the gaze, and it is therefore reasonable to suppose that there is some sort of conjugate innervation of the principal actors in any binocular movement, in the sense that the passage of impulses to one is accompanied by the facilitated passage of impulses to the other. Broadly speaking we may expect that for a conjugate deviation to the right the left medial rectus will be associated with the right lateral rectus; for a movement up and to the right, the right superior rectus (the elevator with the stronger action in abduction) will be associated with the left inferior oblique (the elevator with the stronger action in adduction). The muscles of the two eyes may thus be classified in pairs, or "associates", such that in any general type of movement a pair may be designated as the chief actors. The clinician makes use of this classification in order to assess a muscle deficiency; the "diagnostic directions of gaze", classified below, are chosen so that any defect in binocular fixation may be related to the muscle associates chiefly acting:

Diagnostic Direction	Chief Actors R.E.	L.E.
Right	L. rectus	M. rectus
Left	M. rectus	L. rectus
Up and right	S. rectus	I. oblique
Up and left	I. oblique	S. rectus
Down and right	I. rectus	S. oblique
Down and left	S. oblique	I. rectus

That this linkage is real is proved by the phenomenon of the secondary deviation of a squinting eye. For example, if the left lateral rectus is partially paralysed, when the eyes look left the eyes squint because the left eye fails to make the necessary movement (Fig. 211*b*). If, however, a special effort is made by the left eye to fixate the point, then the right eye, by overaction of its medial rectus, squints, giving what is called a *secondary deviation* (Fig. 211*c*).

Limits of Movements. For practical purposes, the limits of binocular conjugate movements are given by the *field of binocular single vision*, i.e. one determines the position, for any direction of the gaze, at which one eye can no longer keep pace with the other, as evidenced by an insuperable diplopia. Duane's measurements have shown that this field extends to at least 40° in any given direction and usually up to 50° or more, in fact he states that most individuals retain binocular single vision up to the very limits of the excursions of the eyes, so that it may happen that the binocular field of single vision is larger than either monocular field of fixation taken separately. Thus it appears that, no matter what the maximum excursion of each eye separately is, the excursion of both together is, in normal cases, such that one eye keeps pace with the other and diplopia is avoided even in extreme deviations of the eyes.

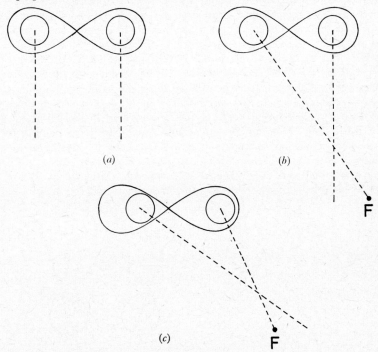

(a)

(b)

(c)

F

F

FIG. 211. Illustrating primary (b) and secondary (c) deviations due to paresis of left lateral rectus.

Disjunctive Movements

Converging Power. When a near object is fixated, the eyes converge as a result of the combined contractions of the medial recti. There is, of course, a limit to which the fixation axes can be deviated inwards so that if, for example, a pencil in the median line is moved closer and closer to the eyes, a point is reached where the two eyes can no longer fixate the pencil and it appears double. This point, the nearest consistent with single vision, is called the *near point of convergence* (Fig. 212); in 80 per cent of normal individuals it lies between 5 and 10 cm in front of the eyes and is not greatly affected by age as with the near point of accommodation. The *far point of convergence* is given by the point of intersection of the fixation axes in the physiological position of rest; since the eyes in this condition are usually divergent, this point is generally behind the eyes (Fig. 212), its distance being given a negative value.

METRE-ANGLE

By the *converging power* of the eyes, for any given point of fixation in the median line, is meant the angle through which *either* fixation axis moves from the primary position when the convergence takes place (Fig. 212); this angle can, of course, be represented in degrees but Nagel introduced the *metre-angle* as a convenient unit. The metre-angle is the angle through which each eye has rotated from the primary position in order to fixate an object 1 metre away in the median line. If the eyes converge on an object 2 metres away, clearly the angle is approximately halved, and the convergence of each eye is said to be 0·5 metre-angle; if the object is 10 cm away the angle is approximately ten times as large and the convergence is 10 metre-angles. In general, the converging power of each eye is given by the reciprocal of the distance of the fixation point in metres. For an average interpupillary distance of 6 cm the metre-angle corresponds to the convergence angle of 1·7°. It will be noted that the accommodative power in dioptres for any point is given by the same reciprocal (p. 570), so that in some respects this is a convenient notation; nevertheless, because the actual angle of convergence depends on the interpupillary distance, this equality can be misleading. Thus for an emmetrope converging on a point ⅓ metre away the convergence and accommodation are respectively 3 metre-angles and 3 dioptres. With interpupillary distances of 5, 6 and 7 cm the actual convergences are, however, 15, 18 and 21 prism-dioptres respectively. This is an important point when considering accommodative convergence (Breinin, 1957).

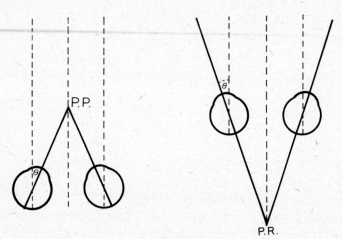

FIG. 212. The near point (P.P.) and far point (P.R.) of convergence.

AMPLITUDE OF CONVERGENCE

The *amplitude of convergence* is given by the difference in the converging powers of the eyes for the near and far points; hence with a near point of +8 cm and a far point of −1 metre, the amplitude is:

$$1/0·08 - 1/-1 = 12·5 + 1 = 13·5 \text{ metre-angles.}$$

A prism alters the apparent direction in which an object is seen (p. 541); if it is base-out, the object appears to be deviated inwards and the eye adducts in order to fixate it (a base-out prism is called an *adducting prism*). Theoretically, then, the near point of convergence may be determined by placing pairs of base-out prisms in front of the eyes until a distant object can no longer be seen single (Fig. 213). The strength of the prisms required to achieve this could be converted into metre-angles of convergence by the formula:

Strength of prism in dioptres = No. of metre-angles × ½ interpupillary distance in cm.

Thus with a near point of 12·5 m.a. and an interpupillary distance of 6 cm the prism strength would be given by:

$$\text{Prism strength} = 12\cdot5 \times 3 = 37\cdot5\varDelta \text{ (or } 18\cdot75^{\circ}\text{d).*}$$

In practice, however, this would be an unreliable measure of convergence-power since the ability to converge in response to prisms varies from moment to moment, and with practice.

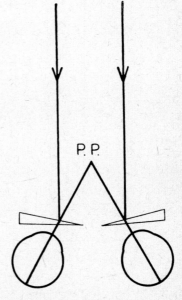

FIG. 213. The use of adducting prisms to cause maximum convergence. (P.P., near point of convergence.)

FUSION SUPPLEMENT

It is stated that the desire for fusion permits a stronger convergence than that obtained without this stimulus; if an object is fixated at the near point of convergence and one eye is then covered up, the other deviates a little; the extra degree of convergence obtained when the stimulus for fusion is present is called the *fusion supplement.*

"See-Saw" Movements. Disjunctive vertical movements—"see-saw" movements—in which one eye moves up and the other down, are not normally produced voluntarily; in the interests of single vision, however, the eyes are

* The terminology of convergence is so varied that it is also necessary to be able to convert a convergence distance into an angle of convergence; the formula supplied by Duane is:

$$\text{Convergence angle} = \frac{\text{I.P.D. (mm)} \times 25}{\text{Convergence distance}} + 1\cdot5^{\circ}.$$

able, on the average, to overcome the effects of a 2Δ prism base-up or base-down in front of one eye, i.e. they can exhibit a "supervergence" of 1°.

Convergence as a Function of the Two Eyes. It must be appreciated that disjunctive movements are a function of both eyes so that if a distant object in the median line is fixated and the strongest tolerated adducting prism is put in front of one eye, on placing a prism, however weak, in front of the other, diplopia results, i.e. no further convergence is possible. The near point of convergence is thus measured by half the strength of the strongest prism tolerated by one eye (when the other has none) or the strength of equal prisms tolerated by both eyes. The explanation for this, according to Hering, was that both eyes shared in the convergence although the direction of one fixation axis might remain finally unchanged. Thus in Fig. 214 the fixation axes of the left and right

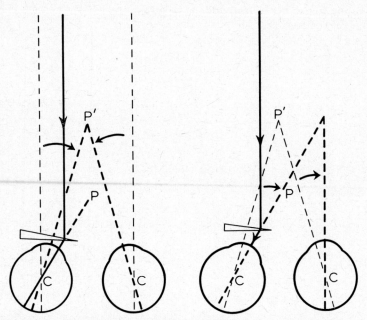

FIG. 214. The fixation axes were originally parallel. A strong adducting prism has been placed in front of the left eye so that it must be directed to P if it is to maintain fixation of the distant point. Both eyes converge to P' and then make a conjugate deviation to the right. As a result, the left eye is directed at P whilst the right eye re-assumes its original direction of fixation.

eyes were initially parallel; a prism has been placed in front of the left eye so that it adducts maximally towards P. The right eye, during this process, was said not to remain fixed and the motions of the two eyes were analysed into an initial convergence to a point P', such that the convergence of each eye is half the angle necessary for the left fixation axis to move through to fixate P, followed by a conjugate deviation to the right through the same angle; this leaves the right fixation axis pointing in its original direction, whilst the left is directed at P.

In a similar way the asymmetrical convergence to a point P, on the fixation axis of one eye when it is looking straight ahead (Fig. 215), was said to be achieved by an initial convergence to a point P′ in the median line, followed by a conjugate deviation. Since convergence is equally shared by the two eyes, we can obtain the convergence of *each* eye for any position of the fixation point by halving the sum of the angles moved through by the two fixation axes from their primary positions.

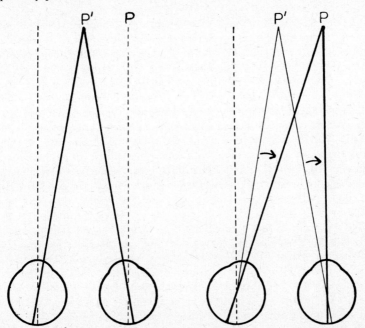

FIG. 215. Illustrating asymmetrical convergence on the point P. The two eyes converge on the point P′ and then make a conjugate movement to the right. The direction of the right fixation axis is thus finally unchanged.

EXPERIMENTAL EVIDENCE

The asymmetrical convergence illustrated by Fig. 215 need not take place in the gross steps illustrated, and the process may consist in a series of convergence and conjugate lateral movements that may escape observation in the absence of an accurate means of measuring either the movements of the eyes or activity in their muscles. Alpern & Wolter, employing the electro-oculogram (EOG, p. 338) to record the eye movements, found that they were, indeed, compound, but apparently the conjugate movement preceded the convergence, i.e. the eyes tended to look to the right of P and subsequently to converge on it. As Alpern & Wolter emphasized, the speed of the conjugate type of movement, namely about 400°/sec, is so much greater than that of convergence (about 25°/sec) that the former might be missed. By measuring the eye movements by photographing the light reflected from the cornea, Westheimer & Mitchell were able to confirm Hering's view exactly, the asymmetrical convergence consisting of a rapid "saccade", i.e. a conjugate version, followed by convergence. Using the electromyograph, Tamler, Jampolsky & Marg (1958) showed that the apparently stationary eye in asymmetrical convergence showed activity in both its medial and lateral rectus; this certainly confirms Hering's view of conjugate innervation of both medial recti, and suggests that either the tendency for the apparently stationary eye to adduct is balanced by the pull of the lateral rectus, or that the eye does adduct and then abducts in a conjugate version as illustrated by Fig. 215.

Rolling

When the eyes are in the primary position they cannot be rolled voluntarily; however, the stimulus for fusion can cause such a movement. It is possible, with the aid of 90° prisms placed in front of one eye, to cause the image of an object on the retina to rotate about the fixation axis, e.g. the image of a cross may be tilted over. The two eyes are thus presented with different images that can only be fused if the eyes are rolled in opposite directions; the stimulus for fusion causes such a rolling, which may amount to as much as 10°–14°, divided between the two eyes, but is generally much smaller.* Convergence is said to be associated with a certain degree of outward rolling of the two eyes, so that their vertical meridians make an angle with each other which, in extreme instances, may amount to as much as 25°. This rolling must produce a rotation of the two retinal images in respect to each other, so that points in space formerly seen single might now be seen double (p. 447); on converging on a line of print, for example, diplopia should result from this asymmetrical torsion; in practice this does not occur so that it would seem, as Harms has pointed out, that when there is a stimulus for fusion, rolling does not take place during convergence; only when the object fixated has adequate symmetry about the fixation axis, so that its retinal image is not seriously modified by rotation about this axis, does rolling occur. For example, if an individual fixated a circle binocularly, rolling of one or both eyes would make no difference to the two retinal images; with a cross, on the other hand, the images would tilt in opposite directions and the individual would have to fuse dissimilar images; the rolling that occurs during convergence is therefore probably largely countered by a corrective fusion movement. As we shall see, the eyes execute a conjugate rolling movement when the head is tilted towards the shoulder; this is a reflex response to the altered pull of gravity on the head, and to the stretch of neck muscles (p. 371).

Analysis of the Movements

It was Dodge in 1903 who made the first serious quantitative attempt to analyse the movements of the eyes; subsequent work has depended for its fruitfulness on improvements in the technique for recording eye movements. Probably the most satisfactory method consists of reflecting light from the cornea, or a contact glass fitting snugly on the eye, and focusing it on to a moving photographic film. By suitable calibration the movements of the reflected light may be converted into angular movements of the eye quite accurately.

Electro-oculogram. When an electrode is placed on the cornea and another on the posterior pole of the eye or elsewhere on the head, a "standing potential" is recorded of about 1 mV, the cornea being positive. Illumination of the eye gives alterations of this standing potential which we call the electroretinogram (ERG); thus the ERG is really superimposed on the "standing potential".

---

* According to Kertesz & Jones (1970), however, when cyclovergence is measured objectively under these conditions, it is negligible so that, in fact, fusion of the rotated images has taken place without rolling of the eyes, i.e. disparate images are fused.

Because of this potential the eye behaves like a dipole orientated along its antero-posterior axis, and if electrodes are placed at right-angles to this axis, e.g. in the two canthi, then movements of the eye will cause changes of potential between the electrodes because the dipole's orientation with respect to these electrodes changes. This is the basis of *electro-oculography* (EOG), an indirect method of recording eye movements, which can be employed even with the eyes closed. Methods specially adapted to record the small involuntary movements that take place during maintained fixation are those of Rashbass (1960) and Ginsborg & Maurice (1959).

In experimental animals, such as the monkey, Fuchs & Robinson (1966) have implanted a fine wire coil under the conjunctiva; this was joined to a connector fixed to the skull. The animal's head was exposed to two alternating magnetic fields in spatial and phase quadrature, and signals could be made to indicate horizontal and vertical eye motions.

Types of Movement

The conjugate movements of the eyes may be classified as *saccades* and *smooth following*, or *tracking*, movements. Saccade was the name originally given by Dodge to the rapid movements between fixation-pauses that occur in reading, but it is now applied more generally to the conjugate shifts of gaze from one fixation point to another. The very small saccades, occurring during steady fixation, as described later, are now called *microsaccades* or *flicks*. The movements of the eyes during convergence take place so much more slowly than the conjugate movements that a separate effector system has been postulated (Alpern & Wolter, 1956).

The Saccade. The mechanics of the saccade were first seriously investigated by Westheimer (1954), the subject being told to fixate a light whose position could be suddenly changed. He found a reaction-time of some 120–180 msec, after which both eyes simultaneously moved; the maximum velocity attained varied with the magnitude of the saccade, being $500°/sec$ for a $30°$ movement and $300°/sec$ for $10°$. Subsequent studies of Robinson (1964), in which he measured both the eye movements and the tensions developed in the extraocular muscles, showed that the eye was impulsively driven in a saccade by a brief burst of force much greater than necessary to maintain the eye in its final position; it is essentially this "pre-emphasis", or excess force, that allows the saccade to be executed so rapidly, permitting a $10°$ saccade to occur within 45 msec after its beginning. The deceleration of the eye that brings it to rest at the end of the saccade is not achieved by any checking action of the antagonistic muscle, but relies entirely on the viscous-elastic damping of the muscles and orbital connective tissue, a finding that has been amply confirmed by EMG and single unit studies of the eye motor nuclei (p. 342). It is this orbital stiffness, of course, that requires that the muscles contract when the eyes deviate from their primary positions, the greater the angular deviation the greater the force of contraction.* In general, the

---

* The mechanical studies of Childress & Jones (1967), in which they caused adduction by a mechanical pull on the eye and observed the responses to quick releases, have generally confirmed Robinson's assessment of the eye in its socket as a high-inertia heavily damped system. Fuchs (1967) has studied the eye movements of the monkey, which are qualitatively similar in so far as the differentiation between saccades and pursuit movements is concerned; quantitatively the monkey differs in being able to follow more rapidly moving targets, and in executing the saccades more rapidly.

saccadic movement matches that of the target to about 0·2° (Rashbass, 1961) but sometimes it does not, in which case it is followed, after a new reaction-time, by a second saccade; in some subjects there may be three or four saccades, each separated by one reaction-time, before the final fixation is reached. Rashbass also observed that there was a threshold target displacement, so that a movement of 0·25–0·5° is usually not accompanied by a response.

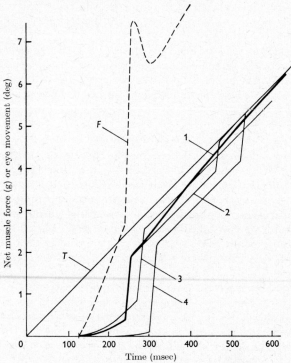

FIG. 216. A composite graph to show the various types of eye-movement responses to a 10 deg/sec ramp target motion (T). Response 1 is the mean of fourteen responses of the most common type and (F) is the net isometric muscle tension in the horizontal recti associated with it. Responses 2, 3 and 4 illustrate variations in overshoot and temporal spacing between smooth and saccadic components. (Robinson, *J. Physiol.*)

CONTROL-MECHANISM

Young & Stark (1963) showed that the control mechanism during the saccadic type of target-following could be described as a "sampled data system", in the sense that the brain made discontinuous samples of the position of the eyes in relation to the target, i.e. of the "error", and corrected this. Proof of such discontinuity was provided by Westheimer who showed that, if the target was caused to step aside and return to its original position within 100 msec, the eye nevertheless made a saccade 200 msec later.*

* Fuchs (1967) has discussed the length of the sampling time; he suggests that the sample may be taken over half a reaction time, a position-correction decided on the basis of the information collected, and the correction effected without change half a reaction time later. Such a system would be described as *semi-discrete*, to distinguish it from smooth pursuit behaviour.

LARGE SACCADES

According to Becker & Fuchs (1969) a saccade of over 15° occurs in two pre-packaged stages; the first saccade brings the eyes short by about 10 per cent of the target, and after a reduced latency (130 instead of 230 msec) a second saccade brings the eyes on target.

Smooth Pursuit or Tracking Movements. Under these experimental conditions the fixated spot of light is caused to move at a constant velocity in a given direction; if this movement does not exceed a certain speed, namely about 30–40°/sec, the eyes move with the same speed, so that the image remains effectively stationary on the retina. As we should expect, the smooth pursuit movement is usually, itself, compound; thus, because of the reaction-time of some 150 msec before the smooth movement begins, and because it takes at least 400 msec for the eye to reach the target speed, the pursuit movement, by itself, would inevitably leave the point of fixation of the eye trailing behind the target unless a saccade intervened; and this in fact happens. Fig. 216 from Robinson (1965) shows the four main types of following movement, that marked 1 being the most common; here, after a delay of 125 msec, the eye accelerates at a mean value of 60°/sec² to reach a velocity of 6·1°/sec and a displacement of 0·38° in 112 msec before the interruption of a saccade; this has an amplitude of 1·24°, so that an error of 0·7° remains. The eye leaves the saccade at a smooth velocity of 12·2°/sec, i.e. it has overshot the target velocity; it maintains this excessive velocity until the error is nearly zero, when the velocity slackens off to become equal to that of the target. Fig. 216 also illustrates the development of tension in the horizontal recti associated with the movement; as with the saccade, there is an initial "pre-emphasis", this time on the rate of change of tension.

CONTROL

Young & Stark considered that smooth pursuit movements were also controlled by a sampled data system, the data utilized being the positions of the retinal image only. The studies of Rashbass and of Robinson, however, indicate that the control is continuous, presumably of the feed-back type, the eye responding to disparity, *per se*, and to the rate of change of position on the retina, i.e. there is response to both position and velocity, the latter being the dominant influence. Thus Rashbass' experiments have provided a number of interesting examples of the independence of the systems of control of saccades and pursuit movements. For example, if a target is given an initial rapid displacement of, say, 3° to the left, followed immediately by movement at uniform velocity to the right (step-ramp stimulus), a saccade is made to the left, after which the pursuit movement to the right occurs. Hence the saccade, because of the reaction-time, begins during the smooth movement of the target in the opposite direction; in this case, then, the position stimulus has taken precedence over the velocity stimulus. If the initial step-displacement is reduced to 1°, then the saccade fails, and the eye starts moving, after a longer-than-normal latency of 150 msec, in the same direction as the target. In this case the eye begins a smooth pursuit movement in response to target velocity even though the movement is away from the target position at this time. Thus the pursuit movement is primarily responsive to target velocity rather than position.

Young & Stark assumed that a retinal image-error exceeding 0·2° was adequate stimulus for a saccade, but the step-ramp experiment of Rashbass shows the inadequacy of this picture; similarly in Curve 1 of Fig. 216, after the initial saccade, an error of 0·7° remained but no saccade was elicited, the velocity over-shoot being adequate for ultimate correction. Thus, the decision as to whether a saccade is to be made, as well as the determination of its magnitude, are functions of both retinal image-error and its rate of change, and it may be that the ganglion cells described by Barlow, Hill & Levick (p. 165), sensitive to direction of image-motion, are involved here.

Disjunctive Movements. The disjunctive movement of convergence is altogether slower than the conjugate movement, although the reaction-time of about 160 msec is of the same order. The movement takes place with a constant velocity, which is a function of the magnitude of the stimulus;* this velocity falls off asymptotically, with a total movement-time of about 800 msec. Even with a sudden change of vergence of the target, it is possible for control to be continuous rather than of the sampled data type; and according to Rashbass & Westheimer (1961a) this is true, so that, when momentary changes in the vergence stimulus are introduced during the execution of a movement, these are responded to after a reaction-time; thus disjunctive eye movements can be modified during their progress, and information concerning disparity can be assimilated during the reaction-time as well as during movement.

When the following of a target involves simultaneous conjugate and disjunctive movements, then, along whatever path the target moves, the movement of the eyes is resolved into two components corresponding to mean target position and target vergence, respectively, and appropriate responses to these components are made by two independent systems of control (Rashbass & Westheimer, 1961b).

Electromyogram

According to Robinson (1966), the electromyographic record of the medial recti during convergence bears little relation to the development of tension; for example, the peak discharge occurs at 200 msec after the onset of movement when the tension has only developed to 37 per cent of its final value. This finding lends support to the suggestion of Alpern & Wolter (1956) that vergence is brought about by the slow fibres of the medial recti, since the electromyographic record is determined by the spikes occurring in fast fibres.

Functions of Slow and Fast Fibres. The presence of two or three types of muscle fibre in the extraocular muscles doubtless has physiological significance, but whether it resolves into the clear-cut distinction between employment in saccades and vergence movements it is not at all easy to say. Jampel (1967) has discussed this point in some detail; there is little doubt that the slow type of movement, as seen in vergence or following, may operate separately from the saccade, as in the pathological "dissociations", but this, of course, may only refer to the neural mechanisms, and it does not follow that a given type of muscle fibre is employed only for certain types of motion; a slow following movement might be brought about by a less powerful and more prolonged discharge in motor neurones to the same muscle fibres as those involved in a saccade. Nemet & Miller (1968) have endeavoured to detect differences in the type of evoked potential in cat extraocular

* Crone (1969) has emphasized the kinetic aspects of the retinal disparity that provokes a convergent fusional movement; the rate at which a fusional movement is brought about is, indeed, a linear function of the magnitude of the disparity.

muscle according as fast movements are evoked by stimulation of the third nucleus and slow by stimulation of the vestibular complex or the reticular formation.

Vision During a Saccade

It was shown by Dodge in 1900 that vision was apparently suppressed during a saccade; a convincing demonstration of this was that of Ditchburn (1955) in which a subject fixated a bright spot on a cathode-ray oscillograph, the position of which moved in advance of the eye movement (positive feed-back); as a result the spot made rapid flicks whenever the eyes moved, and the subject was unable to see these movements although an observer beside him could. Volkmann (1962) presented 20 μsec flashes to a subject who was told to move his eyes at given times; she found that, during a fixation movement, there was a 0·5 logunit rise in the threshold to the flashes, the change beginning some 40 msec before the beginning of the movement and lasting 80 msec after the end. In addition, the finest gratings that could be resolved by the moving eye were about 1·4 times coarser than those resolved by the fixating eye. Beeler (1967), in confirming the 0·5 logunit rise in threshold, pointed out that this decrease in visual sensitivity would not have been sufficient to prevent Ditchburn's subject from seeing the illuminated spot, so that there must be some other factor operating to prevent detection of movement. He presented a subject with a light-spot and made this move sideways 15 minutes of arc, the movement of the spot being arranged to occur at random intervals after an involuntary saccade; the movement of the spot was too small in extent to cause a fixation movement by the subject. Beeler determined the frequency of seeing the spot when it moved at different intervals after the saccade; and this increased from zero, during the saccade, to 93 per cent if the movement was separated from the flick by more than 100 msec. Thus, during a saccade there is apparently a suppression of target information; Beeler suggested that the cortex investigates the oculomotor system when a movement of the retinal images occurs, for example after a flick; during this time there is no awareness of further motions. If the efferent system has operated, i.e. if the retinal image-movement is due to motor activity and not to actual object movement, then all image-motion is ignored.* It is interesting, in this context, that MacKay (1970) showed that the movement of a patch of light caused a reduced sensitivity to flash-stimuli, the time relations being sufficiently similar to those of saccadic suppression to suggest a common mechanism.

Cortical Neurones. So far as the activity of cortical neurones is concerned, there is no doubt that this, in the presence of an appropriate stimulus falling on their receptive fields, may be markedly altered when the eyes make a rapid movement; this may be an increase in unit activity or a suppression; in others, however, there is no alteration at all (Kawamura & Marchiafava, 1968; Wurtz, 1969).

Movements during Steady Fixation

It has been known for a long time that the eyes do not maintain a steady fixation of a point for any length of time; instead, fixation is interrupted by rapid jerks at irregular intervals. The actual movements that take place have been

* Michael & Stark (1967) measured the visually evoked response (VER, p. 519) with scalp electrodes and found that, when saccadic suppression of light-flashes occurred, there was suppression or reduction of some waves on this record. Again, Zuber, Stark & Lorber (1966) found that the reduction in pupillary area in response to a flash could be as little as 10 per cent of the control value if the flash occurred during a saccade. Mitrani, Mateeff & Yakimoff (1970) have demonstrated very neatly that "retinal smear" is a factor of some significance by showing that the depression of visual acuity was large for vertically orientated gratings, when the eye made a horizontal saccade, but small for horizontally orientated gratings. On the other hand, Collewijn (1969) found depression of the VER in rabbit cortex and colliculus evoked by a flash in one, immobile, eye when the saccade occurred only in the other, non-seeing, eye.

analysed by several workers using modern methods of recording, usually employing the reflexion of light from the cornea.* They have been resolved into three types, as follows: (1) Irregular movements of high frequency (30–70/sec) and small excursion, namely some 20 sec of arc; these are described as *tremor*. (2) Flicks or microsaccades of several minutes of arc, occurring at irregular intervals of the order of 1 sec. (3) Between flicks there are slow irregular drifts extending up to 6 minutes of arc. Fig. 217 illustrates the region in which the

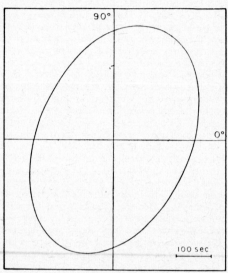

FIG. 217. Indicating the probable range of movements of the eye over a period of steady fixation. The solid line bounds the region within which a point retinal image will be found projected in space 68 per cent of the time during 30-sec. fixation intervals. (After Nachmias, *J. Opt. Soc. Amer.*)

fixation point would be 68 per cent of the time during a 30 seconds fixation period, and it will be seen that the range of excursion is not large, certainly not large enough to bring the image of the fixated point off the fovea. Most workers are agreed that the flick is a corrective movement that brings the image back when, as a result of a drift or previous saccade, it is in danger of moving off the central fovea. This is deduced from statistical studies that show that the chance of a flick occurring in a given direction correlates well with the degree of deviation from central fixation.†

* Of the contributors to this subject mention may be made of Riggs and his colleagues (Ratliff & Riggs, 1950; Cornsweet, 1956; Riggs & Niehl (1960); Krauskopf, Cornsweet & Riggs, 1960); Barlow (1952); Ditchburn & Ginsborg (1953); Drischel & Lange (1956); Nachmias (1959).
† Cunitz & Steinman (1969) have pointed to some remarkable similarities between the flicks that occur during steady fixation and the larger saccades occurring during reading. Thus the intersaccadic intervals are remarkably similar; furthermore, during the reading pauses, the number of flicks is very small compared with a similar period of steady fixation, and they have argued that the flick, like the reading saccade, is a scanning mechanism. They point out that when a subject is asked to "hold" rather than to fixate a target, flicks are very rare (Steinman *et al.*, 1967). Zuber, Stark & Cook (1965) have shown that, when maximum velocity is plotted against amplitude, the values for flicks fall on the same graph as those for the larger saccades in voluntary fixation movements.

Binocular Fixation. When the involuntary movements in the two eyes are studied together during binocular fixation, it is found that the slow drifts take place independently in the two eyes, leading to small degrees of fixation-disparity; i.e. if the eyes are looking at a distant point the fixation axes will tend to converge or diverge at random. On the other hand, the rapid saccades take place simultaneously and in the same direction, and corrections for convergence-errors result from the circumstance that the saccades in the two eyes are not always equal in extent. According to Krauskopf, Cornsweet & Riggs, the eyes apparently behave as independent detectors of fixation error, and in correcting their own errors they cause conjugate movements of the other eye that may not necessarily be in the right direction for correction of this eye's error. Since the chance of initiating a saccade by either eye is proportional to the magnitude of its fixation error, we may expect the eye with the larger error to initiate the corrective saccade; if the movement of the other eye in this saccade is a little smaller than that of the initiator, then on a statistical basis we may expect a reasonably accurate binocular fixation. Thus it appears that it is not the error in convergence that is detected, and corrected for, but simply the error of fixation by either eye acting as an independent unit from the viewpoint of detection.

A more exhaustive study by St.-Cyr & Fender (1969) of both monocular and binocular fixation, in which the corrective effects of flicks and drifts were examined statistically suggests, however, that retinal disparity, as such, can act as a stimulus for corrective movements; furthermore, the drifts seem to be the movements responsible for these vergence corrections. Another point brought out by this work is the large individual variations in the character of the fixation movements, and the dependence of their direction on the type of target fixated; thus a vertical line tended to cause the flicks to take an upward direction.

GENERAL REFERENCES

ALPERN, M. (1969). Movements of the Eyes. In *The Eye*, Ed. Davson, 2nd ed. vol. III. pt. I. London and New York: Academic Press.

COGAN, D. G. (1956). *Neurology of the Ocular Muscles*, 2nd ed. Springfield: C. C. Thomas.

HELMHOLTZ, H. VON (1925). *Physiological Optics*, vol. III, Ed. Southall. Optical Soc. Am.

WHITTERIDGE, D. (1960). Central control of eye movements. In *Handbook of Physiology*, pp. 1089–1109. Ed. Field, Magoun & Hall. Washington: Am. Physiol. Soc.

SPECIAL REFERENCES

ALPERN, M. & ELLEN, P. (1956). A quantitative analysis of the horizontal movements of the eyes in the experiment of Johannes Muller. *Am. J. Ophthal.* **42**, pt. 2, 289–296.

ALPERN, M. & WOLTER, J. R. (1956). The relation of horizontal saccadic and vergence movements. *Arch. Ophthal.* **56**, 685–690.

BACH-Y-RITA, P. & ITO, F. (1966). Properties of stretch receptors in cat extraocular muscles. *J. Physiol.* **186**, 663–688.

BARLOW, H. B. (1952). Eye movements during fixation. *J. Physiol.* **116**, 290–306.

BECKER, W. & FUCHS, A. F. (1969). Further properties of the human saccadic system: eye movements and correction saccades with and without visual fixation points. *Vision Res.* **9**, 1247–1258.

BEELER, G. W. (1967). Visual threshold changes resulting from spontaneous saccadic eye movements. *Vision Res.* **7**, 769–775.

BJÖRK, A. & KUGELBERG, E. (1953). Motor unit activity in the human extraocular muscles. *E.E.G. clin. Neurophysiol* **5,** 271–278.

BJÖRK, A. & KUGELBERG, E. (1953). The electrical activity of the muscles of the eye and eyelids in various positions and during movement. *E.E.G. clin. Neurophysiol.* **5,** 595–602.

BJÖRK, A. & WAHLIN, A. (1960). Muscular factors in enophthalmos and exophthalmos. *Acta ophthal. Kbh.* **38,** 701–707.

BOEDER, P. (1961). The co-operation of extraocular muscles. *Am. J. Ophthal.* **51,** 469–481.

BRANDT, D. E. & LEESON, C. R. (1966). Structural differences of fast and slow fibres in human extraocular muscle. *Am. J. Ophthal.* **62,** 478–487.

BREININ, G. M. (1957). Electromyographic evidence for ocular proprioception in man. *Arch. Ophthal.* **57,** 176–180.

BREININ, G. M. (1957). Quantitation of extraocular muscle innervation. *Arch. Ophthal.* **57,** 644–650.

BREININ, G. M. (1957). The nature of vergence revealed by electromyography. *Arch. Ophthal.* **58,** 623–631.

BRINDLEY, G. S. & MERTON, P. A. (1960). The absence of position sense in the human eye. *J. Physiol.* **153,** 127–130.

CHENG, K. & BREININ, G. M. (1966). A comparison of the fine structure of extraocular and interosseus muscles in the monkey. *Invest. Ophthal.* **5,** 535–549.

CHILDRESS, D. S. & JONES, R. W. (1967). Mechanics of horizontal movement of the human eye. *J. Physiol.* **188,** 273–284.

CLOSE, R. (1967). Properties of motor units in fast and slow skeletal muscles. *J. Physiol.* **193,** 45–55.

COLLEWIJN, H. (1969). Optokinetic eye movements in the rabbit; input-output relations. *Vision Res.* **9,** 117–132.

COOPER, S. & DANIEL, P. D. (1949). Muscle spindles in human extrinsic eye muscles. *Brain* **72,** 1–24.

COOPER, S., DANIEL, P. D. & WHITTERIDGE, D. (1955). Muscle spindles and other sensory endings in the extrinsic eye muscles. *Brain* **78,** 564–583.

COOPER, S. & FILLENZ, M. (1955). Afferent discharges in response to stretch from the extraocular muscles of the cat and monkey and the innervation of these muscles. *J. Physiol.* **127,** 400–413.

CORNSWEET, T. N. (1956). Determination of the stimuli for involuntary drifts and saccadic eye movements. *J. Opt. Soc. Am.* **46,** 987–993.

CRONE, R. A. (1969). The kinetic and static function of binocular disparity. *Invest. Ophthal.* **8,** 557–560.

CUNITZ, R. J. & STEINMAN, R. M. (1969). Comparison of saccadic eye movements during fixation and reading. *Vision Res.* **9,** 683–693.

DANIEL, P. D. (1946). Spiral nerve endings in the extrinsic eye muscles of man. *J. Anat. Lond.* **80,** 189–192.

DIETERT, S. E. (1965). The demonstration of different types of muscle fibers in human extraocular muscle by electron microscopy and cholinesterase staining *Invest. Ophthal.* **4,** 51–63.

DITCHBURN, R. W. (1955). Eye-movements in relation to retinal action. *Optica Acta* **1,** 171–176.

DITCHBURN, R. W. & GINSBORG, B. L. (1953). Involuntary eye movements during fixation. *J. Physiol.* **119,** 1–17.

DRISCHEL, H. & LANGE, C. (1956). Uber unwillkürliche Augapfelbewegungen bei einäugigen Fixieren. *Pflug. Arch.* **262,** 307–333.

DUKE-ELDER, W. S. & DUKE-ELDER, P. M. (1930). The contraction of the extrinsic muscles of the eye by choline and nicotine. *Proc. R. Soc. B.* **107,** 332–343.

FILLENZ, M. (1955). Responses in the brainstem of the cat to stretch of extrinsic ocular muscles. *J. Physiol.* **128,** 182–199.

FUCHS, A. F. (1967). Saccadic and smooth pursuit eye movements in the monkey. *J. Physiol.* **191,** 609–631.

FUCHS, A. F. & LUSCHEI, E. S. (1970). Firing patterns of abducens neurons of alert monkeys in relationship to horizontal eye movement. *J. Neurophysiol.* **33,** 382–392.

Fuchs, A. F. & Robinson, D. A. (1966). A method for measuring horizontal and vertical eye movement chronically in the monkey. *J. appl. Physiol.* **21**, 1068–1070.

Ginsborg, B. L. & Maurice, D. M. (1959). Involuntary movements of the eye during fixation and blinking. *Br. J. Ophthal.* **43**, 435–437.

Hess, A. (1967). The structure of vertebrate slow and twitch muscle fibers. *Invest. Ophthal.* **6**, 217–228.

Hess, A. & Pilar, G. (1963). Slow fibres in the extraocular muscles of the cat. *J. Physiol.* **169**, 780–798.

Irvine, S. R. & Ludvigh, J. E. (1936). Is ocular proprioceptive sense concerned with vision. *Arch. Ophthal.* **15**, 1037–1049.

Jampel, R. S. (1967). Multiple motor systems in the extraocular muscles of man. *Invest. Ophthal.* **6**, 288–293.

Kawamura, H. & Marchiafava, P. L. (1968). Excitability changes along visual pathways during eye tracking movements. *Arch. ital. Biol.* **106**, 141–156.

Kern, R. (1965). A comparative pharmacologic-histologic study of slow and twitch fibers in the superior rectus muscle of the rabbit. *Invest. Ophthal.* **4**, 901–910.

Kern, R. (1968). Uber die adrenergischen Receptoren der extraoculären Muskeln des Rhesusaffen. *v. Graefes' Arch. Ophthal.* **174**, 278–286.

Kertesz, A. E. & Jones, R. W. (1970). Human cyclofusional response. *Vision Res.* **10**, 891–896.

Krauskopf, J., Cornsweet, T. N. & Riggs, L. A. (1960). Analysis of eye movements during monocular and binocular fixation. *J. opt. Soc. Am.* **50**, 572–578.

Krüger, P. (1952). *Tetanus und Tonus der quergestreiften Skelettmuskel der Wirbeltiere und des Menschen.* Leipzig: Acad. Verlag, Geest & Portig.

MacKay, D. M. (1970). Elevation of visual threshold by displacement of retinal image. *Nature* **225**, 90–92.

Manni, E., Bortolami, R. & Desole, C. (1968). Peripheral pathway of eye muscle proprioception. *Exp. Neurol.* **22**, 1–12.

Manni, E., Bortolami, R. & Deriu, P. L. (1970). Superior oblique muscle proprioception and the trochlear nerve. *Exp. Neurol.* **26**, 543–550.

Manni, E., Bortolami, R. & Desole, C. (1967). Relationship of Gasserian cells to extraocular muscle proprioception in lamb. *Experientia* **23**, 230–231.

Manni, E., Desole, C. & Palmieri, G. (1970). On whether eye muscle spindles are innervated by ganglion cells located along the oculomotor nerves. *Exp. Neurol.* **28**, 333–343.

Marg, E., Jampolsky, A. & Tamler, E. (1959). Elements of human extraocular electromyography. *Arch. Ophthal.* **81**, 258–269.

Marquez, M. (1949). Supposed torsion of the eye around the visual axis in oblique directions of gaze. *Arch. Ophthal.* **41**, 704–717.

Matyushkin, D. P. (1961). Phasic and tonic neuromotor units in the oculomotor apparatus of the rabbit. *Sechenov Physiol. J. USSR.* **47**, 960–965.

Matyushkin, D. P. (1964). Motor systems in the oculomotor apparatus of higher animals. *Fed. Proc. Transl. Suppl.* **23**, T 1103–1106.

Mayr, R., Stockinger, L. & Zenker, W. (1966). Elektronen-mikroskopische Untersuchungen an unterschiedlich innervierten Muskelfasern der äusseren Augenmuskulatur des Rhesusaffen. *Z. Zellforsch.* **75**, 434–452.

Merton, P. A. (1951). The silent period in a muscle of the human hand. *J. Physiol.* **114**, 183–198.

Michael, J. A. & Stark, L. (1967). Electrophysiological correlates of saccadic suppression. *Exp. Neurol.* **17**, 233–246.

Miller, J. E. (1958). Electromyographic pattern of saccadic eye movements. *Am. J. Ophthal.* **46**, Pt. 2, 183–186.

Miller, J. E. (1967). Cellular organization of rhesus extraocular muscle. *Invest. Ophthal.* **6**, 18–39.

Mitrani, L., Mateeff, St. & Yakimoff, N. (1970*a*). Smearing of the retinal image during voluntary saccadic eye movements. *Vision Res.* **10**, 405–409.

Mitrani, L., Mateeff, St. & Yakimoff, N. (1970*b*). Temporal and spatial characteristics of visual suppression during voluntary saccadic eye movement. *Vision Res.* **10**, 417–422.

Moses, R. A. (1950). Torsion of the eye on oblique gaze. *Arch. Ophthal.* **44,** 136–139.

Nachmias, J. (1959). Two-dimensional motion of the retinal image during monocular fixation. *J. opt. Soc. Am.* **49,** 901–908.

Nemet, P. & Miller, J. E. (1968). Evoked potentials in cat extraocular muscle. *Invest. Ophthal.* **7,** 592–598.

Page, S. G. (1965). A comparison of the fine structures of frog slow and twitch fibres. *J. Cell Biol.* **26,** 477–497.

Peachey, L. D. (1968). Muscle. *Ann. Rev. Physiol.* **30,** 401–429.

Peachey, L. D. & Huxley, A. F. (1962). Structural identification of twitch and slow striated muscle fibers of the frog. *J. Cell Biol.* **13,** 177–180.

Pilar, G. (1967). Further study of the electrical and mechanical responses of slow fibers in cat extraocular muscles. *J. gen. Physiol.* **50,** 2289–2300.

Quereau, J. V. D. (1954). Some aspects of torsion. *Arch. Ophthal.* **51,** 783–788.

Quereau, J. V. D. (1955). Rolling of the eye around its visual axis during normal ocular movements. *Arch. Ophthal.* **53,** 807–810.

Rashbass, C. (1960). New method for recording eye movements. *J. opt. Soc. Am.* **50,** 642–644.

Rashbass, C. (1961). The relationship between saccadic and smooth tracking eye movements. *J. Physiol.* **159,** 326–338.

Rashbass, C. & Westheimer, G. (1961a). Disjunctive eye movements. *J. Physiol.* **159,** 339–360.

Rashbass, C. & Westheimer, G. (1961b). Independence of conjugate and disjunctive eye movements. *J. Physiol.* **159,** 361–364.

Ratliff, F. & Riggs, L. A. (1950). Involuntary motions of the eye during monocular fixation. *J. exp. Psychol.* **40,** 687–701.

Reinecke, R. D. & Poser, C. M. (1961). Subjective evaluation of ocular proprioception monitored with the use of the electro-oculograms. *Arch. Ophthal.* **66,** 130–132.

Riggs, L. A. & Niehl, E. W. (1960). Eye movements recorded during convergence and divergence. *J. opt. Soc. Am.* **50,** 913–920.

Robinson, D. A. (1964). The mechanics of human saccadic eye movement. *J. Physiol.* **174,** 245–264.

Robinson, D. A. (1965). The mechanics of human smooth pursuit eye movement. *J. Physiol.* **180,** 569–591.

Robinson, D. A. (1966). The mechanics of human vergence movements. *J. Pediat. Ophthal.* **3,** 31–37.

Robinson, D. A. (1970). Oculomotor unit behaviour in the monkey. *J. Neurophysiol.* **33,** 393–404.

Rogers, K. T. (1957). Ocular proprioceptive neurons in the developing chick. *J. comp. Neurol.* **107,** 427–437.

Sanghvi, I. S. & Smith, C. M. (1969). Characterization of stimulation of mammalian extraocular muscles by cholinomimetics. *J. Pharmacol.* **167,** 351–364.

Sears, M. L., Teasdall, R. D. & Stone, H. H. (1959). Stretch effects in human extraocular muscle. *Bull. Johns Hopk. Hosp.* **104,** 174–178.

Shackel, B. (1960). Pilot study in electro-oculography. *Br. J. Ophthal.* **44,** 89, 337, 608.

St. Cyr. G. T. & Fender, D. H. (1969). The interplay of drifts and flicks in binocular fixation. *Vision Res.* **9,** 245–265.

Steinman, R. M., Cunitz, R. J., Timberlake, G. T. & Herman, M. (1967). Voluntary control of microsaccades during maintained monocular fixation. *Science* **155,** 1577–1579.

Tamler, E., Jampolsky, A. & Marg, E. (1958). An electromyographic study of asymmetric convergence. *Am. J. Ophthal.* **46,** pt. 2, 174–182.

Tamler, E., Jampolsky, A. & Marg, E. (1959). Electromyographic study of following movements of the eye between tertiary positions. *Arch. Ophthal.* **62,** 804–809.

Torre, M. (1953). Nombre et dimensions des unités motrices dans les muscles extrinsiques de l'oeil et, en général, dans les muscles squelettiques. *Schweiz. Arch. Neurol.* **72,** 362–378.

VOLKMANN, F. C. (1962). Vision during voluntary saccadic eye movements. *J. opt. Soc. Am.* **52**, 571–578.

WESTHEIMER, G. (1954). Mechanism of saccadic eye movements. *Arch. Ophthal.* **52**, 710–724.

WESTHEIMER, G. (1954). Eye movement responses to a horizontally moving visual stimulus. *Arch. Ophthal.* **52**, 932–941.

WESTHEIMER, G. & MITCHELL, A. M. (1956). Eye movement responses to convergence stimuli. *Arch. Ophthal.* **55**, 848–856.

WHITTERIDGE, D. (1955). A separate afferent nerve supply from the extraocular muscles of goats. *Quart. J. exp. Physiol.* **40**, 331–336.

WHITTERIDGE, D. (1959). The effect of stimulation of intrafusal muscle fibres on sensitivity to stretch of extraocular muscle spindles. *Quart. J. exp. Physiol.* **44**, 385–393.

WOLTER, J. R. (1952). Uber Nervenendigungen in der äusseren Augenmuskulature. *Acta neuroveg.* **4**, 343–353.

WOLTER, J. R. (1955). Morphology of the sensory nerve apparatus in striated muscle of the human eye. *Arch. Ophthal.* **53**, 201–207.

WURTZ, R. H. (1969). Comparison of effects of eye movements and stimulus movements on striate cortex neurons of the monkey. *J. Neurophysiol.* **32**, 987–994.

WURTZ, R. H. (1969). Response of striate cortex neurons to stimuli during rapid eye movements in the monkey. *J. Neurophysiol.* **32**, 975–986.

YOUNG, L. R. & STARK, L. (1963). Variable feedback experiments testing a sampled data model for eye tracking movements. *Int. Elect. Electronics Engineers. Transactions on Human Factors in Electronics.* HFE-4, 38–51.

ZUBER, B. L., STARK, L. & COOK, G. (1965). Microsaccades and the velocity-amplitude relationship of saccadic movements. *Science* **150**, 1459–1460.

ZUBER, B. L., STARK, L. & LORBER, M. (1966). Saccadic suppression of the pupillary light reflex. *Exp. Neurol.* **14**, 351–370.

PART 2

NERVOUS CONTROL OF THE EYE MOVEMENTS

General Considerations

The activities of the muscles of the neck and trunk are ultimately determined by the discharges in the motor neurones of the ventral horns of the spinal cord, which are described as the *final common path*. These neurones are acted upon by sensory neurones, and by interneurones localized in all levels of the central nervous system, from the various segments of the cord to the cerebral cortex. Thus the activity of any given muscle is determined by nervous activity in practically the whole of the central nervous system; nevertheless the activity of an organism may be resolved into certain stereotyped patterns of behaviour— the *reflexes*—which may be evoked independently of many regions of the central nervous system; thus the flexor reflex may operate in the spinal animal, *i.e.* the animal whose spinal cord has been severed from the brain; whilst a large variety of reflex responses may be achieved in the decerebrate animal, the animal whose cerebral hemispheres and basal ganglia have been removed. In general, we may say that these patterns of behaviour may be controlled and modified by activity in brain centres that are located higher in the system than those required for the performance of the reflex activity. This is manifest especially in the voluntary control of muscular activity mediated by the frontal motor cortex and along the pyramidal tracts.

The movements of the eyes form no exception to these general principles; well established reflex patterns of behaviour are present and these may be largely controlled, or at any rate influenced, by voluntary activity initiated, it is presumed, also in the frontal cortex. The final common path is given by motor neurones in the brain, rather than the spinal cord, whilst higher coordinating centres are found in the *superior colliculi* and in the *occipital cortex*. These reflex centres are well established; somewhat more problematical are certain regions of the brain stem, notably the tegmentum of the midbrain, pons and medulla, stimulation of which electrically gives rise to well co-ordinated patterns of eye movement.

In this short, and necessarily somewhat elementary, discussion of the nervous control of the eye movements we shall be concerned with the reflexes and their nervous pathways—the *reflex arcs*. It must be strongly emphasized, however, that most reflexes are only exhibited reproducibly and in the "pure" form under artificial conditions, for example, in the decerebrate animal; any actual movement in the normal individual represents the final outcome of the antagonistic interplay of numerous reflexes which, in their turn, may be modified to a greater or less extent by the voluntary centres. For example, we shall see that a bright light in the periphery of the visual field evokes a *fixation reflex* whereby the eyes are turned towards the light so as to bring its image on the fovea; this does not mean, however, that the eyes are invariably turned to any light in the peripheral visual field. The reflex, defined as the response of the individual to a stimulus, must be considered, especially in man, as a component in behaviour; under special conditions it may be elicited with the utmost regularity, whilst in the normal intact animal the response to the same stimulus may be greatly different according to the conditions prevailing.

THE FINAL COMMON PATH

The motor neurones activating the eye muscles are derived from the cranial nuclei as follows:

N III. (*Oculomotor nerve*). Superior rectus. Inferior rectus. Medial rectus. Inferior oblique.

N IV. (*Trochlear*). Superior oblique.

N VI. (*Abducens*). Lateral rectus.

These nerves represent essentially the axons of cell bodies collected in the grey matter of the mid-brain and pons—the *eye-muscle nuclei*.

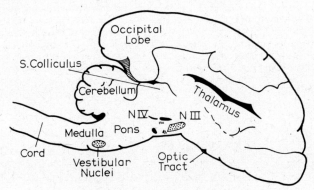

FIG. 218. Longitudinal section of the cat's brain.

The Nucleus of Nerve III. This lies in the tegmentum in the central grey matter in the upper part of the mid-brain below the superior colliculus, beneath and running parallel to the aqueduct of Sylvius (Fig. 218). It is some 5 to 6 mm long and extends between the third and fourth ventricles. It has been conventionally divided into a series of discrete zones each concerned with a pair of muscles; thus according to Brouwer it is divided primarily into a principal (lateral) nucleus

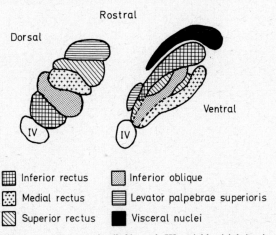

FIG. 219. Illustrating Brouwer's (left) and Warwick's (right) views of the organization of the oculomotor nucleus. (After Warwick, *J. comp. Neurol.*)

providing the motor neurones to the levator palpebrae and four extraocular muscles; the *median nucleus of Perlia*, considered to be a coordinating centre for activating the medial recti during convergence; and the *Edinger-Westphal nucleus*, containing the autonomic motor neurones controlling the ciliary muscle and sphincter pupillae. The principal lateral nucleus was divided into zones from above downwards (rostrocaudally) each zone representing the motor neurones for the levator, superior rectus, medial rectus, inferior oblique and inferior rectus in this order. Subsequent work, notably that of Warwick (1953), has shown that the organization is not so simple. Thus, although the individual eye muscles are indeed operated by localized groups of neurones within the principal nucleus, the simple rostro-caudal arrangement postulated by Brouwer apparently does not pertain. Instead, the arrangement is that shown in Fig. 219 (*right*); it will be seen that the inferior rectus, inferior oblique and medial rectus possess discrete motor pools arranged along the axis of the somatic nucleus in this *dorsoventral* order. The superior rectus, on the other hand, is innervated by neurones scattered along the medial aspect of the caudal two thirds of the main nucleus. Moreover, Warwick (1955) showed that in monkey and man, the two species in which convergence of the eyes is a prominent feature of the movements, the median nucleus of Perlia is difficult to find, and, when present, its neurones are not internuncial, as one would expect of a coordinating centre, but rather they are motor neurones supplying the superior rectus and

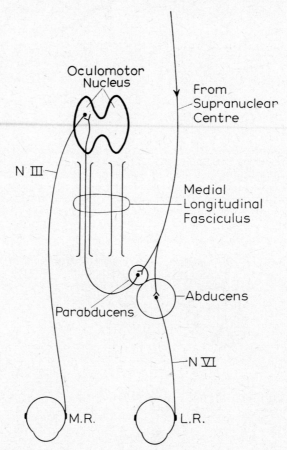

Fig. 220. Illustrating the conjugate innervation of the lateral rectus of one side and the medial rectus of the other.

inferior oblique. Hence we must assume that the act of convergence is organized at a higher level. The Edinger-Westphal nucleus is well established as the autonomic division of the oculomotor nucleus.

The Nucleus of Nerve IV. This is a small clump of cells lying caudally to the nucleus of N III, being situated in the central grey matter in the floor of the aqueduct of Sylvius and at the level of the lower part of the inferior colliculus. The nerve fibres, which represent the motor pathway for the superior oblique, undergo an almost complete decussation in the anterior medullary velum, emerging on the dorsal aspect of the brain stem.

The Nucleus of Nerve VI. This nucleus lies in the pons, in the floor of the fourth ventricle.

Medial Longitudinal Fasciculus. A conjugate lateral movement requires a co-ordinated response on the part of the medial rectus of one eye, innervated by N III, and the lateral rectus of the other, innervated by N VI. This co-ordination is brought about by interneurones running from the nucleus of N VI or a nearby region—the so-called *parabducens nucleus* of Crosby. These interneurones run to the nucleus of N III in the medial longitudinal fasciculus (also called posterior longitudinal bundle), a tract associated with the brain-stem reticular formation which connects the three motor nuclei bilaterally. Thus the initiation of a conjugate lateral movement may be illustrated by Fig. 220, the primary stimulus reaching the parabducens and abducens nuclei; an interneurone in the latter passes up in the median longitudinal fasciculus to activate a motor neurone in N III which passes to the medial rectus.*

Lesions in the median longitudinal fasciculus may well be responsible for the so-called internuclear ophthalmoplegia; the loss of the power to adduct the eye in conjugate lateral movements with its retention in convergence. Certainly this syndrome may be evoked by bilateral section of the fasciculus (Hyde & Slusher).†

CO-ORDINATING CENTRES

So much for the nuclei of the nerves controlling the eye movements; these nuclei may possibly be regarded as "centres" integrating, in a limited way, the activities of the individual muscles so that a given movement is carried out smoothly and efficiently, but it is probably safer to regard them as the final common path for the necessary movements elaborated at higher levels of the nervous system. Certainly the intimate connection of the eye movements with bodily activity, and the dependence of the eye movements on visual impressions, demand centres at a higher level capable of integrating these associated activities on a wider scale. Centres that are considered to be important for these integrative activities are the superior colliculi, certain regions of the brainstem tegmentum, the occipital cortex and the frontal cortex.‡

* Reciprocal innervation demands, of course, that inhibitory impulses pass to the antagonists; thus the parabducens nucleus would have to send inhibitory impulses to N III of the same side besides excitatory impulses to N III of the opposite side. We may note that opinions differ as to whether the fibres related to conjugate deviation cross at once to ascend in the contralateral median longitudinal fasciculus, as indicated in the diagram, or whether they ascend on the same side as the deviation and decussate within the nucleus of N III (see Crosby, 1953, p. 440).

† A possible autonomic control of the extraocular musculature has been discussed by Eakins & Katz (1967); they found that stimulation of the cervical sympathetic, or injection of epinephrine, increased the tension of the superior rectus, an action that was antagonized by α-blockers such as phentolamine. The results were similar to the effects on the nictitating membrane, except that atropine was without effect on the rectus muscle.

‡ The literature contains frequent allusions to *gaze centres*, the one controlling lateral movements and corresponding to the so-called *parabducens nucleus*, the other, near the superior colliculi, supposed to control vertical movements. The parabducens nucleus is best regarded as a relay-station on the motor pathway in conjugate lateral movements, and not as a co-ordinating centre. As Bender & Shanzer (1964) point out, it is too vaguely

Superior Colliculi

The functions of these two bodies, often called the *optic tectum*, are not completely known, especially in man, but there is little doubt that they constitute an important centre in the organization of movement. This is clear from their anatomical connections since they receive afferent impulses from the medial and trigeminal lemnisci, whilst the bulk of their efferent impulses pass to the eye-muscle nuclei by way of the fountain decussation of Meynert and the medial longitudinal fasciculus and to the motor nuclei generally by way of the tectobulbar and tectospinal tracts. In lower animals, such as the cat, moreover, we may conclude that the colliculi are centres for *visually directed movements*. The anatomical basis for this is provided by fibres from the optic tracts that terminate directly in the colliculi, which are thus called *primary optical centres*; in general, the higher the animal phylogenetically the larger the proportion that pass to the lateral geniculate body and thence to the cerebral cortex. In man and the higher apes, therefore, only very few fibres terminate in the colliculi.

Response to Stimulation. Electrical stimulation of the superior colliculi with electrodes implanted in the waking and freely moving cat by Hess' technique causes a movement of the head towards the opposite side; a careful study of the eyes shows that the movement of the head is preceded by a conjugate movement of the eyes (Burgi). Usually there is an upward or downward component in the movement, according to the part of the colliculus stimulated; and in the rostro-median region a pure upward movement can be obtained. In the anaesthetized cat, stimulation of the retina by light does not evoke any eye movements; if, however, the excitability of the superior colliculus is enhanced locally by the application of a crystal of strychnine, then light causes a prompt movement of the eyes directed to a definite spot in space characteristic for the particular region of the colliculus that has been made thus sensitive. By systematically changing the position of the strychninized area, a map of the surface of the colliculus may be plotted indicating regions of the visual field towards which the eyes will be directed when a given point is rendered hypersensitive. Presumably there is a reflex path from the retina through the colliculus that determines that the eyes will move towards the point in the visual field that stimulates the retina; in the anaesthetized animal this reflex is inhibited, but treatment with strychnine removes the inhibition and reveals the reflex (Apter).

Probable Roles. Thus in lower animals we may regard the colliculi as centres that bring about reflex movements of the eyes in response to a visual stimulus. In man and apes the number of visual fibres passing to the colliculi is small, and it is unlikely that these mediate any reflex activity since, in man at any rate, eye movements in response to visual stimuli cannot be evoked in the absence of visual cortex. However, as the number of visual fibres terminating in the

---

defined anatomically and physiologically to be treated as a centre controlling lateral eye movements. The movements evoked by stimulation of the colliculi usually have a strong vertical component and it is doubtless for this reason that the hypothetical gaze centre for vertical movements has been sited near these bodies. According to Szentagothai, the interstitial nucleus of Cajal may be regarded as a vertical gaze centre because of its connections with the eye-muscle nuclei concerned with elevation and depression, whilst a region $\frac{1}{2}$ mm lateral to this nucleus he regards as a horizontal gaze centre by virtue of its connections with the abducens nucleus and that part of the oculomotor nucleus concerned with the medial rectus. Once again, however, Bender & Shanzer (1964) incline to reject these areas as specific centres.

colliculi decreases, there is a corresponding increase in the number of fibres running from the occipital cortex to the colliculi, so that on phylogenetic grounds we may assume that the colliculi retain their function as stations for executing eye movements, but they become subordinated to the occipital cortex (Crosby & Henderson, 1948). According to Pasik, Pasik & Bender (1966), however, the absence of any permanent oculo-motor deficit in monkeys after ablation of the superior colliculi rules these bodies out as subsidiary motor centres, and they consider that their function is to integrate sensory modalities since afferents from almost every sensory system converge here. Let us consider, now, the regions of the cerebral cortex that are concerned with eye movements.

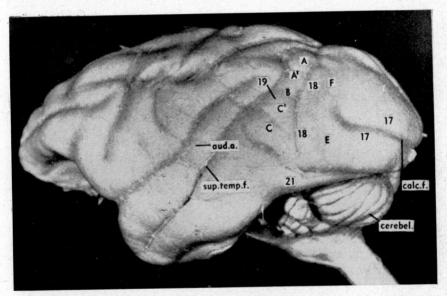

Fig. 221. A photograph of the left side of the brain of *Macaca mulatta* ($\times$ 1·8). On area 19 the various points from which eye movements were elicited are indicated by letters. Stimulation of A produced conjugate upward deviation of the eyes, of A′ conjugate deviation obliquely upward toward the right, of B conjugate horizontal deviation toward the right, of C′ conjugate deviation obliquely downward to the right, and of C, conjugate downward deviation. Stimulation in the region of E (area 17) caused oblique conjugate deviation of the eyes upward and toward the right, and in the region of F, conjugate deviation obliquely downward to the right. (Crosby & Henderson, *J. comp. Neurol.*)

Cortical Centres

Occipital Motor Centre. The occipital lobe is differentiated into three areas on histological grounds; area 17, the *striate area*, is where the visual fibres terminate, whilst areas 18 and 19, the *peristriate* and *parastriate areas*, are intimately associated with area 17, neurones passing from area 17 to area 18, and from area 18 to area 19 and back to area 17. The classical investigators of the cerebral cortex by electrical stimulation observed contralaterally directed conjugate movements of the eyes in response to stimulation of area 17 in primates, the movements being apparently directed to the parts of the visual field that are projected on to the cortex. More recent work has largely confirmed this; Fig. 221 illustrates the regions of the occipital cortex of the macaque that gave movements in the experiments of Crosby & Henderson, whilst Fig. 222 illustrates the results obtained by Wagman, Krieger & Bender. According to these last workers, the occipital cortex may be divided into four quadrants by two planes; a midsagittal vertical plane dividing the cortex into a right and left half, and a roughly horizontal plane through the calcarine fissure. Two of these quadrants are illustrated in Fig. 222. The movements of the eyes were such as to direct the gaze away from the quadrant stimulated.* In man, Foerster obtained eye movements by stimulating area 19. Wagman *et al.* raise the point as to whether the movement of the eyes is the result of stimulating a sensory region, evoking a sensation of light to which the animal responds, or whether the stimulus does, indeed, activate a motor centre. This question is probably somewhat academic since the modern tendency is to treat many areas of the brain as both sensory and motor. The finding that certain occipital lesions may impair motor activity without impairing vision (Bender, Postel & Krieger, 1957), and that in human subjects eye movements that are not always associated with visual sensation may be evoked by electrical stimulation, suggests that there are, indeed, motor regions in the occipital cortex (Bickford *et al.*, 1953).

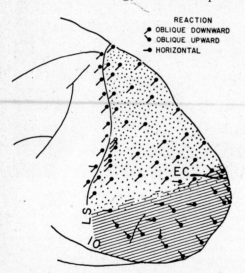

REACTION
OBLIQUE DOWNWARD
OBLIQUE UPWARD
HORIZONTAL

FIG. 222. Contralateral conjugate eye movements elicited upon stimulation of the dorsal, lateral and lateral inferior surfaces of the occipital lobe of *Macaca mulatta*. The solid circles indicate the points that were actually stimulated, whilst the lines originating at these circles indicate the direction of the contralateral eye movement in response to this stimulation. The dots outline the regions from which obliquely downward and contralateral deviations were obtained, whilst the shaded area indicates the region from which obliquely upward and contralateral deviations of the eyes were obtained. (Wagman, Krieger & Bender, *J. comp. Neurol.*)

Frontal Motor Centre. In man and other primates there is a region in the frontal cortex, stimulation of which gives rise to movements of the eyes. In man the region is close to the precentral gyrus, stimulation of which gives rise to discrete movements of the parts of the body as indicated in Fig. 223 (area 8 $\alpha\beta\delta$); in apes the region lies farther forward. Movements of the eyes following stimulation of this area in the chimpanzee were described by Grünbaum & Sherrington in 1903, and similar movements in man by Foerster and Penfield & Boldrey. In the monkey, Graham Brown described movements of

---

* Crosby & Henderson found two separate areas in Area 19, an upper one giving upward movements and and a lower one giving downward movements, but Wagman *et al.* (1958) were unable to confirm this, the upper area giving downward movement as with the rest of the occipital lobe.

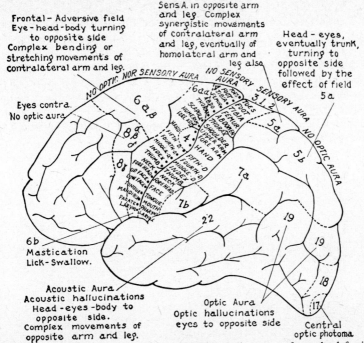

FIG. 223. Diagram of the human cerebral hemisphere seen from the left side illustrating the areas, electrical stimulation of which is followed by movement or sensation. The numbers of the areas are those of Vogt. (Herrick, *Introduction to Neurology*.)

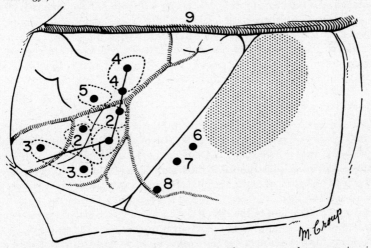

FIG. 224a. Drawing of left human fronto-parietal area exposed at operation in which cortical stimulation resulted in movements at the following points: 1, conjugate deviation of the eyes upward and to the other side; 2 horizontal conjugate gaze to the other side; 3, divergence of the eyes; 4, conjugate deviation of the eyes downward and toward the other side; 5, no response; 6, hand movement and flexion of the fingers on the right side; 7, flexion of the right thumb; 8, retraction of the right lower lip; 9 indicates longitudinal sinus. (Lemmen, Davis & Radnor, *J. comp. Neurol.*)

two kinds; in the first place, stimulation of the middle frontal convolution caused a conjugate movement of the eyes to the opposite side, the head remaining still; in the second place, stimulation of the superior frontal convolution caused a turning of the head, the eyes remaining fixed in their sockets. In both cases, therefore, the animal "looked" to the opposite side, but in the former instance the eyes were employed whilst in the latter the muscles of the neck. Interestingly, if the head was prevented from turning, then the eyes moved. More recently the frontal centre has been studied by Crosby in the macaque and Lemmen, Davis & Radnor in man; comparison of Figs. 224a and b shows that the results are essentially similar.

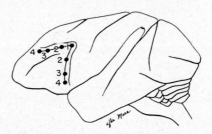

Fig. 224b. Diagram of the left hemisphere of the macaque showing eye deviations from stimulation of the frontal eye fields at the following points: 1, eyes up to the right; 2, deviation to right; 3, divergence; 4, eyes down to right. (After Crosby.)

Finally, we may mention an interesting study of Bizzi (1967) on the unanaesthetized monkey; by inserting an electrode into area 8, records could be obtained from single cortical cells that altered their discharge rate when the eyes moved, the final rate of discharge being a function of the direction of gaze; the interesting feature of these cells was that the change of discharge only occurred after the movement had been initiated, so that Bizzi postulated that this area of the cortex was only concerned with the tonic control of eye position after a movement had been made. He points out that, in the monkey, ablation of area 8 only leads to paralysis of eye movements for a few days (Pasik, & Pasik, 1964).*

VOLUNTARY CENTRE

It is considered that in man, at any rate, the frontal region is the centre for voluntary eye movements in contrast to the occipital regions which are thought to be concerned with purely reflex movements of the eyes in response to a visual stimulus. Thus a stimulus applied to the frontal area is said to be prepotent over a simultaneous one applied to the occipital area; e.g. stimulation of the left occipital lobe causes a conjugate movement to the right; stimulation of the right frontal centre causes a deviation to the left. If both areas are stimulated simultaneously, the eyes move to the left. Surgical removal of the "ocular responsive cortex" in man leads to the failure of the individual to direct his eyes voluntarily to the side opposite the lesion. In monkeys and apes stimulation of the ocular responsive cortex causes head movements, opening of the palpebral fissures, and dilatation

---

* In a later study (1968) these cortical neurones were found to be of two types: Type I, which fired during voluntary saccades or the quick phase of nystagmus in a given direction, whilst they were silent during the slow phase; and Type II, which discharged during fixation of gaze in a specific direction as well as during smooth pursuit movements.

of the pupils, besides eye movements. In general, as with man, the lateral movements are most frequent; if, however, the medial and lateral recti are cut, cortical stimulation leads to upward and downward movements. Simultaneous stimulation of both sides of the cortex can lead to the fixation of the gaze directly forward. As indicated above, however, it may be that area 8 in monkeys plays a more subsidiary role.

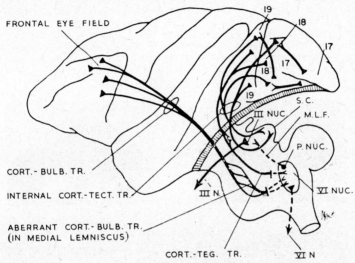

FIG. 225. Pathways for the cortical control of the eye movements. (Slightly modified from Crosby, *J. comp. Neurol.*)

Efferent Pathways. The pathways by which cortical impulses determining eye movements reach the eye-muscle motor neurones are not completely worked out. The pathway from the frontal centre is independent of the occipital centre, since ablation of the latter does not abolish voluntary eye movements. Fig. 225 illustrates Crosby's views. According to this, the pathway from the frontal centre is by way of the corticobulbar tract, i.e. the tract that carries the voluntary impulses to the musculature generally and to which the name *pyramidal* is given in its course through pons, medulla and cord. Thus the fibres run in the internal capsule and cerebral peduncle; at about the midbrain level, fibres to the nucleus of N VI leave the main tract to enter the medial lemniscus, descending in this to the abducens and parabducens nuclei; the remaining eye-muscle nuclei are thought to be indirectly activated by way of the median longitudinal fasciculus. The occipital centres in areas 17, 18 and 19 are thought to activate the abducens and parabducens nuclei indirectly through the colliculi by way of the internal cortico-tectal tract and directly through the corticotegmental tract. The reason for imputing this double pathway is the old observation of Graham Brown that damaging the colliculi left eye movements intact but, as Whitteridge (1960) has pointed out, this evidence is not conclusive since the frontal pathway is known to be independent of the colliculi, and the eye movements observed might have been mediated by the frontal centre.*

* We may note that the eye movements evoked by occipital cortex and superior colliculus may be affected differently by lesions in the brain stem, once again indicating separate pathways (Hyde, 1962) furthermore, Hyde & Slusher found that discrete lesions in the median longitudinal fasciculus could abolish adductions in response to some forms of stimulation but not to others. This looks as though the concept of a parabducens centre, controlling lateral movements through a final common pathway up the median longitudinal fasciculus, postulated by Crosby, is unsound.

Brain Stem Stimulation. A systematic exploration of the brain stem in anaesthetized *encéphale isolé* cats* by Hyde has revealed a number of regions, in particular the tegmentum of the midbrain, pons and medulla, from which conjugate movements may be evoked on electrical stimulation. It is unwise to describe these regions as "centres" controlling the execution of eye movements, however, since they may simply contain the pathways from the cortex to the eye-muscle nuclei. Nevertheless it must be appreciated that, in lower animals at any rate, there are regions of the brain stem in which specific types of movement are organized; these are called by Jung & Hassler *direction-specific movements*, and they cause the animal to turn in a characteristic direction. Thus, stimulation of the interstitial nucleus of Cajal, which is close to the nucleus of N III, causes a rotation of the head about its longitudinal axis as illustrated by Fig. 226. In higher animals, such as the apes and man, turning of the eyes rather than of the head is the more prominent feature of a change of gaze, and it would not be surprising if direction-specific movements of the eyes were organized at midbrain levels.

FIG. 226. Direction-specific types of movement resulting from electrical stimulation in the region of the interstitial nucleus of Cajal by means of electrodes implanted in the otherwise intact brain. (Hassler & Hess, *Arch. Psychiat.*)

THE REFLEXES

Psycho-Optical Reflexes

The most important group of reflexes involving eye movements in man are those in which the incidence of light on the retina acts as the primary stimulus; it is as a result of these reflexes that the eyes are (*a*) directed to objects of interest and (*b*) maintained in the necessary orientation to ensure that the images formed by the two eyes fall on the foveae. We are accustomed to think of reflexes as being executed without the intervention of awareness on the part of the individual; these visual reactions, however, definitely involve awareness and are therefore classed as *psycho-optical reflexes*; as we shall see, there can be no doubt of their essentially involuntary nature.

The Fixation Reflex. All psycho-optical reflexes may be regarded as manifestations of the more general *fixation reflex*—the reflex which brings the images of an object on to the foveae of the two eyes and maintains them there. The

* This preparation is obtained by a section of the brain stem at about the level of C1, leaving intact the nucleus of N V; in this condition the animal has many of the features of the waking state by contrast to the *cerveau isolé* preparation, where the section is higher, at the collicular level.

fixation reflex develops early, since an infant a few days after birth will fixate a bright light; binocular fixation, however, does not develop for some months.

The reflex exhibits itself in two aspects—(a) the response to a peripheral stimulus and (b) the maintenance of fixation; the mechanism involved in the two processes is nevertheless the same. Thus let us imagine that the eyes are stimulated by a bright object in the peripheral field. Attention is aroused and the eyes move so that the images of the object approach the foveae—suppose that the nervous response grows in intensity as the images approach closer and closer to the foveae and reaches a maximum when they fall actually on them. The eye muscles can then be considered to behave as a tuning system, causing movements of the eyes until a maximum nervous response is obtained, just as the knob of a tuning dial is turned to obtain maximum sound from a wireless set. Fixation is thus achieved when maximum nervous discharge is reached. If the head moves slightly, the images are displaced and the nervous response falls off; immediately the tuning mechanism comes into play and the eyes are readjusted to maintain perfect fixation. Thus the same principle, the movement of the eyes to maintain a maximum nervous response, operates both to initiate and to maintain fixation. This principle may also explain the neural basis for the frequent small movements of the eyes that take place even during attempts at steady fixation (p. 343). As studies on the stabilized retinal images have shown (p. 252), the perceived image becomes rapidly indistinct in the absence of a shift of this image on the retina, so that the nervous discharge, postulated above, tends to decrease as the eye remains stationary. The eyes presumably respond to this falling off with a movement which shifts the images on the retinae sufficiently to excite a new set of cones and the discharge again becomes maximal. The constant, small movements of the eyes in fixation may thus be regarded as the restless groping of the oculo-motor centres for a position that gives maximum nervous discharge. *Ocular nystagmus*—a pendular to-and-fro movement of the eyes—may be regarded as an exaggeration of this groping; if fixation is difficult, as in amblyopia, the eyes move excessively in a vain attempt to find a clear image. The nystagmus of miners working for long periods in the dark is a complex syndrome; nevertheless an element in the aetiology must be the necessity for using peripheral vision; under these conditions fixation on the fovea reduces the visual acuity rather than enhances it, so that the "best position" for the retinal image is much more vaguely defined and this might well result in a ceaseless search for it over relatively large areas of retina to give a pendular type of nystagmus. It must be appreciated, however, that this point of view represents a crude simplification of what must be a more complex process; the factors of attention and meaning cannot be ignored; for example, corrective fusion movements, to be described below, are brought about as a result of a changed orientation of the retinal images, e.g. when one eye rolls, and the intensity of the "nervous discharge" remains unaltered.

Other manifestations of the fixation reflex are given by:

(i) the following movements in *optokinetic nystagmus*;
(ii) the *corrective fusion movements* made in the interests of single vision;
(iii) the movements of the eyes concerned with the *visual righting reflexes*.

Optokinetic Nystagmus. This describes the movements of the eyes when the body is in motion and the subject is gazing at the environment, as when he looks out of a railway carriage (train-nystagmus) or alternatively when the subject is still and his environment, or a prominent part of it, moves around him, as in the clinical test when a rotating striped drum is gazed at steadily. Here the stimulus is the movement of the image of the fixation point on the retina, the eye responding, during the slow phase of the nystagmus, by movements designed to keep the image on the fovea. At a certain moment, fixation of this point ceases and the eyes make a rapid saccade in the opposite direction, to fixate a new stripe on the drum, and the slow movement begins again. The slow movement is obviously a following movement, but the rapid movement is difficult to categorize. It was originally thought that it occurred simply because the eyes were unable to maintain fixation on the fixated stripe, i.e. because the stripe had moved out of the field of binocular fixation. In fact, however, the rapid movement usually takes place long before this limit has been reached; moreover, it is not apparently due to a deliberate attempt to fixate a new stripe appearing in the periphery since optokinetic nystagmus occurs in homonymous hemianopia when, because of a lesion in the visual pathway (p. 482), the eye is blind to half the visual field (Fox & Holmes; Pasik, Pasik & Krieger). Thus the eye makes a rapid movement into its blind side. We must conclude, therefore, that the unusual stimulus-condition, namely the apparent rotation of the visual environment, activates a *nystagmus centre* that triggers off these involuntary rhythmical movements (Rademaker & Ter Braak).* The involuntary nature of this movement in man is best revealed when he is surrounded by the moving drum so that practically the whole visual field moves; in this case it is impossible to avoid the nystagmus so long as the eyes are open. When only a small part of the visual field moves, as when the subject looks at a small striped drum, then the subject may avoid the nystagmus by "looking through" the drum, i.e. by refusing to fixate any given stripe (Carmichael, Dix & Hallpike).

Cortical and Subcortical Nystagmus

In animals, such as the rabbit and dog, an optokinetic nystagmus may be evoked even in the absence of cerebral hemispheres, and for this reason it is called *subcortical*. Visual reactions in man, in the absence of a cortex, are, as we have seen, impossible, presumably because of the relative insignificance of the visual projection to the colliculi. It is unwise, therefore, to transpose studies on lower animals to man, and certainly unwise to speak of a *subcortical nystagmus* as such, although the presence of a nystagmus-centre in the lower parts of the brain is not an unreasonable hypothesis. In man this would be activated by the cortical visual stimuli, but in lower animals the colliculi might suffice. Under pathological conditions in man, involving lesions in the cerebral hemispheres, optokinetic nystagmus may be impaired, frequently with the appearance of a *directional preponderance*, in the sense that the reflex is normal when the drum moves one way, and reduced when it moves in the other. In the case of optokinetic nystagmus in man the preponderance is seen when the drum rotates away from the affected

---

* The *rhythm* of the nystagmus will be largely determined by the amplitude of the excursions for a given speed of rotation; the fact that this alters under a variety of experimental conditions is further proof that the initiation of the fast movement, which puts an end to the slow movement, is not due to the eyes reaching the maximal extent of the field of fixation. We may note that, according to Ter Braak, the eyes lag slightly behind the movement of the drum during the slow phase; consequently there is a slight drift of the retinal image over the retina.

side, so that the study of this nystagmus may be a valuable diagnostic tool (Fox & Holmes). According to Carmichael *et al.* the regions of the cortex influencing the optokinetic nystagmus are in the supramarginal and angular gyri of the temporal lobe, the right hemisphere, for example, facilitating nystagmus on moving the drum to the left, and inhibiting that on moving the drum to the right.* A schematic pathway for the nystagmus is illustrated in Fig. 227; it should be noted, however, that Pasik *et al.* (1959) do not agree with the location of the cortical centres in the supramarginal and angular gyri.

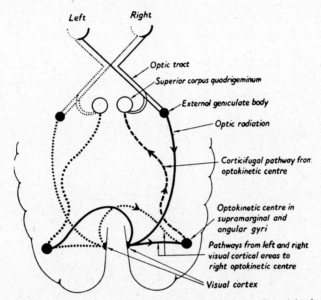

FIG. 227. Suggested schema of optokinetic mechanisms. (Carmichael, Dix & Hallpike, *Brit. med. Bull.*)

SINGLE MOVING POINT

We may note that when a single object moves across the stationary visual field, there is no optokinetic nystagmus unless the field is otherwise empty; thus according to Rademaker & Ter Braak, we may think of two opposing tendencies, the shifting of the images of stationary objects on the retina, tending to inhibit the following movement, and the shifting of the image of the single moving point that incites to movement. According to the relative preponderance of the moving or fixed field, one or other of the tendencies will determine the result, i.e. whether the eye remains fixed or follows the moving point. This was well demonstrated by Ter Braak who immobilized one eye of a rabbit and examined the movements of

* In the macaque Henderson & Crosby abolished optokinetic nystagmus by bilateral destruction of areas 18 and 19; unilateral destruction abolished the nystagmus on rotating the drum away from the injured side, i.e. there was a directional preponderance *towards* the injured side, by contrast with man where it is *away* from the lesion. The frontal eye fields (area 8) exerted an inhibitory action so that on ablation of these the nystagmus was more marked.

We may note the danger of confusion according as the direction of the nystagmus is described in terms of the quick component or the direction of rotation of the drum; thus when Carmichael, Dix & Hallpike (1954) say that nystagmus to the healthy side is suppressed, they are referring to the quick phase of the nystagmus, so that there is suppression of nystagmus when the drum rotates away from the healthy side, i.e. towards the lesion.

the other, which was blinded. Moving a single point across the visual field of the immobilized eye now caused the other eye to give a genuine nystagmus; because the stimulated eye was immobilized, there was no movement of images of stationary objects over its retina, and so the normal inhibition of the nystagmus was released. A similar experiment on a human subject, with one eye immobilized by paralysis of its extraocular muscles, is described by Mackensen (1958).

OPTOKINETIC AFTER-NYSTAGMUS

When a rotating drum is stopped after the subject viewing it has experienced a nystagmus for some time, there is an after-nystagmus lasting for two or three beats only; it has been studied in some detail by Mackensen (1959). It is best evoked if the eyes are presented with a uniform field, i.e. in the dark.

TORSIONAL NYSTAGMUS

Brecher (1934) induced an optokinetic torsional nystagmus by causing a subject to view a rotating sectored disc; the nystagmic jerks were of several degrees in amplitude.

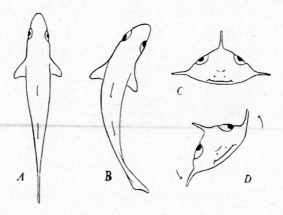

FIG. 228. Showing the rigid relationship between the positions of the eyes and the parts of the body in fishes. A and C illustrate the positions of the fish when at rest. B shows the compensatory movement of the eyes that occurs during swimming or on bending the tail into the position indicated. D illustrates the compensatory movement that occurs on rotating the fish about its longitudinal axis (Bard, *Macleod's Physiology in Modern Medicine*.)

Corrective Fusion Movements. The orientation of both eyes so as to produce images of the same object on corresponding parts of the two retinae requires continuous adjustments, which are brought about by this reflex. Here the effective stimulus is diplopia.

Visual Righting Reflexes. In man and the ape the eyes play a very important part in the maintenance of posture. It is quite clear that the movements of the eyes must be closely attuned to the posture of the animal; in lower forms this is achieved by a rigid association of the eye muscles with the vestibular apparatus so that the position of the eyes is completely determined by the orientation of the body in relation to gravity and the parts of the body in relation to each other (Fig. 228). Here, then, the labyrinth dominates eye movements. The visual impressions alone, however, can clearly provide information as to the orientation of the head, and therefore the body. Thus if a subject is falling forward, his original point of fixation moves upwards and, as a result of the fixation reflex, his eyes move upwards to maintain fixation; the necessary postural adjustments for recovery can be made in the light of impressions provided by the eyes and their adjustments. In this case proprioceptive impulses from the superior recti demand a movement of the head upwards so that the eyes can re-assume the primary position. The proprioceptive impulses from the neck muscles now demand postural changes that will permit the head to re-assume its normal erect position. In man and the ape posture can be maintained quite adequately by means of the visual righting reflexes when the labyrinths have been destroyed.

Vestibular and Proprioceptive Reflexes

The primary stimulus for these reflexes is not visual but arises from the vestibular apparatus or the proprioceptive stretch receptors in the neck muscles. They are classed as *static (or tonic) reflexes* and *kinetic reflexes*; by the first class a definite orientation of the eyes is evoked and maintained for a given position of the head in relation to the trunk and the pull of gravity, whilst by the latter a definite *movement* of the eyes follows a movement or acceleration of the head.

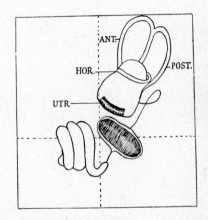

FIG. 229. The positions of the semicircular canals in the head viewed from the side. (After Quix.)

The Vestibular Apparatus. A brief description of the vestibular apparatus would be useful at this point. In man it consists essentially of three semicircular canals in mutually perpendicular planes, and two sacs, the *utricle* and *saccule* (Figs. 229 and 230). Each canal widens at one end into an *ampulla* containing a sense organ—the *crista ampullaris* or *crista acustica*, which bears as an auxiliary structure a jelly-like *cupula* which encloses the sense hairs protruding from the surface of the crista (Fig. 231). The utricle and saccule contain sensory organs called *maculae*, with a similar type of sense cell but with shorter hairs. The sense hairs of the macula are embedded in an otolith membrane which is encrusted with numerous concretions of lime crystals (*otoconia*) (Fig. 232). The membranous labyrinth, the cavities of which are filled with endolymph, is surrounded by cartilage or bone, the space between it and the surrounding tissues being filled with perilymph.

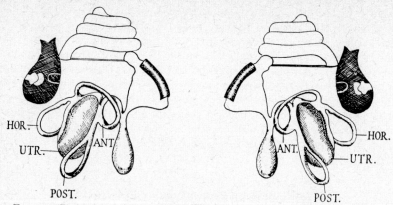

FIG. 230. Positions of the semicircular canals in the head viewed from above. (After Quix.)

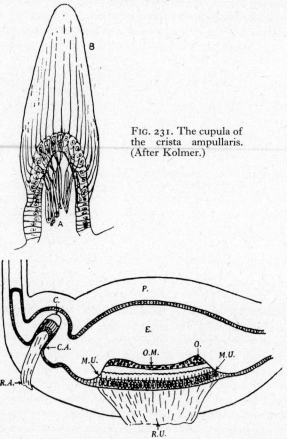

FIG. 231. The cupula of the crista ampullaris. (After Kolmer.)

FIG. 232. Section through the macula utriculi and the ampulla of a vertical semicircular canal in man. C, cupula; C.A., crista ampullaris; E, endolymph; M.U., macula utriculi; O, otoconia; O.M., otolith membrane; P, perilymph; R.A., ramus ampullaris; R.U., ramus utricularis of N VIII. (Lowenstein, after Kolmer, Biol. Rev.)

THE UTRICLE

The macula of the utricle is most probably stimulated by a gravitational pull on the otolith membrane which alters the tension in the hair cells; the macula lies normally in a horizontal plane, and under these conditions the tension is at a minimum; if the head is held upside down the stimulus is at a maximum. The utricle is thus a receptor which can respond to alterations in the orientation of the head in relation to the direction of gravity; it can therefore mediate the static class of reflex, i.e. compensating poses of the eyes, head, etc. It is worth noting, however, that a sudden linear acceleration of the head could exert an inertial pull on the hairs of the macula and give rise to a kinetic type of reflex response.

THE SEMICIRCULAR CANALS

The semicircular canals are receptor organs responding to acceleration of the head; opinion on the manner in which the crista ampullaris is stimulated by an acceleration has fluctuated, but the experiments of Steinhausen and Dohlman have now firmly established the original notion of Mach that it is the inertial movements of the endolymph that deform the cupula and stimulate the hair-fibres embedded in it. As Fig. 233 shows, when the head is accelerated the endolymph within the membranous canal tends to be "left behind" and so produces an inertial force that deflects the cupula, which moves in much the same way as a rotating door. When the acceleration ceases there is an inertial movement of fluid in the opposite direction.

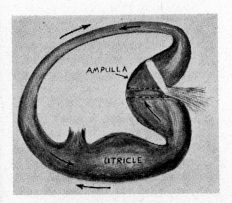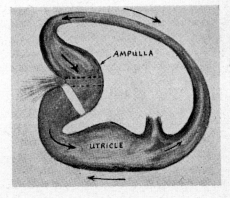

FIG. 233. Illustrating movements of the cupula in the ampullae of the horizontal canals on accelerating the head to the right. In the left canal the inertial flow of fluid is away from the ampulla, whilst in the right canal it is towards the ampulla.

The elements determining the dynamic behaviour of the canals are thus the moment of inertia of the ring of fluid in the canal, the viscous drag of the fluid, and the "spring-stiffness" of the cupula, which always returns to its neutral position in the absence of external forces. It is apparently the viscous drag that is the dominant force; the velocity of flow is proportional to the angular acceleration of the head, so that the position of the cupula, which is the integral of the fluid velocity, is proportional to the angular velocity of the head. Thus the sense organ behaves like an integrating accelerometer, with the result that its neural discharge indicates the angular velocity of the head. If the head movements are too short or too long the viscosity no longer predominates over the inertial or spring forces, and the canal ceases to integrate head acceleration properly (Robinson, 1968). Thus, if the head is accelerated to a definite angular speed and then this speed is maintained, as when a subject is rotated in a chair, at the end of the acceleration the deflexion of the cupula indicates the velocity attained, but now the spring-stiffness asserts itself and the cupula gradually returns to its original position, although the velocity remains the same.

The function of the macula of the saccule in man is unknown; all the evidence suggests that the utricle and semicircular canals are responsible for all the adapting reactions to acceleration and gravity.

Static Vestibular Reflexes. These reflexes, mediated by the utricle, serve to maintain a normally orientated field of vision despite changes in the position of the head. Thus when the body is tilted towards the prone position the eyes move upwards; when the body is inclined sideways, the eyes roll; it should be emphasized, however, that in man the rolling of the eyes by no means compensates entirely for the altered position of the head; psychological adjustments are necessary (p. 338). Movement of the body or the head about a vertical axis will clearly have no influence on the direction of gravitational pull on the otolith membrane, hence any compensatory movement of the eyes that takes place must be mediated by another type of receptor. It is here that the proprioceptive impulses from the neck muscles are involved.

PROPRIOCEPTIVE NECK REFLEXES

Bending the head forward and backwards towards or away from the chest, besides activating the utricle, will stretch the neck muscles, and it may be shown that the compensatory movements of the eyes that are elicited may be caused by both of these receptive mechanisms. Thus movement of the trunk, keeping the head fixed, gives the same reflex, but not by virtue of the stretched neck muscles. Lateral movements in response to moving the head about a vertical axis are mediated exclusively by the neck mechanism. Hence cutting the cord in the cervical region may reduce an animal's ability to maintain orientation (Cohen).

Kinetic Reflexes. Kinetic eye reflexes are compensatory effector movements called forth chiefly by angular acceleration of the head. As illustrated by Fig. 230, the canals are arranged in the head in three mutually perpendicular planes. We may distinguish on each side a horizontal or external canal, both lying in the same plane which, in effect, is not horizontal when the head is erect but slopes backwards at 30°, hence they only come into the horizontal plane when the head is tilted forward through this angle. Rotation of the head about a vertical axis has its main effect on this pair of canals, which we may describe as a *synergic pair*. When the head is rotated about its vertical axis, e.g. when the head looks to the right, the reflex movement of the eyes in the opposite direction that maintains fixation on a point that was originally straight ahead of the eyes will be mediated by the inertial stimuli in the horizontal canals. The anterior vertical canal of one side is in the same plane as the posterior vertical canal of the other, so that we may distinguish two more synergic pairs, namely left anterior-right posterior, and left posterior-right anterior. When the head is moved forwards and backwards, towards and away from the chest, the compensatory rapid movements of the eyes upwards and downwards are mediated by the synergic pair whose planes lie in a sagittal plane, whilst when the head is moved towards and away from the shoulder the other vertical pair, lying in the frontal plane, parallel to the forehead, will mediate the rolling movement that takes place.

ELECTROPHYSIOLOGY

Studies of the action potentials in fibres from the ampullae made by Lowenstein and his collaborators have shown that there is normally a tonic discharge from them all when the head is at rest; on accelerating the head about a vertical axis, for example, the discharge from the ampulla of one horizontal canal is increased, whilst that of the other is inhibited. The inhibited ampulla in this case is the one that "leads" the canal, i.e. on rotating the head to the right the left ampulla is inhibited; with the vertical canals the leading ampulla gives the increased discharge. Thus the message sent to the central nervous system, when the head is accelerated in looking round horizontally, is double, consisting of an increased discharge in some vestibular nerve fibres and a decreased discharge in others. Furthermore, when the movement ceases, the deceleration of the head causes inertial currents in the opposite direction, so that the cupulas in the horizontal canals will be deflected in the opposite way, and this will lead to a new message, in which those fibres that had shown the increased discharge will now show inhibition, whilst the inhibited fibres will give the discharge. It is presumably this second message that brings the compensatory eye-movement to a stop.

Vestibular Nystagmus. The mechanism of the semicircular canals is best investigated in the intact animal or man by the study of the nystagmus that occurs on rotating about a given axis. If a human subject is rotated about a vertical axis when he is seated in a rotating chair with his head bending forward to bring his horizontal canals into the plane of rotation, then his eyes go into a *rotatory nystagmus*; it is as though they were making repeated attempts at maintaining fixation of an object in front, but they are not visual responses since they occur when the eyes are closed, the eye-movements being recorded electrically. The nystagmus consists in a slow movement of the eyes in the opposite direction to that of rotation, followed by a rapid saccade in the direction of rotation. The speed of the slow movement is that of the speed of rotation. The direction of the nystagmus is named after the direction of the fast movement; hence the direction of a rotatory nystagmus is the same as that of the rotation. If the rotation is maintained for about 20 to 30 seconds the nystagmus ceases.

MECHANISM

Why should the eyes go into a nystagmus on rotation? That is, what is there in the stimulus-conditions that results in this repetitive type of motion? We may answer the last question first. It will be recalled that for any abrupt movement of the head there are an acceleration and a deceleration, and therefore there are two sets of messages sent to the central nervous system, determined by two types of inertial fluid movement. The cupula is deflected one way and then the other. By contrast, when the rotation is sustained, the initial acceleration is not followed by a deceleration, and consequently no inertial force is brought into play to return the cupula to its mid-position. Instead the cupula must rely on its own elasticity, and experiments have shown that it behaves as a heavily damped pendulum, requiring some 20 to 30 seconds to return to its mid-position after maximal deflection, i.e. it requires as long as the nystagmus lasts. Thus a sustained deflection of the cupula, with its accompanying pattern of nervous discharges, gives rise to rhythmic nystagmoid movements, whilst a deflection followed by a rapid return gives rise only to a compensatory deflection of the eyes. Experimentally this was demonstrated by Steinhausen who showed that if he applied a transitory pressure to a canal, through a tube inserted into it, the effect was only a simple movement of the eyes; if he maintained a constant pressure, so that the cupula remained deflected, the eyes went into a nystagmus which lasted as long as the pressure was maintained.

Post-Rotatory Nystagmus

We may imagine that our human subject has been rotated at a steady speed for some 20 to 30 seconds; the cupulas have returned to their mid-positions and the rotatory nystagmus has ceased. We now abruptly cease the rotation. There is now an inertial flow of endolymph such as to deflect the cupulas in the opposite direction. Moreover, these deflections will be maintained until the natural elasticity brings the cupulas back to their mid-positions. Hence the conditions are such as to induce another nystagmus, lasting another 20 to 30 seconds, but differing only in the direction; thus the slow component will be in the direction of rotation. It is this nystagmus that is easy to demonstrate; the subject is rotated at an even speed for some 20 to 30 seconds and then the rotation is abruptly stopped and the experimenter observes the nystagmoid movements.

Efficiency

The vestibularly evoked movement serves to keep the image of an external object on the retina stationary, even when the angular velocity is as high as $300°/sec$; the control mechanism may therefore be said to be far more effective than the visual tracking movement, which is only effective up to velocities of $30°/sec$.

The Vertical Canals

The effects of stimulating the vertical canals are more complex. Studies of the discharges in the nerve fibres from the ampullae indicate that, whereas rotation in the plane of the horizontal canals only stimulates the horizontal ampullae, rotation in other planes causes a complex group of inhibitory and excitatory responses in the ampullae of *all* the canals. The final compensatory movement of the eyes, resulting from a rotation in a plane other than that of the horizontal canals, represents the integrated response to stimuli arising in all the ampullae. It must be appreciated, moreover, that the macula of the utricle may be stimulated by these rotatory accelerations, and thus make its contribution to the general response of the eyes.*

* The effects of electrical stimulation of the ampulla of a single canal have been described by Fluur (1959), but as he points out, this is an exceedingly artificial procedure, since under normal conditions of stimulation, there are reactions in both sides of the head simultaneously. Stimulation of the left lateral canal caused a conjugate movement to the right; of the left anterior vertical canal a conjugate movement upwards, and the left posterior vertical canal a conjugate movement downwards. Combined stimulation of the left anterior and posterior canals caused a conjugate counter-clockwise rolling.

Compensatory Rolling. The rolling of the eyes that takes place when the head is tilted towards the shoulder has attracted considerable interest. It is clear that both otolith and semicircular canals may act as receptors, but if the compensatory rolling is to be sustained for any length of time clearly it is only the otolith, or neck proprioception mechanism, that will operate. Merton's (1956) studies have emphasized the much greater effectiveness of the semicircular canals in producing a rolling. When a subject is swung from side to side on a swing hung from a ball-bearing, such that the axis of rotation corresponds to the visual axis of the eye, then it is found that the eyes roll in the opposite direction to that of the movement of the head; the rolling movement lags behind the head movement, so that the compensation is never complete; in other words, the vertical meridians of the eyes do not remain vertical. The compensation amounts to about a quarter of the actual movement of the head. When a head movement is sustained, however, this partial compensation is not maintained, so that the vertical meridian returns almost completely to the position it would have had if no rotation had occurred. According to Woellner & Graybiel, for example, at a tilt of 15° the rolling is only 1·5°, whilst with a tilt of 70° it is only 4°. As Merton has explained, the partial compensatory movements during rotation are of value in reducing the angular movements of the retina, and thereby permitting a reasonable visual acuity even when the head is swinging through quite large angles.

Reaction of Labyrinth to Artificial Stimuli

Caloric Nystagmus. A current of warm water (40°–45°) or cold water (22°–27°) through the auditory meatus causes a nystagmus of vestibular origin; the slow movement is towards the stimulated side, with cold water, and away from this side with warm water. The fact that the nature of the nystagmus depends on the position of the head suggests that the cause is associated with thermal currents in the endolymph.

Galvanic Nystagmus. Passing a direct current from one ear to the other causes a nystagmus, the slow phase being towards the positive pole. The nature of the nystagmus is independent of the position of the head.

Compression Nystagmus. Experimental increase in the pressure in the middle ear causes a nystagmus.

Unilateral labyrinthectomy causes a nystagmus for a few days with the slow phase directed towards the operated side; injection of cocaine has a similar effect. It will be recalled that in the normal animal at rest there are tonic discharges from both ampullae of a synergic pair. Acclerations, requiring compensatory movements, cause an imbalance in this system, the discharge from one ampulla increasing and that from the other being inhibited. Clearly, then, extirpation of one labyrinth creates an imbalance that will be interpreted by the central nervous system as an acceleration, and some compensatory movements of the eyes will be initiated. In this way we may explain the effects of extirpation; with irritative lesions in the vestibular apparatus or in its nervous pathway the balance is upset once again, but this time in the opposite sense (see, for example, Cranmer, 1951).* After a time the necessary central compensations take place and the nystagmus subsides. When, after the nystagmus due to a lesion in the labyrinth has subsided, the labyrinth on the opposite side is injured, then a new nystagmus occurs with the

* Because of the involvement of proprioceptive impulses from the neck muscles in eye movements, we might expect irritative lesions in the cervical cord to cause nystagmus; in fact they do. Presumably impulses pass up the cord in the spino-vestibular pathways which connect the proprioceptive centres of the cord with the inferior vestibular nuclei; within the vestibular areas there is a transfer of impulses to pathways connecting with the oculomotor nuclei (Crosby, 1953).

slow phase now directed towards the newly injured side—the so-called *Bechterew nystagmus*.

Utricular Nystagmus. Gernandt (1970) exposed one utricle in the cat and monkey, and showed that a brief puff of air caused an immediate movement of the eyes in a vertical, or nearly vertical, direction, often combined with a rolling movement; repetitive stimulation led to summation of the responses until, at a limiting frequency of about 5 puffs/sec, the eyes remained fixed in the upward-looking posture. With continuous stimulation, a horizontal nystagmus followed the initial upward movement; the direction of the nystagmus was variable from one animal to another, and when the continuous stimulation ceased it could be followed by an after-nystagmus in the opposite direction.

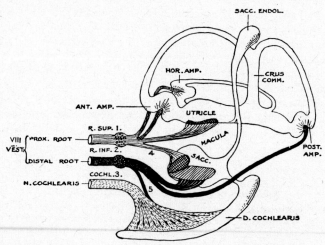

Fig. 234. Innervation of the membranous labyrinth. (De Burlet, *Anat. Anz.*)

The Nervous Pathways in the Labyrinthine Reflexes

Vestibular Reflex Arc. The labyrinth is supplied by the vestibular branch of N VIII; as with other sensory nerves, the fibres have their cell bodies outside the central nervous system, in this case the *ganglion of Scarpa* (Fig. 234); the fibres terminate in four main nuclei in the medulla, namely, *Deiter's nucleus, Bechterew's nucleus, Monakow's nucleus*, and the *triangular nucleus*. Secondary neurones pass:

(*a*) to the cord as the vestibulo-spinal tract,
(*b*) to the cerebellum,
(*c*) to the mid-brain eye-muscle nuclei by way of the median longitudinal fasciculus. Tentatively, then, we might illustrate a vestibular reflex arc by Fig. 235, where the parabducens nucleus is treated as a relay station to bring about coordinated activity between abducens (N VI) and oculomotor (N III) nuclei. However, there is no doubt that some fibres from the vestibular nucleus pass directly up the median longitudinal fasciculus to the eye-muscle nuclei, so that the intervention of the parabducens nucleus is not an anatomical necessity.

Peripheral Mechanism. Yamanaka & Bach-y-Rita (1968) have measured activity in the abducens nerve (N VI) during vestibular nystagmus; the slow phase, corresponding to a pursuit movement, was always associated with nerve fibres conducting slowly (6–40 m/sec) whilst the fast component, a saccade, was associated with activity in the rapidly conducting fibres (41–83 m/sec); if, as seems well established, the more slowly conducting fibres supply slow muscle fibres, then the thesis that separate muscle fibres are used in pursuit movements and saccades is supported. Studies on the mechanical features of the eye muscles during nystagmus generally supported the conclusion from motor unit activity (Yamanaka & Bach-y-Rita, 1970).

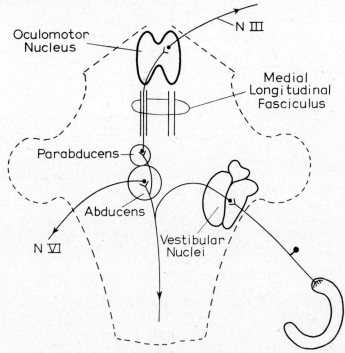

FIG. 235. Possible reflex arc in the response of the eyes to a labyrinthine stimulus. (After Crosby, *J. comp. Neurol.*)

Nystagmus Centres. There is by no means any certainty as to the existence of one or more specific centres controlling nystagmus, either optokinetic or vestibular, or both. According to de Kleyn, the vestibular nystagmus may be elicited in a rabbit from which the cerebral hemispheres and cerebellum have been removed, and the brain stem has been sectioned just anterior to the abducens nucleus, so that the nuclei of N III and IV are out of action, whilst a further section through the medulla, just behind the vestibular nuclei, reduces the amount of the brain involved practically to the vestibular nuclei. Under these conditions the lateral rectus exhibits a slow contraction and quick relaxation on stimulating the labyrinth. For vestibular nystagmus to occur, therefore, little more than a three-neurone arc, involving the vestibular nucleus, seems to be necessary. Recent work, moreover, is inclined to rule out the vestibular nuclei as centres controlling vestibular nystagmus; thus when stimulation of the labyrinth elicited a rhythmical discharge of units belonging to the lateral vestibular nucleus no synchronism was found between the rhythm of the lateral vestibular units and that of the nystagmus (Manni & Giretti, 1968).

CENTRAL NYSTAGMUS

Electrical stimulation of the parieto-temporal region of the guinea pig and the temporo-occipital zone of the rabbit (Fig. 236) induces a central nystagmus, with the quick phase directed to the opposite side; when the stimulus is over, the nystagmus continues for several seconds as an after-nystagmus in the same

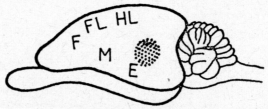

FIG. 236. Lateral view of the cerebral cortex of the rabbit. The nystagmogenic centre is located inside the dotted area. The other motor foci are indicated as follows: F, face; FL, foreleg; HL, hindleg; M, mastication; E, ear. (Manni, Azzena & Desole, *Arch. ital. Biol.*)

direction. The motor pathway uses the same circuitry as that involved in vestibular nystagmus, including the vestibular nuclei, although these are not necessary (Manni & Giretti, 1968). Fibres from the centre apparently pass through the superior colliculi, so that stimulation of these bodies evokes a central nystagmus. By using the pontomesencephalic preparation illustrated by Fig. 237, which

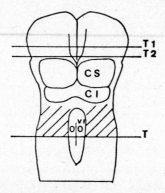

FIG. 237. A schematic dorsal view of the brain stem of the guinea pig, after removal of the cerebellum, showing the superior (CS) and the inferior (CI) colliculi, the transections (T, T1, T2) delimiting the pontomesencephalic preparation and the lesion on the floor of the fourth ventricle (hatched area) in order to destroy the superior vestibular nuclei and the cranial part of the medial and lateral vestibular nuclei. T is just caudal to the abducens nuclei. T1 is at the level of the posterior thalamus and rostral to the posterior commissure region and to the pretectum. T2 is just precollicular, rostral to the oculomotor nuclei and involves the posterior commissure and the pretectum. (Manni & Giretti, *Exp. Neurol.*)

contains the nuclei of N III, IV and VI, part of the trigeminal nuclear complex, and part of the reticular formation, but lacks the cerebellum and vestibular nuclei, Manni & Giretti (1970) showed that the nystagmus evoked by collicular stimulation was organized by the reticular formation lying between and including the oculomotor and abducens nuclei.

VESTIBULAR AND OPTOKINETIC INTERACTION

With optokinetic nystagmus, obviously visual centres are involved, and the question arises as to whether and how the vestibular centre may also become involved in optically induced nystagmus, since there is no doubt that vestibular nystagmus may be influenced by higher centres. According to Ohm, for example, optokinetic and vestibular nystagmus were able to interact in the vestibular nuclei, which were therefore the centre through which both types of nystagmus operated. That the vestibular nuclei are not a necessary stage in the circuitry was suggested by the observation that in cases of streptomycin lesions of the vestibular nuclei, severe enough to abolish caloric nystagmus, optokinetic nystagmus could be elicited normally (Dix, Hallpike & Harrison, 1949). Lachmann, Bergmann & Monnier (1958) discovered a mesencephalic region, stimulation of which would cause a nystagmus; the most sensitive region was near the thalamus and was thus considerably rostral to the vestibular nuclei. By studying the interaction of this "central nystagmus", evoked by electrical stimulation of the brain, with vestibular nystagmus obtained by mechanical stimulation of the labyrinth, these workers came to the conclusion that the two forms of nystagmus had a common meeting ground in the midbrain, since there was no doubt of their mutual reaction, both subtractive and additive effects being obtained (Bergmann et al. 1959). Presumably, then, vestibular nystagmus is mediated through this mesodiencephalic region or one related to it, although de Kleyn's and other experiments show that this is not necessary. As indicated earlier, the cerebral regions that, when stimulated, give a central nystagmus share a common pathway with vestibular nystagmus, and it seems likely that the mesodiencephalic region of Lachmann et al. is closely related to the cortical regions (Manni, Azzena & Atzori, 1965), in which case the locus of interaction is presumably the pontomesencephalic reticular formation.*

CO-OPERATION

So far as their involvement in the ordinary responses to the environment, we may say that usually a movement of the visual field occurs because of a movement of the head, so that both visual and labyrinthine mechanisms are stimulated together. The labyrinthine mechanism responds rapidly, so that its "gain" is initially high by comparison with that of the visual response; the labyrinthine system does not survive conditions of constant velocity, however, so that under these conditions the optokinetic system takes over (Collewijn, 1969).

* When the cortical centre was destroyed the mesodiencephalic centre still gave a nystagmus, so that the effects of stimulation are not merely attributable to the excitation of corticofugal fibres. By the Marchi method, Manni et al. (1965) showed that degeneration extended through the internal capsule, cerebral peduncle, the medial and lateral geniculate bodies and superior colliculus, i.e. to structures surrounding the mesodiencephalic centre. Thus this latter centre is a group of relay neurones intercalated in the neural paths connecting the cortical nystagmogenic area with the oculomotor nuclei.

It would seem from the studies of Bergmann and his colleagues (see, for example, Bergmann, Costin & Chaimowitz, 1970) that electrical stimulation of any region of a single visual pathway in the rabbit will cause a nystagmus, and it is interesting that, just as with a unilateral lesion of the labyrinth, so unilateral injury to the lateral geniculate body causes a nystagmus, with the slow phase directed away from the injured side. This nystagmus ceases after a few days, and now injury to the other lateral geniculate body causes a new nystagmus in the opposite direction, the analogue of the "Bechterew nystagmus".

The Visual, Labyrinthine and Proprioceptive Postural Mechanisms

The visual righting reflex enables the animal to maintain posture independently of the gravistatic labyrinthine mechanism; as we have seen, the primary stimulus is a disorientation of the visual fields from normal. The gravity reflexes tend to maintain normal posture, utilizing as primary stimulus the abnormal gravitational pull on the body; these reflexes also tend to maintain the visual fields normal, independently of the visual stimulus, since they can be evoked in the decerebrate animal. The actual response of man, for example, to a disorientating influence, is therefore the result of these two main groups of reflexes working together, and it is of interest to speculate on their relative contributions to any given postural righting movement.

Suppose then that a human subject is carefully poised on one foot and his equilibrium is suddenly disturbed. The changed pull of gravity in relation to his long axis, and the acceleration of his head, cause rapid labyrinthine reflex adjustments of the body, and also eye movements calculated to maintain a normal orientation of the visual field in the new position of the body. The first phase, then, is the re-establishment of correctly orientated visual fields. This has been done, however, on the basis of crude sensory data—gravitational—and for the eyes, which require a careful adjustment for both to fixate the same point, this reflex compensation is probably grossly inadequate. The fixation reflex now ensures the correct adjustment of the eyes, i.e. use is made of more precise sensory data. We now have the eyes directed accurately so as to maintain a normal orientation of the visual fields in both eyes; the centres controlling postural mechanism can now receive data from three sources—the labyrinth, proprioceptive receptors of the muscles of the body generally and proprioceptive receptors of the eye muscles.* The data from the last-named source, in view of the extreme delicacy of adjustment possible with the eye muscles, are doubtless of major importance and they indicate the exact movements of the head necessary to permit the eyes to return to their primary position and still retain a normally orientated visual field. The head righting response is thus largely controlled with the aid of eye-muscle proprioceptive impulses, although labyrinthine and neck-muscle proprioceptive impulses contribute as well. The visual reflex thus reinforces and gives precision to the much cruder gravitational type of reflex.

Cortical Control of the Eye Movements

The fixation reflex operates by way of the occipital cortex; that this reflex—although an important component in behaviour—does not dominate the eye movements is clear from the fact that an individual does not respond to every peripheral stimulus by a movement of the eyes; similarly, on turning the head, more often than not the eyes move *with* the head and not against as would happen if purely labyrinthine or proprioceptive impulses determined the eye movements. There is thus an over-riding control over the eye movements capable of inhibiting the more primitive reflex patterns; the centre for this control in man is in the frontal lobe of the cortex.

Impairment of Voluntary Control. The importance of the voluntary frontal centre is revealed by the defects in patients suffering from lesions in this area or, more usually, in the projection fibres in the internal capsule or anterior part of the midbrain. As described by Holmes, the patient cannot move his eyes in response to a command although an obvious effort is made to obey, accompanied often by great distress. There is no paralysis of eye movements

---

* The problem of proprioceptive impulses from the extraocular muscles has been discussed earlier (p. 325); the term will be used to avoid laborious circumlocutions, but we need not commit ourselves to the view that such sensory impulses really are utilized in this context; the motor discharge required to achieve and maintain a given fixation point could, of itself, provide the information.

since a slowly moving object may be followed; if the movement is at all rapid, however, following is unsuccessful. Reading may be carried out slowly provided that the letters are not too far apart. When the eyes are deviated by following a moving point, great difficulty may be experienced in bringing them back to their primary position. When the head moves, the eyes generally move in the opposite direction. Clearly the oculomotor behaviour of the patient is now determined by reflex patterns—primarily the fixation reflex. The following movements are responses of the eyes to a movement of the image off the fovea; if the point moves too quickly the eyes become the prey to fixation reflexes initiated from other points in the visual field and so cease to follow—the following movements in the normal human subject therefore represent the combined effects of voluntary control and the fixation reflex; the voluntary centre inhibits other reflexes which, if permitted to operate, would prevent the full execution of the primary intention, and it is thus only when this centre is unimpaired that true purposive movements of the eyes can be carried out.

The powerful nature of the fixation reflex is revealed in lesions of the sort described; the caloric or galvanic nystagmus is a labyrinthine reflex regularly evoked in the normal subject; in these patients, however, such responses can only be elicited when the patient is in the dark—in the light, the fixation reflex, operating without any inhibitory control from the frontal centre, can and does dominate oculo-motor behaviour to the extent of inhibiting vestibular reflexes. Again, when the patient has fixated an object, great difficulty is experienced in removing the gaze from it and many expedients are used to achieve this, such as jerking the head or covering the eyes with the hand.

The fact that a patient can follow a row of letters in reading indicates that the fixation reflex implies not only a visual stimulus but also interest and attention; after reading the first group of letters interest in them ceases and the occipital centre is ready to respond to the letters in the peripheral field; in the normal subject this response is immediate since the primary reflex response to the first group is readily inhibited by the frontal centre; in the sufferer from a frontal lesion the new response is slow and is only brought about if the new stimulus falls close to the fovea. That the patient reads in the right direction shows the importance of attention and interest even when the voluntary centre is not active.

Impairment of Fixation Reflex. A disturbance in the fixation reflex in which vision is unimpaired is rare; however, cases of lesions of the cortico-tectal tract—the efferent pathway in the reflex—have been described. In these patients the movements of fixation are normal but the maintenance of fixation is defective, especially if strong deviation of the eyes is necessary. If the eyes are in their position of rest—and fixation is most easily maintained in this position—the voluntary effort is quite adequate to sustain normal fixation; with the eyes deviated, fixation requires a more complex and continuous muscular effort and it is then that the fixation reflex becomes important—the automatic pilot which keeps the plane on its course while the pilot attends to other things. When the head, or the object of fixation is moving, maintenance of fixation becomes likewise a complex act and the patient fails to follow a moving object and complains of being unable to "focus" objects while he is moving, as in a vehicle. The phenomena of defects in the fixation reflex thus fit in well with the general conception of its function and relation to the voluntary frontal centre.

Allocation of Cortical Centres. Robinson (1968) has suggested that the saccade, pursuit movement and vergence systems are controlled by different cortical regions; he inclines to allocate areas 17–19 to the smooth pursuit system, since smooth movements are evoked by stimulation here, but since most investigators have studied the anaesthetized animal, in which all responses are smooth, further work on the unanaesthetized animal is required. Saccades can be evoked by stimulation of area 8, and only saccades, whilst stimulation of areas 19 and 22 produces vergence movements in the lightly anaesthetized monkey (Jampel, 1959).*

Other Reflex Responses. The abrupt turning of the eyes to an unexpected sound—the *acoustic reflex*—or towards a stimulus applied to the body, primitive as it may appear, involves integrations of a high order since spatial localization enters into the response; it would seem that the frontal cortex is involved (i.e. that this part of the brain is not exclusively concerned with voluntary acts). In other movements implying judgment of the nature of the stimulus, such as blinking and aversion of the eyes as a result of a threatening gesture, much wider regions of the cortex are concerned, and the pulvinar of the thalamus—phylogenetically related to the lateral geniculate body—is regarded as a possible "lower visual centre" intimately associated with these and related responses, since it has close connections with the precuneus, angular gyrus, parieto-occipital and occipito-temporal lobes.

THE CEREBELLUM

This region of the brain is off the main ascending and descending pathways from cord to midbrain and cortex, and may be regarded essentially as a regulating device for the smooth execution of the motor activity of the organism. It may be described as a "feed-back centre", in the sense that it receives information from all regions of the body and from the parts of the brain directly concerned in motor activities; because of its reciprocal connections with these parts, including the regions of the cortex concerned with voluntary movements, it is able to modify their activities continuously in the light of this information. The control of the eye movements, both reflex and voluntary, therefore involves activity of this part of the brain. The direct connections of the flocculonodular lobe with the vestibular nuclei in the medulla suggest a very intimate connection between vestibularly evoked eye reflexes and this, the so-called paleocerebellum. Certainly ablation of this portion of the cerebellum causes nystagmus—of a transient kind, however. When extraocular muscles of the lamb were stretched, evoked responses were recorded from the ipsilateral simplex lobule of the cerebellum (Azzena, Desole & Palmieri, 1970).†

GENERAL REFERENCES

BENDER, M. B. (Ed.) (1964). *The Oculomotor System.* New York: Harper & Row.
CAMIS, M. & CREED, R. S. (1930). *Physiology of the Vestibular Apparatus.* Oxford.
COGAN, D. G. (1956). *Neurology of the Extraocular Muscles,* 2nd ed. Springfield: C. C. Thomas.

* The extensive studies of Pasik & Pasik (1964) on cortical ablation and stimulation have led them to be sceptical about cortical motor centres; they conclude: "We move the eyes with our whole brain. It appears as if oculomotor function is initiated in the entire cortex and uses numerous corticifugal systems that descend and converge toward the tegmentum of the brainstem."
† The role of the cerebellum in the control of the eye movements has been reviewed by Dow & Manni (1964), whilst Collewijn (1970) has made an interesting study of the disturbance in the saccadic type of movement, with normal pursuit movements, in cerebellectomized rabbits.

HOLMES, G. (1938). The cerebral integration of the ocular movements. *Br. Med. J.* **ii**, 107–112.

ROELOFS, C. O. (1954). Optokinetic nystagmus. *Doc. Ophthal.* **7–8**, 579–650.

SZENTAGOTHAI, J. (1943). Die zentrale Innervation der Augenbewegungen. *Arch. f. Psychiat.* **116**, 721–760.

WHITTERIDGE, D. (1960). Central control of the eye movements. In *Handbook of Physiology*, Ed. Field, Magoun & Hall, vol. II, pp. 1089–1109. Washington: Am. Physiol. Soc.

SPECIAL REFERENCES

APTER, J. T. (1945). Projection of the retina on superior colliculus of cats. *J. Neurophysiol.* **8**, 123–134.

APTER, J. T. (1946). Eye movements following strychninization of the superior colliculus of cats. *J. Neurophysiol.* **9**, 73–86.

AZZENA, G. B., DESOLE, C. & PALMIERI, G. (1970). Cerebellar projections of the masticatory and extraocular muscle proprioception. *Exp. Neurol.* **27**, 151–161.

BENDER, M. B., POSTEL, D. M. & KRIEGER, H. P. (1957). Disorders of oculomotor function in lesions of the occipital lobe. *J. Neurol. Neurosurg. Psychiat.* **20**, 139–143.

BENDER, M. B. & SHANZER, S. (1964). Oculomotor pathways defined by electric stimulation and lesions in the brainstem of monkey. In *The Oculomotor System*, Ed. M. B. Bender, pp. 81–140. New York: Harper & Row.

BERGMANN, F., COSTIN, A. & CHAIMOWITZ, M. (1970). Influence of lesions in the lateral geniculate body of the rabbit on optic nystagmus. *Exp. Neurol.* **28**, 64–75.

BERGMANN, F., LACHMANN, J., MONNIER, M. & KRUPP, P. (1959). Central nystagmus. III. Functional correlations of mesodiencephalic nystagmogenic centre. *Am. J. Physiol.* **197**, 454–460.

BICKFORD, R. G., PETERSEN, M. C., DODGE, H. W. & SEM-JACOBSEN, C. W. (1953). Observations on depth stimulation of the human brain through implanted electrographic leads. *Proc. Staff Meetings Mayo Clin.* **28**, 181–187.

BIZZI, E. (1967). Discharge of frontal eye field neurons during eye movements in unanesthetized monkeys. *Science* **157**, 1588–1590.

BIZZI, E. (1968). Discharge of frontal eye field neurons during saccadic and following eye movements in unanaesthetized monkeys. *Exp. Brain Res.* **6**, 69–80.

BRECHER, G. A. (1934). Die optokinetische Auslösung von Augenrollung und rotorischem Nystagmus. *Pflüg, Arch. ges. Physiol.* **234**, 13–28.

BROWN, T. G. (1922). Reflex orientation of the optical axes and the influence upon it of the cerebral cortex. *Arch. néerl. Physiol.* **7**, 571–578.

BURGI, S. (1957). Das Tectum opticum. Seine Verbindungen bei der Katze und seine Bedeutung beim Menschen. *Deutsch. Z. Nervenheilk.* **176**, 701–729.

CARMICHAEL, E. A., DIX, M. R. & HALLPIKE, C. S. (1956). Pathology, symptomatology and diagnosis of organic affections of the eighth nerve system. *Br. med. Bull.* **12**, 146–152.

CARMICHAEL, E. A., DIX, M. R. & HALLPIKE, C. S. (1954). Lesions of the cerebral hemispheres and their effects upon optokinetic and caloric nystagmus. *Brain* **77**, 345–372.

COHEN, L. A. (1961). Role of eye and neck proprioceptive mechanisms in body orientation and motor coordination. *J. Neurophysiol.* **24**, 1–11.

COLLEWIJN (1969). Changes in visual evoked responses during the fast phase of optokinetic nystagmus in the rabbit. *Vision Res.* **9**, 803–814.

COLLEWIJN, H. (1970). Dysmetria of fast phase of optokinetic nystagmus in cerebellectomized rabbits. *Exp. Neurol.* **28**, 144–154.

CRANMER, R. (1951). Nystagmus related to lesions of the central vestibular apparatus and the cerebellum. *Ann. Otol.* **60**, 186–196.

CROSBY, E. C. (1953). Relations of brain centres to normal and abnormal eye movements in the horizontal plane. *J. comp. Neurol.* **99**, 437–479.

CROSBY, E. C. & HENDERSON, J. W. (1948). Pathway concerned in automatic eye movements. *J. comp. Neurol.* **88**, 53–91.

CROSBY, E. C., YOSS, R. E. & HENDERSON, J. W. (1952). The pattern for eye movements on the frontal eye field etc. *J. comp. Neurol.* **97**, 357–383.

DE BURLET, H. M. (1924). Zur Innervation der Macula sacculi bei Säugetieren. *Anat. Anzeiger* **58**, 26–32.

DIX, M. R., HALLPIKE, C. S. & HARRISON, M. S. (1949). Some observations upon the otological effects of streptomycin intoxication. *Brain* **72**, 241–245.

DOHLMAN, G. (1935). Some practical and theoretical points in labyrinthology. *Proc. roy. Soc. Med.* **28**, 1371–1380.

DOW, R. S. & MANNI, E. (1964). The relationship of the cerebellum to extraocular movements. In *The Oculomotor System*, Ed. M. B. Bender, pp. 280–292. New York: Hoeber.

EAKINS, K. E. & KATZ, R. L. (1967). The role of the autonomic nervous system in extraocular muscle function. *Invest. Ophthal.* **6**, 253–260.

FLUUR, E. (1959). Influences of semicircular ducts on extraocular muscles. *Acta otolaryngol. Suppl.* **149**.

FOERSTER, O. (1931). The cerebral cortex in man. *Lancet* (2), 309–312.

FOX, J. C. & HOLMES, G. (1926). Optic nystagmus and its value in the localisation of cerebral lesions. *Brain* **49**, 333–371.

GERNANDT, B. E. (1970). Nystagmus evoked by utricular stimulation. *Exp. Neurol.* **27**, 90–100.

GRÜNBAUM, A. S. F. & SHERRINGTON, C. S. (1903). Observations on the physiology of the cerebral cortex of the anthropoid ape. *Proc. Roy. Soc. B.* **72**, 152–155.

HASSLER, R. & HESS, W. R. (1954). Experimenteller und anatomische Befunde über die Drehbewegungen und ihre nervösen Apparate. *Arch. Psychiat.* **192**, 488–526.

HYDE, J. E. (1962). Effect of hindbrain lesions on conjugate horizontal eye movements in cats. *Exp. Eye Res.* **1**, 206–214.

HYDE, J. E. & SLUSHER, M. A. (1961). Functional role of median longitudinal fasciculus in evoked conjugate ocular deviations in cats. *Am. J. Physiol.* **200**, 919–922.

JUNG, R. & HASSLER, R. (1960). The extrapyramidal motor system. In *Handbook of Physiology*, Ed. Field, Magoun & Hall, vol. II, pp. 863–927. Baltimore: American Physiological Society.

DE KLEYN, A. (1939). Some remarks on vestibular nystagmus. *Confin. Neurol.* **2**, 257–292.

LACHMANN, J., BERGMANN, F. & MONNIER, M. (1958). Central nystagmus elicited by stimulation of the meso-diencephalon in the rabbit. *Am. J. Physiol.* **193**, 328–334.

LACHMANN, J., BERGMANN, F., WEINMAN, J. & WELNER, A. (1958). Central nystagmus. II. Relationship between central and labyrinthine nystagmus. *Am. J. Physiol.* **195**, 267–270.

LEMMEN, L. J., DAVIS, J. S. & RADNOR, L. L. (1959). Observation on stimulation of the human frontal eye field. *J. comp. Neurol.* **112**, 163–168.

LOWENSTEIN, O. (1936). The equilibrium function of the vertebrate labyrinth. *Biol. Rev.* **11**, 113–145.

LOWENSTEIN, O. (1956). Comparative physiology of the otolith organs. *Br. med. Bull* **12**, 110–114.

MACKENSEN, G. (1958). Zur Theorie des optokinetischen Nystagmus. *Klin. Mbl. Augenheik.* **132**, 769–780.

MACKENSEN, G. (1959). Untersuchung zur Physiologie des optokinetischen Nach-nystagmus v. *Graefes' Arch. Ophthal.* **160**, 497–509.

MANNI, E., AZZENA, G. B. & ATZORI, M. L. (1965). Relationships between cerebral and mesodiencephalic nystagmogenic centres in the rabbit. *Arch. ital. Biol.* **103**, 136–145.

MANNI, E., AZZENA, G. B. & DESOLE, C. (1964). Eye nystagmus elicited by stimulation of the cerebral cortex in the rabbit. *Arch. ital. Biol.* **102**, 645–655.

MANNI, E. & GIRETTI, M. L. (1968). Vestibular units influenced by labyrinthine and cerebral nystagmogenic impulses. *Exp. Neurol.* **22**, 145–157.

MANNI, E. & GIRETTI, M. L. (1970). Central eye nystagmus in the pontomesen-cephalic preparation. *Exp. Neurol.* **26**, 342–353.

MERTON, P. A. (1956). Compensatory rolling movements of the eye. *J. Physiol.* **132,** 25–27 P.

OHM, J. (1936). Ueber Interferenz mehrerer Arten von Nystagmus. *Proc. Acad. Sci. Amst.* **39,** 549–558.

PASIK, P. & PASIK, T. (1964). Oculomotor functions in monkeys with lesions of the cerebrum and the superior colliculi. In *The Oculomotor System*, Ed. M. B. Bender, pp. 40–80. New York: Harper & Row.

PASIK, T., PASIK, P. & BENDER, M. B. (1966). The superior colliculi and eye movements. *Arch. Neurol.* **15,** 420–436.

PASIK, P., PASIK, T. & KRIEGER, H. P. (1959). Effects of cerebral lesions upon optokinetic nystagmus in monkeys. *J. Neurophysiol* **22,** 297–304.

PENFIELD, W. & BOLDREY, E. (1937). Somatic motor and sensory representation in the cerebral cortex of man as studied by electrical stimulation. *Brain* **60,** 389–443.

RADEMAKER, G. G. J. & TER BRAAK, J. W. G. (1948). On the central mechanism of some optic reactions. *Brain* **71,** 48–76.

ROBINSON, D. A. (1968). Eye movement control in primates. *Science* **161,** 1219–1224.

STEINHAUSEN, W. (1933). Uber die Beobachtung der Cupula in den Bodengangsampullen des Labyrinths des lebenden Hechts. *Pflüg. Arch.* **232,** 500–512.

TER BRAAK, J. W. G. (1936). Untersuchungen ueber optokinetischen Nystagmus. *Arch. néerl. Physiol.* **21,** 309–376.

WAGMAN, I. H., KRIEGER, H. P. & BENDER, M. B. (1958). Eye movements elicited by surface and depth stimulation of the occipital lobe of macaque mulatta. *J. comp. Neurol.* **109,** 169–193.

WARWICK, R. (1953). Representation of the extra-ocular muscles in the oculomotor nuclei of the monkey. *J. comp. Neurol.* **98,** 449–503.

WARWICK, R. (1955). The so-called nucleus of convergence. *Brain* **78,** 92–114.

WOELLNER, R. C. & GRAYBIEL, A. (1959). Counterrolling of the eyes and its dependence on the magnitude of gravitation or inertial force acting laterally on the body. *J. app. Physiol.* **14,** 632–634.

YAMANAKA, Y. & BACH-Y-RITA, P. (1968). Conduction velocities in the abducens nerve correlated with vestibular nystagmus in cats. *Exp. Neurol.* **20,** 143–155.

YAMANAKA, Y. & BACH-Y-RITA, P. (1970). Relations between extraocular muscle contraction and extension times in each phase of nystagmus. *Exp. Neurol.* **27,** 57–65.

PART 3

THE PUPIL

Sphincter and Dilator Muscles

THE aperture of the refracting system of the eye is controlled by the iris which behaves as a diaphragm, contracting or expanding as a result of the opposing actions of two muscles of ectodermal origin, namely the *sphincter pupillae* and the *dilator pupillae*. As classically described, the sphincter is an annular band of smooth muscle, 0·75–0·8 mm broad in man, encircling the pupillary border, which on contraction draws out the iris and constricts the pupillary aperture. At the pupillary border the muscle is closely associated with the pigment epithelium, so that, on contraction, the latter tends to be drawn on to the anterior surface of the iris. The sphincter is closely adherent to the adjacent connective tissue so that, after an iridectomy, it does not contract up, and the pupil remains reactive to light. Like the sphincter, the dilator muscle is of ectodermal origin, but in the sphincter the ectodermal cells have been transformed into true muscle fibres whereas the cells of the dilator retain their primitive characteristics and are called *myoepithelial cells*. It will be recalled (p. 12) that the ciliary epithelium consists of two layers of cells, an outer pigmented layer and, inside this, the unpigmented layer; these two layers are continued over the posterior surface of the iris, the pigmented layer becoming the layer of myoepithelial cells and the unpigmented layer acquiring pigment and becoming the posterior epithelium of the iris. The myoepithelial cells are spindle shaped with long processes that run radially to constitute the fibres of the dilator, making a discrete layer anterior to that made up by the bodies of the myoepithelial cells; the layer of fibres is called the *membrane of Bruch*. Close to the edge of the pupil the dilator fibres fuse with the sphincter, and at the other end they continue into the ciliary body where they take origin; between the insertion and origin, "spurs" make connection with adjacent tissue. When it contracts, the dilator draws the pupillary margin towards its origin and thus dilates the pupil.

Innervation. The sphincter is innervated by parasympathetic fibres from N III by way of the ciliary ganglion and the short ciliary nerves; the dilator is controlled by the cervical sympathetic, the fibres relaying in the superior cervical ganglion; post-ganglionic fibres enter the eye in the short and long ciliary nerves (Fig. 6, p. 5).

Functions of the Pupil. A pupil of varying size performs three main functions:

(*a*) It modifies the amount of light entering the eye, thus permitting useful vision over a wide range of luminance levels. The amount of light entering the eye is directly proportional to the area of the pupil; under conditions of night vision the absolute amount of light entering the eye is of great importance and a sixteenfold increase of sensitivity of the eye to light should be obtained by an increase of pupil diameter from 2 to 8 mm. At the other end of the scale, excessive luminance reduces visual acuity; the strong pupillary constriction occurring in these circumstances mitigates this condition and thus extends the range of useful vision. In nocturnal animals this aspect of pupil-size is of greater importance and it is probable that the slit of the fully constricted pupil of the cat is an adaptation permitting a greater restriction of the amount of light penetrating the eye in daylight.

(*b*) As a result of constriction, the pencils of light entering the eye are smaller and the depth of focus of the optical system is increased (p. 406). When the eye is focused for distant objects, the depth of focus is large and the pupil-size is relatively unimportant; when the eye is focused for near objects, on the other hand, the depth of focus becomes small and a constricted pupil contributes materially.

(*c*) By the reduction in the aperture of the optical system aberrations are minimized. In the dark these aberrations are unimportant, since form is only vaguely perceived, so that the advantage accruing from the increased light entering a dilated pupil far outweighs the effects of aberrations; in daylight, however, the reverse is true; and we have already seen how, at varying luminances, the size of the pupil is adjusted to give optimum visual acuity (p. 250).

THE PUPILLARY REFLEXES

The accurate adjustment of the pupil-size to the optical requirements of the eye implies a reflex response to illumination; this is the basis of the *light reflex* under which term we may include the constriction of the pupil when the luminance is increased, and the dilatation when the luminance is decreased. The other important reflex, from the optical point of view, is the *near reflex*, the constriction of the pupil that takes place during the focusing on a near object. Additional reflexes are the *lid-closure reflex*, a constriction in response to closure of the lid, and the *psycho-sensory reflex*, a dilatation of the pupil in response to a variety of sensory and psychic stimuli.

Experimental Measurement—Pupillography. The experimental study of the size of the pupil—*pupillography*—has been carried out by a variety of techniques; since the pupil responds to visible light the method of measurement must dispense, if possible, with illumination unless this involves flashlight photography where the duration of the flash is so short that the pupil has no time to respond. For continuous studies flash-photography is obviously of no use, since the flashes have their effects, although delayed; by the use of infrared photography, however, a cinematographic record of the pupil in the dark may be obtained and this is the method that has been most commonly employed. More recently Lowenstein & Loewenfeld have developed a method of "electronic pupillography" based on infrared scanning of the pupil, the reflected light from the scanned pupil being brought on to an infrared sensitive photo-tube. The scanned image is made visible by the usual television principles, if this is required; alternatively the size of the pupil may be recorded less directly on a cathode-ray screen. Thus when the scanning spot crosses from the iris to the pupil there is an abrupt decrease in the reflected light, and this shows up as a sudden fall in the signal on the oscilloscope. The larger the pupil the longer the portion of a given scan will give the low signal.

The Light Reflex

When the light falling on one eye is increased, the pupil of this eye constricts—the *direct light reflex*; at the same time the pupil of the other, unstimulated, eye also contracts—the *indirect* or *consensual light reflex*. A record of the changes in pupil-size in the two eyes during exposure to a bright light is shown in Fig.

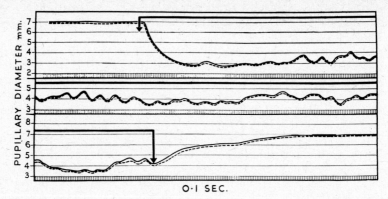

FIG. 238. Effect of exposure of right eye to a bright light. Ordinate: pupillary diameter in mm. Abscissa: time of exposure. Record in full line applies to right eye; in broken line to left eye.

First line: the pupils were large and quiet in darkness. At the arrow the right eye was exposed to a steady illumination.

Second line: this shows the pupillary oscillations after the right eye had been exposed for 3 min to the light.

Third line: when the light was turned off, at the arrow, the pupils dilated and the oscillations disappeared. (Lowenstein & Loewenfeld, *Amer. J. Ophthal.*)

238, where the pupil-diameter is given as the ordinate and is plotted against time of exposure. The record is shown in three parts; at the arrow on the top, the light is switched on to illuminate the right eye, whose pupillary diameter is indicated by the full line. Both pupils contract down to about 3 mm, and the sizes oscillate together whilst, as time progresses, there is an increase in size on which the oscillations are imposed; this is seen by the middle record, the mean pupil-size being about 4 mm. At the arrow in the bottom record the light was turned off, and the pupils return slowly to their original diameters. The increase in diameter that occurs during sustained illumination may be regarded

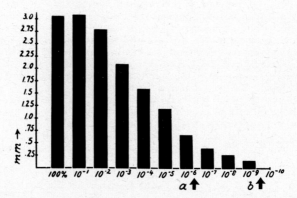

FIG. 239. Average extent of pupillary reflexes elicited by light-stimuli of decreasing intensity. The heights of the columns indicate the average degree of constriction, whilst the abscissa shows the intensities of the light-stimuli (duration 1 second). The arrow *a* shows the threshold for colour vision and the arrow *b* the absolute threshold. (Lowenstein & Loewenfeld, *Amer. J. Ophthal.*)

as an effect of light-adaptation. On flashing a light into the eye, then, the response is a constriction of the homolateral and contralateral pupils, the extent of the constriction depending on the intensity of the light and the state of adaptation of the eye. If the light-stimulus is prolonged, the pupil dilates slowly and reaches a size which, for a given individual, depends on the intensity of the light. When both eyes are stimulated there is a summation of the individual responses so that both pupils constrict rather more than when one alone is stimulated with the same intensity of light; conversely if, when both are being subjected to a light stimulus, one eye is closed, the contralateral pupil dilates slightly.

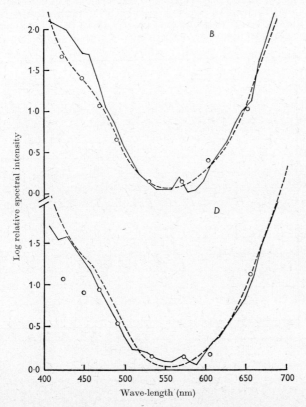

FIG. 240. Mean spectral sensitivity curve for the pupil response of two subjects. O, Differential threshold measurements are plotted for 2 sec flashes of a 2° test patch centrally fixated and seen against a continuous blue background approximately 15° in diameter, which produced a retinal illuminance somewhere between 100 and 200 td. Interrupted line, C.I.E. photopic luminosity curve; solid line, mean results of psychophysical measurements of photopic luminosity (flicker photometry) on the same two subjects with the same apparatus. (Alpern & Campbell, *J. Physiol.*)

Effect of Luminance. The magnitude of the initial constriction caused by a light-stimulus depends, as we should expect, on the intensity of the light. This is demonstrated by Fig. 239 where the constriction, in millimetres, is represented by the heights of the black bars, whilst the relative luminances of the stimuli are plotted logarithmically as abscissae. At the arrow marked *a*, the intensity is some millionth of that required to evoke the 3 mm response, and this corresponds to the threshold for colour vision, i.e. the cone threshold. At the arrow marked b, the intensity is some 10^{-9} to 10^{-10} of that required to produce the 3 mm response; this is the threshold for rod vision, and it will be seen that under the conditions of these experiments the threshold for pupillary response is higher than that for the visual response.

Rods and Cones. The receptors for the pupillary response to light are obviously the rods or cones or both; in fact, many studies have shown that both receptors may bring about the reflex. At threshold intensities for pupillomotor activity the visual sensation is colourless, proving that the rods are mediating the response; on plotting the sensitivity to the different wavelengths a typical scotopic sensitivity curve is obtained with maximum in the region of 5000 Å (Schweitzer, 1956).

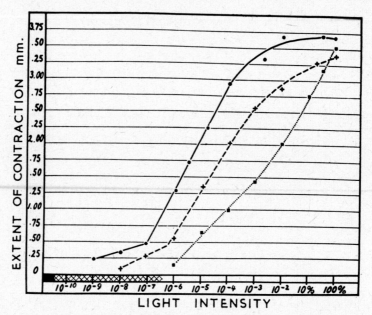

FIG. 241. Pupillary contraction as a function of the intensity of the light-stimulus for three different types of light. Solid line and dots: curve for white light. Broken line and crosses: curve for blue light obtained by placing filter in front of white source; the apparent increase in the threshold is due to the absorption of light by the filter. Dotted line and squares: curve for red light obtained by placing red filter in front of the white source. The black bar above the abscissa indicates values below the absolute threshold; the hatched bar indicates the range of scotopic vision. Note that with red light pupillary responses are only obtained at intensity levels corresponding to photopic vision. (Lowenstein & Loewenfeld, *Amer. J. Ophthal.*)

Spectral Sensitivity. The most thorough investigation of the influence of wavelength on pupillary response is that of Alpern & Campbell (1962), who studied both transient pupillary changes, i.e. the response to rhythmically altered intensities, and steady-state changes. In both cases, the responses to relatively high luminances never gave an uncomplicated photopic luminosity curve comparable with that for vision; instead, the maximum was between 5300 and 5400 Å, suggesting that, even under highly photopic conditions and foveal stimulation, the rods and cones were both operative, due, presumably, to scatter from the fovea. When the effects of rods were excluded by keeping them adapted to a blue light acting as a background upon which the foveal test-stimulus was imposed, then the action-spectrum gave a curve with a peak at about 5600 Å, in reasonable agreement with the C.I.E. standard curve (Fig. 240). Analysis of the curve obtained when the maximum was at 5300–5400 Å, on the basis of the different effects of the Stiles-Crawford phenomenon with red and blue lights, suggested that the rod and cone effects were, indeed, simultaneously present, and that their separate contributions to the final pupillary response could be estimated if the measured sensitivity was treated as the weighted mean of the logarithms of the individual rod and cone sensitivities at a given wavelength.* If the response of the pupil is plotted against intensity of the light-stimulus, as in Fig. 241, there appears an obvious break in the curve when white or blue light is employed, a break corresponding to the transition from rod to cone vision. When red light is employed there is no break, and the threshold is higher because the red light is very ineffective as a rod stimulator (p. 258). In general, the cone response to a flash has a shorter latency (0·2–0·3 second) and is more rapid and longer-lasting than the rod response, which has a latency of 0·45–0·6 second.

Summation. Schweitzer has studied the effects of increasing the area and duration of the stimulus. It will be recalled that the visual stimulus shows a limited amount of spatial summation, in the sense that if the thresholds for a small patch of light and a larger patch are compared, the threshold is lower for the larger patch; this is due, as we have seen, to the convergence of rods and cones on to ganglion cells (p. 216). The area over which summation of the visual response occurs is relatively small by comparison with that for the pupillographic response; thus Schweitzer found complete summation up to an area subtending 4° in the periphery and 0·5° at the fovea, whilst partial summation extended over much wider areas. This means that to make the most efficient use of light in the pupillomotor response we must stimulate a large area of retina; when this is done we find that the visual and pupillomotor thresholds are equal, but Schweitzer was unable to push the sensitivity of the pupillomotor response beyond this point, i.e. he could not obtain a contraction of the pupil in the absence of visual sensation.† As a result of the operation of the light reflex, the size of the pupil tends to adopt a characteristic value for each level of luminance; according to De Groot & Gebhard the diameter, d, in mm, is related to the luminance, B, in mL, by the empirical equation:

$$\log d = 0·8558 - 0·000401 \, (\log B + 8·1)^3$$

In the completely dark-adapted eye the mean pupillary diameter is given variously as 6·6 to 7·6 mm.

The Near Reflex

By the near reflex is meant the constriction of the pupil accompanying the convergence of the eyes on to a near object; associated with these two actions of convergence and miosis, there is, of course, the act of accommodation, all three being mediated by N III. The response is bilateral and the maximal constriction obtained with this reflex is approximately equal to that obtained with the light reflex; the two reflexes may summate their responses.

* Alexandridis & Koeppe (1969) have described the typical Purkinje shift in maximal sensitivity to wavelength on passing from scotopic to photopic stimulation; their sensitivity curves coincided remarkably well with the corresponding sensitivity curves for vision, with maxima at 5000 Å and 5600 Å.

† It is commonly stated that the threshold for pupillomotor activity is higher than that for vision, and this is true for small test-fields; thus if this subtended 1° 50′, the ratio of thresholds varied from 8 to 1000; when the Ulbricht integrating sphere was employed, permitting exposure of a very large field, the thresholds became equal because of the greater degree of summation in the pupillomotor response. Temporal summation, whereby two stimuli, following each other rapidly, are more effective than a single one, depends essentially on the photochemical characteristics of the receptors (Bunsen-Roscoe Law) rather than on any neural organization; it is not surprising, therefore, that temporal summation is the same for the pupillomotor and visual responses (Schweitzer, 1956).

The Lid-Closure Reflex

This has been classically described as a homolateral pupillary constriction associated with closure of the lid; it was said to be evoked if the lid was held open while the effort to close the eye was made. Lowenstein & Lowenfeld (1969) have pointed out that this latter phenomenon is really an example of the near reflex, since it does not occur if the subject is told to look at a far point all the time. As these authors state, it is best to include all pupillary changes associated with lid-closure in the class of "lid-closure reflexes", but the variety of conditions under which the lids close means that the pupillary changes will be different and have different mechanisms. Thus, if the lids close to avoid a noxious stimulus, it is probable that the pupils will dilate because of the psycho-sensory reflex; a brief, naturally occurring, blink is associated with dilatation due to the operation of the light-reflex in reverse, i.e. the response to the short period of darkness.

Psycho-Sensory Reflex

This is a dilatation of the pupils evoked by stimulation of any sensory nerve or by strong psychical stimuli. Pupillary dilatation thus occurs during the induction stage of anaesthesia, in extreme fear and in pain; in sleep, on the other hand, when these stimuli are lacking, the pupils are generally constricted. The mechanism is clearly a cortical one and is apparently mediated both by way of the cervical sympathetic, which activates the dilator pupillae, and by way of N III which inhibits the tonus of the sphincter pupillae. The reaction is not seen in the newborn, but appears in the first few days of life, developing fully at the age of six months.

NERVOUS PATHWAYS

The receptors of the light reflex are the rods and cones; the afferent pathway is therefore through the optic nerves and tracts; the pupillary fibres leave the optic tract before the lateral geniculate body is reached, and for many years it was thought that they relayed in the superior colliculus. Magoun & Ranson showed, however, that in cats and monkeys the fibres pass through the medial border of the geniculate body and then run, by way of the brachium of the superior colliculus, into the *pretectal nucleus*, a group of cells occupying a position at the junction of the diencephalon and the tectum of the mid-brain; here they have their first synapse. In the cat the majority of the secondary fibres cross in the posterior commissure and terminate in the Edinger-Westphal nucleus; in primates the crossing is not so extensive. Destruction of the superior colliculus, therefore, leaves the light reflex intact; stimulation of the pretectal area gives a bilateral constriction of the pupils; conversely, destruction of this area gives a dilated and fixed pupil. From the Edinger-Westphal nucleus the pupillo-constrictor fibres pass with the IIIrd nerve fibres to the ciliary ganglion. The reflex arc for the direct and indirect light reflexes is indicated in Fig. 242; the diagram indicates quite clearly how a light stimulus, falling on one eye, gives rise to pupillary constriction in both eyes.*

* The actual areas of the nucleus of N III concerned with pupillo-constrictor function have been examined recently by Sillito & Zbrożyna (1970); they include a region containing both the cells of the Edinger-Westphal nucleus and the anteromedian nucleus where they adjoin, and the caudal continuation of this nucleus. The authors emphasize

Reflex Inhibition. On the basis of Sherrington's principle of reciprocal innervation we might expect the pupilloconstrictor activity to be accompanied by an inhibition of the normal tonic activity of the dilator. Nisida, Okada & Nakano measured the normal resting activity in the long ciliary nerve of the cat, this being the sympathetic pathway to the eye and thus to the dilator pupillae; the spontaneous discharge in many of the fibres was inhibited when light was shone into the eye. Other fibres were unaffected and were presumably concerned with the control of the blood vessels. According to Okada *et al.*, the central pathway in this reflex inhibition is essentially similar to that illustrated by Fig. 242 although the final motor pathway is by way of the nerve cells of the lateral column of the cervical and thoracic segments of the cord, the so-called *ciliospinal centres of Budge*, between C6 and T4.

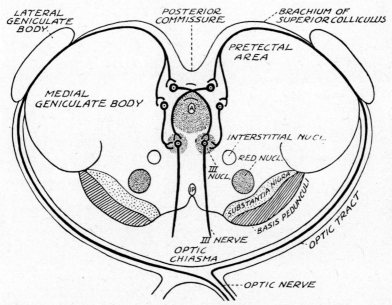

FIG. 242. Diagram of the path concerned in the pupillary light reflex. (Magoun & Ranson, *Arch. Ophthal.*)

Cortical Centres. Since the pupil is subject to psychical influences it is reasonable to expect cortical connections with the pretectal centre; furthermore, the intimate connection between pupillary, ciliary and convergence activities in the "near response" demands the presence of an integrating centre, which, according to Jampel, is unlikely to be at subcortical levels. In the monkey Jampel has described a region at the junction of areas 19 and 22, i.e. in the occipito-temporal region, stimulation of which causes a constriction of the pupils together with accommodation and convergence. It is here, therefore, that the integration of the pupillary, ciliary and convergence mechanisms in the

that this pupillary constrictor nucleus is not a paired structure, as represented in Fig. 242 for example, but is strictly a group of cells with no midline cleavage; in agreement with this, all pupillomotor responses produced by stimulation of the nucleus, or of supranuclear structures, were strictly bilateral.

"near response" probably takes place. The general scheme for pupillary control, postulated by Jampel, is indicated in Fig. 243, where the preoccipital centre is marked PAC; this is influenced by the entire cortex, indicated by X (accounting for psychosensory responses) and by visual stimuli. Impulses pass down to the pretectal nucleus whence they are relayed to the Edinger-Westphal nucleus to activate the sphincter pupillae and ciliary muscle. Influences from the spinal cord are shown by neurones from Z. The pathway for the light reflex has not

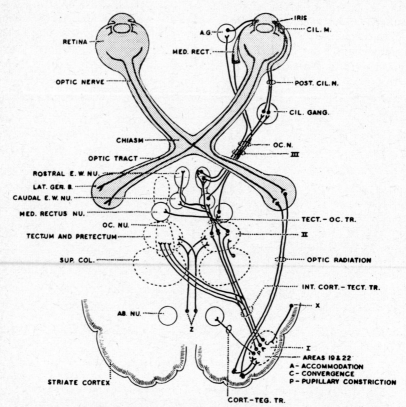

FIG. 243. A diagrammatic scheme, illustrating the three levels of function in the nervous system involved in the near response (see text). (I) This indicates the functionally specific area for the near response in the preoccipital cortex. (II) This indicates the level of the tectum and pretectum. (III) This indicates the final common pathway for impulses to the ciliary muscle, medial rectus, and iris sphincter. (X) This represents the input of nervous activity from the entire cerebral cortex that modifies the output of the functionally specific areas for the near response in the preoccipital cortex. (Z) This represents the input of nervous activity from the spinal cord and brain stem which modifies the impulses discharged from the cerebral cortex into the tectum and pretectum. A.G., accessory ganglion; AB. NU., abducens nucleus; CAUDAL E. W. NU., caudal Edinger Westphal nucleus; CIL. GANG., ciliary ganglion; CIL. M., ciliary muscle; CORT.-TEG. TR., cortico-tegmental tract; INT. CORT.-TECT. TR., internal cortico-tectal tract; LAT. GEN. B., lateral geniculate body; MED. RET., medial rectus; MED. RECTUS NU., medial rectus nucleus; OC. N., oculomotor nerve; OC. NU., oculomotor nucleus; POST. CIL. N., posterior ciliary nerves; ROSTRAL E.W. NU., rostral Edinger Westphal nucleus; SUP. COL., superior colliculus; TECT.-OC. TR., tecto-oculomotor tract. (Jampel, *Amer. J. Ophthal.*)

been incorporated in the diagram, to avoid confusion; presumably the near and light reflexes interact in the pretectal nucleus. The phenomenon of the *Argyll Robertson pupil*—abolition of the direct and indirect light reflexes with retention of the near reflex—suggests that the motor pathways to the iris muscles are different in the two reflexes, and Lowenstein (1956) has presented some experimental evidence in favour of this hypothesis.

Other Pupillomotor Centres. The sphincter pupillae undoubtedly dominates pupil activity so that the pretectal and cortical pupilloconstrictor centres have received most attention. The strong association of dilatation of the pupil with emotional states would indicate the presence of dilator centres in those parts of the brain that have been shown to be strongly associated with emotional states, namely the hypothalamus and the limbic system generally. In fact, electrical stimulation in the hypothalamus, over quite a wide region, causes dilatation of the pupil (Hess). The pathway from the brain in dilator responses involves the sympathetic system, the pupillomotor fibres leaving the cord from the first and second thoracic segments and relaying in the superior cervical ganglion. Stimulation of the cord between the VIth cervical and IVth thoracic segments causes dilatation of the pupil and it is customary to speak of spinal dilator centres, but it seems most likely that the stimulus merely activates the sympathetic motor neurones in the cord.

The Peripheral Mechanisms

Reflex Dilatation. Stimulation of the cervical sympathetic trunk in the neck causes a pronounced dilatation of the pupil. If the sympathetic trunk is cut there is an initial dilatation, due to the irritative effect of the cut, and this is followed by a constriction so that the pupil on the cut side is smaller than the other; this is presumably due to the fact that normally the size of the pupil is determined by a balance of forces between sphincter and dilator; putting the dilator out of action allows the sphincter to take control. The same phenomenon is manifest in the reflex response to light; in the sympathectomized eye the response is generally larger. Reflex dilatation, brought about by some sensory stimulus, is mainly due to active dilatation, mediated by the sympathetic, since when the sympathetic trunk is cut there is only a small dilatation, due to inhibition of sphincter tone. Some time after removing the sympathetic supply to the iris it becomes hypersensitive to adrenaline and this gives rise to what has been called "paradoxical pupillary dilatation". Thus a sensory stimulus causes, in this sympathectomized eye, a small dilatation due to relaxation of sphincter tone, and this is followed by a much larger dilatation due, presumably, to the liberation into the blood of adrenaline as a consequence of the sensory stimulus.

Stimulation of N III, or the ciliary ganglion, causes constriction of the pupil; damage to the ciliary ganglion or IIIrd nerve reduces the reflex responses to light very considerably, such constriction as occurs being due to inhibition of sympathetic tone.* Weak psycho-sensory stimuli result in reflex pupillary dilatation in both normal and ciliary ganglionectomized eyes, the effect being a little larger in the normal eye, as we should expect.

THE EFFECTS OF DRUGS

The main effects of drugs on the pupil are largely what one would expect of a system innervated by both parasympathetic and sympathetic fibres.

Parasympathetic Drugs. It will be recalled that parasympathetic activity

---

* In the cat, in contrast to man and monkey, the sympathetic supply to the iris does not pass through the ciliary ganglion; hence in this animal damage to, or removal of the ganglion does not influence sympathetic activity.

is mediated by the liberation of acetylcholine, so that this substance and drugs with a similar chemical constitution will behave in the same way as parasympathetic stimulation, causing a constriction of the pupil. Thus acetylcholine, pilocarpine, muscarine and many synthetic compounds are *parasympathomimetic*, by virtue of their chemical similarity to the parasympathetic chemical mediator. Another group mimics the parasympathetic by virtue of their inhibitory action on cholinesterase, the enzyme that destroys acetylcholine at its site of action and thus prevents the chemical mediator from acting indefinitely. They are said to *potentiate* the action of acetylcholine. Eserine, physostigmine and many modern synthetic drugs, such as diisopropylfluorophosphate (DFP), act in this way causing a prolonged pupillary constriction by preserving from destruction the acetylcholine, liberated by the tonic and reflex activity of the parasympathetic nerve endings. Another group of substances block the action of acetylcholine by competing for its sites of activity on the muscle; an example of this *parasympatholytic* action is given by atropine, and various synthetic drugs such as homatropine. These drugs, by blocking the tonic parasympathetic activity, leaving sympathetic activity normal, cause a long-lasting dilatation of the pupil, which is now unresponsive to light.

Ganglion Blockers, etc.

Since the parasympathetic innervation operates through the ciliary ganglion, we may expect drugs that act specifically on autonomic ganglia to influence the pupil; in fact, hexamethonium and pentolinium block transmission and cause dilatation of the pupil, due to cutting off of tonic action. Transmission through the ganglion is brought about by the liberation of acetylcholine, hence this transmitter, as well as nicotine and tetramethyl ammonium, causes discharge in the postganglionic fibres and thus brings about constriction of the pupil. However, the sympathetic system, causing dilatation of the pupil, also operates through a ganglion, this time the superior cervical, hence the final effect of a blocking agent on the pupil will depend on which division of the autonomic system is having the dominant action at the time of administration of the drug.

Sympathetic Drugs. The sympathetic chemical mediator is noradrenaline, whilst adrenaline, liberated by the adrenal medulla in response to stress, has an essentially similar action, in so far as it mimics many of the actions of the sympathetic system. These substances cause dilatation of the pupil.

Alpha- and Beta-Actions

Adrenaline has two types of action on tissues, suggesting that there are two types of receptor with which it may react; in accordance with Ahlquist's (1948, 1962) nomenclature, those actions, largely excitatory, that are blocked by such "classical" adrenaline antagonists as ergotoxin, dibenamine and phentolamine, are the result of reaction with the α-receptors; reaction with the β-receptors, largely inhibitory, is blocked by dichloroisoproterenol (DCI) and pronethanol. Noradrenaline, the adrenergic transmitter, has a mainly α-action, whilst isoproterenol (isoprenaline) has mainly β-action. In general, then, when the order of activity of the amines is: Adrenaline > Noradrenaline ≫ Isoproterenol, the influence of α-receptors is involved, whereas the order: Isoproterenol ≫ Adrenaline > Noradrenaline indicates β-activity. If the tissue has both α- and β-receptors, then any of these amines will activate both, and the response will be determined by the dominant one; by the use of the selective blockers, the individual contributions may be assessed.

INTRAOCULAR INJECTIONS

Bennett *et al.* (1961) injected sympathetic amines into the posterior chamber of rabbits, and the anterior chamber of dogs, measuring the pupillary effects. If the dilatation caused by adrenaline was put equal to 100, that due to noradrenaline was 37 and to isoproterenol 19; the action of isoproterenol was not blocked by DCI, so that we may attribute all three effects to α-activity. In these animals, then, pupillary activity seems to be dominated by α-receptors in the dilator.

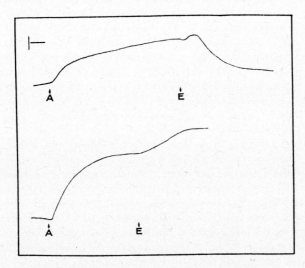

FIG. 244. Showing effects of epinephrine on the iris sphincter:
Top: At A contraction was induced by acetylcholine; at E relaxation is caused by epinephrine.
Bottom: After treatment with DCI, the contracture induced by acetylcholine is stronger and the effect of epinephrine is reversed into an additional contraction. (Van Alphen, Robinette & Macri, *Int. J. Neuropharmacol.*)

ISOLATED MUSCLE

The study of isolated muscle-strips allows the assessment of the effects on dilator and sphincter separately. Van Alphen, Robinette & Macri (1964), working on the cat, found that, with the dilator, the order of effectiveness in causing contraction was: Adrenaline > Noradrenaline ≫ Isoproterenol—. These effects were blocked by phenoxybenzene so that there is little doubt that the main activity is through α-receptors. Some β-activity was demonstrated by the observation that smaller doses of isoproterenol often produced a relaxation, which was blocked by DCI; again, phenoxybenzene not only blocked the contraction due to isoproterenol but could reverse this into a relaxation. With the sphincter, a relaxation of the muscle, previously caused to contract with acetylcholine, was caused by the amines in the order: Isoproterenol ≫ Adrenaline > Noradrenaline, thus indicating mainly β-activity. The action of adrenaline is indicated by Fig. 244; at A, contraction was induced with acetylcholine; at E, adrenaline was applied, which caused a relaxation after a brief contraction. When DCI was added (lower trace) the relaxation failed and adrenaline now caused contraction due to its α-action. Thus the cat sphincter is, indeed, influenced by adrenergic amines, the dominant influence being relaxation through β-receptors. Hence the effects of adrenaline in the cat are synergic, the α-activity on the dilator being assisted by the β-relaxation of the sphincter. Essentially similar conditions were found by Van Alphen, Kern & Robinette (1965) in the rabbit, and by Patil (1969) in the ox; in the monkey, however, α-activity was strongly predominant in both sphincter and dilator, so that under the action of sympathetic amines there should be antagonism.*

Ultrastructure of Nerve Terminals. The nature of nerve terminals on smooth muscle may be ascertained with some degree of probability by the types of vesicles within them, as seen in the electron microscope (see, for example, Richardson, 1964). Nishida & Sears (1969) found two types of axon running in the sphincter muscle of the albino guinea pig; 85 per cent belonged to the type containing agranular vesicles characteristic of cholinergic fibres whilst the remaining 15 per cent contained vesicles with a dense core characteristic of adrenergic fibres; these were morphologically identical with the majority of fibres in the dilator muscle. Twenty four hours after superior cervical ganglionectomy these fibres showed degenerative changes.

Other Drugs. Certain drugs act directly on smooth muscle, and since the sphincter is the stronger of the two iris muscles, their action on the pupil is largely determined by whether they cause this muscle to relax or contract. Thus *histamine* causes a powerful miosis which overcomes the mydriasis of atropine. *Morphine* and veratrine cause a miosis by direct action on the plain muscle; the pin-point pupil of morphia addicts is thought to be due, however, to the action of the drug on the central nervous system. The ions *barium*, *strontium* and *potassium* stimulate smooth muscle and cause miosis. We have seen that mechanical stimulation of the peripheral end of the trigeminal nerve causes constriction of the pupil (p. 47) and this is due to the liberation of one or more prostaglandins previously called *irin* by Ambache.

Pupillary Unrest

It is clear from the above that the state of the pupil at any moment is determined by a variety of synergic and antagonistic nervous influences; in general, external circumstances—light, and proximity of the fixation point—tend to cause constriction whilst the internal factors of sensation, and psychic activity generally, cause dilatation. The continual interplay between these opposing forces results in a constant state of pupillary activity—the pupil is restless—and the size of the pupil under any given conditions is a fluctuating quantity. The fact that both pupils fluctuate in size in harmony shows that the phenomenon of pupillary unrest is determined centrally, and is not merely an example of the independent rhythmic activity so frequently seen in smooth muscle (Stark, Campbell & Atwood). The pathological condition in which this activity is abnormally great is called *hippus*.

---

* Electrical stimulation of the excised cat sphincter causes contraction, abolished by atropine; in the presence of the latter, the contraction is replaced by active relaxation, presumably due to electrical stimulation of the adrenergic fibres (Schaeppi & Koella, 1964).

GENERAL REFERENCES

Loewenfeld, I. E. (1958). Mechanisms of reflex dilatation of the pupil. *Doc. Ophthal.* **12**, 190–448.

Lowenstein, O. & Loewenfeld, I. E. (1969). The Pupil. In *The Eye*, 2nd ed., Ed. Davson, vol. III, pt. 2. London & New York: Academic Press.

SPECIAL REFERENCES

Ahlquist, R. P. (1948). A study of the adrenotropic receptors. *Am. J. Physiol.* **153**, 586–600.

Ahlquist, R. P. (1962). The adrenotropic receptor. *Arch. int. Pharmacodyn.* **139**, 38–41.

Alexandridis, E. & Koeppe, E. R. (1969). Die spektrale Empfindlichkeit der für den Pupillenlichtreflex verantwortlichen Photoreceptoren beim Menschen. *v. Graefes' Arch. Ophthal.* **177**, 136–151.

Alpern, M. & Campbell, F. W. (1962). The spectral sensitivity of the consensual light reflex. *J. Physiol.* **164**, 478–507.

Ambachie, N. (1959). Further studies on the preparation purification and nature of irin. *J. Physiol.* **146**, 255–294.

Bennett, D. R., Reinke, D. A., Alpert, E., Baum, T. & Vasquez-Leon, H. (1961). The action of intraocularly administered adrenergic drugs on the iris. *J. Pharmacol.* **134**, 190–198.

De Groot, S. G. & Gebhard, J. W. (1952). Pupil size as determined by adapting luminance. *J. Opt. Soc. Am.* **42**, 492–495.

Hess, W. R. (1957). *The Functional Organization of the Diencephalon*. New York: Grune & Stratton.

Jampel, R. S. (1959). Representation of the near-response on the cerebral cortex of the macaque. *Am. J. Ophthal.* **48**, pt. 2, 573–582.

Lowenstein, O. (1956). The Argyll Robertson pupillary syndrome. *Am. J. Ophthal.* **42**, pt. 2, 105–121.

Lowenstein, O. & Loewenfeld, I. E. (1958). Electronic pupillography. *Arch. Ophthal.* **59**, 352–363.

Lowenstein, O. & Loewenfeld, I. E. (1959). Scotopic and photopic thresholds of the pupillary light reflex in normal man. *Am. J. Ophthal.* **48**, Pt. 2, 87–98.

Lowenstein, O. & Loewenfeld, I. E. (1959). Influence of retinal adaptation upon the pupillary reflex to light in normal man. *Am. J. Ophthal.* **48**, pt. 2, 536–550.

Magoun, H. W. & Ranson, S. W. (1953). The central path of the light reflex. *Arch. Ophthal.* **13**, 791–811.

Nisida, I. & Okada, H. (1960). The activity of the pupillo-constrictor centres. *Jap. J. Physiol.* **10**, 64–72.

Nisida, I., Okada, H. & Nakano, O. (1960). The activity of the cilio-spinal centres and their inhibition in pupillary light reflex. *Jap. J. Physiol.* **10**, 73–84.

Nishida, S. & Sears, M. (1969). Fine structural innervation of the dilator muscle of the iris of the albino guinea pig studied with permanganate fixation. *Exp. Eye Res.* **8**, 292–296.

Nishida, S. & Sears, M. (1969). Dual innervation of the iris sphincter muscle of the albino guinea pig. *Exp. Eye Res.* **8**, 467–469.

Okada, H., Nakano, O., Okamoto, K., Nakayama, K. & Nisida, I. (1960). The central part of the light reflex via the sympathetic nerve in the cat. *Jap. J. Physiol.* **10**, 646–658.

Patil, P. N. (1969). Adrenergic receptors of the bovine iris sphincter. *J. Pharmacol.* **166**, 299–307.

Richardson, K. C. (1964). The fine structure of the albino rabbit iris with special reference to the identification of adrenergic and cholinergic nerves and nerve endings in its intrinsic muscles. *Am. J. Anat.* **114**, 173–184.

Schaeppi, U. & Koella, W. P. (1964). Adrenergic innervation of cat iris sphincter. *Am. J. Physiol.* **207**, 273–278.

SCHWEITZER, N. M. J. (1956). Threshold measurements on the light reflex of the pupil in the dark adapted eye. *Doc. Ophthal.* **10,** 1–78.

SILLITO, A. M. & ZBROZYNA, A. W. (1970). The localization of pupilloconstrictor function within the mid-brain of the cat. *J. Physiol.* **211,** 461–477.

STARK, L., CAMPBELL, F. W. & ATWOOD, J. (1958). Pupil unrest: an example of noise in a biological servomechanism. *Nature, Lond.* **182,** 857–858.

VAN ALPHEN, G. W. H. M., KERN, R. & ROBINETTE, L. (1965). Adrenergic receptors of the intraocular muscles. *Arch. Ophthal. N.Y.* **74,** 253–259.

VAN ALPHEN, G. W. H. M., ROBINETTE, S. L. & MACRI, F. J. (1964). The adrenergic receptors of the intraocular muscles of the cat. *Int. J. Neuropharmacol.* **2,** 259–272.

PART 4

ACCOMMODATION

THE change in the refractive power of the eye, when the image of a near object is brought into focus on the retina, is described as accommodation; such a process must involve an *increase* in the dioptric power of the system (p. 561).

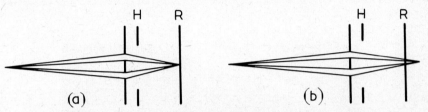

FIG. 245. Scheiner's experiment. (*a*) The eye is focused on a point close to the eye; this point is seen single. (*b*) The eye is focused for infinity; a double blurred image of the point is now formed on the retina.

Scheiner's Experiment. Scheiner, as long ago as 1619, demonstrated that the eye changes its dioptric power on viewing a near point. Fig. 245 illustrates his experiment. A near object was viewed through a screen in which two holes were bored sufficiently close together so that light, passing through both, entered the pupil. The near object was seen single as we should expect (Fig. 245 *a*); on focusing the eye on a distant object, however, the near object immediately appeared double. In this case the image of the near object is formed behind the retina (Fig. 245*b*) and since two separate pencils of light are concerned with each point, a double blurred image is formed.

THEORY OF ACCOMMODATION

According to the modern theory of accommodation, which is largely based on Helmholtz's ideas, the capsule has sufficient elasticity to mould the lens into a more strongly curved system than is necessary for distant vision. The elasticity of the capsule is held in check by the normal tension in the zonule, so that accommodation consists in a relaxation of the tension in the zonule which permits the capsule to mould the lens into a more strongly curved system. The relaxation of the zonule follows a *contraction* of the ciliary muscle.

The Ciliary Muscle. The ciliary muscle is classically described as containing three types of unstriped muscle fibre:

(*a*) *Meridional fibres* arising from the epichoroid and attaching to the scleral spur.

(*b*) *Radial fibres*, situated more internally and anteriorly, arranged in the manner of a fan.

(*c*) *Circular fibres*, continuous with, and inseparable from the radial fibres, running round the free edge of the ciliary body just behind the root of the iris.

However, Fincham has shown that this formulation is rather too rigid; the fibres take origin from the epichoroid, and in the outer part of the muscle (nearest the sclera) they run a a purely meridional course, to be inserted in the scleral spur.

More internally, the fibres make up a branching meshwork and it is this part that has been incorrectly called the radial portion—the fibres are not arranged in a regular radial manner but, as indicated, in a reticular meshwork. As a result of the branching of the bundles of fibres, a number acquire a more or less circular direction, a tendency that becomes more pronounced towards the inner edge of the muscle. Thus the innermost edge consists essentially of circular fibres— the so-called *sphincter muscle of Müller*—but they are not to be regarded as making up a separate muscle. The circular elements give off short tufts of fibres which are inserted into a ring of elastic tissue (A of Fig. 246) at the base of the iris. Another layer of elastic tissue (B) is situated between the muscle and the pigment epithelium of the ciliary body. The elastic elements presumably resist the tension of the zonule.

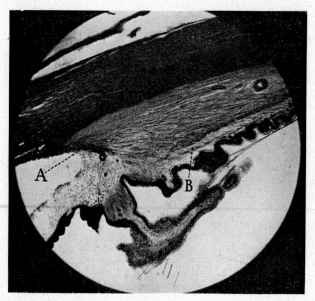

Fig. 246. The ciliary muscle. A and B indicate elastic tissue. (Fincham, *Brit. J. Ophthal.*)

EFFECTS OF CONTRACTION

The general anatomical characteristics of the ciliary muscle indicate that its contraction must relax the zonule; the meridional fibres pull on the epichoroidal tissue and drag the ciliary body forward. The circular fibres, probably the most important in accommodation, must, on shortening, decrease the radius of the circle that they constitute; this can only be achieved by a radial extension of the ciliary body which must reduce the tension in the zonule (Fig. 247). These anatomical considerations are confirmed by the observation that during accommodation the external surface of the ciliary body moves forward by 0·5 mm. Moreover, careful observation of the lens during accommodation shows that it becomes tremulous as though its normal taut suspension had been relaxed; again, the equator of the capsule of the unaccommodated lens shows fine wrinkles, proving that it is held under tension; during accommodation these wrinkles are smoothed out.

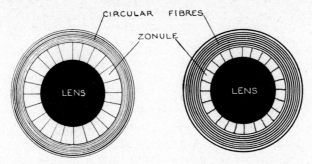

Fɪɢ. 247. Illustrating the action of the circular fibres in relaxing the tension of the zonule.

The Zonular System. The actual mechanism whereby contraction of the ciliary muscle affects the tension in the zonular fibres is not completely clear, since the muscle has no direct connections with the fibres. The structure of the human zonule has been examined in some detail by Rohen & Rentsch (1969) with a view to elucidating this point. According to this study, the fibres fall into two classes, the separation into two groups occurring in the middle third of the ciliary body; behind this, the system is a uniform mat of fibres. As indicated in Fig. 248 the main system of "holding fibres", H, runs forward towards the pars orbicularis where the individual fibres split at the *zonular fork*, ZF, and finally both branches insert into the lens capsule. Branching off from the holding fibres is the second system of *tension fibres*, T, inserting into the ciliary epithelium in the orbicularis region of the ciliary body. It seems that it is because of the close relation of these tension fibres to the ciliary epithelium on the ciliary processes, and in the valleys between, that effects of changes in volume of the ciliary body, due to contraction of the muscle,

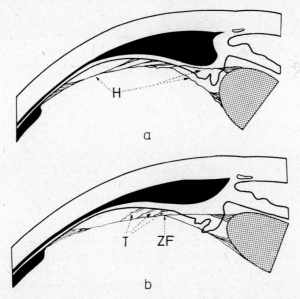

Fɪɢ. 248. Illustrating Rohen & Rentsch' concept of the structure of the zonular fibres. (a) The accommodated state. (b) The relaxed state. H are the *holding fibres* which split at the *zonular fork*, ZF. Branching from the holding fibres are the *tension fibres*, T, inserting into the ciliary epithelium. (After Rohen & Rentsch, *v. Graefes' Arch. Ophthal.*)

are transmitted to the lens. Thus Fig. 248 illustrates the postulated changes; below, the muscle is relaxed in the unaccommodated state, and the holding fibres are stretched, maintaining tension on the capsule. Above, the muscle has contracted and the changed shape of the ciliary body has caused the tension fibres to become stretched, and this has permitted the main holding fibres to relax. It is essentially the tension fibres that cause the important change in tension in the capsule, and they do this by virtue of their attachment to the ciliary epithelium, the changed shape of the ciliary body causing a forward and axial pull.

Changes in the Lens

The actual changes in the lens during accommodation were studied by Helmholtz by means of the Purkinje images (p. 603); during accommodation image III, formed by the anterior surface of the lens, became smaller, indicating an increase in the curvature of this surface (Fig. 249). Studies on individuals with congenital aniridia have shown that the lens does become smaller (1·25 mm) and thicker (1·3 mm) as Helmholtz's theory would demand.

FIG. 249. The change in the size of the Purkinje image, formed by the anterior surface of the lens, following accommodation. The object, whose image was formed by the reflecting surfaces of the eye, consisted of three bright lamps arranged in a triangle. Left: unaccommodated eye. Right: accommodated eye. (Fincham, *Brit. J. Ophthal.*)

Fincham's work suggested that the central zone of the anterior surface became more convex in relation to the peripheral parts, the surface becoming what Fincham called "conoid", but Fisher's (1969) measurements of the lens in both accommodated and unaccommodated states have shown that the anterior surface may be described accurately as that of an ellipsoid; Fisher also calculated that the energy required for this postulated conoid transformation could not be supplied by the capsule.

Whatever the exact changes may be, they must involve a rearrangement of the lens substance since this is quite incompressible; the general increase in curvature must mean a decrease in the surface of the lens and this must mean that the shape approaches more closely to that of a sphere (in the sphere we have the condition of maximum volume for minimum area; if we decrease the surface of a body and maintain its volume the same, we must make it more nearly spherical). The effects of this rearrangement may be observed with the slit-lamp microscope when there are mobile opacities in the lens; the movement consists generally of an axial flow from the periphery, as we should expect.

EXCISED LENS

It follows from the theory of accommodation that the excised lens must be in a state of maximum accommodation; Hartridge demonstrated the truth of this proposition in a remarkably neat manner. He estimated the dioptric power of the cat's eye *in situ* by measuring the magnification of a portion of the fundus observed through a modified ophthalmoscope. On removal of the lens its curvatures were measured and its dioptric power calculated. He showed that the combined power of the cornea and excised lens amounted to 12 dioptres more than the power of the eye *in vivo*, *i.e.* the lens, with the curvatures it possessed in the excised state, would provide the eye with an extra 12 dioptres of refractive power. If these measurements are correct, this indicates that the theoretical amplitude of accommodation of the cat is 12 dioptres; in fact, this is found experimentally to be only 2 dioptres (Ripps, Breinin & Baum, 1961), when the ciliary ganglion is stimulated maximally. Presumably the absence of circular fibres in the cat's ciliary muscle limits the loosening of the zonule.

THE LENS CAPSULE

According to Fincham, the shape of the lens is largely determined by the structure of the capsule; where the capsule is thinnest it might be expected that the lens contents would bulge out, thus giving the larger curvature of the posterior surface where the capsule is, indeed, thinnest (Fig. 250).

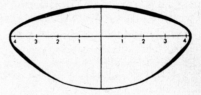

FIG. 250. The lens capsule. (Fincham, *Brit. J. Ophthal.*)

MECHANICAL PARAMETERS

Fisher (1969a) has measured the mechanical parameters of the human lens capsule, in particular Young's modulus, the ratio of *Stress/Strain*, the *stress* being the force applied to unit length of the material to produce a given change in length, the *strain*; its dimensions are dynes/cm $\div$ cm = dynes/cm². Thus the greater the value of Young's modulus the greater the force the capsule can exert when it has been stretched a given amount. The results have been compared with those for collagen and aortic elastic tissue in Table 13; in addition, the maximal percentage elongations of the tissues are also given. It will be seen that the Young's modulus is much less than that of collagen, so that a collagenous capsule would be able to exert much more power on the lens substance; however, it would only stand a 5 per cent elongation before rupturing, whereas it may be calculated that, during accommodation, an elongation of capsular fibres as great as 6·5 per cent could occur. Elastic tissue would allow of far greater elongation, but its low modulus of elasticity would not provide the capsule with adequate energy. Fisher confirmed that the human lens capsule was thicker anteriorly than posteriorly, and since the elastic energy is a function of both modulus of elasticity and thickness, this means that we may expect the anterior capsule to exert the dominant influence during changes of shape; his measurements confirmed that this was, indeed, so.

Lens Substance. The ultrastructure of the lens suggests that its well orientated system of fibres, meeting at suture lines, would give it sufficient coherence that, independently of the capsule, it would have some tendency to resume its original shape after deformation; Fincham remarked on the tendency of the decapsulated lens to resume an unaccommodated shape, and the mechanical studies of Kikkawa & Sato (1963) and Itoi, Itoi & Kaneko (1965) on the excised lens in its normal and decapsulated condition indicate that a part of the visco-elastic property does, indeed, reside in the substance rather than the capsule.

TABLE 13

Young's Modulus and Minimum per cent Elongation of Lens Capsule and other Tissues (Fisher, 1969a).

Tissue	Young's Modulus (Dynes/cm^2)	Maximal Elongation (%)
Human lens capsule	$2 \cdot 0 – 6 \cdot 0 \times 10^7$	29
Rat collagen	300×10^7	5
Aortic elastic tissue (human)	$0 \cdot 3 \times 10^7$	127

The Change in Accommodation with Age

The amplitude of accommodation decreases progressively with age (presbyopia); in a child of 10 years it is, on the average, about 14 D; at 40 it is 6 D; and at 60 it is only 1 D.* The factors determining this reduction in power have been lucidly discussed by Weale (1962) but they are not yet completely understood.†
It is unlikely that alterations in the power of the ciliary muscle to contract are important, and the generally accepted view is that there is a failure of the lens capsule to modify the shape of its contents; thus Fincham observed that the curvature of the excised senile lens was considerably less than that of a juvenile one. This failure could be due to a hardening of the lens material, *sclerosis*, to a decrease in modulus of elasticity or thickness of the capsule, or to all three of these factors.

Capsular Surface Tension. In an attempt to assess the importance of the various factors Fisher (1969b) has measured the capsular surface tension, i.e. the energy that unit area of it can exert, at rest and during accommodation; this was computed from the measured changes in outline of the anterior capsule and the force required to stretch it. In Fig. 251 the effective energy of the capsule is plotted against the amplitude for lenses taken from different aged eyes; there is a remarkable correspondence between the two since the capsule actually increases in thickness with age up to 60 years; this suggests that the modulus of elasticity, which decreases from some 6×10^7 dynes/cm^2 in childhood to $1 \cdot 5 \times 10^7$ dynes/cm^2 in extreme old age, plays an important role. However, this parallelism does not rule out a significant contribution of increased stiffness of the lens material; it

---

* Hamasaki, Ong & Marg (1956) consider that after the age of 52 the amplitude of accommodation is actually zero the apparent ability to accommodate being due to the depth of focus of the eye (p. 405); their objective measurements confirmed this.
† The effects of age on several features of the refractive system of the eye have been discussed by Weale; thus the increase in size of the lens (from about 140 mg at 0–10 years to about 230 mg at 70–80) means that the equator approaches the ciliary body, and thus the zonule will tend to be slack or else the lens will become more spherical. The increase in radius of curvature that also occurs will tend to counteract the optical effects of this, however.

does show, however, that the earlier tendency to attribute presbyopia entirely to lens sclerosis is unjustified.*

When the effects of altered elasticity and lens shape† were taken into account, the measurements indicated that, in the absence of changed power of the ciliary muscle, about 50 per cent of the loss of accommodative power would be due to lens hardening.

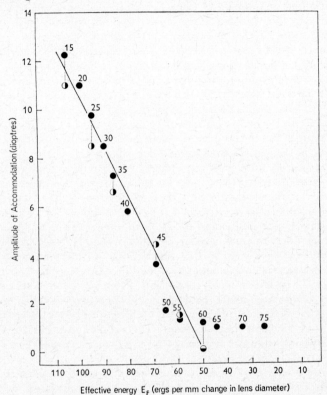

FIG. 251. Showing presbyopic changes in amplitude of accommodation and effective energy of the lens capsule. ● Duane's data; ◑ Bruckner's data mean per decade; ◐ Hamasaki's data. Age in years (15 to 75). (Fisher, *J. Physiol.*)

The Nervous Mechanism

Parasympathetic. Contraction of the ciliary muscle is brought about by the activity of the parasympathetic fibres of N III, with their origin in the Edinger-Westphal nucleus; the fibres relay in the ciliary ganglion. Stimulation of N III causes an increase in the refractive power of the eye (the eye becomes

---

* The difficulty in accepting lens sclerosis as the sole cause of presbyopia is that the condition begins essentially in childhood, when it is unlikely that "senile" changes in the lens fibres would have occurred.

† The effective energy of the capsule, in relation to its power of altering refractive power, depends on the shape of the lens as well as on the modulus of elasticity; the flatter human lens allows the capsule to exert a greater moulding pressure for a given modulus of elasticity than the more spherical lenses of the cat and rabbit; to some extent, then, the differences in accommodative power of these species are attributable to this since their elasticities are not greatly different: 5, 0·9 and 2·5 × 10⁷ dynes/cm² respectively.

myopic), an effect that is inhibited by blocking the ciliary ganglion or by the application of atropine to the eye. Eserine, which potentiates the action of acetylcholine, and pilocarpine, which mimics its action, cause a spasm of accommodation, that by the former being due to failure to hydrolyse the transmitter liberated as a result of tonic action of the nerve.

Sympathetic. Opinions have varied as to whether the sympathetic plays any role in accommodation (see, e.g. Cogan, 1937); the action of the ciliary muscle is antagonized by the elasticity of the capsule, and fine adjustments are therefore possible by means of the interplay of these opposing forces; an antagonistic muscle, innervated by the sympathetic, whose contraction tightens the zonule and therefore flattens the lens, may be thought to be unnecessary in man on this account. Some studies of Olmsted and Morgan and their colleagues have suggested, however, that the sympathetic nerves might cause an active flattening of the lens, insults likely to cause generalized sympathetic stimulation, such as a tap on the nose of a rabbit, being said to cause an increase of the animal's normal hypermetropia by about 1 D; stimulation of the cervical sympathetic in the anaesthetized cat and dog caused large changes in refractive condition, especially if the parasympathetic division to the eye was cut.

In primates Törnqvist (1966, 1967) has shown that stimulation of the cervical sympathetic causes a small decrease in the dioptric power of the eye; this becomes greater if the stimulation is carried out against a background of parasympathetic activity, either by pretreatment of the eye with pilocarpine or by stimulation of the oculomotor nerve; it was never greater than 1·5–2·0 D, however. He pointed out that the responses were so slow, taking some 5–10 seconds to begin and 10–40 seconds to reach a maximum, that they are unlikely to have physiological significance. These effects could, conceivably, be due to the presence of genuine radial fibres, innervated by the sympathetic, whose action would be, in contraction, to tighten the zonule, although it has been argued that the effects are due to constriction of the blood vessels in the eye with a consequent decrease in volume of the ciliary body leading to an increase in the pull on the zonule. Thus Fleming & Olmsted (1955) observed that the transient myopia that occurs on cutting the cervical sympathetic runs a parallel course to the transient dilatation of blood vessels. However, Törnqvist (1966) found that the effects of sympathetic stimulation were not blocked by a typical sympathetic α-blocker, phentolamine, and this would be expected if the action depended on the vascular effects of sympathetic stimulation; again, the effect on accommodation was inhibited by a β-blocker, propanolol, which had no effect on the vasoconstriction due to sympathetic stimulation.*

EXCISED MUSCLE

Studies on excised strips of ciliary muscle leave little doubt that sympathetic amines are active, but the effect varies with the species; in the cat the order of increasing effectiveness in causing relaxation is Isoproterenol ≫ Noradrenaline > Adrenaline, an order indicating a β-action of the sympathetic, especially since the relaxation due to isoproterenol is blocked by DCI. That there are also α-receptors is indicated by the fact that the relaxation caused by adrenaline and noradrenaline is converted to contraction when the β-activity is blocked by DCI (Bennett et al., 1961). In the monkey, sympathetic activity is exclusively β whilst in the rabbit the action is mainly α with a little β.

* Biggs, Alpern & Bennett (1959) found a decrease in accommodative power in humans after subconjunctival injections of adrenaline.

The Stimulus for Accommodation

It might be thought that the effective stimulus for accommodation would be the blurring of the retinal image, but if this were the only clue, the centre responsible would be unable to determine whether the object had come nearer or moved farther away, since in both cases the image would be blurred. We should therefore expect the eye to "grope" for the correct change in power, whereas it was found by Fincham (1951) that when the eye observes an object, and the vergence* of the light is altered, the accommodative effort is correct; that is, if the vergence is changed so that the object is apparently closer, the eye increases its power, and *vice versa*. There is no oscillation, as we should expect in an instrument unable to distinguish the nature of the vergence-change.

Aberrations. What then is the stimulus? Fincham's experiments indicated that in many subjects the chromatic aberration of the retinal image provided the necessary clue, the aberration rings being theoretically different according to the nature of the out-of-focus image. Thus, if the object were too close, the image of a point would be a light disc surrounded by a red fringe, whilst if it were too far away the disc would be surrounded by a blue fringe. By using monochromatic light many subjects were unable to make correct adjustments. A certain number of subjects were able to do so, however, and it would seem that spherical aberration provides a clue (Campbell & Westheimer) since reducing the spherical aberration of the eye reduced the subject's accuracy in accommodating in monochromatic light.†

Fluctuations. When a subject views a near object steadily his accommodation fluctuates through a range of about 0·25 D with a dominant frequency of the order of 2 cycles/sec (Campbell, Robson & Westheimer); this is presumably analogous with pupillary unrest, and with the high-frequency tremor in the eye movements although here the frequency is in the range of 75 cycles/sec. Since the unrest is considerably reduced when the subject views an empty field it is possible that these oscillations are determined by variations in the clarity of the image—retinal feedback.

Depth of Focus of the Eye

In a lens system there is theoretically only one position of the object that will give an image at a definite distance from the principal plane; hence with the eye, where the position of the retina is fixed, there is only one position, for a

---

* By *vergence*, in this context, is meant the degree of divergence or convergence of the rays of light in a pencil as they strike the principal plane of a refracting system; the closer an object, the greater the vergence of its rays, hence the reciprocal of its distance from the principal plane may be used as a measure of vergence; if this is in metres the unit of vergence is the dioptre; thus the vergence of the light from an object is the *dioptric value of the object-distance*. We may note that the term vergence is also applied to the degree of convergence or divergence of the fixation axes; this often leads to ambiguities and confusion. Thus by "vergence-induced accommodation" a writer could mean accommodation induced by convergence or by changed light-vergence.

† Fincham (1951) observed that if a subject, who failed to accommodate correctly in monochromatic light, was allowed to scan the object in view, correct accommodation became possible, and he pointed out that when the eye turns through a small angle from the point of fixation the rays of light at the opposite sides of the blur-circles on the retina acquire different degrees of obliquity to the surface, this difference being opposite in direction when the image is out of focus due to being too close and too far away. The different degrees of obliquity will be translated into different sensations of brightness by the operation of the Stiles-Crawford effect.

given state of the dioptric power, where an object will produce a distinct image. For the emmetropic eye, this is at infinity. When the object is brought nearer to the eye, the image of a point on it becomes a "blur-circle" because the true image is behind the retina. If the blur-circle is not large, however, it is not recognized as differing from a point and the object may thus be brought within a finite distance of the eye before the blur-circle becomes so large that the image is appreciated as "indistinct". The extent to which the object may be moved before the points on it appear out of focus is called the *depth of focus*. As Fig. 252 shows, the depth of focus obviously depends on the pupil-size. It

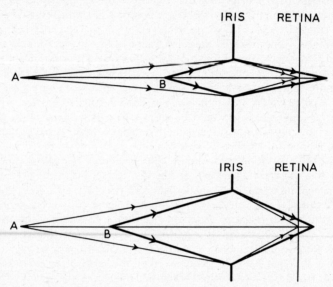

FIG. 252. Illustrating the influence of the size of the pupil on the depth of focus. With the small pupil (above) B produces a blur-circle of fixed size when it is closer to the eye than with the large pupil (below). The depth of focus (AB) is smaller with the larger pupil.

will also depend on the distance of the fixation point from the eye; if this is large (e.g. 100 m) the point may be moved many metres closer to the eye before it goes out of focus; if it is small (e.g. 25 cm) a movement of a centimetre or two is enough to blur the image. For the human eye the figures obtained by Campbell were as follows:

Pupil diameter	Depth at infinity	Depth at 1 metre
1 mm	From infinity to 1·25 m	5·0 m to 56 cm
2 mm	2·33 m	1·8 m to 70 cm
3 mm	2·94 m	1·5 m to 75 cm
4 mm	3·57 m	1·4 m to 78 cm

Visual Acuity. Ogle & Schwartz presented the subject with a chart and changed the vergence of light from it by means of a lens, keeping the image-size constant. At a certain point the visual acuity was decreased and the change of vergence was recorded. This amounted to 0·6 dioptres and Ogle & Schwartz defined the depth of focus in terms of this amount of vergence. The depth of field, on the other hand, could be defined as the equivalent distance through which the chart would have been moved to produce this change of vergence. The *hyperfocal distance* was defined as the distance of the fixation point at which objects beyond this all appear in focus.

Tolerated Change of Vergence. Adamson & Fincham measured the change of vergence tolerated by the eye, i.e. if an object was made to appear closer or farther away, by means of lenses, a point was reached when the eye adjusted its accommodation, and this was indicated by a change of its refracting power as measured with an optometer. The average change in vergence tolerated by the eye was found to be ± 0.25 D. An object 1 m away (vergence -1 D) may be moved so that the vergence changes to -0.75 D or to -1.25 D without blurring of the image; i.e. it can be moved to 1.33 m or 0.80 m, giving a depth of focus of 53 cm. If the object is 25 cm away (vergence -4 D) the limits of tolerance are -3.75 D and -4.25 D, representing a depth of focus of 3.0 cm (26.6—23.6 cm). The difference between the values found by Ogle & Schwartz and Fincham depends on their different criteria.*

GENERAL REFERENCES

ALPERN, M. (1969). Accommodation. In *The Eye*, Ed. Davson, 2nd ed., vol. III, pt. 2. London & New York: Academic Press.

FINCHAM, E. F. (1937). The mechanism of accommodation. *Br. J. Ophthal.*, Mon. Suppl. No. 8.

SPECIAL REFERENCES

ADAMSON, J. & FINCHAM, E. F. (1939). The effect of lenses and convergence upon the state of accommodation of the eye. *Trans. Ophthal., Soc. U.K.* **59**, 163–179.

BIGGS, R. D., ALPERN, M. & BENNETT, D. R. (1959). The effect of sympathomimetic drugs upon the amplitude of accommodation. *Am. J. Ophthal.* **48**, pt. 2, 169–172.

CAMPBELL, F. W., ROBSON, J. G. & WESTHEIMER, G. (1959). Fluctuations in accommodation under steady viewing conditions. *J. Physiol.* **145**, 579–594.

CAMPBELL, F. W. & WESTHEIMER, G. (1958). Sensitivity of the eye to difference in focus. *J. Physiol.* **143**, 18 P.

CAMPBELL, F. W. & WESTHEIMER, G. (1959). Factors influencing accommodation responses of the human eye. *J. opt. Soc. Am.* **49**, 568–571.

COGAN, D. G. (1937). Accommodation and the autonomic nervous system. *Arch. Ophthal.* **18**, 739–766.

FINCHAM, E. F. (1951). The accommodation reflex and its stimulus. *Br. J. Ophthal.* **35**, 381–393.

FISHER, R. F. (1969a). Elastic constants of the human lens capsule. *J. Physiol.* **201**, 1–19.

FISHER, R. F. (1969b). The significance of the shape of the lens and capsular energy changes in accommodation. *J. Physiol.* **201**, 21–47.

FLEMING, D. G. & OLMSTED, J. M. D. (1955). Influence of cervical ganglionectomy on the lens of the eye. *Am. J. Physiol.* **181**, 664–668.

HAMASAKI, D., ONG, J. & MARG. E. (1956). The amplitude of accommodation in presbyopia. *Am. J. Optom.* **33**, 3–14.

* Campbell & Westheimer (1958) prevented their subjects from accommodating with atropine and they were asked to recognize the rhythmic blurring of the image of an object brought nearer and farther from the eye; when this rhythmic excursion was too small, no change was experienced, and on increasing it a point was reached when the rhythmical changes were noted. The average "sensitivity to difference of focus" was 0.2 D. Oshima (1958) found a much higher sensitivity, 0.02—0.06 D.

ITOI, M., ITO, N. & KANEKO, H. (1965). Visco-elastic properties of the lens. *Exp. Eye Res.* **4**, 168–173.

KIKKAWA, Y. & SATO, T. (1963). Elastic properties of the lens. *Exp. Eye Res.* **2**, 210–215.

MORGAN, M. W., OLMSTED, J. M. D. & WATRUS, W. G. (1940). Sympathetic action in accommodation for far vision. *Am. J. Physiol.* **128**, 588–591.

OGLE, K. N. & SCHWARTZ, J. T. (1959). Depth of focus of the human eye. *J. opt. Soc. Am.* **49**, 273–280.

OLMSTED, J. M. D. & MORGAN, M. W. (1939). Refraction of the rabbit's eyes in the unexcited and excited state. *Am. J. Physiol.* **127**, 602–604.

OLMSTED, J. M. D. & MORGAN, M. W. (1941). The influence of the cervical sympathetic nerve on the lens of the eye. *Am. J. Physiol.* **133**, 720–723.

OSHIMA, S. (1958). Studies on the depth of focus of the eye. *Jap. J. Ophthal.* **2**, 63–72.

RIPPS, H., BREININ, G. M. & BAUM, J. L. (1961). Accommodation in the cat. *Trans. Am. Ophthal. Soc.* **59**, 176–193.

ROHEN, J. W. & RENTSCH, F. J. (1969). Der konstruktive Bau des Zonulaapparatus beim Menschen und dessen funktionelle Bedeutung. *v. Graefes' Arch. Ophthal.* **178**, 1–19.

TÖRNQVIST, G. (1966). Effect of cervical sympathetic stimulation on accommodation in monkeys. *Acta physiol. scand.* **67**, 363–372.

TÖRNQVIST, G. (1967). The relative importance of the parasympathetic and sympathetic nervous systems for accommodation in monkeys. *Invest. Ophthal.* **6**, 612–617.

WEALE, R. A. (1962). Presbyopia. *Br. J. Ophthal.* **46**, 660–668.

PART 5

THE NEAR RESPONSE

Convergence, Accommodation and Miosis

WHEN a near object is focused, three events occur simultaneously—convergence, accommodation and pupillary constriction (miosis). The motor pathway for all three activities is through the nucleus of N. III, accommodation and miosis being mediated by the autonomic division of this motor nucleus—the Edinger-Westphal nucleus; and there is no doubt that these pathways, along the oculomotor nerve, are separate since the three reactions may be abolished selectively by appropriate treatment of the nerve or ciliary ganglion. As we have seen, the central coordination of this triad of responses is probably at a higher level of the brain, in the cerebral cortex. The question naturally arises as to the extent to which these responses to the near stimulus are independent of each other. Is it possible to converge without accommodating? Can one accommodate without converging? If the answers to these questions are yes, is pupillary constriction tied preferentially to the convergence or the accommodation? These questions, so important for the theory of squint, have been examined and discussed intensively, whilst in recent years Fincham in England and Alpern in America have submitted them to exact quantitative study.

Accommodation and Convergence. Suppose first we examine a normal young subject and measure his accommodation, by means of an objective optometric test, whilst he converges on an object that is brought closer and closer to his eyes. Clearly, if he is to see single and distinctly, the convergence and accommodation must "match", 1 dioptre of accommodation occurring with 1 metre-angle of convergence. The results of such a measurement are shown by the line marked "normal" in Fig. 253; it will be seen that accommodation does not match exactly; and since diplopia does not occur, we may attribute this to an "accommodation lag", the eye tolerating a certain degree of blurring of the image that is important for depth of focus (p. 405).* The line marked "normal" of Fig. 253 thus gives us a standard of reference, telling us how convergence and accommodation go together when the stimuli for these responses match, and presumably reinforce each other.

Accommodative Convergence. Next we may allow the two eyes to view separate objects independently by a haploscopic device; under these conditions there is no compulsion for fusion and the eyes may adopt any degree of convergence without experiencing diplopia. The apparent nearness of the test-chart, seen by the right eye, is now varied by means of lenses, so that in order for it to be seen distinctly the right eye must accommodate. The line in Fig. 253 marked AC shows that the eyes converge as a result of the accommodation of the right eye; this *accommodative convergence* matches the accommodation over

---

* It has been argued, e.g. by Ogle, that a certain degree of "fixation disparity" is tolerated, in which case, of course, the failure of accommodation to match convergence could be due to errors in binocular fixation, the resultant physiological diplopia (p. 447) being in some way suppressed or tolerated. Not all workers are willing to accept the existence of a significant fixation disparity, however.

a certain range; but not exactly, so that with 6 dioptres of accommodation the convergence is only a little over 4 metre-angles.*

Convergence-Induced Accommodation. Next, we may measure the changes of accommodation that take place when the eyes are *made* to converge whilst the object remains at the same apparent distance—the *vergence* of the light (p. 405) is said to be fixed.† The *convergence-induced accommodation* is indicated in the line of Fig. 253 marked CA. Once again the linkage between the two processes is manifest, and in this case, too, the accommodation lags behind the convergence. This is because of the age of the subject who was 32 years old, and the lag is a sign of presbyopia, subjects under 24 years old showing an almost perfect match between convergence and accommodation. *In a young subject, therefore, the stimulus to convergence automatically induces an accommodation of the right order of magnitude for distinct vision.*

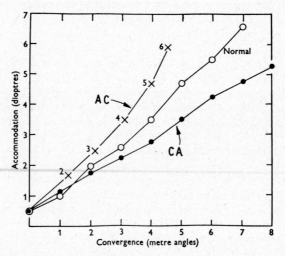

FIG. 253. The relation between accommodation and convergence. The curve marked AC represents accommodative convergence, i.e. the convergence that takes place when the eyes accommodate under conditions where altered convergence does not lead to diplopia. The curve marked CA represents the accommodation that takes place when the eyes converge, the arrangement being such that altered accommodation does not lead to blurring of the image. (Fincham & Walton, *J. Physiol.*)

Conflict between Reflexes. In the experimental arrangements so far employed there has been no conflict between the two stimuli, an alteration in the degree of convergence, for example, being compatible with single vision over the whole range. If we cause a subject to fixate a test-chart some 0·33 metres away, his accommodation for distinct vision is 3 D whilst his convergence is 3 metre-angles. By means of prisms, or other devices, we may make it necessary

---

* The ratio of the accommodative convergence, indicated by AC, over the accommodation, indicated by A, is called the AC/A ratio, and is considered a valuable diagnostic measurement. It has been examined in great detail by Alpern and his collaborators.
† Convergence was stimulated by bringing the object closer, but the effective vergence of the light was held constant by the use of very narrow pencils of light so that the rays entering the eye were effectively parallel for all distances of the object from the eye.

for him to converge further, or diverge, in order to maintain single vision. We may ask: What happens to his accommodation? If the linkage were still maintained, an increase in convergence would be accompanied by an increased accommodation, but since the vergence of the light has not been altered the chart would become blurred by the unnecessary increase in power. There would be thus a conflict between the accommodation reflex, which demands that the accommodation remain unchanged, and the convergence-induced accommodation that operates through the link. Fig. 254 shows that the link has some flexibility.

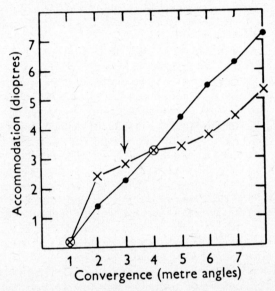

FIG. 254. Illustrating the convergence-induced accommodation that occurs when alterations of accommodation will lead to blurring of the image. The line through the circles shows the normal relationship between accommodation and convergence, whilst the line through the crosses shows the convergence-induced accommodation taking place when the subject views an object 0·33 m away whilst the degree of convergence is varied by a haploscopic device. (Fincham & Walton, *J. Physiol.*)

The line through the circles shows the convergence-induced accommodation that occurred under the earlier conditions where there was no compulsion to maintain any degree of accommodation. The curve through the crosses shows the convergence-induced accommodation during the present experiment; the arrow indicates the initial condition, the subject converging on an object 0·33 m away, showing 3 metre-angles of convergence and 3 D of accommodation. As convergence increases, accommodation does indeed follow it, but to a much smaller extent, an increase of convergence from 3 to 7 metre-angles being associated with an increase of only about 1 D, by contrast with an increase of 3 D that would have occurred under the "free conditions". Thus some blurring of the image has occurred, but this has acted as a "brake" on the convergence-induced accommodation. This experiment thus reveals, once again, the presence of a link between convergence and accommodation, but also demonstrates that the link is not rigid and may be relaxed when its effects militate against distinct vision.

RELATIVE AMPLITUDE OF ACCOMMODATION

The converse experiment is carried out by fixing the convergence and placing lenses in front of the eyes to alter the degree of accommodation. Any accommodative convergence, under these conditions, will cause diplopia, so that the compulsion for fusion operates as an inhibitor of this linked convergence. In fact we do indeed find that convergence is maintained to give single vision over quite a wide range of accommodation, but the conflict between the two requirements leads to a restriction of the amount of accommodation that can be exercised, in other words the *relative amplitude of accommodation* is less than the amplitude when the eyes are free to converge and diverge in accordance with the nearness or farness of the object.

Convergence-Accommodation Link. In general, then, we may conclude that accommodation and convergence are linked together by a common nervous mechanism that activates the two together. Because the stimulus of diplopia is more effective than that of blurred vision, we may consider that the response to nearness is primarily one of convergence; in a young person the associated activation of the accommodation mechanism is accurate enough to make the image fairly distinct, whilst the fine adjustment is provided by the accommodation reflex *per se*, which adjusts the focal power of the eye more exactly, using as clues the alterations of chromatic aberration etc. that take place when the image is not accurately focused. In older persons the convergence-induced accommodation is less adequate and the reflex response must be larger.

Pupillary Response. Finally we may ask how the pupillary reflex is involved. When the eyes are made to converge, with accommodation kept fixed, there is usually some pupillary constriction, but very small by comparison with that which would occur were accommodation permitted to occur to match the convergence (Knoll). Thus the inhibition of the convergence-induced accommodation has been associated with inhibition of the near pupillary reflex. By contrast, the relationship with accommodation is very strong (Marg & Morgan), in fact, according to Alpern, Ellen & Goldsmith, the pupil continues to constrict even when the amplitude of accommodation has been exceeded; thus the increased effort to accommodate is apparently sufficient to cause pupillary constriction. In a similar way, with presbyopic subjects, the stimulus to accommodation is sufficient to cause pupillary constriction, so that presenting a near object, even if accommodation is inadequate, causes a constriction of a magnitude that would be expected had the accommodation taken place (Alpern, Mason & Jardinico).

GENERAL REFERENCES

ALPERN, M. (1969). Movements of the eyes. In *The Eye*, Ed. Davson, vol. III, 2nd ed. pt. 1. London & New York: Academic Press.
BREININ, G. M. (1957). Relationship between accommodation and convergence. *Trans. Am. Acad. Ophthal.* **61**, 375–382.

SPECIAL REFERENCES

ALPERN, M. (1958). Vergence and accommodation. I and II. *Arch. Ophthal.* **60**, 355–357, 358–359.
ALPERN, M., ELLEN, P. & GOLDSMITH, R. I. (1958). The electrical response of the human eye in far-to-near accommodation. *Arch. Ophthal.* **60**, 595–602.

ALPERN, M., MASON, G. L. & JARDINICO, R. E. (1961). Vergence and accommodation. V. Pupil size changes associated with changes in accommodative convergence. *Am. J. Ophthal.* **52**, pt. 2, 762–766.

FINCHAM, E. F. & WALTON, J. (1957). The reciprocal actions of accommodation and convergence. *J. Physiol.* **137**, 488–508.

KNOLL, H. A. (1949). Pupillary changes associated with accommodation and convergence. *Am. J. Optom.* **26**, 346–357.

MARG, E. & MORGAN, M. W. (1950). The effect of accommodation, fusional convergence and the proximity factor on pupillary diameter. *Am. J. Optom.* **27**, 217–225.

PART 6

THE PROTECTIVE MECHANISM

THE eyes are protected from mechanical insults by two mechanisms—blinking and the secretion of tears; by the latter process irritating particles and fumes are washed away from the sensitive cornea and the surface of the globe is maintained in a normally moist condition. Blinking, besides its obvious protective function, also operates to prevent dazzle by a blinding light, to maintain the exposed surface of the globe moist by spreading the lacrimal secretions and, during sleep, by preventing evaporation. Finally the act of blinking assists in the drainage of tears. The pathological condition of *keratitis sicca* may thus result from either defective lid closure or defective lacrimal secretion.

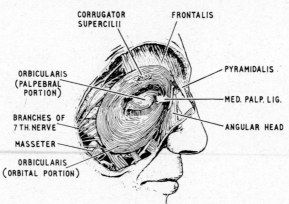

FIG. 255. Illustrating the orbicularis oculi muscle. (After Wolff, *Anatomy of the Eye and Orbit.*)

BLINKING

Orbicularis Oculi. Closure of the eyes is brought about by contraction of the orbicularis oculi muscle, associated with a reciprocal inhibition of the levator palpebri of the upper lid. The orbicularis oculi is an oval sheet of concentric striped muscle fibres extending from the regions of the forehead and face surrounding the orbit into the lids (Fig. 255). Although a single muscle, with its origin mainly in the medial canthal region of the orbit, it is customary to divide it into a number of separate muscles, or parts, the main divisions being the *pars orbitalis* and *pars palpebralis*. The pars palpebralis is divided itself into the upper and lower pretarsal and upper and lower preseptal muscles (Fig. 256). All the palpebral muscles insert into the lateral palpebral raphe and thus, by contracting, act as tensor muscles to the lids, moving the nasal end of each lid towards the nose.

ORBITAL PORTION

The orbital portion of the muscle is not normally concerned with blinking, which may be carried out entirely by the palpebral portion; however, when the eyes are closed tightly, for example in the blepharospasm consequent on a painful stimulus, then the skin of the forehead, temple and cheek is drawn towards the medial side of the orbit; the radiating furrows caused by this action of the orbital portion eventually lead to the "crow's feet" of elderly persons. The two portions can be activated independently; thus the orbital portion may contract, causing a furrowing of the brows that reduces the amount of light entering from above, e.g. in strong sunshine, whilst the palpebral portions remains relaxed and allows the eyes to remain open.

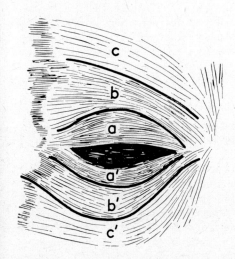

FIG. 256. Illustrating the subdivisions of the orbicularis muscle. *c* and *c'* represent the pars orbitalis; *b* and *b'* are the pretarsal, and *a* and *a'* the preseptal muscles. (After Jones, *Trans. Amer. Acad. Ophthal.*)

Levator Palpebrae. Opening of the eye is not just the result of passive relaxation of the orbicularis muscle; the upper lid contains the *levator palpebrae superioris*, which takes origin with the extraocular muscles at the apex of the orbit as a narrow tendon and runs forward into the upper lid as a very broad tendon, the *levator aponeurosis*, which inserts into the anterior surface of the tarsus and the skin covering the upper lid. Contraction of the muscle will obviously cause elevation of the upper eye-lid, and it is interesting that the nervous connections of this muscle are closely related to those of the superior rectus, required to elevate the eye, so that when the eye looks upwards the upper eyelid moves up in unison.

Müller's Muscle. The orbicularis and levator are striped muscles under voluntary control; the lids contain, in addition, unstriped muscle fibres deep to the septum orbitale; they take origin among the fibres of the levator in the upper lid and the prolongation of the inferior rectus in the lower, whilst they are inserted into the attached margins of the tarsal plates. They are described as the *superior* and *inferior palpebral muscles* respectively. The muscles are activated by the sympathetic division of the autonomic system and tend to widen the palpebral fissure by elevation of the upper, and depression of the lower, lid.

Accessory Muscles of Lid-Closure and -Opening. In addition to the muscles already described, we may note that other, facial, muscles often cooperate in the acts of lid-closure or opening; thus the *corrugator supercilii* muscles pull the eyebrows towards the root of the nose, making a projecting roof over the medial angle of the eye and producing characteristic furrows in the forehead; they are used primarily to protect the eye from the glare of the sun. The *pyramidalis*, or *procerus*, muscles occupy the bridge of the nose; they arise from the lower portion of the nasal bones and, being inserted into the skin of the lower part of the forehead on either side of the midline, they pull the skin into transverse furrows. In lid-opening, the *frontalis muscle*, arising midway between the coronal suture of the scalp and the orbital margin, is inserted into the skin of the eyebrows; contraction therefore causes the eyebrows to rise and opposes the action of the orbital portion of the orbicularis; the muscle is especially used in gazing upwards. It is also brought into action when vision is rendered difficult, either by distance or the absence of sufficient light.

Reflexes. Reflex blinking may be caused by practically any peripheral stimulus but the two functionally significant reflexes are (*a*) that resulting from stimulation of the endings of N V in the cornea, lid or conjunctiva—the *sensory blink reflex*, or *corneal reflex*, and (*b*) that caused by bright light—the *optical blink reflex*. The corneal reflex is rapid (0·1 second reflex time) and is the last to disappear in deepening anaesthesia; it is mediated by N V, impulses being relayed from the nucleus of this nerve to N VII. The reflex is said to be under the control of a medullary centre. The optical reflex is slower and, in man, the nervous pathway includes the visual cortex; the reflex is absent in children of less than nine months.

The blink- or wink-responses to a tap on the face, for example on the supra-orbital region, have been described as a group of *facial reflexes* to which various names, after their discoverers, have been attached (McCarthy, v. Bechterew, etc.). The name *orbicularis oculi reflex* has been suggested to cover them all since their reflex paths are essentially the same, and they apparently consist of two phases—a *proprioceptive* response due to stretch of the facial musculature, followed by a *nociceptive* response which may be regarded as protective in function (Rushworth, 1962).

Normal Rhythm. In the waking hours the eyes blink fairly regularly at intervals of two to ten seconds, the actual rate being a characteristic of the individual. The function of this is to spread the lacrimal secretions over the cornea, and it might be thought that each blink would be reflexly determined by a corneal stimulus—drying and irritation. As a result of extensive studies by Ponder & Kennedy (1928) on human subjects, it would appear that this view is wrong; the normal blinking rate is apparently determined by the activity of a "blinking centre" in the globus pallidus of the caudate nucleus. This is not to deny, however, that the blink-rate is modified by external stimuli, particularly through N V and N II, and also by the emotional state of the individual; Ponder's work shows this most clearly—for example, the blink-rate of an individual in the witness-box went up markedly during cross-examination; that of another decreased on ingestion of a large dose of alcohol, and so on; the fact remains, however, that when any given route for sensory impulses is blocked, e.g. by cocainization of the cornea, the blink-rate is, on the average, unchanged, i.e. characteristic of the individual.

Movements of Eyes. In conclusion, attention should be drawn to the strong association between blinking and the action of the extraocular muscles. A movement of the eyes is generally accompanied by a blink and it is thought that this aids the eyes in changing their fixation point. It will be remembered that in lesions of the frontal motor centre the fixation reflex asserts itself so strongly that a change in the fixation point becomes very difficult but may be brought about by cutting off the visual stimulus by closing the eyes or jerking the head. It is possible that blinking in the normal individual is an aid, although not a necessary one, in inhibiting the fixation reflex preparatory to the adoption of a new point of fixation.

LACRIMATION

The surface of the globe is kept moist by the tears, secreted by the lacrimal apparatus, together with the mucous and oily secretions of the other secretory organs and cells of the conjunctiva and lids.

Lacrimal Glands. The lacrimal gland, proper (Fig. 257), lies in the upper and outer corner of the orbit, just within the orbital margin. It is divided by the lateral horn of the aponeurosis of the levator muscle into two lobes, *orbital* and *palpebral*. About twelve ducts gather the secretion of the entire gland and open through the palpebral conjunctiva, some 4–5 mm above the upper border of the tarsus. It has a typical serous gland structure, and is supplied mainly by the lacrimal branch of the ophthalmic artery, the lacrimal vein draining the blood into the superior ophthalmic vein. The glands are supplied by the *lacrimal nerve*, which is the smallest terminal twig of the ophthalmic division of the trigeminal (Fig. 259, p.422).

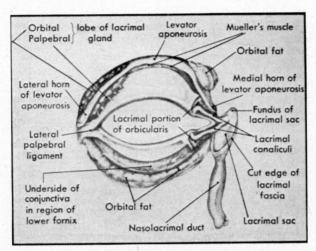

FIG. 257. Tarsal plates and lacrimal system. (Kronfeld, McHugh & Polyak, *The Human Eye in Anatomical Transparencies.*)

Accessory Lacrimal Glands. In addition to the lacrimal gland, there are numerous microscopical *accessory lacrimal glands* in the conjunctiva—the glands of Krause and Wolfring—structurally organized in an essentially similar fashion but on a very much smaller scale.

Sebaceous and Mucous Secretions. An oily fluid is secreted by the glands of Zeis and the Meibomian glands, in the eyelids, whilst goblet cells in the conjunctiva secrete mucus.

Precorneal Film. The fluid layer on the cornea is called the precorneal film and is said by Wolff to consist of an innermost layer of mucus, a middle layer of lacrimal secretions and an outer oil film which serves to reduce evaporation of the underlying watery layer (Mishima & Maurice, 1961) as well as preventing overflow at the lid-margin. Ehlers (1965) absorbed the film on a sponge and estimated, from its weight, that it would be some 8·5 to 4·5 μ thick according to the length of time between blinks; its chemistry suggested that the lipid was

predominantly from the Meibomian glands, containing as it did mainly chole-
sterol esters.

Rate of Evaporation. The importance of the precorneal film in reducing the
evaporation from the surface of the eye was demonstrated by Mishima & Maurice
(1961a) who used, as a measure of evaporation, the thinning of the cornea that
took place as a result of loss of water; in the normal eye no such thinning took
place since the fluid lost from the surface was made good by the aqueous humour;
by replacement of the aqueous humour with oil, however, there was a steady
thinning; the rate was increased greatly by washing the surface of the cornea with
saline, i.e. by removing the precorneal film but this could be brought back to the
normal rate by allowing the rabbit to blink, a procedure that presumably replaced
the lipid precorneal film, since preliminary destruction of the Meibomian glands

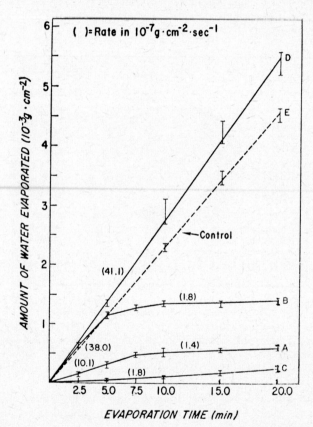

Fig. 258. The weight of water evaporated, and evaporation rates, from the
precorneal and corneal surfaces as measured over a 20 minute period. Line A
shows the results from the intact tear film. Two discrete phases with differing
rates are noted. Line B shows the results from the tear film after the superficial
lipid layer was removed. Again, two discrete phases with differing rates are seen.
The rate of the second phase in Line B parallels that of the second phase in Line
A. Line C shows the results measured from the epithelial surface. Only one phase
with a rate similar to the second phases of A and B is noted. Line D shows the
evaporation from the stromal surface which occurs at a rate similar to that from a
pure water surface (Line E). (Iwata, Lemp, Holly & Dohlman, *Invest. Ophthal.*)

rendered lid-closure ineffective. Removal of the epithelium had no obvious effect so that Mishima & Maurice concluded that it offered no obstruction to evaporation of water. Iwata *et al.* (1969) measured the evaporative rate more directly and confirmed the striking increase in rate when the precorneal film was washed away; as Fig. 258 shows, this rapid loss does not last for long, and the rate settles down to a much slower value similar to that from the intact eye. The initial rapid loss occurs because of the presence of a layer of tears enclosed by the precorneal film; when this has evaporated, the loss is now apparently determined by the rate at which new fluid can pass from aqueous humour to the surface of the eye; thus preliminary blotting off of this layer gives a slow evaporative loss equal in rate to that reached with the lipid film removed. The limiting factor under these conditions seems to be the epithelium, since removal of this allowed the high rate of evaporation to be maintained indefinitely (Fig. 258).

Normal Output. The normal daily output of tears was estimated by Schirmer in 1903 in subjects with their lacrimal ducts and sacs removed; the time for a drop to form was measured; the amount of evaporation that had occurred was estimated by a separate experiment, and knowing the weight of the drop the formation during this period was computed. Schirmer estimated a secretion of 0·5 to 0·75 g in 16 hours. He concluded that there was no secretion during sleep. Schirmer also devised tests for normal lacrimation; these consisted of hooking a strip of filter paper over the lower lid margin and measuring the length of wetting that occurred in a fixed time. In some tests the presence of the filter paper was the stimulus for secretion, whilst in adaptations of this test the nasal mucosa was stroked with a brush, or the subject was asked to look into the sun. Applying this test de Roetth observed a considerable decrease in the magnitude of this lacrimation-response with age, the secretion at 80 years being only about a fifth that at 20 years.

Chemistry of the Lacrimal Secretion

The fluid collecting in the lacus lacrimalis is a mixture of mucus and secretions of the lacrimal glands—the tears proper—and is thus slightly opalescent. According to Krogh, Lund & Pedersen-Bjergaard, the tears are isosmolal with blood plasma, but the precise relationship between the two fluids will obviously depend on the degree of evaporation that the fluid undergoes before collection, and, presumably, on the rate of secretion of the fluid. Thus Giardini & Roberts found the concentration of chloride to fall from 135 mM to 110 mM as the rate of secretion was increased by exposing the human eye to larger and larger concentrations of tear-gas. Essentially similar changes were found by Balik for chloride, sodium, urea and phosphate, whilst Thaysen & Thorn found the concentrations of urea, sodium, potassium and chloride to be independent of rate of secretion.

Whilst there is little doubt that evaporation will tend to increase the concentrations of solutes in the tear-film, and hence the osmolality of the fluid, the extent to which this happens will be governed by the slow evaporative rate, discussed above, and the replacement from the aqueous humour. v. Bahr (1956) observed that the cornea did not change in thickness when a 1 per cent solution of NaCl was placed on it, whereas higher concentrations caused it to come thinner; hence the tear-film probably has an equivalent osmolality; Mishima & Maurice (1961b) found a value of 0·9 per cent NaCl if the eye was closed and 1·0 per cent if open.*

Balik defined a "secretion ratio", F, being the ratio of the concentrations of a given substance in tears over plasma; with urea, sodium and phosphate this was greater than unity with normal rates of flow, suggesting possible active accumulation in the tears, although evaporation as the main cause of these ratios cannot be ruled out. The concentration of glucose in the tears is remarkably low, being only a few per cent of that in plasma (Giardini & Roberts). The pH is about 7·2 becoming more alkaline with continued lacrimation (Swan, Trussell & Allen).

Proteins. The concentration of proteins in the tears is low, being of the order of 0·1 to 0·6 per cent by comparison with 6–7 per cent for plasma; they have been resolved into a number of components by the technique of paper-electrophoresis.†

LYSOZYME

Thus Brunish described an albumin, representing some 39 per cent of the total, 3 globulins representing 45 per cent, whilst the rest was made up of *lysozyme*, an antibacterial substance first described by Fleming in 1922 which is responsible for the low bacterial count of normal human tears. Lysozyme is a strongly basic protein, of molecular weight 14,000–25,000; with an iso-electric point of 10·5 to 11·0 it has a net positive charge at physiological pH and thus migrates to the cathode by contrast with the other proteins which have a net negative charge. It is an enzyme, and owes its bactericidal activity to its power of dissolving the mucopolysaccharidic coats of certain bacteria. It is not peculiar to human tears, being widely distributed in the plant and animal kingdoms, egg-white and fig-tree latex being rich sources, for example.

OTHER FRACTIONS

Although it is usual to designate the globulins as α-, β- and γ-, it is unlikely that the protein fractions of tears are identical with their plasma analogues, so that it is better to indicate them by non-committal names; thus U. Krause isolated six fractions which he denoted by arabic numerals, 1 to 5 in descending order of speed of migration in the electric field. His fraction 1 would correspond to "tear-albumin" of Erickson (1956); fraction 2 to a globulin with a speed of migration close to that of α_2-globulin, and so on.

It would seem that the tear proteins differ from their plasma analogues by virtue of the larger amounts of mucopolysaccharide in combination with them; at any rate, according to McEwan, Kimura & Feeney, they stain strongly with the PAS reagent (p. 72).

* The authors point out that the thinning of the cornea when the eye is open, and its thickening when the eye is kept closed, indicated that under normal conditions the aqueous humour loses water, and they found a difference of 1·5 per cent in the estimated osmolalities of the fluid in the open and closed eyes.
† The literature on the protein fractionation of tears is large and has been summarized by U. Krause (1959) and McEwan & Goodner (1969). The first description is that of Miglior & Pirodda (1954); other outstanding contributions are from the laboratories of McEwen, Erickson and Brunish, to name only a few. We may not that François (1959) has shown that, by contrast with aqueous humour and lens, the use of agar in place of paper for electrophoresis is contra-indicated; at any rate the fractions obtained by this technique are so different as to suggest interaction of the proteins with the agar.

Drainage of Tears

When the secretion of tears exceeds the loss due to evaporation the excess fluid, if it is not to overflow—*epiphora*—must be drained away. The passage for this drainage consists of the *canaliculi*, the *lacrimal sac* and the *naso-lacrimal duct*. The lacrimal *puncta* are the openings of the canaliculi in the lid-borders of the inner canthus; the canaliculi course at first vertically and then more or less horizontally to the lacrimal sac, which is continued inferiorly as the naso-lacrimal duct emptying into the inferior meatus of the nose (Fig. 257). The sac and the duct may be regarded as essentially a continuous structure, the term "sac" really applying to *that part of the duct lying within the orbit*. Any theory of the mechanism of conduction of the tears must take into account the common experience that the act of blinking favours drainage and prevents overflow; certainly the dimensions of the canaliculi and puncta are so small (about 0·5 mm diameter) as to make it unlikely that an adequate drainage could be maintained by the unassisted action of gravity (siphoning would of course be necessary for the upper canaliculus).

The Lacrimal Pump. The mechanism has been analysed most recently by Jones, and is described by him as the *lacrimal pump*. Essentially the pumping action consists in an alternate negative and positive pressure in the lacrimal sac caused by the contraction of the orbicularis muscle. Because of the anatomical relationships of this muscle, in particular that part called Horner's muscle, to the palpebral ligament and lacrimal sac, contraction during blinking causes a dilatation of the sac. At the same time, the muscular contraction also causes the canaliculi to become shorter and broader. Finally contraction of the orbicularis also tends to invert the lower lid, thus ensuring that the punctum dips in the lacus lacrimalis.* Negative pressure in the nose during inhalation, and gravity, are also factors in emptying the sac.

The Nervous Mechanism

Tears are secreted reflexly in response to a variety of stimuli, e.g. irritative stimuli to the cornea, conjunctiva, or nasal muscosa; thermal stimuli, including peppery foods, applied to the mouth and tongue; bright lights, and so on. In addition it occurs in association with vomiting, coughing and yawning. The secretion associated with emotional upset is described as *psychical weeping*. Section of the sensory root of the trigeminal prevents all reflex weeping, leaving psychical weeping unaffected; similarly the application of cocaine to the surface of the eye, which paralyses the sensory nerve endings, inhibits reflex weeping, even when the eye is exposed to potent tear-gases. The afferent path in the reflex is thus by way of N V. The motor pathway to the main lacrimal gland is more complex.

Parasympathetic. The parasympathetic supply is from the *facial nerve* (N VII); preganglionic fibres arise from the *lacrimal nucleus* in the pons; they run in the *nervus intermedius*, through the *geniculate ganglion* (without relaying), to the *sphenopalatine ganglion* as the *greater superficial petrosal nerve*. Post-ganglionic fibres enter the *zygomatic nerve* and thence the *lacrimal* (Fig.

---

* According to Brienen & Snell (1969), however, the punctum always opens into a strip of tear fluid, so that a "dipping into the tear lake" during eye-closure is out of the question. These authors attribute drainage entirely to the pressure developed in the conjunctival sac.

259). The sympathetic pathway is by way of the *deep petrosal nerve* which arises from the cervical sympathetic plexus; it joins the greater superficial petrosal nerve to form the *nerve of the pterygoid canal*, or *Vidian nerve*, which ends in the sphenopalatine ganglion; thence the fibres run with the parasympathetic as described above. Other fibres reach the gland by way of the lacrimal artery. Innervation of the lacrimal gland is not always complete at the time of birth; thus Sjögren found a complete lack of reflex secretion of tears in 13 per cent of normal full-term babies and in 37 per cent of premature babies during the first three days of extra-uterine life. Since this condition was not associated with any dryness of the cornea—keratitis sicca—it might be suspected that normally the secretion of the accessory lacrimal glands, which are not involved in reflex secretion, are adequate to keep the cornea moist. This suggestion is supported by the relative absence of corneal symptoms after complete removal of the lacrimal gland. Thus the reflex secretion may be regarded as an emergency response.

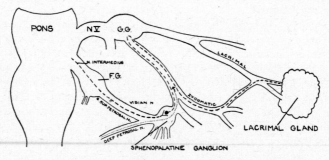

FIG. 259. Innervation of the lacrimal gland. G.G., Gasserian ganglion; F.G., facial ganglion. (After Mutch, *Brit. J. Ophthal.*)

Sympathetic. The function of the sympathetic in the secretion of tears is problematical; section of the superficial petrosal nerve—a procedure often concomitant with operative removal of the Gasserian ganglion—or blocking of the sphenopalatine ganglion, causes a failure of reflex and psychical lacrimation, suggesting that the parasympathetic only is concerned with reflex secretion. Furthermore, there is no apparent interference with weeping as a result of bilateral section of the cervical sympathetic. According to Whitwell (1961), therefore, it is the sympathetic innervation that is responsible for a tonic secretion adequate to keep the eye moist, but the possibility must be considered that the main gland is normally at rest, whilst the accessory glands operate continuously and independently of any innervation. In this event the sympathetic would play no role in the formation of tears.

Drugs. As one would expect of a parasympathetic-dominated process, intravenous pilocarpine produces a large and prolonged increase in flow-rate of tears; according to a study of Goldstein, de Palau & Botelho (1967) there may be an eleven-fold increase, in the rabbit, lasting for 33 minutes. Intravenous noradrenaline increased flow in this animal some threefold for two minutes; it inhibited the increased flow due to pilocarpine, however, perhaps because of its vasoconstrictive action.

Crocodile Tears. Some human subjects lacrimate strongly when eating, and the syndrome has been given the name of "crocodile tears". The mechanism has been discussed by Golding-Wood (1963), and he concludes that the prime cause is a degeneration of the greater superficial petrosal nerve which provokes sprouting of fibres from the lesser superficial that normally subserves salivary secretion; as a result, the salivary secretory fibres stimulate the lacrimal gland. He found that intracranial section of the glossopharyngeal nerve relieved the condition.

GENERAL REFERENCES

McEwan, W. K. & Goodner, E. K. (1969). Secretion of tears and blinking. In *The Eye*, Ed. Davson, 2nd ed. vol III, pt. 3. London & New York: Academic Press.

Mutch, J. R. (1944). The lacrimation reflex. *Br. J. Ophthal.* **28**, 317–336.

Rushworth, G. (1962). Observations on blink reflexes. *J. Neurol. Neurosurg. Psychiat.* **25**, 93–108.

Veirs, E. R. (1955). *The Lacrimal System*. London. & New York: Grune & Stratton.

SPECIAL REFERENCES

Balik, J. (1959). Uber die Ausscheidung von Natrium in die Tränen. *Ophthalmologica, Basel* **137**, 95–102.

Balik, J. (1959). Uber die Ausscheidung von Harnstoff in die Tränenflüssigkeit bei Keratoconjunctivitis sicca. *Klin. Mbl. Augenheilk.* **135**, 533–537.

Balik, J. (1959). Uber die Ausscheidung von Natrium in die Tränen bei trockener Keratoconjunctivitis. *v. Graefes' Arch. Ophthal.* **160**, 633–657.

Balik, J. (1960). Secretion of inorganic phosphorus in tears. *Am. J. Ophthal.* **49**, 941–945.

Brienen, J. A. & Snell, C. A. R. D. (1969). The mechanism of the lacrimal flow. *Ophthalmologica, Basel* **159**, 223–232.

Brunish, R. (1957). The protein components of human tears. *Arch. Ophthal.* **57**, 554–556.

Ehlers, N. (1965). The precorneal film. *Acta Ophthal., Kbh.* Suppl. 81.

Erickson, O. F. (1956). Albumins in lacrimal protein patterns. *Stanf. med. Bull.* **14**, 124–125.

Fleming, A. (1922). On a remarkable bacteriolytic element found in tissues and secretions. *Proc. Roy. Soc. B.* **93**, 306–317.

François, P. (1959). Micro-electrophorese sur gélose des larmes humaines. *Bull. Soc. belge d'Ophthal.* **122**, 343–351.

Giardini, A. & Roberts, J. R. E. (1950). Concentration of glucose and total chloride in tears. *Br. J. Ophthal.* **34**, 737–743.

Golding-Wood, P. H. (1963). Crocodile tears. *Br. med. J.* (i), 1518–1521.

Goldstein, A. M., de Palau, A. & Botelho, S. Y. (1967). Inhibition and facilitation of pilocarpine-induced lacrimal flow by noradrenaline. *Invest. Ophthal.* **6**, 498–511.

Iwata, S., Lemp, M. A., Holly, F. J. & Dohlman, C. H. (1969). Evaporation rate of water from the precorneal tear film and cornea in the rabbit. *Invest. Ophthal.* **8**, 613–619.

Jones, L. T. (1958). Practical fundamental anatomy and physiology (of lacrimal apparatus). *Trans. Am. Acad. Ophthal.* **62**, 669–678.

Krause, U. (1959). A paper electrophoretic study of human tear proteins. *Acta ophthal. Kbh.* **53**, Suppl., pp. 67.

Krogh, A., Lund, C. G. & Pedersen-Bjergaard, (1945). The osmotic concentration of human lacrymal fluid. *Acta physiol. scand.* **10**, 88–90.

McEwen, W. K., Kimura, S. J. & Feeney, M. L. (1958). Filter-paper electrophoresis of tears. III. *Am. J. Ophthal.* **45**, 67–70.

Miglior, M. & Pirodda, A. (1954). Indagini elettroforetiche sulla composizione proteica delle lacrime umane normali. *Giorn. ital. Oftal.* **7**, 429–439.

Mishima, S. & Maurice, D. M. (1961a). The oily layer of the tear film and evaporation from the corneal surface. *Exp. Eye Res.* **1**, 39–45.

MISHIMA, S. & MAURICE, D. M. (1961b). The effect of normal evaporation on the eye. *Exp. Eye Res.* **1,** 46–52.

PONDER, E. & KENNEDY, W. P. (1928). On the act of blinking. *Quart. J. exp. Physiol.* **18,** 89–110.

DE ROETTH, A. (1953). Lacrimation in normal eyes. *Arch. Ophthal.* **49,** 185–189.

SCHIRMER, O. (1903). Studien zur Physiologie und Pathologie der Tränenabsonderung und Tränenabfuhr. *v. Graefes' Arch. Ophthal.* **56,** 197–291.

SJÖGREN, H. (1955). The lacrimal secretion in newborn, premature and fully developed children. *Acta ophthal. Kbh.* **33,** 557–560.

SWAN, K. C., TRUSSELL, R. E. & ALLEN, J. H. (1939). pH of secretion in normal conjunctival sac determined by glass electrode. *Proc. Soc. exp. Biol. Med., N.Y.* **42,** 296–298.

THAYSEN, J. H. & THORN, N. A. (1954). Excretion of urea, potassium and chloride in human tears. *Am. J. Physiol.* **178,** 160–164.

WHITWELL, J. (1961). Role of the sympathetic in lacrimal secretion. *Br. J. Ophthal.* **45,** 439–445.

SECTION IV

Visual Perception

INTRODUCTION

Sensation and Perception

IN Section II we have investigated the "mechanism of vision" in the limited sense of attempting to understand the elementary processes in the retina that result in discharges in the optic nerve, and of attempting to correlate these with the sensations. To this end, very simple visual phenomena were chosen—the sensation of luminosity, the discrimination of flicker, colour and form. In respect to this last category, we confined ourselves to the very element of form discrimination—the resolution of points and lines—and in visual acuity studies we indicated that the recognition of letters was an unsuitable basis for the study of resolution since it involved activities at a higher level than those we were immediately concerned with. In this section we must consider some of the manifestations of this higher form of cerebral activity.

The Perceptual Pattern. Essentially we shall be concerned with the modes in which visual sensations are interpreted; the eyes may gaze at an assembly of objects and we know that a fairly accurate image of the assembly is formed on the retinae of the two eyes, and this is "transferred" by a "point-to-point" projection to the occipital cortex. This is the basis for the visual sensation evoked by the objects but there is a large variety of evidence, derived from everyday experience, which tells us that the final awareness of the assembly of objects involves considerably more nervous activity than a mere point-to-point projection of the retina on the cortex would require. As a result of this activity we appreciate a *perceptual pattern*; the primary visual sensation, resulting from stimulating the visual cortex, is integrated with sensations from other sources presented simultaneously and, more important still, with the memory of past experience. For example, a subject looks at a box; a plane image is produced on each retina which may be represented as a number of lines, but the perceptual pattern evoked in the subject's mind is something much more complex. As a result of past experience the lines are interpreted as edges orientated in different directions; the light and shade and many other factors to be discussed later are all interpreted in accordance with experience; its hardness is remembered and its uses, and the original visual sensation is *interpreted as a box*, not a series of lines. It is impossible, in practice, to separate the sensation of seeing from the perceptual process; the sensation is essentially an abstraction—what we think would be the result of a visual stimulus if this higher nervous integrative activity had not happened. In the case of a flash of light the influence of higher nervous activity is very small and the sensation is virtually indentical with the final perception. With a written word, on the other hand, the final perceptual pattern presented to consciousness is something far different from the bare visual sensation.

We may begin by illustrations of the perceptual process, as revealed by psychophysical studies, and later we shall consider some electrophysiological studies that illustrate the possible neural basis for some of these higher integrative functions.

HIGHER INTEGRATIVE ACTIVITY

Inadequate Representation. *Schroeder's staircase* (Fig. 261) shows how the same visual stimulus may evoke different meanings in the same subject; to most people the first impression created by Fig. 261 is that of a staircase; on gazing at it intently, however, the observer may see it as a piece of overhanging masonry. (If the reader has any difficulty in seeing Fig. 261 as a piece of over-hanging masonry, he has only to turn the page upside down by slowly rotating the book, keeping the eyes fixed on the drawing.) *Rubin's vase* (Fig. 262) may

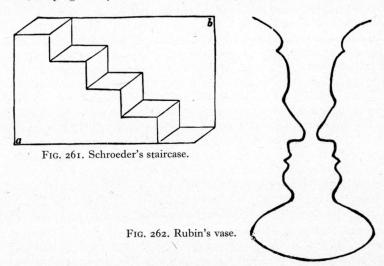

Fig. 261. Schroeder's staircase.

Fig. 262. Rubin's vase.

be seen either as a vase or two human profiles. In general, diagrams of this sort are called *inadequate representations* of the object they are intended to copy, or *ambiguous figures*. The lines, as such, can be interpreted with more or less ease in two different ways and the fact that they are perceived in one or other definite manner emphasizes the arbitrary element of interpretation in the visual process. Helmholtz quotes in this connection the case of a windmill, observed against the night sky; according as the observer fancied that the sails were in front of, or behind, the body of the mill he saw them revolving in different directions. Brown (1962) has described an experimental set-up that simulates the situation described by Helmholtz, the subject observing the shadows cast by a set of six vertical pins arranged in a circle and rotating at constant speed about a vertical axis through the centre of the circle. The stimulus on the screen was two-dimensional, with one set of pins moving left and the other right, but the subject perceived it three-dimensionally, with the apparent direction of rotation of the pins varying. This is because the basic stimulus contains strong depth cues, but without decisive ones concerning the depth relation between the two sets of pins moving in right and left directions.

Careful experimentation with ambiguous figures has shown that, over a period of time, the fluctuations in meaning exhibit a periodicity character-istic of the observer. This is not to deny, however, that with an effort of will, by moving the eyes, blinking, etc., the meaning of the pattern may be made to change. Nevertheless, if the subject merely regards the figure for several minutes and the times during which it is seen in its two meanings are recorded, it is found that the fluctuations are definitely not random in nature. It is interesting that the periodicity may be affected by drugs.

Brown's experimental arrangement described above is particularly suitable for investigation of the rate of change of apparent movement; he found that it increased with viewing time. If this increase had been achieved by monocular observation, then the other eye showed the same increase, so that we may say that there was *interocular transfer* of this adaptational effect, indicating that the process occurs at a higher central level than the retina, namely at some point in the visual pathway common to both eyes.

Reading. The reading process is an instance of the arbitrary attachment of meaning to symbols; with practice the meaning becomes associated not with the complete picture of a word but with its general shape and thus the practised reader can be imposed on easily with mis-spelt words; the context leads him to expect a certain word and he actually sees it even though it is not on the page. Here the perceptual process is imposing a meaning on the visual stimulus that does not correctly belong to it, and many examples of this arbitrary activity of the higher centres may be cited; for example most people read the word "vicegerent" as "viceregent" because the latter is the word they expect even though it does not occur in the dictionary. The filling up of the blind spot and the failure of the hemianopic subject to realize his defect are further instances.

Colour Perception

Colour Constancy. By this term we mean the tendency of the subject to perceive the colour of an object as being the same in spite of large variations in the actual chromaticity of the light emanating from it. The colour becomes one of an object's attributes—its *memory colour*—and there is a strong tendency to perceive this even when the physical conditions have been chosen to make this difficult. For example, a red letterbox may be illuminated with blue light so that far more blue light is reflected off it than red, yet the subject continues to call it red, and his *perception* is that of a red object. If the object, so illuminated, is looked at through a tube, so that its colour may be divorced from its shape, it is said to appear blue. Again, we may place a green filter at arm's length and note that the part of the field of vision seen through the filter acquires a distinct green colour—*colour conversion*—but when the filter is brought close to one eye and the other is closed, the natural colours of the objects viewed are identified readily—*colour constancy*. Thus, as Helmholtz pointed out: ". . . with all coloured surfaces without distinction, wherever they are in the sphere of the coloured illumination, we get accustomed to subtracting the illuminating colour from them in order to find the colour of the object." The basis on which the "discounting of the illuminant colour" is achieved must be an unconscious recognition of the chromaticity of the light reflected off the more highly reflecting objects in the scene, e.g. a piece of white paper.

If we call the colour perceived in an object as the *object colour*, we may refer to these phenomena as examples of *object-colour constancy*. When a coloured light is viewed independently of any object, as through a hole in a screen, then we may call this an *aperture colour*, but it must be appreciated that the perceived aperture colour is by no means invariant for a given chromaticity of its emitted light, and this is largely because of adaptational and contrast effects. Thus, as Helson (1938) showed, it is easily possible to devise a scene in which an object reflecting light of any chromaticity whatsoever is perceived as grey, the main basis for this being the tendency for the subject to see the colour of the illuminant, the colour of the after-image complementary to the illuminant, or achromaticity, depending on the relative reflectances of the surface being viewed and the adaptation reflectance.

These principles, which lead to modifications of the perceived colours of objects, have been formulated quantitatively by Judd (1940), the process of discounting the spectral quality of the illuminant being achieved by defining, on the Maxwellian triangle, a point to correspond with the perception of grey in the object-mode of viewing, this point being close to the chromaticity of the illuminant. The prediction of the hue reported by an observer was given by the vector extending from this achromatic point to the point defining the chromaticity of the light actually emanating from the object.

Two Primary Projections. Interest in colour constancy, and its related phenomena, was revived with the description by Land (1959) of a series of experiments in which he produced, with the aid of only two illuminants, pictures that seemed to give a faithful reproduction of the colours of various scenes— pictures that, on the basis of trichromatic theory, should have required three independent sources of illumination. In general, Land produced two transparent positives of the scene by photographing it first through a red-transmitting filter (5850–7000 Å) and then through a filter transmitting the middle third of the spectrum (4900 to 6000 Å). The image of one of these positives was usually projected on to a white screen with red light (5900–7000 Å), whilst that of the other usually by an incandescent lamp, the two images being adjusted carefully to be in register. Space will not permit of a summary of the findings, but his Experiment 3 may be quoted as an example: By combining the long-wave picture shown in red light with the middle-wave picture shown in incandescent light, the subject perceived the full colour in a portrait of a blonde girl, i.e. blonde hair, pale blue eyes, red coat, blue-green collar, and strikingly natural flesh tones. For adequate colour representation through photography, a three-primary system is ordinarily employed, and the question to be asked is how this apparently true representation of colours is achieved with only two positives. It must be appreciated, however, that the use of white light in the incandescent illuminant provides the viewer with all colours of the spectrum, and it is essentially the fact that the subject ignores the exact chromatic content of the light reflected off the various parts of the screen that permits him to perceive a fairly true representation of the original. According to Judd's analysis, the perceived colours in this, and in other examples described by Land, are, in fact, predictable on the basis of his earlier quantitative formulation, so that there is no reason to abandon the trichromatic theory of colour perception on the basis of these experiments of Land.

Tone Constancy. Essentially similar phenomena are described for the blacks, greys and whites. A grey wall may be compared with a grey piece of paper on a table reflecting the same amount of light; the wall appears white whilst the paper appears grey, presumably because we associate lightness with the walls of a room. These phenomena are classified under the heading of *tone constancy*, but it must be appreciated that, just as with chromatic vision, the final perception depends on inductive effects taking place at the retinal level. For example, it is argued that the black of this type, seen on a fine day, reflects more light than does the white of the paper in dim illumination, and yet the black is seen as black in both circumstances. The psychologist "explains" this as tone constancy; the physiologist, however, would merely state that the black sensation is evoked by a certain area of the retina being stimulated less intensely than another, adjoining, area; he would be the last to suggest that there is an absolute black sensation depending on some critical level of retinal stimulation. We know, moreover, that the sensitivity of the retina is enormously affected by light-adaptation (p. 290), so that it is impossible to expect that the same sensation will be evoked by the same stimulus under widely differing states of adaptation. We must therefore be careful of attributing to the activities of higher centres phenomena that properly belong to the retina.

Having clarified our position in relation to the difference between a perception and a sensation, we can now proceed to analyse some of the simple mechanisms concerned in the perceptual processes.

MONOCULAR PERCEPTION

The Projection of the Retina

Objects are perceived in definite positions in space—positions definite in relation to each other and to the percipient. Our first problem is to analyse the physiological basis for this spatial perception or, as it is expressed, the *projection of the retina into space*.

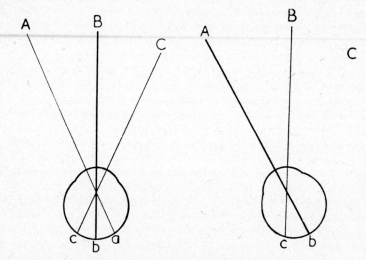

FIG. 263. (*a*) A group of objects, A, B, C, produces images on the retina, *a*, *b*, *c*; the retinal images are projected outwards in space towards the points evoking them. (*b*) The eye has moved to the left so that the image of A falls on *b*. *b* was previously projected to B but it is now projected to A.

Relative Positions of Objects. The perception of the positions of objects in relation to each other is essentially a geometrical problem. Let us confine the attention, for the present, to the perception of these relationships by one eye—*monocular perception*. A group of objects, as in Fig. 263, produces images on the retina in a certain fixed geometrical relationship; for the perception of the fact that A is to to the left of B, that B is to the left of C, etc., it is necessary that the incidence of images at *a*, *b*, *c* on the retina be interpreted in a similar, but, of course, inverted geometrical relationship. The neural requirements for this interpretation are (*a*) that the retina should be built up of elements which behave as units throughout their conducting system to the visual cortex and (*b*) that the retinal elements should have "local signs." The local sign could represent an innate disposition or could result from experience—the association of the directions of objects in space, as determined by touch, etc., with the retinal pattern of stimulation. However they arise, the local signs of the retinal elements provide the complete information as to the relative positions of A, B and C.

The retinal stimuli at *a*, *b* and *c* are appreciated as objects outside the eye— the retina is said to be *projected* into space, and the field of vision is thus the projection of the retina through the nodal point. It will be seen that the geometrical relationship between objects and retinal stimuli is reversed; in the retina *a* is to the right of *b*, and so on; it is customary to speak of the "psychological erection of the retinal image"; this is thoroughly misleading since it suggests an awareness, on the part of the higher centres, of the actual points on the retina stimulated. This of course is untrue; *a* and *c* are projected to the left and right of B, not because the higher centres have learnt the laws of geometrical optics but because such a projection corresponds with experience.

Position in Relation to Observer. The recognition of the directions of objects in relation to the observer is more complex. If the eye (Fig. 263, *b*) is turned to the left, the image of A falls on the retinal point *b*, so that if *b* were always projected into the same direction in space, A would appear to be in B's place. In practice we know that A is perceived as fixed in space in spite of the movements of the eye; hence the direction of projection of the retinal point is constantly modified to take into account movements of the eye—we may call this *"psychological compensation"*. It will be seen that correct projection is achieved by projecting the stimulated retinal point through the nodal point of the eye. Movements of the eye due to movements of the head must be similarly compensated. As a result, any point in space remains fixed in spite of movements of the eye and head. Given this system of "compensated projection," the recognition of direction in relation to the individual is now feasible. B may be said to be due north, or more vaguely "over there"; when the head is turned, since B is perceived to be in the same place, it is still due north or "over there".

False Projection. Under some conditions the human subject will make an error in projecting his retinal image so that the object giving rise to this image appears to be in a different place from its true one; the image is said to be *falsely projected*. If the eye is moved passively, for example by pulling on the conjunctiva with forceps, the subject has the impression that objects in the outside world are moving in the opposite direction to the movement of the eye. The exact nature of this false projection is illustrated in Fig. 264. The eye is fixated on an object F, whose image falls on the fovea, *f*.* The eye is moved forcibly through 30°, so that the image of F falls on a new point, *q* (Fig. 264*b*).

---

* In this section the term "fovea" will be used very loosely to indicate the position of the image of a centrally fixated point; it will be the intersection of the visual axis with the retina.

The correct projection of q would be through the nodal point out to F; however, projection takes place as though the eye had not moved. The image falls 30° to the right of the fovea; with the eye in its original position, such a retinal stimulus would correspond to an object 30° to the left, the point Q in Fig. 264, *a*, so that the stimulus on q, in the new position of the eye, is projected 30° to the left, along the line rR. Consequently, as the eye is moved to the right, the retinal image of F is projected progressively to the left.

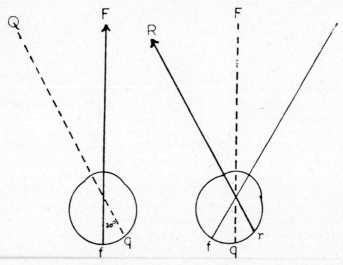

Fig. 264. Illustrating false projection resulting from the forcible movement of the eye. (*a*) The point F is fixated. (*b*) The eye has been moved forcibly 30° to the right. The image of F falls on q, and the projection of q is made on the assumption that the eye has not moved, i.e. towards R.

INFORMATION FROM MOTOR DISCHARGE

Quite clearly, under these abnormal conditions, the necessary information as to the position of the eye is not being sent to what we may call a "*space representation centre*" (Ludvigh), so that interpretation of the retinal image is made on the basis of inadequate information. The tension in the spindles and tendon-organs will undoubtedly be altered as a result of this passive movement, and the fact that no use is made of this proprioceptive information indicates that normally the space representation centre employs a different type of information on which to base its estimates of the position of the eye. This information could be provided by the motor discharge from the voluntary and reflex centres controlling the movements of the eyes, as suggested by Helmholtz; thus when the subject moves his eyes to the left to fixate point A in Fig. 263, the centre that initiates the motor discharges to the medial and lateral recti, besides sending its impulses to the motor neurones of the nuclei of N III and VI, may also send a message to the space representation centre indicating the intensities of these discharges, and thus the approximate extents of the eye movements. The space representation centre thus interprets the shifting images of A, B and C over the retinae not as a movement of these points but as a movement of the eyes to fixate A. The proprioceptive messages from the muscle spindles and tendon-organs operate on a completely unconscious level, simply serving to moderate the strengths of the contractions of individual fibres in accordance with variations in local conditions—the so-called *parametric feedback* (Ludvigh). In the case of the eye that has been moved passively, the space representation centre has obviously not been informed of the eye movement, and the interpretation of the shifting retinal images is now what would be expected had the eyes remained stationary, i.e. the interpretation is that of a moving of the external objects from right to left, with its consequent false projection.

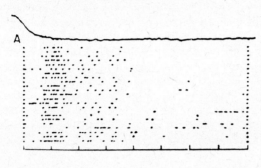

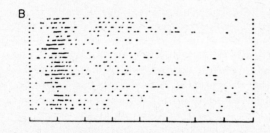

FIG. 265. Response of a cortical neurone to, A, movement of the eye through 20° and, B, movement of the visual field through the same angle. The solid line in A is the electro-oculogram, and the rows of dots are spike responses resulting from a single movement, successive rows corresponding to successive movements. Note similarity in responses in spite of the different ways of stimulating the receptive field of the neurone. (Wurtz, *J. Neurophysiol.*)

COROLLARY DISCHARGE

Helmholtz' postulate of a corollary discharge is an attractive one, but to be confirmed it is desirable to find some difference in the behaviour of cortical neurones according as to whether the retina moves past a stationary image or whether the image moves over a stationary retina. Wurtz (1969) has made such a search among neurones of area 17, but this was unsuccessful; thus in Fig. 265, A illustrates the response of a neurone to a rapid movement of the eye across a visual stimulus; the top line is the electro-oculogram showing a 20° horizontal movement, and the stimulus crossed the receptive field of the unit half-way through the movement. Each line of dots represents unit discharges during one movement, and successive movements are indicated in successive lines. It will be seen that each time the eye moved there was a burst of discharges. In B the eye was stationary, and the stimulus was moved rapidly across the receptive field;

the similarity in responses is remarkable. If a given neurone's activity was suppressed by an eye movement, then image-movement also suppressed it; if there was no influence, then there was no influence of image-movement.*

OCULOGYRAL ILLUSION

Another example of false projection is given by the oculogyral illusion described by Graybiel & Hupp in 1946. A subject is seated in a rotatable chair in the dark, and he fixates a small source of light which is attached to the chair and rotates with it. When the chair is rotated the light appears to move in the same direction. In this case the rotation has stimulated the semicircular canals, and the space representation centre has been informed that the body has rotated. The subject is unaware that the fixation light has also moved, so that the stationary retinal image is reinterpreted; clearly, if the head moves whilst the image of a point remains stationary, the point must have moved to the same angular extent as the head. Hence the retinal image is projected as though the object had moved in relation to the observer. Thus, if the chair is rotated to the right, the light will be falsely projected to the right of the observer, although it has remained fixed in relation to him.

AUTOKINETIC ILLUSION

The explorer Alexander von Humboldt in 1799 noticed that after watching a faint star for a short time it would start to move around in various directions. An obvious explanation for this apparent movement may be the occurrence of involuntary eye movements, the space representation centre being inadequately informed of them whilst the relatively uniform background against which the light is viewed fails to provide adequate retinal information of the eye movements. There is still some disagreement as to whether such movements could always account for the phenomenon, so that other factors have been invoked as well. The subject has been reviewed by Grosvenor (1959), whilst the contribution of eccentric vision to the phenomenon has been described by Crone & Verduyn Lunel (1969).

Apparent Movement of an After-Image. The apparent movement of an after-image, when the eye moves, is an excellent illustration of psychological compensation. A retinal stimulus, being normally projected through the nodal point, is projected into different points in space as the eye moves; an after-image can be considered to be the manifestation of a continued retinal impulse, and its projection changes as the eye moves. The after-image thus appears to move in the same direction as that of the movement of the eye. Whether or not the gradual drift of an after-image across the field of view is entirely due to eye movements, it is difficult to say. One certainly has the impression that the eye is "chasing" the after-image.

Figural After-Effects. This is the term given by psychologists to a variety of illusions having as their basis the adaptation to prolonged observation of a given figure; thus a subject fixating a curved line for some time sees it as less and less

---

* There is no doubt that in the cat's eye movements *per se* may evoke changes in the lateral geniculate body (Bizzi, 1966), probably a presynaptic depolarization of optic tract terminals, but this apparently does not happen in the monkey, at any rate so far as the activity of single units is concerned; Feldman & Cohen (1968) did find monophasic waves as a result of eye movements, however; these occurred late in the eye movement and were recorded in complete darkness and when the eye muscles were paralysed.

curved, whilst a straight line presented immediately after this prolonged inspection of the curved line appears curved in a direction opposite to that of the original curvature (see, for example, Gibson, 1933), Particularly instructive is Fig. 266 from Köhler & Wallach (1944); the subject inspects the fixation cross of the left-hand ensemble for about 40 seconds and then inspects the fixation cross of the right-hand ensemble. As a result of the first period of fixation the right-hand pair of squares look closer to each other in the vertical direction than do the left-hand pair. Thus there is a tendency for the perceived object to be displaced as a result of previous viewing, and the direction of displacement is away from the previously inspected object.

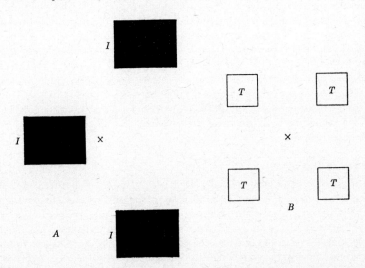

FIG. 266. Demonstrating displacement effects. Regard the fixation cross of A at reading distance for about 40 seconds, and then look at the fixation cross of B. After A has been observed, the right-hand squares of B appear nearer the central horizontal axis of this figure than do the left-hand squares. Inspection of the left-hand I-figure in A causes the left-hand squares of B to "move apart vertically"; and inspection of the right-hand I-figures of A causes the right-hand squares of B to "move together". Subjects say that test figures are displaced from the positions of previous inspection figures. (Graham, *Vision and Visual Perception.*)

Visual Estimates

The Directions of Lines. We have so far considered the problem of estimating the positions of points, in relation to each other and to the percipient. The estimate of the directions of lines involves no really new principles, since if two points, A and B, are exactly localized, the direction of the line AB can be appreciated. The perspective distortion, however, and the torsion of the eye resulting from a change in the direction of fixation, can modify the estimate of the direction of a line, referred to a fixed line in space—e.g. the horizontal or vertical—unless some psychological compensatory process is introduced.

RETINAL MERIDIANS

A horizontal line—fixated in the primary position of the eye—is recognized to be horizontal because its image falls on a set of receptor elements on the retina lying on the horizontal meridian. Similarly a vertical line is recognized because its image falls on the vertical meridian. When the eye moves to the right, into a secondary position, a horizontal line still falls on the meridian because there has been no torsion. On looking up and to the right, however, the image of a horizontal line is inclined at a certain fixed angle, depending on the position of the eye, to the horizontal meridian of the retina. Because of the torsion a psychological compensation must be made so that an image, inclined across the horizontal meridian, is perceived as the image of a truly horizontal line. The full significance of Donders' Law (p. 328) now becomes evident. Movements of the eye involve torsion—this cannot be avoided—and torsion involves a disturbance of orientation of lines unless psychologically compensated. The higher centres, in order to compensate, must know the degree of torsion, and this information becomes physiologically simple to acquire if the torsion is characteristic of the position of the eye. Thus the motor innervation bringing about a given position of the eyes is "reported" to the space representation centre, and this centre, in interpreting the retinal image, works on the assumption that this movement has brought about a characteristic degree of torsion. Donders' Law states that the torsion is a definite quantity for a given position of the eyes; and this regularity is achieved by the eyes rotating about an axis in Listing's plane (Listing's Law).

VESTIBULAR INFLUENCES

Movements of the head are associated with compensatory reflex movements of the eyes, mainly of vestibular origin, designed to keep the visual fields normally orientated. In man, however, the rolling necessary to compensate for an inclination of the head towards the shoulder is quite inadequate, amounting to only one-tenth to one-fifth of the inclination of the head; since horizontal lines appear horizontal even when the head is strongly inclined, psychological compensation must be involved here; presumably the information required is provided by the labyrinth. That this labyrinthine information is not entirely adequate is shown, however, by *Aubert's phenomenon*. A subject regards a bright vertical slit in a completely dark room; on bending the head over to the side, the slit appears to incline in the opposite direction. When the lights are switched on the slit becomes vertical. If, in the dark, the head is inclined suddenly the slit remains vertical for a while but finally tilts over; it would seem that the abrupt movement stimulates the semicircular canals (p. 367) which cause adequate compensation. If the higher centres have to depend on the tonic receptors only—utricle and proprioceptive neck impulses—the compensatory mechanism becomes inadequate (Merton). It would appear, therefore, that the tonic receptors must be reinforced by the visual sensation; the observer knows that the bars of the window, for example, are not usually tilted over and so a compensation is initiated.

ACCURACY OF ESTIMATES

The power of appreciating whether a line is horizontal or vertical is very highly developed in man. Volkmann describes experiments in which he repeatedly set a line to what he considered to be the true horizontal; his mean error was only 0·2° with his left eye, the left-hand end of the line was too low; with his right eye the right-hand end was too low. We may say that Volkmann's *"apparently horizontal meridians"* were slightly inclined to the true horizontal. With vertical lines his error was greater—more than 1°—but it was a constant error indicating an inclination of his *"apparently vertical meridians"*. More recent work, notably that of Gibson, bears out the findings of Volkmann; vertical lines are, on the average, estimated to within 0·28° and horizontal lines to within 0·52°; when an attempt is made to adjust a line to a fixed angle with the vertical, e.g. 60°, the error is much greater (1·74°) a finding which suggests that the vertical and the horizontal act as a frame of reference to which all other directions are referred. A modification of this type of experiment consists in presenting the subject with a short line, AB (Fig. 267) at a given angle of slant, α, from the

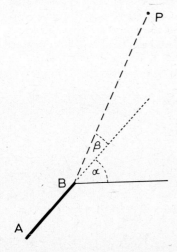

FIG. 267. The subject views the line AB and must adjust the position of P such that it appears to coincide with the prolongation of the line AB. (After Bouma & Andriessen, *Vision Res.*)

horizontal, and asking him to adjust a dot, some distance away, to a position such that an extension of the line would pass throught it. The error is indicated by the angle, β, and the angle (α+β) is called the perceived direction. Bouma & Andriessen (1968) found a tendency for the perceived orientation to be closer to the nearest horizontal or vertical than the geometrical slant, α. It was found that β decreased with the length of the short line, AB, up to a subtense of about 1°, so that a foveal line-segment was employed in the estimate; two dots were as effective as a line-segment.

Comparison of Lengths. The influence of the movements of the eyes in the estimation of length was emphasized by Helmholtz. An accurate comparison of the lengths of the lines AB and CD (Fig. 268) can be made, whereas if an attempt is made to compare the lines A′B′ and C′D′, quite large errors occur. According to Helmholtz, the eye fixates first the point A and the line AB falls along a definite row of receptors, thereby indicating its length. The eye is now moved to fixate C, and if the image of CD falls along the same set of receptors the length of CD is said to be the same as that of AB. Such a movement of the eye is not feasible with the lines A′B′ and C′D′. Similarly the parallelism, or otherwise, of pairs of lines can be perceived accurately because on moving the eye over the lines the distance between them must remain the same.

Fairly accurate estimates of relative size may be made, nevertheless, without movements of the eyes. If two equal lines are observed simultaneously, the one with direct fixation and the other with peripheral vision, their images fall, of course, on different parts of the retina; if the images were equally long it could be stated that a certain length of stimulated retina was interpreted as a certain length of line in space. It is probable that this is roughly the basis on which rapid estimates of length depend; a white line stimulating, say, twenty-five retinal elements being considered to have the same length as another line stimulating the same number of elements in another part of the retina. It is easy to show, however, that the matter is not quite so simple; the retina is curved so that lines of equal length in different parts of the visual field do not produce images of equal length on the retina; the distortion in the image resulting from the use of oblique pencils of light also complicates the matter, so that it is unlikely that such a simple relationship between size of image and projected size is achieved. The problem of the estimation of absolute size is closely linked with that of the estimation of distance and will be discussed under that heading (p. 442).

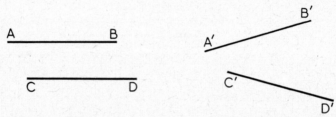

FIG. 268. AB and CD, being parallel, can be compared in length with accuracy whereas A'B' and C'D' cannot.

Estimates of Angles. Two angles may be compared with accuracy if eye movements are permitted and if the sides are parallel. The errors in comparison increase greatly if the sides are not parallel. Essentially we are dealing with the same problem—the superimposition of the retinal pattern on another. In the estimation of right-angles some characteristic errors are encountered. Thus the cross made by the lines *ab* and *cd* (Fig. 269) appeared to Helmholtz to be

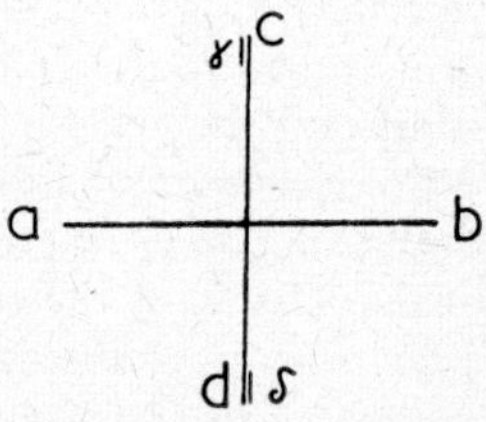

FIG. 269. The cross *abcd* appeared correct to Helmholtz, whereas the true cross is given by the line-segments γ and δ.

truly rectangular, whereas it is correctly given by the line-segments γ and δ. This error is an expression of the fact that the apparently horizontal and vertical meridians of the retina (p. 437) are not exactly at right-angles. When the cross was turned at an angle of 45° the error in assessing the right-angles became much greater; the true right-angles to the right and left appeared to be 92° whilst those lying above and below appeared to be 88°.

Optical Illusions. There are innumerable instances in the psychological literature of well-defined and consistent errors in visual estimates under special

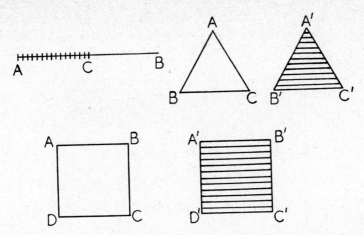

FIG. 270. Some optical illusions. AC appears greater than CB. The angles B′ and C′ of the equilateral triangle A′B′C′ appear to be greater than A′. The sides A′D′ and B′C′ of the square A′B′C′D′ appear to be greater than the sides A′B′ and C′D′.

conditions; there is probably no single hard-and-fast rule by which they can be explained, but an important element may be the tendency for distinctly perceptible differences to appear larger than those more vaguely perceived. Thus the line AB of Fig. 270 has been correctly bisected at C, but because AC is divided into easily perceptible parts, whilst the line CB is not, the former appears longer. The triangles ABC and A′B′C′ are both equilateral but, because of the division of the lines A′B′ and A′C′, the triangle A′B′C′ appears taller and therefore the angles at the base appear greater than 60°. Similarly, ABCD and A′B′C′D′ are squares yet the height of the latter figure appears to be greater than its breadth.

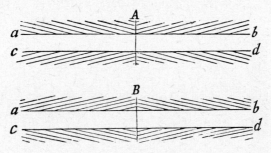

FIG. 271. Hering's illusion.

In Fig. 271 (after Hering) the lines *ab* and *cd* are truly parallel and yet they appear to converge on each other on moving away from A (above) and diverge on moving away from B (below). It may be shown that the illusion results from the tendency of an angle to appear greater when lines are drawn through it parallel to one of the sides (thus the angles B′ and C′ in the triangle A′B′C′ of Fig. 270 appear too large). The illusion created by Fig. 272, in which the letters appear to slope in different directions, is created by the disorientating effect of the background-lines. In Fig. 273 the thin line, although it is continuous and straight, appears to be bent in towards the thick line; no satisfactory explanation of this illusion has so far been given.

Fig. 272. An optical illusion; the letters are in fact all correctly orientated.

Summary. To summarize, we may say that the pattern of objects in the outside world is perceived as a result of the interpretation of the geometrical pattern of the image on the retina. It is not sufficient, however, to know merely the pattern of objects in space, it is useful also to relate *direction* on the retina with certain fixed directions in space. Primarily these are the vertical and the horizontal, the most frequently recurring directions of contours in the outside world. The retina has therefore become specialized by the development of apparently vertical and horizontal meridians; these act as a pair of rectangular co-ordinates through the fovea to which the position of any point or line is referred. In order that the interpretation of the retinal pattern be independent of the movements of the fixation axis, and of the inclination of the apparently horizontal meridian (torsion), a compensated system of projection into space is necessary. As a result of this compensation, the *absolute* position of an object in space is always correctly interpreted and the directions of lines, in relation to the horizontal and vertical, are also accurately estimated.

Fig. 273. The thin line appears to be bent where it joins and leaves the thick line.

There is a great deal of evidence, that the tendency to project the retina is "innate" although few would deny that accurate projection is only acquired with experience. Presumably the development in evolution of an innate tendency to project the retina in accordance with certain geometrical rules had survival value; the "rules" turn out to be similar to those that would be followed if the individual learnt to project the retina in accordance with his own experience.

The Monocular Perception of Depth

From purely geometrical considerations, the estimation of the relative positions of objects in three dimensions by a single eye is not feasible; the points A, B and C in space (Fig. 274) produce images, c, b and a on the retina which are projected outwards along the lines cA, bB, and aC, but there is nothing in the geometry of this projection system to enable it to differentiate between the points B and B' for example. The projection of the retinal image of a single eye is thus two-dimensional and objects in space are presented to consciousness as a plane pattern. Nevertheless, the individual may *perceive* depth with a single eye by making use of a variety of clues, whose significance has been ascertained by accumulated experience. These clues are:

"Aerial Perspective". As a result of the scattering of light by the atmosphere the colours of distant objects give some clue to their distance; the scattered light is predominantly blue and violet and the eye views a distant object through a "wall" of blue light. The thicker this wall, the bluer will the object appear to be; distant mountains thus appear blue, a fact exemplified in the well-known lines:

"T''is distance lends enchantment to the view
And robes the mountain in its azure hue."

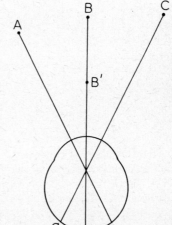

FIG. 274. With uniocular projection the relative positions of B and B' cannot be distinguished.

It must be appreciated, however, that a self-luminous object, such as a lamp, is visible because its own emitted radiations reach the eye; since blue light is scattered preferentially to red, a distant white light appears yellow or red when seen through a hazy atmosphere; the red light penetrates the atmosphere whilst the blue is scattered on the way. It might be objected that a mountain reflects light and that we see it because of this reflected light; this reflected light suffers scattering and so the mountain should appear yellow or red. Distant objects, however, are seen largely through contrast with the brighter sky; it is because they reflect less light than that emitted by the background that they are discriminated; the reflected light is therefore not important in perception and it is the wall of scattered light that determines the colour sensation. (When the brightness of "wall" plus the brightness of the mountain due to reflected light becomes equal to the brightness of the background, the mountain becomes invisible. For this reason, the visibility of distant points in the landscape constitutes a good measure of the atmospheric haze.)

The scattering of light also blurs the outlines of objects and thus gives another clue to their distance from the observer; the apparent closeness of mountains in a very clear atmosphere is an illusion based on the normal association of clearly defined objects with proximity to the observer.

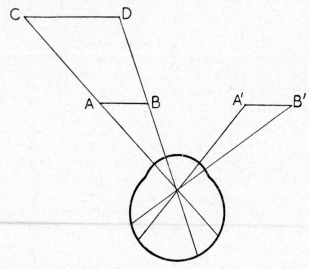

FIG. 275. Illustrating the value of relative size in estimating depth.

Relative Sizes of Objects. The apparent size of an object may be represented by the angle it subtends at the nodal point of the eye; thus AB and A′B′ (Fig. 275) appear to have the same size because they subtend the same angle and they are the same distance away; CD, moreover, appears as big as AB because it subtends the same angle although its real size is twice as great as that of AB. If the observer knows that CD is twice as large as AB, e.g. if AB is a bicycle and CD is a car, he can say that CD is twice as far away from him as AB, i.e. he can estimate their *relative* distances in the third dimension. A fair estimate of the *absolute* distance away can be made if the object viewed is familiar, e.g. a man or a cottage. When a man walks away from an observer the image on the latter's retina becomes smaller and smaller; this is interpreted as a more and more distant, not as a smaller and smaller, man. In gunnery, the visual estimation of range is made almost entirely on this basis.

The association of size with distance can give rise to common illusions; in a thick fog a figure "looming up" looks very large because the figure is assumed to be farther away than it actually is. Similarly objects seen through binoculars appear to be very close whereas their images, seen through the instrument, are at infinity.

ABSOLUTE SIZE

The determination of *absolute size* is thus closely interwoven with that of distance; any rigid formulation of this process on the basis of the number of retinal elements stimulated by the image of an object is therefore doomed to failure. A given retinal distance represents an absolute size of an object in space only for one distance of the object from the observer. A change in this distance must necessitate a revaluation of retinal distance. When an object is recognized to be far away, a small retinal distance is interpreted as a certain unit of "absolute size" in space; if the same object is brought much closer, its image becomes larger so that a larger retinal distance now becomes equivalent to the same unit of "absolute size" in space. The close linkage between this revaluation process and the acts of convergence and accommodation is shown by the condition of "*paresis micropsia.*" As a result of a defective convergence mechanism, for example, a strong effort of convergence is made in order to fixate a near object; this effort, being out of all proportion to the resulting convergence, is associated with much too great a change in retinal dimension values; the object converged upon appears smaller because it is treated as being much closer and the retinal dimension values have been scaled down too far.

THE MOON ILLUSION

The assignment of a size to objects very far away, such as the sun and moon, must be entirely arbitrary, and so it is not surprising that the assigned size varies, although the reasons for the variations are not apparent. Thus the sun and moon, when seen on the horizon appear very much larger than when viewed at the zenith. It is interesting that the magnitude of the discrepancy between the

FIG. 276. The value of the overlapping of contours in the estimation of depth.

apparent sizes of a disc viewed horizontally and vertically is much greater in children than in adults; thus in 4–5-year-olds a 10-inch disc viewed horizontally was estimated to be of the same size as a 20-inch disc seen vertically; in adults it was considered equivalent to a 16-inch disc (Leibowitz & Hartman, 1959).

FIG. 277. Full side view.

FIG. 278. Three-quarters view. The front wheel appears smaller than the back; both appear elliptical.

Form. The contours of objects give some idea as to their relative positions in the third dimension. Thus in Fig. 276 the cottage is in front of the tree which is in front of the windmill which itself is perceived to be in front of the hill, entirely because of the overlapping of the contours.

FIG. 279. Illustrating the value of light and shade in giving the appearance of depth to a sphere.

Perspective. The projected retinal image of an object in space may be represented as a series of lines on a plane, e.g. a box; these lines, however, are not a unique representation of the box since the same lines could be used to convey the impression of a perfectly flat object with the lines drawn on it, or of a rectangular, but not cubical, box viewed at a different angle. In order that a three-dimensional object be correctly represented to the subject on a two-dimensional surface, he must know what the object is, i.e. it must be familiar to him. Thus a bicycle is a familiar object; viewed flat-on it has the appearance of Fig. 277; if it is turned away from the observer it appears as in Fig. 278; the wheels, for example, are elliptical and the front one is smaller than the rear one. Because the observer knows that the wheels are actually circular and of the same size he can state that the bicycle is directed into the third dimension, i.e. he perceives depth in a two-dimensional pattern of lines. The perception of depth in a two-dimensional pattern thus depends greatly on experience—the knowledge of the true shape of things when viewed in a certain way. The limitations of two-dimensional representation become evident when we look at the photograph of an unfamiliar object, such as a complex crystal; the picture is almost meaningless because we have no idea of the actual shapes of its component parts.

FIG. 280. Protective shading of an animal (see text).

Light and Shade. A sphere always appears in a two-dimensional pattern as a circle, however viewed; perspective can therefore give no clue to its solidity. The light and shade on its surface, however, immediately give it a three-dimensional appearance (Fig. 279); in general, it is important that the origin of the source of light be known and then the relations of the shadows of objects to each other indicate their mutual arrangement in space. The protective colouring of a large number of animals is designed to offset the appearance of solidity resulting from uneven illumination. Thus the back of an animal receives more light than the belly and if the back reflected the light just as efficiently as the belly it would appear much brighter and convey the impression of solidity (Fig. 280). However, the white belly of the donkey, for example, increases the amount of light reflected by it so as to make it approximate in brightness to its back, which is grey. As a result the animal appears, at a distance, to be fairly uniformly bright over its whole surface and therefore appears flat (Fig. 280), a characteristic that helps it to merge with its background. Under certain conditions of illumination, light and shade can give valuable clues to the length of a room, for example, in the third dimension; if the source of illumination is close to the observer, the luminance of the walls decreases progressively with their distance away. As a result of the process that results in "tone constancy," the walls generally appear to be evenly illuminated, the actual change in luminance being interpreted as an *increase in distance* from the observer. In this case the phenomenon of "tone constancy" is essentially similar to that of "size constancy"—the tendency to interpret a diminishing retinal image as an object of constant size moving away from the observer.

Parallactic Displacement. On movement of the head from side to side, the objects nearer to the observer appear to move, in relation to more distant points, in the opposite direction. Similarly, when the observer moves forward, the nearer objects appear to move past him whilst the more distant objects appear to move with him. Helmholtz remarks that if one sits in a wood with one eye closed and the head motionless, the picture presented is quite flat; on moving the head, however, the environment springs to life due to the parallactic displacements, which indicate the essentially three-dimensional nature of the scene.

Accommodation. Within a limited range, the accommodative effort required to see an object clearly might be expected to given a clue as to its absolute distance from the observer, but although opinions differ, it seems safe to conclude that the kinaesthetic sensations from the ciliary muscle do not contribute materially to the power to assess distance, and the same would appear to be true, in binocular vision, for the process of convergence (Zajac, 1960). The extreme difficulty in separating the factor of accommodation from all the other factors concerned reveals that depth perception is a complex process whose components are so interwoven as to make it a somewhat academic exercise to attempt to separate them and measure their individual contributions.

This description of the monocular factors in depth perception shows that there are many ways in which the one-eyed person can perceive depth in his environment; nevertheless his perceptions are grossly inadequate when compared with those of a normal subject, and we may now inquire into the general characteristics of binocular vision.

BINOCULAR PERCEPTION

The two eyes behave, in so far as their motor taxis is concerned, as a single unit; binocular vision is likewise integrated, so that in many respects the perception can be described in terms of a single eye.

"The Cyclopean Eye". If its two images on the retinae were projected independently into space, an object would appear to be differently orientated in relation to the individual according as its direction was referred to the primary position of one or the other eye. For example, in Fig. 281, the point A is fixated by both eyes, and it is correctly projected into space at A; to the left eye, however, it is to the right of the fixation axis in its primary position, whilst to the right eye it is to the left. Since it is symmetrically disposed in regard to the two eyes, the object is considered to be straight ahead of the observer— neither to the right nor to the left—so that binocular projection can be represented by the "*Cyclopean eye*", as in Fig. 281, an imaginary eye in the forehead to which all directions are, in fact, referred. The point A is considered to be straight ahead of the observer because it lies on the fixation axis of the Cyclopean eye in its "primary" position.

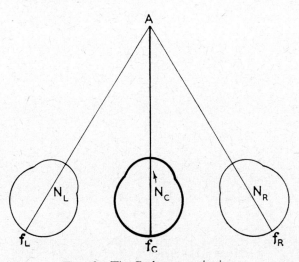

FIG. 281. The Cyclopean projection.

Binocular Double Vision. The Cyclopean projection is represented by transposing the images on the two retinae to the single, median, eye as in Fig. 281. When the images on the two retinae are not symmetrically disposed about the fixation point, however, such a transposition leads to a double projection, and it is found, in fact, that when the same object can be projected in two directions by the Cyclopean eye double vision does actually occur. In Fig. 282, the two eyes fixate the point A, as before. A point B, closer to the eye, on the median line, produces images at b_L and b_R; the problem is to determine how the observer projects these images. In Fig. 282, the Cyclopean eye, occupying a mean position between the two eyes, is drawn in. The image of A falls on the fovea of each eye and hence its projection by the Cyclopean eye is given by projecting f_C through the nodal point, i.e. it coincides with the true direction of A. The image of B on the left retina, b_L, occurs to the left of f_L and must be put in a similar position on the Cyclopean retina; b_L is thus projected to B'. On the right retina, B forms an image, b_R, to the right of f_R, which, when transposed to the Cyclopean eye, gives a projection to B". Since B is projected to B' and B" at the same time, it is seen double; the diplopia is called *heteronymous diplopia* because on closing one eye the image appears to move away from the fixing eye, e.g. on closing the left eye the image of B appears to be at B", i.e. away from the right, fixing, eye, The *homonymous diplopia*, caused by an object farther away from the fixation point, is indicated in Fig. 283. The normal presence of double images in the binocular field of vision is not usually noticed; they are easily demonstrated, however, by placing a pencil close to the eyes in the median line; when a distant object is fixated, the pencil is seen double.

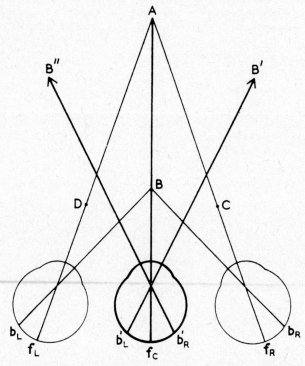

FIG. 282. Heteronymous diplopia. B is closer to the eyes than the fixation point. By the left eye the image of B is projected to B'; by the right to B''.

Binocular Alignment. On a binocular projection, the point B is aligned with the point A, so that if an observer places a pencil in front of him so as to coincide with a distant object when both eyes are open, he really puts the pencil in such a position that its double images lie at equal distances to the left and right of the distant object—B' and B'' of Fig. 282. Such a Cyclopean alignment is, in general, not accurate; the careful alignments necessary for sighting a rifle, for example, are made with one eye; if the alignment were made with the right eye a point C, on the fixation axis, would be correctly aligned with A; for the left eye, the point D (Fig. 282). (The reader should try rapidly pointing with a pencil at a distant object; keeping the pencil fixed, he should inspect the double images of the pencil while he fixates the distant object; if the latter lies between the double images, the pointing has been carried out on a Cyclopean projection; if, on the other hand, one of the double images coincides with the distant object, the pointing has been carried out on the basis of a monocular projection.)

Diplopia due to Squint or Prisms. In Fig. 282 the point A was seen single, on a Cyclopean projection, because the lines joining it to the nodal point of each eye met the retinae in the foveae. B was seen double because the lines from it through the nodal points met the retinae on opposite sides of the fovea. A

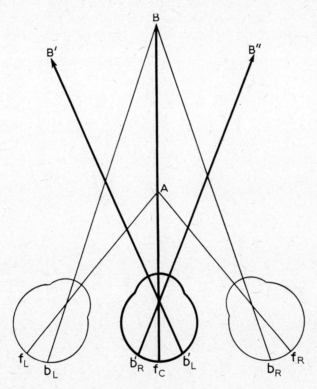

FIG. 283. Homonymous diplopia. B is farther away from the eyes than the fixation point A. By the left eye the image of B is projected to B′; by the right eye to B″.

similar state of affairs occurs in squint. Fig. 284 represents a divergent squint. A is fixated by the right eye, but the left visual axis diverges from A; the image of A on the left retina thus falls at a_L. If the images on the two retinae are transposed to the Cyclopean eye, the image of A is projected towards A and towards A′, to the right of A. A heteronymous diplopia is therefore caused by this divergent squint.

PRISM

A prism may cause diplopia, provided that the eyes do not compensate for the change of apparent direction of the object caused by the prism. In Fig. 285, a base-out prism has been placed before the left eye. The image of A falls on f_R the fovea of the right eye, but in the left eye it falls on a, to the left of the fovea f_L. On the Cyclopean projection, the images of A are projected in two directions, A and A′. In Fig. 286, the eye has compensated for the prism by adducting; both images now fall on the foveae, and the Cyclopean projection indicates single vision. It is interesting that a single projection is obtained in this instance; the left eye has adducted, so that on the basis of our earlier arguments (p. 431) an image falling on the fovea should be projected through the nodal point—to the right of A. If the left eye, only, were being used, such a projection would occur—a *false projection*, in the sense that the image of A is not projected towards A. The fact that with binocular vision the two images of A are correctly projected to a single point indicates that the Cyclopean projection is determined by the projection of the eye that has not been interfered with.

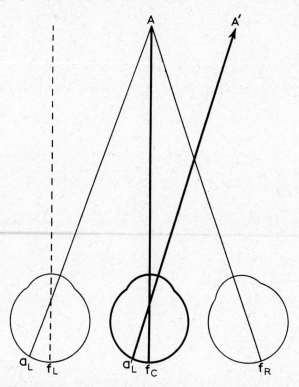

FIG. 284. Heteronymous diplopia due to divergent squint. The image of A is correctly projected by the left, fixating, eye to A; by the right eye it is projected falsely to A'.

When prisms are placed in front of both eyes, if the necessary oculo-motor adjustments are possible, single vision is retained, but this time the retinal images of objects are erroneously projected, either closer to, or farther away from, their true position.

The "False Macula". The diplopia due to squint is an instance of false projection; the squinting eye fails to project a retinal image through the nodal point, e.g. the stimulated point a_L of Fig. 284 is not projected through the nodal point to A but its projection may be described as though the point stimulated were to the left of a_L. The fovea of the squinting eye is likewise projected falsely, an image falling on f_L is projected towards A, the fixation point of the other eye. In many cases the false

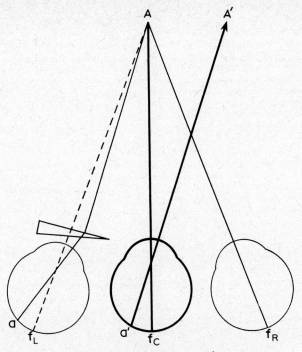

FIG. 285. Diplopia due to a prism.

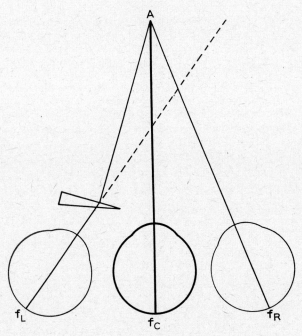

FIG. 286. Single vision with prism before one eye.

projection may be corrected by the development of a *"false macula"*, that portion of the retina on which the image of the fixation point, A, of the other eye falls, acquiring some of the characteristics of the fovea. The change is a cortical one—it is impossible for the anatomical characteristics of the retina to alter; the development of a false macula represents a change in the significance of the stimulus falling on it; the stimulus becomes the centre of attention, and that falling on the true macula is suppressed; moreover, the false macula is correctly projected into space. We have seen that projection is essentially a matter of localization in space of a retinal stimulus; to do this correctly the positions of the eye and head must be known and psychological compensations for movements of the eye and head are an integral part of the projection mechanism. The development of the false macula requires a re-adjustment of this psychological compensation to take into account the abnormal relationship of the visual axes of the two eyes.*

Innate Projection. The existence of false projection following squint reveals that the projection of the eye through the nodal point is an innate characteristic, developed doubtless during the early years of life, but nevertheless an inherited quality. The possibility of a development of a false macula reveals, on the other hand, a certain plasticity in the function of the visual cortex. We have already remarked on the rigidity of the relationship between stimulus and sensation in this area of the cortex, a rigidity which differentiates it from other parts of the cerebral mantle, and it is interesting that the relationship may be modified by experience and is therefore not absolute as with processes occurring at lower levels in the brain.

A condition of *monocular diplopia* may arise from the development of a false macula. Thus the stimulus falling on a_L (Fig. 284) is originally projected to A'; after the development of a false macula it is correctly projected to A. Sometimes both projections take place simultaneously, so that, as a result entirely of cortical activity (there is clearly only one image on the retina), the point A is seen simultaneously at A and A'.

The development of a *"pseudo-fovea"* may result from hemianopia; in this condition one half of the foveal image is seen, a condition so annoying that the affected person tends to use an adjacent portion of the retina for fixation. With time this point acquires the characteristics of the true fovea—it is correctly projected into space and the images of objects falling on it become the centre of attention and interest.

Convergence. With two eyes we have the geometrical basis for an estimate of the absolute distance of a point from an observer, since the converging power necessary for binocular fixation is uniquely determined by its distance away in the third dimension of space. The relative distances of different objects are likewise determined by the relative degrees of convergence necessary for binocular fixation. That variations in convergence may cause three-dimensional illusions was shown long ago by Meyer in the so-called *wallpaper phenomenon.* As described by Helmholtz, the illusion is brought about by looking at a wallpaper with a vertical line pattern; by making his eyes cross, so that the two

---

* There is unfortunately a serious danger of confusion arising out of the different senses in which "false projection" is regarded. To the neurologist the macula of the squinting eye is falsely projected because the individual grasps wide of the object whose image falls on it (he "past-points"). False projection thus means the projection of a retinal image to some other point in space than that occupied by the object evoking it, and this is the sense maintained in this book. To some ophthalmologists, on the other hand, false projection is synonymous with what is called "*abnormal retinal correspondence*". With the development of a "false macula", an image, not on the macula of the squinting eye, is projected to the same point in space as the image on the macula of the normal eye; the retinal correspondence (p. 453) has been readjusted to give single binocular vision of an object whose images do not fall on the innate corresponding points of the two retinae. Thus the neurologist would say that the "false macula" is correctly projected into space, whilst some ophthalmologists would say that it is falsely projected.

eyes were fixating adjacent lines, Helmholtz found that the wallpaper appeared to float out of the wall, whilst if he made his eyes diverge the paper receded into the wall in the region of the fixation points. The crossing or divergence of the eyes caused no apparent diplopia because the fields of view consisted of only vertical lines. As Helmholtz and others have interpreted the phenomenon, the changed convergence *per se* was responsible for the apparent three-dimensionality of the wallpaper; in other words, the act of convergence contributed to the sensation of depth. However, more modern workers are inclined to deny to the convergence *per se* any significant contribution to depth perception (Stevenson Smith, 1946; 1949; Zajac, 1960); it is because alterations in convergence, by altering the fixation point, change the relative sizes and positions of other objects around the fixation point, that the retinal images are modified and thus the appearance of the fused binocular retinal images.

STEREOSCOPIC DEPTH PERCEPTION

We have so far considered the geometrical consequences of the use of the two eyes in estimating distance—the system has been treated as a range-finder—and the problem of the fusion of the two images in the separate eyes has not been considered. Not only does the fusion of the images provide a unitary perception of the binocular field of view, but it presents the objects in this field in a three-dimensional pattern that is quite impossible of achievement with monocular vision. In the following pages we shall inquire into those characteristics of the two retinal images that (*a*) permit or favour fusion and (*b*) contribute to the three-dimensional percept.

Corresponding Points and the Horopter

Corresponding Points. We have seen, in our discussion of binocular projection, that a single object in space is projected in two different directions when its images, formed by the two eyes, do not both fall on the foveae (Fig. 282); this results in diplopia, and single vision is only achieved when the retinal stimuli are projected to the same point in space. The foveae may therefore be described as *corresponding points*; they are points on the two retinae normally projected to the same point in space.

Every point on the retina of each eye is projected into space to give the field of vision; where the projections overlap we have the binocular field of vision; the problem is to determine which points in the binocular field of vision are seen single or, more correctly, which pairs of points in the retinae of the two eyes are projected to the same point in space. The matter is subject to experimental investigation and the following conclusions were drawn by Helmholtz:

(*a*) The apparently horizontal meridians of the two retinae correspond.

(*b*) The apparently vertical meridians correspond.

(*c*) Points on the apparently horizontal meridians, equally distant from the foveae, are pairs of corresponding points.

(*d*) Points equally distant from the apparently horizontal meridians, which are on the apparently vertical corresponding lines, are a pair of corresponding points.

In so many words, points equidistant from the apparently horizontal and vertical meridians of the retinae were said to be corresponding points.

The Vieth-Müller Horopter Circle. Fig. 287 represents a horizontal section through the two eyes, fixating the point F. The points a_L and a_R are equally distant from the vertical meridians of the two eyes and are therefore projected to the same point in space through the respective nodal points, namely, to A. The points b_L and b_R, also symmetrically disposed about the vertical meridians, are projected to B. From the geometry of this binocular projection

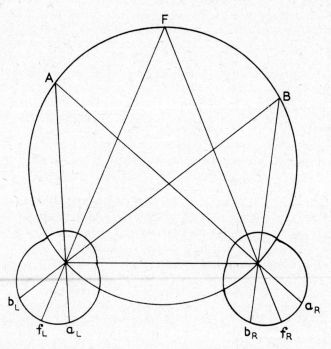

FIG. 287. The Vieth-Müller horopter circle. F is the fixation point. If corresponding points are symmetrically distributed about the foveae the points in space, in the fixation plane, whose images fall on corresponding points lie on the circle.

it is evident that A, F and B lie on a circle passing through the fixation point, F, and the nodal points of the eyes. All points in space, lying on this circle, so long as they produce images on the two retinae, produce them on corresponding points provided that the symmetry, postulated by Helmholtz, exists. The circle is called the *Müller* or *Vieth-Müller horopter circle*: it is the locus of points, in a horizontal plane, that produce images falling on corresponding points of the two retinae when the fixation axes are horizontal, and are directed on the point F. Moreover, if the eyes may be considered to rotate about the nodal points instead of their centres of rotation, the fixation point may move to any point on the circle and all other points will still produce images on corresponding points. With this approximation the Müller horopter circle becomes the locus of points, in a horizontal plane, that produce images on corresponding points for a given degree of convergence of the eyes when the retinal horizons are horizontal. The Müller circle only indicates the points in space *in a single plane* that produce images on corresponding points. If the circle is rotated about a line joining the nodal points, the surface so produced gives the locus of all points in space whose images fall on corresponding points, for a fixed degree of convergence of the eyes. This is only true if the two retinae possess strict spherical symmetry and if the corresponding points are symmetrically disposed around the foveae. As the apparently vertical meridians of the retinae are not parallel to each other (p. 438) there is no such perfect symmetry and this Müller horopter surface, as we may call it, is only an approximation.

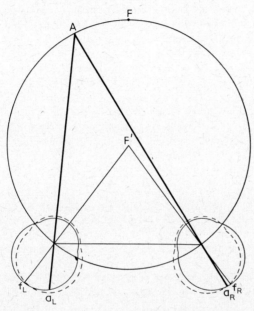

FIG. 288. When F was the fixation point, A lay on the horopter-circle; when F' becomes the fixation point A no longer produces images on corresponding points, so that A is not on the more generally defined horopter.

The Horopter. The *horopter* as originally defined is the *locus of points in space whose images fall on corresponding points of the two retinae*, i.e. no proviso is made as to the fixation point or the degree of convergence. The Müller horopter circle must therefore be regarded as a special case of the horopter when the fixation point has been defined. The number of points in space that will produce images on corresponding points, *however the eyes move*, is strictly limited; thus in Fig. 288, when the eyes move to a new fixation point not lying on the original Müller circle, e.g. the point F', the images of A do not lie symmetrically with regard to the foveae and are therefore not on corresponding points. Hence, although A lies on the Müller horopter circle when the eyes fixate F, it does not lie on the horopter. The horopter was calculated by Helmholtz from experimental data on the positions of corresponding points and the laws of the eye movements; it is a curve of the third degree (a curve that cuts a plane in three points). Under certain conditions the horopter acquires a simpler form; e.g. if the eyes are looking straight ahead in the primary position, the horopter becomes a horizontal plane passing through the feet, all points on this plane being seen single. When the point of fixation is either in the median plane of the head or in a horizontal plane with the fixation axes horizontal, the horopter becomes a Müller circle and a vertical line, intersecting in a point.

Longitudinal Horopter

For practical purposes, the so-called *longitudinal horopter* is the most useful. For a given fixation point it is the locus of points in a horizontal plane whose images fall on corresponding points of the two retinae. Thus, if Müller's assumptions were correct, it would be his horopter circle. In fact, however, the postulated symmetry about the vertical meridians does not hold good, so that the longitudinal horopter is not a circle.

Hering-Hillebrand Deviation. Thus if we wish to find the point corresponding to a_L (Fig. 287) it is, in general, not at a_R such that $f_L a_L = f_R a_R$, but closer to, or farther away from f_R. This so-called *Hering-Hillebrand deviation* causes a characteristic change in the longitudinal horopter, so that not only does it deviate from a circle but its shape changes with the fixation distance. With strong convergence it was found to be concave to the face; at about 2 metres it became approximately a straight line in the frontal plane, whilst beyond this distance it was a curve, convex to the face.

Modern Analysis

Interest in the horopter, up to 1932, was largely academic; in this year, however, Ogle and his associates drew attention to the implications of the horopter from the point of view of the relative sizes of the retinal images. On the basis of certain assumptions, Ogle showed that the shape of the longitudinal horopter should be an index to the degree of discrepancy between the sizes of the retinal images. His interest in the problem of aniseikonia led to a very thorough investigation of the horopter, both from a practical and theoretical aspect; it is impossible in a brief space to do justice to the results or the conclusions of this study. The mathematical analysis confirmed the general conclusions of Hillebrand on the variation of the shape of the horopter with distance. Thus at a certain distance it coincided with the frontal plane whilst at other distances it was curved. The significance of altered image-size, i.e. of *aniseikonia*, will be discussed later (p. 469).

The Stereoscope and Depth Perception

The essential feature of stereoscopic perception is the existence on the two retinae of *different images of the same object*; the object is seen in two aspects because of the different viewpoints of the eyes. Thus Fig. 289 represents two photographs of a book as it would be seen by the left and right eyes; if these photographs are viewed separately by the two eyes and their images fused, the book is perceived as a three-dimensional object.

The Stereoscope. The fusion of the two images can be carried out, with a little practice, by holding them directly in front of the eyes, a piece of cardboard in the median plane being used to keep the fields separate. The conditions are, however, artificial in that the eyes normally converge when near objects are viewed, and a convenient instrument, which permits this convergence when the objects viewed are close to the eyes and directly in front of them, is the

FIG. 289. Stereoscopic photographs of a book.

stereoscope, the optical properties of which are shown in Fig. 290. Two pictures, representing different aspects of the same object, AB and CD, are placed about 10 cm from the eyes. They are viewed separately through base-out prisms, so that symmetrically disposed points on the stereograms are projected to the same point, e.g. *a* and *b* are projected to *c*.

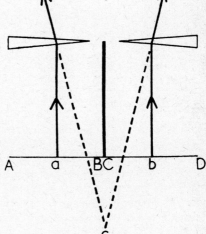

FIG. 290. Optical principles of the Brewster stereoscope.

Discrepant Images. If the two identical circles of Fig. 291*a* are placed in the stereoscope, their images are fused but no sensation of depth is achieved precisely because the pictures are identical and are such as would, in effect, be presented to the eyes were a flat circle viewed. If the dots are placed in the circles, as in Fig. 291*b*, fusion of the images creates the impression that the dot is floating in space in front of the circle, whereas if the dots are placed as in

Fig. 291c, the dots appear to be behind the plane of the circle. The stereograms imitate the appearances to the two eyes of a three-dimensional pattern; thus let us imagine that there is a dot on the surface of a sphere placed symmetrically with regard to the two eyes; viewed with the left eye, the dot appears to the right of the centre of a circle—the outline of the sphere—whilst viewed by the right

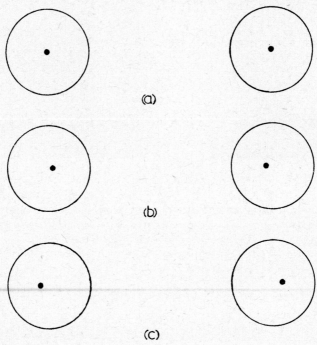

FIG. 291. (a) The spots are in the centres of the two circles; the fused image is two-dimensional. (b) The spots are de-centred nasally. The spot appears, on fusion of the images, to be nearer the observer than the plane of the circle. (c) The spots are de-centred temporally. The spot appears, on fusion of the images, to be farther from the observer than the plane of the circle.

eye it is to the left. The combined image, as seen in the stereoscope, gives the impression of a point on the surface of the sphere. It will be noted that a sphere, viewed from any angle, appears as a circle; the two aspects of a sphere are therefore identical circles and binocular perception of a uniformly bright sphere can give no impression of solidity. It is essentially because a sphere is normally unevenly illuminated, and has light and dark spots on it, that the aspects presented to the eyes are different; the fusion of these aspects places the lights and shades in their true relationships in three dimensions and so creates the appearance of solidity. The importance of light and shade in stereoscopic vision may be seen by examining photographs of unfamiliar objects which reflect light to varying extents on different parts; stereoscopic vision puts the various high lights in their correct positions in the third dimension and so contributes greatly to the intelligibility of the presentation.

Parallax. The difference in the two aspects of the same object (or group of objects) is measured as the *instantaneous parallax*. In Fig. 292, B is closer to

the observer than A; the fact is perceived stereoscopically because the line AB
subtends different angles at the two eyes, and the instantaneous parallax is
measured by the difference between the angles α and β. The *binocular parallax*
of any point in space is given by the angle subtended at it by the line joining
the nodal points of the two eyes; hence the binocular parallax of A is α; that
of B is β; the instantaneous parallax is thus the difference of binocular parallax
of the two points considered.

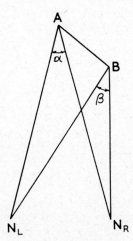

FIG. 292. Binocular and
instantaneous parallax.

Accuracy and Limits of Stereoscopic Perception. The accuracy of
stereoscopic perception is remarkable. Andersen and Weymouth and later,
Ten Doesschate, placed three wires in front of an observer; the distance the
middle wire had to be moved to appear out of the plane of the other two was
measured; the parallax was of the order of four seconds of arc, corresponding
to a disparity of the retinal images far smaller than the diameter of a single
cone. Other values in the literature range from 1·6 to 24 seconds.

With two editions of the same book, it is not possible, by mere inspection, to
detect that a given line of print was not printed from the same type as the same
line in the other book. If the two lines in question are placed in the stereoscope,
it is found that some letters appear to float in space, a stereoscopic impression
created by the minute differences in size, shape, and relative position of the
letters in the two lines. The stereoscope may thus be used to detect whether
a bank-note has been forged, whether two coins have been stamped by the
same die, and so on.

It may be shown, on geometrical grounds, that beyond a certain distance
from the eyes, the difference of parallax of objects becomes effectively zero;
beyond this point, then, objects in the environment appear to be in a single
plane; the distance has been computed to be about 450 metres but of course
it depends on the figure taken for the smallest perceptible instantaneous parallax.

Effect of Distance. Several studies of the effects of distance on relative
depth discrimination have been described. Under these conditions the observer
is presented, in the distance, with two similar objects, separated laterally but
initially at the same distance from him; one object is moved nearer or farther

away until the observer states that he can detect the fact; and the distance between the two objects represents the *linear* depth limen, which may be converted to an angular disparity or parallax. According to Teichner, Kobrick & Wehrkamp, there is an actual improvement in the power to discriminate as the distance increases, although on theoretical grounds we should not expect any change, when the limen is expressed in terms of an angle of parallax. Ogle (1958) has pointed out, however, that the monocular clue given by difference in perceived size, when the two objects are not equidistant from the observer, becomes relatively more important in depth discrimination as distance increases; and it is for this reason that there is an improvement of depth discrimination, the eye being more sensitive to changes of size than to changes of parallax at large distances. This view was confirmed by Jameson & Hurvich.

Duration of the Stimulus. Dove in 1841 observed that stereopsis could be experienced by viewing pictures in a stereoscope when they were illuminated by an electric spark, proving that the experience did not depend on movements of the eyes, such as those involved in convergence. A more systematic study by Ogle & Weil has shown that the accuracy* of stereopsis increases by a factor of four on increasing the exposure-time from 4 to 1000 msec. This suggested to Ogle & Weil that small nystagmoid movements, by contributing to visual acuity, improved stereoscopic acuity, since there is no doubt that these two parameters are related. That this is not the correct explanation, however, is shown by Keesey's study on visual acuity with different exposure-times; this did, indeed, improve as the time increased from 20 to 200 msec, but the same improvement was found when the image on the retina was "stabilized" in such a way that it always remained on the same portion of the retina. Thus the small involuntary scanning movements do not contribute to visual acuity, and the improvement of both visual acuity and stereoscopic acuity with viewing-time implies only that the neural processes concerned with the discrimination require a certain minimum time for their best performance.

Disparate Points and Stereoscopic Perception

Stereoscopic perception results from the fusion of dissimilar images; it follows from this that the fused percept is made up by the integration of stimuli falling on disparate points on the two retinae, i.e. of points which, on a Cyclopean projection, would be seen double. Thus in Fig. 293, A'B'C'D' and A''B''C''D'' are the two aspects of a slab, ABCD, inclined to the median plane. If the lines B'D' and B''D'' correspond, it is quite clear that the lines A'C' and A''C'' *cannot*, the line XY corresponding approximately with A'C'. Hence, on a purely Cyclopean projection, if the edge of the slab, BD, is seen single, the near edge, AC, is seen double. The fused percept is, nevertheless, that corresponding to a single entity—a slab "floating in space." The existence of images of the same point in space on disparate points of the two retinae is, therefore, just as much a characteristic of stereoscopic perception as the existence of images on corresponding points; in fact, if all points on an object produced images on corresponding points, there would be no stereoscopic perception.

---

* The accuracy of stereopsis is determined by measuring the threshold, i.e. the smallest instantaneous parallax that is just detected with stereoscopic vision. As with other thresholds, of course, there is a range of uncertainty over which we must speak of a *probability of detection*, and the threshold must be assessed on a statistical basis (Ogle, 1950; 1962).

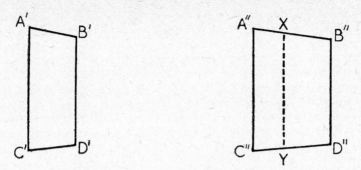

FIG. 293. Illustrating the importance of the incidence of images on non-corresponding points in stereoscopic perception.

Corresponding Points Defined. The facts of stereoscopic perception tell us, therefore, that a pair of corresponding points on the two retinae is not completely defined by the statement that their images are seen single, since disparate points may also be seen single. Let us consider the points A and B of Fig. 294.

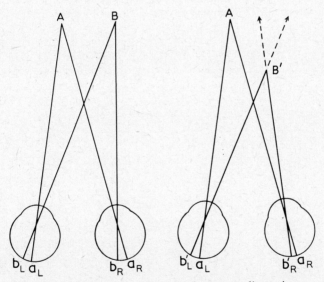

FIG. 294. Illustrating the definition of corresponding points.

They are on the horopter and their images fall on corresponding points, a_L and a_R; b_L and b_R. Both points are seen single in binocular vision. Let us now consider the points A and B'; if A is the fixation point, B' is not on the horopter and, on a Cyclopean projection, would be seen double. If, however, A and B' are points on an object, seen single as a stereoscopic percept, B' must be seen single although it is not on the horopter; we must find what it is that differentiates its images b'_L and b'_R from the pair b_L and b_R which enables us to say that they are not corresponding points. It will be seen that when A and B are on the horopter, closing either eye fails to alter their apparent relative positions;

on the other hand, when A is on the horopter but B′ is not, B′ is projected in different directions in relation to A according as one or other eye is closed. Thus with the left eye closed, B′ appears closer to A than when the right eye is closed. The actual binocular, or stereoscopic, projection of B′ in relation to A is, therefore, different from either of the separate projections. It is essentially this difference that is made use of in the most reliable experimental method for assessing the positions of corresponding points (e.g. the so-called *nonius* method). If the fixation point is used for reference, another point will produce images on corresponding points if its binocular projection is identical with the separate projections of the two eyes.

NONIUS METHOD

Experimentally the points on the longitudinal horopter may be determined by an apparatus illustrated schematically by Fig. 295. The observer views a fixation point and a vertical horopter-rod through an arrangement of screens that permits both eyes to fixate the fixation point, F, whilst the right eye is able to see only the upper half of the vertical rod, and the left eye only the bottom half. The position of this rod is adjusted so that when the eyes are fixating F, upper and lower halves of the rod appear above and below each other. Under these conditions the retinal direction values of the points stimulated by the rod are the same.

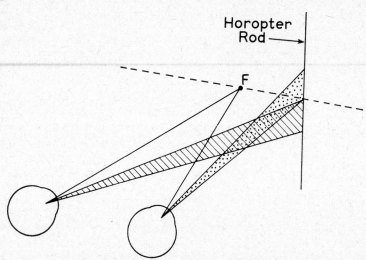

FIG. 295. Nonius method for determining corresponding points. Both eyes are able to fixate F, the fixation point, whilst the right eye sees only the upper half of the horopter rod and the left eye only the bottom half. When the rod is on the horopter its upper and lower halves appear in line. (After Ogle.)

Panum's Area. If the stereograms of Fig. 293 are modified so that the lines A′C′ and B′D′ progressively get closer, whilst the equivalent lines of the right stereogram get progressively farther apart, a point is reached where fusion becomes impossible and double vision takes the place of a stereoscopic percept. There is thus a limit to the disparity of retinal points, beyond which they cannot be fused. Panum attempted to define the limiting degree of disparity beyond which fusion becomes impossible. He stated in general that "contours resembling each other, depicted on approximately corresponding points, must be fused together". The circumference that contained those points on the other retina that could be fused with a given point on the first, he called the *corresponding circle of sensation*. This formulation is probably rather too rigid; the ability to fuse contours is determined largely by the possibility of interpreting them as the right- and left-eyed views of a three-dimensional object, and thus the meaning of the final fused percept contributes to some extent to the ability to fuse disparate images. Nevertheless there is reason to suppose that there is a certain contour of finite size such that any point within it will invariably fuse with a single point on the other retina. It is to be noted that the two aspects of any object in space, as seen by the left and right eyes, differ principally in respect to their width, their height being very nearly the same; it is not surprising therefore that it is possible to fuse pairs of vertical parallel lines with quite large discrepancies in their separation, whereas only very small discrepancies in the separation of horizontal lines are tolerated. Panum's "circle" should thus be described as an ellipse with its long axis horizontal. According to Ogle (1950) the length of the horizontal axis of this ellipse varies from one subject to another and, in a given subject, with the position on the retina at which the measurement is made. In one subject it increased from 5 to 30 minutes of arc on passing from central to peripheral vision, and in another from 10 to 40 minutes. Mitchell (1966), using central vision and flash-presentation of the stimuli so as to preclude fusional eye movements, found that horizontal disparities of 13–23 minutes of arc were necessary if two points were to be seen as double; the required vertical disparities were 8.9 to 15.9 minutes of arc.

Double Images in Stereoscopic Perception. If a pencil is placed vertically at arm's length, an inch or two to the side and in front of a thin bar, a unitary stereoscopic percept of the two is obtained, with no double images. The images would be said to fall within the Panum circles. On bringing the pencil closer to the eyes, with the bar as a fixation point, a stage is reached at which the double images of the pencil can be distinguished; nevertheless the pencil and the bar may still be perceived as a unitary pattern because one of the double images is ignored. When the pencil is brought very close to the eyes, no accurate perception of it in relation to the bar may be obtained because the double images have now become too obvious. In this experiment the pencil was not on the horopter, which passes through the fixation point; the farther removed from the horopter the pencil becomes, the less accurate and well defined is the stereoscopic percept; fortunately, however, for the unitary perception of the outside world, the more distant an object from the horopter, the more vaguely is it seen, owing to the limitations in the depth of focus of the eye. Thus objects in the environment may produce retinal images of such gross disparity that, if they were distinctly perceived, a unitary perception would be impossible. The vagueness of the double images, however, permits the ready suppression of one, so that all objects in the binocular field of view appear normally single and it requires a careful examination to reveal the actual existence of double images.

In general, the closer the position of an object to the horopter, the more accurate is depth perception; for example if one pin is placed on the horopter and another very close to it, but off the horopter, the eyes can estimate with

extraordinary accuracy the displacement in the third dimension. The farther
the pins are off the horopter, the grosser the estimate, a fact following from
Weber's psychophysical law.

Range and Scope of Binocular Depth Discrimination. We may ask whether
stereoscopic depth perception is achieved when the disparity of retinal images is
greater than Panum's area so that an object closer to the fixation point is presented
as a pair of double images. Helmholtz was in no doubt that objects could be quite
accurately located as being nearer to, or farther from, a given point by virtue of
their double images; and it was presumably the crossed disparity in the one
case (cf. Fig. 282) and the uncrossed disparity in the other (cf. Fig. 283) that
permitted this. Ogle (1962) investigated the effects of greater and greater disparities in the retinal images on the perception of depth; he confirmed the efficacy
of double images in permitting estimates of depth, and divided stereoscopic depth
into several parts; for small retinal disparities where fusion takes place, the sensation of depth is compelling and the subject is able to make accurate quantitative
estimates of relative depth in accordance with the degree of retinal disparity;
beyond this region, where double images may be shown to occur, i.e. where
fusion is not real, the experience of depth remains, but quantitative estimates of
relative depth are poor or non-existent. Finally, when the disparity is still larger,
the sense of depth disappears and the two images are indefinitely localized.

More recently Blakemore (1970) has made a thorough investigation of the
limits of retinal disparity within which estimates of depth can be reliably made,
and these turned out to be very large; thus with objects immediately in front of

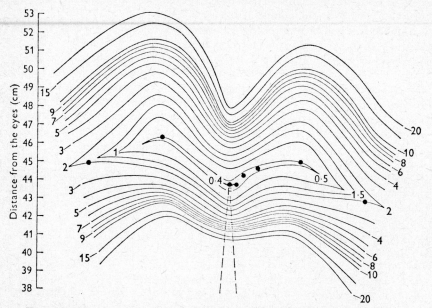

FIG. 296. Illustrating retinal disparities that may be fused, as a function of position
of the images in relation to the fixation point, F. The disparities are represented on
ordinates as distances of the object point from the eyes, causing the disparities.
Each smooth line is a contour of iso-threshold-disparity, and the magnitude of that
threshold, in minutes, is shown at the end of the contour line. (Blakemore, *J.
Physiol.*)

the observer, a convergent disparity as high as 7° could be interpreted as nearness to the observer, and a divergent disparity of 9° could be interpreted as farther awayness. In the periphery the same observer perceived depth with disparities as high as 13–14°. Blakemore confirmed that the accuracy of stereopsis, as indicated by the threshold disparity required to permit discrimination of depth, was greatest at the horopter; this is illustrated by Fig. 296, where lines indicate contours of isothreshold disparity, the actual values in minutes of arc being indicated by the figures. The filled circles represent the subject's longitudinal horopter, and it is interesting how the contours follow the shape of this horopter and how the range is compressed in the middle of the visual field.

BINOCULARLY STABILIZED VISION

Fender & Julesz (1967) have emphasized the dynamic character of Panum's fusional area; thus when fusion of disparate images was brought about using stabilized images, these images could be pulled apart very much farther than the limits of Panum's fusional area, without the appearance of double images. The fact that the images were stabilized means that no disjunctive movements of the eyes occurred, as would have happened in normal vision. During this pulling apart of the retinal images there was no change in the stereoscopic perception, so that if the images were of a vertical line, its localization in space was unaltered. Thus Fender & Julesz speak of first, a labelling process, operative in Panum's fusional region, which establishes corresponding areas in the left and right images having various disparities, and second a cortical registration process that preserves the labelling of the retinal images of a given point on an object in spite of considerable movements of these images in relation to each other, i.e. in spite of considerable variations in their disparity.

Cyclopean and Stereoscopic Projection. The study of stereoscopic perception has revealed a defect in the Cyclopean system of representing the projection of the two eyes in binocular vision. According to this, the point B of Fig. 282 (p. 448) is seen double however close it is to A, whereas we know that, if the separation is not too large, B is seen single. On a *stereoscopic* projection system, therefore, the points b_L and b_R are correctly projected through their respective nodal points to B. The Cyclopean projection of the images of B is unreal, in the sense that it projects the images of a single object to two different points in space simultaneously; the stereoscopic visual process imposes a truer meaning on the two retinal stimuli, not by suppressing one or the other of the two projected images, but by establishing a single projection that corresponds to the true position of B in relation to A. This process extends to retinal images that fall outside Panum's circle, and, even when double images become apparent, it is found that they tend to be projected into space in a direction much closer to the true position of the point giving rise to them than would be expected on a Cyclopean projection.

The Cyclopean projection is not, nevertheless, to be treated as a useless and academic mode of representing the projections of the two eyes; it tells us how the eyes would combine their separate projections if this combination were a simple additive process; thus in Fig. 283 (p. 449) the right eye alone would project its image of B to the right of A, whilst the left eye would project it to the left. The Cyclopean projection is nothing more than an addition of these two projections. By comparing the Cyclopean with the actual stereoscopic projection we can form an idea of the fundamental revaluation of retinal points that must take place to permit single vision as a result of the stimulation of disparate points.

EXPERIMENTAL DEMONSTRATION

An interesting demonstration of this shift in visual directions has been given by Burian and is illustrated in Fig. 297; the subject views a stereogram consisting of a set of three vertically arranged dots. The images of the upper and lower dots fall on corresponding points, whilst those of the middle dots, 2 and 2', do not, being displaced nasally to give a crossed disparity. When viewed, the appearance will be that of the dots 1 and 3 lying in a frontal plane with dot 2 closer to the observer (cf. Fig. 291b). What is remarkable, is that the three dots will all appear to be in the same vertical line; consequently the visual directions of the retinal points, on which the points marked 2 fall, have been altered because of the stereoscopic effect, since we know that if, in the stereograms, the points 2 were placed immediately under points 1 and 3 they would still appear in the same vertical line, but this time, of course, also in the same fronto-parallel plane.

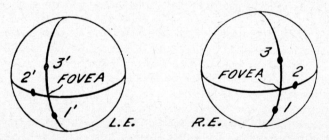

FIG. 297. Illustrating the shift in visual direction of a point by virtue of viewing it stereoscopically. The subject views the stereogram consisting of the sets of dots; on fusion the dots appear in a vertical line. (Burian, *Documenta Ophthalmologica*.)

Stereopsis without Familiarity Cues. It has been argued that retinal disparity is not the only, or even the determining, factor in stereopsis; thus with most stereograms familiarity with the figure represented is thought to be an important element, whilst in addition there are often monocular depth cues in the individual stereograms. However, Julesz (1964) has shown that it is possible for the monocularly viewed stereograms to be such that the object, seen in stereopsis, is completely invisible in the individual stereograms. He devised stereograms composed of random dots such that neither had any significance on its own, yet, on viewing these in the stereoscope, the observer saw a central square which could be in front of, or behind, the rest of the field; in Fig. 298 it is in front of the background.* The manner in which Fig. 298 was constructed is too elaborate to permit a detailed description here; suffice it to say that the stereograms are divided into 9 × 10 picture-elements composed of dots; some of these elements occupy corresponding points whilst others have disparity. It is the organized disparities of certain of these picture-elements that allow the stereoscopic appearance on fusion. This ingenious technique of Julesz allows, now, an unequivocal verification of the interpretation of Dove's experiments with flash-illumination of the stereograms, since it had been argued that fusion under Dove's conditions was only possible because the subjects knew what they were expected to see; with Julesz' stereograms there is "nothing to see" except a random array of dots. The experiments also dispose of Hering's claim that it is the awareness of double images that determines stereopsis; even if the random elements in Fig. 298 were giving double images there could be no awareness of this. Finally, the experiment shows that discrete contours in the left- and right-eye images are not a necessary feature of stereopsis.

FIG. 298. Random-dot stereograms constructed so that the individual pictures reveal no form, yet when viewed in the stereoscope the fused percept is one of a square in front of the background. (Julesz, *Science*.)

Stereopsis and Perception

It has been customary to treat stereopsis as being the result of a higher perceptual process that interprets the separate fields of view of the two eyes. Thus it has been argued that the two fields "are presented separately to consciousness and that the final percept results from a psychic act". As Ogle has emphasized, however, the experience resulting from fused disparate images is so immediate and compelling that it is difficult to separate it from a visual sensation, as such, and it might well be more correct to describe the experience simply as the response to the falling of disparate images on the two retinae. Interpretation, in the sense of conscious analysis of the two retinal images, is of course not an element in the experience. The same may be said of stereoscopic lustre

* When the pictures are rotated through 90°, so that the disparities become vertical, stereopsis is lost but the perception of the central square remains, although unrecognizable in either monocular presentation. Thus we must distinguish between the perception of form in depth and form without depth (Kaufman, 1964a).

(p. 474); for explanatory purposes we may say that the experience results because it is the only way of interpreting the two images of different luminosity, yet once again the experience is so compelling and immediate that we must treat it as the "response to retinal images of different luminosity".

Anomalous Contour. The perception of depth created by observing a distant object binocularly, when nearer objects are interposed between the distant object and the observer, provides an example of what the psychologist has called *anomalous contour*; and its contribution to depth perception, apart from the fusion of disparate images, may be investigated by constructing appropriate stereograms that simulate the actual objects viewed. Thus Fig. 299 from Lawson & Gulick (1967), when viewed in the stereoscope, gives rise to the impression of a white square in front of the dots, which lie in a single plane behind. The two pictures are identical except that the right vertical row of dots in the left picture was omitted and the left vertical line of the right picture. The real situation corresponding to the left- and right-eye views could consist of a square frame of dots with a white opaque screen interposed between it and the observer, as illustrated below, and this is the "interpretation" put on the left- and right-eye views. In this case there is no fusion of disparate images of dots since these all appear to lie in a single plane.

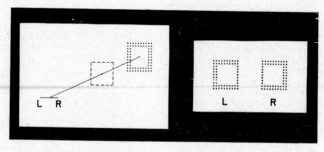

FIG. 299. Illustrating depth perception with anomalous contours. The stereogram shown at the right represents the actual situation illustrated at the left, where the view of the squares of dots is interrupted by the white square. (After Lawson & Gulick, *Vision Res.*)

Reversal of Relief

It will be noted by comparing the stereograms of Fig. 291*b* and *c*, that the dot in the circle appears to be in front or behind according as it is displaced to the right or left of the centre as viewed by the left eye, and according as the dot is displaced to the left or right of the centre as viewed by the right eye. Thus the pair of stereograms, *c*, is really obtained from the pair, *b*, by interchanging the pictures presented to the left and right eyes. In any pair of stereograms a *reversal of relief* may be obtained by interchanging the pictures presented to the right and left eyes. Such a reversal implies a knowledge by the higher perceptual centres of the contributions of each retina to the final fused image, and yet it is interesting that we are normally not aware as to which eye sees any object not seen by both eyes in the binocular field of view. Thus Rogers describes the following experiment. A tube, about 2 inches in diameter, is held in front of the right eye and directed towards the far left corner of the room. A piece of white paper is placed in front of the left eye so as to screen this eye from the part of the room seen by the right eye. The impression is obtained of seeing the objects in the room with the *left eye* through a hole in the paper.

Size Effects

Horizontal Magnification. The importance of the sizes of the retinal images in determining space perception can best be described with the aid of an example, in which the size of one retinal image is increased with the aid of a lens that modifies the image-size but leaves the dioptric power of the eye almost unchanged (an aniseikonia glass). In Fig. 300 (*left*) we have a line on the longitudinal horopter; it appears to be in the frontal plane and the images on the two retinae are of the same size. In Fig. 300 (*right*) a lens has been put in front of the right eye; the lens magnifies in the horizontal meridian only, i.e. it is a cylinder with axis vertical. The retinal image is now formed at a_R and b'_R and is projected through the nodal point. The intersections of the projections of both eyes give the apparent positions of A and B in stereoscopic single vision, namely, at A and B'. It is clear that a considerable distortion of space values has resulted, AB appearing to lie along AB'. AB therefore no longer lies on the "horopter" of the individual with the aniseikonia glass in front of one eye, and the horopter has thus changed its shape. It has been found that the eye is sensitive to as little as a 0·25 per cent change in the size of one retinal image. This spatial distortion can only be evoked by a magnification in the horizontal meridian; if an ordinary spherical magnifying lens is employed, magnifying in both horizontal and vertical meridians, then the distortion disappears.

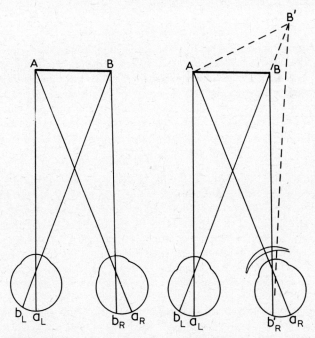

FIG. 300. The effect of an aniseikonia glass on depth perception. *Left:* AB is in the frontal plane and may be said to lie on the longitudinal horopter. *Right:* AB is now projected to AB'.

Vertical Magnification. If the magnification is confined to the vertical meridian, then we have what Ogle calls the *induced size effect*. The space dimensions are changed, but in the sense that would have been produced by horizontal magnification of the image in the other eye. It is for this reason that simple over-all magnification of the image in one eye, by a spherical lens, has no effect with regard to spatial distortion. Thus the horizontal magnification in the right eye would give the distortion illustrated by Fig. 300, with B appearing at B′; the vertical magnification, by its induced effect, would cause A to move back into the plane of B′, and so the distortion disappears.

INDUCED SIZE EFFECT

The induced effect, which is essentially the effect of producing *vertical disparities* in the retinal images, is a most striking phenomenon. The interpretation put forward by Ogle is difficult to present in a short space and this will not be attempted here; suffice it to say that, as with so many stereoscopic and other illusions, the experience may be regarded as a matter of putting the best interpretation on an anomalous set of stimuli. The experience in ordinary viewing conditions where vertical disparities become prominent is in asymmetrical convergence, one eye fixating a point immediately in front of it, whilst the other converges (see, for example, Fig. 215, p. 337). Under these conditions certain readjustments in spatial orientation are made; if, now, the vertical disparities are introduced artifically, then the adjustments in spatial orientation that would have been appropriate to asymmetrical convergence are carried out, leading to distortion of space values.

The Pulfrich Phenomenon

If a swinging pendulum with a luminous bob is observed simultaneously through a green glass by one eye and a dark red glass by the other, instead of moving in a plane the pendulum appears to move through an elliptical path. Thus if the left eye wears the red glass, the pedulum on its left-to-right swing appears to move away from the observer and on its reverse swing to move towards him. If the colours are interchanged, the direction of apparent rotation is reversed. Different coloured glasses are unnecessary to evoke the effect, and it is sufficient if the light entering one eye is reduced by placing a neutral filter before it.

Interpretation. The cause of the illusion is the difference in reaction-times of the two eyes in response to the different colours or different luminances. In Fig. 301 let us assume for simplicity that the right eye responds instantaneously to the image of the bob at any point in its path, whilst the left eye responds after a delay. In effect, therefore, when the two eyes are fixating the bob, indicated in its true position in black at A, the left eye is seeing it as though it were at A′, i.e. as though its image on the left retina were at a instead of at f_L. The disparate images are fused to give a stereoscopic effect, so that the bob is actually projected to a point outside the plane of the pendulum. The disparity is uncrossed, so that the bob on its left-to right swing appears farther from the observer. The degree of disparity will obviously depend on the speed of the bob, and this passes from zero at the extreme of its swing through a maximum at the bottom of its swing; hence the bob will appear in the plane of the pendulum at the extremes of its swing, and closest to and farthest from the observer at the bottom of its right-to-left and left-to-right swings respectively. In other words, its motion will be elliptical. Careful experimental studies of the Pulfrich phenomenon, for example those of Lit, have confirmed this explanation.

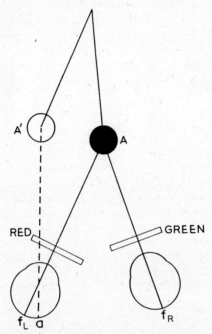

FIG. 301. The Pulfrich phenomenon.

FURTHER ASPECTS OF BINOCULAR VISION

Retinal Rivalry and Ocular Dominance

Stereoscopic vision is essentially the response to disparate images on the retinae of the two eyes; if the images are too disparate, as under the artificial conditions of the stereoscope, great difficulty is experienced in obtaining a single fused percept, the double images being so far apart as to force themselves on the awareness and so destroy the unity of the percept. The importance of the *meaning* of the final percept is evident when one attempts to fuse two pictures in the stereoscope; it may be found impossible for a time, but if fusion once takes place to give a three-dimensional representation of some object, then the significance of the two aspects has become clear and the stereograms may be fused rapidly at will.

Retinal Rivalry. If two pictures that cannot possibly be related as two aspects of the same three-dimensional object are presented to the two eyes, single vision *may*, under some conditions, be obtained but the phenomenon of *retinal rivalry* enters. Thus the two letters **F** and **L** of the stereogram in Fig. 302, when presented to the two eyes separately, can be fused to give the letter **E**; the letters **F** and **L** cannot, however, by any stretch of the imagination be regarded as left and right aspects of a real object in space, so that the final percept is not three-dimensional and, moreover, it is not a unitary percept in the sense used in this discussion; great difficulty is experienced in retaining the appearance of the letter **E**, the two separate images, **F** and **L**, tending to float apart. We are here dealing with a mode of binocular vision that may be more appropriately called *simultaneous perception*; the two images are seen simultaneously, and it is by *superimposition*, rather than fusion, that the illusion of the letter **E** is created.

FIG. 302. The letters **F** and **L** in the stereoscope give simultaneous perception to create the impression of a letter E.

With the stereograms of Fig. 303, if both patterns were fused or superimposed, the result would be a criss-cross pattern; in actual fact, however, no such perfect superimposition is achieved and the phenomenon of retinal rivalry becomes strongly evident. In different parts of the field one or other set of lines predominates whilst the opposing set is suppressed; sometimes for a brief interval one set of lines will appear by itself over the entire field; in general the impression is one of fluctuation and Helmholtz stated that the appearance

FIG. 303. The stereograms do not give a regular criss-cross pattern in the stereoscope.

at any moment was largely determined by the attention paid to any aspect, e.g. if the observer started to count the lines of one set the others, in the opposite direction, tended to disappear. Helmholtz deduced from this observation that "the content of each separate field comes to consciousness without being fused with that of the other field by means of organic mechanisms, and therefore the fusion of the two fields in one common image, when it does occur, is a psychic act." More recent studies of retinal rivalry, however, indicate that there is a periodicity in the recurrence of one or other field in awareness which appears to be independent of volition; the distinction between fusion by a "psychic act" and an "organic mechanism" is, moreover, not one that would be countenanced by the modern physiologist; as we shall see, the fusion of retinal images, when these occur within Panum's area, is brought about by the projection of the evoked nervous discharges from the retinae on to the same cortical neurones. When retinal disparity is so large as to evoke rivalry this projection on to single cortical cells presumably does not occur.

Ocular Dominance. Retinal rivalry may be viewed as the competition of the retinal fields for attention; such a notion leads to the concept of *ocular dominance*—the condition when one retinal image habitually compels attention at the expense of the other. Whilst there seems little doubt that a person may use one eye in preference to the other in acts requiring monocular vision, e.g. in aiming a rifle, it seems doubtful whether, in the normal individual, ocular dominance is really an important factor in the final awareness of the two retinal images. Where the retinal images overlap, stereoscopic perception is possible and the two fields, in this region, are combined into a single three-dimensional percept. In the extreme temporal fields, entirely different objects are seen by the two eyes and as to which dominates the awareness at any moment depends largely on the interest it arouses; as a result, the complete field of view is filled in and one is not aware of what objects are seen only by one eye. Where the fields overlap, and different objects are seen by the two eyes, e.g. on looking through a window the bars may obscure some objects as seen by one eye but not as seen by the other, the final percept is determined by the need to make something intelligible out of the combined fields. Thus the left eye may see a chimney-pot on a house whilst the other eye sees the bar of a window in its place; the final perceptual pattern involves the simultaneous awareness of both the bar and the chimney-pot because the retinal images only have meaning if both are present in consciousness. So long as the individual retinal images can be regarded as the visual tokens of an actual arrangement of objects, it is possible to obtain a single percept, and there seems no reason to suppose that the final percept will be greatly influenced by the dominance of one or other eye. When a single percept is impossible, retinal rivalry enters; this is essentially an alternation of awareness of the two fields—the subject apparently makes attempts to find something intelligible in the combined presentation by suppressing first one field and then the other—and certainly it would be incorrect to speak of ocular dominance as an absolute and invariable imposition of a single field on awareness, since this does not occur.*

When the field of one eye is consistently dominant over that of the other, this generally occurs as a result of a general law that the predominant field along an edge is that field in which the edge lies. Thus if a vertical bar is presented to the left eye and a horizontal bar to the right eye the combined impression may be represented by the cross in Fig. 304, although it must be appreciated that the result is by no means so clear-cut as a simple diagrammatic representation would indicate. At the extremes of the fused cross one eye sees only white whilst the other sees the contour, or edge; this eye is "dominant" and the black bar is seen in preference to the white space. Where the bars overlap there is a mixed state of affairs. To the right eye the field above and below the overlap is white and, since this eye is dominant along the horizontal edge, we may expect white above and below the bar; to the left eye the field to the left and right of the overlap is white and since, in this region, the left eye is also "dominant" we expect white to the left and right.

Experimentally the subject of ocular dominance has received more attention

* An interesting observation of Creed's should be recorded here; in studying the retinal rivalry caused by observation in the stereoscope of two postage stamps, differing in colour as well as slightly in design, he noted that the dominance of one colour was not necessarily associated with the dominance of the design belonging to it—thus at a given moment it was quite possible for the design of one stamp to prevail with the colour of the other.

than it probably deserves, and attempts have been made to relate it to "handed-ness". Dolman's peephole test consists of a 13×20 cm card with a round central hole in it of 3 cm diameter. A target 6 metres away is looked at through the hole held at arm's length. Under these conditions only one eye can actually see the target, and the one that is found to be doing so is adjudged the dominant eye. Walls has summarized the various tests and placed them in five categories and has stated that in effect they reduce to only two, namely those based on motor control of the eyes and those based on the assessment of subjective visual direction. According to Walls, for instance, the motor taxis of the eyes may be likened to the steering apparatus of a motor car; one wheel is directly attached to the steering rod whilst the other is guided by the track rod. It is difficult to translate this analogy into physiological terms, however; we know that the muscles to both eyes are innervated at the same time in any movement, and it would be quite incorrect to think of the one eye dragging the other along with it, as this view of ocular dominance would require.

FIG. 304. Schematic representation of the simultaneous perception of a vertical and horizontal bar in the stereoscope. (Helmholtz.)

Stereoscopic Lustre

If the stereograms of Fig. 305 are fused, the impression of lustre is created. By lustre is meant the appearance of a surface when some of its reflected light seems to come from within it; it is as though one sees some of the light from the object *through* the rest of the reflected light. The stereograms of Fig. 305 represent the two appearances of a crystal when viewed by the two eyes separately, but the two differ also in being bright and dark at corresponding places. Stereoscopic lustre is therefore the result of stimulating corresponding points of the retinae to different extents, and the phenomenon suggests an explanation for

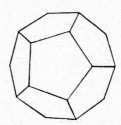

FIG. 305. Stereoscopic lustre.

the lustre in real objects. If the two stereoscopic pictures are to be interpreted as tokens of the same object, the surface of the latter must be such that a small change in viewpoint (from left eye to right eye) makes a profound change in the amount of light reflected into the eye. An object with a diffusely reflecting surface could not produce such a change and it therefore appears "flat"; a strongly reflecting surface, on the other hand, would provide the reflected light with a directional quality such that the amount of light entering the eye varied considerably with the viewpoint; such a surface is, in general, lustrous. The chromatic lustre of a bird's plumage, of a thin film of oil, and so on, is due to the variation in the quantity and *quality* of the light entering the eye with change of viewpoint. It must be pointed out, however, that lustre in objects can be appreciated by *one eye only*; Helmholtz has shown, nevertheless, that the *illusion* of lustre may likewise be produced with a single eye; in these circumstances the illusion is created by rapid changes in the brightness of the points in the retinal image, associated with either movements of the eye or of the object; thus on Helmholtz's view essentially the same impression can be gained in one eye by successive stimuli as is gained in two eyes by simultaneous stimuli. Bartley has confirmed this viewpoint; he has shown that successive monocular presentation of two stereograms of different luminance creates the sensation of lustre. The psychological issues involved in such an equivalence of time and space need not concern us here; the important feature of stereoscopic lustre resides in its demonstration that, wherever this is possible, the higher perceptual centres will read such a meaning into two retinal images as to make them the tokens of a single object in space; the simultaneous presentation to the two eyes of a light and a dark view of an object is *interpreted* as the single presentation of a lustrous object in three dimensions, and this happens even though both the light and dark views reflect light diffusely. If the two views cannot be regarded as the tokens of a single object, retinal rivalry takes the place of this perceptual process; in the case of stereoscopic lustre this can be shown by presenting differently coloured views of the same object; if the difference in colour is not very pronounced the impression of lustre is gained on fusion; when the difference is made great lustre gives way to retinal rivalry.

The Essential Basis of Stereopsis

The studies of Julesz have prompted several investigations into the nature of stereopsis; for example, the question has been asked whether fusion of disparate images is necessary; whether contours are "irrelevant", and so on. Thus Kaufman (1964, 1965), in a series of papers has described stereoscopic effects using, instead of random dots, rectangles composed of randomly selected typewriter letters; in one stereogram a block of these letters is shifted to the left or right so as to be out of phase with the identical block in the other stereogram, and as a result this central block appears in depth in relation to the background of surrounding letters. Kaufman concluded that retinal rivalry was present using this type of pattern as well as Julesz' dot patterns, since the binocular percept did not give rise to the density of dots or letters that would have been expected by a simple physical superposition of the two monocular stereograms; instead, the density corresponded with that of a single stereogram, as though the percept of depth had been evoked by alternate suppression of one or other stereogram. Kaufman (1965) generalizes correctly when he says that stereopsis will occur when correlated stimuli are out of spatial phase with respect to some reference system. The reference system can be another set of correlated objects or point-arrays, or it may be the edge of the over-all half-fields. Any mechanism that can detect the correlation between the binocular stimuli, and also detect a difference in their phase, can yield a representation of depth.

Visual and Objective Unreality. When attempting to steer a way through the rather complex arguments in this field, it is best to keep clearly in mind that the binocular percept, resulting from the presentation of separate and different pictures to the two eyes, will be governed by the relation of these two pictures to reality. How can they be interpreted as left- and right-eyed views of the external world? As indicated earlier, rivalry is the response to *visual* unreality; no object when viewed by left and right eyes simultaneously can appear as a vertical bar and a horizontal bar. *Objective* unreality is given by the formation of double images, e.g. of a pencil held close to the face when viewing a distant point; the double images are real visually in the sense that real objects in space have provoked them, but they are objectively unreal because our sense of touch provides contradictory evidence. Stereopsis is achieved under these conditions, and it is assisted by the suppression of objectively unreal parts of the retinal images. It is probably this ability to suppress parts of the retinal images that permits the development of stereopsis when differently coloured monocular stereograms are presented to the two eyes; according to Treisman (1962), stereopsis is achieved although the colour of one of the fields may be suppressed; thus the brain, under these conditions, was accepting the contours of the stereograms but suppressing the colour of one. Similarly, with many of the random dot and letter stereograms devised by Julesz and Kaufman, the situations presented are often so unreal visually that stereopsis is only achieved by virtue of suppression of parts of the visual fields.

Binocular Aspects of Some Visual Functions

Stereoscopic vision reveals the existence of processes integrating the responses in the two eyes to give a unitary percept quite different from that predictable on the basis of a summation of the separate impressions. Retinal rivalry, on the other hand, provides a condition in which the separate impressions appear, at certain moments, to be unaffected by each other, whilst between these two states we may have a more or less perfect overlap of the visual impressions with, however, dominance of one or other field in certain parts. It is worth inquiring to what extent the simpler aspects of vision discussed in Section II are affected by the use of two eyes. There is, unfortunately, disagreement on many points so that the whole subject merits further investigation.

Light Threshold. If there is the possibility of summation between the two eyes, we may expect that the absolute light threshold, i.e. the smallest light stimulus necessary to evoke the sensation of light in the completely dark-adapted eye, will be smaller when the two eyes are employed to view the stimulating patch. There are conflicting claims regarding this possibility; Pirenne has shown that the threshold is definitely lower in binocular vision, but this *does not* indicate a summation between the two eyes. The use of two eyes increases the chance of one eye receiving the requisite number of quanta from a flash of light to stimulate vision. The decrease in threshold, to be expected on this basis of quantum fluctuations, can be calculated; it corresponds to that actually found.

Brightness Sensation. At higher levels of luminance the position is more complex; when the fields presented to the two eyes are equal, there is no obvious summation—this is evident from everyday experience, since the closing of one eye does not make a previously binocularly viewed piece of white paper appear significantly darker. When there is a gross discrepancy in the luminance of the two fields, retinal rivalry appears, although it may most frequently manifest itself as ocular dominance, the bright field determining the binocular sensation.

(Under special conditions, it will be remembered, the sensation of lustre may be created.) Under intermediate conditions there is a definite fusion of the separate sensations, so that the binocular sensation is compounded of the separate monocular sensations or, to speak more precisely, the binocular sensation is different from that evoked by either eye separately, and is intermediate between these two.

WEIGHTING COEFFICIENTS

Recent experimental studies have been devoted to quantifying the relation between the monocular and binocular stimuli. Thus Levelt (1965) caused subjects to view binocularly two equally illuminated fields; this was the *comparison stimulus*. The sensation of brightness evoked in this way was compared with the *test stimulus*, where the subject once again viewed two monocular fields binocularly, but this time the luminance of one was fixed by the experimenter and that of the other was adjusted by the subject until the binocular sensation of brightness equalled that evoked by the comparison stimulus. An example of the results is given in Fig. 306, where the luminance of the left field is plotted as ordinate against that of the right field, the plotted points representing the left and right luminances which, on fusion, gave a binocular sensation of brightness equal to that obtained when both eyes viewed fields of the same luminance of 30 cd/m². Over part of the range there is a linear relation between the two field luminances, indicating a weighted averaging of the two; e.g. if the left field was 40 cd/m², the right was 16; if the left was 36 the right was 24; if the left was 12 the right was 60, and so on. When the discrepancy between the luminances presented to the two eyes was large, the relation became non-linear and provides the basis for some well known paradoxes. Thus in *Fechner's*

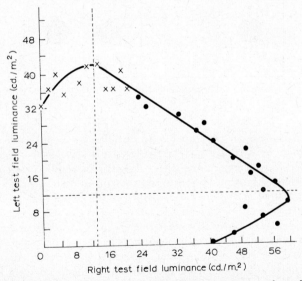

FIG. 306. The subject views test fields of different luminances through left and right eyes separately; one field is fixed and the other is varied by the subject so that the combined impression is one of equality with a standard binocularly viewed luminance of 30 cd/m². Thus a left field of 40 cd/m² viewed in conjunction with a right field of about 16 cd/m² gave the same subjective sensation as that obtained by viewing fields of 30 cd/m² through both eyes. (After Levelt, *Brit. J. Psychol.*)

paradox: a monocularly viewed white surface appears brighter than when it is viewed binocularly in such a way that one eye views it directly and the other through a dark glass. Thus if we look at Fig. 306 we see that if the subject was presented through his left eye with a field of about 40 cd/m² it would provoke a considerably brighter sensation than if this field was seen binocularly with a field of about 12 cd/m², since the combined sensation for binocular vision is equivalent to that evoked by 30 cd/m². In general, Levelt showed that over the linear range, the brightness sensation could be predicted from the weighted mean of the left and right stimuli:

$$W_R E_L + W_R E_R = C$$

where the weighting coefficients, W_R and W_L, add up to unity and are constant for a given observer. If a contour was present in one of the binocular fields and not in the corresponding region of the other, then the weighting coefficient for the eye with the contour increased at the expense of that for the other eye. This is demonstrated by viewing Fig. 307 stereoscopically; when fused, A appears much brighter than C, although at the centres of the discs the luminances are the same; on the other hand no difference between B and C is observed. Usually, contours in non-corresponding parts of the field give rise to rivalry, and this means, essentially, that the contours of one eye dominate the weighting coefficient of this eye when dominant; at another moment, the contour of the other field will become dominant.*

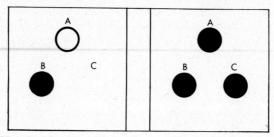

Fig. 307. When the left and right figures are fused in the stereoscope, A appears brighter than C although no difference between B and C is noticed. (Levelt, *Brit. J. Psychol.*)

Flicker. The classical experiments of Sherrington would indicate a more or less complete independence of the two eyes in respect to the discrimination of flicker. Thus under conditions where flicker is perceptible at 20 cycles/sec, the presentation of separate flickering patches to the two eyes in such a way that the dark period of one eye corresponded to the bright period of the other, did not give fusion although a flicker rate of 40 cycles/sec in one eye would have done. Smith has shown that when one eye views a steady light source and the other an intermittent one, the sensation is one of binocular flicker, i.e. it is impossible to discriminate between the sensation so evoked and that which would be produced by two flickering sources presented separately to the two eyes. In the region of the critical fusion frequency, however, retinal rivalry occurs,

* Engel (1967) has examined in some detail the manner in which the sensations are combined; his formulation carries that of Levelt farther, invoking a cross-correlation process between the two eyes that allows of an improved prediction of the effects of combining monocular fields.

so that the flicker sensation alternates with the impression of a stationary source. The critical fusion frequency, under these conditions, is not entirely determined by the luminance of the flickering patch but on the ratio of this luminance to that of the stationary source. The experimental conditions for the study of binocular flicker are clearly essentially the same as those under which retinal rivalry and ocular dominance are studied; where a unitary percept is possible, as in the special case of the two flickering patches being synchronous, we may expect the binocular flicker sensation to be identical with that evoked in either eye separately. When conditions are different in the two eyes, but not too discrepant, retinal rivalry becomes obvious. When the discrepancy is very strong, rivalry gives place to the dominance of one eye. A study of the tendency for dominance, when different pictures are presented in the stereoscope, shows that the picture with the most contours tends to become dominant; in the case of flicker, it is the flickering field that imposes itself on awareness at the expense of the stationary.

Binocular Colour Mixing. When a pair of stereograms consists merely of two coloured pieces of paper, the condition most frequently encountered is that of retinal rivalry, one or other of the colours dominating awareness for a time to give place to the other. A true binocular mixing can be obtained, however, conforming approximately to the laws of monocular colour mixing, if the conditions are suitable. According to some studies by Johansen, conditions favouring mixture are low absolute luminance and a small separation of the two colours in the spectrum. When rivalry occurs, the longer wavelength tends to dominate, and where there is a difference in luminosity between the colours, dominance occurs with the brighter of the two. The fact that so careful an experimenter as Helmholtz denied the existence of binocular colour mixing reveals that there are large individual differences in the capacity to mix colours binocularly. When red and green stereograms of a three-dimensional object are presented to the eye, it is seen in depth but one or other colour is usually suppressed (Treisman, 1962).

Colour Stereoscopy. The stereoscopic appearance obtained by regarding two differently coloured, but otherwise identical, plane pictures with the two eyes separately, is probably due to chromatic differences of magnification. If the left eye, for example, views a plane picture through a red glass and the right eye views the same picture through a blue glass, an illusion of solidity results.

Adaptation. It would seem from the work of Mandelbaum that the adaptive state of one eye is unaffected by that of the other; thus the absolute threshold of one eye was not altered by adapting the other to a variety of luminance levels. This is an important finding, since it justifies the use of one eye as a control in adaptation studies, e.g. those of Wright (p. 146).

THE PERCEPTION OF MOTION

The perception of motion has two elements in it; there is the direct appreciation of movement in consequence of the gliding of the image over the retina when the eye is still (or as a result of the movement of the eye when the image of the object fixated remains stationary on the retina), and there is the more "intellectual" recognition of the fact of movement deduced from the observation that an object is projected to a certain point in space at one moment and at another point after a certain interval of time. Thus we recognize that the large hand of a watch is in motion because at one minute it points at "ten" and at the next minute at "eleven"; there is, however, no real perception of motion, the threshold rate of movement being too low. In experimental studies the measurements are concerned with the direct perception of movement. The sensitivity of the retina to movement is highest at the fovea, as one might expect, since it is here that discrimination, generally, reaches its acme; nevertheless the periphery shows a high sensitivity in comparison with the low order of visual acuity attainable in this region; in the peripheral retina, so great is the difference between visual acuity and motion acuity, that the existence of objects may often only enter awareness when they are moved; as soon as they become stationary they are no longer perceived.*

Thresholds. The threshold sensitivity to movement can be measured in a number of ways as follows:

(a) *The minimum excursion of a moving body necessary for its being in motion to be appreciated.* This minimum depends on the speed of movement, but under favourable conditions it is of the order of 10 seconds of arc. This is an extremely small displacement, so small that if two stationary points were separated by this amount they could not be discriminated as separate. The image of the moving point thus need not move over a whole cone before its movement is appreciated. As a result of diffraction and chromatic aberration, the image of a point covers several cones; these are stimulated to different degrees, since the brightness of the image falls off as the periphery of the blur-circle is approached; a very slight movement of the whole image modifies the distribution of light on the several cones and thus modifies the pattern of discharge in their nerve fibres. A movement much smaller than the diameter of a cone could modify the distribution of light on the cones and so provide the peripheral basis for this high degree of acuity. According to McColgin, the threshold increases with increasing distance from the fovea.

(b) *The minimum angular speed of a point.* In the absence of stationary reference points, this is of the order of one to two minutes of arc per second.

(c) *The minimum duration for a given excursion.* If a point is moved between two stationary points, there is a speed above which the motion cannot be appreciated, i.e. the observer sees the point at all points in its path simultaneously; the moving luminous point becomes a "streak". With an excursion of 10°, the minimum duration is of the order of 50 msec, corresponding to a speed of movement of 200°/sec.

Apparent Movement

A common experimental method of inducing the appearance of movement is to present the subject with a given figure or spot for a short period of time and then to present the same figure at a different position in space; if the times and distances fall within certain ranges, the subject reports the perception of movement. This

---

* This is a somewhat loose way of stating the fact; one becomes aware of an object when it moves in the peripheral field, but this may be an expression of visual acuity rather than motion acuity; owing to adaptation, the awareness of a stationary object in the peripheral field soon ceases, and it requires the stimulation of a new set of receptors to come again into consciousness. The sudden awareness of this fresh stimulation is *interpreted* as a movement of the object, yet experiments seem to show that if the movement of the object is continued there need be no real perception of motion. According to Warden, the direct comparison of visual with movement acuity is beset with difficulties.

tachistoscopic type of presentation is, of course, the basis of motion pictures. According to Wertheimer's (1912) classical study, when the intervals are less than about 30 msec the two stimuli are reported as occurring simultaneously so that no motion is perceived; with 60 msec, movement is optimal whilst with intervals of 200 msec or more the two stimuli are perceived in succession.

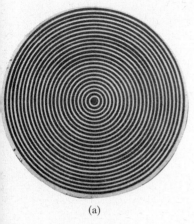
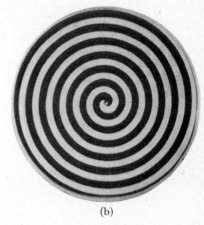

(a)   (b)

FIG. 308. Illustrating illusions of movement. For details see text. (After Bowditch & Hall, *J. Physiol.*)

Illusions. A variety of illusions of movement have been described; for example, the apparent movement of the moon behind fast-flying clouds; the apparent backward motion of the water near the shores of a large river if the current is rapid at the middle; the apparent ducking forward of the masts of a ship in passing under a bridge from which it is observed, and so on. If the series of concentric lines in Fig. 308a is subjected to a slight but rapid circular motion, similar to that used in rinsing a circular dish half-filled with water, an interesting "propeller-like" motion will appear. Again, if Fig. 308b is rotated about an axis through its centre, then, if the direction of rotation is that in which the spiral line approaches the centre, the entire surface appears to expand during rotation and to contract after it has ceased, and *vice versa* if the rotation is in the opposite direction (Bowditch & Hall, 1880–2). Alternatively, the prolonged viewing of an object in motion gives rise to a movement-after-image in the opposite direction— the so-called *waterfall phenomenon*. Barlow & Brindley (1963) have shown that there is interocular transfer of the adaptation effect, since if only one eye is exposed to the moving figure, and this eye is subsequently blinded temporarily by exerting pressure on the globe, the after-effect appears in the other eye. The possible retinal basis for the phenomenon will be mentioned later (p. 503 footnote).

NEUROPHYSIOLOGY OF PERCEPTION

The Visual Pathway

Optic Nerve and Tract. The ganglion cells of the retina (p. 117) are the second-order neurones in the transmission of retinal impulses to the cerebral cortex; the axons of these cells constitute the *optic nerves* (N II) and *optic tracts*, passing either to the *lateral geniculate bodies* or the *superior colliculi* (the pupillary fibres, it will be remembered, pass to the pretectal centre of the midbrain). The general plan of the visual pathway is indicated in Fig. 309; the fibres from the nasal halves of the retinae decussate in the *optic chiasma*, so that each optic tract contains fibres from the temporal half of one retina and the nasal half of the other. A visual impulse, arising from the right half of the field, is therefore conveyed exclusively along the left optic tract and thus to the left cerebral hemisphere. Section of the optic tract thus gives rise to what is called *homonymous hemianopia*, or half-blindness—one half of the visual field of each eye, the temporal half of the opposite side and the nasal half of the other, being obliterated. The degree of decussation varies from species to species and is obviously a function of binocular vision; thus in the guinea pig some 99 per cent of the optic nerve fibres cross to the opposite side, leaving only 1 per cent uncrossed; in the rat some 10 per cent remain uncrossed; in the opossum 20 per cent, whilst in the cat and primates, with frontally directed eyes, about equal numbers of fibres are crossed and uncrossed.

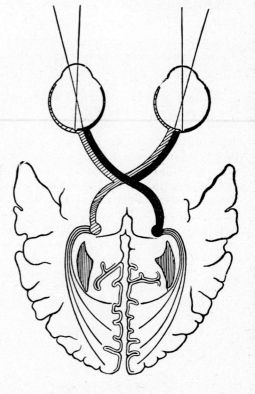

FIG. 309. The visual pathway. Fibres from the nasal retina, including the nasal half of the macula, decussate in the chiasma to join the uncrossed fibres of the temporal half of the retina. (Wolff, after Traquair.)

Phylogenetically considered, then, the uncrossed fibres are the "younger" ones and may be expected to subserve primarily stereoscopic perception of depth (Blakemore & Pettigrew, 1970). Thus in an animal with frontally developed eyes, as Fig. 309 shows, the temporal half of the retina of one eye is receiving an image of the same region of space as does the nasal half of the other. Hence with both laterally and frontally directed eyes, each cerebral hemisphere receives information from the contralateral half of the visual world.

Arrangement of Fibres. The fibres are grouped in the nerve in three bundles according to the retinal fields from which they originate as follows:

(*a*) Uncrossed temporal fibres.
(*b*) Crossed nasal fibres.
(*c*) Macular fibres.

Half of the macular fibres undergo decussation in the chiasma, so that half of the macula is represented in each tract. The relative positions of the bundles in the optic nerve and tract are indicated in Fig. 310.

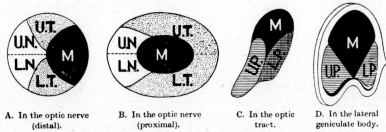

A. In the optic nerve   B. In the optic nerve   C. In the optic   D. In the lateral
    (distal).   (proximal).   tract.   geniculate body.

The crescents below U.P. and L.P. are the uniocular fibres.

FIG. 310. Distribution of the fibres in optic nerve, optic tract, and lateral geniculate body. U.T., upper temporal; U.N., upper nasal; L.N., lower nasal; L.T., lower temporal; U.P., upper peripheral; L.P., lower peripheral. (Wolff, *Anatomy of the Eye and Orbit.*)

The fibres vary in size, being divided roughly into coarse and fine; since the speed of conduction in a nerve fibre increases with its diameter, this variation in size must be associated with different conduction rates in the visual pathway. It is interesting that in the cat the fine, slowly conducting, fibres pass to the midbrain visual centres whilst the coarse fibres go to the visual cortex. So great is the difference in calibre of the fibres, that a visual impulse can be transmitted to the lateral geniculate body, relayed here to the cortex, and the cortical response may pass to the midbrain before the impulse reaches this region by way of the fine fibres directly. It is theoretically possible, therefore, for midbrain reflex activities to be anticipated and modified by cortical influences initiated by the same stimulus. In fact Doty states that the evoked potential recorded from the midbrain, in response to a light stimulus, appears some 12 msec after that recorded from the cerebral cortex of the cat.*

Non-Geniculate Pathways

Collicular Fibres. In all species, including man, a portion of the optic tract fibres avoids the lateral geniculate body and goes to the superior colliculus in what has been called the *mesencephalic root of the optic tract*, to distinguish it from the *diencephalic root* running to the lateral geniculate body. According to the

---

* Studies with the electron microscope have shown that the number of small unmyelinated fibres in the optic nerve is vastly greater than had been estimated on the basis of light-microscopical measurements, so many of these fibres being beyond the limits of resolution of the light-microscope. For example, Maturana has counted some 470,000 unmyelinated fibres in the frog's optic nerve and 12,000 thicker myelinated fibres, whereas earlier counts had shown a total of only about 14,000.

position of the animal in the phylogenetic scale, the proportion of optic fibres travelling to the midbrain station changes; in birds there is very little cortical representation of retinal impulses, the vast majority of the optic fibres running to the optic tectum or superior colliculus; in rodents the collicular projection is less, but still very significant, whilst in apes and man it is very small. The fact that there is a projection of retinal fibres on the superior colliculus in man might suggest that visual responses to light would occur in the absence of an occipital cortex; in fact, however, only pupillary responses, which, as we have seen, are mediated by fibres relaying in the pretectal nucleus, are obtained in the absence of a functioning cortex.

Basal Optic Root. A further projection of the optic tract is given by the *posterior accessory optic tract*; it has been described in all species, including man, although not all workers are agreed as to its existence. The fibres end in grey matter (called the *nucleus opticus tegmenti* in the rabbit) at the mesial end of the substantia nigra ventral to the red nucleus. The function of this projection may conceivably be to activate Magoun's reticular activating system in the so-called *arousal response* (Moruzzi & Magoun). Thus sensory stimuli, besides having specific effects mediated by the thalamus and cortex, have also what has been called a non-specific activity that apparently prepares the brain for the reception of impulses by the specific pathway. This non-specific pathway makes use of certain regions of the brain included under the general term of the reticular formation. Visual stimuli are certainly capable of eliciting responses in the reticular formation (Ingvar & Hunter), but it is not certain that these responses arise from a direct projection from the optic tract, since they may be evoked secondarily from the occipital cortex, this region being known to send corticofugal fibres to the reticular formation (Jasper, Ajmone-Marsan & Stoll).

Lateral Geniculate Body

In primates including man the cells of the dorsal nucleus of the lateral geniculate body are arranged in six laminae, and the fibres of the optic tract make synaptic connections with these cells. It was shown by Glees & LeGros Clark that the crossed optic fibres end in layers one, four and six, whilst the uncrossed, temporal, fibres synapse with cells in layers two, three and five (Fig. 311). There thus seems to be a rigid separation of the fibres arriving from the two eyes, so that the fusion of the two retinal images required by binocular vision cannot apparently take place at this level. In this respect, then, we may regard the lateral geniculate body as a sorting centre that rearranges the crossed and uncrossed fibres that had become quite mixed in the optic tract. By noting the distribution of transneuronal atrophies* following small retinal lesions, Penman and LeGros Clark showed that there is a virtual "point-to-point" representation of the retina in the lateral geniculate body, thereby ensuring that the one-to-one relationship between cones of the fovea, bipolar and ganglion cells is carried through as far as this stage in the visual pathway.

Each nerve fibre of the optic tract, after entering its lamina, breaks up into a number of terminals, each ending in a "bouton" applied to a lateral geniculate body cell; each cell has only one bouton applied to it, so that there is no overlapping of impulses from different retinal receptors. (In the cat, on the other hand, there are a large number of boutons to each cell—in this animal visual acuity is subservient to sensitivity to light.) The terminal branches of each

---

* When a neurone is injured, the degeneration is not necessarily confined to this cell; thus the destruction of primary visual neurones in the retina causes the cell bodies of geniculate cells to degenerate, a phenomenon described as *transneuronal degeneration*; apparently the geniculate neurone depends for its viability on receiving impulses from the ganglion cells of the retina.

optic tract fibre end on up to thirty lateral geniculate cells. Since each geniculate cell receives a bouton from only one optic tract fibre, there should be some 10–30 times as many geniculate cells as optic tract fibres, yet the numbers appear to be about equal.

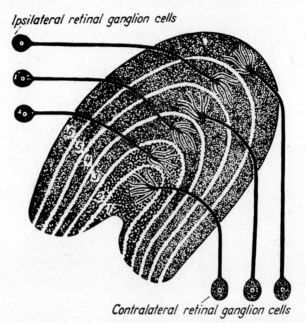

FIG. 311. Illustrating the segregation of the optic tract fibres into the separate laminae of the lateral geniculate body in primates. Fibres from the contralateral retina, i.e. the crossed fibres, relay in layers 1, 4 and 6. (Glees, In *The Visual System*.)

Feedback. Modern neurophysiology has brought out the reciprocal nature of the connections between any two stations in the central nervous system. Traditionally the lateral geniculate body has been regarded as a relay-station from retina to cerebral cortex, and, as such, doing little more than pass the visual messages on. However, there is some evidence that the message is modified, and this modification is effected not only by activity on the part of geniculate cells, but also by *corticofugal fibres*, i.e. by fibres originating in the occipital cortex which, by either excitatory or inhibitory activity, modify the responses of the geniculate cells to the retinal impulses (Guillery, 1967). By this feedback arrangement the cortex is able to modify and control the messages it receives. Thus Hull (1968) found that the responses of some geniculate neurones to photic stimulation of the retina were facilitated by reversible ablation of the cortex through cooling; others were inhibited. Inhibition of geniculate cells following light stimuli may also be achieved by activity on the part of the geniculate cells themselves. According to Vastola, the mechanism might be that illustrated by Fig. 312; here the principal cell, activated by an optic tract fibre, sends its message up to the cerebral cortex but also, by a collateral fibre, it activates a short-axon neurone in the lateral geniculate body, and this in turn inhibits the activities of neighbouring principal geniculate cells. According to Vastola, the cells that are most strongly inhibited are those that receive the weakest excitatory stimuli from the optic tract. Such an inhibition of the weakly stimulated cells by the more strongly excited ones is of obvious value in sharpening up the retinal image, as we have seen when considering the basis of visual acuity. The lateral geniculate body thus continues a process that has already been begun in the retina itself, namely, a process of accentuating the differences in response to light taking place in neighbouring receptors.

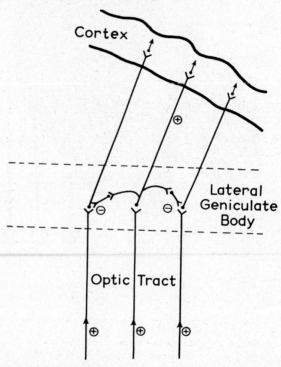

FIG. 312. Illustrating inhibition within the lateral geniculate body. The middle one of the three geniculate cells relays an excitatory influence to the cerebral cortex and also to short-axon inhibitory cells which, as a result of this activation, inhibit the neighbouring geniculate cells.

Significance of Lamination. The segregation of the cells of the lateral geniculate body into six layers has aroused some speculation as to its significance; the lamination seems to be a measure of the position of the animal in the phylogenetic scale since in rodents, such as the rabbit, it is either absent or, at any rate, very difficult to see. In the cat there are three obvious laminae, called, A, A_1 and B; fibres from the contralateral eye relay in A and from the homolateral eye in A_1, whilst fibres relaying in B probably do not lie on the cortical pathway, the neurones in this layer probably running to the thalamus (Bishop & Clare). Recent studies indicate, however, that the homolateral and contralateral fibres are not rigidly separated; thus Fillenz describes two interlaminar layers, A_1B and A_1A, in which fibres from both eyes synapse; moreover, since degenerating and normal synapses may be seen in a single geniculate cell in these layers after cutting one optic nerve, it would seem that fibres from both eyes may converge on a single geniculate cell, i.e. interaction between the eyes is possible at this level in the cat.

In primates, as we have seen, there are six layers, three for the contralateral eye and three for the homolateral. LeGros Clark pointed out that the smallest retinal lesion always caused degeneration in three layers of the lateral geniculate body, suggesting that the unit of conduction to the cortex is a three-neurone one, and he thought that this might be related to the trichromatic nature of vision (p. 263) whereby colour responses are determined by mixtures of three primary sensations. Certainly the electrophysiological studies of DeValois (1960) and Hubel & Wiesel (1961) lend support to the notion of a functional separation of neurones according to layer. Thus De Valois found that the responses of single cells of the monkey's lateral geniculate body depended on the layer in which they were situated; thus the dorsal pair of layers gave responses when the light was flashed on to the retina (ON-responses, p. 282), whilst the cells of the intermediate layer gave responses either at ON or at OFF, according to the wavelength of the light employed; the ventral pair of layers gave responses only when the light was switched off (OFF-responses), so that light projected on to the retina caused only a cessation of spontaneous activity followed by a spike discharge when the light was switched off.

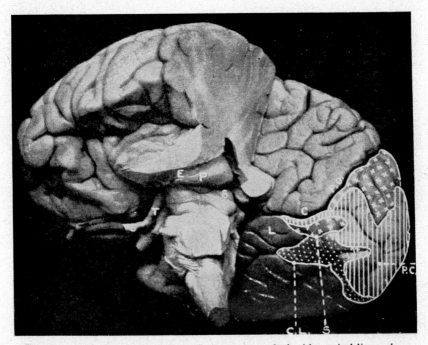

FIG. 313. The cortical visual areas. Striate area marked with vertical lines; the parastriate area with crosses and the peristriate area with dots. C, calcarine sulcus; C.L., collateral sulcus; S, sagittal sulcus of the gyrus lingualis (L); O, optic nerve; T, optic tract; E, lateral geniculate body; P, pulvinar; S, superior colliculus; i, inferior colliculus. (Whitnall, *Anatomy of the Human Orbit.*)

As we shall see, the separation of the afferent impulses from the eyes in the lateral geniculate body means that the combination of the visual fields, required for simultaneous use of the two eyes in binocular vision, is deferred to the cerebral cortex. It follows, then, that it is the function of cortical cells to integrate messages from the separate eyes; given this postponement, it is reasonable that the messages should be kept separate in the lateral geniculate body.

Ventral Nucleus. The significance of the unlaminated, ventral or *pregeniculate* nucleus of the lateral geniculate body is not clear; it has been stated that its prominence decreases as we ascend the phylogenetic scale, so that in primates it is only vestigial, but Polyak (1957) disagrees, showing that many optic tract fibres relay here in the monkey, and pass thence to the midbrain, relaying in the Edinger-Westphal nucleus and thus providing a second pathway, in addition to the pretectal fibres already described, mediating pupillary responses to light. As indicated earlier, Wiesel & Hubel (1966), in their examination of colour-coded geniculate units in the monkey, found all their three types of unit in the dorsal nucleus, and an additional (type IV) unit in the ventral nucleus.

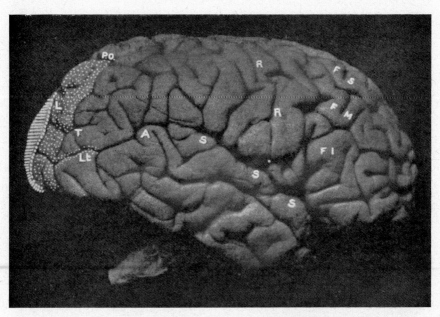

FIG. 314. Lateral view of the right hemisphere. Striate area marked with cross lines; the parastriate and peristriate areas by large and small dots respectively. PO, parieto-occipital; L, lunate; T, transverse occipital; Lt, lateral sulcus. A is placed on the angular gyrus, which surrounds the upturned end of the superior temporal sulcus. S, S, S is the lateral cerebral fissure (fissure of Sylvius). R, R is the central sulcus (fissure of Rolando). FS, FM, FI are placed on the posterior parts of the superior, middle and inferior frontal convolutions respectively; between them and the sulcus are the pre-central sulcus and convolution. (Whitnall, *Anatomy of the Human Orbit.*)

The Visual Cortex

From the lateral geniculate body, the third-order neurones pass as the *optic radiation* to the surface of the occipital pole, relaying with the cortical neurones in Brodman's Area 17, or the *striate area*, so called because the axons of the optic radiation form a well-defined white line of Gennari on their way to their synapses with the subjacent layer IV, together with axons of association fibres passing to adjacent cortex. The region over which the striate area extends is illustrated in Figs. 313 and 314.

Retinal Projection on the Cortex

Small lesions in the striate area of man lead to blindness in well-defined regions of the visual fields, and a systematic study of the effects of gun-shot wounds in this region of the cortex has led to the definite picture of the projection of the retina on the visual cortex reproduced in Fig. 315 from Gordon Holmes. In general the projection may be represented by picturing one half of the retina spread over the surface of one striate area, the macular region being placed posteriorly, the periphery anteriorly, the upper margin along its upper edge and the lower on its inferior border. The macula, as one might expect, is projected over an area out of all proportion to its size, just as with the motor representation of Area 4, where the cortical centres concerned with the hand are much larger than those connected with movements of the proximal segments of the arm. More recent studies of head injuries by Spalding have largely confirmed the picture presented by Gordon Holmes.

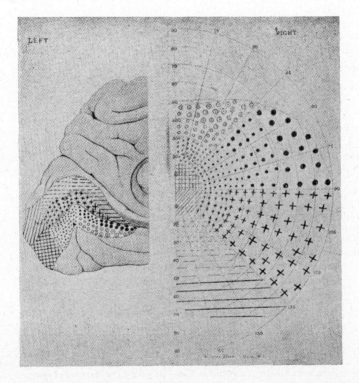

FIG. 315. The projection of the retina on the visual cortex. On the left the striate area of the left hemisphere of the brain is shown, the calcarine fissure being opened up to reveal that portion of it which lines its walls. On the right is the right half of one visual field. The correspondence of the markings indicates the representation of different segments of the visual field on the cortex. (Gordon Holmes, *Proc. Roy. Soc.*)

"Point-to-Point Projection". Anatomical and physiological investigation has confirmed this general plan of the retinal projection and has brought out the existence of a virtual "point-to-point" relationship between retinal stimulus and cortical representation. For example, in the monkey, a definite projection map can be plotted by stimulating the retina with a point source of light and determining the exact locus of the electrical potential evoked in the cortex (Talbot & Marshall). From a strictly anatomical point of view this point-to-point projection, in the sense of a single cone being connected to a single bipolar cell, which is connected to a single ganglion cell, and so on, is not possible; in the retina the one-to-one relationship between cone and optic nerve fibre only holds good in the central fovea, and even here it is only a functional relationship since impulses from a single cone have subsidiary paths which permit a spread of their responses outwards to several ganglion cells. Similarly, as we have just seen, in the lateral geniculate body, although there are no internunciary neurones, each fibre of the tract ends on a number of geniculate cells, and a further dispersion is possible in the cortex by intracortical association cells. The point-to-point relationship is therefore a functional one, in that there is a preferential path from one cone, say, to one cortical cell. This point to-point relationship between retinal light stimulus and cortical evoked response may be found, moreover, only under highly artificial conditions, involving deep anaesthesia. In the intact animal, with implanted electrodes in different regions of the cortex, it is found that the evoked response to discrete stimuli is by no means so highly localized. In a similar way, the more elaborate study of visual sensitivity in human subjects with lesions in their visual pathway has shown that portions of the visual field that are apparently "spared", in the sense that they respond to light, are not necessarily completely normal (Battersby *et al.* 1960).

Cortical Magnification Factor. If the retina is projected on to the visual cortex in the way these studies suggest, we may speak of a *cortical magnification factor*; for example, if 1 mm of retina were represented by 3 mm of cortex, the magnification would be 3, or, since 1 mm of retina corresponds to a visual field of about 3·3°, the magnification would be about 1 mm/degree. Daniel & Whitteridge (1959) have extended Talbot & Marshall's studies on visual cortex of the monkey and baboon and have shown that the magnification decreases evenly from a value of 5·6 mm/degree at the fovea to about 0·1 mm/degree at a visual angle of 60°. Having measured the magnifications along different meridians of the visual field, Daniel & Whitteridge made a map of the striate cortex by applying these magnifications; it was obviously not the simple surface of a sphere, because of the distortion imposed by the variation in magnification. The area of the calculated surface came out at 1320 mm². If this was folded twice, once along the horizontal meridian for its anterior half, and vertically at the junction of the posterior and middle thirds, the resulting folded surface corresponded with the actual surface of the calcarine cortex in monkeys and baboons.

Geniculate Maps. In a similar way Choudhury & Whitteridge (1965) have plotted a map of retinal projection on the rabbit's lateral geniculate body; in this animal there is a retinal region—the streak—in which the density of ganglion cells is very high, and it probably corresponds with the fovea of primates. This streak was "magnified" in the lateral geniculate body in relation to other retinal areas. In the rabbit, a small number of optic nerve fibres project ipsilaterally, and it was found that the ipsilateral and contralateral projection areas overlapped, indicating the possibility of binocular interaction at this level in this non-laminated relay station.

Cortical Lesions and Lateral Geniculate Body. Lesions in the visual cortex are reflected in regions of degeneration in the lateral geniculate body of the same side; the smallest cortical lesion, leading to degeneration, always involves cells in all six layers of the geniculate body; the conducting unit of the optic radiation is therefore apparently six fibred, and impulses from corresponding regions of the six cell layers are brought into relation in the cortex; fusion of the temporal field of one eye with the nasal field of the other is therefore possible at this level.

The effects of discrete lesions in specific layers of the lateral geniculate body are revealed in degeneration of geniculate fibres and terminal boutons in the striate cortex; recent studies by Hubel & Wiesel (1969) on this aspect will be discussed later (p. 509).*

The "Sparing of the Macula". The fact that lesions in the optic radiation, or in the visual cortex, frequently do not involve the macula (the "sparing of the macula") has led many to suppose that fibres from the lateral geniculate body do not pass exclusively to the cortex of the same side, but that some macular fibres cross in the corpus callosum, or alternatively by way of mesencephalic connections, and therefore give rise to a bilateral cortical representation of the macula. In primates and man, however, complete destruction of one occipital lobe is followed by complete cellular atrophy of the homolateral geniculate body, hence the cells of this body project exclusively on to the occipital cortex of the same side. Moreover, such damage leads to loss of half the retinal fields including half of each macula (Spalding; LeGros Clark). The apparent sparing of the macula is probably essentially a result of its large cortical representation by comparison with that of the peripheral retina; moreover, the part of the cortex involved in this projection lies in the margins of distribution of the middle and posterior cerebral arteries so that, when one of these is occluded, a portion of it, at least, may receive sufficient blood from the other to remain functional. The possibility that macular sparing may be due to a crossing of macular fibres in the chiasma is rulled out by Gordon Holmes on the grounds that lesions in the optic tract cause a complete hemianopia extending into the fovea.

We shall see that the integration of the responses of the two eyes in stereoscopic binocular vision demands that some cortical cells of a given hemisphere receive impulses from the "wrong side" of its appropriate retina; this could be achieved by the failure of all the fibres from the nasal half to cross, or by the crossing of some temporal fibres; and this may well happen in the cat, but in man such integration of the nasotemporal region of the retina is almost certainly brought about by neurones transmitting influences from one side of the cortex to the other in the corpus callosum (Hubel & Wiesel, 1967; Berluchi, Gazzaniga & Rizzolatti, 1967).

Areas 18 and 19. Sharply differentiated from the primary visual projection area cytoarchitectonically, are the two neighbouring Areas, 18 and 19, called the *parastriate* and *peristriate* areas respectively. Cells of Area 17 connect by short axons to Area 18, whilst from this area connections are made to Area 19 and back to the striate area. These areas have been described as "visuopsychic" in contrast with the "visuosensory" striate area, and they have been considered to be concerned with the elaboration of visual impulses received first by the striate area. In order to determine how these areas were interrelated, v. Bonin, Garol & McCulloch (1942) applied strychnine to localized portions

---

* The projection of the lateral geniculate body on the cortex has been re-examined by Wilson & Cragg (1967) in both cat and monkey, following up degeneration in the cortex resulting from lesions in the lateral geniculate body. In the monkey they confirmed that the projection was confined to Area 17, but in the cat degeneration extended to Area 18 (Visual II); and this agrees with the finding of short-latency responses to light recorded from this area by Berkley, Wolf & Glickstein, responses that survived cutting of the corpus callosum and removal of Area 17. A further projection was found on the suprasylvian gyrus. Lesions medial to the lateral geniculate body (pulvinar) caused some degeneration in Area 19 (Visual III), and this may correlate with Suzuki & Kato's (1969) description of cells in the posterior thalamus of the cat responding to photic stimulation of the retina. The claim of Glickstein, Miller & Smith (1964) that there was a crossed projection of the geniculate through the corpus callosum to the contralateral visual cortex was disproved.

of any one area of the monkey's cortex and studied the "strychnine spikes" evoked in other regions.* Applied to Area 17, there was remarkably little spread within this area; spikes appeared in Area 18, but not in 19 and never in the contralateral hemisphere, so that the fusion of the two halves of the visual field in any one eye does not take place by direct interhemispheric connections between Areas 17. When applied locally to Area 18, the parastriate area, there was widespread activity within this area and also in Areas 19 and 17. Moreover, activity spread to symmetrical loci in Area 18 of the opposite lobe, and also to the temporal lobe (middle and inferior convolutions). Strychninization of Area 19 caused only local activity and apparently caused a spreading depression so that spontaneous activity in large areas of the cortex was finally suppressed. Thus Area 18, the parastriate area, stands out as the part of the cortex most closely concerned with integrating visual activity in the two hemispheres, and also with activity in the temporal lobe. By applying more modern techniques to the cat and baboon, Choudhury, Whitteridge & Wilson (1965) have confirmed the importance of Area 18 in the bilateral response to a visual stimulus; section of the corpus callosum, or cooling of the contralateral projection point on the cortex, extinguished the ipsilateral cortical response in animals with one optic tract cut.

CORTICAL MAP

We have seen that the retina may be projected topographically on the striate area of man (Fig. 315, p. 489) and other animals. In a similar way Areas 18 and 19 in the cat may be shown to have a topographical localization; Fig. 316 is the map derived by plotting the responses to photic stimuli recorded from the three visual areas I, II and III by Bilge *et al.* (1967). Of great interest is the circumstance that the vertical meridian, indicated by the heavy line marked 270°–90°, separates Visual I from Visual II; it is essentially in the vertical meridian that integration between the two eyes is so important for stereopsis (p. 516).

CORTICAL ASSOCIATIONS

The higher visual functions involving complex association, such as integrating hearing with vision or the visual guiding of motor activity by the hands, must, in the first place, involve the striate area, since in man cortical visual impulses pass to this region alone; it was originally thought that there were long association tracts connecting the visual cortex with the more remote parts of the cerebral mantle; as indicated above, however, in man there are no long-fibre connections made with the striate area, the cells of this area being connected only by short neurones to the immediately adjoining Area 18 (peristriate area). From Area 18 short fibres relay to Area 19 (parastriate area) and back to the striate area.

Effect of Extirpation on Learning. In the monkey, bilateral extirpation of Areas 18 and 19 caused the irrecoverable loss of a learned habit. Thus Ades

---

* Strychnine applied locally to the cortex makes the underlying neurones highly sensitive to any afferent impulse; as a result, the neurone discharges, apparently spontaneously, and the occurrence of "strychnine-spikes" at a distance from the region at which the strychnine was applied indicates the likely point of termination of these artificially activated neurones. Thus the appearance of spikes in Area 18 and not in Area 19, after strychninization of Area 17, suggests that neurones of Area 17 make direct connections with Area 18 but not with Area 19. It was considered earlier that the action of strychnine in the cortex was to abolish inhibition, by analogy with its action in the cord, but studies on cortical neurones rather oppose this view (Krnjević, Randić & Straughan, 1966).

& Raab (1949) presented monkeys with two doors leading to chambers in one of which was placed a pellet of food. By placing a letter F on one door, which led to the compartment containing food, and an inverted letter F on the other, which did not contain food, the monkeys could be trained to distinguish between the upright and inverted F. In other words they were taught to discriminate visual form. After bilateral extirpation of Areas 18 and 19 the habit was lost, but could be re-learned in about the same time as it was originally acquired. If one area only was removed, there was no loss; moreover, if the second area was removed after a delay of two weeks, there was also no loss. This indicates that the learned exercise had become bilaterally organized between operations, the initial excision having initiated some compensatory activity in other parts of the cortex that not only compensated for the loss of the cortex removed on the one side but "in anticipation" for the removal on the other. A similar bilateral compensation for extirpation of the motor area has already been described.

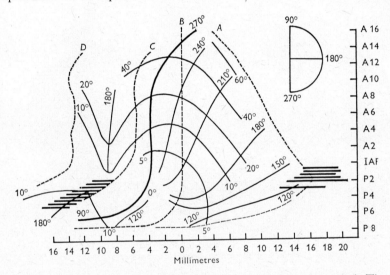

Fig. 316. An extended cortical surface to show the visual areas of the cat. A, The bottom of the splenial sulcus; B, the medial edge of the left hemisphere continuing as a section 1 mm in front of the posterior pole; C, the bottom of the postlateral sulcus; D, the bottom of the lateral sulcus. The heavy line 270°—90° is the vertical meridian of the visual field and separates Visual I medially from Visual II laterally. Between C and D the 180° meridian separates Visual II from Visual III. Meridians are shown 30° apart. Incomplete semicircles of latitude are shown 5°, 10°, 20°, 40° and 60°. Heavy horizontal lines indicate areas in which folding of cortex makes reliable representation of the cortical surface impossible. I.A.P., interaural plane. Ordinates, Clarke-Horsley anterior and posterior planes; Abscissae, extended cortical surface in millimetres. (Bilge *et al.*, *J. Physiol.*)

TEMPORAL LOBE

If the temporal lobes were also removed, all possibility of learning, or re-learning, was lost although removal of the temporal lobes alone was without influence either on the learned habit nor yet on the power of the animal to acquire it. Thus the temporal lobes, in some way, are able to act for the visual "association areas" in the learning of visual discriminations involving form. It is worth noting that none of the animals with temporal lobectomy showed the signs of psychic blindness described by Klüver & Bucy (1939); in other words, they were quite capable of selecting edible from inedible articles from a miscellaneous collection. Since Area 17, the striate area, is not directly connected with the temporal lobe, the learning in the absence of Areas 18 and 19 is presumably brought about by the mediation of subcortical links, e.g. by way of corticofugal fibres from the striate area to the pulvinar of the thalmus, and thence from the pulvinar to the temporal cortex; at any rate Jasper, Ajmone-Marsan & Stoll (1952) have shown that cortical stimulation leads to evoked activity in the pulvinar and also the laterate geniculate body.

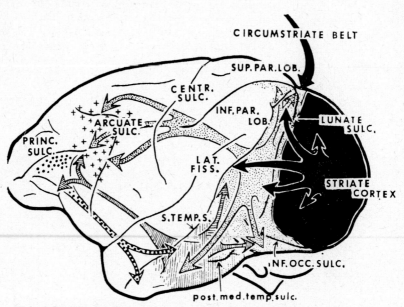

F IG. 317. Diagrammatic representation of some cortical interconnections with special reference to the striate area. (Kuypers *et al.*, 1965.)

When visual input to one side of the brain was cut off, by section of the left optic tract, then learning could be impaired by removing the right temporal lobe; further impairment could be achieved by section of the corpus callosum (Ettlinger, 1958). It would appear, then, that co-operation between the visual cortex of one side and the temporal lobe of the other, in the performance of learned tasks, may take place through the corpus callosum, but is not adequate to replace the co-operation between visual and temporal areas of the same side.

The connections between the visual areas and the temporal lobe have been examined in the rhesus monkey most recently by Kuypers *et al.* (1965), and their conclusions are illustrated by Fig. 317. The striate area projects on to what they call a *circumstriate belt*, which has been mottled in the Figure; this is considerably more extensive than the peri- and parastriate areas. From here fibres relay to the temporal lobe and also to parietal and frontal regions. No evidence for a circumstriate projection on to the precentral gyrus was obtained, however, so that the mechanism for visual guidance of motor behaviour is not so far clarified. A projection to the frontal regions concerned with eye movements was demonstrated, and it is by way of this, presumably, that the visually determined movements of the eyes take place.

Lower Visual Centres

In man removal of the striate areas causes complete blindness; thus awareness of light, and the ability to respond to it by motor activity,* are cortical functions. In lower animals, however, this is not true; for example, in reptiles and birds vision is barely affected by removal of the cortex, so that a pigeon will fly and avoid obstacles as well as a normal one. In rodents, such as the rabbit, removal of the occipital lobes causes some impairment of vision, but the animal can avoid obstacles when running, recognize food by sight, and so on. In monkeys it was thought for many decades that the animal was totally and permanently blind after bilateral occipital lobectomy, and, as Klüver pointed out, this conclusion was easily reached from casual observation of the animal provided it was in unfamiliar surroundings; under these conditions the monkey bumps into walls or objects in the room and makes no attempt to avoid obstacles. It does not seem to be able to find food or other objects except by sense of touch, and it sits for hours in its cage with the door wide open. By contrast, when the monkey is in familiar surroundings its movements are so quick and efficiently performed that on casual observation it passes for normal, and it requires careful experimental study to assess the degree of impairment of visual function. Such a monkey is said to exhibit an optokinetic nystagmus some months after the operation (Pasik & Pasik, 1964), and it may be trained to respond to light and to changes in the total luminous flux entering the eye; up till recently, it was considered that this was the total basis of its visual discrimination. However, Humphrey & Weiskrantz (1967) have shown that, under the proper conditions, a monkey with bilateral striate removal can not only detect objects but locate these accurately in space, i.e. is able to discriminate on the basis of position as well as luminous flux. Thus they found that, in order that monkeys could be made to notice an object, e.g. a cube, it had to be moved, either from side to side or by rotation; once this condition for notice had been fulfilled, the monkey could be trained easily to grasp the object accurately when it appeared at a given point in the visual field, the accuracy being greatest when it was in the central field and poor at 40° of eccentricity. Eventually it became difficult to hold an object still enough for the monkey not to perceive the natural tremor of the experimenter's hand, and finally, after a longer period, stationary objects could be noticed and the monkey trained to localize them accurately.

In the dark, a stationary luminous object was noticed, but in ordinary illumination a neon light, for example, was not, unless it was flashed repetitively; at the frequency required for human fusion the monkey failed to notice the light.†

Subcortical Mechanisms. As to the mechanism for these discriminations, it seems that these must be subcortical since the cortical projection of the retina, at any rate in primates, is exclusively to Area 17. It will be recalled that there is a considerable projection of the optic tract on the superior colliculus, and the early

---

* Except, of course, the pupillary response to light, mediated by optic tract fibres that pass to the pretectal nucleus.
† Pasik, Pasik & Schilder (1969) have shown that monkeys, with striate cortex removed, can discriminate different sized figures illuminated in such a way that the total light-flux entering the eye is the same; they thus apparently were able to discriminate on the basis of size.

study of Apter (1945) showed that this projection was essentially point-to-point; more recently Garey (1965) has shown, moreover, that the projection of the visual cortex on the superior colliculus has the same topological feature, so that every point on the retina is related to a particular part of the colliculus, directly, and indirectly by way of the lateral geniculate body and cortex. Removal of the superior colliculus on one side in the cat causes an obvious visuo-motor defect, so that visually guided activity in the contralateral field is impaired, and the animal exhibits circling movements; in a few weeks, however, compensation is complete. In the past there has been a tendency to regard the colliculus as a purely motor centre but Sprague & Meikle (1965), on the basis of their studies of lesions in the colliculus and in the afferent and efferent pathways, argue that the body must be regarded as an integrating centre rather than as a simple relay station for cortically directed eye movements. In the rabbit, Hill (1966) has examined the receptive fields of single collicular neurones; these are similar to those observed in ganglion cells, and he suggests that, because the colliculus is a layered structure, there may be a columnar ordering of cells with similar receptive fields analogous with the cortical arrangement described by Hubel & Wiesel (p. 509). Finally, we may note that Sprague (1966) has revealed an interesting interaction between the superior colliculus and cortex of the cat; by large cortical lesions on one side he was able to cause a permanent loss of visually guided activity in the contralateral field; on removal of the superior colliculus of the opposite side, visually directed responses returned, so that the responses in both fields were the same, activity in one field being determined by the tectum and in the other by the cortex. Sprague asked: Why, after removal of the cortex on one side, the superior colliculus of the same side did not take over? And, Why did the opposite colliculus have to be removed to allow it to do so? Clearly the opposite colliculus was inhibiting the colliculus on the same side as the cortical lesion, and this was proved by making a commissural incision separating the two colliculi; as a result, the colliculus on the same side as the cortical lesion was now able to take over.*

Corpus Callosum

This is an interhemispherical tract that provides connections between a given point on one hemisphere with a symmetrical point on the other. By stimulating discrete points on the cortex, Curtis (1940) recorded electrical responses on symmetrical points of the cortex of the other hemisphere. In primates, the only area that did not give a response was Area 17, showing the probable absence of interhemispherical connections between these two areas, and the anatomical studies of Myers (1962) and Ebner & Myers (1965) leave no doubt that it is at the junction of Areas 18 and 17 that the great majority of the posterior callosal fibres project.† In the cat Choudhury, Whitteridge & Wilson

---

* Urbaitis & Meikle (1968) found dark-light discrimination in cats with their cortical visual areas removed as well as their superior colliculi, and concluded that the *pretectal* region was capable of organizing visually directed activity.
† Myers (1962) examined the degeneration of fibres in the opposite cortex when the occipital lobe of one side was removed; in addition to Area 18, Areas 7 and 23 of the parietal lobe and Area 20 of the temporal lobe, in the region where they border Area 19, showed degeneration; thus, the striate cortex, receiving no afferents from the corpus callosum, is completely surrounded by cortex that does. We may note that Area 19 was free of degeneration as well as Area 17; in the cat, Ebner & Myers (1965) found essentially the same picture, earlier claims of callosal connections between Areas 19 being due to the inclusion of part of Area 18 in Area 19.

(1965) found that the receptive fields of cortical units in the medial edge of Area 18, just adjacent to Area 17, were within a few degrees of the vertical meridian; these responses were recorded when the optic tract of the same side was cut, so that a callosal route from the opposite hemisphere was probably involved, and this was confirmed by cooling the corresponding point on the opposite hemisphere or by complete section of the callosum.

Callosal Neurones. Again, Berlucchi, Gazzaniga & Rizzolatti (1967) inserted electrodes into the posterior callosum of cats and found axons that gave discharges when the eyes were stimulated with patterns corresponding to those described by Hubel & Wiesel, so that they obtained simple, complex and hyper-complex units (p. 504). In general, it was only when the stimulus was presented close to the vertical meridian that callosal units were excited; three out of seventeen units were driven by both eyes by stimuli located in the same part of the visual field; nine were driven by one eye, and with the remaining five a clear response was obtained with one eye and a weak one with the other. Thus, as Berlucchi *et al.* say, the representation of the visual world is on a continuum, the neurones associated with the vertical meridian and their callosal connections being the hyphen necessary to bring together the two half-fields. Thus the cortical cells connected in this way are at the junction of Areas 17 and 18 and it is here that the vertical meridian projects (Choudhury, *et al.*, 1965; Bilge *et al.*, 1967), whilst it is here, too, that the great majority of the callosal fibres project (Ebner & Myers, 1965).

BINOCULAR FIELDS

When the chiasma of the cat was split, so that the visual input from a given eye was entirely homolateral, Berlucchi & Rizzolatti (1968) found 9/70 units driven by both eyes in the visual cortex at the boundary of Areas 17 and 18; both receptive fields were very similar and covered the vertical meridian. Thus the representation of the visual fields of the two eyes may be indicated by Fig. 318; the central 20° is bilaterally represented through callosal connections; the remaining parts of the binocular field, making up a total of 120°, are represented by cortical neurones driven by both eyes, but a given visual stimulus in this field will only drive cortical cells in one hemisphere. Fields of 80° on each side are uniocular, whilst a final 80° is blind.

Callosal Transfer of Visual Habit. If a cat is placed in a box from which there are two exits, it may be trained to discriminate between two signs, for example between a vertical and a horizontal bar, if these are placed over the respective exits and the animal is rewarded with food when choosing the one exit and not when choosing the other. If one eye is covered while the animal is trained in this pattern-discrimination, it is found that when the animal is made to use the other eye for the discrimination it succeeds as well as if both eyes had been open during the training. In other words, we may say that there has been transfer of the learned habit from one eye to the other; if the left eye is trained the right eye can also execute the task, and vice versa. If the optic chiasma is sectioned by a sagittal incision, so that all visual impulses from one eye are only carried to one hemisphere, it is found that the animal is still able to make the transfer, in the sense that the task learnt with the left eye can still be executed by the right eye. Finally, if both the chiasma and corpus callosum are separated by sagittal incisions, the animal must now learn this habit separately with each eye.

Thus, one cat, operated on in this way, required respectively 80, 170 and 850 trials to learn three different types of visual discrimination with the left eye, whilst on covering this and testing it with the other eye open it required 70, 160 and 830 trials in order to relearn the discriminations. Furthermore, the animals could be taught to make contradictory discriminations with the two eyes; for example, the left eye could be trained to respond positively to the vertical bar whilst the right eye could be trained to respond negatively to this, and positively to the horizontal bar. Hence its choice of exit depended on which eye it was using (Sperry, Stamm & Miner).

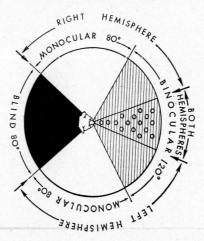

FIG. 318. The visual field of the cat and its relationship with the eyes and the cerebral hemispheres. The visual callosal mechanisms allow both hemispheres to receive projections from an area of the visual field extending 20° on either side of the vertical meridian. The extension of the bihemispheric region of the visual field is inferred from earlier results of Berlucchi et al. (1967) showing that some visual receptive fields of callosal units are as large as 20°. (Berlucchi & Rizzolatti, *Science*.)

However, if the split-brain cats were trained using a shock-avoidance technique instead of a food-reward one, pattern discrimination was successfully transferred, and it was suggested that this was made possible by the meeting of visual and pain stimuli in the superior colliculi (Sechzer, 1963). In these experiments the cats were discriminating patterns; when the simpler discrimination of light-intensity was required, it was found that interocular transfer occurred in the presence of split chiasma and corpus callosum (Meikle & Sechzer, 1960), and only when additional commissures were cut, namely the anterior and posterior, the massa intermedia, habenular, hippocampal and superior collicular commissures, was the interocular transfer prevented (Meikle, 1964). Finally we may train a cat to respond positively to two sets of lines with their directions at right-angles, i.e. to ||| ≡, and negatively to two sets with their directions parallel, i.e. to ||| |||; if the stimuli are presented in such a way that one eye receives the vertical, and the other the horizontal, lines of the positive stimulus, then if the chiasma and the dorsal two-thirds of the corpus callosum are split, the power to discriminate is grossly impaired; only if the rostral third of the callosum is also cut does the performance degenerate to chance level (Voneida & Robinson, 1970).

In monkeys, too, with their optic chiasma sectioned, there is interhemispheral transfer of learned visual discrimination, which is abolished by section of the commissures (Downer 1959). This author studied the "handedness" of his experimental monkeys under these conditions and showed that, according to which eye was open, they employed the contralateral arm, i.e. the arm controlled by that side of the brain to which the visual information was restricted by the section of the chiasma. If they were prevented from doing this, by binding the arm, then the animal groped with its free arm as though it were blind. Unlike cats, split-brain monkeys could not make an interocular transfer of a discrimination based on brightness (Hamilton & Gazzaniga, 1964).*

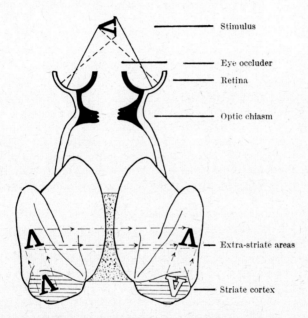

FIG. 319. Diagram to explain mirror-image reversal. With right eye occluded, the input to left striate cortex is propagated to extrastriate areas (Λ) and is laterally reversed in crossing the midline via the forebrain commissures (stippled). When the left eye is occluded the input to the right striate area (W) does not match the stimulus (Λ) transposed from the left hemisphere; it does, however, match its mirror-image. (Noble, *Nature*.)

MIRROR-IMAGE STIMULI

A very revealing study on transfer of learned habit in the monkey is that of Noble (1966) who used as positive and negative stimuli objects that were not bilaterally symmetrical; thus the usual stimuli would be + and o or ||| and ≡, the choice of one being rewarded and the other not. When a split-chiasma animal was trained to react positively to an object and negatively to its mirror-image, e.g. to ⌐ and ⌐, with one eye occluded, it was found that when the animal was tested for transfer to the opposite eye it consistently chose the negative stimulus.

* The normal rabbit behaves similarly to the split-brain split-chiasma cat so far as inability to transfer pattern discrimination is concerned; there is no doubt, however, that the two visual areas are linked by callosal connections (Van Hof, 1970).

Thus, if it had been trained to react positively to ⌈ with its right eye, then it reacted positively to ⌋ with its left eye. Noble considered that this was to be expected on the basis of a mirror-image reversal of the projection of the retinal image on the opposite cortex through Area 18 and the corpus callosum (Fig. 319).

MECHANISM OF TRANSFER

Berlucchi & Rizzolatti asked whether the learned process was coded and transferred to the opposite hemisphere, or whether the information necessary for the learning process was received in both hemispheres, through the corpus callosum if necessary, and the information was subsequently processed into memory patterns separately in the two halves of the brain. Certainly their studies on information from the vertical meridian of the visual field indicates that primary information passes virtually simultaneously to the two hemispheres.

Phylogenetic Significance of Visual Areas

The phylogenetic significance of the striate and associated cortical areas has been well summarized by Holmes:

"In the evolution of the brain, vision is the first sense organ to obtain a representation in the cortex of the forebrain. Even in birds, fibres carrying retinal impulses reach what is regarded as the neopallium; it is only later in evolution that other sense organs attain cortical connections and it is still later that so-called 'motor centres' appear. With vision, too, localization in the cortex reached its acme, for every point of the retinae is represented rigidly in it; then the brain, if we may speak teleologically, decided that such a rigid machine was unsuitable to its further development and adopted a more plastic organization for its later-evolved functions. It is consequently not surprising that cortical visual reactions are more highly organized, and possibly organized on a simpler plan, than other cerebral activities. Even in birds, the original visual cortex is connected with the roof of the mid-brain by efferent fibres— the *tractus occipito-mesencephalicus* of Kappers—which are probably the homologue of the occipito-mesencephalic fibres of mammals, and by means of these the cortex may be able to influence reactions, including ocular movements and postures, in response to retinal stimuli. As the cortex became the main, and finally the exclusive organ for visual perceptions, the primary visual reactions— as reflex direction of the eyes to light, accommodation, fusion when binocular vision was acquired, and fixation—were transferred to it till in man they can be evoked through it only. The next step in development was the appearance of associational areas around the visual cortex which enable it to co-operate with other sense impressions and to elaborate further the faculties of spatial perception, discrimination, and recognition."

Electrophysiology of the Visual Pathway

Ganglion Cells. The organization of the receptive fields of the ganglion cells of the retina has been described earlier; to recapitulate, the basic organization is circular, with a central spot and a surrounding annulus responding in opposite fashions; thus we may have a field in which an ON-response occurs when the light stimulus is projected onto this region of retina whilst the surrounding annulus gives an OFF-response, i.e. switching the light on does not excite, in fact it causes inhibition of the background spontaneous discharge if this is present. On switching the light off, there is a prominent OFF-discharge. This unit would be called an ON-centre unit. Fig. 320, A and B, illustrates both ON- and OFF-centre fields. Quite clearly, then, by the time the responses of the cones or rods have been transmitted through the ganglion cells there has been a considerable elaboration, of which a prominent feature is inhibition. We have seen how this type of unit, acting in conjunction with others, can signal movement and even the specific direction of the movement; in addition, too, the signals of many of the units have become colour-coded in a different manner from the coding found in the receptors themselves, so that a given "dominator-type" for example, can indicate by its frequency of discharge the prevailing luminosity independently of the colour of the light, whilst others, modulators, will respond only to a relatively narrow band of wavelengths.

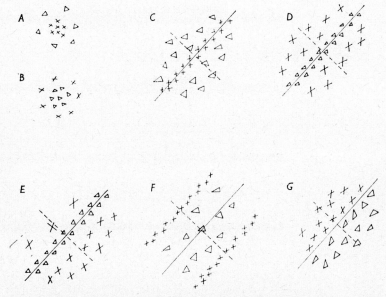

FIG. 320. Types of receptive field. Crosses indicate excitatory or ON-responses; triangles indicate inhibitory or OFF-responses. A: ON-centre response typically seen in retina and lateral geniculate body. B: OFF-centre response. C-G: Various arrangements of "simple" cortical fields. (Hubel & Wiesel, 1962.)

Geniculate Neurones. A study of lateral geniculate neurones, e.g. by Hubel & Weisel (1961), has revealed essentially the same basic organization of receptive fields, characterized by ON-centre—OFF-surround, or *vice versa*; this does not represent a mere copy of the ganglion cell input, however, since a given geniculate cell will have been influenced by many hundreds of ganglion cells excited within the receptive field; moreover, there is considerable accentuation of the inhibitory actions of the centre and surround of any given field. This is illustrated by Fig. 321 where the microelectrode was recording from an optic tract fibre simultaneously with a geniculate cell. A shows the responses to both units when a small light-spot falls on the centre of the receptive field; the optic tract fibre, being an ON-centre one, gives a burst of spikes which are distinguishable from those of the geniculate fibre by their being monophasic and small. The response of the geniculate fibre appears at OFF as large diphasic spikes. Increasing the size of spot, so that the stimulus now falls on the surrounding annulus of opposite response characteristics, has almost completely inhibited the geniculate response, but had only slight effect on the tract fibre (B). In C the spot has been made much larger, and the geniculate response is virtually abolished and the tract response is reduced. In D only the surround has been stimulated, and now the large diphasic spikes appear first at ON, and the small tract spikes at OFF. Thus the geniculate cells differ quantitatively in the degree of "surround antagonism", to such an extent indeed does this occur that diffuse illumination, because it stimulates both centre and surround of the receptive field, has little effect on geniculate cells.

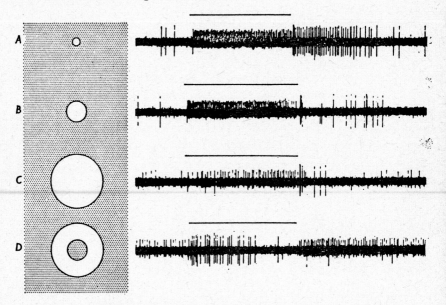

Fig. 321. Simultaneous recording from optic tract fibre and from geniculate neurone; the optic tract fibre gives small monophasic spikes.

A. The small light-stimulus falls on the centre of the receptive field; the optic tract fibre is excited at ON (it is ON-centre) whilst the geniculate one is excited at OFF (it is OFF-centre).

B. The larger stimulus, covering both central region and surrounding region of opposite characteristics, causes almost complete inhibition of the geniculate response.

C. The stimulus is larger and has now reduced the optic tract response as well.

D. Only the surround is stimulated, and the geniculate neurone has fired at ON and the tract neurone at OFF. Horizontal lines above records indicate duration of light stimulus. (Hubel & Wiesel, 1961.)

Thus, by application of inhibition, Nature has reduced the tremendous bombardment of the visual cortex that would occur continuously as long as the eyes were open in the light; instead, many neurones remain silent and are ready to respond to what is important in the environment, namely the constrast involved in contours and the effects of a moving stimulus, be it light or dark, which, passing across the concentric receptive fields, has a definite effect.

DIRECTIONAL SENSITIVITY

In many species, e.g. the frog and rabbit, retinal ganglion cells show directional specificity, responding to movement of a spot of light in the "preferred" direction and showing inhibition of spontaneous discharge in the "null" direction (p. 165). In other species, this phenomenon cannot be detected in the retina, and it is in the lateral geniculate body, or superior colliculus, that such directional-sensitivity is elaborated. In the rat lateral geniculate body, for example, Montero & Brugge (1969) have described neurones with essentially the features described by Barlow *et al.* (1964) for retinal neurones; they showed that the basis of this discrimination was both inhibition of adjacent receptors in the null direction and facilitation in the preferred direction. About 80 per cent of the cells were sensitive to up-and-down movement, and most were concentrated in the area of central vision.*

Striate Cortex. When recordings are taken from neurones in the striate cortex, an entirely different organization of receptive fields appears, the dominant feature being the linear organization of the ON- and OFF-regions, as opposed to the concentric circular organization of the lower neurones. The types of field that are described as "simple" are illustrated in Fig. 320 C to G, A and B being the geniculate fields described earlier. Here the ON or excitatory responses are indicated by crosses and the OFF by triangles. The receptive field represented by C consists of a central slit of retina orientated in a definite direction; light falling on this slit, or a spot of light moving along it, produces ON-responses; light falling on the narrow areas to each side gives OFF-responses, whilst light falling on both central slit and the parallel surrounds leads to complete cancellation, so that diffuse illumination of the retina produces no response in these cortical neurones. Summation occurs along the slits, in the sense that a slit of half the length gives a smaller response than one occupying the whole length of the field.

Illumination of both flanks simultaneously produced a larger response than separate illumination of either. The necessary orientation of the slit stimulus to obtain maximal response was quite critical, so that a variation of more than 5–10° had a measurable effect. We may assume that these cortical neurones receive their inputs from a number of geniculate cells, each with a circular field, and that the effect of this combination is to give a linear array of these fields that would behave effectively as a unit with a linear receptive field as illustrated in Fig. 322. By appropriately arranging geniculate cells the different receptive fields of other types of "simple" cortical neurones, illustrated in Fig. 320, E–G, may be constructed.

To return to Fig. 320, D illustrates the mirror-image of field C, the outer flanks being excitatory and the inner slit inhibitory. In both C and D the flanks of the field were more diffuse than the central regions. F illustrates a cell with a more diffuse centre and narrow flanks, and the optimal response was obtained by simultaneously illuminating the two flanks with parallel slits. In G, only two

* Barlow & Brindley (1963) studied the effects of repeatedly stimulating a direction-sensitive ganglion cell by moving a spot of light over the retina in the preferred direction; before stimulation, there was a maintained discharge of 7/min; the discharge rose to 60/min and finally settled down to 25/min. When the stimulation ceased, the discharge fell to zero and gradually rose, over a period of a minute, to the steady level of 7/min. They consider that this fall to zero and slow return of the background discharge give rise to the illusion of movement, the normal balance of steady discharges between ganglion cells of opposite directions of preferred movement being upset.

regions could be distinguished; the most effective stimulus, here, was two areas of different luminance placed so that the line separating them fell exactly over the boundary of the fields; this was called by Hubel & Wiesel an "edge stimulus". With all these "simple" units a moving stimulus was always very effective, but it is important that the correct orientation of the slit be maintained, the movement being executed at right-angles to the direction of the slit. Usually the response was the same with forward or backward movement, but in others the responses were very unequal, due to an asymmetry in the flanking regions; thus with unit G the difference was very marked.

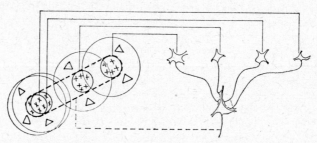

FIG. 322. Scheme to explain the organization of the linear cortical receptive field on the basis of summation of circular fields at the geniculate level. (Hubel & Wiesel, 1962.)

The fields of these units were described as *simple* because, like retinal and geniculate fields, they were subdivided into distinct excitatory and inhibitory regions; there was summation within the separate excitatory and inhibitory parts; there was antagonism between excitatory and inhibitory regions; and finally because it was possible to predict responses to stationary or moving stimuli of varying shapes from a map of the excitatory areas.

COMPLEX UNITS

Another class of units showed similarities with the simple ones, but they were called *complex* because their responses to variously shaped stationary or moving forms could not be predicted from a map made with circular spot-stimuli. If separate ON- and OFF-regions were found, antagonism between these did not always hold. Four types of complex unit were described by Hubel & Wiesel (1962); as with the simple units, the orientation of a slit-stimulus was of the utmost importance for obtaining maximal response, but in general the position of the slit on the retina was not critical, by contrast with the simple unit. The unit was making abstractions of a higher order, responding to the direction of the stimulus and ignoring differences in its position. It is this type of neurone that would be concerned, presumably, with determining the verticality and horizontality of lines in space. Movement of a slit across the visual field of this unit does not have the phasic effects seen with the simple unit; instead, there is a continuous discharge throughout the movement. Space will not permit the description of more than one complex unit in detail; in general they were of four types, namely those activated by a slit with either non-uniform or a uniform field; those activated by an edge, and those activated by a dark bar; each could be treated as analogous with one of the classes of simple units. Fig. 323 illustrates the

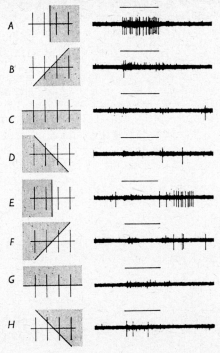

Fig. 323. Responses of a cell with a large (8 × 16°) complex receptive field to an edge projected on the ipsilateral retina so as to cross the receptive field in various directions. At the time of the stimulus, indicated by the upper line of each record, the illumination of the screen to one side of the edge was increased suddenly; thus in A this was to the left of the edge and in E to the right. (Hubel & Wiesel, *J. Physiol.*)

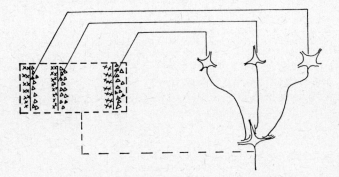

Fig. 324. Possible scheme for explaining the organization of complex receptive fields. A number of cells with simple fields, of which three are shown schematically, are imagined to project to a single cortical cell of higher order. Each projecting neurone has a receptive field arranged as shown to the left: an excitatory region to the left and an inhibitory region to the right of a vertical straight-line boundary. The boundaries of the fields are staggered within an area outlined by the interrupted lines. Any vertical-edge stimulus falling across this rectangle, regardless of its position, will excite some simple-field cells, leading to excitation of the higher-order cell. (Hubel & Wiesel, 1962).

response of an "edge-type"; when a vertical edge was projected on to the retina, by making half the field bright and leaving the other half dark, there was a strong ON-response. When the orientation of the edge was altered the response diminished. Exchanging the stimulus for its mirror image reversed the response giving inhibition at ON and excitation at OFF.

Fig. 324 illustrates the way in which simple cortical neurones can be connected to a higher order cortical neurone to give the latter a complex field; here the cortical simple units are of Type G, and it can be seen that any vertical edge-stimulus, falling across the field, regardless of its position, will excite some simple units provided, only, that the direction is adequate. In a similar way other complex fields can be built up from groups of simple cortical cells. On this basis we should expect the latency of the complex cell to be longer than that of the simple cell; in fact Denney, Baumgartner & Adorjani (1968) found values of 2·27 and 3·84 for simple and complex cells respectively.*

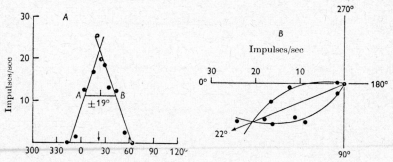

FIG. 325. Illustrating sensitivity of a cortical cell to direction of movement of a grating.

A. Linear plot of response amplitude (of cortical cell) versus the direction of movement (angle) of the stimulus grating. Their point of intersection is taken as the best estimate of the maximum response amplitude and the preferred angle (arrowed). The horizontal line is drawn at half the maximum response amplitude. Spatial frequency: 0·24 c/deg. Drift frequency: 3·8 c/s. Contrast: 0·7.

B. Polar plot of results in A. The arrowed line indicates the preferred angle. 0° = a movement from right to left in the visual field; 90° = a downward movement. (Campbell, Cleland, Cooper & Enroth-Cugell, *J. Physiol.*)

SPATIAL AND ORIENTATIONAL SELECTIVITY

It will be recalled that when a grating is allowed to drift over the receptive field of a ganglion cell, a characteristic response, in phase with the drift, is produced, and the sensitivity of the cell to the contrast of the grating is expressed as a function of the grating-frequency, to give the so-called *contrast-sensitivity function* (p. 247). When geniculate and cortical cells of the cat were examined

* Spinelli & Barrett (1969) have measured the receptive fields of a large number of cortical cells in the cat, using moving spot instead of linear stimuli to plot them. A remarkably large number of these fields were circular, and thus reminiscent of geniculate fields, but the fact that most of the units could be excited by either eye precluded their being fibres of the optic radiation. The units exhibited a very strong direction-sensitivity that would make them valuable as detectors of both movement and of direction of movement. Spinelli & Barrett have discussed the role of single cortical units as "detectors" of a specific feature of the environment, e.g. a contour; clearly such units would be of value, but against this must be weighed the loss of flexibility, so that a more economical system might be obtained by retaining a large number of units uncommitted to any specific feature.

(Campbell, Cooper & Enroth-Cugell, 1969), characteristic contrast-sensitivity functions were obtained with sharp cut-off points such that a small change in spatial frequency of the grating caused an abrupt decrease in sensitivity; the "attenuation points" varied from about 0·2 to 1·5 cycles/degree for cortical cells and 0·3 to 4 cycles/degree for geniculate cells. By human standards the grating discrimination displayed by these cells, is low, indicating a low visual acuity in the cat.

Orientational selectivity in the cat was examined using the same technique, the angle made by the pattern with the optimum, or *preferred*, direction being varied and the mean spike-frequency provoked by moving this pattern over the field being determined. As Fig. 325 shows, the impulse-frequency decreases linearly with the obliquity of the grating from the preferred orientation. The angular selectivity of a unit was defined by drawing the line AB through the value of half-maximum amplitude of response. Bisecting the angle determined by the length of AB gives the angular selectivity, which in this unit was ± 19°. Angular selectivities ranged from 10° to 50°, but it is interesting that there was no preference for the horizontal or vertical as might be suggested by the finding that grating-acuity is highest when the lines are either vertical or horizontal (p. 246).

Binocular Fields. The visual fields of animals with frontally placed eyes overlap, and the phenomena of binocular vision indicate that at some stage, or stages, in the visual neuronal pathway, from retina to striate and adjacent cortical areas, there is interaction between the impulses arriving from the separate eyes. When lateral geniculate neurones were examined by Hubel & Wiesel (1961), they found units that could be driven by one or other eye, but they never found one that could be activated by both eyes. With striate neurones the position was different, some 85 per cent responding to stimulation of either eye (Hubel & Wiesel, 1962); furthermore, as we might expect, the receptive fields in the two eyes that stimulated a single cortical neurone were similar in shape and orientation, and they occupied symmetrical positions in the two retinae, i.e. what are called corresponding areas. Summation and antagonism occurred between the receptive fields of the two eyes. For example, if the ON-area of the left eye and the OFF-area of the right eye were stimulated together, the effects tended to cancel each other. In some cases a cell could only respond if both eyes were stimulated together by an appropriately orientated slit.

Dominance. We may speak of the *dominance* of one or other eye, according to its power of exciting a given cortical neurone by comparison with the other eye; thus a neurone excited only by the left eye, i.e. showing no measurable binocular interaction, would be said to be completely left-dominant; between this condition and right-dominance there were many stages. In general, there was a preponderance of contralateral dominance, presumably corresponding with the preponderance of crossed over uncrossed fibres in the cat's optic tracts. An illustration is in Fig. 326, the response being heavily dominated by the contralateral eye. Each cell, then, receives a somewhat greater excitatory input from the contralateral eye than from the ipsilateral eye. The significance of this will be discussed in connection with stereoscopic vision.

Cortical Architecture. Mountcastle (1957) discovered that the somatosensory cortex was divided by a system of vertical columns extending from surface to white matter, with cross-sectional widths of some 0·5 mm, the responses to

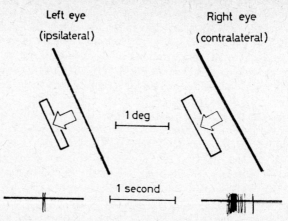

Left eye
(ipsilateral)

Right eye
(contralateral)

1 deg

1 second

FIG. 326. The binocular neurone whose responses are illustrated here belongs to Hubel and Wiesel's eye-dominance group 2: it is heavily dominated by the input from the contralateral eye. The receptive fields for the left and right eyes are shown as rectangles on the left and right, respectively. Below each field is an oscilloscope record of the action potentials typically evoked from the neurone by a single, slow sweep of a thin bright slit across the receptive field in that eye. Clearly the unit is far more strongly activated by a stimulus to the right (contralateral) eye than by one to the left. (Blakemore & Pettigrew, *Nature*.)

neurones in these columns being similar with respect to modality of stimulation, the one responding to superficial and the other to deep sensation. In the visual area, too, Hubel & Wiesel (1963) found that their units responding to a given orientation of a slit were grouped together in columns; when units were classified in accordance with eye-dominance, a similar vertical distribution of units was found, overlapping with those based on line preference. In the monkey, too,

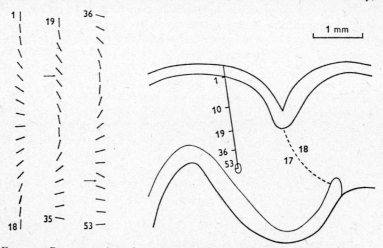

1 mm

FIG. 327. Reconstruction of a penetration through striate cortex about 1 mm from the border of Areas 17 and 18, near the occipital pole of the spider monkey. To the left of the figure the lines indicate orientations of columns traversed; each line represents one or several units recorded against a rich unresolved background activity. Arrows indicate reversal of direction of shifts in orientation. (Hubel & Wiesel, *J. Physiol.*)

Hubel & Wiesel (1968) showed that the units were organized on a vertical basis in so far as orientation of receptive field and eye preference, or ocular dominance, was concerned. When a slightly oblique penetration was made it was interesting to observe the step-by-step changes in orientation of the receptive fields as the electrode passed through and along adjacent orientation columns (Fig. 327). In addition, a horizontal organization became apparent; thus simple cells were highly concentrated in the deep part of Layer III and in Layer IV; it is here that the geniculate fibres relay with the cortical cells. The cortical cells showing simple behaviour in these layers were grouped in accordance with the eye by which they were driven (they were mostly monocularly driven). Layers above and below contained high proportions of complex and hypercomplex cells, which presumably owe their more complex behaviour to connections with simple neurones. The earlier study on the cat had indicated that hypercomplex cells were confined to Areas 18 and 19, but the present work showed that these cells were also frequent in Area 17. The anatomical demonstration of these columns, based on eye-representation, i.e. contralateral or ipsilateral, is illustrated by Fig. 328 from Hubel & Wiesel (1969), who made use of the fact that the lateral geniculate cells were carrying messages from either the contralateral eye (Layers 6, 4, 1) or the ipsilateral eye (Layers 5, 3, 2). Thus a lesion in Layer 6 of the left lateral geniculate body should represent a lesion in the right eye, and the distribution of degenerating axons and terminals in the left hemisphere should indicate how messages from the right eye were being organized. It was found, from the degenerating fibres, that they passed mainly to Layer IV of the cortex, where they relayed with cortical neurones; some passed vertically up into Layer III also. Fig. 328 shows the

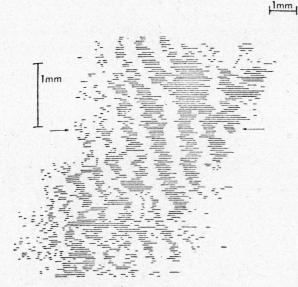

FIG. 328. Reconstructed map of regions of fourth layer degeneration, following localized injury to the lateral geniculate body, as they would appear from above. Each horizontal line represents areas of terminal degeneration from one para-sagittal section. Shaded areas thus represent those parts of Layer IV receiving input from the contralateral eye; unshaded areas, ipsilateral eye. (Hubel & Wiesel, *Nature*.)

pattern of degeneration, which occurred in patches, as viewed from above, so that shaded areas represent those parts of Layer IV receiving input from the contralateral eye, and unshaded portions from the ipsilateral eye. This evidence, together with the results of electrode penetrations, indicates that the cortex is divided into columns, or slabs, with walls perpendicular to the surface and long narrow cross-sections, determined by this Layer IV mosaic.

Eye Dominance

All, or nearly all, of the cells in these columns represent the dominant eye so far as electrophysiological recordings are concerned. The mechanism for this is that a higher order complex cell, lying in a layer superficial or deep to the Layer IV cell (which receives the direct input from the eye), receives its relayed input from the Layer IV patch immediately above or below, and a subsidiary input from adjacent patches, which represent the other eye. Such a cell can thus be influenced from the two eyes, but it will respond much more strongly to the eye that corresponds to the nearest Layer IV patch. Hence the cells of this Layer IV patch, together with all cells directly above or below in the column, constitute a group dominated by one eye. The stripes for eye-preference are about 1 mm wide, but those for direction-preference are probably considerably finer (Hubel & Wiesel, 1968).

Areas 18 and 19. Hubel & Wiesel (1965) considered that visual association areas would be rich in neurones with complex receptive fields, since these areas would consist of areas receiving projections from cortical, rather than geniculate, neurones. On carrying their explorations to regions more and more lateral to the striate area, Hubel & Wiesel found two additional areas, Visual Areas II and III, corresponding architectonically with Areas 18 and 19, as revealed by histological studies; both these areas apparently received projections from Area 17. When the receptive fields of neurones in these two areas were examined, they were found to be entirely of the complex type, as we should expect of neurones receiving projections from cortical neurones. Furthermore, new degrees of complexity in the receptive fields emerged, so that a further class of "hypercomplex" cells had to be made, and within this group "higher order" and "lower order" cells were described. Visual Area II contained almost entirely complex cells similar to those found in Area I, the remainder being hypercomplex, whilst in Area III the majority were hypercomplex. An example of a hypercomplex cell was one that, like the complex cells, could be excited by a line of specific orientation, but this line had to be limited in length in one or both directions to obtain maximal effect. The adequate stimulus was thus a critically orientated line falling within a given region of retina (the activating region), provided that a similarly orientated line did not fall over an adjacent, antagonistic, region. This hypercomplex cell behaved as though it were activated by two complex cells, one excitatory to the cell with a receptive field occupying the activating portion and one, inhibitory to the cell, having its field in the antagonistic portion. The majority of neurones in Areas II and III were driven by both eyes, with a distribution of dominance similar for all three visual areas.

Retinal vs Cortical Cells. The general feature manifest in these studies of neurones of ever higher order is the increase in the number of stimulus parameters that must be fulfilled to fire a cell as we progress from the receptor to the hypercomplex cortical neurone. To obtain an optimum response from a retinal ganglion cell, for example, it is generally sufficient to specify the position, size and luminance of the spot; enlarging the spot beyond the size of the field-centre raises the threshold, but even with diffuse light a response can be obtained at high enough intensity. As Fig. 320 shows, the penalty for increasing the size of the spot is more severe in the geniculate cell. In the cortex, the driving of any cell can only be obtained by restricted stimuli whose position, shape and orientation are specific for the neurone. Diffuse light is a poor stimulus and usually ineffective at any intensity. On passing to the complex cell, however, the specificity is reduced, and now orientation rather than exact position is often the dominant influence. The value of this is doubtless in the recognition of form, as such; as Hubel & Wiesel point out, it represents an economy that permits the use of relatively few cortical cells of high order that will recognize form independently of position on the retina.

RESPONSE TO MOVEMENT

The ability of some complex cells to respond in a sustained manner to a stimulus as it moves over quite a wide field of the retina marks them off as those specially suited for the perception of movement; they adapt rapidly when the image is stationary; presumably the simple field cells, with which the complex cell is connected, also adapt rapidly, but because of their staggered fields the moving stimulus excites them in turn and the higher order complex cell is thus continuously bombarded. The complex cells would also respond continuously to the changes in the retinal images caused by the rapid small eye movements; when these are abolished we have seen that awareness of the image fades, and it would seem that the rapid discontinuous movements of contours would excite complex cells continuously in the same way that we have seen them to be excited by movement of the image across the field.

Neurophysiological Basis of Stereopsis

In looking for electrophysiological correlates of binocular vision involving stereopsis, we must expect to find, as we have indeed, cortical neurones that respond to stimuli in both eyes. Furthermore, as Barlow, Blakemore & Pettigrew (1967) have put it, the neurone must select those parts of the two images that belong to each other, in the sense that they are images of the same point; secondly, for stereopsis, they must assess the small displacements from exact symmetry that give the binocular parallax.

Minimum Response Fields. Barlow *et al.* recorded from neurones in Area 17 of the cat and found, as Hubel & Wiesel had earlier, that binocular stimulation was more effective in stimulating many units than uniocular; if the positioning of the targets was wrong, however, one eye could veto the response of the other as in Fig. 329; here a bar moved over the receptive field of the left eye and right eye separately; acting together there was powerful facilitation. When the separate fields were mapped out for each eye, it was found that these "minimum response fields" did not occupy exactly symmetrical positions in the two eyes, i.e. they were not corresponding areas; instead, for maximal binocular facilitation there had to be a discrepancy; for the unit illustrated by the left part of Fig. 329 this was 5·7°; in this case the cat's eyes were divergent by 6·4°, so that the actual

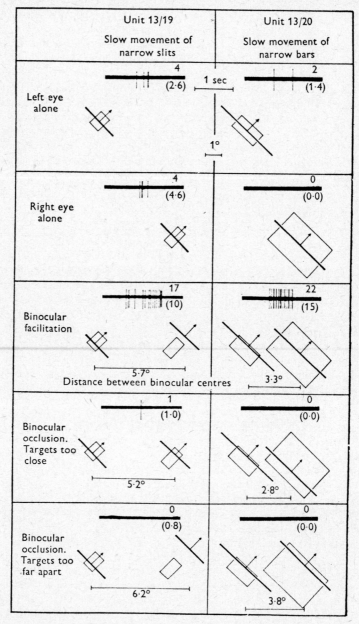

Fig. 329. Illustrating responses of two binocularly driven units. Maximal facilitation in binocular stimulation is obtained only when there has been a certain disparity between the stimulated regions in the two retinae. (Barlow, Blakemore & Pettigrew, *J. Physiol.*)

disparity corresponded to a convergence of 0.7° for the left unit and 3.1° for the right unit. In general, different neurones required different disparities to obtain maximal facilitation between the two eyes.*

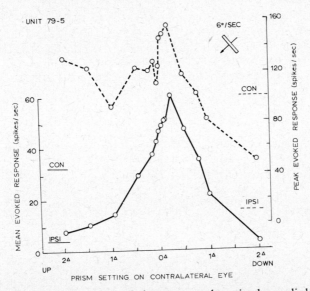

FIG. 330. Responses of a cat's cortical neurone to a bar-stimulus applied to both eyes when the degree of retinal disparity was varied by placing prisms in front of the contralateral eye. The bars marked IPSI and CON indicate the responses of ipsi- and contralateral eyes when stimulated alone. Full lines indicate mean evoked response and dashed lines the peak evoked response. (Pettigrew, Nikara & Bishop, *Exp. Brain Res.*)

Disparity. To estimate the disparities of the fields the optimum orientation of a slit was first found, and this was the same for both eyes. The slit was moved over the retina and the limits beyond which no response was obtained were drawn on a screen in front of the eye. This gave the primary borders of the field; the lateral borders were determined by moving the continuously oscillating slit, maintaining the axis of orientation, until the end of the slit had moved out of the field and no constant response could be elicited. This is called the *minimum response field*, and is indicated in the rectangles of Fig. 329. The fields for the two eyes were not necessarily of the same size, the dominant eye having the larger. It seems likely that the centres of these fields represent the centres of retinal areas connecting to a single cortical neurone. For obtaining the binocular disparities the two fields must be examined together; usually the centres of their separately determined fields give a good clue to this, but to find the exact measure the two slits are oscillated in synchrony over their respective fields to obtain maximal facilitation. At a critical separation this is obtained and the positions of the optimally separated slits are marked with a line on the respective monocularly determined fields, as in Fig. 329. The point on this line where the normal to it would pass through the centre of the monocular minimum response field is called the *binocular centre*. It is the angle between these centres that is used as measure of retinal disparity.

* Nikara, Bishop & Pettigrew (1968) have made comparable studies in the cat, studying the vertical and horizontal disparities between receptive fields of binocularly represented units of Area 17. The disparities were in the range of ±1.2° in both horizontal and vertical directions. The significance of vertical disparity is that it permits single vision of points when the directions of regard of the two eyes are in error due to random movements of the fixation axes in relation to each other.

Detection of Disparity. Pettigrew, Nikara & Bishop (1968) recorded the summated action potentials from binocularly driven cortical units of the cat when the disparity between left and right-eyed fields was changed by means of a prism. Fig. 330 shows the evoked response of a unit as a function of prism setting, 0° corresponding to perfect superimposition. The responses of the ipsi- and contra-lateral eyes are indicated by horizontal bars, and it will be seen that over a certain range of disparities there is facilitation through binocular stimulation. Over the remainder of the range of disparities there is mutual inhibition. The sensitivity of the unit to disparity is high, of the order of 3 minutes of arc (0·1Δ).

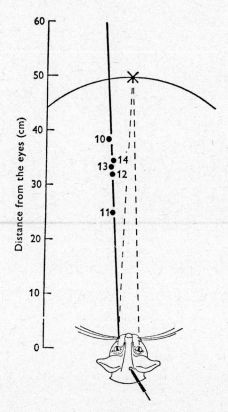

FIG. 331. Illustrating the positions in space of points that would give rise to the observed retinal disparities of cortical neurones in a constant direction column. (Blakemore, *J. Physiol.*)

Depth and Direction Columns. It will be recalled that Hubel & Wiesel found that the neurones of the striate area were organized in columns according to eye-dominance and according to the direction-preference of the slits required to excite; Blakemore (1970) examined the effects of recording from striate neurones at successive depths in a column, and he found that, in certain columns, the neurones all had optimal retinal disparities of about the same direction and magnitude; all these neurones were responding, then, to a strip of space at a definite distance from the fixation point. Adjacent columns had disparities differ-ing by about o·6°. Blakemore called these *constant depth* columns to distinguish them from *constant direction* columns whose neurones showed enormous variations in disparity from unit to unit at successive depths. When the left- and right-eye fields of these units were compared, it was found that for one eye, e.g. the left, the fields of successive units were all superimposed, so that the variation in disparity was due to the variations in the positions of the fields of the right eye. In Fig. 331 a spatial plot of the positions of points in space that would given rise to these dispar-ities is shown, and it becomes clear that these units have the common parameter of constant direction in relation to the left eye. Thus the column of neurones is "peer-ing along a tube of visual space that is lined up along one oculocentric visual direc-tion". Examination of several direction-columns gave a similar pattern, namely that the superimposable response fields were found in the hemisphere contralateral to the retina in which they were evoked, so that those responsible for the disparities were derived from the ipsilateral retina. It has been pointed out earlier that the ipsilateral projection of the retina is the phylogenetically newer projection since in lower vertebrates the decussation tends to be complete. It is interesting therefore that it is the ipsilaterally projecting neurones that show the binocular disparities that are essential for stereoscopic vision.* Thus in non-stereoscopic vision it would be the direction-columns that were of absolute importance; with the requirements of stereopsis information both with respect to direction and dis-parity of retinal images is required, and this is provided by the constant dis-parity (depth) columns.

CORTICAL NEURONES REQUIRED

Blakemore estimated an average of o·6° difference of horizontal disparity between neighbouring depth-columns, and about 4° of oculocentric visual direction between direction-columns. If the mosaic were maintained across the whole binocular field, and if the total range of disparities were 6°, this would require 500 columns for one orientation of the target. If there were ten to fifteen optimal orientations of the target, this would make a grand total of 8000 to 12,000 columns to encode every orientation and every spatial position, a number that is not impossible anatomically (Blakemore, 1970).

Stereopsis in the Monkey. By the use of random-dot patterns, as illustrated in Fig. 298, p. 467, such that the individual pictures presented to the eyes in a stereoscope are indistinguishable, it is possible to study the stereoscopic vision of the monkey, training it to respond only when the two pictures give a required stereoscopic depth perception on fusion. In this way Bough (1970) demonstrated that monkeys could discriminate between pictures that, when fused, would give no stereopsis, and those that would. Hubel & Wiesel (1970) have examined the cortical cells of Areas 17 and 18 of macaque monkeys and have shown that in Area 18 binocularly responding cells are found reacting similarly to complex and hyper-complex cells in Area 17; these they called "ordinary cells of Area 18"; they had receptive fields in anatomically corresponding areas; they responded to separate stimulation of the eyes and showed moderate summation when stimulated together. The other half were called "binocular depth cells", the most common of which

* Blakemore & Pettigrew (1970) have shown that the receptive fields of the ipsilaterally projecting neurones are much more scattered than those of the contralaterally, when penetrations to successive depths in a column are made. Thus the ipsilaterally projecting neurones can give less precise information regarding actual direction in space, but of course the stereoscopic information they can convey is what is important.

gave no response to stimulation of the separate eyes but a brisk response to simultaneous stimulation. Some responded well to stimuli in corresponding points of the retinae, and others exhibited some disparity in their receptive fields. By contrast with the cat, no convincing evidence was found of binocular depth cells in Area 17. Thus Area 18 has two functions in the monkey, namely linking the two half-fields (as also in the cat), and the elaboration of stereoscopic depth perception.

Bilateral Representation of Retina. Blakemore (1969) has emphasized that if the decussation of the optic nerve fibres splits the visual field exactly into two halves across the vertical meridian, so that all stimuli from the temporal side of the field remain ipsilateral, and all on the nasal side cross over, then, when an object produces disparate images on the retinae, as in Fig. 288, a_L and a_R, the images can fall on the nasal halves of each retina and so their messages are carried to opposite hemispheres of the brain. In this event, the single cortical neurones that receive messages from both of these points on the retinae will have to be connected to a pair of cortical neurones in opposite hemispheres, perhaps through the corpus callosum. Alternatively, of course, we may postulate that over the small regions adjacent to the vertical meridians, where retinal disparities are utilized in binocular stereoscopic vision, the decussation may be incomplete, so that this part of the fovea will be represented, in effect, bilaterally. This might account for the "sparing of the macula" discussed earlier, but in the primate the evidence against such a bilateral representation through incomplete decussation is preponderant, and callosal connections mediate the necessary integration between the two hemispheres. Hubel & Wiesel (1967) recorded from separate units simultaneously in the two hemispheres of the cat, and found some that had receptive fields that overlapped in the middle; these were invariably in the boundary between Areas 17 and 18 where the representations of the central vertical meridian in the primary and secondary visual areas are side by side (Fig. 316, p. 493). Blakemore has examined the response fields of binocularly driven units at the junctions of Areas 17 and 18, and has confirmed that some of these overlap, in fact their binocular centres (p. 513) could be on the "wrong" side of the estimated vertical meridian by a degree or two, as illustrated by Fig. 332, where the centres of the response-fields of right eyes have been plotted against azimuth from the vertical meridian. The filled circles indicate responses in the right, or ipsilateral, hemisphere which, in the absence of bilateral representation, should be confined to the left of the Figure, and the open circles indicate responses in the left, contralateral, hemisphere and should be confined to the right of the Figure. The "wrong fields" are indicated by the line joining the points, and this indicates the degree of overlap. This central strip of retina, with a width of about 1·5° and equivalent to horizontal disparities of about twice this, doubtless is the region of bilateral representation that is required for stereopsis in the vertical meridian.

CORPUS CALLOSUM

The mechanism of this overlap could be through the corpus callosum, in which event monocularly driven callosal units should be found, as they apparently are (Berlucchi, Gazzaniga & Rizzolatti 1967), whilst cutting the corpus callosum abolishes stereopsis (Mitchell & Blakemore, 1970), of objects in the midline but not, of course, in the periphery since in this event both images are projected to the same hemisphere. The situation is illustrated by Fig. 333, which describes

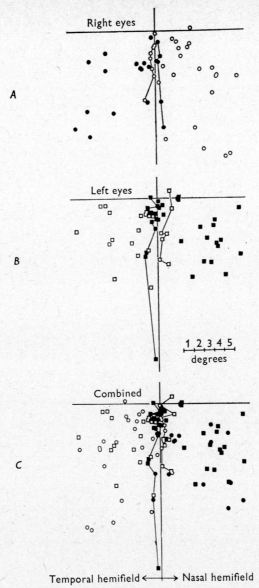

Temporal hemifield ←——→ Nasal hemifield

FIG. 332. A. Response field centres for four right eyes are pooled by super-imposing the true vertical through the middle of the overlap and the horizontal through the estimated area centralis. The open circles are the centres for units recorded in the left (contralateral) hemisphere and the filled circles for units in the right (ipsilateral) hemisphere.

B. The field centres for five left eyes are pooled in the same way. The open squares are centres for right (contralateral) hemisphere units and the filled squares for neurones in the left (ipsilateral) hemisphere.

C. The data from A and B are combined by lateral inversion of the right eyes' field-centres and the superposition of the verticals and horizontals. Now temporal visual hemifield lies to the left of the vertical meridian and nasal hemifield to the right. The symbols are those used in A and B. All field-centres lying on the "wrong" side of the vertical meridian (i.e. in the ipsilateral hemifield), are joined to give an indication of the width of the strip of bilateral projection. (Blakemore, *J. Physiol.*)

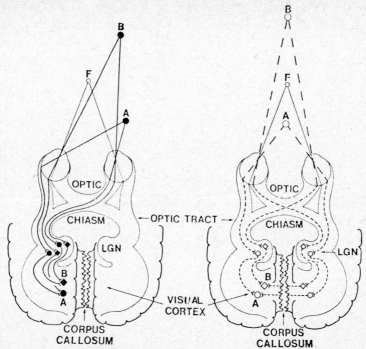

FIG. 333. The possible neural system for binocular depth perception in the split-brain human. On the left is the arrangement of neurones which enabled the subject to recognize the depth of the peripheral objects A and B, relative to the fixation point, F. For both objects the image in the right eye falls on nasal retina and that in the left upon temporal, and all the information projects to the left optic tract. The messages from the two eyes remain segregated at the lateral geniculate nucleus (LGN). Solid circles represent neurones within whose receptive field the image of A falls. Solid diamonds are cells to which object B projects. The two binocular cells, A and B, shown as large symbols in the cortex, encode the disparities of the objects. In this case the sectioned corpus callosum is of no hindrance. The binocular information is processed entirely in the left hemisphere and the judgements can therefore be vocalized.

In the right diagram objects A and B lie directly in the midline and, therefore, their images fall on temporal or nasal retina, respectively, in *both* eyes. The interrupted lines and open symbols in the visual pathway show the neurones normally responsible for the recognition of the disparity of these objects. The binocular cells A and B in the cortex receive one of their inputs from a fibre from the opposite visual cortex, crossing in the corpus callosum. Section of the commissure has severed this link and binocular integration is impossible. (Mitchell & Blakemore, *Vision Res.*)

the neural arrangements in a human subject with sectioned corpus callosum.* To the left, the subject, fixating F, will appreciate the stereopsis created by the positions of A and B because the images fall on the temporal and medial retinae, cortical neurones being binocularly driven by disparate points on the retinae. To the right, the disparities of A and B are such that both the images fall on nasal or temporal halves of the retinae, so that a cortical neurone must receive messages from the contralateral hemisphere to integrate the disparate messages. Mitchell

* This operation is carried out to prevent spread of an epileptic focus from one hemisphere to the other.

& Blakemore found that, whereas the corpus callosum-sectioned subject could discriminate the peripherally placed points A and B in depth, he could not do so when they were in the midline (Fig. 333 *right*).

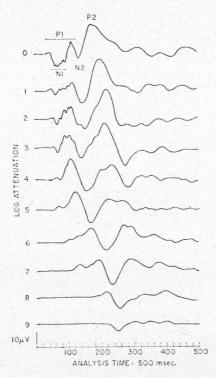

FIG. 334. VER as a function of stimulus intensity. 4° stimulus, 10 μsec duration, 2·5 + 10⁷ mL. peak luminance (0 log attenuation). Electrode placement: mid-occipital to linked ears. (Vaughan in *Clinical Electroretinography*.)

The Visually Evoked Response

The arrival of afferent impulses at the sensory cortex gives rise to a series of electrical changes recorded from the surface of the exposed cortex—the electro-corticogram (ECoG). These are of sufficient magnitude to be distinguished from the background activity, especially when this has been depressed by anaesthesia. In man, attempts have been made to record similar changes through scalp electrodes, but their attenuation through this method of recording demands the use of special "averaging techniques" that allow a separation of the evoked response from the background, i.e. the signal-to-noise ratio must be increased. The summation technique is based on the principle that the average amplitude of a signal, time-locked to a reference point in time, increases in direct proportion to the number of samples, whilst recurrent activity randomly related in time to the reference increases as the square root of the number. The principle, then, is to stimulate with repeated flashes of light and to add the responses by a computer of average transients (CAT).

Amplitude and Latency. Fig. 334 illustrates schematically the cycle of changes of potential that may be obtained by this technique; as will be seen, as the intensity of the stimulus is decreased, the latency of the response increases and the amplitudes of the different waves change in a complex manner, reflecting the algebraic summation of positive and negative components; thus the increase in amplitude of the first positive wave, P_1, as intensity of stimulus decreases, is due to the failure of the negative wave, N_1. Consequently latency is more closely related to intensity than amplitude. The fact that the Stiles-Crawford effect is prominent in the VER indicates that cones have a dominant influence, and when a luminosity curve was constructed, on the basis of the reciprocal of the latency versus wavelength of stimulating light, this had the typical photopic maximum (De Voe, Ripps & Vaughan, 1968).*

Application. The VER has been employed in the study of a number of phenomena of human vision; for example, in retinal rivalry it seems that the amplitude of the flash-evoked VER recorded from a given side of the head is depressed if the contralateral visual field is being suppressed during rivalry (Lansing, 1964; Lawwill & Biersdorf,† 1968); we have already seen how the presence of orientation- and size-selective neurones in the human cortex has been surmised on the basis of studies of the VER (p. 248).

Visual Deprivation

The consequences of failure to use an eye in childhood are well known as clinical manifestations of squint or of the operated congenital cataract (Von Senden, 1960). In the human infant, the deprivation must apparently extend for several years, and the loss of vision exhibits some degree of reversibility if the disuse has not lasted too long. In the cat the subject has been examined by Hubel & Wiesel‡ starting from the basis that closure of the eyes for several weeks after birth leads to complete blindness. Under these conditions, it was found that only one out of eighty-four cortical cells was driven by stimuli applied to the previously occluded eye, and since newborn kittens gave normal responses, it was concluded that deprivation caused disconnection of synapses that were present at birth. It was interesting that the effects of bilateral lid-closure were less than uniocular, so that, although the animals were behaviourally blind, large numbers of cortical cells showed normal responses to light; thus when one eye only is kept closed from birth it seems that the other eye pre-empts cortical cells, actively disrupting the connections of the other eye with them; when both eyes are kept closed, this active process is prevented, and the innately formed connections are retained. When squints were induced from birth by section of a rectus muscle, there was no obvious visual defect, but when the distribution of ocular dominance amongst

---

* Ennever *et al.* (1967) have shown that the potential recorded from the human scalp can be seriously contaminated by non-cortical activity; thus the act of blinking produces a record very similar to the VER, but this is due to modulation of the electric field over the scalp caused by the corneo-retinal potential, movements of scalp, neck and face muscles altering the electrical resistance of the scalp and so altering the recorded potential. We must note, on the other hand, that Cobb & Dawson's (1960) studies of the latencies of the ERG and VER have shown quite conclusively that the VER is not merely the electrotonically transmitted record of the ERG.
† Donchin & Cohen (1970) question this interpretation of the depression of the VER in rivalry; they point out that if the subject's attention is drawn to the flash rather than the rivalry patterns a prominent VER is obtained, whereas the same flash gives no VER if the subject concentrates on the rivalry pattern. This supports the contention of Ennever *et al.* that the VER is a record of eye-movement rather than of cortical responses.
‡ Wiesel & Hubel (1963, a, b; 1965); Hubel & Wiesel (1963; 1965).

cortical cells was examined, this was found to be grossly abnormal; thus although the afferent influx to cells of the cortex had not been altered, a change in their synaptic connections was induced by altering the relation between the afferent impulses from the two eyes.

Eye-Dominance Distribution. The distribution of the eye-dominance amongst randomly chosen cortical cells was, in general, quite characteristic for the normal kitten, and proved to be sensitive to uniocular closure from birth, the distribution becoming highly lopsided in favour of the non-occluded eye. By closing the eye for differing periods, and beginning the closure at different ages, Hubel & Wiesel (1970) were able to define quite precisely the period during which the central visual pathways were susceptible to visual deprivation; this susceptibility began suddenly, near the start of the fourth week, and disappeared around the end of the third month, so that monocular closure in the adult cat, even for very long periods, was completely without effect. The actual period of closure, during the susceptible period, required to produce measurable effects was only three to four days, whilst after a six-day closure the number of cortical cells that could be influenced by the occluded eye became negligible. The effect of deprivation seems, therefore, to be to inhibit connections between the retina and the cerebral cortex; the point at which this failure occurs is apparently at the synapse between geniculate and cortical cells since, although the geniculate cells show microscopically observable abnormalities leading to thinning of the layers, their responses seem quite normal (Wiesel & Hubel, 1963).*

GENERAL REFERENCES

BARTLEY, S. H. (1941). *Vision. A Study of its Basis.* New York: Van Nostrand.
BURIAN, H. M. & JACOBSON, J. H. (1966). Eds. *Clinical Electroretinography.* New York: Pergamon.
GRAHAM, C. H. (1965). Ed. *Vision and Visual Perception.* New York: Wiley.
HELMHOLTZ, H. VON. (1925). *Physiological Optics* (translated J. P. C. Southall). Optical Soc. Am.
OGLE, K. N. (1950). *Researches in Binocular Vision.* Philadelphia: Saunders.
OGLE, K. N. (1958). Present state of our knowledge of stereoscopic vision. *Arch. Ophthal.*, **60**, 755–774.
OGLE, K. N. (1962). The optical space sense. In *The Eye*, Ed. Davson, vol. IV, chapters 11–18. London & New York: Academic Press.
POLYAK, S. L. (1957). *The Vertebrate Visual System.* Chicago: University of Chicago Press.

SPECIAL REFERENCES

ADES, H. W. & RAAB, D. H. (1949). Effect of preoccipital and temporal decortication on learned visual discrimination in monkeys. *J. Neurophysiol.* **12**, 101–108.
ANDERSEN, E. E. & WEYMOUTH, F. W. (1923). Visual perception and the retinal mosaic. *Am. J. Physiol.* **64**, 561–594.
BARLOW, H. B., BLAKEMORE, C. & PETTIGREW, J. D. (1967). The neural mechanism of binocular depth perception. *J. Physiol.* **193**, 327–342.
BARLOW, H. B. & BRINDLEY, G. S. (1963). Evidence for a physiological explanation of the waterfall phenomenon and figural after-effects. *Nature* **200**, 1345–1347.

* Van Hof (1969) found no abnormality in the ability of young rabbits to discriminate patterns when they were deprived of vision for six weeks after birth. Ganz & Fitch (1968) have examined visual function in deprived kittens, and the partial recovery after the period of deprivation. It is interesting that there was no interocular transfer from the deprived to the experienced eye during training of monocularly deprived animals.

BARLOW, H. B. & BRINDLEY, G. S. (1963). Inter-ocular transfer of movement after-effects during pressure blinding of the stimulated eye. *Nature* **200**, 1347.

BATTERSBY, W. S., WAGMAN, I. H., KARP, E. & BENDER, M. E. (1960). Neural limitations of visual excitability: alterations produced by cerebral lesions. *Arch. Neurol.* **3**, 24–42.

BERKLEY, M., WOLF, E. & GLICKSTEIN, M. (1967). Photic evoked potentials in the cat: evidence for a direct geniculate input to visual II. *Exp. Neurol.* **19**, 188–198.

BERLUCCHI, G., GAZZANIGA, S. & RIZZOLATTI, G. (1967). Microelectrode analysis of transfer of visual information by the corpus callosum. *Arch. ital. Biol.* **105**, 583–596.

BERLUCCHI, G. & RIZZOLATTI, G. (1968). Binocularly driven neurons in visual cortex of split-chiasm cats. *Science* **159**, 308–310.

BILGE, M., BINGLE, A., SENEVIRATNE, K. N. & WHITTERISGE, D. (1967). A map of the visual cortex in the cat. *J. Physiol.* **191**, 116–117P.

BISHOP, G. H. & CLARE, M. H. (1955). Organization and distribution of fibres in the optic tract of the cat. *J. comp. Neurol.* **103**, 269–304.

BIZZI, E. (1966). Changes in the orthodromic and antidromic response of optic tract during eye movements of sleep. *J. Neurophysiol.* **29**, 861–870.

BLAKEMORE, C. (1969). Binocular depth discrimination and the nasotemporal division. *J. Physiol.* **205**, 471–497.

BLAKEMORE, C. (1970). The representation of three-dimensional visual space in the rat's striate cortex. *J. Physiol.* **209**, 155–178.

BLAKEMORE, C. (1970). The range and scope of binocular depth discrimination in man. *J. Physiol.* **211**, 599–622.

BLAKEMORE, C. & PETTIGREW, J. D. (1970). Eye dominance in the visual cortex. *Nature* **225**, 426–429.

V. BONIN, G. N., GAROL, H. W. & MCCULLOCH, W. S. (1942). The functional organization of the occipital lobe. *Biol. Symp.* **7**, 165–192.

BOUGH, E. W. (1970). Stereoscopic vision in the macaque monkey: a behavioural demonstration. *Nature* **225**, 42–44.

BOUMA, H. & ANDRIESSEN, J. J. (1968). Perceived orientation of isolated line segments. *Vision Res.* **8**, 493–507.

BOWDITCH, H. P. & HALL, G. S. (1880–82). Optical illusions of motion. *J. Physiol.* **3**, 297–307.

BROWN, K. T. (1962). Complete interocular transfer of an adaptation process responsible for perceptual fluctuations with an ambiguous visual figure. *Vision Res.* **2**, 469–475.

BURIAN, H. M. (1951). Stereopsis. *Doc. Ophthal.* **5–6**, 169–183.

CAMPBELL, F. W., CLELAND, B. G., COOPER, G. F. & ENROTH-CUGELL, C. (1968). The angular selectivity of visual cortical cells to moving gratings. *J. Physiol.* **198**, 237–250.

CAMPBELL, F. W., COOPER, G. F. & ENROTH-CUGELL, C. (1969). The spatial selectivity of the visual cells of the cat. *J. Physiol.* **203**, 223–235.

CHOUDHURY, B. P. & WHITTERIDGE, D. (1965). Visual field projections on the dorsal nucleus of the lateral geniculate body. *Quart. J. exp. Physiol.* **50**, 104–111.

CHOUDHURY, B. P., WHITTERIDGE, D. & WILSON, M. E. (1965). The function of the callosal connections of the visual cortex. *Quart. J. exp. Physiol.* **50**, 214–219.

CLARK, W. E. LEGROS. (1959). The anatomy of cortical vision. *Trans. Ophthal. Soc. U.K.* **79**, 455–461.

COBB, W. A. & DAWSON, G. D. (1960). The latency and form in man of the occipital potentials evoked by bright flashes. *J. Physiol.* **152**, 108–121.

CREED, R. S. (1935). Observations on binocular fusion and rivalry. *J. Physiol.* **84**, 381–391.

CRONE, R. A. & VERDUYN LUNEL, H. F. E. (1969). Autokinesis and the perception of movement: the physiology of eccentric fixation. *Vision Res.* **9**, 89–101.

CURTIS, H. J. (1940). Intercortical connections of corpus callosum as indicated by evoked potentials. *J. Neurophysiol.* **3**, 407–413.

DANIEL, P. M. & WHITTERIDGE, D. (1959). The representation of the visual field on the calcarine cortex of baboons and monkeys. *J. Physiol.* **148**, 33–34P.

DE VALOIS, R. L. (1960). Colour vision mechanisms in the monkey. *J. gen. Physiol.* **43**, No. 6 (Suppl.), 115–128.

DE VOE, G. G., RIPPS, H. & VAUGHAN, H. G. (1968). Cortical responses to stimulation of the human fovea. *Vision Res.* **8**, 135–147.

DENNEY, D., BAUMGARTNER, G. & ADORJANI, C. (1968). Responses of cortical neurones to stimulation of the visual afferent radiations. *Exp. Brain Res.* **6**, 265–272.

DONCHIN, E. & COHEN, L. (1970). Evoked potentials to stimuli presented to the suppressed eye in a binocular rivalry experiment. *Vision Res.* **10**, 103–105.

DOTY, R. W. (1961). Remarks on the optic-tectum. In *The Visual System.* Ed. Jung & Kornhuber, pp. 215–217. Berlin: Springer.

DOWNER, J. L. DE C. (1959). Changes in visually guided behaviour following midsagittal division of optic chiasma and corpus callosum in monkey. *Brain* **82**, 251–259.

EBNER, F. F. & MYERS, R. E. (1965). Distribution of corpus callosum and anterior commissure in cat and raccoon. *J. comp. Neurol.* **124**, 353–365.

ENGEL, G. R. (1967). The visual process underlying binocular brightness summation. *Vision Res.* **7**, 753–767.

ENNEVER, J., GARTSIDE, I. B., LIPPOLD, O. C. J., NOVOTNY, G. E. K. & SHAGASS, C. (1967). Contamination of the human cortical evoked response with potentials of intra-orbital origin. *J. Physiol.* **191**, 5–6 P.

ETTLINGER, G. (1958). Visual discrimination following successive temporal excisions in monkeys. *J. Physiol.* **140**, 38–39P.

FELDMAN, M. & COHEN, B. (1968). Electrical activity in the lateral geniculate body of the alert monkey associated with eye movements. *J. Neurophysiol.* **31**, 455–466.

FENDER, D. & JULESZ, B. (1967). Extension of Panum's fusional area in binocularly stabilized vision. *J. opt. Soc. Am.* **57**, 819–830.

FILLENZ, M. (1961). Binocular interaction in the lateral geniculate body of the cat. In *The Visual System.* Ed. Jung & Kornhuber, pp. 110–117. Berlin: Springer.

GANZ, L. & FITCH, M. (1968). The effect of visual deprivation on perceptual behavior. *Exp. Neurol.* **22**, 638–660.

GAREY, L. J. (1965). Interrelationships of the visual cortex and superior colliculus in the cat. *Nature* **207**, 1410–1411.

GLEES, P. (1961). Terminal degeneration and trans-synaptic atrophy in the lateral geniculate body of the monkey. In *The Visual System.* Ed. Jung & Kornhuber, pp. 104–110. Berlin: Springer.

GLEES, P. & CLARK, W. E. LEGROS. (1941). The termination of optic fibres in the lateral geniculate body of the monkey. *J. Anat.* **75**, 295–308.

GLICKSTEIN, M., MILLER, J. & SMITH, O. A. (1964). Lateral geniculate nucleus and cerebral cortex: evidence for a crossed pathway. *Science* **145**, 159–161.

GRAHAM, C. H. (1965). Perception of movement. *In Vision and Visual Perception,* pp. 575–588. Ed. C. H. Graham. New York: Wiley.

GRAYBIEL, A. & HUPP, D. (1946). The oculogyral illusion: a form of apparent motion which may be observed following stimulation of the semicircular canals. *J. aviat. Med.* **17**, 3–27.

GROSVENOR, T. (1959). Eye movements and the autokinetic illusion. *Am. J. Optom.* **36**, 78–87.

GUILLERY, R. W. (1967). Patterns of fiber degeneration in the dorsal lateral geniculate nucleus of the cat following lesions in the visual cortex. *J. comp. Neurol.* **130**, 197–213.

HAMILTON, C. R. & GAZZANIGA, M. S. (1964). Lateralization of learning of colour and brightness discriminations following brain bisection. *Nature* **201**, 220.

HELSON, H. (1938). Fundamental problems in color vision. I. *J. exp. Psychol.* **23**, 439–476.

HILL, R. M. (1966). Receptive field properties of the superior colliculus of the rabbit. *Nature* **211**, 1407–1409.

HOLMES, G. (1945). Organization of the visual cortex in man. *Proc. Roy. Soc. B* **132**, 348–361.

HUBEL, D. H. & WIESEL, T. N. (1961). Integrative action in the cat's lateral geniculate body. *J. Physiol.* **155**, 385–398.

HUBEL, D. H. & WIESEL, T. N. (1962). Receptive fields, binocular interaction and functional architecture in the cat's visual cortex. *J. Physiol.* **160**, 106–154.

HUBEL, D. H. & WIESEL, T. N. (1963). Receptive fields of cells in striate cortex of very young, visually inexperienced kittens. *J. Neurophysiol.* **26**, 994–1002.

HUBEL, D. H. & WIESEL, T. N. (1965). Binocular interaction in striate cortex of kittens reared with artificial squint. *J. Neurophysiol.* **28**, 1041–1059.

HUBEL, D. H. & WIESEL, T. N. (1967). Cortical and callosal connections concerned with the vertical meridian of visual fields in the cat. *J. Neurophysiol.* **30**, 1561–1573.

HUBEL, D. H. & WIESEL, T. N. (1968). Receptive fields and functional architecture of monkey striate cortex. *J. Physiol.* **195**, 215–243.

HUBEL, D. H. & WIESEL, T. N. (1969). Anatomical demonstration of columns in the monkey striate cortex. *Nature* **221**, 747–750.

HUBEL, D. H. & WIESEL, T. N. (1970). Stereoscopic vision in macaque monkey. *Nature* **225**, 41–42.

HUBEL, D. H. & WIESEL, T. N. (1970). The period of susceptibility to the physiological effects of unilateral eye closure in kittens. *J. Physiol.* **206**, 419–436.

HULL, E. M. (1968). Corticofugal influence in the macaque lateral geniculate nucleus. *Vision Res.* **8**, 1285–1298.

HUMPHREY, N. K. & WEISKRANTZ, L. (1967). Vision in monkeys after removal of the striate cortex. *Nature* **215**, 595–597.

INGVAR, D. H. & HUNTER, J. (1955). Influence of visual cortex on light impulses in the brain stem of the unanaesthetized cat. *Acta. physiol. scand.* **33**, 194–218.

JAMESON, D. & HURVICH, L. M. (1959). Note on factors influencing the relation between stereoscopic acuity and observation distance. *J. opt. Soc. Am.* **49**, 639.

JASPER, H., AJMONE-MARSAN, C. & STOLL, J. (1952). Corticofugal projections to the brain stem. *Arch. Neurol. Psychiat.* **67**, 155–171.

JUDD, D. B. (1940). Hue saturation and lightness of surface colors with chromatic illumination. *J. opt. Soc. Am.* **30**, 2–32.

JUDD, D. B. (1960). Appraisal of Land's work on two-primary color projections. *J. opt. Soc. Am.* **50**, 254–268.

JULESZ, B. (1964). Binocular depth perception without familiarity cues. *Science* **145**, 356–362.

KAUFMAN, L. (1964a). Suppression and fusion in viewing complex stereograms. *Am. J. Psychol.* **77**, 193–205.

KAUFMAN, L. (1964b). On the nature of binocular disparity. *Am. J. Psychol.* **77**, 393–402.

KAUFMAN, L. (1965). Some new stereoscopic phenomena and their implications for the theory of stereopsis. *Am. J. Psychol.* **78**, 1–20.

KAUFMAN, L. & PITBLADO, C. (1965). Further observations on the nature of effective binocular disparities. *Am. J. Psychol.* **78**, 379–391.

KEESEY, U. T. (1960). Effects of involuntary eye movements on visual acuity. *J. opt. Soc. Am.* **50**, 769–774.

KLUVER, H. & BUCY, P. C. (1939). Preliminary analysis of functions of the temporal lobes in monkeys. *Arch. Neurol. Psychiat.* **42**, 979–1000.

KRNJEVIĆ, K., RANDIĆ, M. & STRAUGHAN, D. W. (1966). Pharmacology of cortical inhibition. *J. Physiol.* **184**, 78–105.

KUYPERS, H. G. J. M., SZWARCBART, M. K., MISHKIN, M. & ROSVOLD, H. E. (1965). Occipito-temporal cortico-cortical connections in the rhesus monkey. *Exp. Neurol.* **11**, 245–262.

LAND, E. H. (1959). Color vision and the natural image. Pts. I and II. *Proc. Nat. Acad. Sci. Wash.* **45**, 115–129, 636–644.

LANSING, R. W. (1964). Electroencephalographic correlates of binocular rivalry in man. *Science* **146**, 1325–1327.

LAWSON, R. B. & GULICK, W. L. (1967). Stereopsis and anomalous contour. *Vision Res.* **7**, 271–297.

LAWWILL, T. & BIERSDORF, W. R. (1968). Binocular rivalry and visual evoked responses. *Invest. Ophthal.* **7**, 378–385.

LEIBOWITZ, H. & HARTMAN, T. (1959). Magnitude of the moon illusion as a function of the observer. *Science* **130**, 569–570.

LEVELT, W. J. M. (1965). Binocular brightness averaging and contour information. *Brit. J. Psychol.* **56**, 1–13.

LIT, A. (1960). The magnitude of the Pulfrich stereophenomenon as a function of target velocity. *J. exp. Psychol.* **59**, 165–175.

LUDVIGH, E. (1952). Control of ocular movements and visual interpretation of environment. *Arch. Ophthal.* **48**, 442–448.

MANDELBAUM, J. (1941). Dark adaptation. Some physiologic and clinical considerations. *Arch. Ophthal.* **26**, 203–239.

MATURANA, H. R. (1959). Number of fibres in the optic nerve and the number of ganglion cells in the retina of anurans. *Nature, Lond.* **183**, 1406–1407.

MCCOLGIN, F. H. (1960). Movement thresholds in peripheral vision. *J. opt. Soc. Am.* **50**, 774–779.

MEIKLE, T. H. (1964). Failure of interocular transfer of brightness discrimination. *Nature* **202**, 1243–1244.

MEIKLE, T. H. & SECHZER, J. A. (1960). Interocular transfer of brightness discrimination in "split-brain" cats. *Science* **132**, 734–735.

MERTON, P. A. (1956). Compensatory rolling movements of the eye. *J. Physiol.* **132**, 25–27 P.

MITCHELL, D. E. (1966). Retinal disparity and diplopia. *Vision Res*, **6**, 441–451.

MITCHELL, D. E. & BLAKEMORE, C. (1970). Binocular depth perception and the corpus callosum. *Vision Res.* **10**, 49–54.

MONTERO, V. M. & BRUGGE, J. F. (1969). Direction of movement as the significant stimulus parameter for some lateral geniculate cells in the rat. *Vision Res.* **9**, 71–88.

MORUZZI, G. & MAGOUN, H. W. (1949). Brain stem reticular formation and activation of the EEG. *EEG clin. Neurophysiol.* **1**, 455–473.

MOUNTCASTLE, V. B. (1957). Modality and topographic properties of single neurones of cat's somatic sensory cortex. *J. Neurophysiol.* **20**, 408–434.

MYERS, R. E. (1962). Commissural connections between occipital lobes of the monkey. *J. comp. Neurol.* **118**, 1–10.

NIKARA, T., BISHOP, P. O. & PETTIGREW, J. D. (1968). Analysis of retinal correspondence by studying receptive fields of binocular single units in cat striate cortex. *Exp. Brain Res.* **6**, 353–372.

NOBLE, J. (1966). Mirror-images and the forebrain commissures of the monkey. *Nature* **211**, 1263–1266.

OGLE, K. N. (1953). Precision and validity of stereoscopic depth perception from double images. *J. opt. Soc. Am.* **43**, 906–913.

OGLE, K. N. (1958). Note on stereoscopic acuity and observation distance. *J. opt. Soc. Am.* **48**, 794–798.

OGLE, K. N. & WEIL, M. P. (1958). Stereoscopic vision and the duration of the stimulus. *Arch. Ophthal.* **59**, 4–17.

PANUM, P. L. (1858). *Physiologische Untersuchungen uber das Sehen mit zwei Augen.* Kiel.

PASIK, P. & PASIK, T. (1964). Oculomotor functions in monkeys with lesions in the cerebrum and the superior colliculi. In *The Oculomotor System*, pp. 40–80. Ed. M. B. Bender. New York: Harper & Row.

PASIK, P., PASIK, T. & SCHILDER, P. (1969). Extrageniculostriate vision in the monkey: discrimination of luminous flux-equated figures. *Exp. Neurol.* **24**, 421–437.

PENMAN, G. G. (1934). The representation of the areas of the retina in the lateral geniculate body. *Trans. Ophthal. Soc. U.K.* **54**, 232–270.

PETTIGREW, J. D., NIKARA, T. & BISHOP, P. O. (1968). Binocular interaction on single units in cat striate cortex: simultaneous stimulation by single moving slit with receptive fields in correspondence. *Exp. Brain Res.* **6**, 391–410.

SECHZER, J. A. (1963). Successful interocular transfer of pattern discrimination in "split-brain" cats with shock-avoidance motivation. *J. comp. physiol. Psychol.* **58**, 76–83.

SMITH, S. (1946). The essential stimuli in stereoscopic depth perception. *J. exp. Psychol.* **36**, 518–521.

SMITH, S. (1949). A further reduction of sensory factors in stereoscopic depth perception. *J. exp. Psychol.* **39**, 393–394.

SPALDING, J. M. K. (1952). Wounds of the visual pathway. I. The visual radiation. *J. Neurol.* **15**, 99–107.

SPALDING, J. M. K. (1952). Wounds of the visual pathway. II. The striate cortex. *J. Neurol.* **15**, 169–183.

SPERRY, R. W., STAMM, J. S. & MINER, N. (1956). Relearning tests for interocular transfer following division of optic chiasma and corpus callosum in cats. *J. comp. physiol. Psychol.* **49**, 529–533.

SPINELLI, D. N. & BARRETT, T. W. (1969). Visual receptive field organization of single units in the cat's visual cortex. *Exp. Neurol.* **24**, 76–98.

SPRAGUE, J. M. (1966). Interaction of cortex and superior colliculus in mediation of visually guided behaviour in the cat. *Science* **153**, 1544–1547.

SPRAGUE, J. M. & MEIKLE, T. H. (1965). The role of the superior colliculus in visually guided behaviour. *Exp. Neurol.* **11**, 115–146.

SUZUKI, H. & KATO, H. (1969). Neurons with visual properties in the posterior group of the thalamic nuclei. *Exp. Neurol.* **23**, 353–365.

TALBOT, S. A. & MARSHALL, W. H. (1941). Physiological studies on neural mechanisms of visual localization and discrimination. *Am. J. Ophthal.* **24**, 1255–1264.

TEICHNER, W. H., KOBRICK, J. L. & WEHRKAMP, R. F. (1955). The effects of terrain and observation distance on relative depth discrimination. *Am. J. Psychol.* **68**, 193–208.

TEN DOESSCHATE, G. (1955). Results on an investigation of depth perception at a distance of 50 metres. *Ophthalmologica, Basel* **129**, 56–57.

TREISMAN, A. (1962). Binocular rivalry and stereoscopic depth perception. *Quart. J. exp. Psychol.* **14**, 23–37.

URBAITIS, J. C. & MEIKLE, T. H. (1968). Relearning a dark-light discrimination by cats after cortical and collicular lesions. *Exp. Neurol.* **20**, 295–311.

VAN HOF, M. W. (1969). Discrimination of striated patterns of different orientation in rabbits deprived of light after birth. *Exp. Neurol.* **23**, 561–565.

VAN HOF, H. W. (1970). Interocular transfer in the rabbit. *Exp. Neurol.* **26**, 103–108.

VASTOLA, E. F. (1960). Monocular inhibition in the lateral geniculate body. *E.E.G. clin. Neurophysiol.* **12**, 399–403.

VAUGHAN, H. G. (1966). The perceptual and physiologic significance of visual evoked responses recorded from the scalp in man. In *Clinical Electroretinography*, pp. 203–223. Ed. H. M. Burian & J. H. Jacobson. New York: Pergamon.

VOLKMANN, A. W. (1863). *Physiologische Untersuchungen in Gebiet der Optik.* Leipzig.

VONEIDA, T. J. & ROBINSON, J. S. (1970). Effects of brain bisection on capacity for cross comparison of patterned visual input. *Exp. Neurol.* **26**, 60–71.

WALLS, G. L. (1951). A theory of ocular dominance. *Arch. Ophthal.* **45**, 387–412.

WERTHEIMER (1912). Quoted by Graham (1965).

WIDÉN, L. & AJMONE-MARSAN, C. (1961). Action of afferent and corticofugal impulses on single elements of the dorsal lateral geniculate nucleus. In *The Visual System*. Ed. Jung & Kornhuber, pp. 125–133. Berlin: Springer.

WIESEL, T. N. & HUBEL, D. H. (1963a). Effects of visual deprivation on morphology and physiology of cells in cat's lateral geniculate body. *J. Neurophysiol.* **26**, 978–993.

WIESEL, T. N. & HUBEL, D. H. (1963b). Single cell responses in striate cortex of kittens deprived of vision in one eye. *J. Neurophysiol.* **26**, 1003–1017.

WIESEL, T. N. & HUBEL, D. H. (1965). Comparison of the effects of unilateral and bilateral eye closure on cortical unit responses in kittens. *J. Neurophysiol.* **28**, 1029–1040.

WIESEL, T. N. & HUBEL, D. H. (1966). Spatial and chromatic interactions in the lateral geniculate body of the rhesus monkey. *J. Neurophysiol.* **29**, 1115–1156.

WILSON, M. E. & CRAGG, B. G. (1957). Projections from the lateral geniculate nucleus in the cat and monkey. *J. Anat.* **101**, 677–692.

WURTZ, R. H. (1969). Comparison of effects of eye movements and stimulus movements in striate cortex of the monkey. *J. Neurophysiol.* **32**, 987–994.

ZAJAC, J. L. (1960). Convergence, accommodation and visual angle as factors in perception of size and distance. *Am. J. Psychol.* **73**, 142–146.

Visual Optics

VISUAL OPTICS

THE science of geometrical optics consists in the study of the passage of light through various systems; it assumes as a fundamental proposition that light travels in straight lines. This, of course, is only partially true, the phenomena of diffraction providing instances of the non-rectilinear propagation of light. From the point of view of physiological optics, however, we need only consider these deviations from the point of view of the effect of pupil size on the resolving power of the eye, all other aspects being sufficiently accurately dealt with on the basis of this primary assumption—the rectilinear propagation of light. Light, passing through a medium, may be treated as an infinite number of rays which may be represented as straight lines; accurately defined, a ray of light is the straight line between two points in a uniform medium along which light is propagated.

REFLEXION

Plane Mirrors

When light strikes a surface bounding two media, a part is reflected at the surface. The Law of Reflexion states that the reflected ray lies in the same plane as the incident ray and the normal to the reflecting surface, and that the two rays make equal angles with the normal, i.e. the *angle of incidence*, i, is equal to the *angle of reflexion*, r (Fig. 335).

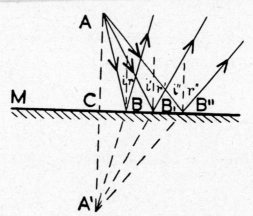

FIG. 335. Reflexion at a plane surface. Point object.

If we consider a luminous point, A, in front of a plane mirror, we may represent the passage of light in the plane of the paper as in Fig. 335, where M is a section of the mirror made by a plane through A at right angles to it (the mirror would thus project out of the paper at right-angles). Rays of light from A may be represented as straight lines, striking the mirror at B, B′, B″, etc. To determine the course of the reflected rays we must erect normals at B, B′, B″, etc. The angles i, i', i'', etc. are the angles of incidence and, by drawing reflected rays to make angles of reflexion r, r', r'', etc. equal to the angles of incidence, the course of the reflected light is obtained.

It will be seen that the reflected rays diverge from each other, and that if they are produced backwards through the mirror they meet at a point A'; to an eye placed in front of the mirror it appears that the reflected rays are actually emanating from this point so that A' is called the *image* of A. It can be shown geometrically that A' is as far behind the mirror as A is in front of it, i.e. AC = A'C. This last fact gives us a rapid means of constructing the image of an object formed by a plane mirror, and of tracing the pencils of light during reflexion. Let us consider an object, the arrow AB of Fig. 336, *a*, in front of a plane mirror,

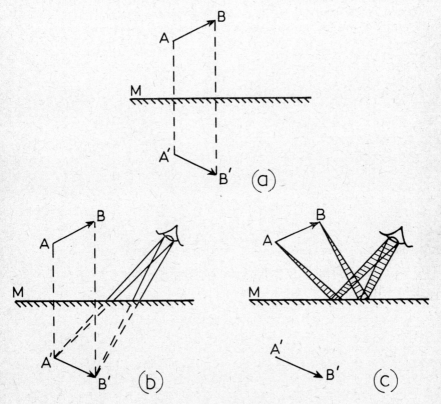

FIG. 336. Reflexion at a plane surface. Extended object. Construction of the pencils of light entering the eye.

M. The position of the image is obtained at once by drawing perpendiculars from A and B to the mirror and producing them behind an equal distance. To the eye, in front of the mirror, all the rays from A and B appear to diverge from A' and B' so that the reflected rays entering the eye may be obtained by drawing lines from A' and B' to the extremes of the pupil of the eye (Fig. 336, *b*); the actual course of the light from A and B to the eye may then be drawn in (Fig. 336, *c*).

A' is said to be the image of A because *all* rays from A appear to emanate from A'. It is referred to as a *virtual image* because the rays only *appear* to diverge from it—the eye is unable to distinguish between a real object at A' or a virtual image, since in both cases the rays of light striking the eye diverge in an identical manner. It will be seen that the image of AB is laterally inverted; thus an observer looking directly at AB would see the head of the arrow on his right; on looking at the image he would see the tail on his right.

Rotation of the Mirror. It is important in retinoscopy to know the effects of rotating the mirror on the position of the image. If M is the mirror (Fig. 337) and A a point source of light, the image A' is obtained as before by dropping a perpendicular and measuring off an equal distance behind. The mirror is now tilted upwards through an angle θ, and the new position of the image, A'', is obtained in the same way. It is easy to show that the line joining the image to the centre of rotation moves through twice the angle of rotation of the mirror.

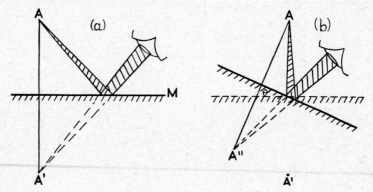

FIG. 337. Effect of rotation of the mirror.

Spherical Mirrors

Construction of the Image. A reflecting surface having the form of a portion of a sphere is termed a spherical mirror (Fig. 338). The *centre of curvature*, C, is the centre of the sphere of which the mirror forms a part; the *pole*, P, is the middle point of the reflecting surface. To represent the path of rays in a single plane we must consider a section of the mirror, and this is chosen so that it passes through the pole and the centre of curvature; it is called a *principal section*. The straight line passing through the centre of curvature and the pole is termed the *principal axis*, to distinguish it from any other line passing through the centre of curvature and striking the mirror, which may be called a *subsidiary axis*.

In Fig. 338, *a*, M is the principal section of a concave mirror, P is the pole, C the centre of curvature, and NP the principal axis. Any ray from a luminous point, O, on the axis, e.g. OA, strikes the mirror and is reflected in accordance with the fundamental law of reflexion. The normal at A is given by the line CA, which is a radius of the sphere of which the mirror forms a part, so that the angle OAC equals *i*, the angle of incidence. The reflected ray is given by drawing AI such that the angle CAI equals OAC; it strikes the principal axis in I. Now it can be proved that all rays from O (with certain reservations to be discussed later) pass through I after reflection at the mirror; consequently, to an eye in front of the mirror, the reflected rays appear to come from I, the image of O. Since the rays after reflexion *do* actually pass through I, the image is said to be real.

A similar construction may be applied to a convex mirror as in Fig. 338, *b*; the reflected rays, in this case, diverge from the axis instead of cutting it, and it may be shown that they all diverge from the same point I behind the mirror. I is thus the virtual image of the point O.

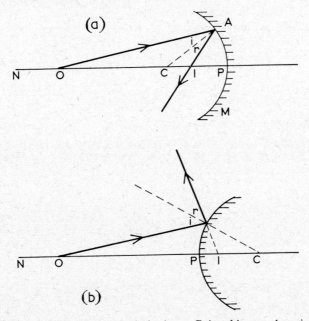

FIG. 338. Reflexion at spherical mirrors. Point object on the axis.

Position of the Image. For any position of the object, the position of its image formed by a spherical mirror can be calculated with the aid of the formula:

$$1/v + 1/u = 2/r \qquad (1)$$

where u is the distance of the object from the mirror,
   v is the distance of the image from the mirror,
   r is the radius of curvature of the mirror.

In applying this formula, a convention as to the signs to be applied to the various distances must be adopted; the convention in common use is the following. All distances are measured from the pole of the mirror to the point in question; if, on passing from the pole to the point, one moves *with* the direction of the incident light the distance is said to be positive; if one moves *against* the direction of the incident light the distance is negative. Thus in Fig. 338, *a*, the radius of curvature, r, is measured from P, the pole, to the centre of curvature, C; the distance is measured against the direction of the incident light and r is given a negative value, e.g.—10 cm. Similarly the distances of the object and image, u and v, are negative. With the convex mirror it will be seen that r

is positive. Conventionally, light is made to come from the left, so that all distances to the right of the mirror are positive and all to the left are negative.

Example. An object is placed 20 cm in front of a concave mirror of radius 10 cm. Where is the image?

Here u is —20 cm and r is —10 cm.

$$1/v - 1/20 = -2/10$$
$$\text{Whence } v = -6.66 \text{ cm}$$

i.e. the image is 6·66 cm in front of the mirror.

Principal Focus. If the luminous object is placed on the principal axis, as before, but very far away indeed from the mirror, i.e. at infinity, the position of the image may be found by substituting in Equation (1):

$$1/v + 1/\infty = 2/r$$
$$\text{i.e. } 1/v = 2/r$$
$$v = r/2.$$

In words, the distance of the image from the mirror is half the radius of curvature. This position of the image is called the *principal focus* of the mirror, and its distance from the pole of the mirror is said to be the *focal length*, f, which is equal to half the radius.

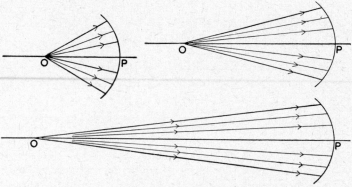

FIG. 339. Illustrating the decreasing divergence of the rays from a point O as the latter is moved farther and farther away from the mirror.

If we consider the rays emanating from a luminous point at various distances from a mirror, as in Fig. 339, we note that when they reach the mirror they diverge from each other, and from the axis, to different extents; the farther the object from the mirror the less the divergence, so that in the limiting case of an object infinitely far away the rays do not diverge but are parallel. Now we have just defined the principal focus of the mirror as the position of the image of a point, infinitely far away on the axis, hence *all rays parallel to the axis pass through the principal focus.* In Fig. 340 the reflexion of rays from a point infinitely far away on the axis, by a convex and a concave mirror, is shown; it will be noted that, in the case of the convex mirror, the reflected rays *appear* to come from the principal focus.

Object Not on the Axis. The position of the image of a point off the axis may be calculated with the mirror formula, but in this event the distances

must be measured along a subsidiary axis. Thus if O is the object (Fig. 341, *a*)
we may draw an axis through O and C, the centre of the mirror; the image
will be formed, say, at I on the axis. When the object is at infinity, but in this
case not on the principal axis, the rays will be parallel with each other but not
parallel to the axis. The distance of the image will be given by the same formula,
i.e. it will be $r/2$, but it will be on the subsidiary axis and so not at the principal
focus. It will be in the *focal plane* of the mirror, i.e. a plane through the principal
focus at right angles to the principal axis. In general we may say that parallel
rays converge on the focal plane of a concave mirror (or diverge from the focal
plane of a convex mirror); when the rays are parallel with the principal axis
they converge on the principal focus; when the rays are parallel, but inclined
to the principal axis, they converge on a point in the focal plane, and this point
is given by the intersection of the subsidiary axis with this plane (Fig. 341, *b*).

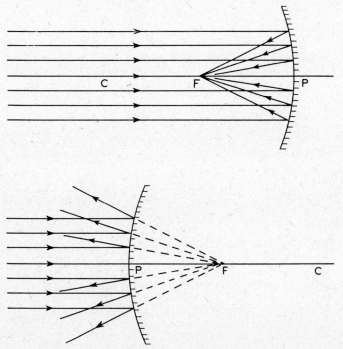

FIG. 340. The principal focus.

Example. A concave mirror of radius 100 cm is placed vertically; a star is
reflected in the mirror. What is the position of the image of the star, if a line
joining the centre of curvature of the mirror to the star makes an angle of 30°
with the horizontal?

The state of affairs is similar to that represented in Fig. 341, *b*. The star is
infinitely far away for optical purposes, on the subsidiary axis OCB, making an
angle of 30° with the principal axis, which is horizontal. The rays of light from
the star are parallel and come to a focus at a distance of $r/2$ from the mirror
on the subsidiary axis, i.e. at I, 50 cm from the mirror.

Finite Object. So far we have considered the images of point objects; an object of finite size can be treated as a series of points, and its image may be found as above, but it is simpler to make use of certain principles already derived. Thus in Fig. 342, *a*, AB is the object, and we wish to find the position of the image. It has already been shown that all rays from any point on the object will meet at a single point on the image; hence if we can follow out two rays from a point on the object, and find their point of intersection, we can find the position of the image of this point. Now a ray from A parallel to the principal axis is reflected back through the principal focus (ADF). Another ray from A, through the centre of curvature of the mirror, is reflected back along its own path, since

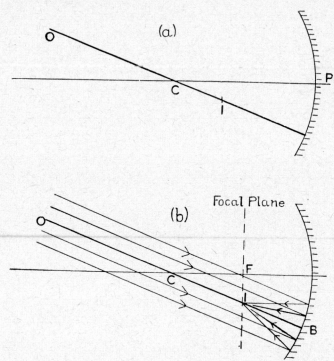

FIG. 341. Reflexion at concave mirror. (*a*) Point object on a subsidiary axis, OC. (*b*) Point object infinitely far away on the same subsidiary axis.

it strikes the mirror normally. It intersects the first ray at A′ which is thus the image of A. Similarly by drawing rays parallel to the axis and through the centre of curvature from B, the position of B′ can be established, and thus the image of the object AB. The image in this case is inverted, real, and diminished. The construction for a convex mirror is indicated in Fig. 342, *b*; here the two rays diverge after reflexion and must be produced back to give the position of the virtual image.

The *linear magnification* for spherical mirrors is given by the ratio: *Size of Image/Size of Object*, and it is equal to *v/u*. If we extend our sign convention so that distances measured above the axis are positive and distances below negative, then the ratio: *Size of Image/Size of Object*, will be positive in the case of an erect image and negative in the case of an inverted one. To make the magnifi-

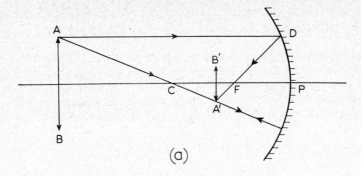

(a)

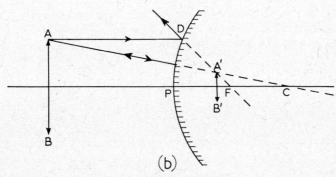

(b)

FIG. 342. Image of a finite object formed by reflexion at concave (a) and convex (b) mirrors.

cation formula fit in with this convention, a negative sign must be put in front of the ratio v/u:

$$Magnification = i/o = -v/u \qquad (2)$$

A Note on Real and Virtual Images. The image formed by a plane or convex mirror is defined as a virtual image because the reflected rays from the mirror *appear* to come from it. The image, A'B', formed by the concave mirror of Fig. 342, *a*, is defined as real because the reflected rays actually meet in a

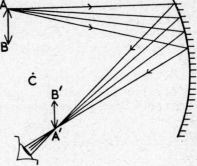

FIG. 343. Showing how a real image may be viewed by the eye.

point at A′ and at B′ and, of course, at intermediate points on the line A′B′. If a screen is placed at A′B′ a clear image of AB will be formed on it because all light from A, for example, reflected by the mirror, comes to a point at A′ and no light from other parts of the object overlaps it. If the image is viewed directly, it will be seen distinctly, as with a virtual image, provided that the eye is not too close; this is because the rays, converging on A′ for example, cross and start to diverge as in Fig. 343, so that the eye is presented with rays diverging from A′ and treats the latter as a definite luminous point in space.

REFRACTION AT PLANE SURFACES

By refraction is meant the change in direction of light (or bending of the rays) which takes place when it passes from one transparent medium to another. Thus if PQ (Fig. 344, *a*) is the surface separating air above it from water below, the ray AB is refracted, or bent towards the normal at B; we may note, incidentally, that the ray AB gives rise to two rays on striking the surface, a reflected ray BD and the refracted ray BC. The change in direction of the ray on refraction takes place at the surface of separation.

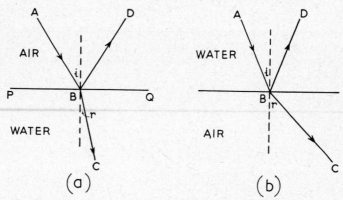

FIG. 344. Refraction and reflexion at an air-water (*a*) and a water-air interface (*b*).

According to whether the refracted ray is bent towards or away from the normal, the medium in which refraction occurs is said to be more or less *optically dense* than the medium in which the incident ray lies. Fig. 344, *a*, represents the passage of light from a medium of less to a medium of greater density, and Fig. 344, *b*, represents the reverse case, where the refracted ray is bent *away* from the normal. The angles made by the incident and refracted rays with the normal are called respectively the *angles of incidence* (*i*) and *refraction* (*r*).

The law defining refraction is called *Snell's Law*. It states that the refracted ray lies in the plane containing the incident ray and the normal, and that the angles of incidence and refraction are connected by the equation:

$$\text{Sin } i/\text{Sin } r = n \tag{3}$$

where *n* depends on the optical characteristics of the two media through which the light passes and is called the *index of refraction*. The index of refraction at the air-water interface should be written: $_{air}n_{water}$ to indicate that the light passes from air to water.

The term *refractive index* is used for the special case where the first medium is a vacuum, i.e. the index of refraction at a vacuum-water interface is called the refractive index of water. The index of refraction is thus a characteristic of two media in contact, whilst the refractive index is a characteristic of a single medium. The optical density of air is very little different from that of a vacuum, so that the index of refraction of an air-water interface, for example, is generally treated as though the air were a vacuum, whence the index of refraction is taken as the refractive index of water.

It is often important to be able to derive the index of refraction of an interface from the refractive indices of the two media, e.g. if we have crown glass of refractive index 1·52 and flint glass of refractive index 1·66, we want to know how light will behave on passing from crown to flint glass, i.e. we want to known the index of refraction: $_{Crown}n_{Flint}$

The answer is given by the ratio:

$$\frac{n_{Flint}}{n_{Crown}}$$

where n_{Flint} and n_{Crown} are the refractive indices of the respective glasses. More generally, on passing from medium (1) to medium (2) the index of refraction: $_1n_2$ is given by the ratio: n_2/n_1. It follows that if we are given the index of refraction for passing from one medium to another, the index of refraction for passing in the reverse direction is the reciprocal of this value; i.e. $_2n_1 = 1/_1n_2$.

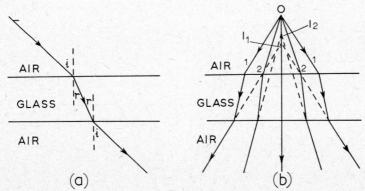

FIG. 345. (*a*) Behaviour of a single ray on passing through a parallel-sided slab of glass. (*b*) Behaviour of several rays.

Parallel-Sided Plate. In Fig. 345, *a*, is shown the passage of a ray of light from air through a parallel-sided plane slab of glass. On passing into the glass the ray is bent towards the normal in accordance with Snell's Law; on emerging it is bent away from the normal to an equal extent, so that it emerges parallel with its original direction. In Fig. 345, *b*, the refraction of a pencil of rays from a luminous point is shown; it will be seen that there is no perfect image of a point, the rays (1) diverge, after refraction, from I_1 and the rays (2) from I_2. These effects of refraction are, however, not large, and with a window-pane, for example, an almost perfect image is obtained which appears to be at the same position as that of the object.

To an eye viewing the object obliquely, there is a lateral displacement of the image; this may be shown experimentally by viewing an iron bar through a thick glass plate so that part of the bar may be seen as an image through the plate whilst the rest is seen directly. If the line of vision is normal to the surface of the plate, the bar appears continuous—the image and object coincide in direction. With oblique vision, however, that part of the bar seen through the plate glass is shifted laterally (Fig. 346).

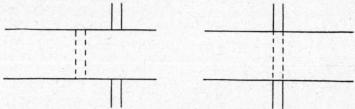

FIG. 346. Displacement of bar by parallel-sided plate glass with oblique vision.

When the eye observes an object *in* a different medium, the image appears to be closer to the eye than the object actually is, if the medium surrounding it is more dense than that surrounding the eye; e.g. a coin at the bottom of a tank of water appears closer to the eye than it actually is; the formation of the image in this case is shown in Fig. 347. If the pencil of rays is small it appears, after refraction, to diverge from a single point.

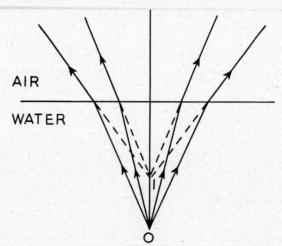

FIG. 347. The object O, viewed by an eye in air, appears to be at I.

Total Internal Reflexion. If the refraction, at a water-air interface, of rays emanating from a luminous point in the water is constructed as in Fig. 348, it is found that as the angle of incidence is made greater and greater the emergent rays are inclined less and less to the surface until a point is reached where the emergent ray just grazes it; if the angle of incidence is increased beyond this

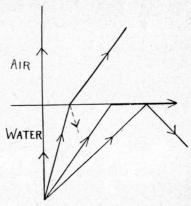

FIG. 348. Total internal reflexion.

value no emergent ray can be constructed, and the ray is said to be *totally internally reflected*; in other words, when this limiting value of the angle of incidence is exceeded, all the light obeys the simple law of reflexion instead of a portion of it being refracted and a portion reflected. The limiting angle of incidence is called the *critical angle*, which is related to the index of refraction of the surface by the equation:

$$\text{Sin } i = {}_{water}n_{air}$$
$$= 1/{}_{air}n_{water}$$
$$= 1/n_{water} \qquad (4)$$

i.e. the sine of the critical angle is equal to the reciprocal of the refractive index of the medium.

The critical angle for a water-air interface is about $49°$; for crown glass and air it is about $42°$.

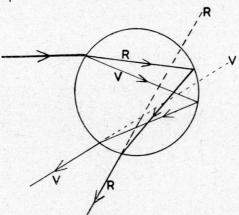

FIG. 349. Illustrating the formation of the primary rainbow.

The formation of rainbows results from total internal reflexion in raindrops; they are seen when the sun shines on falling rain and the observer must have his back to the sun; in favourable circumstances several bows may be seen—the brightest, the primary bow, is red on its outer edge and violet on its inner, and results from rays that have been refracted into and out of a drop of water, suffering a total internal reflexion between the two refractions as in Fig. 349. Because water has a different refractive index for each wavelength of light, the emergent rays from the raindrops make different angles to the eye according to their wavelength. It can be shown that, although the light emerges from many different raindrops in different positions, the final effect must be to make the red rays, for example, appear to come from a different point in the sky than that from which the violet rays appear to come. The secondary bow is formed by rays that have been twice internally reflected by the raindrops.

Prisms

A prism is defined for the purposes of geometrical optics as a figure bounded by three planes intersecting in three parallel straight lines.

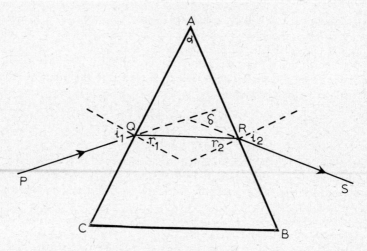

FIG. 350. Passage of a ray through a prism. α is the refracting angle; A the refracting edge; δ the angle of deviation.

The passage of a ray of light through a prism may be constructed on the basis of Snell's Law as in Fig. 350 where the triangle ABC represents a section through a prism; the edge A is called the *refracting edge* and the angle α the *refracting angle*. It is clear that the ray emerging from the prism has been altered in direction, having been bent towards the *base* of the prism, BC. The extent of this change of direction is measured by the *angle of deviation*, δ, formed by producing the incident ray, PQ, forwards and the emergent ray, RS, backwards.

The angle of deviation is determined by three factors:

(*a*) The refractive index of the material of the prism.
(*b*) The angle of the prism (α).
(*c*) The angle of incidence of the ray considered.

When the angle of the prism and the angle of incidence are small, the deviation is given by the equation:

$$\delta = (n-1)\alpha \tag{5}$$

If the prism is made of glass, of refractive index about 1·5, *the deviation is equal to half the refracting angle.*

For a given prism, the smallest value that the deviation may have is called the *angle of minimum deviation*; it can be shown that a ray suffering minimum deviation makes equal angles of incidence and emergence ($i_1 = i_2$; Fig. 350).

Formation of the Image. For a well-defined image of a point object, O, to be seen (Fig. 351) the rays after refraction must appear to diverge from a single point, I. With a prism of finite angle this is theoretically impossible; the nearest approach to perfect image formation is given by choosing the conditions such that the central ray of the pencil of light forming the image is deviated to a minimum. Ophthalmic prisms are designed so that, as far as is possible, the central rays of pencils are minimally deviated.

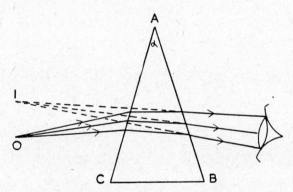

FIG. 351. Image formation by a prism.

Ophthalmic Use of Prisms. A prism may be used to correct a pathological deviation of one eye. Thus if, when the right eye (Fig. 352, *a*) converges on the point O, the left eye looks straight ahead, the images of the object O do not fall on the foveae of the two eyes and diplopia results. The interposition of a prism between the object and the left eye, however, deviates the rays from O so that they fall on the fovea, F. A prism used in this way to counteract a divergent squint is referred to as a "*base-in*" prism. A "*base-out*" prism is required to correct a convergent squint as in Fig. 352, *b*.

Notation of Prisms. The employment of prisms in ophthalmology has led to the use of a number of different methods of indicating the strength of a prism. From a practical point of view we are interested in the apparent displacement of an object caused by the prism; thus in Fig. 353 the object at O appears to be at I; it has been displaced through an angle δ and this angle can be used to indicate the prism's strength; alternatively the linear displacement, *a*, at a fixed distance may be used. On the *prism dioptre* or *tangent centune* system, a prism of one prism dioptre strength (1Δ) displaces an object, 1 metre away, through 1 cm. This corresponds approximately to an angle, θ, of $\frac{1}{2}°$ hence on the tangent centune system the angle of apparent displacement is measured in units of half a degree.

Ophthalmic prisms are used under conditions of minimum deviation, i.e. the central ray in the formation of the image is minimally deviated; the angle of minimum deviation can therefore be used as an index to the strength of an ophthalmic prism (expressed as $°d$), and it will be seen from Fig. 271 that δ and θ are approximately equal when the distance of the object from the prism is large (as occurs in practice). Since, however, the prism dioptre, used as the unit for measuring θ, is equivalent to half a degree, a prism of $5°d$ is equivalent approximately to one of $10\varDelta$.

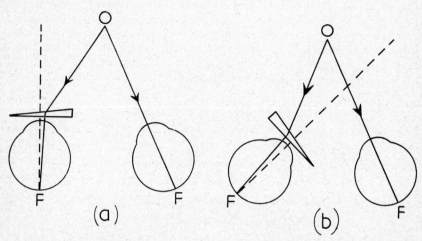

FIG. 352. (a) Use of base-in prism to compensate divergent squint. (b) Use of base-out prism to compensate convergent squint.

Under conditions of ophthalmic practice, equation (5) relating the refracting angle, α, to the deviation, δ, applies; consequently with glass prisms the strength can be indicated by the refracting angle, which is equal to twice the deviation. Thus a prism of refracting angle $10°$ (referred to as a ten-degree-prism) is equivalent to a $5°d$ prism, which, as we have seen, is equal to a $10\varDelta$ prism.

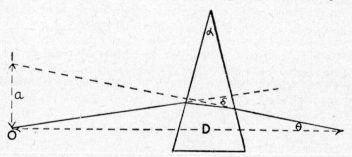

FIG. 353. The notation of prisms. α, refracting angle; δ, angle of minimum deviation; θ, angle of apparent deviation.

To make the confusion even worse the arc centune system is also employed; on this system the angle θ is indicated in units of a hundredth of a radian (∇); this unit, however, is approximately equal to half a degree so that the difference between the prism dioptre and arc centune, or *centrad*, as it is called, is negligible.

Example. An object at 6 metres appears to be displaced 5 cm on placing a prism in front of the eye. What is the strength of the prism in $°d$?

1 cm subtends 1 prism dioptre at 1 metre.

5 cm subtend 5 prism dioptres at 1 metre.

5 cm subtend 5/6 prism dioptres at 6 metres.

The apparent deviation is 5/6 prism dioptres.

The strength in $°d$ is thus 5/12.

REFRACTION AT A SINGLE SPHERICAL SURFACE

The behaviour of rays of light on passing from one medium to another, when the surface of separation is spherical, is of great importance not only for the understanding of the properties of a lens, which, as we shall see, constitutes a special case, but for the study of the eye as an optical instrument.

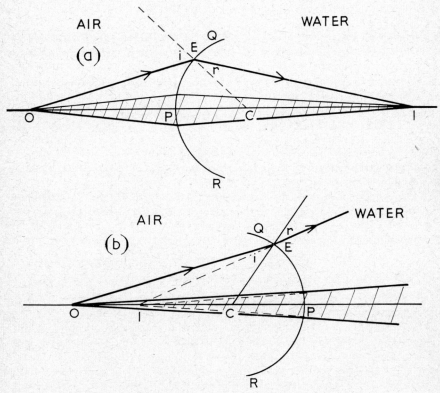

FIG. 354. Refraction at a spherical refracting surface.

In Fig. 354, *a*, PQR represents the spherical boundary between air and a more dense medium, e.g. water. The index of refraction of the surface, $_{air}n_{water}$, can be equated to the refractive index of water, n_{water}, or simply, n.

We may define the *centre*, C, *radius* of *curvature*, CP, *principal axis*, OPC, *pole*, P, and *principal section* in exactly the same way as for spherical mirrors.

If O is a luminous point on the axis, OE can be considered to be a ray emanating from it. This ray strikes the surface making an angle of incidence, i, with the normal (CE being a radius and therefore normal to the surface at the point E). The ray is refracted at E, being bent towards the normal, and strikes the axis at I. Now it may be shown by geometrical reasoning that, provided its angle of incidence is small, *any* ray from O will pass through the same point I; hence I is the real image of the point O provided that the rays from O make a small angle of incidence; this restriction may be conformed with by confining the rays crossing the surface to a thin, axial, pencil as illustrated in Fig. 354, *a*.

If the surface of separation is made concave to the air, the refracted ray OE diverges away from the axis; all other rays behave in the same manner, but so long as the pencil of rays is small they diverge from a definite point, I, the virtual image of the point O (Fig. 354, *b*).

For both convex and concave surfaces, the relative positions of image and object are given by the general formula:

$$n/v - 1/u = (n-1)/r \qquad (6)$$

where u is the distance of the object from the surface, v the distance of the image, and r is the radius of curvature; the sign convention must be observed as with spherical mirrors.

More generally, if the refractive index of the object-space is n_1 and that of the image-space is n_2, the formula becomes:

$$n_2/v - n_1/u = (n_2 - n_1)/r \qquad (6a)$$

Example. An object is placed 25 cm in front of a glass block which presents a convex spherical surface to the object. If the radius of curvature of the surface is 5 cm and the refractive index of the glass is 1·5, find the position of the image of the object formed in the glass block.

The distance of the object, u, is measured against the direction of the incident light and is therefore —25 cm. r, the radius of curvature, is measured with the direction of the incident light and is +5 cm.

Substituting in Equation (6) we get:

$$1·5/v - 1/\text{-}25 = 0·5/5$$
or
$$1·5/v + 1/25 = 0·5/5$$

Whence $v = +25$ cm, i.e. the image is real and 25 cm away from the surface, in the direction of the incident light.

The points O and I in Fig. 354 are described as *conjugate foci* and in any optical system any two points, chosen such that if the object is placed at one the image of it appears at the other, are called conjugate foci. If, in the case of the convex system in Fig. 354, *a*, we imagine that the object is placed at I, its image will be formed at O, since in geometrical optics we can always reverse a ray along its original path. We can thus imagine a divergent beam of rays from I striking this time a *concave* surface of separation and being made, by refraction, to *converge* on O. A convex surface thus causes divergent rays to converge on the axis when the rays pass from a less dense medium to a more dense one; a concave surface causes convergence if the rays pass from a more dense medium to a less dense. In dealing with spherical surfaces, therefore, we cannot state generally that a surface of a particular type, e.g. convex, will cause the incident rays to converge on the axis, unless we specify the change in the medum on passing from one side of the surface to the other.*

Principal Foci. With spherical mirrors the position of the principal focus was defined as the position of the image when the point object was infinitely far away on the axis, i.e. when the rays emanating from it could be considered parallel to the axis. A similar point can be defined in relation to the spherical refracting surface, and it is called the *second principal focus*; in the case of the convex system in Fig. 355, *a* (the rays are passing from a less dense medium to a more dense), the second principal focus represents a real point image; with the concave system, on the other hand, the parallel rays after refraction diverge from the second principal focus which is thus on the "object side" of the surface.

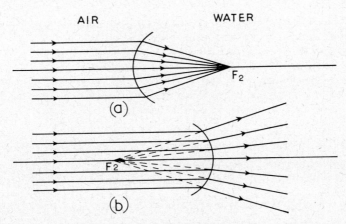

FIG. 355. The second principal focus.

The *first principal focus* is defined as the position a point object must have on the axis such that the rays, after refraction, run parallel to the axis; in this case the image is said to be at infinity. In the case of the convex system (Fig. 356, *a*) the first principal focus represents a real position of an object, whilst with the concave system (Fig. 356, *b*) the position of the object necessary to make the rays from it run parallel to the axis is virtual, in the sense that the rays before refraction must *converge on* the first principal focus. The reader should familiarize himself with this concept of a *virtual object*; the rays from any real point object on the axis will always diverge from this line, hence in order to make them run parallel with the axis after refraction they must meet a converging system as in Fig. 356, *a*; a diverging system can only make the rays diverge even more; in order that rays after refraction by such a divergent system may run parallel with the axis, they must clearly be converging on the refracting surface; the point to which they are converging is called the position of the

---

* In the concave system represented by Fig. 354, *b*, O and I are conjugate foci, hence an object at I produces an image at O. This does not mean, however, that a *real* object at I will produce an image at O. O and I are conjugate foci because the ray from O may be reversed along its path; when we say that the object is at I, light is considered to be coming from the right, i.e. I is the point to which rays *would have* converged in the absence of the spherical surface; they are bent away from the axis and meet at O, a real image. I is called a *virtual object*.

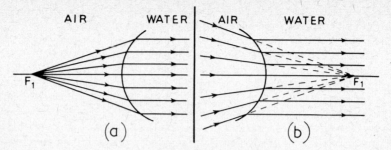

FIG. 356. The first principal focus.

virtual object. With the concave system of Fig. 356, *b*, the first principal focus is therefore on the "image side" of the refracting surface.

The distances of the first and second principal foci from the pole of the refracting surface are called the *first* and *second focal lengths* respectively, and are represented by the symbols f_1 and f_2. They may be evaluated in terms of *r* and *n* by substitution in Equation (6). Thus when *u* is equal to infinity, the distance of the image becomes the second focal length, f_2, i.e.:

$$n/f_2 - 1/infinity = (n-1)/r$$

whence
$$f_2 = nr/(n-1) \tag{7}$$

Similarly, f_1 is given by the equation:

$$f_1 = -r/(n-1) \tag{8}$$

Inspection of Equations (7) and (8) tells us that f_1 and f_2 are related as follows:

$$f_2 = -nf_1$$

i.e. they are of opposite sign and therefore on opposite sides of the refracting surface, and for the special case considered, namely with the light passing from a less dense to a more dense medium, f_2 is greater than f_1 numerically.

The formula relating object- and image-distances becomes:

$$n/v - 1/u = 1/f_1$$
$$= n/f_2$$

or, more generally, if n_1 is the refractive index of the object-space and n_2 that of the image-space:

$$n_2/v - n_1/u = (n_2 - n_1)/r$$
$$= P' \tag{9}$$

where P' is defined as the *surface power* and is equal to $(n_2 - n_1)/r$.

Object Off the Axis. As in the case of reflexion by spherical mirrors, the construction for determining the position of the image of a point object, not on the principal axis, is carried out by drawing a subsidiary axis through C, the centre of curvature of the surface, and the object. A narrow pencil of rays from the object converges to a point image (Fig. 357) whose distance from the surface may be calculated from Equation 6.

Finite Object. The position and nature of the image of a finite object may be determined by drawing a ray, AQ, parallel to the axis, which converges on the second principal focus, F_2, in the case of the convex surface (Fig. 358, *a*), or diverges from this point in the case of the concave surface. A second ray, through the centre of curvature, C, strikes the surface normally and hence passes through undeviated. The point of intersection of the rays, A_1, represents the image of A.

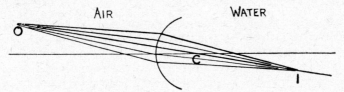

FIG. 357. Formation of the image of a point off the axis.

In the case of the convex surface the image is real, inverted, and diminished; with the concave surface the image is virtual, erect, and diminished.

The magnification produced by a spherical refracting surface may be calculated from the formula:

$$i/o = v/n \cdot u \qquad (10)$$

The position, nature, and magnification of the image depend on the position of the object; they may be determined by construction or by substitution in Equations (6) and (10); the construction of the image for typical positions of the object represents a useful exercise for the student.

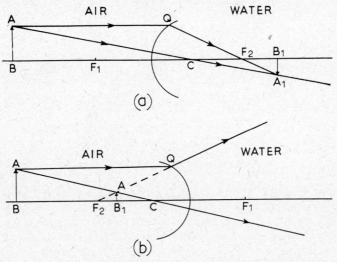

FIG. 358. Formation of the image of an extended object.

The Aphakic Eye. We are now in a position to investigate the refraction of light by the eye from which the lens has been removed (*aphakic*). The cornea and aqueous humour have almost identical refractive indices, consequently we may treat the aphakic eye optically as a homogeneous medium of refractive index 1·3365, separated from air by a spherical surface of separation (Fig. 359). The radius of curvature of the human cornea may be taken as 7·8 mm.

f_1, the first focal length, is given by:

$$f_1 = -r/n-1 = 1-7\cdot8/0\cdot3365 = -23\cdot2 \text{ mm.}$$

f_2, the second focal length, is given by:

$$f_2 = -n \cdot f_1 = 1\cdot336 \times 23\cdot2 = 31\cdot2 \text{ mm.}$$

The first principal focus is thus 23·2 mm in front of the cornea whilst the second principal focus is 31·2 mm behind; this latter point is the position where parallel rays from a point infinitely far away on the axis form an image. The length of the eye, from the anterior to the posterior poles, is given as 24·4 mm, consequently a distant object forms an image some 6·8 mm behind the retina. The image of a distant object on the retina is thus badly blurred in the aphakic eye; this is best demonstrated by considering a pencil of light from a distant object as in Fig. 359; the intersection of the pencil with the retina constitutes a "blur circle" and this circle of finite size represents the image of a point on the object. A distinct image of an object is only obtained if all the rays from any point on it meet in a single point on the image; this is clearly not the case here and the image on the retina is blurred. It will be noted that the size of the blur circle depends on the size of the pencil of light entering the eye, and this depends on the size of the pupil; diminishing the pupil size thus tends to make an out-of-focus image less blurred.

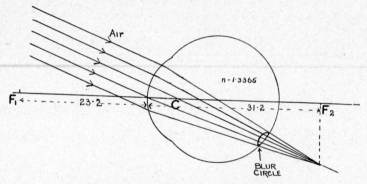

FIG. 359. Image formation by the aphakic eye.

Apparent Position of the Pupil. The pupil, when viewed in the intact eye, is seen as an image formed by refraction at the surface of the cornea. The rays of light, in these circumstances, pass from the more dense medium into the less dense and we must proceed carefully lest we become confused. In Fig. 360, QPR is the surface of the cornea and F_a and F_p are the focal points as calculated by treating the cornea as a spherical refracting surface of radius 7·8 mm.[*] F_a is thus 23·2 mm in front of the cornea and F_p 31·2 mm behind. A ray of light parallel to the axis, passing from air into the eye, converges on to the second principal focus, F_p; on the principle of the reversibility of rays it follows that a ray, passing *out of the eye* such that it appears to have come from F_p, is refracted parallel to the axis; i.e. the ray from A, the top of the pupil, which strikes the surface at D, emerges parallel with the axis. Another ray from A, appearing

[*] See footnote p. 561.

to come from C, the centre of curvature, strikes the surface normally and passes out undeviated. These two emergent rays diverge from the point A′, obtained by producing them backwards; a similar construction gives the image of B at B′; A′B′ is thus the erect, magnified, virtual image of the pupil, AB.

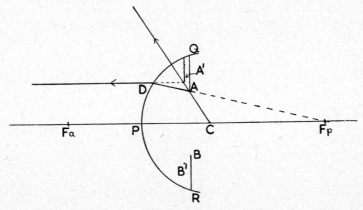

FIG. 360. Apparent position of the pupil. The pupil, AB, is seen as a magnified erect, virtual image, A′B′, closer to the cornea.

It will be seen later that the formation of the image of an object within the eye, as in Fig. 360, has some analogy with the formation of the image of an object within the first principal focus of a convex lens, i.e. the images are in both cases magnified and virtual. There is a fundamental difference, however, in that the image formed by the eye is closer to the surface than the object, whereas the reverse happens with the convex lens.

REFRACTION BY SPHERICAL LENSES

Thin Lenses

A spherical lens is a portion of a refracting medium bounded by two spherical surfaces (if one of the surfaces is plane, e.g. in a plano-convex lens, it is regarded as part of the surface of an infinitely large sphere). The various possible types of lens are shown in Fig. 361. The refractive properties of the spherical lens may best be deduced by applying the simple principles of ray construction used with a single spherical surface.

The Lens Formula. The simple bi-convex lens of Fig. 362 is bounded by two spherical surfaces of radii r_1 and r_2. The rays of light from a luminous point O, on the common principal axis, are refracted at the first surface and converge towards the point I′, the image of O which would be formed if no further refraction took place. At the second surface, however, the rays are made to converge further (they meet a concave surface and are passing from a more dense to a less dense medium), and they meet at the point I, the image of the object O formed by the lens as a whole. Now I may be considered as the image, formed by the second surface of the lens, of the virtual object, I′; the position of I′ can be calculated simply from the formula for a spherical surface

(Eqn. 6) provided the refractive index of glass and the radius of curvature, r_2, are known; similarly, the position of I, the image of I', can be calculated, this time making use of r_1. The position of the image O is thus determined by the radii of curvature and the refractive index of the lens. It should be noted

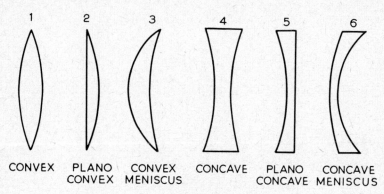

FIG. 361. Types of lens.

that, for an accurate treatment, the thickness of the lens should also be known; if the lens is thin, its thickness is ignored and a simple formula relating the position of the object with the position of the image can be derived:

$$1/v - 1/u = 1/f_2$$

where u and v have the same meaning as before and f_2 is the second focal length of the lens. The second principal focus is defined, as before, as the position of the image of an infinitely distant point on the axis; its distance from the lens is f_2.

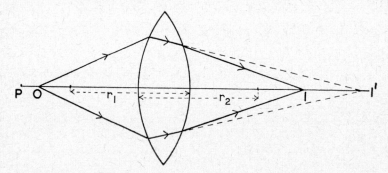

FIG. 362. The formation of the image of a point O by a biconvex lens. I' is the image that would have been formed if refraction had occurred only at the first surface; I is the image resulting from refraction at both surfaces.

The bi-convex lens of Fig. 362 is a "converging lens" in the sense that it tends to bend incident rays, arising from a point object on the axis, towards the axis. The second principal focus is thus a real image of a distant point, and occurs on the right of the lens (Fig. 363, a). The first principal focus is the position of a point object on the axis such that its image is at infinity; it is on the left side of the lens and it may be shown that the first focal length is equal to the second focal length, i.e. $f_2 = -f_1$.

The bi-concave lens in Fig. 363, *b* is a "diverging lens"; parallel rays from an object infinitely far away on the axis thus diverge from the axis after refraction so that the second principal focus is a virtual image on the *left* of the lens; the first principal focus is the position of a virtual object, to the right of the lens, such that the rays, after refraction, emerge parallel to the axis.

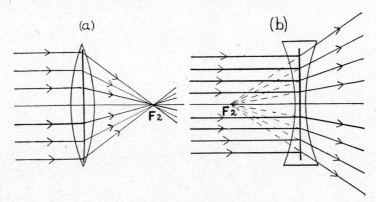

FIG. 363. The second principal focus.

For a converging lens the second focal length is thus, according to our sign convention, a positive quantity; with a diverging lens, on the other hand, it is a negative quantity. *It is customary to characterize a lens by the second focal length*; a converging lens is thus a positive or "plus" lens, a diverging one a negative or "minus" lens. The lens formula is generally written:

$$1/v - 1/u = 1/f \qquad (11)$$

where *f* implies the second focal length.

Example. A converging lens of focal length 10 cm is placed 5 cm in front of an object. Where is the image?

Here,
$$u = -5 \text{ cm}$$
$$f = +10 \text{ cm}$$
$$1/v - 1/-5 = 1/10$$
$$1/v = 1/10 - 1/5$$
$$= -1/10$$
i.e.
$$v = -10 \text{ cm.}$$

The image is ten centimetres away from the lens, on the same side as the object, i.e. it is virtual.

It is a simple matter to calculate, from the curvature of its surfaces, whether a lens of a given shape will act as a converger or diverger; in general, any lens whose thickness diminishes on passing from the centre to the periphery behaves as a converging lens, whilst a lens whose thickness increases on passing in this direction behaves as a diverging lens; thus lenses 1, 2 and 3 are converging (Fig. 361) whilst 4, 5 and 6 are diverging.

Construction of the Image. The image of a finite object may be constructed by the methods of ray tracing described for mirrors and for a single spherical refracting surface. In this case, however, the undeviated ray is the ray passing through the *centre of the lens* (the ray AC of Fig. 364); a ray AQ, parallel to the axis, passes through the second principal focus of a converging lens, and diverges from that of a diverging lens.

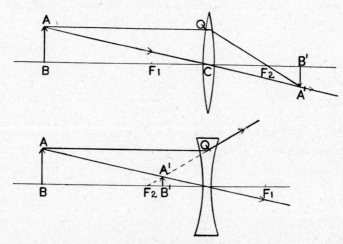

Fɪɢ. 364. Formation of the image of an extended object by a convex and a concave lens.

The linear magnification is given by the ratio:

$$\frac{Size\ of\ Image}{Size\ of\ Object} = \frac{i}{o} = \frac{v}{u} \tag{12}$$

Relative Positions of Object and Image. The position and nature of the image, formed by a converging or diverging lens, for different positions of the object may be deduced quite simply from the lens formula. The results of a systematic calculation are given in Table 14.

TABLE 14

Nature of Lens	Nature and Position of Object	Nature and Position of Image
Converging.	Real. Outside $2f$.	Real. Inverted. Diminished. Between f and $2f$.
,,	Real. At $2f$.	Real. Inverted. Same size as object. At $2f$.
,,	Real. Between f and $2f$.	Real. Inverted. Magnified. Outside $2f$.
,,	Real. At f.	Virtual. Erect. Magnified. At infinity.
,,	Real. Inside f.	Virtual. Erect. Magnified. Farther from lens than object.
,,	Virtual. Any position.	Real. Erect. Diminished. Inside focus.
Diverging.	Real. Any position.	Virtual. Erect. Diminished. Inside focus.
,,	Virtual. Outside focus.	Virtual. Inverted. Outside focus.
,,	Virtual. Inside focus.	Real. Erect. Magnified. Further from lens than object.

Parallel Rays. The same arguments regarding parallel rays, brought forward earlier, may be applied to lenses; in view of the practical importance of the results we may briefly recapitulate the argument.

The second principal focus of a lens is the point to which rays from an infinitely distant point object, on the axis, converge (in the case of a converging lens) or from which they diverge (in the case of a diverging lens). Rays from an infinitely distant point are parallel with themselves and, if the point is on the principal axis, parallel with the latter. If the distant point object is not on the principal axis, the rays reaching the lens are parallel with themselves but are inclined to the principal axis. They come to a focus, F'_2, on a subsidiary axis drawn through the centre of the lens and the distant object, and since the same formula applies to determining the relative positions of object and image on this subsidiary axis, F'_2 is in the same plane as F_2, so that the plane through F'_2 perpendicular to the principal axis is called the *focal plane* of the lens, and is the plane in which all rays from distant objects come to a focus (Fig. 365).

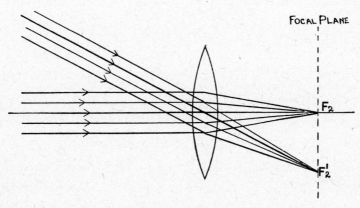

FIG. 365. The second focal plane.

It will be seen that if the bundle of rays parallel to the principal axis are considered to emanate from the bottom point of a distant object, and the bundle of parallel rays, inclined to the axis, to emanate from the top of the same object, $F_2F'_2$ represents the image of the distant object; e.g. if the lens is directed at the sun so that the principal axis coincides with the bottom of the disc, F'_2 corresponds to the image of a point on the top of the disc.

Conversely, if we have an object at the first principal focus of a converging lens, the rays in any pencil, after refraction, all emerge parallel to one another. Thus if AB is the object, the image, as seen by the eye, is infinitely far away since the pencils from A and B are pencils of parallel rays (Fig. 366).

The Magnifying Glass. Fig. 366 illustrates a special instance of the use of a convex lens as a magnifying glass—when the image is at infinity. If the object is within the first principal focus, as in Fig. 367, the virtual erect image, A'B', may be measured, and it is quite clear from the construction that it is magnified. This may be verified by calculation, since the magnification is given by:

$$\frac{\textit{Size of Image}}{\textit{Size of Object}} = \frac{v}{u}.$$

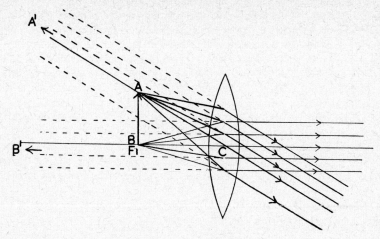

FIG. 366. The magnifying glass. The object AB is at the first principal focus of a convex lens. The image is at infinity.

Here v is greater than u, so that the magnification is greater than unity. When the object is at the first principal focus (Fig. 366) the ratio v/u becomes ∞/f, which is infinite, and it is clear that when dealing with parallel rays, the *linear magnification*, given by the ratio of the lengths of image and object, cannot be used to indicate the effective magnification. The apparent absurdity is due to the fact that the ratio indicating magnification only takes into account the lengths of object and image and not their distances from the eye; thus a

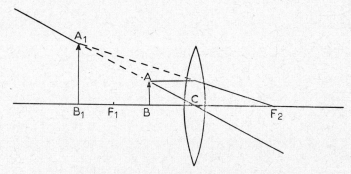

FIG. 367. The magnifying glass. The object AB is within the first principal focus.

very large object like the sun does not appear very large to the eye, because of its distance away. From the point of view of an image viewed by the eye, therefore, we must choose a more satisfactory criterion of size, the *apparent size*, given by the ratio of its length by its distance away. Thus AB, A′B, A″B″ in Fig. 368 all have the same apparent size because the ratios: AB/BN, A′B′/B′N′, A″B″/B″N are equal (this amounts to the statement that AB, A′B′, etc. subtend the same angle at the eye).

This definition of apparent size permits the assignment of a definite magnitude

to the size of an image at infinity. Thus the image of A (Fig. 366) lies on a
subsidiary axis through A and the centre of the lens, infinitely far away to the
left of the lens. The image thus subtends an angle ACB at the centre of the lens;
if the eye is very close to the lens, the image subtends approximately the same
angle at it. It will be seen that the *object* subtends the same angle at the eye,
namely, ACB. Thus the apparent sizes of object and image are the same and
if the magnification were to be expressed as the ratio:

$$\frac{Apparent\ Size\ of\ Image}{Apparent\ Size\ of\ Object}$$

it would be unity. A convex lens is, nevertheless, a useful magnifying instrument
when the image is at infinity, since it permits the perception of details indistinguishable by the naked eye. The reason for this resides in the fact that the lens
permits the eye to approach very close to an object and still obtain a distinct
visual image. If the eye, without the lens, could see an object 1 cm away
distinctly, placing a convex lens in front would not give any magnification.
For comfortable observation an object is viewed at a distance of about 25 cm,
consequently the *magnifying power* of the lens is defined by the ratio:

$$\frac{Apparent\ Size\ of\ Image}{Apparent\ Size\ of\ Object\ 25\ cm\ from\ Eye}$$

The apparent size of the object at 25 cm is given by the ratio: AB/25 cm
and the apparent size of the image by the ratio: AB/BC (the image, even though
infinitely far away, lies on the line AC, and by similar triangles the ratio of its
height by its distance from the lens is always given by AB/BC).

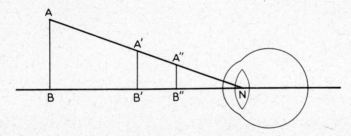

Fig. 368. Illustrating apparent size.

The magnifying power of a convex lens depends on the position of the image;
if this is 25 cm away, it is given by the expression: $1+25/f$, where f is the
focal length in cm. If the image is at infinity it becomes: $25/f$.

Thus a convex lens, of focal length 5 cm, has a magnifying power of five when
the object viewed is at the first principal focus.

Dioptric Power. The shorter the focal length of a lens the more strongly
are rays made to converge or diverge after refraction; the reciprocal of the
focal length is therefore the best measure of the refracting power of a lens.
When the focal length is expressed in metres, the reciprocal of this value gives
the *power* of the lens in *dioptres*. Thus a lens of $+5$ cm focal length has a power
of $1/0.05 = 100/5 = +20$ *dioptres*.

When it is desired to express the power of a single spherical refracting surface in dioptres, we must remember that the first and second focal lengths are not equal. The dioptric power in these circumstances is given by: $P = n/f_2$ or $P = -1/f_1$ where n refers to the index of refraction of the surface; when the object is in air, n becomes the refractive index of the medium separated from air by the curved surface.

When two thin lenses are placed in contact, their combined focal length, F, is given by:

$$1/F = 1/f_1 + 1/f_2 \tag{13}$$

or their combined power by: $P = P_1 + P_2$ $\tag{14}$

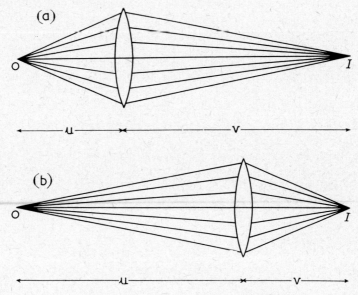

FIG. 369. Illustrating the concept of vergence.

The Concept of Vergence. The concept of power, as a reciprocal of the focal distance, may be extended to other distances in optical systems. For example, the reciprocal of the radius of curvature, $1/r$, may be called the *curvature* of the surface, $1/r = C$ dioptres when the radius is expressed in metres. The distance of the object or image from the refracting surface is a measure of the degree of divergence or convergence of the rays in relation to this surface; thus in Fig. 369, a, where the object is close to the convex lens, the divergence of the rays from the object is high whilst the convergence of the rays to the image is small, by contrast with the opposite relationship in Fig. 369, b. The reciprocals of the object- and image-distances may be defined as the *vergence* of the rays belonging to the object and image respectively; thus $1/u = L$, the object vergence, and $1/v = L'$, the image vergence, once again expressed in dioptres when the distances are expressed in metres. The thin lens formula becomes:

$$L' - L = P \tag{11a}$$

REDUCED VERGENCE

Where the refractive indices of the object- and image-spaces are different the concept of reduced vergence, introduced by Gullstrand, becomes useful. If n_1 is the refractive index of the object-space, the reduced vergence of the object is $n_1/u = L$, whilst the reduced vergence of the image is $n_2/v = L'$ when the refractive index of the image-space is n_2. The formula for the spherical refracting surface (Eqn. 9, p. 546) takes the form:

$$L' - L = P' \qquad\qquad (9a)$$

The Thick Lens

The formula relating the positions of object and image for a thin lens was obtained by considering the refraction at the first surface and treating the image so formed as the object for the next surface. The thickness of the lens was ignored. For most ophthalmic lenses the thin lens formula is quite adequate; however, in treating the eye as an optical instrument, where there are three main refracting surfaces, namely, the cornea and the anterior and posterior surfaces of the lens, the problem is very much simplified if we make use of concepts derived from the study of the thick lens.

The thick lens formula is:

$$1/V - 1/U = 1/F_2 \qquad\qquad (15)$$
$$= -1/F_1$$

where the distances U and V do not mean quite the same thing as before. V is the distance of the image from a certain point, the *second principal point*, whose exact position can be calculated from the curvatures of the surfaces, the thickness of the lens and the refractive index of its material. U is the distance of the object from another point, the *first principal point*, whose position can be similarly calculated. F_2 is the distance of the second principal focus from the second principal point whilst F_1 is the distance of the first principal focus from the first principal point. The principal foci have the same meaning as for a thin lens, and the two focal lengths are equal but of opposite sign. The principal points and foci of a thick lens are shown in Fig. 370.

A plane at right-angles to the principal axis through H_1, the first principal point, is called the *first principal plane*; a similar plane through H_2 is called the *second principal plane*. These planes have the important property that any ray directed towards a point on the first, at a height h above the axis, leaves the second at the same height, h, above the axis. With this knowledge we can construct the image of an object formed by a thick lens, given the positions of the principal points and foci.

In Fig. 370, AB is the object. A ray parallel to the axis strikes the first principal plane at P and leaves the lens as if it had come from Q, on the second principal plane at the same height above the axis as P. It passes through F_2. A ray through F_1 strikes the first principal plane at R and leaves the lens as from S. The two rays intersect at A', which is the position of the image of A. A'B' is thus the image of AB.

Nodal Points. Listing introduced two other points, the *nodal points*, corresponding to the centre of a thin lens. Any ray directed towards the first point, N_1, leaves the system as though it had come from the second point, N_2, and parallel with its original direction. In the case so far treated, where the thick lens is surrounded by the same medium on both sides, the nodal points coincide with the principal points (thus in Fig. 370 AH_1 is parallel with H_2A'). When the media on opposite sides of the lens are different, the principal and nodal points do not coincide.

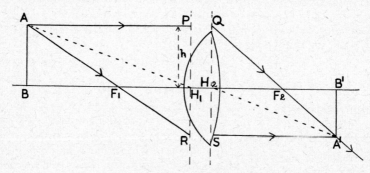

FIG. 370. Formation of the image by a thick lens. H_1 and H_2 are the principal points; F_1 and F_2 the principal foci.

It will be noted that we are able to treat a thick lens, which consists of two refracting surfaces separated by a finite distance, as a single lens of focal length F only if distances are measured from certain calculated points. This really means that a single refracting surface cannot be found which will behave as a combination of two surfaces separated by a finite distance; instead we have to imagine that any ray is refracted at the first principal plane and then refracted again at a new surface, the second principal plane. It must be emphasized that these surfaces are entirely fictitious; refraction takes place at the surfaces of the lens, but the result of this refraction is the same as if it had taken place at these two surfaces with their special property described above.

REFRACTION BY THE EYE

The thick lens is only a special case of the more general Gaussian treatment of any number of spherical refracting surfaces with a common axis, separated from each other by finite distances. Each surface, separating two media of different refractive index, has its own principal plane, nodal point, and first and second principal foci; by a simple calculation these so-called *cardinal points* can be combined to give a single set, no matter how many surfaces there are; thus if we had six separate surfaces, their refracting properties could be described in terms of a pair of principal points, a pair of nodal points, and a pair of principal foci. It may be asked, what are the principal point and nodal point of a single spherical refracting surface, since we must know them in order to obtain the cardinal points of the combined surfaces. The nodal point is a point such that a ray, incident on the refracting system and directed towards this point, emerges after refraction parallel with its original direction; with a single spherical surface this point is equivalent to the centre of curvature; similarly, the principal point of a single surface is the pole and the principal plane corresponds to the surface itself.*

* It is, of course, a contradiction in terms to describe a spherical surface as a "plane"; the curvatures, however, are in practice sufficiently small to justify this idealization.

The Schematic Eye

Refraction by the eye may be approximately described as the effect of the bending of light at three spherical surfaces, refraction at the posterior surface of the cornea being ignored since the difference in refractive index of cornea and aqueous humour is not large. To obtain by the Gaussian treatment the cardinal points of the combined system of surfaces, we require to know their radii of curvature, indices of refraction, and distances apart. The experimental determination of these quantities, the *optical constants of the eye*, will be described later.

TABLE 15. *Cardinal Points of the Eye* (Gullstrand)

Cardinal Point	Distance from Ant. Surface of Cornea
First principal point	1·35 mm
Second principal point	1·60 mm
First nodal point	7·08 mm
Second nodal point	7·33 mm
First focal point	−15·7 mm
Second focal point	24·4 mm

The positions of the cardinal points, calculated from Gullstrand's measurements of the optical constants,* are given in Table 15, where all positions are referred to the anterior surface of the cornea. As we should expect, the second principal focus, 24·4 mm behind the cornea, coincides with the retina of the normal eye.

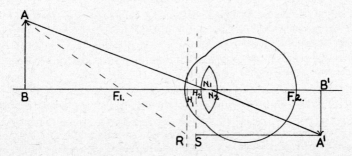

FIG. 371. The schematic eye. Formation of the image of a near object.

The first focal length, F_1, is the distance of the first focus from the first principal point, i.e. $(-15.7 - 1.35)$ or -17.05 mm. The second focal length, F_2, is the distance of the second focus from the second principal point, i.e. $(24.4 - 1.6)$ or $+22.8$ mm. The refractive power of the eye is given by:

$$P = -1/F_1 = n/F_2 = 58 \text{ dioptres}$$

The construction of the image of an object close to the eye is indicated in Fig. 371; a ray through the first focus strikes the first principal plane at R and leaves the second principal plane at the same height, parallel to the axis.

---

* For the purposes of this discussion it is sufficient to retain the classical figures of Gullstrand; the results of more recent and accurate determinations are described later.

A ray through the first nodal point emerges through the second, parallel with its original direction. The rays intersect behind the retina.

The formula relating positions of object and image is of the same type as for a single spherical refracting surface (Eqn. 6, p. 544), namely:

$$n/V - 1/U = n/F_2$$
$$= -1/F_1 \qquad (16)$$

where distances are measured from the principal points.

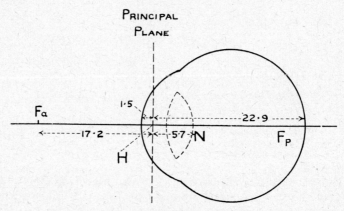

FIG. 372. The reduced eye.

The Reduced Eye

The first and second principal and nodal points of the schematic eye are quite close to each other; Listing simplified the treatment of the eye as a refracting system by choosing a single principal point, lying midway between the two actual points, and a single nodal point likewise situated between the real nodal points. This gives a principal point, H (Fig. 372), 1·5 mm behind the anterior surface of the cornea, and a nodal point, N, 7·2 mm behind the surface. The first and second focal lengths are now referred to a single principal point (as are also U and V) and have values of −17·2 and +22·9 mm respectively.

This simplification of the schematic eye corresponds, in effect, to the treatment of the eye as a *single spherical refracting surface*, the single nodal point representing the centre of curvature whilst the surface itself is a plane, passing through H 1·5 mm behind the cornea. The radius of curvature of this surface is clearly 7·2−1·5 = 5·7 mm. Thus the reduced eye of Listing may be defined as a fictitious spherical surface, separating two media of refractive indices 1 and 1·336, situated 1·5 mm behind the cornea. Its radius of curvature is 5·7 mm and its anterior and posterior focal lengths are −17·2 and 22·9 mm respectively.

The dioptric power of the eye is about +58 D; the first focal length of the aphakic eye was found to be −23·2 mm, corresponding to a dioptric power of +43 D; consequently the lens contributes some 15 dioptres to the total power of the eye.*

---

* This does not mean that the *power of the lens* is +15 D, since we cannot obtain the dioptric power of a refracting system by a mere summation of the powers of its constituents. Its power is given by Gullstrand as +19·1 D (p. 609).

Donders' Reduced Eye. Donders, in the interests of simplicity, treated the eye as a single surface 2 mm behind the cornea, with a radius of 5 mm and anterior and posterior focal lengths of 15 and 20 mm respectively. This is a gross simplification, but has the advantage that the appropriate figures can be remembered easily.

Image of Distant Object. Fig. 373 represents the normal reduced eye, refraction taking place at the plane H. The eye is looking at a distant object such that the principal axis passes through its mid-point. A parallel bundle of rays, (b), from the centre of the object is refracted at H and comes to a point focus on the retina, R, which corresponds to the second focal plane. Another bundle, (a), from the top of the object is inclined to the principal axis and forms an image on the subsidiary axis through the top of the object and the nodal point, N. Since the rays are parallel they come to a focus at the intersection of this subsidiary axis with the retina, at A'. In a similar way a bundle (c) from the bottom of the object comes to a focus at C'. The image is thus inverted.

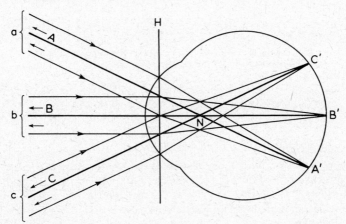

FIG. 373. Formation of the image of a distant object by the reduced eye.

Image of Near Object. The construction is identical with that described for a spherical refracting surface (p. 547); the image of AB (Fig. 374) falls behind the retina; for the image to fall on the retina the refracting power of the eye would have to be modified in the sense that the rays would have to be refracted more strongly; thus if the second principal focus,[*] F_p, were brought closer to the cornea (to F'_p) the ray AP would be bent in more strongly and would intersect the ray through the nodal point[†] sooner, and thus produce a point image of A on the retina. This shortening of the focal length, or increase in the dioptric power, is called *accommodation*.

From Fig. 373 it will be clear that the size of the image, i, of a distant object of size o is given by substituting in the equation: $i/o = $ B'N/BN, where B represents

---

[*] In future the first and second principal foci of the eye will be referred to as the *anterior* (F_a) and *posterior* (F_p) foci respectively. This will save a great deal of confusion since the point in front of the eye we call the first principal focus may be regarded as the second principal focus if light is considered to be passing out of the eye. The same applies *mutatis mutandis* to the retina of the normal eye.

[†] A shortening of the focal length would be associated with a shift of the nodal point towards Fp.

the position of the distant mid-point of the object. B'N is the distance of the retina from the nodal point, namely, $24 \cdot 4 - 7 \cdot 2 = 17 \cdot 2$ mm. Hence if d is the actual distance of the object from the nodal point, the size of the retinal image is given by: $i = \dfrac{17 \cdot 2 \times o}{d}$, remembering that d must be sufficiently large that rays from the object are, in effect, parallel.

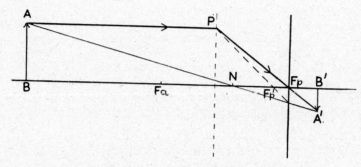

FIG. 374. Formation of the image of a near object by the reduced eye. F_p must come forward to F_p' if the image is to be formed on the retina.

AMETROPIA AND ACCOMMODATION

Ametropia

An unaccommodated eye, in which the image of a distant object falls on the retina, is called a normal or *emmetropic eye*. If the image of a distant object falls behind the retina, the eye is "long-sighted" or *hypermetropic*; if the image falls in front of the retina, the eye is "short-sighted" or *myopic*. The three conditions are indicated in Fig. 375 for the case of a point object infinitely far away on the axis.

We may thus regard the myopic eye, for example, as one that is too long for its refracting system, or alternatively as an eye of normal length but with an abnormally high dioptric power. Myopia can be due to either of these causes but it is thought that the more common condition is an abnormal length of the eye. Hypermetropia may be regarded as the condition in which the eye is too short or in which the dioptric power is inadequate.

Ametropia due to a defect in the axial length of the eye is referred to as *axial ametropia*, whereas ametropia due to a defective refractive power is called *refractive ametropia*.

The Far Point. In the myopic eye, parallel rays are brought to a focus in front of the retina. Our consideration of lenses has shown us that a real image of a real object is closest to the refracting surface when the incident rays are parallel, i.e. when the real object is at infinity. To produce an image farther away from the refracting surface we must bring the object closer, i.e. the eye must be presented with *divergent*, not parallel, rays. Hence if an object is brought from a large distance closer and closer to a myopic eye, a point is reached—the *far point*, or *punctum remotum*—at which the object is seen distinctly.

The distance of the far point from the eye is generally symbolized by the small letter r; r is thus the distance of an object from the completely relaxed eye at which it can be seen distinctly.

For the emmetrope, the far point is at infinity; for the myope it is a finite distance in front of the eye (Fig. 376, *a*), and in accordance with the sign convention its distance from the principal plane has a *negative* value.

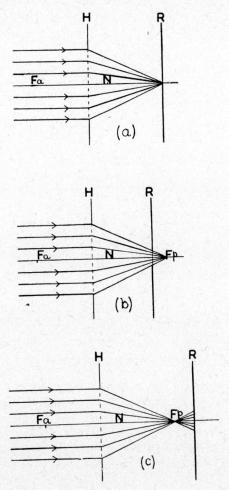

FIG. 375. (*a*) Emmetropic, (*b*) hypermetropic, and (*c*) myopic eye.

In the hypermetropic eye, parallel rays are brought to a focus behind the retina, i.e. the rays, after refraction, do not converge sufficiently to produce an image on the retina. To correct this, we must present the eye with *converging* rays. This cannot, of course, be achieved by moving a real object towards, or away from, the eye, since rays from any object are always divergent. A lens must therefore be used to impose the necessary convergence on the rays from an object if the latter is to be seen clearly by the completely relaxed hypermetropic eye. This is equivalent to presenting the eye with a *virtual object*, and the far point of a hypermetropic eye is, consistently with our definition, the position of a virtual object such that its image falls on the retina in the absence of accommodation. The position of the far point is shown in Fig. 376, *b*; the rays falling on the eye are convergent and would, in the absence of the eye, meet at P.R. After refraction by the eye they come to a focus on the retina. The distance of the far point, P.R., from the principal point of the eye is indicated by *r* once again. This time it has a positive value.

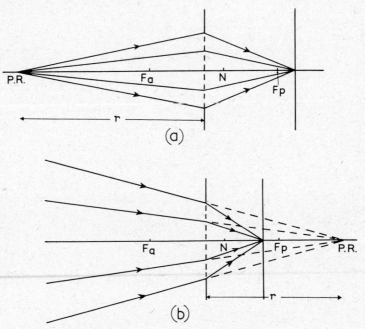

FIG. 376. The far point. (*a*) Myopic eye. (*b*) Hypermetropic eye.

The Ametropic Error. The conception of the far point provides an easy method of assessing the strength of lens required to correct an ametropic condition. Thus, if an individual has a far point of −2 metres, our problem is to permit him to see infinitely distant objects as though they were 2 metres distant, since they would then produce clear images on the retina. We require, therefore, a lens which, when placed in front of the eye, will give an image of distant objects at 2 metres from the principal point, i.e. in the lens formula:

$$1/v - 1/u = 1/f$$

$$u = \infty$$
$$v = 2 \text{ m. Hence } -\tfrac{1}{2} = 1/f$$
$$\mathbf{f = -2 \text{ metres}, \ P = -\tfrac{1}{2} \text{ Dioptres.}}$$

A concave lens of focal length 2 metres, placed at the principal point of the eye, would produce an image of a distant object at the far point.

In general, if *r* is the distance of the far point in metres, *r* is the focal length of the lens which must be placed at the principal point of the eye to correct it for distant vision, and 1/*r* is the dioptric power of the lens.

With a hypermetrope, as we have seen, it is not possible to find a real position of the object to give a distinct image on the retina; the far point can therefore only be found by placing convex lenses in front of the relaxed eye until the necessary convergence has been imposed on the rays, emanating from a distant object, to permit them to come to a focus on the retina. The distance of the far point from the principal point of the eye, r, is thus given by the focal length of the necessary correcting lens (Fig. 377, a).

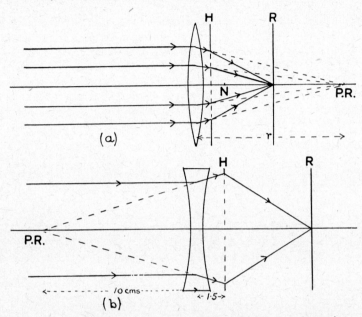

FIG. 377. (a) Correction of hypermetropia. (b) Correction of myopia.

The reciprocal of r, expressed in metres, is symbolized by R; it is the inverse distance of the far point from the principal point of the eye, expressed in dioptres, or the *dioptric value of the far point distance*. R is thus the measure of the refractive error of the eye, and is referred to as the *static refraction* or the *ametropic error*.

Spectacle Point. In the treatment of ametropia the distances have been correctly referred to the principal point of the eye, and the refractive error has been assessed in terms of the correcting lens that must be placed at the principal point. Whereas this gives a true picture of the refractive error, it is inadequate as a measure of the necessary strength of spectacles, owing to the fact that spectacles are worn at a distance in front of the eye, at the so-called *spectacle point*, which is, on the average, 12–13 mm in front of the cornea.

Thus let us suppose that the far point of a subject is 10 cm in front of the principal point; we require the strength of lens, placed at the spectacle point (say, 1·5 cm in front of the principal point) which will make the rays from a distant object appear to diverge from the far point.

The state of affairs is illustrated in Fig. 377, b; the far point, P.R., is at the second principal focus of the concave lens whose focal length is clearly: $-(10 -1·5) = -8·5$ cm, or $-0·085$ m. The dioptric power of the correcting lens is thus: $-1/0.85 = -11.75$ D. The ametropic error is given by $1/r$ in metres, i.e. $-1/0.1 = -10$ D. The strength of the correcting lens, worn at the spectacle point, is called the *spectacle correction*; in this case it is -11.75 D.

Effectivity. The general effects of moving a lens farther away from, or closer to, the eye may now be deduced. In Fig. 378, we may consider a concave lens, d metres in front of the principal point of the eye. A parallel bundle of rays from an infinitely distant object is made to diverge from the far point of the myopic eye, P.R., and thus comes to a focus on the retina, R. The power of the lens is clearly $1/(r-d)$.* If the lens is now moved a distance q away from the eye, the rays after refraction by the lens diverge from a point Q, more distant than the far point, so that the eye fails to bring them to a focus on the retina. The eye is "under-corrected" and it is said that the *effectivity* of the lens has been reduced by moving it away from the eye. A new lens of focal length equal to the distance $(r-d-q)$, is necessary. If the lens is brought a distance q *closer* to the eye, it will be seen, on the same reasoning, that the effectivity is increased.

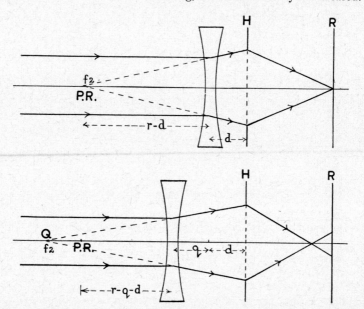

FIG. 378. Illustrating the change in effectivity of a concave lens on moving it away from the eye.

The convex correcting lens is treated in Fig. 379; a parallel bundle of rays is caused to converge on the far point, P.R., by the convex correcting lens of focal length equal to $(r-d)$.† Moving the lens a distance q from the eye causes the rays to converge on a point Q, closer to the retina, R, than P.R.; as a result, the image formed by the eye falls too close to the cornea and the eye is over-corrected. A lens of focal length $(r-d-q)$ is now sufficient for correction, i.e.

---

* The distances r, d, and q are all negative; $(r-d)$ is therefore a negative quantity less than r.

† Here, r is positive whilst d and q are negative; $(r-d)$ is therefore a positive quantity greater than r.

one of less power. The effectivity of a convex lens is therefore increased by moving it away from the eye, and conversely, it is decreased by bringing it closer in.

In general, if a lens of focal length f, at a given position in front of the eye, is required to correct it, a lens of focal length $(f-q)$ is required when the lens is moved a distance q either towards or away from the eye, the value of q being given a negative sign if the direction of movement is away from the eye.

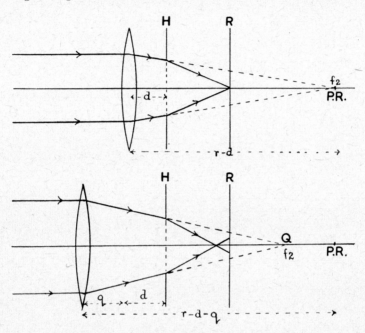

FIG. 379. Illustrating the change in effectivity of a convex lens on moving it away from the eye.

Size of Retinal Image in Corrected Ametropia. We have so far considered the use of correcting lenses from the point of view of moving the image, formed by the eye, forwards or backwards; it is important to know whether the size of the image on the retina is affected by the use of a correcting lens; when the refractive error is not the same in the two eyes (*anisometropia*) this consideration becomes important.

Fig. 380, *a* represents an axially hypermetropic eye; a parallel bundle of rays from the top of a distant object is brought to a focus at A′; a bundle from the bottom of the object (on the axis) comes to a focus at B′. B′A′ is the inverted image of the distant object; it falls behind the retina, R_H, in the plane R_E, i.e. where the emmetropic retina would be. The size of the image of A′B′ is clearly determined by the ray through F_a (the anterior focus of the eye) which, after refraction, runs parallel with the axis.

A convex correcting lens is now placed at F_a; as a result, the image is brought forward into the plane of the hypermetropic retina, R_H. The size of the image is determined, once again, by the ray through F_a, and since this passes through the correcting lens undeviated, the size of the image formed on the hyper-metropic retina is the same as that which would have been produced on the emmetropic retina. By a similar construction it can be shown that the image formed in corrected myopia is of the same size as that formed in the emmetropic eye, provided that the correcting lens is at the anterior focus of the eye.

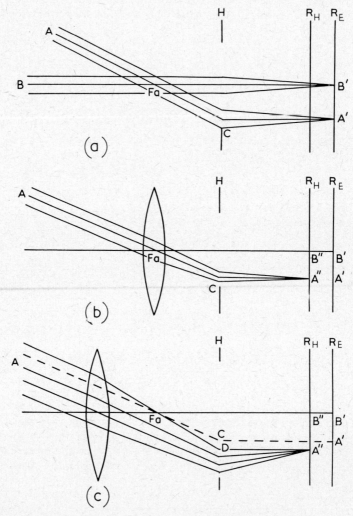

Fig. 380. The size of the image in corrected hypermetropia: (a) The image of a distant object is formed behind the hypermetropic retina. (b) A convex lens at F_a brings the image into the plane of the retina keeping the size the same. (c) A correcting lens beyond F_a produces a larger image.

In Fig. 380, *c* the correcting lens has been placed in front of F_a. In the absence of the lens the dotted ray through F_a would strike the principal plane of the eye at C and would determine the size of the image, B'A', i.e. the size of the emmetro-pic image. In the presence of the lens, the ray through F_a converges more strongly on the axis, as a result of refraction, and consequently strikes the principal plane of the eye at D, lower down than C; it runs parallel to the axis of the eye and strikes the retinal plane, R_H, at A''. All other rays meet here, since the eye is corrected. B''A'' is therefore the size of the image when the correcting lens is in front of the anterior focus of the eye; it is larger than the image formed in the emmetropic eye. Moving the lens forwards therefore increases the size of the retinal image; moving the lens backwards decreases it. By a similar construction it can be shown that the reverse effects are obtained with a concave correcting lens.

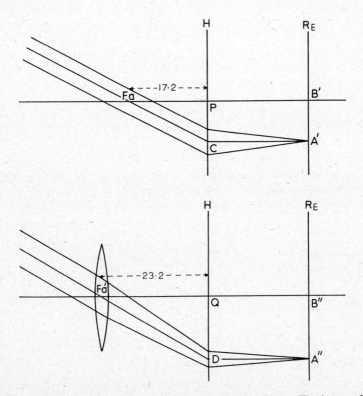

FIG. 381. The size of the image in corrected aphakia. *Above:* The image of a distant object formed by the normal emmetropic eye. *Below:* The image of the same object formed by the corrected aphakic eye, the lens being at F_a', the anterior focus.

Size of Image in Corrected Aphakia. Aphakia is a condition of refractive hypermetropia. In Fig. 381 we have, as before, the passage of a bundle of rays giving the image, A'B', of a distant object on the emmetropic retina. The size is given by the length B'A' or by PC. In the same figure the image of the object is formed by the aphakic eye, with an appropriate correcting lens at its anterior focus, F_a'. Because the anterior focal length of the aphakic eye is greater than that of the normal ($-23\cdot2$ mm compared with $-17\cdot2$ mm) the size of the image B''A'', which is equal to QD, is clearly larger than that in emmetropia,

PC. A comparison shows that the sizes are proportional to the anterior focal lengths:

$$\frac{Size\ of\ Image\ in\ Corrected\ Aphakia}{Size\ of\ Image\ in\ Emmetropia} = \frac{23.2}{17.2} = 1.35$$

i.e. the image in corrected aphakia is 1·35 times as big as that in the emmetropic eye.

Accommodation

Amplitude of Accommodation. We have seen (p. 562) that an increase in the dioptric power of the eye is necessary to bring near objects in focus; as an object is moved closer and closer to the eye, at a certain point—the *near point of accommodation*—the eye can no longer increase its dioptric power and the object becomes indistinct. It will be remembered that the far point is the position of an object such that its image is formed on the retina of the completely relaxed eye. We may define the *amplitude of accommodation*, therefore, as the difference in refracting power of the eye in the two states of complete relaxation and maximal accommodation. Thus if Q is the refracting power of the eye in maximum accommodation, and S its power when completely relaxed, the amplitude, A, is given by:

$$A = Q - S \qquad (17)$$

Since the amplitude is a *difference* in refracting power of the eye in two conditions, it is not necessary to know its actual power in either; the amplitude may be calculated from the near and far points quite simply, as follows.

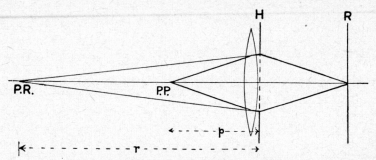

FIG. 382. The amplitude of accommodation.

In Fig. 382 the near and far points of a myopic eye are shown. It is required to find how much the power of the eye is increased when an object, formerly at the far point, is seen distinctly at the near point. To do this we can imagine that the eye, instead of accommodating, has remained in the relaxed condition and that the increased dioptric power to permit distinct vision of the object at the near point has been provided by a converging lens. The power of this lens would represent the increase in dioptric power of the eye in maximum accommodation. The completely relaxed eye sees distinctly an object at the far point; the lens must be chosen, therefore, of such a strength that it will form an image of an object, situated at the near point, at the far point. In the lens formula:

$$1/v - 1/u = 1/f$$

$v = r$, where r is the distance of the far point in metres,
$u = p$, where p is the distance of the near point in metres.

Thus $$1/r - 1/p = 1/f.$$

If we let $1/r$ in metres be R; $1/p$ be P; the power of the lens, A, is given by:

$$R - P = A \qquad (18)$$

R may be called the *dioptric value of the far point distance*; P the *dioptric value of the near point distance*, and A is the amplitude of accommodation, in dioptres.

This relationship between R, P, and A has been derived from the lens formula, consequently it will only hold if the usual sign convention is applied. Thus in the case of the myope R and P are both negative.

Near Point in Different Refractive Conditions. As examples of the application of this formula, we may choose three subjects with the same amplitude of accommodation, namely, 12 D; the first is an emmetrope, the second a hypermetrope of +6 D and the third a myope of −8 D.

Emmetrope. The far point is at infinity, whence R = 0. From Equation (18) we have:

$$0 - P = 12$$

i.e. P $= -12$ D; $p = -1/12$ metres $= -8.33$ cm.

The near point is **8.33 cm** in front of the eye.

Hypermetrope. The dioptric value of the far point distance is the ametropic error, hence R $= +6$ D.

From Equation (18) we have: $6 - P = 12$,

i.e. P $= -6$ D; $p = -1/6$ metres $= 16.7$ cm.

The near point is **16.7 cm** in front of the eye.

Myope. Here R $= -8$ D.

From Equation (18) we have: $-8 - P = 12$

i.e. P $= -20$ D; $p = -1/20$ metres $= -5$ cm.

The near point is **5 cm** in front of the eye.

Thus the myope has the shortest near point distance; this is to be expected since no accommodative effort is required until the object has been brought in to his far point; the hypermetrope, on the other hand, in order to see distant objects clearly, must accommodate; for example, in the case considered, he uses 6 dioptres of his total amplitude in order to bring images of distant objects on his retina.

Hypermetropia and Accommodation. It will be evident that the hypermetrope can see distant objects by accommodating, provided that his amplitude of accommodation is adequate to overcome his refractive error. If the error is sufficiently great, or if the amplitude is sufficiently restricted, distant objects cannot be seen clearly without a positive correcting lens, but so long as he has some accommodative power, the hypermetrope can see distant objects clearly with a *weaker* lens than that necessary to correct his hypermetropia fully. It is therefore customary to divide the hypermetropia into several parts.

The weakest positive lens, required to give distinct vision of distant objects, is said to measure the *absolute hypermetropia* (*Ha*); it is that part of the hypermetropia the subject is unable to correct by accommodation.

The strongest positive lens tolerated for distant vision measures the extent to which the hypermetrope can relax his accommodation, and therefore indicates his *total hypermetropia*, (H). It is found in general, however, that the hypermetrope cannot relax his accommodation completely without the use of a cycloplegic, i.e. that there is a certain amount of *latent hypermetropia* (Hl). The total hypermetropia is thus only given by the strongest positive lens tolerated when the accommodation is completely relaxed by a cycloplegic. The strongest positive lens tolerated in the absence of a cycloplegic is called the *manifest hypermetropia* (Hm); the difference between the total and the manifest hypermetropia is the latent hypermetropia. Finally, the difference between the manifest and absolute hypermetropia is called the *facultative hypermetropia* (Hf), it is the hypermetropia that can be voluntarily overcome.

Thus if the strongest positive lens tolerated by a hypermetrope with a cycloplegic is $+9$ D, his total hypermetropia is $+9$ D. If, without a cycloplegic, it is $+8$ D, the manifest hypermetropia is $+8$ D, and the latent hypermetropia is $9-8 = +1$ D. If the weakest positive lens tolerated without a cycloplegic is $+2$ D, the absolute hypermetropia is $+2$ D, and the facultative hypermetropia is $8-2 = +6$ D.

Range of Accommodation. The *range of accommodation* is defined as the difference between the near and far point distances:

$$p-r = a \tag{19}$$

For the myope, considered above, p was found to be -5 cm. R, the dioptric value of the far point distance, was -8 D, whence $r = -1/8$ metres, or $-12 \cdot 5$ cm. The range is thus:

$$-5+12 \cdot 5 = 7 \cdot 5 \text{ cm}$$

In the case of the emmetrope the far point is at infinity, so that the range is infinite; with the hypermetrope the far point is behind the eye, i.e. a certain "distance beyond infinity" and the range must likewise be considered to be infinite. The conception of the range of accommodation is thus not a very useful one, and the accommodative power should always be indicated by the *amplitude* of accommodation.

NON-SPHERICAL LENSES AND ASTIGMATISM

The Cylindrical Lens

Asymmetrical Refraction. The curved refracting surfaces, so far considered, have been portions of spheres so that refraction has been symmetrical about the principal axis. By this we mean that if we work out the refraction of a ray in any one plane, we know that the refraction of a similar ray in any other plane will be identical. Thus in all constructions we have confined attention to the rays in the plane of the paper; the rays from a point, striking a lens, actually form a cone in three dimensions, and it is only because we know that the rays not in the plane of the paper will behave in the same way as similar rays in the plane that we can state that they will all meet in the image. The same thing is meant when we state that the curvature of a spherical lens is the same in all meridians; any plane through the optic axis intersects the lens in the arc of a circle of the same curvature; this is indicated in Figs. 383 and 384, where a vertical and horizontal plane intersect a spherical lens.

A cylindrical lens is one whose surfaces may be represented as portions of cylinders (Fig. 385). The axis of a cylinder is a line running through its centre, and the axis of a cylindrical lens may be taken as a line running through it, parallel to the axis of the cylinder of which it forms a part (Fig. 385). A principal optic axis of a cylindrical lens is a line, perpendicular to the axis, passing through the centre of curvature. Since a cylinder can be considered to be made up of a large number of circles, one on top of the other, we can have any number of principal optic axes.

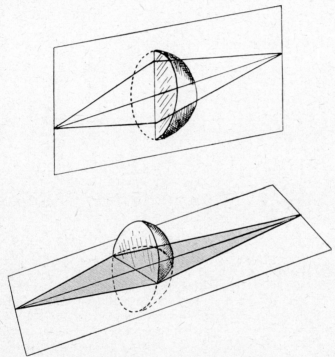

FIGS. 383 and 384. A vertical plane and a horizontal plane intersect a spherical lens in the arc of a circle.

A cylindrical lens is clearly not symmetrical about its optic axis; a plane through the axis of the cylinder, as in Fig. 385, intersects the surface in a straight line; rays in this plane, therefore, meet a linear, as opposed to a curved surface, and therefore pass through the lens finally undeviated. We may say that there is no curvature in the meridian parallel to the axis. A plane at right angles to the axis of the cylinder intersects the latter in the arc of a circle (Fig. 386) so that rays in this plane are deviated as by a spherical lens. We may say that the lens has curvature in the meridian at right-angles to its axis.

Image Formation. To understand the formation of the image of a luminous point by a cylindrical lens, we have only to consider the behaviour of rays in these two mutually perpendicular planes; rays in all other planes may be considered to be partly in one, and partly in the other of these two, and will therefore not modify the shape of the image formed by the rays in these mutually perpendicular planes. The rays in the vertical plane through the axis (Fig. 385), emanating from A, pass through undeviated; the rays in the horizontal plane (Fig. 386) are brought to a point focus at a distance determined by the curvature in this meridian, namely at A'. If Figs. 385 and 386 are combined mentally, it will be seen that the image of A at A' becomes a vertical straight line, i.e. a line parallel with the axis of the cylinder.

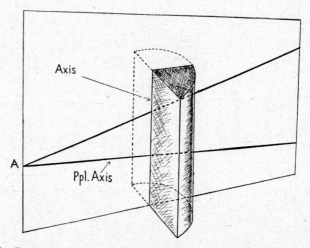

FIG. 385. Rays in the vertical plane through the axis of the cylindrical lens are undeviated.

This is made clearer by Fig. 387; here the refraction in two planes is shown in perspective. The lower plane is horizontal through A, and the rays are brought to a point focus at A' on the principal axis through A; the upper plane is inclined upwards, and the rays in this plane are bent inwards, but not downwards as they would be by a spherical convex lens. They thus converge to the point A" vertically above A'. A'A" is therefore a part of the line image of A.

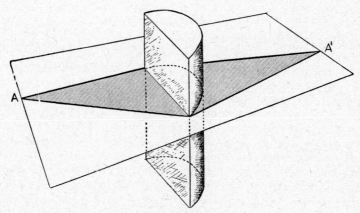

FIG. 386. Rays in the plane at right-angles to the axis are made to converge on the point A'.

Cylindrical lenses may be plano-convex, convexo-convex, plano-concave, etc. The important feature of a cylindrical lens is that it modifies the degree of convergence or divergence (the "vergence") of rays striking it, *only in the plane at right-angles to its axis.* The vergence in the plane parallel with the axis remains unchanged.

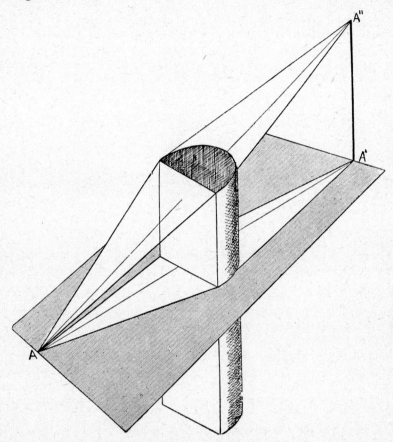

FIG. 387. Illustrating the formation of a line image of a point object by a cylindrical lens.

The Toric Lens

The cylindrical lens is a special case of an *astigmatic lens*—a lens which fails to give a point image of a point object; any lens in which the curvatures in different planes are not equal will be astigmatic, but the only other case with which we need concern ourselves here is the lens of *toric surface*—a lens in which the curvatures in two meridians at right-angles are different. A toric surface is defined as the surface formed by the revolution of a circle about a line parallel with a diameter. For example, the surface of a tyre inner-tube could be formed by rotating a circle of diameter d, equal to the cross-sectional diameter of the tube, about a line distant R from the circle. If we hold the inner-tube horizontally, the curvature in the vertical meridian is given by the curvature of the small circle whereas the curvature in the horizontal meridian is given by that of the large circle of radius R.

We may deduce the refracting properties of a toric surfaced lens in the same way as for a cylinder; instead, however, of treating the components of rays in one plane as being undeviated, we must imagine that rays in all planes are deviated, but that their components in two mutually perpendicular planes converge on (or diverge from) two different points.

Image Formation. Let us consider a plano-toric convex lens with a radius of curvature in the horizontal plane, R_H, and a radius of curvature in the vertical plane R_V, and let us assume that R_H is greater than R_V, i.e. that the *curvature* or *dioptric power* in the horizontal meridian is less than that in the vertical (Fig. 388).

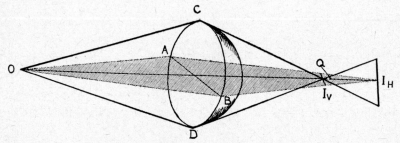

FIG. 388. Refraction by a toric lens.

OC and OD are two rays in the vertical plane through the optic axis; they meet at I_V. A pair of rays in the horizontal plane, OA and OB, meet at I_H.

Since the dioptric power is greater in the vertical meridian, the rays OC and OD converge to the point I_V sooner than the rays OA and OB converge to I_H. The refraction of rays in intermediate planes may be treated by resolving them into the two planes we have just considered, and the result is that at I_V a horizontal line is formed, due to the fact that rays in the vertical plane have come to a focus whilst those in the horizontal plane have not. At I_H the rays in the horizontal plane have come to a focus whilst those in the vertical plane have diverged from I_V, so that at I_H a vertical line is formed.

At the point Q, on the axis of the lens, the degree to which the rays in the vertical plane have diverged from the axis is equal to the degree to which the rays in the horizontal plane have converged on it; the image of O on a screen placed at Q is therefore a circle, referred to as the *circle of least confusion*; it is the nearest approach to a point image of the point object O. If the screen is moved closer to, or farther from, the lens, the image becomes oval. It is not difficult to envisage the shape of the cross-section of the beam at any point; at I_V it is a line; beyond I_V in the direction of Q the rays in the horizontal plane converge, whilst those in the vertical plane diverge, so that the cross-section is an ellipse with its long axis horizontal; at Q the cross-section is a circle; beyond Q it is an ellipse with its long axis vertical; at I_H it is a vertical line. The shape of the emergent beam through a toric lens is called *Sturm's conoid*. It should be noted that the first line-image (I_V of Fig. 388) is parallel to the meridian of least curvature whilst the second line-image is at right-angles to this meridian.

A toric lens may be made up to two lenses in contact, where one is a cylinder and one a sphere, or alternatively of two cylinders of unequal power with their axes at right-angles. Thus if we take a convex cylinder of $+5$ D and a convex sphere of $+3$ D and place them in contact, with the axis of the cylinder vertical, the power of the combination in the horizontal meridian will clearly be $+8$ D; in the vertical meridian the cylinder has no refracting power (its axis is vertical) and the power of the combination is only $+3$ D. We thus have a toric, or astigmatic, combination with a curvature in the horizontal meridian greater than that in the vertical. An infinitely distant point object will produce a line-image at the first line-focus and this line will be vertical (parallel to the meridian of least curvature) at a distance of $100/8 = 12 \cdot 5$ cm. A second, horizontal, line-image will be formed at a distance of $100/3 = 33 \cdot 3$ cm.

Astigmatism

Astigmatism in the eye is a condition of *refractive ametropia*, consisting of a difference in the refracting power of the eye in two planes or meridians; the astigmatism is called *regular astigmatism* if the meridians of extreme curvature are at right-angles to each other—as in the toric surfaces considered—and we need only concern ourselves with this form of astigmatism here, If the meridian of least curvature is horizontal* the astigmatism is described as *direct* or "*with the rule*" whilst if the meridian of least curvature is vertical it is called *inverse* or "*against the rule*". An astigmatic individual does not see a point object distinctly, the best image obtainable on the retina being, theoretically, the circle of least confusion; extended objects are thus blurred. Under certain conditions, however, a clearly defined image of a line can be formed on the retina as the following will make clear.

Image of a Line. In Fig. 389, L is an astigmatic convex lens and O'OO'' is a distant vertical line-object. Let us assume that the astigmatism is such that refraction in the vertical meridian is greater than that in the horizontal, i.e. the astigmatism is regular and direct. A point O on the optic axis forms two line-images at the first and second line-foci, I_V and I_H respectively; since refraction is greater in the vertical meridian, rays in this plane come to a focus first and I_V is a horizontal line; I_H is vertical. Other points, O' and O'' form horizontal and vertical lines I'_V and I''_V, I'_H and I''_H. The appearance of the image of the vertical line O'OO'' formed on a screen placed at I_V therefore consists of a series of short parallel lines placed one above the other and gives the impression of a blurred vertical line (Fig. 389; *inset*). The image of O'OO'' formed at I_H, on the other hand, consists of a series of vertical lines superimposed one above the other and the appearance on a screen is that of a well-defined vertical line (Fig. 389; *inset*).

With a vertical line-object, therefore, we have a clear line-image at I_H, the focus of the horizontal meridian, and we may state that the *meridian in focus is at right-angles to the clearly defined line.*

If we apply the same argument to the images of a horizontal line formed by the same toric surface, we find that the one formed at the first line-focus, I_V,

---

* The meridians of differing curvature must be at right-angles in regular astigmatism; they need not, of course, be absolutely vertical or horizontal. So long as the meridian of least curvature makes an angle of less than 30 degrees with the horizontal plane the astigmatism is said to be "with the rule." If the meridian of least curvature is within 30 degrees of the vertical plane the astigmatism is "against the rule."

consists of overlapping horizontal lines, i.e. it is a clearly defined horizontal line, whilst the image formed at the second line-focus, I_H, consists of a series of short parallel vertical lines, and has the appearance of a blurred horizontal line. Thus with a horizontal line-object, we have a clear line-image at the point I_V, the focus of the vertical meridian, and once again the *meridian in focus is at right-angles to the clearly defined line.*

Hence if the object consists of a cross with arms vertical and horizontal, a screen at the first line-focus receives a clearly defined horizontal line and a blurred vertical, the blurred line being parallel with the meridian in focus (vertical); a screen at the second line-focus receives a clearly defined vertical line and a blurred horizontal, the latter being parallel with the meridian in focus.

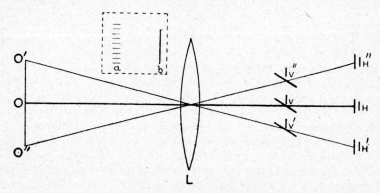

FIG. 389. Refraction by a toric lens. A vertical line O′O″ gives a clear line image at the focus of the horizontal meridian.

We may thus make the generalization: if a screen intercepts the image of two lines at right-angles formed by a toric surface, at such a point that the image of one of the lines is clear and the other blurred, *the screen is at the focus of the meridian parallel to the blurred line.*

Estimation of Astigmatic Correction. This principle—that, when a line is clearly defined whilst a line at right-angles is blurred, the retina is at the focus of the meridian parallel to the blurred line—provides the mean for determining the degree of astigmatism of an eye and its meridians of greatest and least curvature. To the astigmatic eye no point is seen distinctly but a line may be, provided that this line is at right-angles to the meridian in focus at the retina. Thus a series of radiating lines are presented to the subject (an *astigmatism chart*); if all the lines are clearly defined there is no astigmatic error; if regular astigmatism is present one line is sharper than the rest, whilst the line at right-angles to this is most blurred. The retina is thus at the focus of the meridian parallel to the blurred line and the eye requires no correction in this meridian. The meridian at right-angles, on the other hand, is out of focus so that a correcting cylinder is necessary. It will be remembered that a cylinder modifies the vergence of rays in the meridian at right-angles to its axis, consequently, in order that the cylinder may modify the refraction in any meridian, its axis must be at right-angles to this meridian. We must therefore place the axis of the correcting cylinder parallel with the direction of the most blurred line or at right-angles to the direction of the sharply defined line.

Let us suppose that we are confronted with a subject who is unable to see clearly, and that this defect cannot be remedied with the aid of spherical lenses only. We may place the subject before the astigmatism chart and put up spherical lenses until one of the lines, say a vertical one, appears distinct whilst a horizontal line is blurred. Suppose that this is achieved with a +6 D sphere. We now know that the subject is astigmatic and that with a correction of +6 D the horizontal meridian is in focus on the retina. In this meridian, therefore, a correction of +6 D *cyl.* is necessary and the axis must be vertical. More spheres are put up until the horizontal line is distinct and the vertical blurred; suppose a +8 D sphere is required. The retina is now at the focus of the vertical meridian and a correction of +8 D *cyl.*, axis horizontal, is required. The eye may thus be fully corrected by two cylinders: +6 D, axis vertical, and +8 D, axis horizontal; their combined effect is clearly the same as that of: +6 D *spher.*, and +2 D *cyl.* axis horizontal.

NEUTRALIZATION AND DECENTRING

Neutralization of Lenses

Two lenses of the same power, but of opposite sign, when placed in contact have no dioptric power and behave as a piece of plate glass; the power of a lens may thus be estimated by determining the power and sign of the lens necessary to make its combination with the unknown lens behave as a piece of plate glass, i.e. by *neutralizing* it.

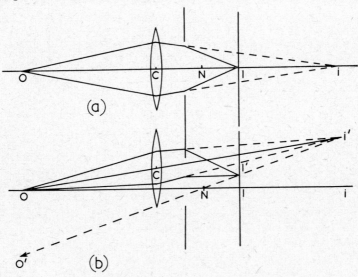

FIG. 390. The effect of decentring a convex lens.

If a lens is moved in front of the eye, objects seen through it appear to move, whereas objects appear to remain stationary when viewed through a thin piece of plate glass under the same conditions. A lens is thus neutralized when moving the combination in front of the eye leads to no apparent movement of objects in the field of view.

In Fig. 390, *a*, a converging lens is placed in front of the eye so that the optic axes coincide. The lens would form an image of O at *i*; this image behaves as a virtual object for the eye, which produces an image I, on the common optic axis.

In Fig. 390, *b* the lens has been moved to the right; an image of O is formed by the lens at *i'* on the optic axis *of the lens*; *i'* serves as a virtual object for the eye and the position of the image of this object on the retina is given, as always, by drawing a line from *i'* through the nodal point; it cuts the retina at I'. The image of I on the retina is projected along I'O' so that moving a convex lens in front of the eye causes an apparent displacement of objects *in the opposite direction*.

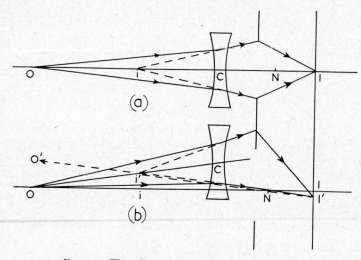

FIG. 391. The effect of decentring a concave lens.

Fig. 391 demonstrates the effects of moving a concave lens, and it is clear that the apparent displacement of objects is *in the same direction* as that of the movement.

An inspection of Fig. 390, for example, shows that the greater the strength of the lens, the greater the apparent displacement; thus if the power of the lens were increased, *i'* would be closer to C, and the line *i'*N would be more strongly inclined to the eye's optic axis, thereby making the excursion II' larger.

There is an exception to the rule which states that the apparent movement of objects viewed through a converging lens is "against" the movement of the lens. When the image of the object viewed is formed in front of the eye's principal plane, the apparent movement is in the same direction as the movement of the lens (Fig. 392).

Decentring of Lenses

Spherical Lenses. A spectacle lens whose optic axis does not coincide with that of the eye is said to be decentred; it is clear from the foregoing discussion that a decentred spherical lens has a prismatic effect in addition to its modification of the vergence of incident rays. Thus in Fig. 390, *b*, an object O appears

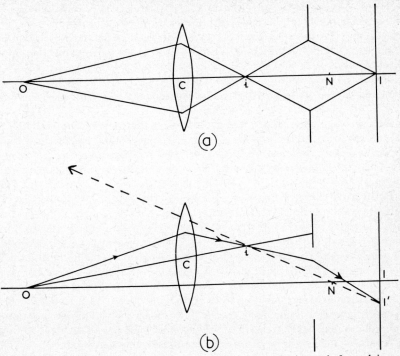

FIG. 392. The effect of decentring a convex lens when the image is formed in front of the eye. The image moves *with* the direction of decentration.

to be at O′; if the eye is imagined to be a right eye, it must be adducted if it is to form an image on the fovea, I. A converging lens whose centre has been moved outwards thus behaves as an adducting, or base-out, prism (p. 542). From Fig. 391, *b*, it is seen that decentring a concave lens outwards causes abduction.

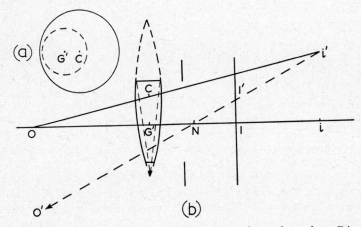

FIG. 393. Illustrating the cutting of a decentred lens from a larger lens. C is the optical centre of the lens but it does not coincide with the geometrical centre G′.

In the sperical lenses so far considered, the optical centre corresponded with the geometrical centre, so that decentration had to be achieved by moving the lens bodily; the same prismatic effect can, however, be obtained by making use of a lens whose geometrical and optical centres do not correspond. If a circular piece of the large lens in Fig. 393, *a* is cut out asymmetrically, the geometrical centre of the piece is at G', whereas the optic centre is at C. Such a lens, when mounted so that its geometrical centre corresponds with the optic axis of the eye, shows a prismatic effect (Fig. 393, *b*).

The prismatic effects are greater, the greater the degree of decentration and the power of the lens; in general, a displacement of the image equal to one prism dioptre is obtained for each centimetre of decentration of a 1 D lens; thus half a centimetre of decentration of a 6 D lens corresponds to 3 prism dioptres.

Cylindrical Lenses. A study of Fig. 395 (p. 574) will show us that, if a cylindrical lens is decentred in the direction of its axis, there is no prismatic effect, in fact it would be incorrect to speak of decentring in this special case, since the optic axis of the eye would still coincide with a principal axis of the cylinder. Moving the lens in a direction at right-angles to its axis must, on the other hand, have a prismatic effect, since now the optic axis of the eye is inclined to all the principal optic axes of the cylinder. The direction of the axis of a cylinder may therefore be obtained by finding the direction in which it must be moved across the eye to give no apprent displacement of the image. When this has been found, the cylinder may be neutralized by placing other cylinders in contact, with their axes parallel, until there is no apparent displacement on moving the lens at right-angles to the direction of the axis.

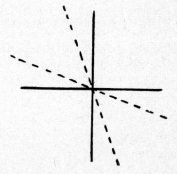

FIG. 394. The scissors effect.

A cylinder modifies the vergence of light in one meridian but not in another at right-angles; the magnification in the two meridians is therefore different, with the result that a distorted image is seen; thus a convex cylinder with its axis vertical may magnify dimensions in the horizontal meridian and leave them unchanged in the vertical; if a cross is regarded through a cylinder, one arm corresponding in direction with the axis, this distortion is not evident since one line is only made a little longer; on rotating the cylinder about the optic axis of the eye, however, a characteristic "scissors effect" is obtained (Fig. 394) since the greatest magnification is obtained, now, along a line between the two arms of the cross.

Diagnosis of a Lens

A spectacle lens of unknown characteristics may be completely diagnosed by the following procedure. A cross is viewed through the lens, care being taken to make the geometrical centre coincide with the optic axis of the eye. Any prismatic effect is revealed by a bodily shift of the centre of the cross (Fig. 395, *a*); when this has been neutralized by a prism, the combination may be rotated until any "scissors" distortion disappears; the axis of the cylindrical component now coincides with one of the arms of the cross. The combination is now moved up and down; dioptric power in the vertical meridian will cause the centre of the cross to move vertically up or down, and this is revealed by an apparent movement of the horizontal line (Fig. 395, *b*). This movement may be neutralized by a cylinder or sphere; thus when the horizontal line remains stationary on moving the combination vertically, the vertical meridian is neutralized. The combination is now moved horizontally and new spheres or cylinders are added until apparent movement of the vertical line (Fig. 395, *c*) ceases. In this way the prism strength, and the dioptric powers in the vertical and horizontal meridians, are found.

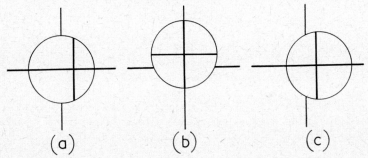

FIG. 395. (*a*) The effect of a prism (or decentred lens) on the appearance of a cross. (*b*) The effect of curvature in the vertical meridian on the appearance of the cross on moving the lens upwards. (*c*) The effect of curvature in the horizontal meridian on the appearance of the cross on moving the lens sideways.

CONTACT LENSES

The contact lens consists of a glass shell which fits in close apposition to the cornea and sclera, and accompanies the eye in its movements. We need only concern ourselves here with the *corneal segment*, i.e. the portion fitting over the cornea, the function of which is to correct optical defects. This is achieved by eliminating the cornea as a refracting surface and, in effect, replacing it by a perfectly spherical one.

The general principle of the use of a contact lens is indicated in Fig. 396. The corneal segment is separated from the cornea by a thin film of saline. Refraction of light takes place as indicated, i.e. the ray is deviated by the anterior surface of the contact glass and also by the posterior surface; since saline has approximately the same refractive index as that of the cornea, no further refraction takes place until the lens is reached. Thus the cornea has been replaced as a refracting surface, and if it contains irregularities (astigmatism, etc.) they are unable to influence the refraction of light since the layer of saline fills in these irregularities and gives a surface corresponding exactly in shape with that of the posterior surface of the contact glass. If the eye is emmetropic apart from these corneal defects, the contact glass need only be designed so that it contributes nothing to the focal power of the system, modification of the vergence of incident rays being, in effect, carried out by the spherical surface of saline. This is achieved by making the curvatures of the two surfaces of the contact glass equal to that of the cornea; in this event the glass is *afocal*—it is equivalent to a parallel-sided thin plate glass; any influence one surface of the glass may have on the dioptric power of the system is cancelled out exactly by the effect of the other surface.

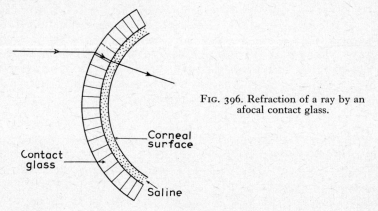

Fig. 396. Refraction of a ray by an afocal contact glass.

If a spherical refractive error is present, it can be corrected by modifying the curvature of the anterior surface of the contact glass. If this curvature is made greater, the contact glass behaves as a convex meniscus lens and corrects hypermetropia.

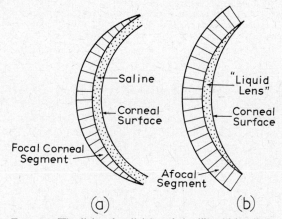

Fig. 397. The "glass lens" (*a*) and the "liquid lens" (*b*).

The Glass and Liquid Lens Systems. When the spherical correction in a contact lens is achieved by modifying the curvature of the anterior surface, as in Fig. 397, *a*, the system is referred to as a *"glass lens,"* the increased dioptric power being provided by the extra curvature of the anterior surface of the glass.

If, on the other hand, the glass is kept afocal by making its surfaces concentric, corrections can still be made by modifying the effective curvature of the cornea, i.e. a so-called *"liquid lens"* is formed (as in Fig. 397, *b*) where the two glass surfaces are concentric but the curvature of the glass is different from that of the cornea; a meniscus-lens shaped space is thus left between glass and cornea; this is filled with saline and the curvature of the cornea has, in effect, been modified.

The term "liquid lens," as opposed to a glass lens, is somewhat of a misnomer; the liquid between the cornea and the glass is lens-shaped, but since the refractive index of the liquid is taken as being equal to that of the cornea, the liquid behaves optically, not as a lens, but as a *single* spherical refracting surface—a modified cornea.

Advantages of the Contact Lens

We have seen that a decentred lens behaves as a prism; when the eyes look obliquely through spectacles the lenses are, in effect, decentred and the consequent prismatic effect can be annoying, especially when the corrections applied to the two eyes are of opposite sign. The contact lens, by moving with the eye, does not suffer from this defect.

The retinal image in corrected ametropia is equal in size to that obtained in the emmetropic eye if the correcting lens is placed at the anterior focus (p. 567). In anisometropia, therefore, the images in both eyes may be made equal with correcting lenses at the anterior foci. When the correcting lens is brought closer to the eye, the image becomes larger, if the correction is for myopia, and smaller for hypermetropia. In anisometropia, therefore, there may be a considerable disparity in the sizes of the corrected images when contact lenses are worn. Contact lenses, from this point of view, are a positive disadvantage. However it has been found that the discomfort associated with the prismatic effect of spectacles is greater than that due to the *"aniseikonia"* of contact glasses in corrected anisometropia; the contact glass is therefore often chosen as the lesser evil.

In corrected unilateral aphakia, however, the condition is different; here the image in the corrected aphakic eye is some 33 per cent larger than that in the emmetropic eye, By moving the correcting lens closer to the eye, the size of the image is diminished. In the case of a $+15$ D correction, the size is reduced by 20 per cent by using a contact glass, so that the disparity is diminished sufficiently to permit fusion of images in binocular vision.

OPHTHALMOSCOPY

Ophthalmoscopy is concerned with the observation of the interior of the eye, principally the fundus; this is viewed by an observer as an image formed by the dioptric system of the eye. We shall be concerned here with the optical principles governing the appearance of the image under different refractive conditions; since the light normally falling on the fundus is entirely inadequate for the purposes of observation, we shall also be concerned with the optical principles involved in the illumination of the eye.

The Direct Method

In Fig. 398 R and H represent the retina and principal plane of an emmetropic eye; N is the nodal point. A point O, on the optic axis, emits a pencil of rays whose size is determined by the pupil; the rays emanate from the posterior focus and consequently run parallel to the axis after refraction at H. A point O′, on the retina but a little way off the axis, gives rise to another parallel bundle of rays, inclined to the first. In order that the two points may be visible to an observer in front of the eye, at least some of the rays from each point, O and O′, must enter the observer's eye. If the eye is close to the subject's principal plane this occurs, but when it is far away the beam from O′, owing to its deflexion downwards, misses the observer's pupil entirely; in this event O′ is *outside the field of view*, the limit being given by a point closer to O. Thus the *field of view* of an observer, looking directly into the eye, decreases with his distance away; it will be clear from Fig. 398 that the field increases with the size of the subject's and observer's pupil (sight-hole). In this, the *direct method of ophthalmoscopy*, the observer's eye must be as close as feasible to the subject's if a reasonable field of view is to be obtained.

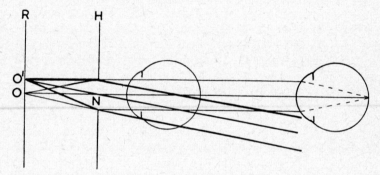

Fig. 398. The direct method of ophthalmoscopy. Showing that the field of view depends on the distance between observer and subject and the sizes of their pupils.

We may now consider the passage of light from the subject's fundus to the retina of the observer. In Fig. 399 R_s is the retinal plane, H_s the principal plane, N_s the nodal point, and F_a the anterior focus of the subject's eye; corresponding positions for the observer are R_o, H_o, and N_o.* The subject is emmetropic, so that O corresponds to the posterior focus. A ray from a point O′ on the subject's retina passes through the nodal point undeviated. A ray from O′ parallel with the optic axis passes through F_a; since the eye is emmetropic all rays from O′ leave the eye as a parallel bundle. To determine the passage of this bundle through the observer's eye we proceed as follows: A ray through the observer's nodal point, N_o, parallel to the bundle, would strike the observer's retinal plane at I′. If the observer is emmetropic, parallel rays meet in the retinal plane, consequently the bundle from O′ comes to a focus at I′.

* In this, and subsequent diagrams, the observer's anterior focus is not assigned any particular position; the constructions are thus independent of the position of the observer's eye provided, of course, that light from the fundus may enter it. The student may work out for himself the special case where the two anterior foci coincide.

Rays from O emerge parallel to the axis and come to a focus at I, so that
II′ is the image, on the observer's retina, of the portion OO′ of the subject's
fundus. It is an inverted image on the retina, and the fundus is therefore seen
erect.

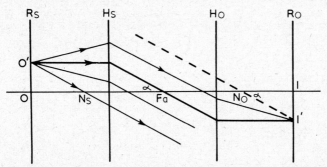

FIG. 399. The direct method. Passage of rays from emmetropic fundus to
observer's retina.

The apparent size of OO′ is determined by the size of II′, and this may be
represented by the angle II′ subtends with the nodal point, N_o, i.e. the angle α.

The formation of the image of a hypermetropic fundus may be tackled in
the same way, but here the emergent beam is not parallel and to trace it we
must know the degree of hypermetropia and thus the position of the far point.

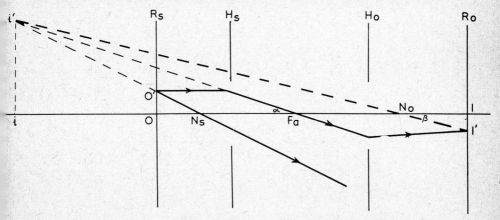

FIG. 400. The direct method. Passage of rays from hypermetropic eye.

The construction is shown in Fig. 400; the far point is the position of an
object such that its image falls on the retina of the unaccommodated eye; the
far point and the retina are thus conjugate foci, and the image of OO′, formed by
the dioptric system of the eye, is therefore at the far point, i.e. behind the eye
of a hypermetrope. Let us construct this image. A ray parallel to the axis passes
through the anterior focus, F_a. A ray through the nodal point passes on undeviated. These two rays diverge and, when produced backwards, meet at $i′$,
the image of O′. $ii′$ is the image of OO′, and all rays emanating from O′, for
example, appear, after refraction by the subject's eye, to diverge from $i′$.

The diverging rays from O′ are brought to a focus on the observer's retina, to give an image I′; the position of this point is given by considering the ray through the nodal point, N_o; this ray appears to diverge from $i′$; all other rays converge on I′, the point of intersection of this ray with the observer's retina.

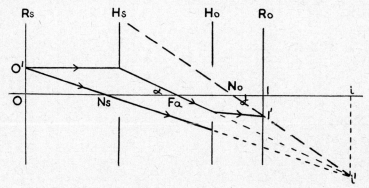

FIG. 401. The direct method. Passage of rays from myopic eye.

With the myope, the far point is in front of the eye, e.g. at i (Fig. 401). The image of OO′ is thus $ii′$ and may be derived by the appropriate construction. Rays from O′ therefore converge, after refraction, on $i′$ and thus provide a virtual object for the observer's eye. The image of $i′$, formed by the dioptric system of the observer's eye, is given, once again, by the ray through the nodal point, N_o.

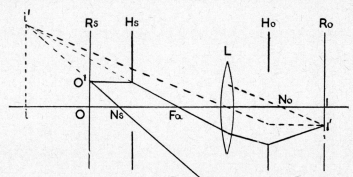

FIG. 402. The direct method. Passage of rays from hypermetropic eye; the observer is using a correcting lens.

Magnification. If the images II′ are compared in Figs. 399–401, it will be seen that their apparent sizes are determined by the angle α for the emmetrope, the angle β for the hypermetrope, and the angle γ for the myope. Since β is less than, whilst γ is greater than α, the apparent size of the subject's fundus is greatest in myopia and least in hypermetropia. The magnification can be calculated quite simply; in the emmetropic eye it is 15.

Correcting Lens. When the subject is hypermetropic, the emergent rays from the fundus diverge and the observer must either accommodate or put

up a correcting lens; the construction of the image in the presence of a correcting lens is shown in Fig. 402; the emergent beam may be constructed as before;
after striking the correcting lens the rays emerge parallel, with their direction
determined by the ray from i' through the centre. Parallel rays now strike the
observer's principal plane and converge on I' whose position is determined by the
ray in this parallel bundle passing through N_o.

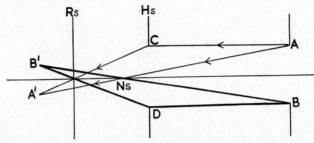

FIG. 403. A'B' is the image of the sight-hole, AB, formed by the refracting
system of the subject's eye.

Field of View. In Fig. 403 R_s and H_s have the same meaning as before;
CD is the subject's pupillary aperture and AB that of the observer or, to be
more accurate, the sight-hole. Let us construct the image of AB formed by the
dioptric system of the subject's eye, assuming that this remains unaccommodated.
A ray through the nodal point, N_s, passes on undeviated; a ray parallel to the
axis passes through the posterior focus. The image of the top of the aperture,
A, is thus formed at A', behind the retina; the image of B is formed at B'.

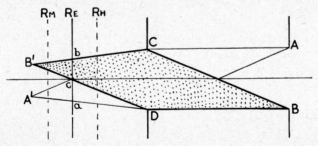

FIG. 404. The projected image of AB on the retina, acb, represents the field of
view.

Having established the position of this image, we may now consider the paths
of the complete pencils of light from A and B, as in Fig. 404; it will be seen that
the pencils cut the subject's fundus in two blur-circles, cb and ca. We may
now show that ab represents the total field of view available to an observer
looking through the aperture AB. This is best done by imagining that the rays
from the image of the sight-hole are reversed; thus the ray bC, on emerging
from the eye, strikes the bottom of the sight-hole. All other rays from b will,
after emerging from the eye, run parallel with this ray and will therefore fail
to get through the sight-hole. The point b thus represents a point on the fundus,

the top of whose emergent pencil just grazes the sight-hole; it is consequently the farthest point above the optic axis consistent with permitting any light from it to enter the sight-hole and is thus the upper limit to the field of view.

The same argument may be applied to the ray aD from a; it strikes the top of the sight-hole and other rays emerge parallel with it and fail to get through. ab is thus the field of view of the subject's fundus; it is, in fact, the *projected image of the sight-hole on the fundus*. We have thus far considered that the subject was emmetropic; the statement that the field of view is the projected image of the sight-hole on the fundus can be shown to be quite general and independent of the refractive condition; we can therefore derive an estimate of the qualitative effects of ametropia on the field of view by simply imagining that the fundus is nearer to, or farther away from, the principal plane. Thus in Fig. 404 it is evident that the field of view is greatest in hypermetropia and least in myopia.

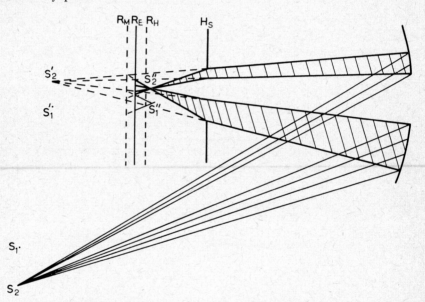

FIG. 405. Illumination with a concave mirror in the direct method.

Method of Illumination. The classical method of illumination of the fundus by means of a concave mirror is indicated in Fig. 405 (not drawn to scale). The mirror is tilted so that the image of the source $S_1 S_2$ lies symmetrically about the optic axis of the eye. In the absence of the eye an inverted, diminished, real image of $S_1 S_2$ would be formed by the mirror at $S_1' S_2'$; this behaves as a virtual object for the dioptric system of the eye which produces an image $S_1'' S_2''$ in front of the retina. The retina is thus illuminated by blur-circles formed by the diverging rays from $S_1'' S_2''$ and from intermediate points on the image. It is clear that the field of illumination is greatest in myopia and least in hypermetropia. If a plane or convex mirror is used, the image of the source, formed by the eye, is behind the retina, since the rays entering the eye are divergent, The fundus is illuminated by blur-circles formed by rays converging on points behind the fundus and the order of field size is reversed, the greatest field being in hypermetropia.

SELF-LUMINOUS INSTRUMENT

In modern practice, of course, the separate source and mirror have been substituted by a compact instrument, the *self-luminous ophthalmoscope*. Fig. 406 illustrates the essential principles; the light from a lamp, L, is brought to a point-focus on the edge of the mirror, so that the effective source is very close to the subject's eye. The rays are so strongly divergent, therefore, that the fundus is illuminated with a large out-of-focus patch. In order to reduce the nuisance caused by the reflexion off the cornea, the *corneal reflex*, the axis of the sight-hole, C, is above that of the illuminating beam.

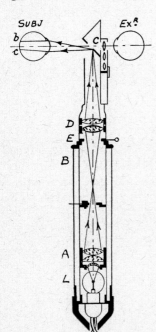

FIG. 406. Modern self-luminous ophthalmoscope of the Fincham type. (Emsley *Visual Optics*.)

The Indirect Method

With the direct method of ophthalmoscopy, the effective field of view in emmetropia is about 6·5°, i.e. little more than the optic disc; it is considerably more restricted in myopia. A glance at Fig. 398 shows us that the restriction in the field is determined, other things being equal, by the deflexion of the pencils of light as they leave the eye; if these pencils could be deflected again towards the observer's eye, the field could be increased. In the *indirect method* this is brought about by interposing a convex lens of about 13 D between the subject's and observer's eyes.

The passage of light from an illuminated point O′ on the fundus, under these conditions, is illustrated in Fig. 407. The total separation between the eyes of subject and observer is now larger, say 60 cm. The condensing lens is placed at a point, C, such that the sight-hole and the subject's pupil are conjugate foci, i.e. the sight-hole is imaged in the subject's pupillary plane; for a 13 D lens the necessary position is about 9 cm from the subject's eye. f_1 and f_2 are respectively the first and second principal foci of the condensing lens.

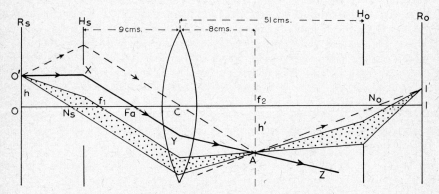

FIG. 407. The indirect method of ophthalmoscopy. The passage of rays from the fundus of an emmetropic eye to the observer's retina.

A ray O'X, parallel to the axis, passes through F_a and strikes the lens in Y. If the eye is emmetropic, all other rays from O' emerge parallel with this ray and after refraction by the condensing lens they converge on to its second focal plane through f_2. The actual point to which they converge may be found by drawing in the ray through the centre of the lens; it is parallel with XY and strikes the focal plane in A. All rays leaving the eye therefore meet in A, which is the image of O' formed by the refracting system of the subject's eye and the condensing lens. It is this *real* image that the observer sees; the image

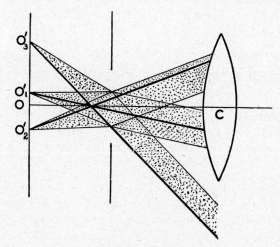

FIG. 408. The indirect method. The field of view is determined by the size of the condensing lens.

of A on the observer's retina is given in the usual way by drawing a ray through the nodal point, N_o. If the observer accommodates, all the rays from A, entering his eye, meet at I'. The actual pencil of light entering the observer's eye is thin compared with that obtained in the direct method; this follows from the fact that the sight-hole and the pupillary plane are conjugate foci; only light from the image of the sight-hole in the pupil can pass through the sight-hole itself; the size of the image of a 4 mm sight-hole is about 0·7 mm.

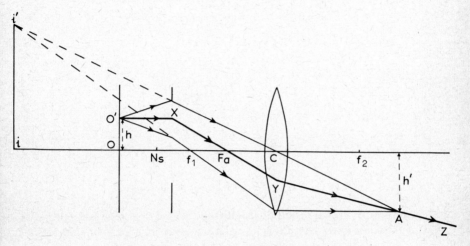

FIG. 409. The indirect method. Hypermetropic eye.

It will be noted that the image of the fundus, as seen in the indirect method, is inverted. The magnification for the particular set-up shown in Fig. 407 amounts to about 2·7, being much less than in the direct method.

Field of View. It will be evident from Fig. 407 that, so long as an emergent pencil of rays strikes the condensing lens, it will be bent back towards the optic axis to converge on the focal plane of the condensing lens, producing an image which the observer can see. Thus the two points O_1' and O_2' (Fig. 408) on the subject's fundus are both visible because pencils from them strike the condensing lens; the point O_3' is not seen. The size of the condensing lens therefore determines the field of view in the indirect method. For manipulative purposes a large condensing lens is used, but owing to aberrations, the effective useful aperture is about 4 cm. The field of view is easily calculated and amounts to about 25°, i.e. it is about four times larger than the effective field in the direct method.

In myopia it may be shown that the angular field of view is greater than in emmetropia, whilst the reverse relationship is found with hypermetropia.

Path of Rays in Ametropia. Exactly the same principles of ray tracing may be applied to determining the image of the fundus of hypermetropic and myopic eyes in the indirect method. Thus, as with the direct method, the position of the image of the fundus at the far point is first determined (Figs. 409 and 410). The emergent pencil from O', in the case of hypermetropia, diverges from its image i'; the course of the rays after striking the condensing lens is determined by considering the ray through its centre. It appears to come from i' and passes through undeviated. Another ray from i', through the first principal focus of the condensing lens, runs parallel to the axis after refraction. These two rays intersect at A, the image of O' seen by the observer. The further passage of the rays to the observer's retina may be drawn in as before.

The constructions show quite clearly that the size of the image of the fundus, seen by the observer, is greatest in hypermetropia, intermediate in emmetropia, and least in myopia.

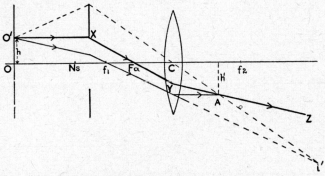

FIG. 410. The indirect method. Myopic eye.

Estimation of the Refractive Condition. We may analyse the effects of moving the condensing lens nearer to, or farther away from the subject by deriving the essential features of the three preceding constructions (Figs. 407, 409 and 410). There is a ray common to all three conditions, namely, the ray O'X, parallel to the optic axis of the eye. This passes through the anterior focus of the subject's eye, F_a, in each case and, since it strikes the lens at the same point, Y, and at the same angle, it is refracted in the same direction (along YZ). We may thus draw a general diagram to embrace the three refractive conditions, as in Fig. 411, the ray O'XYZ being considered to have emanated from a point h mm above the pole of the fundus in each condition.

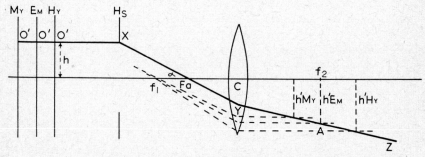

FIG. 411. The indirect method. First focus of the condensing lens is within the anterior focus of the eye. The ray O'XYZ diverges from the axis and the image, h', is greatest in hypermetropia.

The position of the image of O', formed by the eye's refractive system and the condensing lens, is, of course, different in the three conditions. In emmetropia it is determined by a ray through f_1 (the first focus of the lens) which, belonging to a parallel pencil, meets the ray O'XYZ in the second focal plane of the condensing lens. In hypermetropia, this second ray through f_1 diverges from XY and therefore meets the ray XYZ beyond f_2 and, because YZ slopes away from the axis, the size of the image, h', is greater than in emmetropia. In myopia, the second ray, through f_1, converges towards XY and therefore meets this ray at a point within the second principal focus, f_2, and the size of the image, h', is smaller.

The relative sizes of the images in different refractive conditions are thus determined by the fact that the ray YZ slopes away from the axis.

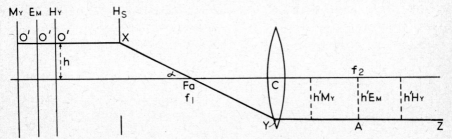

FIG. 412. The indirect method. First focus of the condensing lens coincides with the anterior focus of the eye. The ray O'XYZ runs parallel with the axis and the images are of the same size in all refractive conditions.

If the condensing lens is brought farther from the subject, so that f_1 coincides with F_a, the determining ray, O'XYZ, after refraction at the condensing lens, now runs parallel to the optic axis, since it has passed through the first principal focus of this lens (Fig. 412). The image of O' is given as before by the intersection with any other ray from O', but clearly the size is now independent of the position of the point of intersection, so that we may conclude that, in this special case, the size is the same in all three refractive conditions. A comparison of Figs. 411 and 412 shows, moreover, that the size of the image is the same as for an emmetrope when the condensing lens was closer to the subject.

If, now, the condensing lens is moved so that its first focus, f_1, is farther from the eye than F_a (Fig. 413), the determining ray O'XYZ converges on to the optic axis; it is converging because it comes from a point on the axis, F_a, outside

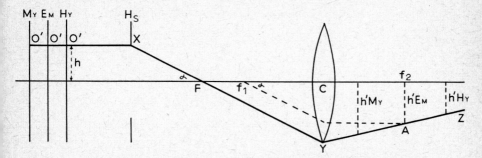

FIG. 413. The indirect method. First focus of the condensing lens is outside the anterior focus of the eye. The ray O'XYZ converges on the axis and the image, h', is greatest in myopia.

the first focus of the lens; the point to which it converges is found by considering a ray, parallel to it, passing through f_1; the two rays must meet in the second
focal plane of the condensing lens, at A. The point A corresponds to the position
of the image of O′ in the emmetropic eye, since it is the point to which parallel
rays converge; the height h' is the same as before and consequently the size
of the image of the emmetropic fundus is still unchanged. With the hypermetrope, as before, the image is formed farther away along the ray O′XYZ,
and since this converges on the optic axis, it will now be smaller than in emmetropia. In myopia the image, being formed closer to the lens, will be larger than in
emmetropia.

To take a numerical example, suppose that with the condensing lens close
to the subject (Fig. 411) the sizes in the refractive conditions are:

Hypermetropia	*Emmetropia*	*Myopia*
3	2	1

When the lens is brought out so that the first principal focus corresponds
with the anterior focus of the subject's eye (Fig. 412) the sizes are:

Hypermetropia	*Emmetropia*	*Myopia*
2	2	2

being equal and of the same size as that in emmetropia under the first conditions.
When the first focus of the lens is outside the anterior focus of the eye we have:

Hypermetropia	*Emmetropia*	*Myopia*
1	2	3

Thus on moving the condensing lens from a point close to the subject to a
point far away, if the size of the image remains constant the eye is emmetropic;
if the size diminishes, the eye is hypermetropic; whilst if it increases, the eye is
myopic.

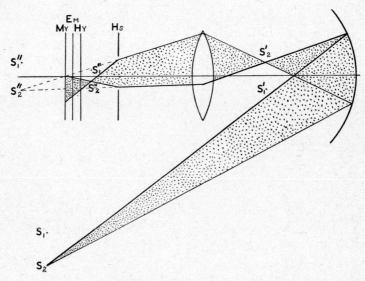

FIG. 414. The indirect method. Illumination of the eye.

Illumination

The illumination in the indirect method is complicated by the presence of the condensing lens but otherwise the optical principles are the same as those involved in the direct method. The source of light, $S_1 \ldots S_2$ (Fig. 414) is imaged by the mirror at $S_2' \ldots S_1'$ in front of the subject's eye. The condensing lens forms an image $S_1'' \ldots S_2''$ behind the subject's eye; this behaves as a virtual object for the refracting system of the eye which produces a real image, $S_1''' \ldots S_2'''$ in front of the fundus. A pencil of rays from S_2, for example, after reflexion at the mirror converges on to S_2' and after passing through this point diverges until it strikes the lens; it then converges towards S_2'' until it meets the principal plane of the eye, when it converges on to S_2'''. A blurred image of $S_1''' \ldots S_2'''$ is formed on the fundus and it is evident from Fig. 414 that the field of illumination is the greatest in myopia and least in hypermetropia.

RETINOSCOPY

By the aid of ophthalmoscopy, particularly with the indirect method, an estimate of the subject's refractive condition can be made, but by no means accurately. *Retinoscopy*, or *skiascopy*, is the name given to a more accurate objective method of determining refractive errors. By means of a mirror, as in ophthalmoscopy, the fundus is illuminated; the study of the movements of the image of the fundus, when the mirror is tilted, permits an accurate estimate of the subject's refractive condition.

In Fig. 415, *a* the point O on the optic axis and fundus is illuminated; the

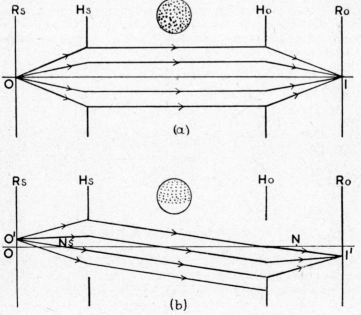

FIG. 415. Retinoscopy. Illustrating movement of the reflex.

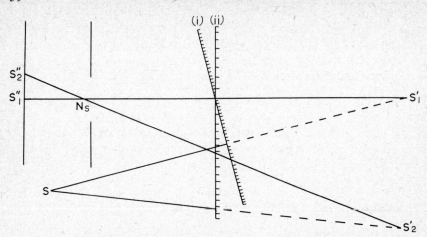

FIG. 416. Effect of rotation of a plane mirror. An upward movement of the mirror causes the illuminated point to move from S_1'' to S_2''; the reflex therefore moves upwards, i.e., it moves "with" the mirror.

pencil of light from O fills the pupil; since it emerges symmetrically disposed about the optic axis, it enters the pupil (or sight-hole) of an observer's eye placed so that its optic axis coincides with that of the subject. The point O is illuminated as in ophthalmoscopy, with a mirror, and the reflected light from the fundus is called the *reflex*. The observer views this reflex in relation to the subject's pupil, i.e. he "projects" it on to the pupil. It will be seen that light from all points on the subject's pupil enters the sight-hole, with the result that the reflex fills the pupil. If, now, the illumination of the fundus is altered so that the point O′ is illuminated, the light leaving the eye no longer travels symmetrically about the axis but is deviated downwards; rays from the lower part of the pupil thus fail to enter the sight-hole and this part of the pupil appears dark. To the observer, then, the reflex now occupies only the upper part of the pupil, and it is said to move upwards (Fig. 415, *b*; *inset*).

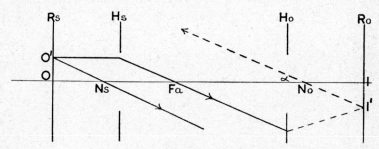

FIG. 417. Retinoscopy. Emmetropia.

The change in the illumination of the fundus from O to O′ is achieved by moving the mirror; if the latter is plane, it must be tilted upwards; the movement of the reflex is thus "with" the movement of the mirror (Fig. 416).

We may now consider how the movement of the reflex varies with the refractive condition of the subject's eye.

Emmetropia. In Fig. 417 R_s and R_o are the retinal planes of subject and observer, H_s and H_o their principal (pupillary) planes. The image of O is formed at I on the common optic axis. If the mirror is moved so that O' is illuminated, the image of O' on the observer's retina is at I', the point of intersection of the ray, in the parallel emergent beam, through the nodal point, N_o, with the observer's retina. I' is below I and the movement of the reflex is interpreted therefore, as an upward movement through an angle α, the angle subtended by II' at the observer's nodal point.

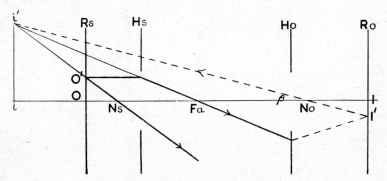

FIG. 418. Retinoscopy. Hypermetropia.

Hypermetropia of Low Degree. Fig. 418 represents the condition in hypermetropia. The image of O falls, as before, at I. The image of O' is at I', this position being obtained, as in the construction in the direct method of ophthalmoscopy, by considering the image of OO' formed by the subject's eye at its far point, namely, ii'. The image on the observer's retina is obtained by drawing a ray from i' through the nodal point, N_o. I' is below I, and the movement of the reflex is therefore upwards, through the angle β, which is less than α; consequently the movement appears less rapid.

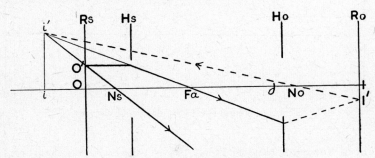

FIG. 419. Retinoscopy. Hypermetropia of high degree.

Hypermetropia of High Degree. Here the image of OO' at the far point is smaller, so that the angle γ, formed at the observer's nodal point, is smaller than either α or β; the movement of the reflex is in the same direction as before, but slower (Fig. 419).

Myopia of Low Degree. In myopia the far point is in front of the eye; if it is beyond the nodal point of the observer's eye, as in Fig. 420, the image of O' formed on the observer's retina is still such that I' is below I. Thus the image of OO' formed by the subject's dioptric system is at *ii'*; this acts as a virtual object for the observer's eye, and the image is constructed by drawing a ray from *i'* through the nodal point, N_o; this strikes the retina below the optic axis. The movement of the reflex is thus, once again, upwards or "with," and the angle δ measures the speed of movement. δ is clearly greater than α.

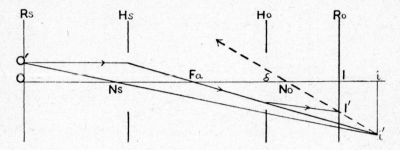

FIG. 420. Retinoscopy. Myopia of low degree.

Myopia of High Degree. If the far point falls between the subject and observer, as in Fig. 421, the image of O' falls at I', *above* the optic axis, i.e. the reflex appears to move downwards or "against" the movement of the plane mirror. The angle ε measures the rate of movement and this will clearly decrease, the greater the degree of myopia.

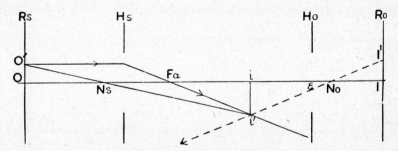

FIG. 421. Retinoscopy. Myopia of high degree.

Point of Reversal. Neutral Point. Between the condition of myopia, with the far point behind the observer's nodal point, and that in which it is in front, the movement of the reflex changes sign; the actual point at which this change takes place is called the *point of reversal* or *neutral point*; it is the point at which the observer's nodal point coincides with the subject's far point. Thus in Fig. 422 the far point, *i*, coincides with N_o. The image of OO' formed by the dioptric system of the subject's eye is thus *ii'* in the plane of the nodal point, N_o. If O is illuminated, it forms an image at N_o and the reflex is seen by the observer.

O′, however, forms an image, i, vertically below N_o, and to construct the image of $i′$ formed by the observer's dioptric system we would have to draw a line through $i′$ and N_o, i.e. a line vertically up and down. Under these conditions, therefore, there is no image of O′ formed by the dioptric system of the observer's eye. Moreover, as soon as we pass from the point O, on the axis, to any other point very close, the image is vertically above or below N_o and consequently the smallest movement of the mirror makes the reflex invisible. We may therefore say that at this, the point of reversal, the movement of the reflex is infinitely rapid and is measured by the angle 90°.

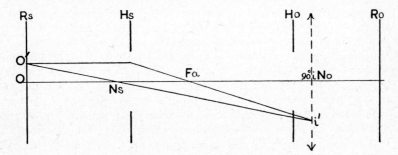

FIG. 422. Retinoscopy. Point of reversal.

A review of Figs. 417–422 shows us that as we pass from extreme hypermetropia, through emmetropia, to myopia of low degree the movement of the reflex is "with" the movement of a plane mirror, and increases progressively in speed. At the point of reversal the speed is infinite. When the myopia is greater than this, the movement is reversed and decreases in speed with increasing degree of myopia.

The refractive condition of a subject may be determined quantitatively by observing the direction and speed of movement of the reflex, obtained by directing light on to the fundus with a mirror. If the subject is myopic, the far point lies at a finite distance in front of the eye; the observer may find its position by moving the mirror and his eye backwards and forwards until reversal is obtained. In other refractive conditions, the subject may be made artificially myopic by putting up positive lenses, and the point of reversal then determined. Thus if the point of reversal is found to be 20 cm in front of the spectacle point when a $+5$ D lens is put up, the subject is emmetropic because a $+5$ D lens in front of an emmetropic eye gives a far point of 20 cm. If the point of reversal is 20 cm in front of the eye with a $+7$ D lens, the refractive error is now hypermetropia of $+2$ D, 5 D of the lens strength being necessary to bring the far point from infinity to 20 cm and the remaining 2 D to correct the actual hypermetropia. In general, if R′ is the dioptric value of the distance of the point of reversal from the spectacle point, when a lens of D dioptres is placed at this point, the spectacle refraction is given by $D+R′$, attention being paid to the signs; R′ is in front of the eye and is given a negative value in accordance with the sign convention.

Illumination. In general, a plane mirror is used and the principles involved are similar to those in the direct method of ophthalmoscopy.

OPTICAL CONSTANTS OF THE EYE

In order to elucidate the refracting properties of a spherical optical system we must know:

(*a*) The index of refraction of the various surfaces, i.e. the refractive indices of the different phases.

(*b*) The radii of curvature of the surfaces.

(*c*) The relative positions of the surfaces.

Refractive Index

The best known instrument for the measurement of the refractive index of a liquid is the *Abbé refractometer*. If ABC (Fig. 423) is the section of a prism, the liquid whose refractive index is required is spread over the surface AC. The surface AB, which is ground, is illuminated; some of the rays, emanating from a point S, are both refracted at the surface AC and reflected; the ray SP strikes the surface at the critical angle of incidence (p. 539) and thus just grazes the surface, whilst a ray ST, making a greater angle of incidence, is totally internally reflected. A ray SX, on the other hand, is partly reflected and partly refracted. To an observer looking towards the face BC through a telescope, the field of view is sharply divided into a bright area, corresponding to rays of light that have not suffered refraction, and a darker area corresponding to rays that have suffered both refraction and reflexion. By measuring the position at which this division of the field occurs, the critical angle can be calculated and hence the refractive index of the liquid.

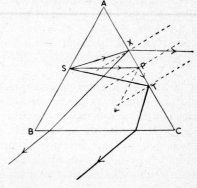

FIG. 423. Principle of the Abbé refractometer.

The figures given by Gullstrand for the refractive indices of the transparent tissues of the eye are:

Cornea	1·376
Aqueous	1·336
Vitreous	1·336
Lens	1·386 (Cortex)
Lens	1·406 (Core)

Radii of Curvature

A refracting surface does not transmit all the incident light, but a certain proportion is regularly reflected, the greater the index of refraction the greater the proportion reflected. A spherical surface may thus be treated as a mirror and its radius of curvature may be deduced from the magnification of the image. The magnification is given by:

$$i/o = -v/u$$

If u, the distance of the object, is made large, v, the distance of the image, is approximately equal to the focal length, f, which is half the radius of curvature.

Hence $$i/o = -R/2u$$
And $$R = -2i \cdot u/o$$

The quantities i, u, and o can all be measured, and R can be calculated.

Purkinje-Sanson Images. The image (catoptric) formed by reflexion at the eye surfaces are called *Purkinje-Sanson images*; the principal ones are four in number, formed respectively by the anterior and posterior surfaces of the cornea and lens; their mode of formation is indicated in Fig. 424 from Emsley.

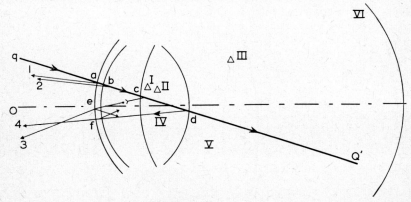

FIG. 424. Formation of the Purkinje-Sanson images. (Emsley, *Visual Optics.*)

It should be noted that images II, III and IV are formed by reflexion followed by refraction; thus image III is formed by reflexion at the anterior surface of the lens followed by refraction at the posterior and anterior surfaces of the cornea; allowance must be made for these refractions when calculating the radius or curvature of the reflecting surface from the size of image III.

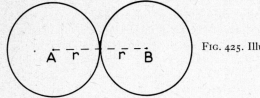

FIG. 425. Illustrating the principle of doubling.

The Ophthalmometer. The main problem in estimating the radius of curvature of a refracting surface thus resolves into measuring the size of the catoptric image; the difficulty in the way results from the movements of the eye and to overcome this Helmholtz made use of the principle of "*doubling*". If we know by how much the centres of two identical objects must be separated so that their edges just touch, we know the size of the objects; *e.g.* in Fig. 425

the centres of two identical circles of radius *r* are separated so that their circumferences just touch; the diameter of the circles is given by the separation of their centres.

In the *Helmholtz ophthalmometer*, the image, I, of an object in front of the corneal surface, is observed through the optical system indicated by Fig. 426; two inclined glass plates cause the rays from I to diverge from I′ and I″, and thus the image I is doubled. The two lenses behave as a telescope for the observation of I′ and I″, real images X and Y being formed by L_1; these are viewed through the eye-piece lens, L_2. The separation of I′ and I″ is determined by the inclination of the glass plates to the vertical, so that the two images can be made just to touch by varying this inclination. The separation of the images, under these conditions, may be calculated and thence the size of the image I. Small involuntary movements of the eye during the measurement do not disturb the observer since both images move together. Helmholtz found a value of 7·83 mm for the radius of curvature of the anterior surface of the cornea.

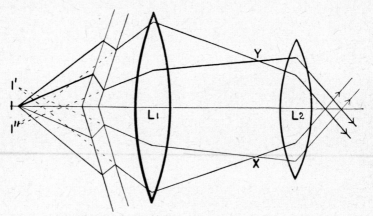

FIG. 426. The Helmholtz ophthalmometer.

In the Javal-Schiøtz instrument, the inclined plates are replaced by a Wollaston prism, consisting of two rectangular quartz prisms cemented together, with their optic axes at right-angles, to form a plate. Such a prism is doubly refracting (p. 625) and gives two images of a single object. With this system, the separation of the images is fixed, and the size of the object in front of the cornea is varied until its doubled images I′ and I″ just touch. To facilitate the variation in object-size, so-called "*mires*" are used mounted on a circular frame. The mires are pieces of opal glass illuminated by electric lamps (Fig. 427) and are treated as the two ends of the object; by sliding them closer together or farther apart, the size of the object is decreased or increased.

The mires, A and B, are shaped as in Fig. 428 (*inset*), and the object is considered to extend from the point *a* to the point *b*, i.e. it may be treated as the line *ab*. Two images of AB, XY and X′Y′, are formed by the doubly refracting system, and the appearance seen through the telescope will depend on the distance apart of A and B. Let us suppose that the appearance is that shown in Fig. 428, *a*, A giving the two images X and X′ and B the images Y and Y′. The object was considered to be the line *ab*, and its images are *xy* and *x′y′*; it is clear that these images are *too small* to permit the lower edge of the one (*y*) to coincide with the upper edge of the other (*x'*).

In Fig. 428, *b* the distance *ab* has been increased, so that *xy* and *x'y'* are greater but still not large enough to give exact abutment. In Fig. 428, *c* the distance *ab* has been adjusted to give exact abutment, so that *y* coincides with *x'*. In Fig. 428, *d* the condition in which the object is *too large* is shown.

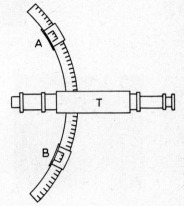

Fɪɢ. 427. The Javal-Schiøtz ophthalmometer.

The ophthalmometer may be used for measuring the curvature of the cornea in different meridians; thus the circular frame may be placed vertically in front of the eye and the mires A and B adjusted to give exact abutment. The frame is then turned through 90° and if the curvature is different in the horizontal meridian, coincidence of the points *x'* and *y* will no longer be obtained and the appearance will be as in Fig. 428, *b* or *d*. If the latter, it can be said that the

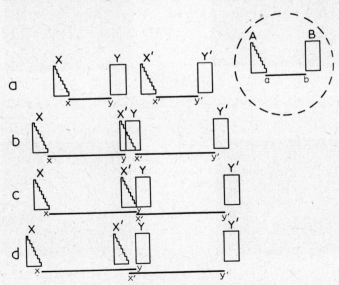

Fɪɢ. 428. Appearances of the images of the mires when the object-size is too small (*a* and *b*), correct (*c*), and too large (*d*) for abutment.

FIG. 429. The keratoscopic disc.

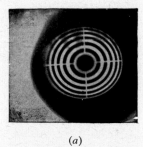

(*a*)

Regular astigmatism of 5·5 D with the rule

(*b*)

Regular astigmatism against the rule.

(*c*)

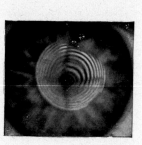

(*d*)

Irregular astigmatism. Eczematous keratitis; course of affection within fourteen days.

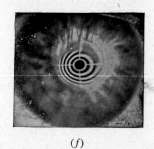

(*e*)

Conical cornea—base of cone.

(*f*)

Conical cornea—apex of cone.

FIG. 430. Appearances of photo-keratoscopic images. (Emsley, *Visual Optics*.)

image size, *xy*, is too large and since the object size, *ab*, has been kept constant, it can be deduced that the curvature is less in the horizontal meridian than in the vertical, i.e. that the astigmatism is with the rule. The mire A is designed so that an overlap of one step corresponds to a difference in refracting power of 1 D.

The Keratoscopic Disc. It should be noted that the ophthalmometer utilizes only a small portion of the cornea (1 to 2 mm) and the figures obtained for the radius of curvature apply to the central area only; more extensive studies with the keratoscopic disc have shown that it is only a central zone of about 4 mm diameter that can be treated as a spherical surface; peripherally the cornea is considerably flattened.

The *keratoscopic disc* (Fig. 429) is a card on which concentric circles are drawn; it may be used as an object in the ophthalmometer with the doubly refracting prism removed. The images of the circles would be perfect circles if the cornea were a perfectly spherical surface (apart from aberrations, p. 612); if the curvature were variable, the circles would be deformed, becoming elongated along the meridian of least curvature. Some typical appearances are shown in Fig. 430. Even in the optical zone of the normal cornea a certain amount of astigmatism is generally present ("physiological astigmatism"). On the average, the difference in refracting power in two meridians at right-angles amounts to 0·5–0·74 D in 89·4 per cent of individuals, the meridian of least curvature being horizontal (with the rule). The keratoscopic disc has revealed a considerable peripheral flattening of the normal cornea.

For routine measurements of the curvatures of both corneal and lens surfaces Sorsby, Benjamin & Sheridan photographed the Purkinje images and with the aid of nomograms, computed by Bennett, the necessary parameters were quickly calculated from the measured sizes of these images.

The results of Gullstrand's measurements of the curvatures of the central regions of the refracting surfaces of the eye may be summarized as follows:

Anterior surface of cornea	7·7 mm
Posterior surface of cornea	6·7 mm
Anterior surface of lens	10·0 mm
Posterior surface of lens	6·0 mm

Positions of the Surfaces

The apparent positions of the refracting surfaces of the eye can be measured with the slit-lamp microscope; the instrument is arranged so that the microscope is directed constantly towards the focus of the illuminating system. By moving the system towards the eye the anterior surface of the cornea can be brought into focus. The instrument is then racked forward until the surface in question is in focus, and the focusing scale on the slit-lamp is read. In routine examinations it is more convenient to photograph the optical section of the anterior segment of the eye and to measure the required distances afterwards, applying the necessary corrections for angle of view, apparent position, etc. (Bennett, 1961). The positions of the refracting surfaces are shown in Table 11.

The Lens

If the lens consisted of a homogeneous medium, its optical properties could be easily determined from its refractive index and curvatures; in fact, of course,

TABLE 16

Positions of the Refracting Surfaces of the Eye, referred to
the Anterior Surface of the Cornea (Gullstrand)

Cornea. Ant. surface	o	mm
Cornea. Post. surface	0·5	,,
Lens. Ant. surface	3·6	,,
Lens. Post. surface	7·2	,,

the refractive index varies from cortex inwards. Gullstrand treated the lens schematically as a "core" of refractive index 1·406 surrounded by material of refractive index 1·386 (Fig. 431), the so-called *"equivalent core lens."* In order that this schematic lens may have the dioptric power actually found, the core must have entirely fictitious curvatures (anterior and posterior radii of 7·91 and 5·76 mm respectively). It should be noted that the refracting power of such a system is greater than that of a homogeneous lens of the same dimensions, even if the refractive index throughout were as high as that of the core. As the lens ages the cortex hardens and the whole lens tends towards homogeneity; this may lead to a diminished dioptric power in spite of the over-all increase in refractive index. For this reason the myope in old age may tend to become emmetropic.

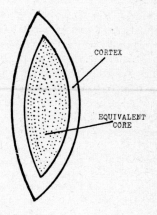

CORTEX

EQUIVALENT CORE

FIG. 431. The equivalent core lens.

Length of Eye

Up till recently the figures accepted for the dimensions of the eye have been obtained by measurements on the excised organ. With the aid of X-rays, however, an accurate estimate of the length of the eye in the living subject can be made (Rushton). A beam of X-rays at right-angles to the optic axis strikes the dark-adapted eye; as a result, a circle of light is seen by the subject, corresponding to an optical section of the globe. If the slit through which the X-rays pass is moved posteriorly by an apparatus orientated in relation to the cornea, the circle decreases in size and ultimately disappears. At this position of the slit, the beam of X-rays just grazes the pole of the retina. By measuring the distance of the slit from the cornea, the length of the optic axis could be determined to within 0·1 mm. The average length of the emmetropic globe was found by Goldmann and Hagen to be 23·4 mm, comparing with Gullstrand's value of 24·4 mm. In myopic eyes the lengths were greater.

Dioptric Power

The X-ray technique has been applied to determining the size of the image of an external object formed on the retina; the X-ray tube, just below the line of sight, projected two beams through the eye; X-rays are not significantly refracted by the media of the eye so that the two beams appeared to the subject as two lines separated by a definite distance. This distance was determined by causing the subject to gaze at a wall on which lines of known distance apart were projected with a lantern. The image on the retina was projected on to this wall and the pair of lines whose separation corresponded with that of the projected image was found. The X-ray beams struck the retina 5 mm apart; consequently the pair of lines on the screen formed an image on the retina this distance apart. The magnification of the eye was thus determined and the refractive power could be calculated. The figure obtained by Goldmann and Hagen was 59·2 D, comparing with Gullstrand's value of 58·6 D. The corneal refraction was measured in the usual way, when the dioptric power of the lens was determined; it came out at 19·2 D comparing with Gullstrand's value of 19·1 D.

Emmetropia and Ametropia

Strenström published the results of a study of the optic elements of the eyes of 315 human subjects chosen without reference to their refractive condition, the X-ray technique being used for determining the axial length. His results may be summarized as follows:

Corneal radius	7·86 ± 0·008 mm
Depth of anterior chamber	3·68 ± 0·009 mm
Power of cornea	42·84 ± 0·044 D
Power of lens	17·35 ± 0·044 D
Total power of eye	58·13 ± 0·056 D
Length of optic axis	24·00 ± 0·035 mm

The extreme limits of variation in the length of the eye were 20 and 29·5 mm. Although there was a strong correlation between axial length and refractive error—myopes having longer axial lengths usually than emmetropes—Strenström's results showed that it would be quite wrong to speak of a fixed "normal" axial length to give the condition of emmetropia; both the axial length and refractive power of the emmetropic eye varied over quite a large range, the larger emmetropic eye having the weaker power, and *vice versa*. Emmetropia is therefore achieved rather by a balance of refractive power and axial length. This is in some measure understandable since the larger the eye the smaller will be the curvature of the cornea, other things being equal.

A later study by Sorsby *et al.* (1957) reemphasized that emmetropia was essentially the result of a correct balance between the optic elements of the eye, especially between axial length and corneal curvature. Thus, in the emmetropes, the axial length varied between 22 and 26 mm, the condition of emmetropia resulting from a correlation between corneal power and axial length, a strongly curved cornea being accompanied by a short axial length and *vice versa*. Moreover, with hypermetropia and myopia up to 4 D, the axial lengths varied over the same range, so that the ametropia within this range must be regarded as the result of a faulty correlation of axial length with corneal power, rather than as a result of an "abnormal" axial length. With refractive errors above 4 D the correlation between axial length and refractive error was strong.

Changes During Growth. Sorsby, Benjamin & Sheridan studied some
1432 children aged between 3 and 15 years in which the refractive error, corneal
and lens curvatures, and depth of the anterior chamber were measured by
photographic techniques. As Fig. 432 shows, children at the age of 3 years
have an average hypermetropia of 2 D and an axial length of 23 mm. With
increasing age the ametropia decreases whilst the axial length increases to
reach a probable limiting length for boys of 24·1 mm and 23·7 mm for girls
(Sorsby *et al.* 1957). The increase in axial length from the age of 3 to 14, whilst
it is continuous, is remarkably small, only about 1 mm, or some 4–5 per cent,
compared with an increase in body-length of some 60 per cent. After adolescence

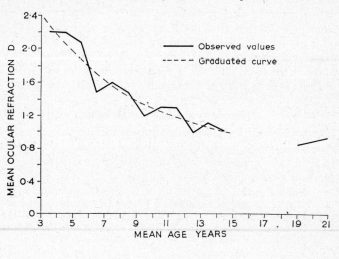

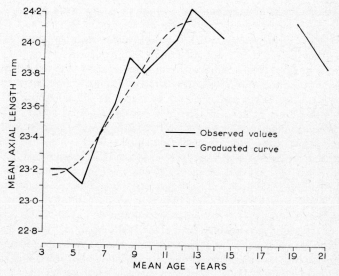

FIG. 432. *Upper.* Variation of ocular refraction with age. *Lower.* Variation of
axial length with age. (Sorsby, Benjamin & Sheridan. *M.R.C. Sp. Rep.* By
permission of Her Majesty's Stationery Office.)

little further change occurs. The changes in corneal power with age are slight, so that the decrease in power of the optic elements necessary to compensate for the increase in axial length is due mainly to a fall in lens-power from 21·8 to 19·8 D. The depth of the anterior chamber increased by about 0·1 to 0·2 mm. By following up some 386 children over a period of 2–6 years, some of the changes with age, already manifest by the initial cross-sectional study just described, could be seen more definitely, and it became clear that the increase in axial length that usually takes place during normal growth is largely compensated by a reduced dioptric power. For example, although 32 per cent of cases showed an increase of length during the period of study that would have led to a myopia of 2 D if uncompensated, only 5 per cent actually showed this amount of increase in refractive error. During growth, therefore, the axial elongation, with its tendency to myopia, is compensated by a decreased power of cornea and lens, in particular of the lens.

Studies on children younger than 3 years are not easy, so that the changes taking place between birth and this age have not been so accurately ascertained. Sorsby & Sheridan give a mean axial length of 17·9 for full term male infants, so that by the age of 3 years the axial length of the eye has increased by 5 mm, comparing with only 1 mm for the remainder of its growth. According to Cook & Glasscock, who refracted the eyes of 1000 newborn babies under atropine, there is a wide scatter of refractive errors at birth, ranging between 7 D hypermetropia to 7 D of myopia, whilst the mean refraction was one of 1·5 D of hypermetropia. Thus the classical view that all newborn babies are strongly hypermetropic is not born out, some 25 per cent of the total examined being myopic. The actual refractive error that an adolescent or adult has may usually be regarded as a failure of coordination of axial length with the powers of the optic elements; it seems likely that the large errors, however, are due to defective growth in the 0–3 years-old period, so that the child begins the "definitive stage" of development with too high an error to permit adequate compensation.

DEFECTS IN THE IMAGE

The laws relating to the formation of images by mirrors and refracting surfaces were derived on the assumption that the rays concerned made small angles with the axis (so-called *paraxial rays*), in which event it could be shown that all rays from a point on the object would meet at a point in the image. If this condition is not fulfilled, the light from any point on the object occupies a definite area in the image, with the result that overlapping of the images of adjacent points on the object gives rise to blurring of the over-all image. In practice it is often difficult, or disadvantageous, to limit the light to paraxial beams and we must now determine the consequences of permitting non-paraxial or *marginal* rays to enter into image formation.

Points on the Axis

Spherical Aberration. It may be shown that marginal rays are brought to a focus closer to the refracting or reflecting system than paraxial rays; the consequence of this variation in focal length is called *spherical aberration*. In Fig. 433 the rays from O marked (1), are brought to a focus at a point I_1, whilst the paraxial rays, marked (2), are brought to a focus at I_2, the so-called *paraxial* *image of* O. Rays intermediate between these extremes come to foci between I_1 and I_2. The distance I_1I_2 is the *axial spherical aberration*; a screen placed at I_2 receives a circle of light instead of a point image; as the screen is moved towards, I_1 the circle diminishes in size and passes through a minimum at c, the *circle of least confusion*. The diameter of this circle is called the *lateral spherical aberration*.

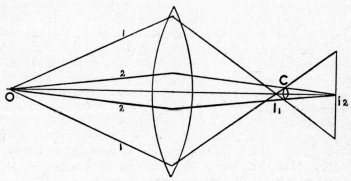

Fig. 433. Spherical aberration.

If a series of rays are drawn as in Fig. 434, it will be will be seen that in approaching their respective foci they cross; the envelope of these rays is called a *caustic*, and constitutes a bright line because of the concentration of rays.

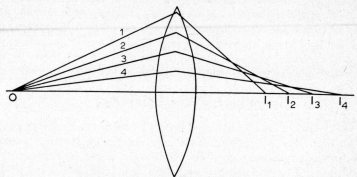

Fig. 434. Spherical aberration. The caustic.

Spherical aberration can be partially remedied by placing a stop in front of the lens so as to permit only paraxial rays to enter, but this is disadvantageous in that it restricts the field of view and reduces the brightness of the image. It can be shown that the spherical aberration of a lens is considerably reduced if the bending of rays by each surface is the same; spherical aberration thus depends on the shape of a lens and on which surface faces the incident light; in Fig. 435 different types of converging lens are shown; the spherical aberration decreases on passing from left to right to reach a minimum between types 3 and 4, if the incident light comes from the left of the lens.

The influence of shape on the deviations at the two surfaces is made clearer by considering the plano-convex lens in Fig. 436; rays parallel to the axis,

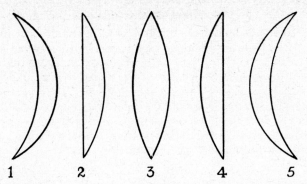

FIG. 435. Spherical aberration in different types of lens. The aberration is at a minimum with a type intermediate between (3) and (4).

incident on the plane surface, pass through undeviated, since they are normal; at the second surface they are refracted away from the normal, i.e. bent towards the axis. All the bending is thus done at the second surface. If the convex surface faces the incident light, the rays are bent towards the normal at the first surface and away from it at the second; both refractions cause a bending of the ray towards the axis, and if spherical aberration is reduced by dividing up the deviations between the two surfaces, clearly it will be less when the convex surface faces the incident light.

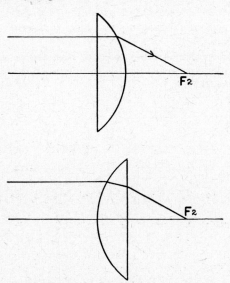

FIG. 436. Illustrating the importance of lens shape. *Above:* All the bending occurs at the second face. *Below:* The bending is divided between the two faces.

For a certain special position of object and image a single spherical refracting surface may produce no spherical aberration; under these conditions the surface is said to be *aplanatic*, and the positions of object and image are *aplanatic points*.

Points off the Axis

Spherical aberration represents a failure of all rays from a point object to meet at the same point in the image, due to the variation of the focal length of the lens according as the rays are marginal or paraxial. If we wish to consider the possible aberrations in the image of a finite object we must study the images of points off the axis, and we must consider, moreover, the *spatial relationships* between the points on the object and the points on the image. Failure to bring all rays from a point on the object to a point on the image leads to a loss of definition, whereas a disturbance in the spatial relationships of the points in the image leads to distortion of some sort, e.g. curvature.

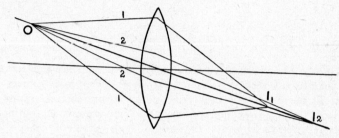

FIG. 437. Spherical aberration with a point off the axis.

The image of a point object, not on the principal axis, suffers spherical aberration as might be expected, but in addition two other aberrations enter, due to the asymmetrical position of the object in relation to the lens; these are *coma* and *astigmatism*.

The spherical aberration for the point off the axis is shown in Fig. 437; this error can be reduced, as before, by a stop to cut out marginal rays, or by the design of the lens shape.

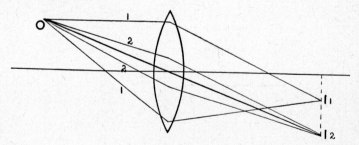

FIG. 438. Coma.

Coma. In Fig. 438, the lens is assumed to have been corrected for spherical aberration, so that the focal length is the same for all rays. However, the paraxial rays meet at the paraxial image, I_2, on the central ray through the lens, whereas the marginal rays meet at a point I_1, at the same distance from the lens as I_2, but closer to the axis. Rays intermediate between (1) and (2) meet at points between I_1 and I_2, and thus the image of O is far from being a point, its shape being similar to the tail of a comet, from which the aberration takes its name.

Spherical aberration and coma are related defects, which can be abolished by a suitable design of the lens surfaces and choice of the position of the object.

Astigmatism. Because of its asymmetrical position in respect to the lens, the curvatures in two planes at right angles through the point object are not the same. If the point is vertically above the optic axis in the plane of the paper, the *meridional plane* of the system, defined as the plane containing the object and the axis, is the plane of the paper (Fig. 439); the curvature of the lens in the meridional plane is greater than that in the plane at right-angles, the so-called *sagittal plane*. Rays in the meridional plane from O (rays marked 1) thus come to a focus before those in the sagittal plane, and form a *tangential line focus*, T; the rays in the sagittal plane, coming to a focus later, form a vertical *sagittal line focus*, S. Consequently, in a lens in which both spherical aberration and coma are corrected, the best image of a point object, off the axis, is a circle of least confusion; hence the image of an extended object will have poor definition unless this error, too, is corrected. This is generally done by combining a converging and diverging lens, with focal lengths and refractive indices conforming to the relationship:

$$n_1 f_1 + n_2 f_2 = 0 \qquad\qquad (10)$$

This is known as the *Petzval condition*.

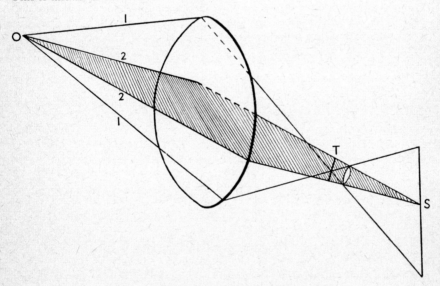

FIG. 439. Astigmatism due to oblique pencils.

Curvature and Distortion. Spherical aberration, coma, and astigmatism are defects in a lens that prevent a point object from producing a point image. Even when these three aberrations are remedied, a perfect image of an extended object is not formed, owing to the defects of *curvature* and *distortion*.

Let us imagine that spherical aberration, coma, and astigmatism have been corrected in a lens, as in Fig. 440. Curvature is due, principally, to the fact that the focal lengths along the subsidiary axes are shorter, the more the pencils are inclined to the lens; thus although all rays from any point on the object come to a point focus in the image, the image of O is at I, whilst the images of O_1, O_2, O_3, etc. are at I_1, I_2, I_3, etc. progressively closer to the lens; the image of OO′ is thus a curved line, II′, concave to the object. With a concave lens the curvature is in the opposite sense.

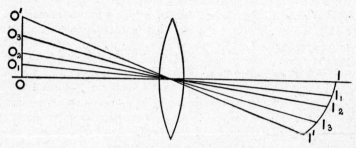

FIG. 440. Curvature of field.

Curvature and astigmatism are related defects, and both are abolished on applying the Petzval condition, the curvature due to the convex lens of the combination being cancelled by that due to the concave lens.

When all the preceding aberrations are cancelled out, one final defect can be present if a stop is placed on the optic axis of the lens, so as to prevent the principal ray from any point (i.e. the central ray of a pencil) from passing through the centre of the lens.

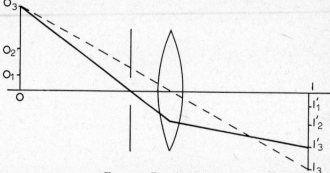

FIG. 441. Barrel distortion.

In the absence of the stop, the image of the point O_3 (Fig. 441) on the object would be at I_3 on the ray that would pass through the centre of the lens; however, in the presence of a stop a finite distance from the lens, the central ray from O_3 does not pass through the centre of the lens, but is refracted as shown, and all rays from O_3 come to a focus on this ray and produce an image at I'_3. The image of OO_3 is thus II'_3, smaller than if the stop were absent. Similarly, the image of O_2 falls at I'_2 and II'_2 is smaller than if the stop were absent, but not so much smaller in proportion; O_1, close to the axis, produces an image, I'_1, such that II'_1 is of approximately the theoretical size. Thus the magnification of different parts of the image varies, being least on the parts farthest from the principal axis. The appearance of the image of a gauze is consequently that indicated in Fig. 442, the distortion being called *barrel distortion*. When the stop is on the other side of the lens, the magnification is greatest for points farthest off the axis, and a gauze gives an image of the shape shown in Fig. 443, the distortion being called *pin-cushion distortion*.

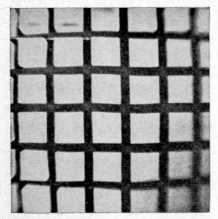

FIG. 442. Image of a gauze showing barrel distortion. (Barton, *Text-book on Light*. Photo by Mitchell.)

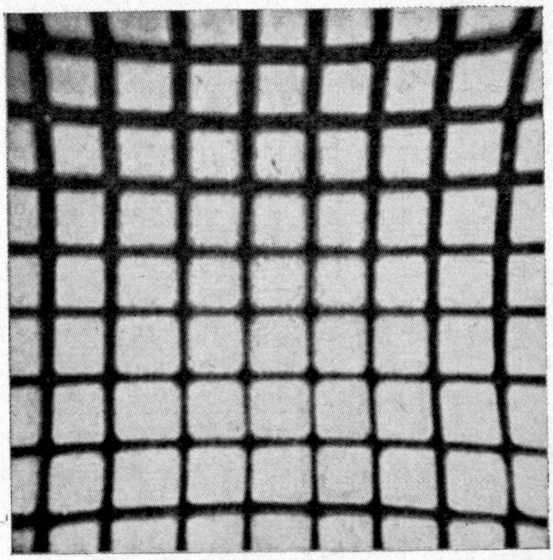

FIG. 443. Image of a gauze showing pin-cushion distortion. (Barton, *Text-book on Light*. Photo by Mitchell.)

Chromatic Aberration

The aberrations considered above are called *monochromatic aberrations*, since they occur when light of a single wavelength is used; chromatic aberration results from the use of mixed wavelengths, e.g. white light, and is due to the dependence of the refractive index on the wavelength of light; with a long wavelength like red, a material exhibits a smaller refractive index than with a short wavelength like violet. In a convex lens, uncorrected for chromatic aberration, the red rays, being refracted least strongly, come to a focus F_r (Fig. 444)

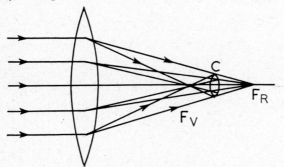

FIG. 444. Chromatic aberration.

farther away from the lens than the violet rays with a focus at F_v. The best image of a point is thus a circle of least confusion. Chromatic aberration is a far more serious defect than spherical aberration. The variation of the refractive index with wavelength of light for any given material is indicated by its *dispersive power*; this characteristic does not run strictly parallel with the refractive index when different materials are compared, so that it is possible to make a combination of a convex and concave lens such that the chromatic aberration of the one is cancelled by that of the other, their dioptric powers being only partially cancelled. Such a combination is called an *"achromatic pair"*. Since the diameter of the circle of least confusion in chromatic aberration depends on the diameter of the lens, the defect can be reduced by the presence of a stop.

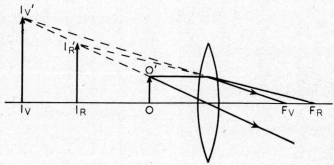

FIG. 445. Chromatic aberration and the magnifying glass.

Chromatic Aberration and the Magnifying Glass. It is worth noting that when a convex lens is used as a magnifying glass, chromatic aberration is not apparent, owing to the fact that the different coloured images subtend the same angle at the eye. Thus in Fig. 445, F_v and F_r are the second principal foci for violet and red light respectively. The virtual images of OO′ lie on the ray through the centre of the lens and, although they are at different distances from the lens, they have the same apparent size to an eye close to the lens; their images on the retina thus overlap exactly.

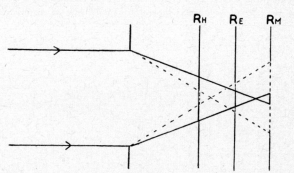

FIG. 446. Chromatic aberration in the eye. Red rays in full lines; violet rays broken.

Aberrations in the Eye

The eye is not a perfect optical instrument and exhibits the aberrations described above to a greater or less extent. Chromatic aberration can be demonstrated by looking at a distant luminous point through a cobalt glass which transmits the red and violet ends of the spectrum and cuts out the intermediate band. The appearance varies with the refractive condition of the eye; to the myope it is that of a red spot surrounded by blue; to the hypermetrope it is that of a blue spot surrounded by red, and to the normal eye the spot appears purple with perhaps a blue border due to the partial overlapping of the blue and red blur-circles (Fig. 446).

Hartridge has calculated that the difference in the focal lengths of the eye for red and violet lights amounts to 0·47 mm. As a result, a point source of white light on the optic axis of the eye forms an image consisting of concentric fringes of colour. Generally the eye focuses for yellow light, in which event the fringes have the following diameters (calculated by Hartridge):

> Red (6,500 Å) 0·0216 mm (6·4 cones).
> Yellow (6,000 Å) 0·0078 mm (2·3 cones).
> Green (5,500 Å) 0·0090 mm (2·7 cones).
> Blue-green (5,000 Å) 0·03 mm (9 cones).
> Blue (4,500 Å) 0·0588 mm (17·5 cones).

the number of foveal cones covered by the fringes being indicated in brackets. A bright point, viewed on a dark background, should thus appear as a central yellow spot merging, through yellowish white, greenish white, bluish white and purple, with blue. In actual fact a white point source appears white; moreover Hartridge has shown that even when the chromatic aberration of the normal eye is doubled by the use of appropriate lenses, it is not normally detected; only when the aberration is quadrupled are the blue and yellow fringes distinctly visible. Hartridge has suggested that there is a suppressor mechanism in the retina or brain which depresses the sensitivity to both blue and yellow. In evidence he adduces the fact that a small point source of yellow light appears white, whilst a similar blue source appears black. The chromatic fringes on the retina are small; the suppressor mechanism operates to convert the yellow to white and the blue to black, that is, to make the image resemble the object. When the images of coloured objects formed on the retina are large, the suppressor mechanism does not operate and the colours are correctly appreciated.

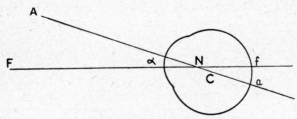

FIG. 447. The angle alpha.

It should be noted that the chromatic aberration so far considered has been a defect observable with a point object on the optic axis; with a point off the axis, the asymmetry introduced causes a chromatic difference of magnification; i.e. the lens is effectively decentred and has a prism effect causing a spectrum to be formed as the image of a point.

Spherical aberration is present in the eye and may be demonstrated subjectively. Aberrations due to obliquity of points from the optic axis must enter, even with foveal fixation, because the optic axis does not coincide normally with the visual axis. Thus in Fig. 447 the point F is fixated so that its image falls on the fovea, *f*. FN*f* represents the *visual axis*. The optic axis, AN*a*, diverges from the visual axis, the angle between the two being called the *angle alpha*; it amounts on the average to 5°; when the visual axes are parallel, e.g. when viewing an infinitely distant object, the optic axes of the two eyes thus diverge by 10°.

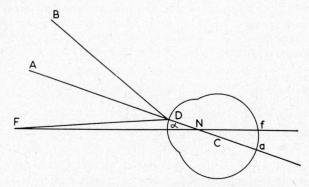

FIG. 448. Measurement of the angle alpha.

The angle alpha may be approximately determined on the perimeter. The subject looks at a light at the centre of the arc so that F*f* (Fig. 448) represents the visual axis. The point B, on the perimeter, is determined at which the reflex image of the light on the cornea appears to an observer to occupy the centre of the pupil. Since $\angle$ FDA = $\angle$ BDA, the angle read off on the perimeter represents twice the angle FDA; this is approximately equal to $\angle$ FNA, the angle between the visual axis and the optic axis. (Actually the line ADN, through the centre of the pupil, does not coincide exactly with the optic axis— it is called the *central pupillary line* and the angle ADF is called the *angle kappa*.) The fixation axis, it will be remembered, is the line joining the fixation point with the centre of rotation of the eye; the angle made by the fixation axis and the optic axis is called the *angle gamma*.

Thus a centrally fixated point is actually viewed obliquely, so far as the optic axis of the eye is concerned, and we must expect coma and astigmatism to be present. Coma and spherical aberration are reduced by the pupil and also by the peripheral flattening of the lens and cornea. Curvature of the field is not a defect, owing to the curvature of the retina. The pupil does not coincide with the nodal point of the optical system so that some barrel distortion should be present.

The image of a point, formed by the eye, is thus very far from being a point, as the geometrical constructions might lead us to believe, the most serious defect being chromatic aberration, but, as we have seen, this appears to be compensated by a physiological mechanism; in a similar way it seems likely that, even with monochromatic errors, inductive processes, which accentuate the response of the most strongly stimulated cone and reduce those of nearby elements, generally sharpen the image. It will be appreciated that, as a result of spherical and chromatic aberrations, there is no single focal length of the eye;

this permits a certain tolerance in the focusing of the dioptric system, so that small variations in the axial length of the eye need not necessarily be associated with a significant degree of ametropia; moreover, small changes in the vergence of the light entering the eye need not necessarily be followed by a change in accommodation. The aberrations thus give the eye a certain depth of focus, or tolerance to a change in vergence, which would be absent were the formation of the image to follow strictly the "paraxial laws." It seems probable that the tolerance of ± 0.25 D, observed by Fincham (p. 407), is due to these causes. The retinoscopy test for ametropia is said to be accurate to within $\frac{1}{8}$ D, but it seems doubtful whether such precision is necessary, when the aberrations are taken into account.

INTERFERENCE AND OTHER PHENOMENA

Wave-Motion

The passage of light through a medium may be regarded as a series of waves or "ripples"; these waves are made up of the movements of "particles" in the medium in a direction at right-angles to that of propagation. When a wave-motion is established in any medium the position of any given particle goes through a cyclical change, as in Fig. 449, *a* where the displacement of the particle from its position of rest is plotted against time. The time required for a complete cycle is called the *period*, T, of the vibration. The *amplitude* of the vibration, which measues its energy,* is the height of a crest.

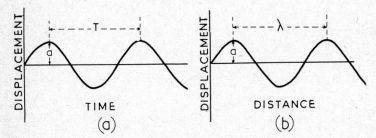

Fig. 449. Wave-motion.

Fig. 449, *a* represents the displacement of a single particle at different times; since wave-motion is due to the propagation of a disturbance by one particle to another, points adjacent to the one just considered likewise pass through the cyclical change depicted in Fig. 449, *a* but they are "out of phase" in the sense that they reach their crests at different times. Instead of plotting the displacement of one particle against time, we can plot the displacements of all the particles in a plane at a given moment against the distance from a given point, as in Fig. 449, *b*. The curve has the same shape as the first, and as before the height of the crest is the amplitude; the distance between two crests, however, represents an interval of space, and is called the *wavelength*, λ.

* The energy of the vibration varies as the square of the amplitude. If, as a result of interference, the amplitude at a point is doubled the brightness at this point will be quadrupled.

The time taken by any particle to go through a complete cycle, T, also represents the time taken for a disturbance to be propagated through a distance λ, the wavelength. Hence the velocity of light is equal to λ/T. If $1/T$ is called the *frequency*, n, we have:

$$V = n \cdot \lambda \tag{21}$$

Interference

If a continuous series of waves is propagated through a medium from a point, the motion of any particle is that represented by Fig. 449; if now a new set of waves is propagated from another point, the motions of the particle are modified; this modification is called *interference*, and its nature depends on the difference in phase of the disturbances due to the two wave-motions. Thus in Fig. 450 the displacements of a given particle due to the first wave-motion are represented by the curve A; those due to the second motion by the curve B. The waves are out of phase and tend to cancel each other in some parts and to accentuate each other elsewhere; for example at t_1 the particle would have been at its maximum height due to the first wave; the second wave, acting alone, would have raised it, but not so much, and the resultant position is given by the algebraic sum of the two displacements. It is easy to see that if the second wave is half a wavelength out of phase with the first the result will be complete suppression of particle movement; if the two waves are completely in phase, the amplitude of the motion will be doubled.

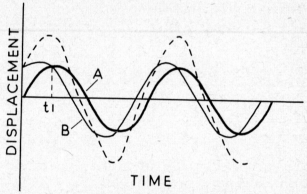

FIG. 450. Illustrating interference.

If light is to be regarded as a wave-motion we must expect interference, resulting in light and dark patches of illumination, when two sources of light illuminate the same screen. That this phenomenon is not normally observed is due to a number of factors. In the first place the sources must be "*coherent*," i.e. they must retain the same phase difference continuously, otherwise the points in space at which interference takes place will vary, and a point of darkness at one moment will become a point of maximum brightness at another. Sources of light commonly used vary in phase continuously, so that we cannot expect interference to be observed with two lamps, for example. Secondly, the sizes of the dark and bright patches, the so-called *fringe-widths*, decrease with increasing distance apart of the sources; unless the sources are very close together, the interference pattern consists of innumerable very thin lines of black and white which are unresolvable by the eye. Thirdly, the fringe widths decrease with decreasing wavelength.

Young's Slits. Interference may be demonstrated quite simply by the experimental set-up depicted in Fig. 451, known as *Young's slits*. Light is focused on to a slit S; it diverges from S and meets two other slits S_1 and S_2 which behave as two separate sources of light; they are coherent because they both arise from S, any variation in phase in this source being reflected in similar changes in both S_1 and S_2. Where the two beams from S_1 and S_2 overlap, interference occurs and a series of black and white lines are seen on a screen placed in this region. The wavelength of the light can be calculated quite simply from the fringe width.

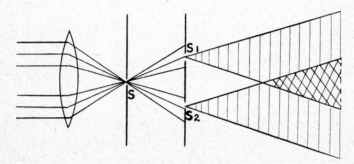

FIG. 451. Young's slits.

Diffraction

If vibrations are set up at a point in a homogeneous medium they spread in all directions; the locus of all points suffering the same displacement at any given moment is the surface of a sphere; it is called the *wave-front*. (The ripples in water are wave-fronts; here the disturbance is propagated in two dimensions and the front appears as a circle.) In Fig. 452, *a* each of the circles represents a wave-front due to successive disturbances propagated from the same point O. The direction, OP, in which any very small portion of a wave-front travels, is normal to the wave-front and is called a *ray*. As the wave gets farther from its point of origin, the surface of its front gets flatter and flatter so that when the

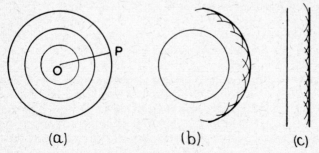

(a)   (b)   (c)

FIG. 452. (*a*) Successive wave-fronts. OP is a ray. (*b*) Huygens' construction for circular wave-front. (*c*) Huygens' construction for plane wave-front.

disturbance arises at a point infinitely far away the wave-front becomes a plane, and the rays are parallel.

According to Huygens' Principle, the position of a wave-front, t sec after its present position, is found by regarding each point of the front as the source of secondary wavelets, travelling at the speed of light in the same manner as the original wave. Hence the new position of the wave-front is given by drawing spheres of radius $c \times t$ (where c is the velocity of light) round each point on the front; the envelope of these spheres is the new position (Fig. 452, b and c). A plane wave-front clearly propagates as a plane.

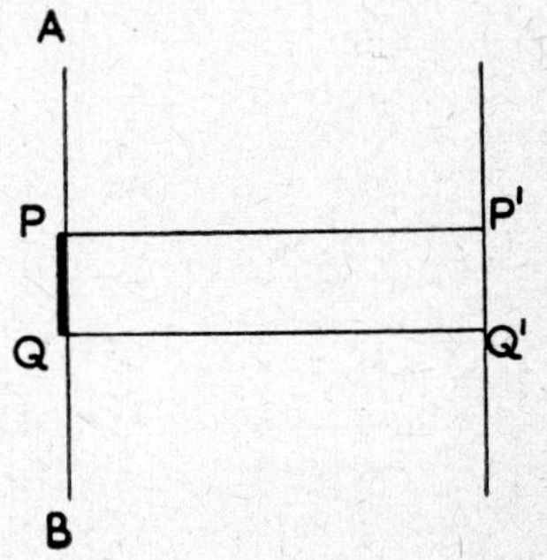

FIG. 453. The plane wave-front, AB, meets an obstacle PQ. The geometrical shadow is P'Q'.

This mode of viewing the propagation of light raises the problem as to why reasonably clear-cut shadows of objects are seen; thus in Fig. 453 the plane wave-front AB strikes an object PQ which casts a shadow P'Q' on the screen; this shadow, according to the laws of geometrical optics, should be of the same size as PQ since the rays are parallel. However, if the propagation of the wave-front at P can be considered to be due to the setting up of wavelets, the point P, on the edge of the object, must give rise to disturbances reaching all parts of P'Q', i.e. light must travel round corners and P'Q' must be illuminated in spite of the obstruction PQ. Such a deviation from the rectilinear propagation of light would be called *diffraction*, and it is easy to show that this does occur. We may reasonably ask, however, why PQ casts a shadow at all, and, moreover, why the size of the shadow is such as to make us think that light *does* travel in straight lines. In general terms the explanation lies in the fact that although the wavelets from P and Q do strike P'Q', as a result of interference their effects are largely cancelled out and P'Q' is a well-defined shadow. The edges of the shadow are, however, not perfectly defined since the cancellation of wavelets is not complete and it is possible to show, with a narrow obstacle such as a pin, that the shadow consists of a central bright line and, on its edges, a series of black and white lines.

As a result of diffraction, light passing through a small circular aperture, e.g. the pupil of the eye, is not a perfect circle of white; in general it consists of a central bright spot with concentric rings of dark and light. Consequently the image of a luminous point formed by the dioptric system of the eye is not a point but a series of rings of finite magnitude. Thus although decreasing the aperture of the eye reduces the aberrations discussed earlier, it tends to blur the image as a result of diffraction effects. It is found that, with pupil sizes above 3 mm, visual acuity is independent of pupil size, a decrease in diffraction being counterbalanced by an increase in aberrations.

Polarization and Double-Refraction

A doubly-refracting material produces two images of a point object because a part of the light is refracted in accordance with Snell's Law, giving rise to the "*ordinary ray*," whilst the remainder gives a value of the ratio: Sin i/Sin r, varying with the angle of incidence; it is thus generally refracted to a different extent and gives rise to the "*extraordinary ray*."

It is easy to show that the light emerging from a doubly refracting crystal, such as tourmaline or calcite, is modified during its passage; thus on passing light through two flat tourmaline crystals (which absorb the ordinary ray and only transmit the extraordinary ray) the amount of light passing through the second crystal depends on the position of its axis in relation to that of the first; if the axes are parallel, all the light emerging from the first is transmitted by the second, whilst at intermediate angles smaller amounts pass through and, when the axes are at right-angles, no light at all is transmitted. The light emerging from a doubly refracting crystal is therefore said to be *polarized*; non-polarized light consists of transverse vibrations in all planes at right-angles to the direction of propagation of the light, whereas in polarized light the transverse vibrations are confined to a single plane; the light is said to be *plane polarized*.

The process of polarization of light may be likened to passing the vibrations through a slot; if the slot is vertical, only vibrations in a vertical plane parallel with the direction of propagation can pass through, so that the slot acts as a sorter or polarizer. If it is desired to determine in what plane the vibrations are polarized, another slot is used, the direction in which the slot must be placed, to permit the light to pass through, giving the plane of polarization. Thus if a beam of polarized light is viewed through a tourmaline crystal, it may appear quite bright at one position of the axis, but in another position at right-angles to this, no light at all may be transmitted.

Various devices are used for obtaining and analysing polarized light. In the Nicol prism the ordinary ray is got rid of by causing it to be totally internally reflected at a Canada-balsam surface. "Polaroid" is made up of a large number of very small crystals of a substance, herapathite, which behave effectively as a single crystal. The substance is similar to tourmaline in that it absorbs the ordinary ray. The crystals are embedded in nitrocellulose sheets.

GENERAL REFERENCES

BARTON, A. W. (1944). A Textbook on Light. London: Longmans Green.
BENNETT, A. G., & FRANCIS, J. L. (1962). Visual optics. In *The Eye*. Ed. Davson. Vol. IV. Pt. 1. London & New York: Academic Press.
DUKE-ELDER, W. S. (1954). The Practice of Refraction. London: Churchill.
EMSLEY, H. H. (1944). Visual Optics. London: Hatton Press.
TREISSMAN, H. & PLAICE, E. A. (1946). Principles of the Contact Lens. London: Kimpton.

SPECIAL REFERENCES

BENNETT, A. G. (1961). The computation of optical dimensions. Appendix D in Sorsby, Benjamin & Sheridan (1961).

Cook, R. C. & Glasscock, R. E. (1951). Refractive and ocular findings in the newborn. *Amer. J. Ophthal.*, **34**, 1407–1413.

Goldmann, H. & Hagen, R. (1942). Zur direkten Messung der Totalbrechkraft des lebenden menschlichen Auges. *Ophthalmologica, Basel*, **104**, 15–22.

Hartridge, H. (1947). The visual perception of fine detail. *Phil. Trans.*, **232**, 516–671.

Rushton, R. H. (1938). Clinical measurement of the axial length of the living eye. *Trans. Ophthal. Soc. U.K.*, **58**, 136–142.

Slateper, F. J. (1950). Age norms of refraction and vision. *Arch. Ophthal.*, **43**, 466–481.

Sorsby, A., Benjamin, B., Davey, J. B. Sheridan, M. & Tanner, J. M. (1957). Emmetropia and its aberrations. *Med. Res. Counc. Sp. Rep. Ser.* No. 293.

Sorsby, A., Benjamin, B. & Sheridan, M. (1961). Refraction and its components during the growth of the eye from the age of three. *Med. Res. Counc. Sp. Rep. Ser.* No. 301.

Sorsby, A. & Sheridan, M. (1960). The eye at birth: measurement of the principal diameters in 48 cadavers. *J. Anat. Lond.*, **94**, 193–197.

Stenström, S. (1946). Untersuchungen über die Variationen und Kovariationen der Elemente des menschlichen Auges. *Acta. ophthal. Kbh. Suppl.*, **26**.

INDEX

PRINTED IN GREAT BRITAIN AT ABERDEEN UNIVERSITY PRESS